Nanotechnology: Technological Progress

Nanotechnology: Technological Progress

Editor: Andrew Green

NYRESEARCH
P R E S S

New York

Published by NY Research Press
118-35 Queens Blvd., Suite 400,
Forest Hills, NY 11375, USA
www.nyresearchpress.com

Nanotechnology: Technological Progress
Edited by Andrew Green

International Standard Book Number: 978-1-63238-550-5 (Hardback)

The publisher's policy is to use permanent paper from mills that operate a sustainable forestry policy. Furthermore, the publisher ensures that the text paper and cover boards used have met acceptable environmental accreditation standards.

Trademark Notice: Registered trademark of products or corporate names are used only for explanation and identification without intent to infringe.

Cataloging-in-Publication Data

Nanotechnology : technological progress / edited by Andrew Green.
    p. cm.
Includes bibliographical references and index.
ISBN 978-1-63238-550-5
1. Nanotechnology. 2. Nanostructured materials. I. Green, Andrew.
T174.7 .N36 2017
620.5--dc23

Printed in the United States of America.

Contents

Preface..IX

Chapter 1 **Ionic polarization-induced current–voltage hysteresis in CH₃NH₃PbX₃**
perovskite solar cells...1
Simone Meloni, Thomas Moehl, Wolfgang Tress, Marius Franckevičius,
Michael Saliba, Yong Hui Lee, Peng Gao, Mohammad Khaja Nazeeruddin,
Shaik Mohammed Zakeeruddin, Ursula Rothlisberger & Michael Graetzel

Chapter 2 **Deciphering the origin of giant magnetic anisotropy and fast quantum**
tunnelling in Rhenium(IV) single-molecule magnets.. 10
Saurabh Kumar Singh & Gopalan Rajaraman

Chapter 3 **High-efficiency electrochemical thermal energy harvester using carbon nanotube**
aerogel sheet electrodes...18
Hyeongwook Im, Taewoo Kim, Hyelynn Song, Jongho Choi, Jae Sung Park,
Raquel Ovalle-Robles, Hee Doo Yang, Kenneth D. Kihm, Ray H. Baughman,
Hong H. Lee, Tae June Kang & Yong Hyup Kim

Chapter 4 **Nanocaged enzymes with enhanced catalytic activity and increased stability**
against protease digestion...27
Zhao Zhao, Jinglin Fu, Soma Dhakal, Alexander Johnson-Buck, Minghui Liu,
Ting Zhang, Neal W. Woodbury, Yan Liu, Nils G. Walter & Hao Yan

Chapter 5 **Ferroelastic switching in a layered-perovskite thin film** 36
Chuanshou Wang, Xiaoxing Ke, Jianjun Wang, Renrong Liang, Zhenlin Luo,
Yu Tian, Di Yi, Qintong Zhang, Jing Wang, Xiu-Feng Han, Gustaaf Van Tendeloo,
Long-Qing Chen, Ce-Wen Nan, Ramamoorthy Ramesh & Jinxing Zhang

Chapter 6 **4Pi-RESOLFT nanoscopy**..45
Ulrike Böhm, Stefan W. Hell & Roman Schmidt

Chapter 7 **Structural semiconductor-to-semimetal phase transition in two-dimensional**
materials induced by electrostatic gating...53
Yao Li, Karel-Alexander N. Duerloo, Kerry Wauson & Evan J. Reed

Chapter 8 **Visualizing the orientational dependence of an intermolecular potential**............... 61
Adam Sweetman, Mohammad A. Rashid, Samuel P. Jarvis, Janette L. Dunn,
Philipp Rahe & Philip Moriarty

Chapter 9 **Pure and stable metallic phase molybdenum disulfide nanosheets for hydrogen**
evolution reaction...68
Xiumei Geng, Weiwei Sun, Wei Wu , Benjamin Chen, Alaa Al-Hilo,
Mourad Benamara, Hongli Zhu, Fumiya Watanabe, Jingbiao Cui & Tar-pin Chen

Chapter 10   **Exceptional damage-tolerance of a medium-entropy alloy CrCoNi at cryogenic temperatures**... **75**
Bernd Gludovatz, Anton Hohenwarter, Keli V.S. Thurston, Hongbin Bei,
ZhenggangWu, Easo P. George & Robert O. Ritchie

Chapter 11   **Temperature-feedback upconversion nanocomposite for accurate photothermal therapy at facile temperature**.. **83**
Xingjun Zhu, Wei Feng, Jian Chang, Yan-Wen Tan, Jiachang Li, Min Chen,
Yun Sun & Fuyou Li

Chapter 12   **Frequency comb transferred by surface plasmon resonance**................................. **93**
Xiao Tao Geng, Byung Jae Chun, Ji Hoon Seo, Kwanyong Seo, Hana Yoon,
Dong-Eon Kim, Young-Jin Kim & Seungchul Kim

Chapter 13   **Arbitrary cross-section SEM-cathodoluminescence imaging of growth sectors and local carrier concentrations within micro-sampled semiconductor nanorods**.. **100**
Kentaro Watanabe, Takahiro Nagata, Seungjun Oh, Yutaka Wakayama,
Takashi Sekiguchi, János Volk & Yoshiaki Nakamura

Chapter 14   **Realization of mid-infrared graphene hyperbolic metamaterials**.................... **109**
You-Chia Chang, Che-Hung Liu, Chang-Hua Liu, Siyuan Zhang, Seth R. Marder,
Evgenii E. Narimanov, Zhaohui Zhong & Theodore B. Norris

Chapter 15   **Seamless growth of a supramolecular carpet**...**116**
Ju-Hyung Kim, Jean-Charles Ribierre, Yu Seok Yang, Chihaya Adachi,
Maki Kawai, Jaehoon Jung, Takanori Fukushima & Yousoo Kim

Chapter 16   **A light-driven three-dimensional plasmonic nanosystem that translates molecular motion into reversible chiroptical function**...................................**125**
Anton Kuzyk, Yangyang Yang, Xiaoyang Duan, Simon Stoll,
Alexander O. Govorov, Hiroshi Sugiyama, Masayuki Endo & Na Liu

Chapter 17   **Selectively enhanced photocurrent generation in twisted bilayer graphene with van Hove singularity**..**131**
Jianbo Yin, HuanWang, Han Peng, Zhenjun Tan, Lei Liao, Li Lin, Xiao Sun,
Ai Leen Koh, Yulin Chen, Hailin Peng & Zhongfan Liu

Chapter 18   **Creating single-atom Pt-ceria catalysts by surface step decoration**.............**138**
Filip Dvořák, Matteo Farnesi Camellone, Andrii Tovt, Nguyen-Dung Tran,
Fabio R. Negreiros, Mykhailo Vorokhta, Tomáš Skála, Iva Matolínová,
Josef Mysliveček, Vladimír Matolín & Stefano Fabris

Chapter 19   **Micro-total envelope system with silicon nanowire separator for safe carcinogenic chemistry**..**146**
Ajay K. Singh, Dong-Hyeon Ko, Niraj K. Vishwakarma, Seungwook Jang,
Kyoung-Ik Min & Dong-Pyo Kim

Chapter 20   **Switchable friction enabled by nanoscale self-assembly on grapheme**...........**153**
Patrick Gallagher, Menyoung Lee, Francois Amet, Petro Maksymovych,
Jun Wang, Shuopei Wang, Xiaobo Lu, Guangyu Zhang, Kenji Watanabe,
Takashi Taniguchi & David Goldhaber-Gordon

Chapter 21   **Topological phase transitions and chiral inelastic transport induced by the squeezing of light**.. 160
Vittorio Peano, Martin Houde, Christian Brendel, Florian Marquardt &
Aashish A. Clerk

Chapter 22   **Multiscale deformations lead to high toughness and circularly polarized emission in helical nacre-like fibres**.. 168
Jia Zhang, Wenchun Feng, Huangxi Zhang, Zhenlong Wang, Heather A. Calcaterra,
Bongjun Yeom, Ping An Hu & Nicholas A. Kotov

Chapter 23   **Three-dimensional structural dynamics and fluctuations of DNA-nanogold conjugates by individual-particle electron tomography**.................................. 177
Lei Zhang, Dongsheng Lei, Jessica M. Smith, Meng Zhang, Huimin Tong,
Xing Zhang, Zhuoyang Lu, Jiankang Liu, A. Paul Alivisatos & Gang Ren

Chapter 24   **A highly active and stable hydrogen evolution catalyst based on pyrite-structured cobalt phosphosulfide**... 187
Wen Liu, Enyuan Hu, Hong Jiang, Yingjie Xiang, Zhe Weng, Min Li, Qi Fan ,
Xiqian Yu, Eric I. Altman & Hailiang Wang

Chapter 25   **Multifunctional hydrogel nano-probes for atomic force microscopy**............................ 196
Jae Seol Lee, Jungki Song, Seong Oh Kim, Seokbeom Kim, Wooju Lee,
Joshua A. Jackman, Dongchoul Kim, Nam-Joon Cho & Jungchul Lee

Chapter 26   **Flexible single-layer ionic organic–inorganic frameworks towards precise nano-size separation**.. 210
Liang Yue, Shan Wang, Ding Zhou, Hao Zhang, Bao Li & Lixin Wu

Chapter 27   **Photoresponse of supramolecular self-assembled networks on graphene–diamond interfaces**.. 220
Sarah Wieghold, Juan Li, Patrick Simon, Maximilian Krause, Yuri Avlasevich,
Chen Li, Jose A. Garrido, Ueli Heiz, Paolo Samorí, Klaus Müllen, Friedrich Esch,
Johannes V. Barth & Carlos-Andres Palma

            Permissions

            List of Contributors

            Index

Preface

The world is advancing at a fast pace like never before. Therefore, the need is to keep up with the latest developments. This book was an idea that came to fruition when the specialists in the area realized the need to coordinate together and document essential themes in the subject. That's when I was requested to be the editor. Editing this book has been an honour as it brings together diverse authors researching on different streams of the field. The book collates essential materials contributed by veterans in the area which can be utilized by students and researchers alike.

Nanotechnology is an emerging field of study which has significantly contributed towards the growth of several other disciplines such engineering, medicine, etc. it is concerned with the manipulation of matter at a miniscule level like atomic or supramolecular level. There has been rapid progress in this field and its applications are finding their way across multiple industries. This book presents the complex subject of nanotechnology in the most comprehensible and easy to understand language. The various sub-fields of this discipline along with technological progress that have future implications are glanced at in this book. As this field is emerging at a fast pace, this book will help the readers to better understand the concepts of nanotechnology.

Each chapter is a sole-standing publication that reflects each author's interpretation. Thus, the book displays a multi-facetted picture of our current understanding of applications and diverse aspects of the field. I would like to thank the contributors of this book and my family for their endless support.

Editor

Ionic polarization-induced current–voltage hysteresis in $CH_3NH_3PbX_3$ perovskite solar cells

Simone Meloni[1,2], Thomas Moehl[3], Wolfgang Tress[3,4], Marius Franckevičius[3,5], Michael Saliba[4], Yong Hui Lee[4], Peng Gao[4], Mohammad Khaja Nazeeruddin[4], Shaik Mohammed Zakeeruddin[3], Ursula Rothlisberger[1,2] & Michael Graetzel[3]

$CH_3NH_3PbX_3$ (MAPbX$_3$) perovskites have attracted considerable attention as absorber materials for solar light harvesting, reaching solar to power conversion efficiencies above 20%. In spite of the rapid evolution of the efficiencies, the understanding of basic properties of these semiconductors is still ongoing. One phenomenon with so far unclear origin is the so-called hysteresis in the current–voltage characteristics of these solar cells. Here we investigate the origin of this phenomenon with a combined experimental and computational approach. Experimentally the activation energy for the hysteretic process is determined and compared with the computational results. First-principles simulations show that the timescale for MA^+ rotation excludes a MA-related ferroelectric effect as possible origin for the observed hysteresis. On the other hand, the computationally determined activation energies for halide ion (vacancy) migration are in excellent agreement with the experimentally determined values, suggesting that the migration of this species causes the observed hysteretic behaviour of these solar cells.

[1] Laboratoire de Chimie et Biochimie Computationnelles, ISIC, FSB-BCH, École Polytechnique Fédérale de Lausanne (EPFL), Lausanne CH-1015, Switzerland. [2] National Competence Center of Research (NCCR) MARVEL—Materials' Revolution: Computational Design and Discovery of Novel Materials, Lausanne CH-1015, Switzerland. [3] Laboratory of Photonics and Interfaces, ISIC, Swiss Federal Institute of Technology (EPFL), Lausanne CH-1015, Switzerland. [4] Group for Molecular Engineering of Functional Materials, ISIC-Valais, Swiss Federal Institute of Technology (EPFL), Lausanne CH-1015, Switzerland. [5] Center for Physical Sciences and Technology, Savanorių Avenue 231, Vilnius LT-02300, Lithuania. Correspondence and requests for materials should be addressed to T.M. (email: Thomas.moehl@epfl.ch) or to U.R. (email: Ursula.roethlisberger@epfl.ch).

The perovskite MAPbI$_3$ (methylammonium lead triiodide) and its halide analogues have emerged as one of the new and very interesting absorber materials for highly efficient solar cells[1-5]. Because of the ease of processing and wide range of applications (solar cells, photodetectors[6-10] and lasing[11-14]) they have attracted intense attention in the research community. Currently, the increase of efficiency of the solar cell devices is one of the main focuses[15,16], but the understanding of the basic operation principles of devices containing this material is also slowly evolving[17-19]. One of the main open questions is the origin of the observed hysteresis of the current-voltage (JV) curve of MAPbI$_3$-based solar cells, first reported by Dualeh et al.[18], and investigated in more detail by Snaith et al.[20], Tress et al.[17], O Regan et al.[21] and other authors[22-24]. This effect complicates the determination of the 'real' solar-to-electrical power conversion efficiency of such devices and can make 'bad cells look good' as presented in a recent publication by Christians et al.[25] Moreover, hypotheses were put forward suggesting a fundamental link between hysteresis and the limited long-term stability of halide perovskites[17]. Several ideas for the peculiar phenomenon underlying hysteresis have been proposed, the main advocated views involving either a ferroelectric effect[26-30] or ionic (vacancy) movement[17,31] inside the perovskite as probable cause.

Possible ferroelectric effects can be caused by the orientation of the organic (dipolar) cations, namely MA, or induced by deformation of the inorganic framework. In both cases the crystal acquires a net dipole moment. Under the effect of the external potential the MA molecules align with the external electric field, and the dipole moment of aligned MA molecules produces a balancing field lowering the effective field acting on the charge carriers. If the characteristic timescale necessary for the alignment is of the same order as the scanning time of the external potential, the effective potential acting on the charge carriers depends on the 'history' of the experiment, resulting in the hysteresis. Although several theoretical[27,30,32,33] and experimental[26,34-36] studies have been undertaken in this direction, a recent investigation of Fan et al.[35] showed that no room temperature ferroelectric behaviour could be measured.

An alternative hypothesis is that the hysteresis is due to the movement of ionic species[17,24,31]. It is well known from the literature that inorganic perovskites, like CsPbBr$_3$ or CsPbCl$_3$, are excellent halide conductors[37,38]. Moveable ions (or their vacancies) will induce a retarded reaction towards the change of electronic charge distribution of such a device under operation that could explain the so-called slow time component. Under the influence of an external biasing field or the 'built-in' potential of the device, the migration of ionic vacancies may result in a net charge accumulation in certain regions of the MAPbX$_3$ or at its contacts. The change of the concentration profile of the vacancies produces a balancing internal counterfield acting on the electronic charge carriers. Again, if the characteristic time of polarization of the sample is associated to the migration of ionic species is of the same order of magnitude as the potential scanning time, this phenomenon will result in a hysteresis of the JV curve.

Previous experimental studies[39,40] have estimated the phenomenon at the origin of the observed hysteresis to take place on a timescale of microseconds to seconds. The relatively long timescale of this phenomenon suggests that it is a thermally activated process characterized by an activation energy sizably higher than the thermal energy available to the system. Here we have used a combined experimental and theoretical approach to determine this activation energy, and to identify the causative process. Experimentally we have determined the activation energy of the hysteretic process from the temperature-dependent measurement of the JV curve for MAPbI$_3$ and MAPbBr$_3$. By density functional based simulations we have determined the activation energy of different ion (vacancy) migrations in the crystal lattice as well as the characteristic rotational time of the MA ions. This combined approach allowed us to establish the general nature of the phenomenon and its activation energy. Our results support the hypothesis that hysteresis is due to halide ion (vacancy) migration induced polarization of the perovskite layer and exclude a ferroelectric effect due to the alignment of the MA ions.

Results

Experimental determination of the activation energy. To determine the activation energy, E_a, of the process underlying the hysteretic behaviour, we have chosen a simple approach consisting in analysing the effect of the temperature on the JV curve of perovskite solar cell devices under illumination. A typical plot of different JV curves under illumination is presented in Fig. 1. The efficiencies of the devices with iodide were in the range of 10–14% PCE. Bromide devices had lower efficiencies, mainly because of the low J_{sc} (about 3–5 mA cm^{-2}), resulting in a power conversion efficiency (PCE) of 2–4%.

The general protocol for most of the measurements was (if no other conditions are indicated): (i) 50 s waiting at reverse potential (-0.5 V); (ii) scanning the JV curve with 50 mV s^{-1} to 1.1 V forward bias (in the following denoted as 'forward scan'); (iii) 50 s waiting at 1.1 V forward bias; and (iv) scanning the JV curve with 50 mV s^{-1} back to -0.5 V reverse bias ('reverse scan').

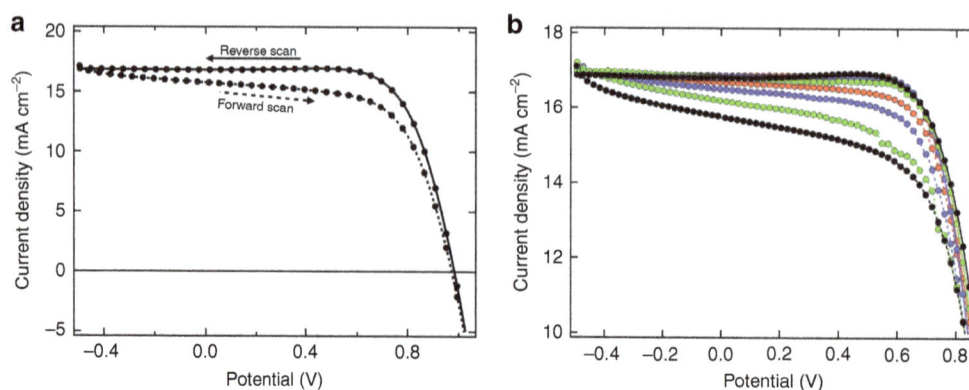

Figure 1 | JV curves of an iodide-based device. (a) At 1 sun at -15° C **(b)** at different temperatures (for a better comparison all curves are scaled to reach the same J_{sc} for the reverse scans). The voltage scan rate is 50 mV s^{-1} between -0.5 to 1.1 V. Between each forward and backward scan the potential was kept constant for 50 s to let the device equilibrate (please see text for further explanation). Red (20 °C); blue (5 °C); green (-5 °C); and black (-15 °C).

The measurements have been performed at several different temperatures (normally between 20 and $-20\,^{\circ}\mathrm{C}$). Figure 1b shows the dependence of the JV curve on the temperature. The measurements were started at $20\,^{\circ}\mathrm{C}$. After each measurement cycle, the JV curve of the device was re-measured at $20\,^{\circ}\mathrm{C}$ to ensure that the device did not degrade during the measurements. Several features can be observed in the JV curve in Fig. 1b. The reverse scans show minor dependence on the temperature exhibiting generally a similar shape. Furthermore, one can observe a 'bump' immediately before reaching a plateau of the current. This bump appeared for most of the devices though shape and change with temperature was not further investigated (please, see also text below regarding the phenomenon). The forward scans show a stronger dependence on the temperature. The slope of the overall current increases just as if the shunt resistance decreases with temperature. In other words, at room temperature the hysteresis between forward and reverse scan is small. Upon decreasing the temperature, the hysteresis between forward and reverse scan increases. To extract the E_a, we measured the hysteresis as the difference of the current at a given voltage along the backward and forward scan, $\Delta I = J_B(V) - J_F(V)$, and studied its dependence on the temperature.

The simultaneous presence of several types of charge carriers (electrons, holes and ionic defects) in the system, the possibility that polarization affects the carrier dynamics in the perovskite layer or at the interface with the contacts, altering absorption, transport and recombination properties, makes it impossible to derive an analytical expression of the dependence of ΔI as a function of the temperature. A numerical solution, on the other hand, requires the determination of the dependence of the generation rate, diffusion coefficient, recombination rate and surface recombination velocity on temperature and the polarization of the perovskite layer. Moreover, in view of simulating hysteretic behaviour as a function of the temperature and sweeping rate, the solution of the time-dependent transport-reaction problem is required, and not the simpler steady-state solution of the time-independent problem. All this renders also the numerical solution option for interpreting experimental results out of reach at the moment. Thus, similarly to other authors (see, for example, Eames et al.[24] and Yang et al.[41]), we have used an empirical relation between ΔI and T:

$$\frac{1}{\Delta I} = A \times e^{-\frac{E_a}{k_B T}} + C \qquad (1)$$

with A as prefactor, k_B as Boltzmann's constant, T as temperature and C as a constant. The reason for using the inverse of the difference between the backward and forward photocurrent is

that this current difference is reduced when the process generating the hysteresis relaxes more quickly to the stationary condition during the voltage scanning. In other words, ΔI is expected to have an inverse proportionality to the ionic current ($\Delta I \sim 1/\Delta I_{\mathrm{ionic}}$), which depends on the corresponding diffusion coefficient, typically having an Arrhenius-like dependence on the temperature. Intuitive arguments supporting the empirical relation above between ΔI and E_a will be further elaborated in the Discussion section. In the following, we show that this relation is, indeed, obeyed by the experimental data. In addition, we will use results of atomistic simulations to show that the experimentally determined activation energy is, indeed, associated to a microscopic diffusion process.

Figure 2a shows ΔI as a function of the potential for an iodide-based device at different temperatures. Generally ΔI increases at lower temperature. One can also observe the strong increase of ΔI in the high forward bias region originating from the already mentioned 'bump' often observed in the reverse scan of perovskite solar cells. In Fig. 2b $\ln(1/\Delta I)$ is plotted against $1/T$ for selected potentials in the range of -0.2 to $0.4\,\mathrm{V}$. By fitting these data with the Arrhenius-like equation, E_a is determined from the slope of $\ln(1/\Delta I)$ vs $1/T$ (similar plots for the bromide-based devices are shown in Supplementary Fig. 1).

The activation energy determined at high forward ($>0.4\,\mathrm{V}$) and reverse ($<-0.2\,\mathrm{V}$) bias range showed the highest deviation from the E_a values at potentials in the medium range. This has two-fold reasons. One is the already mentioned bump in the reverse scan, as also observed by Tress et al. The origin of this bump is, presumably, the initially low field in the device for the reverse scan and, therefore, the low driving force for any kind of process which then sets in towards lower forward voltages. Therefore only the E_a values determined in the bias range -200 to $400\,\mathrm{mV}$ have been used for the calculation of the averaged E_a of Table 1. The hysteresis effect of the devices with bromide is generally less pronounced, which also leads to a higher error in the estimation of the activation energy from experimental data (vide infra). As a result, some Br-based devices did not seem to show the expected dependence on the temperature, for example, showing very-low or negative activation energies.

The current density may also have an impact on the hysteretic behaviour. To compare the hysteresis at similar J_{sc} in the two halide systems we have measured the iodide-based devices also at low light intensities (0.1–0.2 sun, see Supplementary Fig. 2). We did not find any significant differences due to different light intensities in the E_a's shown in Fig. 3. In this figure, one can also clearly observe the lower activation energy for the devices with bromide. The different activation energies for the different

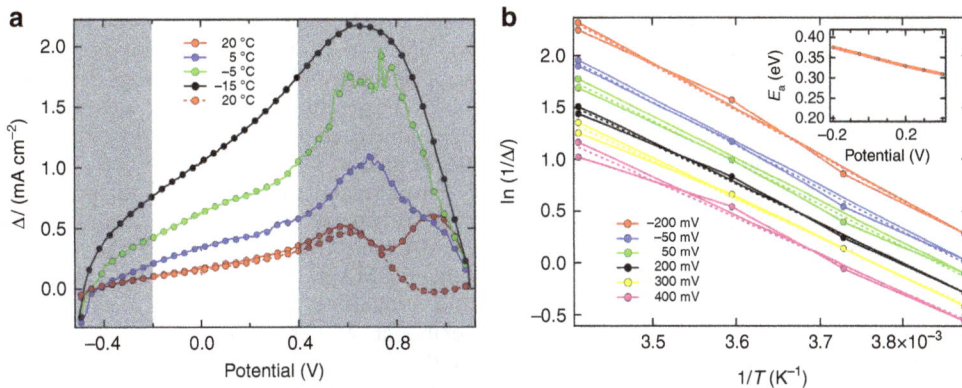

Figure 2 | ΔI and potential dependent activation energies. (a) The current difference between forward and reverse voltage scan and the potential window (0.2–0.4 V) used for the fitting by the Arrhenius-type equation. (**b**) Plot of $\ln(1/\Delta I)$ vs the inverse of the temperature for selected values of the potential (inset shows the determined activation energy independence of the potential).

Table 1 | Averaged activation energies.

Perovskite	Light intensity	E_a (meV)
MAPbI$_3$	High (~1 sun)	314 (±48)
	Low (0.1-0.2 sun)	341 (±42)
	All iodide devices	333 (±47)
MAPbBr$_3$	~1sun	168 (±43)

Determined average activation energies for iodide and bromide-based MAPbX$_3$ devices. Values averaged between −200 and 400 mV.

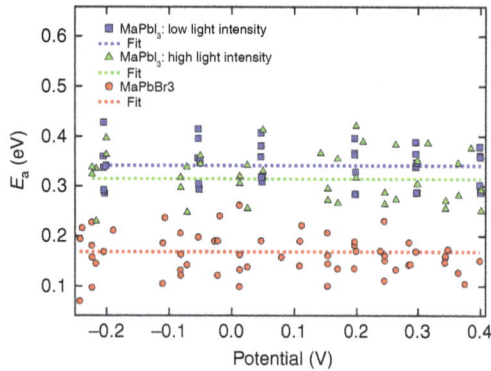

Figure 3 | Potential dependent activation energies of different samples. Collection of the activation energies at different potential for the different samples measured.

devices with iodide (high and low light intensities) and with bromide are given in the Supplementary Fig. 3.

ΔI might also depend on the scan velocity and on the scan bounds of the voltage interval next to the already mentioned light intensity. It is clear that hysteresis can only be observed if the scan rates are performed within times similar to the characteristic timescale of the underlying phenomenon. In fact, it was shown by Tress et al. that if the scan velocities are too fast ($\geq 100,000 \, \mathrm{mV \, s^{-1}}$) or too slow ($< 10 \, \mathrm{mV \, s^{-1}}$), no hysteresis is observed[17]. Therefore, we have tested different measurement conditions—different scan rates (50, 100 and 200 mV s^{-1}, see Supplementary Fig. 4a) and scan bounds (−0.5, −0.2 and 0 V, Supplementary Fig. 4b)—but could not observe any significant changes in E_a.

Noticing that the major difference between the forward and backward scan of the JV curves at different temperatures is the slope of the JV curve along the forward scan, we also computed the activation energy associated to this resistance-like term at 0 V. The barriers obtained from ΔI and the slope (in principle equal to an Arrhenius-type relation of $\ln(1/R)$ vs $1/T$) are consistent (see Supplementary Figs 1d, 2d and 5a). This justifies the approach of taking either ΔI or the slope to measure the activation energy of the slow process as hysteresis in this voltage range is mainly governed by a rate limited process that does not strongly depend on the actual voltage applied during the scan. Thus, the determined E_a directly reflects the activation energy of this slow process, which is reacting retarded to the change of the applied voltage (as will be explained in more detail in the discussion section and in the gedankenexperiments in the last part of the SI).

Extracting the activation energies for the different devices by the procedure(s) described yields a clear trend (see Supplementary Figs 3 and 6). The iodide-based devices show activation energies of in average 333 ± 47 meV, and the bromide-based devices of about 168 ± 43 meV. When using the

determination over the slope of the forward scan (at 0 V), the tendency is similar with E_a being 275 ± 19 meV for the iodide-based perovskite devices and 176 ± 43 meV for the bromide-based perovskite devices (see Supplementary Fig. 5b). As mentioned above, the relative error on the estimation of E_a is larger for the bromide devices but the tendency is clear. The E_a dependence on the halide rules out that E_a describes a temperature-activated transport in the contacting materials, which are the same for all devices. In addition, changed transport properties of the contacting materials should take effect as series resistance under high forward bias reducing the fill factor. The results reported above clearly indicate that the nature of the halide significantly affects the hysteresis via the barrier of the associated thermally activated process, which is lower for bromide than for iodide. This suggests that the process underlying hysteresis involves movements of halide ions (or their vacancies).

Simulations of the ferroelectric effect. Present experiments show that hysteresis is due to a 'thermally activated' process, with an associated barrier in the range of ~0.1–0.4 eV, depending on the type of perovskite used in the device and the conditions of the experiment. To identify what is the microscopic process causing hysteresis, in particular, whether it is due to ferroelectricity or ionic polarization, we performed two types of simulations. First, we performed ~30 ps long first-principles (on the basis of density functional theory) molecular dynamics simulations (MD) at various temperatures ($T = 100, 200, 300$ and 400 K) starting from the tetragonal MAPbI$_3$ crystal phase. The computational sample consisted of a $2 \times 2 \times 2$ supercell of the simple tetragonal analogue of the experimental body-centered tetragonal crystal[42], containing 32 stoichiometric units (384 atoms).

The system equilibrated at different temperatures is able to assume different crystalline phases with a trend consistent with experimental results[42,43]. At 100 and 200 K, the atoms arrange in an orthorhombic-like phase with non-negligible values of all the three tilting angles. At 300 K, two of the three tilting angles are reduced and the structure becomes tetragonal-like. Finally at 400 K, all the three tilting angles are approximately 0° and the structure is cubic-like. More details on the temperature-dependent simulations are also provided in the Supplementary Notes 1 (Supplementary Figs 7–12).

Albeit the relatively short simulation times of 30 ps, the computational results suggest that a ferroelectric origin of hysteresis in unlikely. Hysteresis induced by ferroelectricity might be due to either a break of symmetry of the PbI$_3$ lattice, perhaps induced or enhanced by the lack of inversion symmetry in the crystal due to the MA cation, or by the alignment of the polar MA molecules[27,44]. A break of symmetry in the PbI$_3$ lattice should result in a histogram of the Pb–I bond distances (< 3.2 Å) with multiple maxima. The computed distribution ($g_{PbI}(r)$, Supplementary Fig. 7) shows no evidence of this feature at any of the temperatures investigated, suggesting that there is no break of symmetry in the PbI$_3$ framework whatever the crystal phase of the sample is.

Furthermore, we investigated the possibility that ferroelectricity originates from a persistent preferential alignment of MA ions. Under the effect of the external bias plus built-in potential MA molecules might align with the overall electric field and produce a counterfield. To determine the characteristic rotational reorientation time, the time correlation function of the C–N unit vector, $\langle d(t) \cdot d(0) \rangle$ ($d = r_N - r_C / |r_N - r_C|$), is computed, and it is fitted with a double exponential decay, $\langle d(t)d(0) \rangle = a_1 e^{-t/\tau_1} + a_2 e^{-t/\tau_2}$. The time for MA molecules to loose memory of their initial orientation, τ_1, is on the picoseconds timescale at 200–400 K (see Fig. 4). At 100 K the reorientation time is much longer, probably because of the fact that complete rotation of MA ions is hindered at this temperature. These results are consistent with previous first-principles calculation using a different, more qualitative, approach to determine the characteristic reorientation time[39,45], recent classical MD simulations[46] and experimental data[47–50]. The consistency with classical MD results on bigger samples suggest that no relevant finite size effects affect the estimated reorientation times. From the dependence of τ_1 on the temperature in the range 200–400 K, we estimated an activation energy for MA reorientation of 0.042 eV (inset of Fig. 4), consistent with experimental results[50].

If the dynamics of the MA molecules is uncorrelated, that is, if they rotate independently from each other, the characteristic orientational correlation time determined in the simulations is also the time the sample takes to polarize under the action of a bias. This leads to the conclusion that polarization due to dipole alignment takes place on the picosecond timescale. This is too short a time to account for the JV hysteresis, that is associated to a process with a characteristic time in the milliseconds-to-seconds range[39,51]. To estimate the amount of correlation between MA molecules, we computed the spatial correlation function, $\langle d_i \cdot d_j \rangle$ with d_i and d_j unit C-N vectors of molecules i and j (Supplementary Fig. 12). The spatial correlation function is 1 if two MA ions have the same orientation during the simulation, −1 if they have opposite orientation and 0 if the orientation of one is independent of that of the others. MD simulations results show that at room temperature the correlation between MA ions is small. This, indeed, is consistent with the fact that all the MA ions undergo a quick decorrelation, as shown by the time autocorrelation curves of individual molecules, $\langle d_i(t) \cdot d_i(0) \rangle$ (see Supplementary Fig. 10). Thus, we expect that the alignment of the sample takes place on the timescale of the rotational reorientation time of a single MA ion, that is, on the picosecond timescale estimated above. Further information about these simulations are available in the SI (Supplementary Figs 9, 10 and 11).

Recent Monte Carlo simulations[32] on a lattice model of MAPbI3 have estimated the alignment under the action of an electric field to take place in 10^4 spin steps. Making an accurate estimation of the reorientation time from Monte Carlo steps is not simple, as there is no one-to-one correspondence between a Monte Carlo step per MA molecule and the time MA ions would take to cover the corresponding reorientation. However, an approximate upper limit estimation of the polarization time of

the sample can be obtained by assuming that a global Monte Carlo step, that is, one step for each molecule in the sample, costs a time corresponding to the rotational reorientation time, τ_1. This would yield an alignment time of ~ 50 ns (10^4 steps $\times \sim 5$ ps), still too short for MA-reorientation-related ferroelectricity to be responsible for hysteresis.

Simulations of the ionic migration. To probe the second hypothesis, that is, that a bias-induced stepwise migration of ions might result in a change of their local concentration and then in a balancing internal counterfield, we performed MD simulations on systems containing a single MA^+, Pb^{2+} or I^- vacancy, respectively. The reason for focusing on vacancies is that previous experiments have shown that ionic transport in Br and Cl perovskites is most likely assisted by this type of defect.

During the 10 ps of first-principles MD, no spontaneous ionic vacancy migration was observed. This is not surprising as it is known that ionic/vacancy migration in MAPbX3-related materials is slower than the timescale of our simulations[37,52,53]. The longer timescale of ionic migration processes makes this second phenomenon a more plausible candidate as source of the observed JV hysteresis. To assess this, we performed string simulations[54] aimed at computing the vacancy-driven ionic migration path and the associated energy barrier, E_a, for MAPbI3 and MAPbBr3. In the case of MAPbBr3, for which the stable phase at room temperature is cubic, we considered both tetragonal and cubic structures. This allows to distinguish between the effect of halide substitution and phase change on the migration barriers. In the case of cubic MAPbBr3 we observed no qualitative changes with respect to the tetragonal case, and the activations energies typically are between the values estimated for the corresponding tetragonal system.

Moreover, to estimate the possible effect on E_a arising from the arbitrary choice of the initial orientation of MA, with its high orientational mobility, we also investigate two systems containing the spherically isotropic Cs^+ monovalent cation, namely CsPbI3 and CsPbBr3. For these systems we considered the tetragonal phase. Because of the crystal symmetry, MA^+/Cs^+ and Pb^{2+} ionic migration can take place either along or perpendicularly to the tetragonal axis, and the barriers of these two processes can differ. Here we consider both cases: axial (a) and equatorial (e) migration (see Fig. 5). It is also important to remark that tetragonal perovskite crystals contain two non-equivalent I/Br sites. The first, denominated axial in the following, is the one in which the halide ion forms the Pb–X–Pb triplex oriented along the tetragonal axis (see Fig. 5). The second, the equatorial site, is the one in which the halide forms Pb–X–Pb laying on the plane orthogonal to the tetragonal axis. Halide migration can take place from an equatorial to another equatorial site, e2e, or from an equatorial to an axial site, e2a (or vice versa, a2e). Stroboscopic images of the migration processes mentioned above for the case of MAPbI3 are shown in Fig. 6a–f.

The migration barriers for all the processes and systems mentioned above are reported in Table 2. The comparison of the migration paths of the X^-, Pb^{2+} and A^+ ions easily explains the difference in the migration energy of the various species. Figure 6a,b shows that a very-small distortion of the crystal structure accompanies halide migration, while migration of A^+ (Fig. 6c,d) and Pb^{2+} (Fig. 6e,f) requires a significant even though local rearrangement of the crystal structure.

The migration of halide ions essentially affects only the PbX6 units involved in the event, with negligible distortions of the rest of the lattice. The migration barrier does not change significantly replacing MA^+ with Cs^+. In particular, the trend of the migration barrier with the chemical nature of the halide is

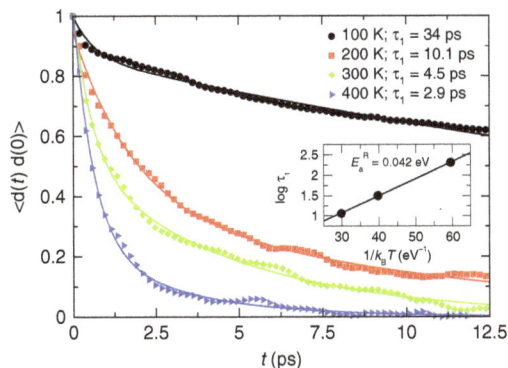

Figure 4 | Time-dependent autocorrelation function of the unit C–N vectors. The autocorrelation function has been computed averaging over the microcanonical ensemble sampled by constant number of particles, volume and energy (NVE) first-principles MD simulations performed at four different energies corresponding to average temperatures of 100, 200, 300 and 400 K. In the inset log τ_1 vs $1/_{k_B T}$ is shown, together with the linear fitting from which the rotational reorientation activation energy is obtained.

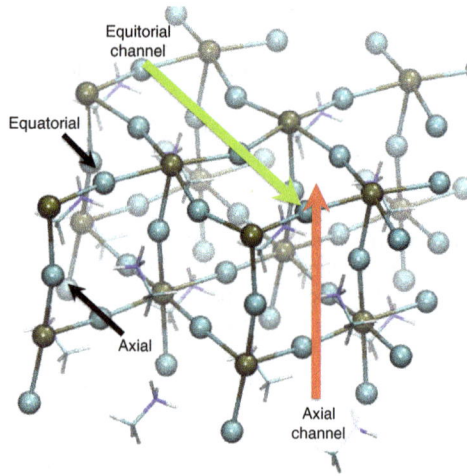

Figure 5 | Propagation channels. Periodic crystalline (defect free) MAPbI$_3$ sample. Cyan and brown spheres represent I and Pb ions, respectively. MA is shown as sticks. Periodic boundary conditions are applied in all the crystallographic directions. Black arrows point to equatorial and axial I$^-$ ions. Red and green arrows indicate the axial and equatorial channels for MA$^+$ and Pb^{2+} ion migration.

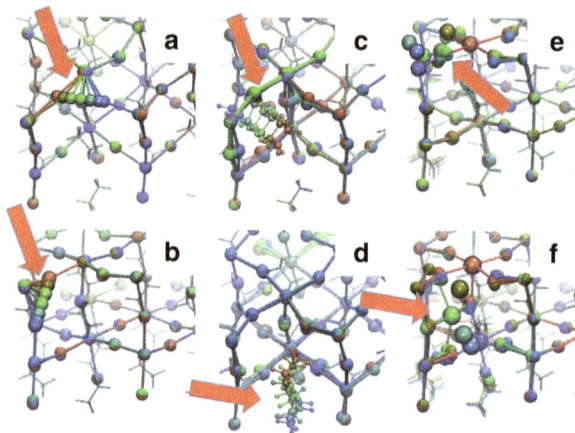

Figure 6 | Stroboscopic images of the ionic migration paths in MAPbI$_3$. (a–f) Paths for the corresponding migration events in MAPbBr$_3$ are analogous. The colours of the frames go from blue (initial states) to green (intermediate states) to red (final states). (a,b) Migration path of I$^-$ along the equatorial-to-equatorial and equatorial-to-axial channels, respectively. (c,d) Migration paths of MA$^+$ along the equatorial and axial channels, respectively. (e,f) Migration paths of Pb^{2+} along the axial and equatorial channels. Both paths are rather complicated. The details are described in the main text.

confirmed. This suggests that our results for MAPbX$_3$ samples are not significantly biased by the arbitrary choice of the initial orientation of the MA ions. The axial vacancy is significantly less stable than the equatorial one ($\Delta E \sim 0.1$–0.2 eV), and a vacancy in the axial state is expected to migrate towards an equatorial one. Thus, a complete migration event along the axial channel, bringing the vacancy from an equatorial site to another, requires two steps: an e2a migration followed by an a2e one. Thus, this channel requires the crossing of two barriers. In the case of MA-perovskites, the first barrier, e2a (450–460 meV), is sizably higher than the single barrier of the equatorial channel, e2e (200–280 meV), and the one-step e2e migration channel will give the major contribution to the halide transport in perovskites. It is, then, the activation energy of this channel determining the halide transport in perovskites.

The migration of the monovalent cation requires the opening of the PbX$_3$ framework separating the initial and final A$^+$ sites. The shape of this framework is different in the axial and equatorial directions. Thus, the activation energy and the migration path, including the orientation of the MA ions along it, depend on the channel, whether equatorial or axial. Given the high orientation disorder of MA$^+$, it is expected that under experimental conditions this ion migrates following paths characterized by different orientations of the C–N bond and, possibly, different migration barriers. We expect that the migration barrier of Cs$^+$ represents a lower bound for that of MA$^+$, consistent with the fact that Cs$^+$ is smaller than MA$^+$ (as suggested by the larger size of the lattice of MA-perovskites with respect to Cs-perovskites computed in simulations).

Also the migration of Pb^{2+} requires a significant distortion of the crystal structure. This is necessary to let the Pb^{2+} ion leave its original site and enter into the new one. This explains the high barrier for the migration of Pb^{2+}.

Walsh et al.[55] have shown that the defects in MAPbI$_3$ are formed according to the Schottky mechanism, in which the amount of iodide and cation defect concentrations have the same order of magnitude. In particular, according to Walsh et al. the most probable point defect formation reaction is

$$\text{nil} \rightarrow V'_{MA} + V^\circ_I + \text{MAI} \qquad (2)$$

with an associated formation enthalpy of 0.08 eV. These results, taken together with the calculated activation barriers for defect migration, show that not only V°_I is the most mobile defect, but it is also present in similar concentrations with the other defects considered in the present work, and that, indeed, the ionic mobility is the limiting step in the vacancy-driven polarization of the perovskite sample. Thus, considering the migration barriers of all the ionic species shown in Table 2, at room temperature the timescale of the migration of the MA$^+$ and Pb^{2+} ions is $\sim e^{(\Delta E^*_{A/Pb} - \Delta E^*_X)/k_B T} \geq e^{10}$ larger than the one for X$^-$ ions. This is consistent with the experiments by Yang et al.[41] with solid-state electrochemical cells of MAPbI$_3$, in which it has been shown that the active mobile ion is iodide.

Indeed, the I$^-$ and Br$^-$ computational migration barriers along the e2e channel are in very good agreement with the experimental activation energies. In addition, the effect of halide substitution follows the experimental trend discussed in the previous section, with a lower barrier for halide migration in Br than in I perovskites. The good match between experimental and computational results, both in the absolute value of the migration barriers and in the trend with halogen substitution, together with experimental results reported previously[17,39], strongly support the hypothesis that hysteresis is due to the polarization of the perovskite layer associated to halide-vacancy migration. As proposed in refs 17,39, halide ions/vacancies migrate in the same direction as the corresponding charge carriers. Since ions are not extracted at the contacts, they accumulate at the interface of the electrodes producing a balancing potential that reduces the efficiency of collection of charge carriers. In extreme cases, and in the absence of the compact-TiO$_2$ hole-blocking layer, the internal balancing potential can significantly reduce the V_{oc} (ref. 17).

Discussion

Before discussing experimental and computational results, it is worth giving the intuitive arguments at the basis of the empirical relation between ΔI and T (Eq. (1)). A first observation is that polarization of the sample due to ionic movements, or to the alignment of MA molecules, produces a counterfield opposing to

Table 2 | Activation barriers for ionic migration.

	E_a (meV)	
	$A^+ = Cs^+$	$A^+ = MA^+$
APbI3		
Vacancy I$^-$		
e2e	360	**280**
e2a	290 (170)	450 (130)
Vacancy A$^+$		
e	590	1,120 (880)
a	1,160	700 (600)
Vacancy Pb^{2+}		
e	810	1,390
a	990	1,780

		Tetragonal	Cubic
APbBr3			
Vacancy Br$^-$			
e2e	270	**200**	220
e2a	290 (160)	460 (90)	
Vacancy A$^+$			
e	700	1,130 (950)	1,010
a	1,200	800 (700)	
Vacancy Pb^{2+}			
e	940	1,350	1,620
a	1,220	1,800	

In the tetragonal phases, there are different possible migration channels. For the e2a case of halide migration, where the initial and final states have a different energy, the barriers of the inverse process, a2e, are reported in parentheses. The e2e halide migration channel (reported in bold in the table) is expected to be the most efficient ionic transport path, and it is the corresponding activation energy that must be compared with the experimental values (168 meV for MAPbBr$_3$ and 333 meV for MAPbI$_3$). The initial and final states of the MA$^+$ migration are non-equivalent, due to non-equivalent MA orientation in the initial and final states. Also in this case we report in parentheses the activation energy of the reverse process.

the external bias plus built-in voltage. This results in a reduction of the measured current. A possible explanation is that reducing the overall field for extracting charge carriers increases electron–hole recombination. A second observation is that in this work we focus on a temperature range, in which hysteresis shrinks with the temperature (see below). In this case, the faster is the relaxation to the stationary condition from the preset state along the voltage scanning, the lower is the hysteresis. Thus, we assume that hysteresis is in inverse proportion to the relaxation rate. The relaxation rate is related to the diffusion coefficient, in the case of ionic polarization, or the reorientation time, in the case of ferroelectric polarization. In both cases, the dependence on the temperature is via an Arrhenius-like relation characterized by an activation energy. These arguments are summarized in Eq. (1). The values determined for the activation energy in our experiments are in close agreement with the values reported by several other authors using different experimental approaches[24,37,38,41,56], supporting the validity of such an empirical $\Delta I - E_a$ relation.

The comparison between experimental and theoretical results suggests that hysteresis is due to ionic movements rather than ferroelectricity. Here we propose a possible mechanism explaining the experimental observations based on this hypothesis. The effect of ionic displacement is the polarization of the perovskite layer at the contacts, which eventually changes the characteristic properties of the device. A similar effect has been exploited in the field of piezophototronics[57,58]. It has been shown that the accumulation of cations or anions at the interface of, for example, p–n junctions can significantly alter the band bending and can also affect the conduction and valence band edge positions of the involved semiconductors (or work functions of the contacting materials). In perovskite-based solar cells, mobile ions can

similarly accumulate at the interface with the contacting materials under the action of the electric field generated by the space charge region at the contacts. The effect is minimized close to V_{oc}, where the built-in potential is approximately balanced by the external bias and a minimum force is acting on the ions. Thus we can assume that at V_{oc} ions are distributed almost uniformly in the perovskite layer. In contrast, when the external bias is zero (and the internal field is high), ions accumulate at the contact(s). The switching from the polarized to unpolarized state is not instantaneous, and this results into the hysteresis.

However, considering a broader temperature range, the dependence of the hysteresis with temperature is more complex: In the temperature regime below $\sim 180\,K$, an increase in the hysteresis can be observed as shown by Zhang et al.[59] Since the perovskite is in its ferroelectric crystal phase, with the J_{sc} being very low and the overall solar cell efficiency going under 0.1% the reasons for the hysteretic behaviour are more complex and its interpretation goes beyond the scope of this manuscript. At 180 K, nearly no hysteresis is visible in the JV curve. At this temperature the ions do not move significantly on the timescale of the scanning plus dwell time (~ 60–80 s, depending on the scanning rate). In absence of polarization, hysteresis cannot be observed as the state of the system at the given voltage is independent of scan direction. Increasing the temperature increases the mobility of ions and, thus, the polarization during the dwell time at the starting point of the voltage scanning. During the voltage scan, which lasts for maximum $\sim 30\,s$ in the case of the slowest scan rate (50 mV s^{-1}), the system cannot reach the stationary condition at each voltage and hysteresis increases with T (in the temperature regime with $180\,K < T < \sim 240\,K$). There exists a temperature, $T > \sim 240\,K$, at which the system is able to approach a stable polarization during the dwell time. Thus, any further increase of the temperature does not significantly increase the initial polarization before the scan. However, mobility keeps increasing with the temperature and the system can more closely approach the voltage-dependent stationary ionic distribution during the scan. This results in a reduction of the hysteresis. This behaviour is shown in Supplementary Figs 1a and 2a, which show a slight decrease of the measured current with increasing temperature in the backward scan, and a complementary increase in the forward scan. Summarizing, a complex dependence of the hysteresis on temperature is observed, and this is due to the interplay of the effect of the increased mobility of the ions on the polarization during the dwell time at the starting points of the scanning, and the relaxation during the scanning. The effect of the temperature between 80 and 360 K has been investigated in more detail by Zhang et al.[59]

The outlined model leading to the observed hysteresis can explain the characteristics of perovskites solar cells measured in this and other recent works. Tress et al.[17] have observed a shift in the forward and backward JV curves in the dark by changing the pre-conditioning potentials. This device was made without a TiO$_2$ blocking layer. In this case, the charge polarization can act as a term balancing the built-in potential and/or changing the work function of the fluorine-doped tin oxide (FTO) due to the different ionic distribution at this interface, and, thus, shifting the dark current curves. Almora et al.[60] have reported JV curves in the dark presenting a capacitive loop. The (retarded) variation of ionic polarization during the voltage scans can result into a change in the extension or charging of the space charge zone, which is equivalent to a varying capacitance along the backward and forward scan. Our experiments and calculations, as well as the above mentioned experiments, interpreted in light of the model described here, suggest that the JV characteristics of perovskite solar cells under illumination are strongly influenced through a modulation of the current by ionic polarization. We have made two gedankenexperiments that are presented in the Supplementary Notes 2 highlighting in more detail the thoughts we have presented in this paragraph (see also Supplementary Fig. 13).

The possibility that the width of the space charge region, edge positions of the semiconductor bands (which also includes the contacting semiconductors, for example, the TiO_2) or the work function of the conductive substrate (for example, FTO as shown by Tress et al.) are altered by changes of the ionic environment, leads to a complicated interplay of the different materials. We want to stress that the qualitative model outlined above and in the SI requires additional investigation to identify the detailed effect of ion accumulation on band bending and band edge position, and ultimately on charge separation and collection.

In summary, here we have shown that the hysteresis observed in JV curves of different $MAPbX_3$ perovskite solar cells is due to a thermally activated process and we have determined the associated activation energy. The variation in the processing parameters for the preparation of the perovskite layer, as well as different measurement conditions (JV-scan rate and value of the potential at which hysteresis is measured) did not significantly affect the determined values for the activation energy, demonstrating the independence of the origin of the hysteretic effect on the processing and measurement conditions. Consistently, the activation energy for the bromide- and iodide-based devices showed an average of 168 ± 43 and 333 ± 47 meV, respectively.

We paralleled the experimental investigation with first-principles MD simulations to determine the characteristic time for (re-)orienting MA molecules. We could show that this characteristic time, in the picosecond range, is too short for the process being associated with JV hysteresis. Furthermore, we determined the activation energy of the migration of vacancies of the various ionic species forming the perovskite and show that the lowest activation energy for vacancy migration is the one for halides. Its values matches well with the experimentally determined activation barriers involved in hysteresis. The dependence of this barrier on the type of halide computed in the simulations is also in agreement with the experimental trend. Present experimental and computational results strongly support the hypothesis put forward in ref. 17 that hysteresis is due to polarization of ionic charges in the perovskite layer under the influence of the built-in and applied potential. The mobility of the other possible ionic species (MA^+ and Pb^{2+}) than the halides is much lower, and we do not expect them to give any significant contribution in the polarization of devices in experiments with a scanning rate in the range of 10 to 10,000 mV s^{-1}.

Methods

General methodological information. Refer to the Supplementary Information for experimental section, figures on E_a for different scan velocity and scan bounds, different preparation methods and light intensities. Additional computational details and results, and further discussion of the outlined model is available free of charge via the Internet at http://pubs.acs.org.

References

1. Kojima, A., Teshima, K., Shirai, Y. & Miyasaka, T. Organometal halide perovskites as visible-light sensitizers for photovoltaic cells. J. Am. Chem. Soc. **131**, 6050 (2009).
2. Im, J. H., Lee, C. R., Lee, J. W., Park, S. W. & Park, N. G. 6.5% Efficient perovskite quantum-dot-sensitized solar cell. Nanoscale **3**, 4088–4093 (2011).
3. Kim, H. S. et al. Lead iodide perovskite sensitized all-solid-state submicron thin film mesoscopic solar cell with efficiency exceeding 9%. Sci. Rep. **2**, 591 (2012).
4. Burschka, J. et al. Sequential deposition as a route to high-performance perovskite-sensitized solar cells. Nature **499**, 316 (2013).
5. Lee, M. M., Teuscher, J., Miyasaka, T., Murakami, T. N. & Snaith, H. J. Efficient hybrid solar cells based on meso-superstructured organometal halide perovskites. Science **338**, 643–647 (2012).
6. Moehl, T. et al. Strong photocurrent amplification in perovskite solar cells with a porous TiO_2 blocking layer under reverse bias. J. Phys. Chem. Lett. **5**, 3931–3936 (2014).
7. Lee, Y. et al. High-performance perovskite-graphene hybrid photodetector. Adv. Mater. **27**, 41–46 (2015).
8. Hu, X. et al. High-performance flexible broadband photodetector based on organolead halide perovskite. Adv. Funct. Mater. **24**, 7373–7380 (2014).
9. Dou, L. T. et al. Solution-processed hybrid perovskite photodetectors with high detectivity. Nat. Commun. **5**, 5404 (2014).
10. Xia, H. R., Li, J., Sun, W. T. & Peng, L. M. Organohalide lead perovskite based photodetectors with much enhanced performance. Chem. Commun. **50**, 13695–13697 (2014).
11. Deschler, F. et al. High photoluminescence efficiency and optically pumped lasing in solution-processed mixed halide perovskite semiconductors. J. Phys. Chem. Lett. **5**, 1421–1426 (2014).
12. Dhanker, R. et al. Random lasing in organo-lead halide perovskite microcrystal networks. Appl. Phys. Lett. **105**, 151112 (2014).
13. Sutherland, B. R., Hoogland, S., Adachi, M. M., Wong, C. T. O. & Sargent, E. H. Conformal organohalide perovskites enable lasing on spherical resonators. ACS Nano **8**, 10947–10952 (2014).
14. Xing, G. C. et al. Low-temperature solution-processed wavelength-tunable perovskites for lasing. Nat. Mater. **13**, 476–480 (2014).
15. Jeon, N. J. et al. Compositional engineering of perovskite materials for high-performance solar cells. Nature **517**, 476–480 (2015).
16. Jeon, N. J. et al. Solvent engineering for high-performance inorganic–organic hybrid perovskite solar cells. Nat. Mater. **13**, 897–903 (2014).
17. Tress, W. et al. Understanding the rate-dependent J-V hysteresis, slow time component, and aging in $CH_3NH_3PbI_3$ perovskite solar cells: the role of a compensated electric field. Energy Environ. Sci. **8**, 995–1004 (2015).
18. Dualeh, A. et al. Impedance spectroscopic analysis of lead iodide perovskite-sensitized solid-state solar cells. ACS Nano **8**, 362–373 (2014).
19. Gonzalez-Pedro, V. et al. General working principles of $CH_3NH_3PbX_3$ perovskite solar cells. Nano Lett. **14**, 888–893 (2014).
20. Snaith, H. J. et al. Anomalous hysteresis in Perovskite solar cells. J. Phys. Chem. Lett. **5**, 1511–1515 (2014).
21. O'Regan, B. C. et al. Opto-electronic studies of methylammonium lead iodide perovskite solar cells with mesoporous TiO_2; separation of electronic and chemical charge storage, understanding two recombination lifetimes, and the evolution of band offsets during JV hysteresis. J. Am. Chem. Soc. **137**, 5087–5099 (2015).
22. Azpiroz, J. M., Mosconi, E., Bisquert, J. & De Angelis, F. Defect migration in methylammonium lead iodide and its role in perovskite solar cell operation. Energy Environ. Sci. **8**, 2118–2127 (2015).
23. Haruyama, J., Sodeyama, K., Han, L. & Tateyama, Y. First-principles study of ion diffusion in perovskite solar cell sensitizers. J. Am. Chem. Soc. **137**, 10048–10051 (2015).
24. Eames, C. et al. Ionic transport in hybrid lead iodide perovskite solar cells. Nat. Commun. **6**, 7497 (2015).
25. Christians, J. A., Manser, J. S. & Kamat, P. V. Best practices in perovskite solar cell efficiency measurements. avoiding the error of making bad cells look good. J. Phys. Chem. Lett. **6**, 852–857 (2015).
26. Wei, J. et al. Hysteresis analysis based on the ferroelectric effect in hybrid perovskite solar cells. J. Phys. Chem. Lett. **5**, 3937–3945 (2014).
27. Frost, J. M. et al. Atomistic Origins of high-performance in hybrid halide perovskite solar cells. Nano Lett. **14**, 2584–2590 (2014).
28. Yuan, Y. B., Xiao, Z. G., Yang, B. & Huang, J. S. Arising applications of ferroelectric materials in photovoltaic devices. J. Mater. Chem. A **2**, 6027–6041 (2014).
29. Juarez-Perez, E. J. et al. Photoinduced giant dielectric constant in lead halide perovskite solar cells. J. Phys. Chem. Lett. **5**, 2390–2394 (2014).
30. Stroppa, A. et al. Tunable ferroelectric polarization and its interplay with spin-orbit coupling in tin iodide perovskites. Nat. Commun. **5**, 5900 (2014).
31. Xiao, Z. et al. Giant switchable photovoltaic effect in organometal trihalide perovskite devices. Nat. Mater. **14**, 193–198 (2015).
32. Frost, J. M., Butler, K. T. & Walsh, A. Molecular ferroelectric contributions to anomalous hysteresis in hybrid perovskite solar cells. Appl. Mater. **2**, 081506 (2014).
33. Liu, S. et al. Ferroelectric domain wall induced band gap reduction and charge separation in organometal halide perovskites. J. Phys. Chem. Lett. **6**, 693–699 (2015).
34. Chen, H. W., Sakai, N., Ikegami, M. & Miyasaka, T. Emergence of hysteresis and transient ferroelectric response in organo-lead halide perovskite solar cells. J. Phys. Chem. Lett. **6**, 164–169 (2015).
35. Fan, Z. et al. Ferroelectricity of $CH_3NH_3PbI_3$ Perovskite. J. Phys. Chem. Lett. **6**, 1155–1161 (2015).

36. Kutes, Y. *et al.* Direct observation of ferroelectric domains in solution-processed $CH_3NH_3PbI_3$ perovskite thin films. *J. Phys. Chem. Lett.* **5**, 3335–3339 (2014).

37. Mizusaki, J., Arai, K. & Fueki, K. Ionic conduction of the perovskite-type halides. *Solid State Ionics* **11**, 203–211 (1983).

38. Kuku, T. A. Ionic transport and galvanic cell discharge characteristics of CuPbI3 thin films. *Thin Solid Films* **325**, 246–250 (1998).

39. Mosconi, E., Quarti, C., Ivanovska, T., Ruani, G. & De Angelis, F. Structural and electronic properties of organo-halide lead perovskites: a combined IR-spectroscopy and ab initio molecular dynamics investigation. *Phys. Chem. Chem. Phys.* **16**, 16137–16144 (2014).

40. Zhao, Y. *et al.* Anomalously large interface charge in polarity-switchable photovoltaic devices: an indication of mobile ions in organic-inorganic halide perovskites. *Energy Environ Sci.* **8**, 1256–1260 (2015).

41. Yang, T.-Y., Gregori, G., Pellet, N., Grätzel, M. & Maier, J. The significance of ion conduction in a hybrid organic–inorganic lead-iodide-based perovskite photosensitizer. *Angew Chem. Int. Ed. Engl.* **54**, 7905–7910 (2015).

42. Stoumpos, C. C., Malliakas, C. D. & Kanatzidis, M. G. Semiconducting tin and lead iodide perovskites with organic cations: phase transitions, high mobilities, and near-infrared photoluminescent properties. *Inorg. Chem.* **52**, 9019–9038 (2013).

43. Trots, D. M. & Myagkota, S. V. High-temperature structural evolution of caesium and rubidium triiodoplumbates. *J. Phys. Chem. Solids* **69**, 2520–2526 (2008).

44. Frost, J. M., Butler, K. T. & Walsh, A. Molecular ferroelectric contributions to anomalous hysteresis in hybrid perovskite solar cells. *Appl. Mater.* **2**, 081506 (2014).

45. Carignano, M. A., Kachmar, A. & Hutter, J. Thermal effects on $CH_3NH_3PbI_3$ perovskite from *ab initio* molecular dynamics simulations. *J. Phys. Chem. C* **119**, 8991–8997 (2015).

46. Mattoni, A., Filippetti, A., Saba, M. I. & Delugas, P. Methylammonium rotational dynamics in lead halide perovskite by classical molecular dynamics: the role of temperature. *J. Phys. Chem. C* **119**, 17421–17428 (2015).

47. Poglitsch, A. & Weber, D. Dynamic disorder in methylammoniumtrihalogenoplumbates(Ii) observed by millimeter-wave spectroscopy. *J. Chem. Phys.* **87**, 6373–6378 (1987).

48. Jamnik, J. & Maier, J. Treatment of the impedance of mixed conductors—equivalent circuit model and explicit approximate solutions. *J. Electrochem. Soc.* **146**, 4183–4188 (1999).

49. Leguy, A. M. A. *et al.* The dynamics of methylammonium ions in hybrid organic-inorganic perovskite solar cells. *Nat. Commun.* **6**, 7124 (2015).

50. Chen, T. *et al.* Rotational dynamics and its relation to the photovoltaic effect of $CH_3NH_3PbI_3$ perovskite. Preprint at http://arXiv:150602205 (2015).

51. Zhao, Y. *et al.* Anomalously large interface charge in polarity-switchable photovoltaic devices: an indication of mobile ions in organic-inorganic halide perovskites. *Energy Environ. Sci.* **8**, 1049 (2015).

52. Hoshino, H., Yokose, S. & Shimoji, M. Ionic conductivity of lead bromide crystals. *J. Solid State Chem.* **7**, 1–6 (1973).

53. Schoonman, J. The ionic conductivity of pure and doped lead bromide single crystals. *J. Solid State Chem.* **4**, 466–474 (1972).

54. E, W., Ren, W. & Vanden-Eijnden, E. Simplified and improved string method for computing the minimum energy paths in barrier-crossing events. *J. Chem. Phys.* **126**, 164103 (2007).

55. Walsh, A., Scanlon, D. O., Chen, S., Gong, X. G. & Wei, S.-H. Self-regulation mechanism for charged point defects in hybrid halide perovskites. *Energy Environ. Sci.* **8**, 995–1004 (2015).

56. Hoke, E. T. *et al.* Reversible photo-induced trap formation in mixed-halide hybrid perovskites for photovoltaics. *Chem. Sci.* **6**, 613–617 (2015).

57. Pan, C. *et al.* Enhanced Cu2S/CdS coaxial nanowire solar cells by piezo-phototronic effect. *Nano Lett.* **12**, 3302–3307 (2012).

58. Wang, Z. L. in *Microtechnology and MEMS* (Springer, 2012).

59. Zhang, H. *et al.* Photovoltaic behaviour of lead methylammonium triiodide perovskite solar cells down to 80 K. *J. Mater. Chem. A* **3**, 11762–11767 (2015).

60. Almora, O. *et al.* Capacitive dark currents, hysteresis, and electrode polarization in lead halide perovskite solar cells. *J. Phys. Chem. Lett.* **6**, 1645–1652 (2015).

Acknowledgements

We greatly acknowledge Guido Rothenberger, Alessandro Mattoni and Carlo Massimo Casciola for valuable hints and discussions during the preparation of the manuscript. M.G. acknowledges a Sciex fellowship under Project Code 13.194. Financial support from SNF-NanoTera (SYNERGY), Swiss Federal Office of Energy (SYNERGY) and CCEM-CH in the 9th call proposal 906: CONNECT PV is gratefully acknowledged. M.G. and T.M. acknowledge the European Research Council (ERC) for an Advanced Research Grant (ARG no. 247404) funded under the 'Mesolight' project. M.K.N. thank the Swiss Federal Office for Energy (Energy center special funds) and the European Union Seventh Framework Program under grant agreement numbers 604032 of the MESO, 308997 of the NANOMATCELL. U.R. acknowledges funding from the Swiss National Science Foundation via individual grant No. 200020-146645, the NCCRs MUST and MARVEL, and support from the Swiss National Computing Center (CSCS) and the CADMOS project for computing resources. We acknowledge PRACE for awarding us access to computer resources at Supermuc at LRZ, Germany.

Author contributions

S.M., U.R., T.M. and M.G. conceived, designed and led the study. S.M. carried out the computational calculations. M.F., Y.L. and M.S. prepared devices. T.M. carried out the measurements and did the data analysis with the help of W.T. and P.G. synthesized the methyl ammonium halides. S.M., T.M. and W.T. wrote the manuscript with the help of U.R. and M.G., S.M.Z. and M.K.N. contributed to the discussions.

Additional information

2

Deciphering the origin of giant magnetic anisotropy and fast quantum tunnelling in Rhenium(IV) single-molecule magnets

Saurabh Kumar Singh[1] & Gopalan Rajaraman[1]

Single-molecule magnets represent a promising route to achieve potential applications such as high-density information storage and spintronics devices. Among others, $4d/5d$ elements such as Re(IV) ion are found to exhibit very large magnetic anisotropy, and inclusion of this ion-aggregated clusters yields several attractive molecular magnets. Here, using *ab intio* calculations, we unravel the source of giant magnetic anisotropy associated with the Re(IV) ions by studying a series of mononuclear Re(IV) six coordinate complexes. The low-lying doublet states are found to be responsible for large magnetic anisotropy and the sign of the axial zero-field splitting parameter (D) can be categorically predicted based on the position of the ligand coordination. Large transverse anisotropy along with large hyperfine interactions opens up multiple relaxation channels leading to a fast quantum tunnelling of the magnetization (QTM) process. Enhancing the Re-ligand covalency is found to significantly quench the QTM process.

[1] Department of Chemistry, Indian Institute of Technology, Bombay Powai, Mumbai 400076, India. Correspondence and requests for materials should be addressed to G.R. (email: rajaraman@chem.iitb.ac.in).

I n the quest of single-molecule magnets (SMMs)[1–6] with enhanced magnetic properties, magnetic anisotropy is found to be the most influential parameter, which governs the barrier height for slow relaxation of magnetization[7–11]. Owing to inherently large magnetic anisotropy, lanthanide-based complexes are promising candidate for single-ion magnets[2,3,5,12–17] and mononuclear SMMs based on transition metal ions are relatively scarce in the literature, as stronger ligand field interactions suppress the orbital contributions to the anisotropy and hence the barrier heights (U_{eff})[18–25]. In the past few years, late-transition metal ions have gained much attention in the area of SMMs. The diffused magnetic orbitals of the $4d/5d$ ions translate stronger magnetic exchange, whereas larger spin-orbit coupling constants (SOCs) exhibited by these ions[26], often lead to highly anisotropic ground state (highly anisotropic g-tensors with an unusually large zero-field splitting values (ZFS)). These two essential conditions along with a possibility of exhibiting anisotropic/anti-symmetric exchange makes this class of molecules ideal for observing SMM behaviour[27–33]. Owing to these advantages, these $4d/5d$ ions show better SMM behaviour compared with their $3d$ congeners at many occasions[27,28,31–33]. Among the $4d/5d$ ions, the chemistry of Re(IV) metal ion is very rich as they have been successfully used to isolate several single-chain magnets (SCMs)/SMMs with an attractive U_{eff} values[26,34–41]. Apart from rich magnetic studies, these Re(IV) complexes are also explored in the development of the new anticancer drugs[42].

Despite several years of comprehensive experimental efforts in designing Re(IV) ion-based SMMs/SCMs, the origin of giant magnetic anisotropy is not well understood[39]. As in most of the cases, the axial ZFS (D) values are extracted using magnetization measurements, which are known to be insensitive to the sign and strength of the D parameter[41,43–51]. On other hand, the most promising high frequency-electron paramagnetic resonance (HF-EPR) technique has its own limitation, where the sign can be accurately determined but such large magnitude of D are often difficult to estimate[39]. An alternate solution to resolve the ambiguities in the sign/magnitude of D value is to analyse the ZFS parameter using *ab initio* calculations[9,22,39,52–57], which has been widely used in this respect. Moreover, strategic designing of new generation SMMs based on Re(IV) ions requires a thorough understanding of the nature ZFS and how the magnitude and the sign of the D and E vary depending on the ligand field environment. As the magnitude of E and the hyperfine interactions are correlated to the quantum tunnelling of magnetization (QTM)[58], the possibility to fine tuning these values is of paramount importance in this area.

The goal of the present communication is to gain a thorough understanding of the magnetic anisotropy in six coordinate Re(IV) complexes using state-of-the-art *ab initio* calculations. By modelling structurally diverse 13 mononuclear six coordinate Re(IV) complexes[39,41,43–51], we aim to answer the following intriguing questions (i) What is the suitable theoretical methodology to compute ZFS parameters in $5d$ transition metal ions such as Re(IV) complexes? (ii) What is the origin of giant D values and is there a correlation between the nature of the donor atoms and the sign of the D values in these complexes? (iii) What is mechanism of magnetic relaxation in Re(IV) single-ion magnets and how this is influenced by the metal–ligand covalency?

Results

Magnetic anisotropy and spin-Hamiltonian. The free Re(IV) ion is a d^3 Kramers ion with a ^4F ground state term, which splits into three states $^4A_{2g}$, $^4T_{2g}$ and $^4T_{1g}$ with the $^4A_{2g}$ being the ground state in an octahedral environment. Due to perfect cubic symmetry, pure octahedral complexes do not possesses any ZFS,

however, any distortions from the octahedral geometry are expected to yield large D values via mixing of the subsequent excited states because of very large SOC $(\lambda \sim 1,000\,cm^{-1})$. To begin with, we have studied the homoleptic $[ReCl_6]^{2-}$ model complex in tetragonal environment to analyse the origin of ZFS. Abragam and Bleaney have proposed a qualitative equation to predict the sign of the D values of a tetragonally distorted d^3 ion.

$$D = -\frac{4\lambda^2}{\Delta_0} + \frac{4\lambda^2}{\Delta_1} \qquad (1)$$

where λ is the SOC and Δ_0, Δ_1 is the tetragonal distortion parameter. If $\Delta_0 < \Delta_1$, the sign is predicted to be negative, whereas for $\Delta_0 > \Delta_1$, the sign is predicted to be positive. To corroborate this qualitative analysis, we have performed *ab initio* calculations, using complete active self consistent field (CASSCF) and CASPT2 methods incorporating spin-orbit effects with the RASSI-SO module in MOLCAS[59,60]. A positive D value of $+19.3\,cm^{-1}$ has been obtained for axially elongated model, whereas a negative D value of $-24.3\,cm^{-1}$ has been observed for axially compressed D_{4h} $[ReCl_6]^{2-}$ model complex (see Supplementary Fig. 1 and Supplementary Table 1 for details). The energy splitting pattern of the first three same spin-free states $(^4T_{2g}(F))$ are arranged as expected based on the ligand field theory and the sign of the D values computed using CASPT2 are in line with the expected values based on the equation (1). Although CASSCF calculations predict a similar splitting pattern of first three same spin-free states, it fails to reproduce the correct sign of the D values compared with CASPT2 methods for both elongated and compressed geometries. This suggests that spin-flip states rather than same spin-free states govern the sign as well as magnitude of D values for $5d$ elements such as Re(IV) ion. Hence, here after all the results discussed are performed at CASPT2 level of theory (vide infra).

To further understand the nature of D and E values, we have selected 13 mononuclear Re(IV) complexes and classified into three categories type-I: $[ReX_4(L)]$ (where L = a bidentate ligand on the equatorial plane), type-II: $[ReX_4(L)_2]$ (where L = monodentate ligand in the axial positions) and type-III $[ReX_5(L)]^-$ (where L = monodentate ligand; see Figs 1 and 2 for details). Continuous symmetry measure analysis (SHAPE)[61] of the X-ray structures reveals that all the complexes are in the distorted octahedral geometry (see Supplementary Fig. 2 and Supplementary Tables 2 and 3 for details).

Sign and magnitude of ZFS parameter for 1–13. Calculations reveal that eight spin-free states corresponding to ^2G states are found to be low-lying and thus are expected to contribute significantly to the D values via spin-flip excitations in all complexes **1–13** studied (see Supplementary Figs 3 and 4, Supplementary Tables 4–16 and Supplementary Note 1 for details). The MS-CASPT2 + RASSI-SO computed D, E and the first spin-free excitation energies for all complexes are depicted in Table 1. For complexes **1–6**, large negative D and significantly large |E/D| values, with D as high as $\pm 132\,cm^{-1}$ (for **4**) have been

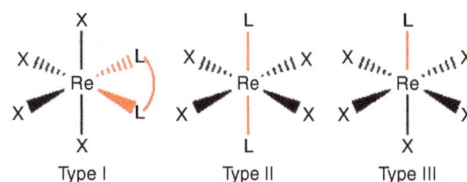

Figure 1 | Structural topology. Classifications of substituted hexa halo Re(IV) complexes (where X = Cl, Br and L = coordinating ligand).

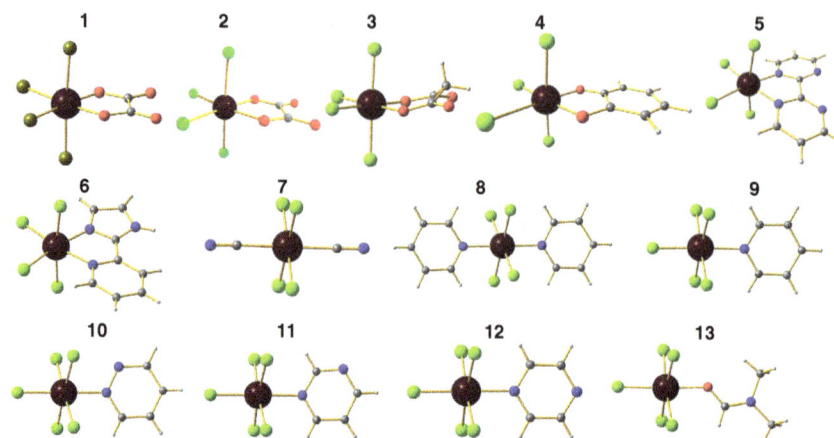

Figure 2 | X-ray crystal structures. Crystal structure of Re(IV) mononuclear complexes. Colour code: dark brown, Re; pale brown, Br; light green, Cl; blue, N; grey, C; white, H. [ReBr$_4$(ox)]$^{2-}$ (1); [ReCl$_4$(ox)]$^{2-}$ (2); [ReCl$_4$(mal)]$^{2-}$ (3); [ReCl$_4$(cat)]$^{2-}$ (4); [ReCl$_4$(bpym)] (5); [ReCl$_4$(pyim)] (6); [ReCl$_4$(CN)$_2$]$^{2-}$ (7); [ReCl$_4$(py)$_2$] (8); [ReCl$_4$(py)]$^{-}$ (9); [ReCl$_5$(pyz)]$^{-}$ (10); [ReCl$_5$(pyd)]$^{-}$ (11); [ReCl$_5$(pym)]$^{-}$ (12); [ReCl$_4$dmf]$^{-}$ (13). bpym, 2,2'-bipyrimidine; cat, catechol; dmf, dimethyl formamide; mal, malonato; ox, oxalato; pyim, 2-(2'-pyridyl)imidazole; py, pyridine; pyd, pyridazine; pym, pyrimidine; pyz, pyrazine.

Table 1 | MS-CASPT2 + RASSI-SO computed D and $|E/D|$ value for all studied Re(IV) mononuclear complexes along with first spin-free excitation energy.

Complex	D_{cal}	$|E/D|_{cal}$	D $(|E/D|)_{exp}$	ΔE	References
1 [ReBr$_4$(ox)]$^{2-}$	−93.0	0.18	−73 (0.20)*	8,079.2	39
2 [ReCl$_4$(ox)]$^{2-}$	−85.0	0.24	−57 (0.26)*	8,873.5	39
3 [ReCl$_4$(mal)]$^{2-}$	−61.6	0.15	55†	8,442.8	44
4 [ReCl$_4$(cat)]$^{2-}$	±132.6	0.30	95†	7,526.7	50
5 [ReCl$_4$(bpym)]$^{2-}$	−47.9	0.23	—	7,616.5	46
6 [ReCl$_4$(pyim)]$^{2-}$	−34.7	0.17	—	7,804.7	48
7 [ReCl$_4$(CN)$_2$]$^{2-}$	+16.2	0.23	+11 (0.29)*	7,377.3	41
8 [ReCl$_4$(py)$_2$]	+55.6	0.18	9.56†	5,950.5	43
9 [ReCl$_5$(py)]$^{-}$	±32.7	0.31	3.52†	7,841.7	51
10 [ReCl$_5$(pyd)]$^{-}$	+41.0	0.24	14.1†	7,385.9	47
11 [ReCl$_5$(pym)]$^{-}$	+24.6	0.17	6.2†	8,300.7	47
12 [ReCl$_5$(pyz)]$^{-}$	+32.5	0.20	9.4†	7,630.2	49
13 [ReCl$_5$(dmf)]$^{-}$	+18.5	0.18	10.1†	8,467.4	45

bpym, 2,2'-bipyrimidine; cat, catechol; dmf, dimethyl formamide; mal, malonato; ox, oxalato; pyim, 2-(2'-pyridyl)imidazole; py, pyridine; pyd, pyridazine; pym, pyrimidine; pyz, pyrazine.
All the D values and spin-free excitations energies are provided in cm^{-1}.
*HF-EPR reported values.
†Obtained from magnetic susceptibility measurements, no sign convention has been used.

---

witnessed. On other hand, complexes of type II and type III categories (7–13) found to posses positive ZFS parameter with D as high as $+55$ cm^{-1} has been noticed (for 8; see Supplementary Fig. 5 and for the orientation of D tensor). The magnetic susceptibility and powder magnetization data computed for 1–13 reproduces nicely the experimental behaviour, adding confidence to the computed values (see Supplementary Figs 6–10 and Supplementary Note 2 for details). Simulation of HF-EPR spectra reported earlier[35,39] confirm the negative sign of the D with a large E/D values for complexes 1 and 2, with the D estimated to be ca -73 and -57 cm^{-1}, respectively. Calculations yield D value of -93 and -85 cm^{-1} for complexes 1 and 2, where both the sign as well as the magnitude of the D values are correctly reproduced compared with the experimental values. More importantly, the magnitude of the $|E/D|$ values and g-tensors (both pseudo-spin 1/2 and 3/2), which are precisely estimated from the experiments are very well reproduced in our calculations (see Supplementary Tables 17 and 18 for details).

For complex 6, the magnetization data[41] yield an estimate of D as -14.4 cm^{-1}, which is in agreement with CASSCF results

(-21 cm^{-1}, see Supplementary Table 17) but in disagreement with MS-CASPT2 values ($+16.2$ cm^{-1}, see Table 1). However, HF-EPR experiments performed lately[35], where both the magnitude as well as the sign of the D value is estimated accurately, places the D value to be $+11$ cm^{-1}. This highlights the issue of obtaining the sign/magnitude of the D value from the magnetization data and also emphasise the need for CASPT2 approach and hence incorporation of dynamic correlation to correctly reproduce the sign of the D values[62,63]. Inclusion of dynamic correlation on CASSCF computed wavefunctions drastically stabilizes the doublet states compared with the computed CASSCF states, leading to a pronounced contribution from these states to the D values as discussed earlier (see Supplementary Fig. 3 for details). Moreover, either pseudo-spin or effective Hamiltonian approach[62] needs to be employed to extract the ZFS parameters as other theoretical methodologies found to yield ambiguous sign and magnitude of D values (see Supplementary Tables 19 and 20 for details).

The SINGLE_ANISO computed orientations of the main anisotropic axes (D_{XX}, D_{YY} and D_{ZZ}) and main magnetic axes (g_{XX}, g_{YY} and g_{ZZ}) for all the complexes 1–13 are provided in

Fig. 3 and Supplementary Fig. 5. It is evident from the figures that, the principal anisotropic axes (D_{ZZ}) are oriented towards the L_{ax}–R–L_{ax} (molecular $-z$ axis) direction, however, a significant tilt from this axis is witnessed[64]. Moreover, the main magnetic axes (g_{XX}, g_{YY} and g_{ZZ}) and main anisotropic axes (D_{XX}, D_{YY} and D_{ZZ}) do not coincide with each other and such non-coincidence has been previously noticed by Askevold et al.[65] The orientation of the D_{ZZ} axis is tilted by 28.8°, 36° and 33.7° from their molecular $-z$ axis for complexes 1, 7 and 10, respectively. Large structural distortions and the associated large $|E/D|$ values are responsible for such deviations. Larger $|E/D|$ values detected in complex 7 has larger tilt compared with complex 1, where smaller $|E/D|$ value leads to smaller tilt. This trend is visible also for other structures. However, in the rhombic limit, the nature of the easy axis is ambiguous, as even the small structural distortions flip the Eigen values and hence the orientation of the easy axis of magnetization. This can be better visualized for complex 4, where presence of large $|E/D|$ value of 0.30 causes the flipping of D_{ZZ} axis to the equatorial plane.

Origin of ZFS in complexes 1–13. To shed light on the sign of ZFS, we have analysed the molecular orbitals (MOs) of complexes 1 and 7 and 10 as a representative examples for type I, II and III

defined earlier (see Fig. 4 for details). For 1, presence of unsymmetrical ligand in the equatorial position leads to the d_{xz} orbital being the lowest lying in energy followed by degenerate d_{yz} and d_{xy} orbitals. The $\pi^{\star}$ orbitals of the oxygen donors interact rather strongly with the d_{yz} and d_{xy} orbitals because of acute $\angle$ O-Re-O bite angle. The d_{xz} orbital on the other hand faces less repulsion leading to a slight stabilization. Besides, stronger $\sigma^{\star}$ interaction by the oxalate ligand lead to destabilization of the $d_{x^2-y^2}$ orbital compared with the d_{z^2} orbital. This orbital ordering has the following consequences to the D values: (i) all spin-conserved excitations from the d_{xy}, d_{xz} and d_{yz} orbitals to the vacant d_{z^2} orbital contribute to positive D values. This is affirmed by an additional calculation incorporating only the quartet states in the estimation of D and this yield a positive D value ($+13$ cm^{-1}). (ii) Spin-flip excitations from the d_{xz} to the d_{yz} orbitals contribute to negative D values (excitation between same $|m_l|$ levels). As the gap between these two orbitals is very small (667 cm^{-1}), this transition governs both the sign and the magnitude of the D value for this complex. A similar pattern predicted for complexes 2–6 rationalize the observed negative sign for these complexes.

Strong π acceptor cyanide ligand stabilizes the d_{xz} and d_{yz} orbitals via $p\pi$–$d\pi$ interactions compared with the d_{xy} orbital. Stronger σ-donation along the axial directions destabilizes the d_{z^2}

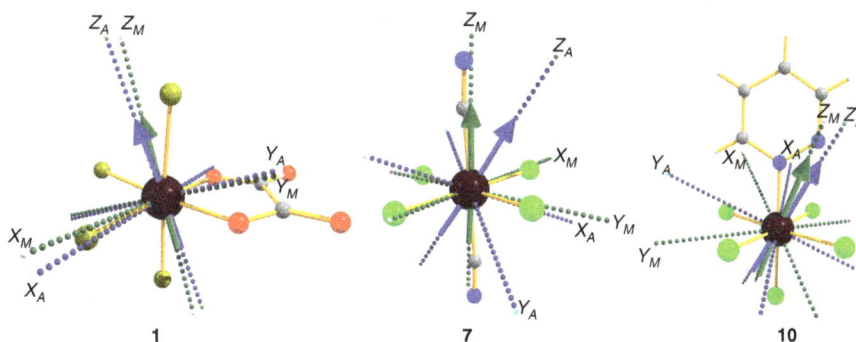

Figure 3 | Orientation of g and D tensors. SINGLE_ANISO computed main magnetic (X_M, Y_M and Z_M) axes representing g-tensors orientation and main anisotropic axes (X_A, Y_A and Z_A) representing D tensors orientation for complexes 1, 7 and 10.

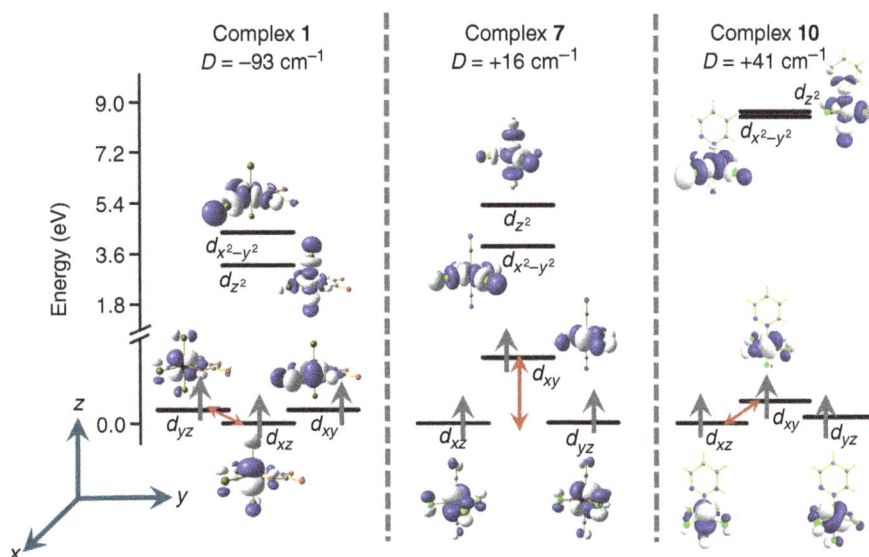

Figure 4 | Molecular orbital analysis and nature of excitations. Computed d-orbital ordering for complexes 1, 7 and 10. The iso-density surface plotted with the iso-value of 0.02 e$^-$/bohr^3. The double headed arrow represents the gap between the orbitals, which are contributing significantly to the D value. The orbitals which are appeared as degenerate in the figure are not strictly degenerate due to symmetry arguments.

orbital leading to different orbital arrangement compared with complex **1**. This orbital arrangement has the following consequences to the D value: (i) the spin-conserved $d_{xy} \rightarrow d_{x^2-y^2}$ transition contributes to the negative D value, affirmed by an additional calculation incorporating only the quartets in the estimation of D and this yields $-26\,cm^{-1}$ as the D value, (ii) the spin-flip excitations from the degenerate $d_{xz}-d_{yz}$ orbitals to the d_{xy} orbital contribute to the positive D value (the gap here is $3,123\,cm^{-1}$). Here also the later term dominates leading to an overall positive D value for **7**.

For complex **10**, positive D value is expected as the splitting pattern is found to be very similar to that of complex **7** (Fig. 4). However, the presence of weak π-donor pyridazine ligand on the axial position reduces the splitting between the $d_{xz}-d_{yz}$ orbital and the d_{xy} orbital ($1,044\,cm^{-1}$). This suggests that much less energy is required to flip the spin in complex **10** compared with complex **7**, thus the D value is expected to be large in this case. A similar splitting pattern is predicted for complexes **8**, **9** and **11–13** rationalize the observed positive D values for these complexes.

Rationale for the observed variation in the ZFS parameter. Among **1–6**, the equatorial positions are occupied by π-donor ligand, except in case of complex **5** where 2,2'-bipyrimidine (bpym) ligand serves as a weak π-acceptor ligand. Independent of the nature of the ligand (π-donor versus π-acceptor), in complexes **1–6**, the $d_{yz} \rightarrow d_{xz}$ transition dominates the D value over other transitions, leading to a large negative D values. To understand the large difference in the D values of complexes **1** and **2**, we have performed additional calculations on model complexes where Cl in complex **1** is modelled as Br maintaining Re–Br distance same as that of complex **2**. For this model, the D is estimated to be $-86\,cm^{-1}$ compared with $-85\,cm^{-1}$ for the Cl analogue and this suggest that apart from the spin-orbit coupling of Br, the structural distortion such as –*cis* angles play an important role in determining the strength of the D value (see Supplementary Table 3 for selected structural parameters of complexes **1** and **2**)[55].

The strength of the donor–acceptor abilities significantly affects the magnitude of the D values (see Fig. 5 and Supplementary Table 21 for further details)[20,62]. In complexes **1–6**, larger charge on the donor atoms are found to yield large D values (see Fig. 4 and Supplementary Figs 11 and 12 for quantitative charges computed). Larger charges on the donor atoms stabilizes the d_{xz} orbitals compared with the d_{xy}/d_{yz} orbitals leading to different transition energies (see Table 1 for details) and thus the computed charges are found to strongly correlated to the magnitude of the D values. This striking observation offers a rational approach to fine tune the magnitude of the negative D value in this set of complexes. Moreover, stabilization of d_{xz} orbital also affects the E values, as it increases the difference between the D_{XX} and D_{YY} contributions leading to a larger E with large charge on the ligand (see equation 2 and Supplementary Fig. 12 for details).

$$D = D_{zz} - \tfrac{1}{2}(D_{XX} + D_{YY}); E = (D_{XX} - D_{YY})$$
$$D_{ZZ} \sim \left[E(d_{xz}) - E(d_{yz})\right]^{-1}$$
$$D_{XX} \sim \left[E(d_{xz}) - E(d_{xy})\right]^{-1} \quad (2)$$
$$D_{YY} \sim \left[E(d_{xz}) - E(d_{z^2})\right]^{-1}$$

In contrast to complex **7**, where axial positions are occupied by two strong π-acceptor ligands, complex **8** posses two weak π-donor (pyridine) ligands on the axial position and therefore the $d_{xz}-d_{yz}$ orbital to d_{xy} orbital is found to be ($1,360.7\,cm^{-1}$) much smaller than that of $3,123\,cm^{-1}$ observed in case of complex **7**. Moreover, the first spin-flip-excited state is found at $5,950.5\,cm^{-1}$ (in case of complex **8**), which is again much

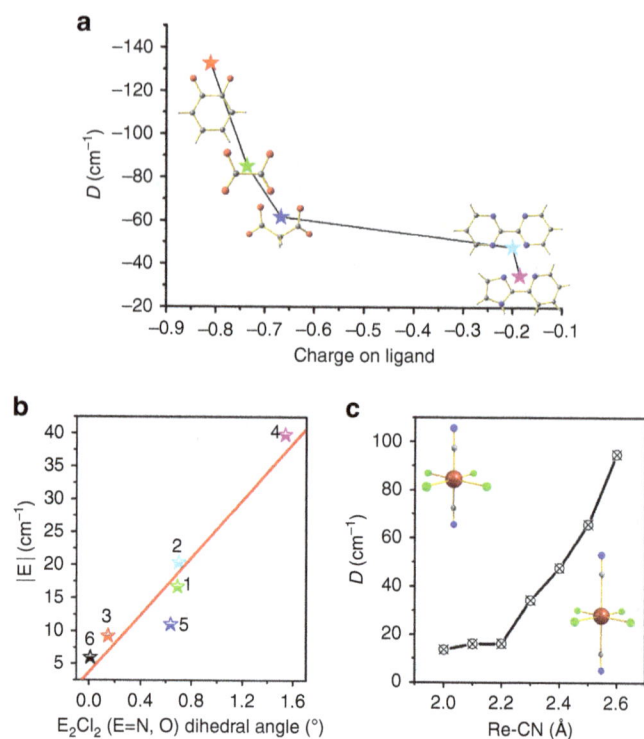

Figure 5 | Impact of structural distortions on magnetic anisotropy. (**a**) Plot of computed D value versus charge of the coordinated ligand atoms. (**b**) Plot of the computed $|E|$ value versus Re(O/N)$_2$Cl$_2$ dihedral angle; (**c**) magneto-structural correlation by varying Re–CN bond distances of complex **7**.

smaller than the $7,377.3\,cm^{-1}$ gap observed for complex **7** (see Table 1 for details). This leads to larger D value for complex **8** compared with complex **7**.

Complexes **9–13** possess positive D values ranging from $+18\,cm^{-1}$ (complex **13**) to $+41\,cm^{-1}$ (complex **10**). As the structural parameters across the series are very similar, the differences in the magnitude of the D values are expected to arise from the donor strength of the ligand. To affirm this point, we have analysed the donor–acceptor interactions using second-order perturbation theory natural bonding orbitals (NBO) analysis for complexes **10** and **13**. NBO analysis suggests a significant σ-donation from lone pair of nitrogen to Re d_{z^2} orbital and this strength is estimated to be $19.6\,kcal\,mol^{-1}$ for complex **10**, whereas $22.6\,kcal\,mol^{-1}$ for complex **13** (see Supplementary Fig. 13 for details). Larger σ-donation in complex **13** leads to smaller D value compared with complex **10**. A similar analogy can be drawn also for other complexes.

Here independent of the nature of π-donor/acceptor ligands, type I complexes found to yield negative D values, whereas type II and III complexes found to yield positive D values. This is in stark contrast to the earlier observations where lighter transition metal d^3 ions found to switch the sign of ZFS parameter by changing the nature of the ligand donor atoms (π-donor ligands found to yield $+D$ values, whereas π-acceptor ligands yield $-D$ values)[66]. This is essentially due to the fact that spin-flip doublet transitions are the dominating factor to the D values in Re(IV) complexes, whereas in lighter elements due to smaller crystal-field splitting, the spin-allowed transition dominates the D value.

Magneto-structural D-correlations. To probe how the axial bond length influences the D value, we have developed a magneto-structural D correlations on complex **7** (see Fig. 5c for details), where the axial -CN bonds are varied from 2.0 to 2.6 Å (compression and elongation of axial bonds). As $[ReCl_4(CN)_2]^{2-}$ unit has been employed as a building block for synthesis of polynuclear SMMs/SCMs[34–36,41], magneto-structural correlation developed on this model will serve the purpose of obtaining qualitative single-ion Re(IV) anisotropy in diverse polynuclear framework. This Re–C bond distance is found to vary significantly among structures, particularly when the -CN ligands are found to bind to other metal ions. In our correlation, the magnitude of the D value found to drastically increase (from $+13.6$ to $+94.7 \, cm^{-1}$) as the Re–C bond length increases to 2.7 Å. As the metal–ligand interactions are weaker at longer distances, the transition energies are further lowered leading to larger D values for axially elongated structures[62,63]. Besides our results reveal that tetragonal distortions does not alter the sign of D values and this suggests that the $[ReCl_4(CN)_2]^{2-}$ unit unlikely to offer negative single-ion D value in any polynuclear framework.

Mechanism of magnetic relaxation. Complex 1 exhibits field-induced SMM behaviour with a barrier height of $9.6 \, cm^{-1}$ at higher temperatures and $1.5 \, cm^{-1}$ at lower temperatures. The experimental relaxation observed at higher temperature (up to 3.5 K) is unlikely due to Orbach process as the first excited Kramer's doublet (KD) is estimated to lie at $195 \, cm^{-1}$. Thus, the relaxation is expected to be a multi-phonon Raman process. The fast relaxation observed at lower temperature is essentially due to QTM process, which is facilitated due to the presence of transverse anisotropy, hyperfine interactions and external perturbations such as internal magnetic field provided by surrounding molecules. To gain insights into the QTM process, we have analysed the wavefunction of the ground-state KD and our analysis suggests that the ground-state KD comprised of 44% of $|3/2, \pm 3/2\rangle$ and 47% of $|3/2, \pm 1/2\rangle$. As the D value is very large $(D >> kT)$, the $|3/2, \pm 1/2\rangle$ KD will be completely depopulated and the ground state can be treated as a pseudo spin 1/2 system. The presence of large E term offers a strong mixing between the $|3/2, \pm 3/2\rangle$ and $|3/2, \pm 1/2\rangle$ components, which allows QTM to facilitate at low temperatures. To qualitatively analyse the mechanism of magnetic relaxation, we have computed the matrix elements between the connected KDs (see Supplementary Fig. 14 and Supplementary Note 3 for details)[67]. Our calculations predict very large tunnelling probability between the ground-state KDs and this is in line with the analysed wave function analysis. Such a prominent QTM process expected to quench the magnetization completely and this is consistent with the absence of zero-field SMM behaviour[56]. On the other hand, application of external d.c. field lifts the degeneracy and suppresses this fast relaxation. However, the QTM process cannot be ignored even under the applied field conditions as hyperfine interactions and intermolecular dipolar couplings facilitate this process. For diluted samples where intermolecular interactions are negligible, hyperfine interactions are the only factor which governs the resonant QTM process[58,68]. To gain further insights, we have computed hyperfine interactions of the Re(IV) ions as it has two dominant isotopes ^{185}Re and ^{187}Re ($I = 5/2$) with a significant natural abundance (see Supplementary Table 22). Particularly, the transverse component of this internal nuclear spin of Re(IV) (measured as $|A_X|$ and $|A_Y|$ hyperfine tensors) give rise to a small internal magnetic field inside the molecule facilitating the QTM in zero external field. The hyperfine interactions computed for complexes 1 and 2 are found to be significantly large (1,661, 1,668, 1,669 MHz for complex 1 and 1,962, 1,967 and 1,969 MHz for complex 2) leading to fast QTM both in the presence and absence of magnetic field. A detailed experimental characterization on the diluted sample needs to be studied to verify the proposed mechanism and to gain further understanding on the relaxation process.

Role of metal–ligand covalency on magnetic anisotropy. With an aim to analyse the role of metal–ligand covalency on the D value of Re(IV) complexes, here we have modelled complexes, $[ReCl_4(E_2C_6H_4)]^{2-}$ (here $E = O$ (complex 4), S (4a) and Se (4b)). For the optimized geometries of 4a and 4b, the computed D values are $+112$ and $+114 \, cm^{-1}$, respectively. Although the strength of the D values are only moderately affected, the sign can be predicted unambiguously here as there is a significant drop in the E/D values (0.30, 0.23 and 0.24 for 4, 4a and 4b, respectively). Besides, the NBO analysis reveals that the Re–S/Se bond is more covalent than Re–O bond (for the Re–O, the Re contribution is 18.4%, whereas oxygen contribution is 81.6% and for Re–S(Se) bond, Re contribution is 32%(36%), whereas the S(Se) contribution is 67% (63%); see Supplementary Figs 15 and 16 and Supplementary Tables 23–25 for details). This difference in covalency lead to larger orbital splitting within t_{2g} sub-shell for S/Se analogues[22]. The larger orbital splitting are compensated by the large SOC associated with S and Se atoms leading to rigorous mixing of excited states with the ground state yielding similar strength of D values compared with the oxygen analogue. Interestingly, the π-interactions in 4a and 4b stabilize the d_{xy} and destabilize the d_{xz}/d_{yz} orbitals leading to a similar strength of D_{XX} and D_{YY} contributions. This lead to a decrease in the E values for complexes 4a and 4b compared with complex 4.

Discussion

Here we have probed the origin of contrasting behaviour observed for Re(IV) SMMs where both the giant magnetic anisotropy and fast QTM found to co-exist. The low-lying doublet states are found to govern the sign and magnitude of ZFS parameters in this class of complexes. Our method assessment reveals that pseudo spin approach or effective Hamiltonian approach coupled with CASPT2 calculations needs to be employed to correctly reproduce the sign and magnitude of ZFS parameters. Quite interestingly, the sign of D values are found to be predictable based on the coordination mode of the ligands in these complexes where type I complexes found to possess larger negative D values, whereas type II and III found to possess a positive D values. Nature of the donor ligands as well as charge on the coordinated atoms found to influence only the strength but not the sign of D values. Very large hyperfine interactions (both transverse and axial) and rhombic anisotropy computed on these system found to govern the QTM process. By performing additional calculations and by developing magneto-structural correlations, we offer a way to enhance (diminish) the negative D ($|E|$) value in these classes of complexes. Some interesting observations are noted where the metal–ligand covalency found to govern the transverse anisotropy, offering a way to quench the inherent fast QTM process in this class of complexes.

Methods

Ab initio **calculations.** We have performed the *ab initio* calculations based of wave function theory approach to compute the ZFS in these set of mononuclear complexes. All the calculations have been performed using MOLCAS 7.8 suite of programme[69]. Here we have employed the state average-CASSCF method to compute the ZFS. The active space comprises of three active electrons in five active orbitals (CAS(3,5)). With this active space, we have computed all the 10 quartets and 40 doublet states in the configuration interaction procedure. On top of the converged CASSCF wave function, we have performed MSCASPT2 calculations to treat the dynamical correlations. We have employed ionization potential electron affinity (IPEA) shift of 0.25 to avoid the intruder states problem in CASPT2 calculations. The MS-CASPT2 computed states were further treated in RASSI-SO module, which explicitly computes the spin-orbit states. Furthermore, SINGLE_ANISO module has been utilized on top to compute the reliable spin-Hamiltonian (D and g-tensor, orientation of main magnetic axes and main anisotropic axes and local magnetic susceptibility) for each complex. The following ANO-RCC basis sets were used: [8s7p5d3f2g1h.] for Re, [ANO-RCC...5s4p2d.] for Cl, [ANORCC...6s5p3d.] for Br, [ANO-RCC...4s3p2d.] for O, C and N and [ANO-RCC...2s1p] for H during the calculations. The Cholesky decomposition for two electron integral is employed throughout the calculations to save the disk space. Moreover, additional ZFS calculations have been performed using two different techniques: (i) effective Hamiltonian approach and (ii) second-order perturbation method to check out the robustness of reported theoretical methods in predicting correct sign and magnitude of D values.

DFT calculations. Hyperfine interaction of the Re(IV) nuclei were computed within DFT framework, using electron paramagnetic resonance/nuclear magnetic resonance (EPR/NMR) module in the ORCA code[70]. We have employed meta-GGA TPSSH functional along with SARC basis set for the Re, which is much more flexible at core region to estimate all the components of the A-tensors (Fermi Fermi Contact, Spin-dipolar and Spin-orbit coupling) along $-x$, $-y$ and $-z$ directions. A very tight self consistent field (SCF) (1×10^{-8} E_h) has been kept throughout the calculations.

References

1. Sessoli, R., Gatteschi, D., Caneschi, A. & Novak, M. A. Magnetic bistability in a metal-ion cluster. *Nature* **365**, 141–143 (1993).
2. Blagg, R. J., Muryn, C. A., McInnes, E. J. L., Tuna, F. & Winpenny, R. E. P. Single pyramid magnets: Dy-5 pyramids with slow magnetic relaxation to 40 K. *Angew. Chem. Int. Ed.* **50**, 6530–6533 (2011).
3. Blagg, R. J. *et al.* Magnetic relaxation pathways in lanthanide single-molecule magnets. *Nat. Chem* **5**, 673–678 (2013).
4. Leuenberger, M. N. & Loss, D. Quantum computing in molecular magnets. *Nature* **410**, 789–793 (2001).
5. Woodruff, D. N., Winpenny, R. E. P. & Layfield, R. A. Lanthanide single-molecule magnets. *Chem. Rev.* **113**, 5110–5148 (2013).
6. Wernsdorfer, W. & Sessoli, R. Quantum phase interference and parity effects in magnetic molecular clusters. *Science* **284**, 133–135 (1999).
7. Neese, F. & Pantazis, D. A. What is not required to make a single molecule magnet. *Faraday Discuss.* **148**, 229–238 (2011).
8. Ruiz, E. *et al.* Can large magnetic anisotropy and high spin really coexist? *Chem. Commun.* 52–54 (2008).
9. Singh, S. K. & Rajaraman, G. Probing the origin of magnetic anisotropy in a dinuclear {MnIIICuII} single- molecule magnet: the role of exchange anisotropy. *Chem. Eur. J* **20**, 5214–5218 (2014).
10. Waldmann, O. A criterion for the anisotropy barrier in single-molecule magnets. *Inorg. Chem.* **46**, 10035–10037 (2007).
11. Cirera, J., Ruiz, E., Alvarez, S., Neese, F. & Kortus, J. How to build molecules with large magnetic anisotropy. *Chem. Eur. J* **15**, 4078–4087 (2009).
12. Langley, S. K. *et al.* A {(Cr2Dy2III}-Dy-III} single-molecule magnet: enhancing the blocking temperature through 3d magnetic exchange. *Angew. Chem. Int. Ed.* **52**, 12014–12019 (2013).
13. Liu, J. L. *et al.* A heterometallic Fe-II-Dy-III single-molecule magnet with a record anisotropy barrier. *Angew. Chem. Int. Ed.* **53**, 12966–12970 (2014).
14. Rinehart, J. D. & Long, J. R. Exploiting single-ion anisotropy in the design of f-element single-molecule magnets. *Chem. Sci* **2**, 2078–2085 (2011).
15. Meihaus, K. R. & Long, J. R. Magnetic blocking at 10 K and a dipolar-mediated avalanche in salts of the Bis(eta(8)-cyclooctatetraene) complex [Er(COT)(2)](-). *J. Am. Chem. Soc.* **135**, 17952–17957 (2013).
16. Zhang, P. *et al.* Equatorially coordinated lanthanide single ion magnets. *J. Am. Chem. Soc.* **136**, 4484–4487 (2014).
17. Rinehart, J. D., Fang, M., Evans, W. J. & Long, J. R. Strong exchange and magnetic blocking in N-2(3-)-radical-bridged lanthanide complexes. *Nat. Chem* **3**, 538–542 (2011).
18. Chen, L. *et al.* Slow magnetic relaxation in a mononuclear eight-coordinate Cobalt(II) complex. *J. Am. Chem. Soc.* **136**, 12213–12216 (2014).
19. Colacio, E. *et al.* Slow magnetic relaxation in a Co-II-Y-III single-ion magnet with positive axial zero-field splitting. *Angew. Chem. Int. Ed.* **52**, 9130–9134 (2013).
20. Harman, W. H. *et al.* Slow magnetic relaxation in a family of trigonal pyramidal iron(II) pyrrolide complexes. *J. Am. Chem. Soc.* **132**, 18115–18126 (2010).
21. Jurca, T. *et al.* Single-molecule magnet behavior with a single metal center enhanced through peripheral ligand modifications. *J. Am. Chem. Soc.* **133**, 15814–15817 (2011).
22. Vaidya, S. *et al.* A synthetic strategy for switching the single ion anisotropy in tetrahedral Co(II) complexes. *Chem. Commun.* **51**, 3739–3742 (2015).
23. Zadrozny, J. M. & Long, J. R. Slow magnetic relaxation at zero field in the tetrahedral complex [Co(SPh)4]2-. *J. Am. Chem. Soc.* **133**, 20732–20734 (2011).
24. Zadrozny, J. M. *et al.* Magnetic blocking in a linear iron(I) complex. *Nat. Chem* **5**, 577–581 (2013).
25. Atanasov, M., Zadrozny, J. M., Long, J. R. & Neese, F. A theoretical analysis of chemical bonding, vibronic coupling, and magnetic anisotropy in linear iron(ii) complexes with single-molecule magnet behavior. *Chem. Sci* **4**, 139–156 (2013).
26. Wang, X. Y., Avendano, C. & Dunbar, K. R. Molecular magnetic materials based on 4d and 5d transition metals. *Chem. Soc. Rev.* **40**, 3213–3238 (2011).
27. Dreiser, J. *et al.* Three-axis anisotropic exchange coupling in the single-molecule magnets NEt4[MnIII2(5-Brsalen)2(MeOH)2MIII(CN)6] (M = Ru, Os). *Chem. Eur. J* **19**, 3693–3701 (2013).
28. Dreiser, J. *et al.* Frequency-domain fourier-transform terahertz spectroscopy of the single-molecule magnet (NEt4)[Mn-2(5-Brsalen)(2)(MeOH)(2)Cr(CN)(6)]. *Chem. Eur. J* **17**, 7492–7498 (2011).
29. Magee, S. A. *et al.* Large zero-field splittings of the ground spin state arising from antisymmetric exchange effects in heterometallic triangles. *Angew. Chem. Int. Ed.* **53**, 5310–5313 (2014).
30. Mironov, V. S. Strong exchange anisotropy in orbitally degenerate complexes. A new possibilityfor designing single-molecule magnets with high blocking temperatures. *J. Magn. Magn. Mater.* **272**, E731–E733 (2004).
31. Mironov, V. S. Trigonal bipyramidal spin clusters with orbitally degenerate 5d cyano complexes [Os-III(CN)(6)](3-), prototypes of high-temperature single-molecule magnets. *Dokl. Phys. Chem* **415**, 199–204 (2007).
32. Pedersen, K. S. *et al.* Enhancing the blocking temperature in single-molecule magnets by incorporating 3d-5d exchange interactions. *Chem. Eur. J.* **16**, 13458–13464 (2010).
33. Singh, S. K. & Rajaraman, G. Can anisotropic exchange be reliably calculated using density functional methods? A case study on trinuclear Mn-III-M-III-Mn-III (M = Fe, Ru, and Os) cyanometalate single-molecule magnets. *Chem. Eur. J* **20**, 113–123 (2014).
34. Feng, X. W., Harris, T. D. & Long, J. R. Influence of structure on exchange strength and relaxation barrier in a series of (FeReIV)-Re-II(CN)(2) single-chain magnets. *Chem. Sci* **2**, 1688–1694 (2011).
35. Feng, X. W., Liu, J. J., Harris, T. D., Hill, S. & Long, J. R. Slow magnetic relaxation induced by a large transverse zero-field splitting in a (MnReIV)-Re-II(CN)(2) single-chain magnet. *J. Am. Chem. Soc.* **134**, 7521–7529 (2012).
36. Harris, T. D., Coulon, C., Clerac, R. & Long, J. R. Record ferromagnetic exchange through cyanide and elucidation of the magnetic phase diagram for a (CuReIV)-Re-II(CN)(2) chain compound. *J. Am. Chem. Soc.* **133**, 123–130 (2011).
37. Martinez-Lillo, J. *et al.* A heterotetranuclear [(NiRe3IV)-Re-II] single-molecule magnet. *J. Am. Chem. Soc.* **128**, 14218–14219 (2006).
38. Martinez-Lillo, J. *et al.* Metamagnetic behaviour in a new Cu(II)Re(IV) chain based on the hexachlororhenate(IV) anion. *Chem. Commun.* **50**, 5840–5842 (2014).
39. Martinez-Lillo, J. *et al.* Highly Anisotropic Rhenium(IV) Complexes: New Examples of Mononuclear Single-Molecule Magnets. *J. Am. Chem. Soc.* **135**, 13737–13748 (2013).
40. Pedersen, K. S. *et al.* [ReF6](2-): A robust module for the design of molecule-based magnetic materials. *Angew. Chem. Int. Ed.* **53**, 1351–1354 (2014).
41. Harris, T. D., Bennett, M. V., Clerac, R. & Long, J. R. [ReCl4(CN)(2)](2-): A high magnetic anisotropy building unit giving rise to the single-chain magnets (DMF)(4)MReCl4(CN)(2) (M = Mn, Fe, Co, Ni). *J. Am. Chem. Soc.* **132**, 3980–3988 (2010).
42. Martinez-Lillo, J. *et al.* Rhenium(IV) compounds inducing apoptosis in cancer cells. *Chem. Commun.* **47**, 5283–5285 (2011).
43. Mroziński, J., Kochel, A. & Lis, T. Synthesis, structure and magnetic properties of trans-tetrachloro-bis-(pyridine)-rhenium(IV). *J. Mol. Struct.* **610**, 53–58 (2002).
44. Cuevas, A. *et al.* Rhenium(IV)-copper(II) heterobimetallic complexes with a bridge malonato ligand. Synthesis, crystal structure, and magnetic properties. *Inorg. Chem.* **43**, 7823–7831 (2004).
45. Martinez-Lillo, J. *et al.* Ligand substitution in hexahalorhenate(IV) complexes: synthesis, crystal structures and magnetic properties of NBu4[ReX5(DMF)] (X = Cl and Br). *Inorg. Chim. Acta* **359**, 3291–3296 (2006).
46. Chiozzone, R. *et al.* A novel series of rhenium-bipyrimidine complexes: synthesis, crystal structure and electrochemical properties. *Dalton Trans.* 653–660 (2007).
47. Arizaga, L. *et al.* Synthesis, crystal structure, electrochemical and magnetic properties of (NBu4)[ReCl5(L)] with L = pyrimidine and pyridazine. *Polyhedron* **27**, 552–558 (2008).
48. Martinez-Lillo, J., Armentano, D., De Munno, G. & Faus, J. Magneto-structural study on a series of rhenium(IV) complexes containing biimH(2), pyim and bipy ligands. *Polyhedron* **27**, 1447–1454 (2008).
49. Martinez-Lillo, J. *et al.* Pentachloro(pyrazine)rhenate(IV) complex as precursor of heterobimetallic pyrazine-containing Re-IV M-2(II) (M = Ni, Cu) species: synthesis, crystal structures and magnetic properties. *Dalton Trans.* 4585–4594 (2008).
50. Cuevas, A. *et al.* Synthesis, molecular structure and magnetic properties of a rhenium(IV) compound with catechol. *J. Mol. Struct.* **921**, 80–84 (2009).

51. Kochel, A. Solvothermal synthesis, characterization and properties of [ReCl5(py)] — complex with T (Néel) of 5.5 K. *Trans.Met. Chem* **35**, 1–5 (2010).

52. Maganas, D., Sottini, S., Kyritsis, P., Groenen, E. J. J. & Neese, F. Theoretical analysis of the spin hamiltonian parameters in Co(II)S4 complexes, using density functional theory and correlated *ab initio* methods. *Inorg. Chem.* **50**, 8741–8754 (2011).

53. Zadrozny, J. M. *et al.* Slow magnetization dynamics in a series of two-coordinate iron(II) complexes. *Chem. Sci* **4**, 125–138 (2013).

54. Gomez-Coca, S., Cremades, E., Aliaga-Alcalde, N. & Ruiz, E. Mononuclear single-molecule magnets: tailoring the magnetic anisotropy of first-row transition-metal complexes. *J. Am. Chem. Soc.* **135**, 7010–7018 (2013).

55. Singh, S. K., Gupta, T., Badkur, P. & Rajaraman, G. Magnetic anisotropy of mononuclear Ni-II complexes: on the importance of structural diversity and the structural distortions. *Chem.Eur. J* **20**, 10305–10313 (2014).

56. Ungur, L., Thewissen, M., Costes, J. P., Wernsdorfer, W. & Chibotaru, L. F. Interplay of strongly anisotropic metal ions in magnetic blocking of complexes. *Inorg. Chem.* **52**, 6328–6337 (2013).

57. Ye, S. & Neese, F. How do heavier halide ligands affect the signs and magnitudes of the zero-field splittings in halogenonickel(II) scorpionate complexes? A theoretical investigation coupled to ligand-field analysis. *J. Chem. Theo. Comput* **8**, 2344–2351 (2012).

58. Wernsdorfer, W., Bhaduri, S., Boskovic, C., Christou, G. & Hendrickson, D. N. Spin-parity dependent tunneling of magnetization in single-molecule magnets. *Phys. Rev. B* **65**, 180403 (2002).

59. Chibotaru, L. F. & Ungur, L. Ab initio calculation of anisotropic magnetic properties of complexes. I. Unique definition of pseudospin Hamiltonians and their derivation. *J Chem Phys* **137**, 064112–064122 (2012).

60. Ungur, L. & Chibotaru, L. *Lanthanides and Actinides in Molecular Magnetism* (Wiley-VCH Verlag GmbH & Co. KGaA, 2015).

61. Pinsky, M. & Avnir, D. Continuous symmetry measures. 5. The classical polyhedra. *Inorg. Chem.* **37**, 5575–5582 (1998).

62. Maurice, R. *et al.* Universal theoretical approach to extract anisotropic spin hamiltonians. *J. Chem. Theo. Comp* **5**, 2977–2984 (2009).

63. Ruamps, R. *et al.* Giant ising-type magnetic anisotropy in trigonal bipyramidal Ni(II) complexes: experiment and theory. *J. Am. Chem. Soc.* **135**, 3017–3026 (2013).

64. Carmieli, R. *et al.* The catalytic Mn2 + sites in the enolase-inhibitor complex: crystallography, single-crystal EPR, and DFT calculations. *J. Am. Chem. Soc.* **129**, 4240–4252 (2007).

65. Askevold, B. *et al.* Square-planar ruthenium(II) complexes: control of spin state by pincer ligand functionalization. *Chem. Eur. J* **21**, 579–589 (2015).

66. Goswami, T. & Misra, A. Ligand effects toward the modulation of magnetic anisotropy and design of magnetic systems with desired anisotropy characteristics. *J. Phys. Chem. A* **116**, 5207–5215 (2012).

67. Ungur, L. & Chibotaru, L. F. Magnetic anisotropy in the excited states of low symmetry lanthanide complexes. *Phys. Chem. Chem. Phys.* **13**, 20086–20090 (2011).

68. Gomez-Coca, S. *et al.* Origin of slow magnetic relaxation in Kramers ions with non-uniaxial anisotropy. *Nat. Commun* **5**, 4300 (2014).

69. Aquilante, F. *et al.* MOLCAS—a software for multiconfigurational quantum chemistry calculations. *WIREs Comput. Mol. Sci* **3**, 143–149 (2013).

70. Neeese, F. The ORCA program system. *WIREs Comput. Mol. Sci* **2**, 73–78 (2012).

Acknowledgements

G.R. acknowledges DST (EMR/2014/000247), INSA, DST Nanomission (SR/NM/NS-1119/2011) for funding. S.K.S. thanks Department of Chemistry, IITB, for Research Associate position. The authors also thank the anonymous reviewer for his constructive comments.

Author contributions

S.K.S and G.R. designed the project and S.K.S performed all the calculations. S.K.S and G.R analysed the results and wrote the manuscript.

Additional Information

3

High-efficiency electrochemical thermal energy harvester using carbon nanotube aerogel sheet electrodes

Hyeongwook Im[1,*], Taewoo Kim[1,*], Hyelynn Song[1], Jongho Choi[1], Jae Sung Park[2], Raquel Ovalle-Robles[3], Hee Doo Yang[4], Kenneth D. Kihm[5], Ray H. Baughman[6], Hong H. Lee[7], Tae June Kang[8] & Yong Hyup Kim[1,9]

Conversion of low-grade waste heat into electricity is an important energy harvesting strategy. However, abundant heat from these low-grade thermal streams cannot be harvested readily because of the absence of efficient, inexpensive devices that can convert the waste heat into electricity. Here we fabricate carbon nanotube aerogel-based thermo-electrochemical cells, which are potentially low-cost and relatively high-efficiency materials for this application. When normalized to the cell cross-sectional area, a maximum power output of $6.6\,W\,m^{-2}$ is obtained for a $51\,°C$ inter-electrode temperature difference, with a Carnot-relative efficiency of 3.95%. The importance of electrode purity, engineered porosity and catalytic surfaces in enhancing the thermocell performance is demonstrated.

[1] School of Mechanical and Aerospace Engineering, Seoul National University, Seoul 151-742, South Korea. [2] Institute of Advanced Machinery and Design, Seoul National University, Seoul 151-742, South Korea. [3] Nano-Science & Technology Center, Lintec of America, Inc., Richardson, Texas 75081, USA. [4] Department of NanoMechatronics Engineering, College of Nanoscience and Nanotechnology, Pusan National University, Busan 609-735, South Korea. [5] Department of Mechanical, Aerospace and Biomedical Engineering, University of Tennessee, Knoxville, Tennessee 37996, USA. [6] Alan G. MacDiarmid NanoTech Institute, University of Texas at Dallas, Richardson, Texas 75080, USA. [7] School of Chemical and Biological Engineering, Seoul National University, Seoul 151-744, South Korea. [8] Department of Mechanical Engineering, INHA University, Incheon 22212, South Korea. [9] Institute of Advanced Aerospace Technology, Seoul National University, Seoul 151-742, South Korea. * These authors contributed equally to this work. Correspondence and requests for materials should be addressed to T.J.K. (email: tjkang@inha.ac.kr) or to Y.H.K. (email: yongkim@snu.ac.kr).

Harvesting energy from waste heat has received much attention due to the world's growing energy problem[1-4]. Critical needs for harnessing waste heat are to improve the efficiency of thermal energy harvesters and decrease their cost[5]. Solid-state thermoelectric devices have been long investigated for the direct conversion of thermal energy to electrical energy, and many exciting advances have been made[6,7]. However, device performance relative to cost has so far limited application for waste heat recovery[8]. Thermal electrochemical energy harvesters[9] might have major advantages, as suggested in a very preliminary way by previous comparisons of Wh/dollar of solar and electrochemical thermocells[10]; however, they presently have no commercial applications because of their low energy conversion efficiencies and low areal output power. The goal of the present work is to increase the obtained energy conversion efficiencies and areal output power of thermocells to the point where they outperform thermoelectrics in energy output per device cost during device lifetime for low-grade thermal energy harvesting.

To increase the energy conversion efficiency of thermocells, carbon nanomaterials have been introduced as cell electrodes to take advantage of the fast redox processes, high thermal and electrical conductivities, and high gravimetric surface areas that these materials can provide[11]. Hu et al.[10] reported an energy conversion efficiency as high as 1.4%, relative to Carnot cycle efficiency, when carbon multi-walled nanotube (MWNT) buckypaper was used for thermocell electrodes. This efficiency was raised to 2.6% by introducing a carbon single-walled nanotube (SWNT)/reduced graphene oxide (rGO) composite electrode[12]. Improved mass transport due to enhanced porosity of the optimized SWNT/rGO composite was found responsible for the efficiency enhancement. Despite recent advances in thermocell technology, a significant efficiency increase is required for thermocells to become commercially attractive, considering that the Carnot-relative efficiency has to be 2–5% for commercial viability[13]. While efficiency is undoubtedly a key factor, another important quantity is the specific power the device can generate.

Here we exploit planar and cylindrically wound carbon nanotube (CNT) aerogel sheets as thermocell electrodes and devise various additional ways to optimize Carnot efficiency. The deployed optimization strategies to improve thermocell performance involve the use of CNT aerogel sheets as electrodes, removal of low activity carbonaceous impurities that limit electron transfer kinetics, decoration of CNT sheets with catalytic platinum nanoparticles, mechanical compression of nanotube sheets to tune conductivity and porosity, and the utilization of a cylindrical cell geometry. The output power density generated by a described cylindrical thermocell reaches $6.6\,\mathrm{W\,m^{-2}}$ for a 51 °C inter-electrode temperature difference, which corresponds to a Carnot-relative efficiency of 3.95% (that is, 3.95% of the maximum energy conversion efficiency possible for a heat engine operating between two given temperatures).

Results

CNT aerogel sheets as high-performance electrodes. A generic aqueous electrolyte thermocell, which utilizes the $Fe(CN)_6^{4-}$/$Fe(CN)_6^{3-}$ redox couple and K^+ as the counter ions, is schematically illustrated in the inset of Fig. 1a. An inter-electrode temperature difference causes a difference in the redox potentials of the electrolyte at the electrodes. This thermally generated potential difference drives electrons in the external circuit and ions in the electrolyte, thereby enabling electrical power to be generated. Continuous operation of the thermocell requires transport of the reaction products formed at one electrode to the other electrode. If either electrode is not furnished with the redox

molecules needed for electron generation or consumption, then power production will cease.

The thermoelectric potential in a thermocell is generated by the temperature dependence of the free energy difference between reactant and product of a reaction taking place at the electrolyte–electrode interface[9,13]. Minor effects on the cell potential arise from the thermal diffusion (Soret effect)[13-16] and the transport entropy of ions[17]. However, for most systems of interest, the electrode potentials dominate and the minor effects on potential can be neglected for practical purposes[13,16].

A planar-type thermocell, shown in Fig. 1a, was used to investigate the potential of CNT aerogel sheets as high-performance electrodes. For preparation of the thermocell electrode, CNT aerogel sheet was drawn from a CNT forest and laid on a rectangular tungsten frame that is connected to a motor. Then, the motor was rotated at 10 r.p.m. to simultaneously draw the sheet from the forest and warp it onto the frame (Fig. 1b). The thickness and area of the CNT electrode can be easily controlled by the number of motor rotations and the frame size, respectively. A 100-μm-thick CNT sheet electrode with an area of $1.0 \times 1.0\,\mathrm{cm^2}$ was used to evaluate thermocell performance. All thermocells and all electrochemical impedance measurements (to determine equivalent series resistance (ESR) and charge transfer resistance (R_{ct})) used as electrolyte a 0.4 M aqueous solution of potassium ferro/ferricyanide. Cyclic voltammetry (CV) measurements used an aqueous solution of 10 mM $K_3Fe(CN)_6$ and 0.1 M KCl.

Fast transport of the redox mediator ions into electrodes is required to obtain high-areal power generation from CNT thermocells. Randomly oriented CNT sheets, such as CNT buckypaper, may impede ion transport into electrode depths, due to the high tortuosity of the pore structure. On the other hand, the well-aligned CNTs in the aerogel sheets might result in faster ion transport deep within the electrodes, as schematically illustrated in Fig. 1c. The effectiveness of ion transport in CNT aerogel sheets was characterized by measuring the mass transport coefficient using an electrolytic flow cell (see Methods and Supplementary Fig. 1). Exploiting the limiting-current method, the mass transfer coefficient (k_c) was estimated from the dependence of limiting current on electrolyte concentration by using the following equation[18,19]:

$$k_c = \frac{i_L}{nFAC_\infty} \qquad (1)$$

where i_L is the limiting current, n is the number of moles of electrons transferred, F is the faradaic constant, A is the electrode area and C_∞ is the bulk species concentration.

Various aqueous 1:2 concentrations of $Fe(CN)_6^{3-}$/$Fe(CN)_6^{4-}$ redox couple in 0.5 M NaOH were used for the experiments, by varying the $Fe(CN)_6^{3-}$ concentration from 20 to 80 mM. The concentration of $Fe(CN)_6^{4-}$ was twice the $Fe(CN)_6^{3-}$ concentration for each experiment to ensure a limiting reaction rate at the cell cathode (that is, $Fe(CN)_6^{3-} + e^- \rightarrow Fe(CN)_6^{4-}$). The electrolyte stored in a glass container was circulated through the cell by a peristaltic pump. The flow rate was kept low at 6.6×10^{-6} $\mathrm{m^3\,s^{-1}}$ for allowing laminar flow (Reynolds number, Re ~ 650, Supplementary Note 1 and Supplementary Table 1) in the cell.

The limiting-current method is based on driving an electrochemical reaction to the maximum possible reaction rate, which is limited by the mass transport of redox ions. The reaction rate limit is indicated by a current plateau on a polarization curve plot. Polarization curves of the thermocells based on CNT buckypaper and CNT aerogel sheet electrodes are shown in Fig. 1d,e, respectively. A linear relationship between limiting current and reactant concentration is evident from the insets of the figure panels. Using equation (1), the mass transfer coefficient of CNT

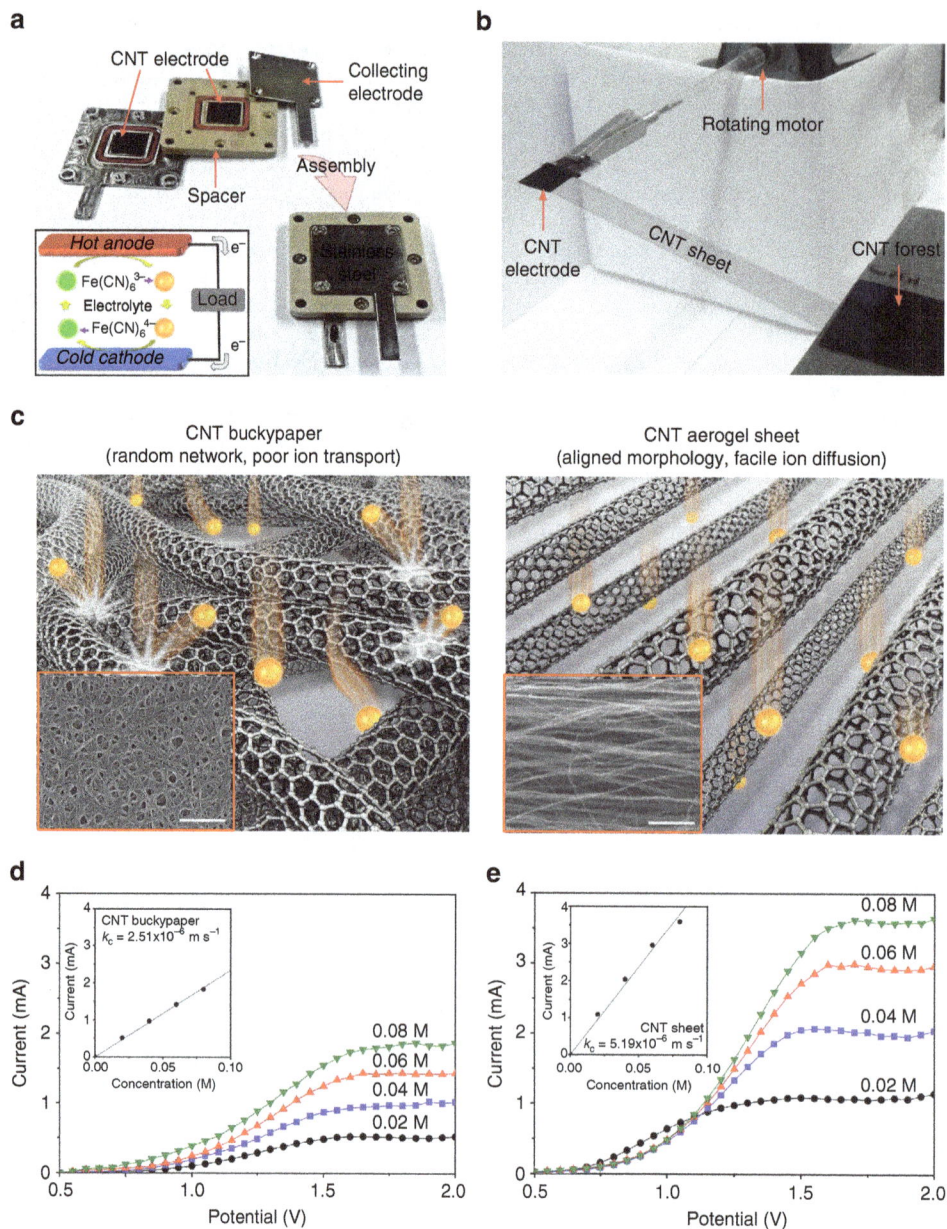

Figure 1 | Carbon nanotube aerogel sheets as high-performance electrodes. (a) Photographs of cell components and their assembly into a planar thermocell (inset: schematic drawing of thermocell operation). **(b)** An apparatus for continuously drawing a carbon nanotube (CNT) aerogel sheet from a CNT forest, and wrapping it around a metal frame to form an electrode for a planar thermocell. **(c)** Illustrations and SEM micrographs (insets) comparing CNT buckypaper and CNT aerogel electrodes and the relationship of these morphologies to ion transport (scale bars in the insets, 1-μm). MWNT bundling is not shown and only MWNT outer walls are pictured. Polarization curves for **(d)** CNT buckypaper and **(e)** CNT aerogel electrodes. The insets show the dependence of limiting current on ferrocyanide concentration.

aerogel sheet (5.19×10^{-6} m s^{-1}) is estimated to be twice that of CNT buckypaper (2.51×10^{-6} m s^{-1}). The transfer coefficient of CNT aerogel sheet electrode approaches the theoretically limiting mass transfer coefficient (5.76×10^{-6} m s^{-1}, see Supplementary Note 1), which corresponds to the unobstructed transport of reactants to a flat plate. These electrochemical results and the SEM micrographs suggest that the low tortuosity of the pore structure in the CNT aerogel, compared with that of CNT buckypaper, results in faster ion diffusion to deep within the electrode and corresponding higher limiting currents at a given redox concentration.

Optimization of thermocell performance. It is well-known that impurities, such as carbonaceous byproducts, are introduced during CNT synthesis[20]. A coating of these impurities on CNTs

could restrict charge exchange with redox ions in the electrolyte, thereby degrading cell performance[21]. Various efforts have been made to remove undesirable amorphous carbon from CNT surfaces, for the purpose of improving performance for diverse applications of CNTs[21,22]. We take advantage of the difference in oxidation rate in air between CNT and byproducts[23] to purify forest-drawn CNT sheets, which are ultimately used as electrodes.

To find the optimum heating schedule, the oxidation temperature was first varied in the range of 300–400 °C, while the oxidation time was held constant at 5 min. The purity of the CNT sheets was characterized using Raman spectroscopy for 514 nm excitation. The effect of oxidation temperature on the Raman intensity ratio of D band to G band (I_D/I_G) is shown in Fig. 2a, where the D band peak corresponds to defective carbon (like in amorphous carbon or carbon in defect sites) and the

G band arises from ordered sp^2 carbons[24]. The minimum ratio I_D/I_G occurs at anneal temperatures between 325 and 350 °C, signifying that higher anneal temperatures should not be used for annealing times as short as 5 min. For further optimization of CNT aerogel sheet quality, the anneal temperature was fixed at 340 °C and the heating time was varied between 0 and 60 min to further minimize the I_D/I_G ratio (Fig. 2b). The Raman spectra for an as-drawn sheet and for the optimally annealed sheet are shown in the figure inset. The merit of this method for removing CNT impurities was confirmed by high-resolution transmission electron microscopy (HR-TEM) images of MWNTs in the annealed sheets. The non-annealed MWNT (Fig. 2c) is covered with a carbonaceous coating, which is reduced by the 5-min anneal at 340 °C (Fig. 2d), and eliminated by a 15-min anneal at 340 °C (Fig. 2e), which corresponds to the minimum obtained ratio of I_D/I_G in Fig. 2b.

Thermocell performance was strongly affected by the annealing time at 340 °C, as shown in Fig. 3a,b. In this experiment, the hot plate temperature was maintained at 25 °C and the cold plate temperature was at 5 °C. While this temperature difference (ΔT) of 20 °C was applied between heating and cooling plates, the actual ΔT between the electrodes, which was calculated using the open-circuit voltage and the observed thermo-electrochemical Seebeck coefficient ($1.43\,mV\,K^{-1}$)[16], was smaller (17.5 °C). This difference is caused by thermal resistances (and corresponding temperature drops) at the interfaces between hot and cold electrodes and the respective heating and cooling plates. The curves of cell voltage versus areal current density and the corresponding power density curves obtained for various thermal oxidation times (shown in Fig. 3a,b) reveal that a lower I_D/I_G ratio, indicating that a cleaner CNT sheet, yields a higher maximum output power. The maximized areal power density (P_{MAX}), normalized to the square of the inter-electrode

temperature difference (ΔT^2), is shown as a function of thermal oxidation time at 340 °C in Fig. 3c. A 186% increase (from 0.07 to $0.13\,mW\,m^{-2}\,K^{-2}$) in the $P_{MAX}/\Delta T^2$ is shown to result from removing carbonaceous impurities from the surfaces of CNTs and CNT bundles. This optimum thermal oxidation condition (15 min at 340 °C) was used for further experiments.

To explain this major improvement in performance, we used electrochemical impedance measurements to characterize the three primary internal resistances of the thermocell (that is, the activation, ohmic and mass transport resistances)[16,25]. The activation resistance is the loss incurred in overcoming the activation barrier associated with reactions at the electrodes. The ohmic resistance is mainly due to the series resistances of electrode and electrolyte and the mass transport resistance is associated with the kinetics of ion diffusion and convection in the thermocell. The changes in these internal resistances with respect to heating time during sheet oxidation at 340 °C are shown in the inset of Fig. 3c. This figure shows that a major reduction in the activation resistance is realized by removal of carbonaceous impurities, while the decrease in the combined ohmic and mass transport resistances is much smaller. This result indicates that amorphous carbon covering the surface of CNT hinders electrochemical reaction, resulting in performance degradation.

The fact that the electrode is an aerogel implies that increased contact between CNTs (and correspondingly decreased inter-electrode electrical and thermal resistances) could be realized by compressing the electrode. However, the correspondingly increased density can restrict ion transport into the interior of the electrode (see Supplementary Fig. 2). To evaluate these opposing effects, mechanical compression of a planar electrode was accomplished by pressing CNT sheets (wrapped on the tungsten frame used for electrode fabrication), using two silicon wafers that were coated with gold films (which provide low

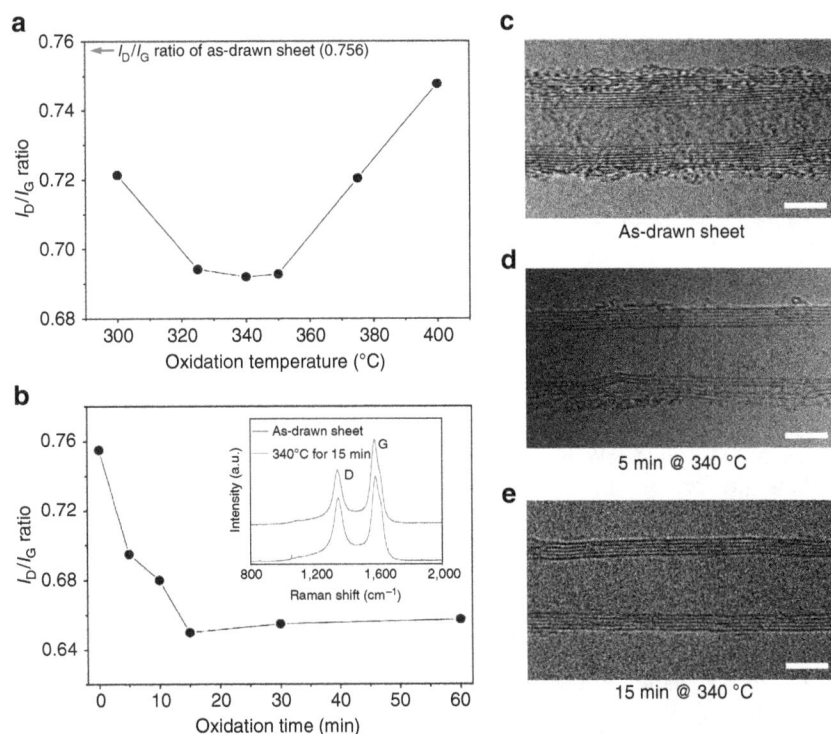

Figure 2 | The effects of thermal oxidation temperature and time. (a) The dependence of Raman I_D/I_G ratio on oxidation temperature for a 5-min anneal of CNT aerogel sheets in ambient air. **(b)** The dependence of the I_D/I_G ratio for CNT aerogel sheets on oxidation time in ambient air at 340 °C. The inset shows the Raman spectra for an as-drawn CNT sheet and for a CNT sheet that has been thermally oxidized in air for 15 min at 340 °C. **(c-e)** High-resolution transmission electron microscope (HR-TEM) images of **(c)** as-drawn, **(d)** 5-min oxidized, and **(e)** 15-min oxidized CNT aerogel sheets (all scale bars, 5-nm).

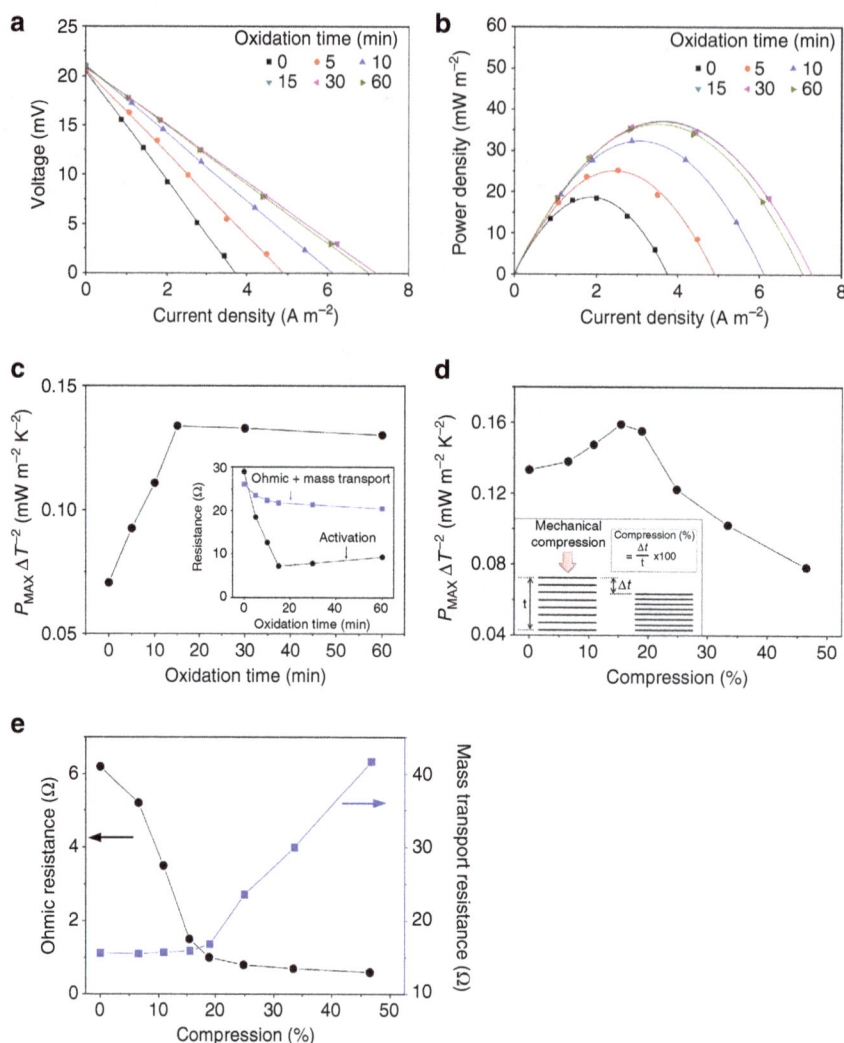

Figure 3 | The effect on thermocell performance of oxidation and compression of the CNT aerogel. (**a**) Cell voltage versus current density and (**b**) cell power density versus current density for samples having different thermal oxidation times. (**c**) Maximum power density normalized to the inter-electrode temperature difference ($P_{MAX}/\Delta T^2$) as a function of thermal oxidation time. (Inset: the dependence of activation resistance and the sum of ohmic and mass transport resistance on thermal oxidation time). (**d**) $P_{MAX}/\Delta T^2$ generated by the thermocell as a function of the per cent mechanical compression of cell electrodes. The inset illustrates the mechanical compression of a planar CNT aerogel electrode. (**e**) The dependence of ohmic and mass transport resistance on the compressive strain shown in **d**.

adhesion to the CNT sheet stack). The maximum obtained power density, normalized to ΔT^2, is plotted in Fig. 3d as a function of the per cent compression in the sheet thickness direction. These results show that $P_{MAX}/\Delta T^2$ is maximized for ∼15% sheet compression, and that this compression provides a 23% increase in this performance metric (from 0.13 to 0.16 mW m^{-2} K^{-2}).

The data in Fig. 3e show the origin of this effect of sheet compression. The mass transport resistance increases with increasing level of compression, whereas the ohmic resistance decreases with increasing compression. Ohmic resistance decreases with increasing compression, because a shorter electron pathway to the external circuit results from an increased number of contacts between CNTs and a decreased electrode thickness. On the other hand, mass transport resistance, typically affected by ion accessibility into electrodes, is increased by compression, indicating sluggish ion diffusion within the compressed CNT network (Supplementary Fig. 2). Therefore, $P_{MAX}/\Delta T^2$ is maximized at an intermediate compression (15%), where mass transport resistance is not much affected by compression, but ohmic resistance is dramatically decreased.

To further improve electrode performance, Pt nanoparticles were deposited on the thermally oxidized CNT aerogel sheet electrodes by chemical reduction of a platinum salt solution (see Methods and Supplementary Fig. 3). This Pt nanoparticle deposition increases the electrode surface area for the thermocell reaction, and the catalytic activity of platinum reduces the charge transfer resistance[26]. The HR-TEM image of Fig. 4a and the statistical size analysis of deposited Pt nanoparticles in Supplementary Fig. 3b show that platinum nanoparticles with an average diameter of 2.4 nm are uniformly deposited on the surfaces of individual CNTs and nanotube bundles to provide 86 wt% of Pt in the electrodes.

CV measurements were used to obtain the changes in electroactive surface area (ESA)[27] and redox potential difference between oxidation and reduction that Pt deposition provided. The CV curves of Fig. 4b indicate an increase in the Faradaic peak current for the thermally oxidized sheet electrode and Pt-decorated thermally oxidized sheet electrode (Pt-sheet) compared with that for the as-drawn sheet electrode. Moreover, the Pt-sheet has the highest faradaic current. This increased current can be attributed to the increased ESA, which according

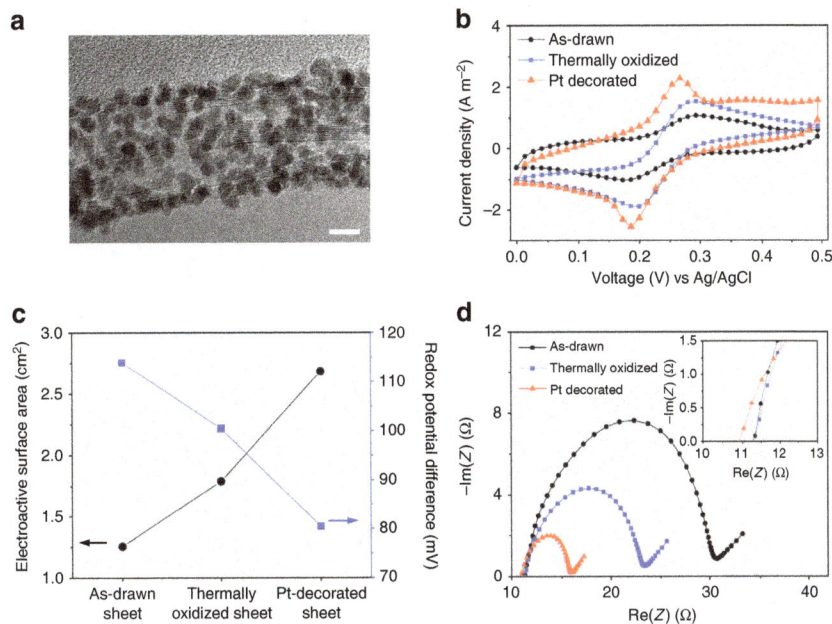

Figure 4 | Pt nanoparticle deposition for improving electrode performance. (**a**) HR-TEM image of individual CNT decorated by Pt nanoparticles. Scale bar, 5 nm. (**b**) Cyclic voltammograms obtained at a 100 mV s^{-1} scan rate using as-drawn, thermally oxidized, and thermally oxidized and Pt-deposited CNT electrodes. (**c**) Electroactive surface area and redox potential difference for the above various CNT electrodes. (**d**) Nyquist impedance plots for various electrodes. The inset shows a close-up of the high frequency region of the curves.

to the Randles–Sevcik equation[28] is given by:

$$I_p = 2.69 \times 10^5 \cdot \text{ESA} \cdot D^{1/2} \cdot n^{3/2} \cdot \upsilon^{1/2} \cdot C \qquad (2)$$

where I_p is the faradaic peak current, D is the diffusion coefficient, n is the number of electrons transferred during the redox reaction, υ is the potential scan rate, and C is the concentration of probe molecule.

The redox potential differences and ESAs determined by CV are shown in Fig. 4c for electrodes that are as-drawn, thermally oxidized and platinum-deposited after chemical cleaning by thermal oxidation. This figure shows that the cleaned CNT electrode has a higher ESA and a smaller redox potential difference than the as-drawn CNT electrode, which led to the improved performance shown in Fig. 3c. The largest ESA and the lowest potential difference resulted (Fig. 4c) when Pt nanoparticles were deposited onto the cleaned surface of CNTs. These results indicate that the Pt-decorated sheet electrodes provide the highest performance in thermocells, because they yield the largest effective surface area and the greatest reduction in the potential difference between peaks for oxidation and reduction. Electrochemical impedance spectroscopy analysis was also performed to support the performance improvement. The ESR (the intercept of the curve with the x-axis of the Nyquist plot[29] in the inset of Fig. 4d) is slightly reduced after Pt decoration. The Nyquist plot in Fig. 4d shows that the charge transfer resistance (the diameter of the semicircle)[30] is lower for the thermally oxidized sheet than for the as-drawn sheet and that it is lowest for the Pt-decorated, thermally oxidized sheet. These observations from electrochemical impedance spectroscopy measurements agree with the CV results.

Cylinder-type CNT thermocells and their performance. Building on our results for planar electrodes, we further optimized thermocell performance by using the cylindrical electrode configuration of Fig. 5a (while deploying the same 0.4 M ferro/ferricyanide aqueous electrolyte as for the flat thermocells).

The cylindrical electrodes pictured in this figure were fabricated by first using the apparatus of Fig. 5b to wind a forest-drawn CNT sheet onto a 300-µm diameter tungsten wire (which facilitates current collection) until a 3.0- to 3.5-mm outer scroll diameter was obtained. The SEM images of Fig. 5c show the realized highly uniform structure of the electrode sidewall and the orientation of the CNTs around the electrode circumference. Using the results of Fig. 3d on the optimal degree of electrode compression to produce densification, a cleaned CNT sheet electrode (wound to a 3.5 mm diameter) was compressed to 3.0 mm diameter by using two grooved templates (shown in Supplementary Fig. 4). Following deposition of Pt nanoparticles by chemical reduction of a platinum salt solution, two resulting cylindrical CNT sheet electrodes were inserted into a 3 mm inner diameter, cylindrical glass tube to form the thermocell of Fig. 5a.

The performance of the above thermocells containing differently prepared cylindrical CNT aerogel electrodes is presented in Fig. 5d,e for $\Delta T \approx 51\,°C$. The cylindrically configured as-drawn sheet electrode generates a P_{MAX} of 2.0 W m^{-2}. These results show that removal of carbonaceous impurities by the described optimized thermal oxidation increases the power output to 3.7 W m^{-2}, which is further increased for the cylindrical electrode thermocell to 6.0 W m^{-2} when Pt nanoparticles are deposited within the cylindrical electrodes. Finally, the maximum power density was further enhanced to 6.6 W m^{-2} when the cleaned, Pt-sheet electrode was laterally compressed. This output corresponds to an energy conversion efficiency of 3.95%, relative to the theoretically limiting Carnot cycle efficiency, as shown in Fig. 5f.

The energy conversion efficiency (η) of the thermocells was calculated using

$$\eta = \frac{(1/4)V_{OC} \cdot I_{SC}}{A_c \cdot \kappa (\Delta T/d)} \qquad (3)$$

where V_{OC} is the open-circuit voltage, I_{SC} is the short-circuit current, A_c is the cross-sectional area of the cell, κ is the thermal conductivity of the electrolyte, ΔT is the temperature difference

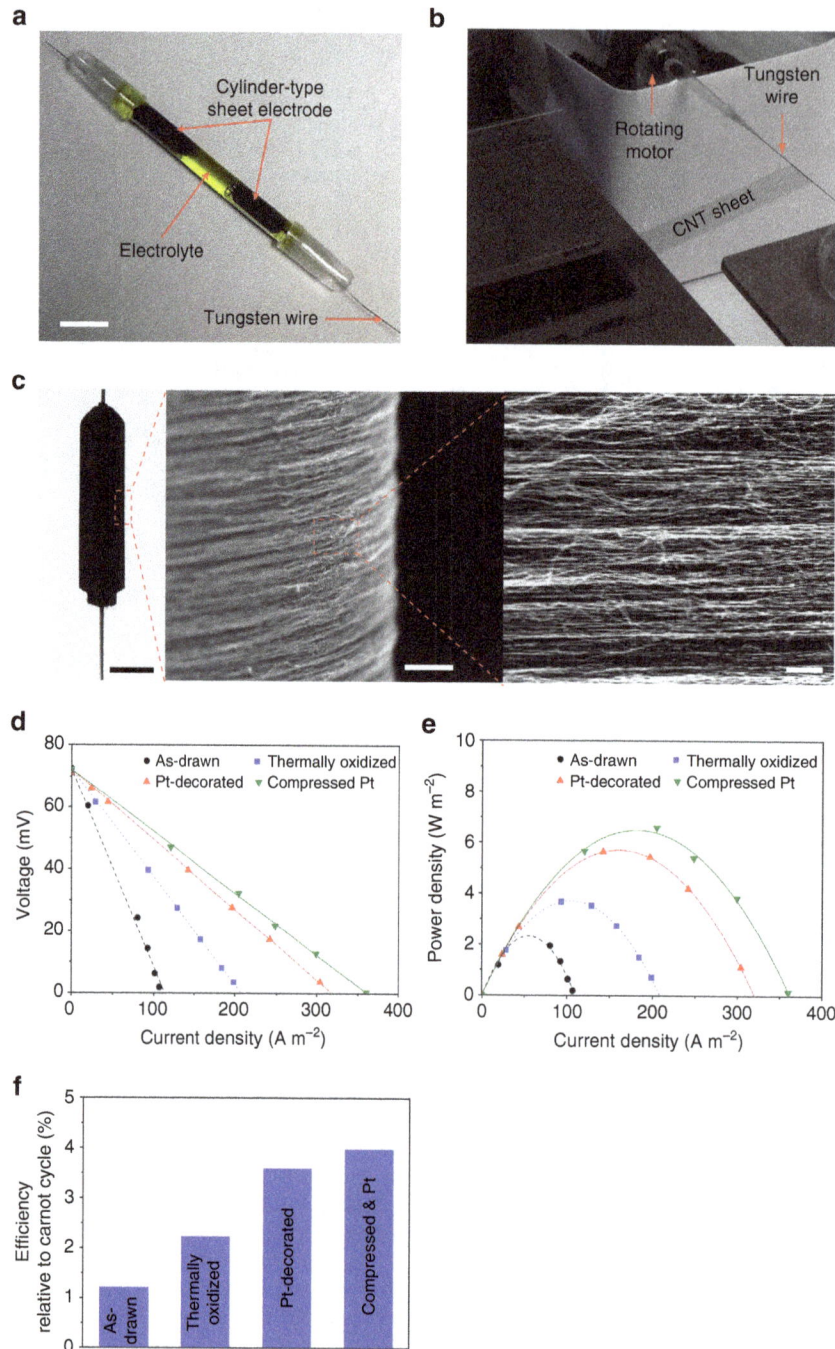

Figure 5 | Fabrication of cylinder-type CNT thermocells and their performance for $\Delta T \approx 51\,°C$. (a) Photograph of an assembled thermocell (scale bar, 2 cm) and **(b)** apparatus for fabrication of cylindrical CNT thermocell electrodes. **(c)** Photograph and SEM images of a cylindrical CNT electrode. Scale bars, 3 mm (left of panel), 30 μm (middle of panel) and 1 μm (right of panel). **(d)** Cell voltage and **(e)** power density versus current density for variously treated cell electrodes. **(f)** Energy conversion efficiency relative to Carnot cycle efficiency for various CNT electrodes in the cylindrical cell configuration. All sheet samples (except the as-drawn) were purified by thermal oxidation before assembly into cylindrical electrodes.

between the electrodes and d is the electrode separation distance (see Supplementary Note 2).

The open-circuit voltage was $V_{OC} = 72$ mV and the ΔT thereby calculated (from this voltage and the Seebeck coefficient of 1.43 mV K^{-1}, Supplementary Note 3, and Supplementary Fig. 5b) was 51.4 °C for all electrode types in the cylindrical thermocell configuration. Thermal conduction dominates the heat transfer through the electrolyte between the hot and cold electrodes (see Supplementary Note 4, Supplementary Fig. 6 and Supplementary Table 2). The cell area and the thermal

conductivity[31] are 7.1×10^{-6} m^2 and 0.57 W m^{-1} K^{-1}, respectively. The electrode separation distance is 2.5 cm, corresponding to the closest approach distance between the hot and cold electrodes shown in Fig. 5a. The short-circuit current obtained from as-drawn, thermally oxidized, thermally oxidized and Pt-decorated; and thermally oxidized, compressed and Pt-decorated sheets in the cylindrical cell configuration are 0.8 mA, 1.5 mA, 2.3 mA and 2.6 mA, respectively. Using equation (3), η values of 0.17%, 0.32%, 0.51% and 0.56% are attained through the use of cylindrical thermocell configuration for these cylindrical thermocells from differently processed CNT aerogel sheets, respectively.

The energy conversion efficiency (η_r), relative to the Carnot efficiency limit of a heat engine, is

$$\eta_r = \frac{\eta}{(\Delta T/T_H)} \tag{4}$$

where T_H is the hot temperature. We calculated the Carnot-relative energy conversion efficiency using equations (3) and (4) (See Supplementary Table 3 for parameters used in the equations), the measured $\Delta T = 51.4\,°C$, and an estimated T_H for the hot side electrode temperature of $90.7\,°C$ (estimated by considering symmetric geometry of cylindrical cell and electrodes arrangement, and ΔT). The resulting calculated Carnot-relative efficiency was $\eta_r = 3.95\%$ for the optimized cylindrical thermocell (based on cylindrical electrodes made of thermally purified, optimally compressed, Pt-decorated CNT sheets). This thermocell efficiency is substantially higher than the previously reported record Carnot-relative efficiency for a thermocell, which is 2.63% (ref. 12).

Discussion

The ability to simultaneously obtain a record Carnot-relative efficiency (3.95%) and a record areal current density (6.6 W m^{-2} for a temperature difference of only 51.4 °C) bodes well for eventual practical deployment of thermo-electrochemical cells for harvesting low-grade heat as electrical energy. This work demonstrates the importance of electrode purity, engineered porosity and catalytic surfaces on thermocell performance and the ways that each predominately affects thermocell parameters. It is doubtful that thermal electrochemical cells will ever match the efficiencies obtained for costly thermoelectrics. However, potentially low cost (using future low-cost catalyst) and convenient deployability (like the demonstrated wrapping of flexible CNT sheet thermocells about hot pipes[10]) might eventually lead to their importance in the menagerie of means for practically harvesting thermal energy.

Methods

Mass transfer coefficients of CNT electrodes. The mass transport characteristics of CNT buckypaper and CNT aerogel sheet electrodes were investigated using the electrolytic flow cell of Supplementary Fig. 1. CNT buckypaper was prepared by vacuum filtration of MWNTs (CM-95, Hanwha Nanotech) suspended in anhydrous N,N-dimethylformamide (N,N-DMF, Sigma-Aldrich) onto a membrane filter (Millipore PTFE filter, 0.2 μm pore size, 47 mm diameter), washing with deionized (DI) water and methanol, drying in vacuum and removal of the formed sheet from the filter. CNT aerogel sheet electrodes were prepared by sheet draw from a CNT forest. The electrolytic flow cell shown in Supplementary Fig. 1 consists of two parallel collecting electrodes that are separated by 10 mm. For both the CNT buckypaper and CNT aerogel sheet evaluations, 100-μm-thick CNT electrodes having an identical area of $1.0 \times 1.0\,cm^2$ are attached on the centre of both collecting electrodes using carbon paste. Other exposed areas of the collecting electrodes were covered with 100-μm-thick polyethylene terephthalate film. Various aqueous 1:2 concentrations of $Fe(CN)_6^{3-}/Fe(CN)_6^{4-}$ redox couple in 0.5 M NaOH were used for the experiments by varying the $Fe(CN)_6^{3-}$ concentration from 20 to 80 mM. The concentration of $Fe(CN)_6^{4-}$ was twice as high as the $Fe(CN)_6^{3-}$ concentration for each experiment to ensure a limiting reaction rate at the cell cathode (that is, $Fe(CN)_6^{3-} + e^- \rightarrow Fe(CN)_6^{4-}$). The electrolyte stored in a glass container was circulated through the cell by a peristaltic pump (Longer pump, BT100-2J), and the flow rate was kept low at $6.6 \times 10^{-6}\,m^3\,s^{-1}$ for allowing laminar flow in a cell channel. A power supply (Keithley, 2400 Source-meter) was used to drive electrochemical reactions.

Preparation and post-treatment of CNT aerogel sheet. Vertically aligned MWNT arrays were grown on an iron-catalyst-coated silicon (Si) substrate by chemical vapour deposition of acetylene gas[32]. The diameter and height of MWNTs are ~10 nm and ~200 μm, respectively. CNT sheet was continuously drawn from a sidewall of the MWNT forests using a dry spinning process[32,33] and wrapped around a rectangular tungsten frame for planar-type electrodes and tungsten wire for cylinder-type electrodes, using a connected rotating motor at 10 r.p.m. The planar electrode has a thickness of 100 μm and area of $1.0 \times 1.0\,cm^2$

and the cylinder electrodes had a diameter and length of 3 mm and 2.0 cm, respectively. Thermal annealing of the CNT sheets to remove carbonaceous impurities was performed in ambient atmosphere using a halogen-lamp-heated quartz tube furnace, which had a ramp time to target temperatures (from 300 to 400 °C) of 1 min. For deposition of Pt nanoparticles by chemical reduction of a platinum salt solution[34], 3.75 mg of K_2PtCl_4 (Aldrich) was dissolved in 50 ml of diluted ethylene glycol solution (3:2 by volume ethylene glycol:DI water). Afterwards, the described thermally oxidized CNT sheet electrodes were immersed in the platinum salt solution and Pt nanoparticle deposition was permitted for 3 h at 110 °C with weak stirring to prevent mechanical damage to CNT sheet. After the reaction, the pH of the solution was decreased to 2 using HCl and the electrode was washed with DI water several times to remove excess ethylene glycol. To avoid liquid-based aerogel densification during liquid evaporation, the electrodes were not dried before use in the thermocells.

Thermocell testing and materials characterization. The electrolyte used in all tests was a 0.4 M aqueous solution of $Fe(CN)_6^{4-}/Fe(CN)_6^{3-}$. The ionic conductivity of the electrolyte at different temperatures was measured using a conductivity meter (Metter Toledo S-230), as shown in Supplementary Fig. 7. For evaluation of planar thermocells, the cell (Fig. 1a) was placed between two fluid-heated plates that were connected to hot and cold thermostatic baths to provide ± 0.1 °C control of plate temperatures. For cylindrical thermocell testing, the glass wall surrounding each cell electrode (which are separated by 2.5 cm) was wrapped and bonded with a TYGON tube, through which cooling fluid or heating fluid was passed. The inter-electrode direction of the thermocell was oriented horizontally during measurements. While the hot and cold fluid temperatures were 100 °C and 30 °C, respectively, thermal resistances between the heating and cooling sources and the respective electrodes reduced the inter-electrode temperature difference from 70 °C to $\Delta T = 51.4\,°C$. This ΔT of 51.4 °C, obtained by dividing the measured V_{OC} by the thermo-electrochemical Seebeck coefficient, would correspond to a temperature difference between the cold and hot electrodes averaged over the radial shape of CNT electrode. A voltage-current meter (Keithley, 2000 Multimeter) was used for characterizing cell voltage versus cell current for different external resistive loads, and thereby determining power output. The glass tube having 3 mm inside diameter was used for the cylinder thermocell experiments (Fig. 5a). Raman spectra as a function of the thermal oxidation times and the temperature used for removing carbonaceous impurities were recorded for 514 nm excitation using a Renishaw inVia Raman Microscope. Sample structure was further characterized by field-emission scanning electron microscopy (FE-SEM, Hitachi-S4800) and transmission electron microscopy (JEM-300F). Electrochemical impedance measurements were conducted in the frequency range between 10 kHz and 50 mHz using a commercial instrument (Zahner, IM6ex). CV (using a Digi-Ivy, DY2100 instrument) used 10 mM $K_3Fe(CN)_6$ with 0.1 M KCl as the supporting electrolyte in aqueous solution and a scan rate of 100 mV s^{-1}. Platinum and Ag/AgCl electrodes were used as counter and reference electrodes, respectively, for the CV measurements.

References

1. Lan, Y., Minnich, A. J., Chen, G. & Ren, Z. Enhancement of thermoelectric figure-of-merit by a bulk nanostructuring approach. *Adv. Funct. Mater.* **20**, 357–376 (2010).
2. Biswas, K. *et al.* High-performance bulk thermoelectrics with all-scale hierarchical architectures. *Nature* **489**, 414–418 (2012).
3. Zhao, D. Waste thermal energy harvesting from a convection-driven Rijke–Zhao thermo-acoustic-piezo system. *Energy Convers. Manage.* **66**, 87–97 (2013).
4. Hochbaum, A. I. *et al.* Enhanced thermoelectric performance of rough silicon nanowires. *Nature* **451**, 163–167 (2008).
5. Hall, P. J. & Bain, E. J. Energy-storage technologies and electricity generation. *Energy Policy* **36**, 4352–4355 (2008).
6. Kraemer, D. *et al.* High-performance flat-panel solar thermoelectric generators with high thermal concentration. *Nat. Mater.* **10**, 532–538 (2011).
7. Hewitt, C. A. *et al.* Multilayered carbon nanotube/polymer composite based thermoelectric fabrics. *Nano Lett.* **12**, 1307–1310 (2012).
8. Vining, C. B. An inconvenient truth about thermoelectrics. *Nat. Mater.* **8**, 83–85 (2009).
9. Kuzminskii, Y. V., Zasukha, V. A. & Kuzminskaya, G. Y. Thermoelectric effects in electrochemical systems. Nonconventional thermogalvanic cells. *J. Power Sources* **52**, 231–242 (1994).
10. Hu, R. *et al.* Harvesting waste thermal energy using a carbon-nanotube-based thermo-electrochemical cell. *Nano Lett.* **10**, 838–846 (2010).
11. Nugent, J. M., Santhanam, K. S. V., Rubio, A. & Ajayan, P. M. Fast electron transfer kinetics on multiwalled carbon nanotube microbundle electrodes. *Nano Lett.* **1**, 87–91 (2001).
12. Romano, M. S. *et al.* Carbon nanotube—reduced graphene oxide composites for thermal energy harvesting applications. *Adv. Mater.* **25**, 6602–6606 (2013).
13. Quickenden, T. I. & Mua, Y. A review of power generation in aqueous thermogalvanic cells. *J. Electrochem. Soc.* **142**, 3985–3994 (1995).
14. Eastman, E. D. Thermodynamics of non-isothermal systems. *J. Am. Chem. Soc.* **48**, 1482–1493 (1926).

15. Eastman, E. D. Theory of the Soret effect. *J. Am. Chem. Soc.* **50**, 283–291 (1928).

16. Kang, T. J. *et al.* Electrical power from nanotube and graphene electrochemical thermal energy harvesters. *Adv. Funct. Mater.* **22**, 477–489 (2012).

17. deBethune, A. J., Licht, T. S. & Swendeman, N. The temperature coefficients of electrode potentials: the isothermal and thermal coefficients—the standard ionic entropy of electrochemical transport of the hydrogen ion. *J. Electrochem. Soc.* **106**, 616–625 (1959).

18. Wragg, A. A. & Leontaritis, A. A. Local mass transfer and current distribution in baffled and unbaffled parallel plate electrochemical reactors. *Chem. Eng. J.* **66**, 1–10 (1997).

19. Cañizares, P., García-Gómez, J., Fernández de Marcos, I., Rodrigo, M. A. & Lobato, J. Measurement of mass-transfer coefficients by an electrochemical technique. *J. Chem. Educ.* **83**, 1204 (2006).

20. Moon, J.-M. *et al.* High-yield purification process of singlewalled carbon nanotubes. *J. Phys. Chem. B* **105**, 5677–5681 (2001).

21. Fang, H.-T. *et al.* Purification of single-wall carbon nanotubes by electrochemical oxidation. *Chem. Mater.* **16**, 5744–5750 (2004).

22. Rinzler, A. G. *et al.* Large-scale purification of single-wall carbon nanotubes: Process, product, and characterization. *Appl. Phys. A* **67**, 29–37 (1998).

23. Dementev, N., Osswald, S., Gogotsi, Y. & Borguet, E. Purification of carbon nanotubes by dynamic oxidation in air. *J. Mater. Chem.* **19**, 7904–7908 (2009).

24. Dresselhaus, M. S., Dresselhaus, G., Jorio, A., Souza Filho, A. G. & Saito, R. Raman spectroscopy on isolated single wall carbon nanotubes. *Carbon* **40**, 2043–2061 (2002).

25. Im, H. *et al.* Flexible thermocells for utilization of body heat. *Nano Res.* **7**, 1–10 (2014).

26. Rubi, J. M. & Kjelstrup, S. Mesoscopic nonequilibrium thermodynamics gives the same thermodynamic basis to butler – volmer and nernst equations. *J. Phys. Chem. B* **107**, 13471–13477 (2003).

27. Pacios, M., del Valle, M., Bartroli, J. & Esplandiu, M. J. Electrochemical behavior of rigid carbon nanotube composite electrodes. *J. Electroanal. Chem.* **619–620**, 117–124 (2008).

28. Papakonstantinou, P. *et al.* Fundamental electrochemical properties of carbon nanotube electrodes. *Fuller. Nanotubes Carbon Nanostruct.* **13**, 91–108 (2005).

29. Park, S.-M. & Yoo, J.-S. Peer reviewed: electrochemical impedance spectroscopy for better electrochemical measurements. *Anal. Chem.* **75**, 455 A–461 A (2003).

30. Randles, J. E. B. Kinetics of rapid electrode reactions. *Discuss. Faraday Soc.* **1**, 11–19 (1947).

31. Romano, M. *et al.* Novel carbon materials for thermal energy harvesting *J. Therm. Anal. Calorim.* **109**, 1229–1235 (2012).

32. Zhang, M. *et al.* Strong, transparent, multifunctional, carbon nanotube sheets. *Science* **309**, 1215–1219 (2005).

33. Im, H. *et al.* Enhancement of heating performance of carbon nanotube sheet with granular metal. *ACS Appl. Mater. Interfaces* **4**, 2338–2342 (2012).

34. Jining, X., Nanyan, Z. & Vijay, K. V. Functionalized carbon nanotubes in platinum decoration. *Smart Mater. Struct.* **15**, S5 (2006).

Acknowledgements

This research was supported in Korea by the National Research Foundation of Korea (Grants 2009-0083512, 2014R1A2A1A05007760 and 2014R1A1A4A01008768). Financial support at the University of Texas at Dallas was from Air Force grants FA9550-13-C-0004, FA9550-15-1-0089 and Robert A. Welch Foundation grant AT-0029.

Author contributions

H.I. and T.K. contributed to experiment design, measurements, data analysis and manuscript preparation. H.S., J.C., J.S.P. and H.D.Y contributed to experimental measurements and data analysis. R.O.-R. made and characterized nanotube samples. K.D.K., R.H.B., H.H.L., T.J.K. and Y.H.K. contributed to planning experiments, data analysis and manuscript preparation.

Additional information

Nanocaged enzymes with enhanced catalytic activity and increased stability against protease digestion

Zhao Zhao[1,2,*], Jinglin Fu[3,*], Soma Dhakal[4], Alexander Johnson-Buck[4], Minghui Liu[1], Ting Zhang[3], Neal W. Woodbury[2,5], Yan Liu[1,2], Nils G. Walter[4] & Hao Yan[1,2]

Cells routinely compartmentalize enzymes for enhanced efficiency of their metabolic pathways. Here we report a general approach to construct DNA nanocaged enzymes for enhancing catalytic activity and stability. Nanocaged enzymes are realized by self-assembly into DNA nanocages with well-controlled stoichiometry and architecture that enabled a systematic study of the impact of both encapsulation and proximal polyanionic surfaces on a set of common metabolic enzymes. Activity assays at both bulk and single-molecule levels demonstrate increased substrate turnover numbers for DNA nanocage-encapsulated enzymes. Unexpectedly, we observe a significant inverse correlation between the size of a protein and its activity enhancement. This effect is consistent with a model wherein distal polyanionic surfaces of the nanocage enhance the stability of active enzyme conformations through the action of a strongly bound hydration layer. We further show that DNA nanocages protect encapsulated enzymes against proteases, demonstrating their practical utility in functional biomaterials and biotechnology.

[1] Center for Molecular Design and Biomimetics, the Biodesign Institute at Arizona State University, Tempe, Arizona 85287, USA. [2] School of Molecular Sciences, Arizona State University, Tempe, Arizona 85287, USA. [3] Department of Chemistry, Center for Computational and Integrative Biology, Rutgers University-Camden, Camden, New Jersey 08102, USA. [4] Department of Chemistry, Single Molecule Analysis Group, University of Michigan, Ann Arbor, Michigan 48109, USA. [5] Center for Innovations in Medicine, the Biodesign Institute at Arizona State University, Tempe, Arizona 85287, USA. * These authors contributed equally to this work. Correspondence and requests for materials should be addressed to H.Y. (email: hao.yan@asu.edu) or to J.F. (email: jinglin.fu@rutgers.edu) or to N.G.W. (email: nwalter@umich.edu).

Common micro- and nanoscale subcellular compartments are formed from either lipids or proteins and include mitochondria, lysosomes, peroxisomes, carboxysomes and other metabolosomes, as well as multi-enzyme complexes[1–5]. Compartments increase the overall activity and specificity of the encapsulated enzyme pathways by maintaining a high local concentration of enzymes and substrates, promoting substrate channelling and protecting their content from damage, as well as by segregating potentially damaging reactions from the cytosol. Spatial confinement is also an important aspect for chaperone-assisted folding of linear polypeptides into active tertiary and quaternary conformations, as well as for preventing proteins from aggregating under cellular stress conditions[6]. A better understanding of the effects of spatial confinement on protein function will not only enhance our fundamental knowledge of cellular organization and metabolism but also increase our ability to translate biochemical pathways into a variety of noncellular applications, ranging from diagnostics and drug delivery to the production of high-value chemicals and smart materials[7–11]. Over the past few decades, artificial enzymatic particles have been created using compartmentalization by virus-like protein particles[7], liposomes or polymersomes[5] and chemical crosslinking[8]. However, severe obstacles to a broader application remain, including low encapsulation yield of large proteins because of steric hindrance[12], insufficient access of substrates to the encapsulated enzymes, aggregation of vesicle shells[13] and limited control over the spatial arrangement of proteins within the compartments[7,10].

Recently, DNA nanostructures have started to emerge as promising molecular scaffolds to organize biomolecules at the nanoscale based on their programmable, sequence-driven self-assembly[14–20]. For example, multi-enzyme cascades have been assembled on DNA nanostructures with precise control over the spatial arrangement to enhance catalytic activity by substrate channelling[19,21]. Conversely, self-assembling DNA nanoboxes and -cages have shown promise in the delivery of macromolecular payloads such as antibodies[22,23] and enzymes[24]. Tubular DNA nanostructures have also been used to construct efficient enzyme cascade nanoreactors[25,26]. Here, we describe a simple and robust strategy for the DNA nanocage-templated encapsulation of metabolic enzymes with high assembly yield and controlled packaging stoichiometry. With such an approach in hand, we sought to test the hypothesis that the recently described, chaperone-like stabilizing impact of polyphosphate on metabolic protein enzymes[27] together with the cryptic RNA binding properties of many enzymes[28] may lead to beneficial effects when enzymes are surrounded by DNA nanocages.

Results

Enzyme encapsulation strategy. As shown in Fig. 1a, our approach for enzyme encapsulation within DNA nanocages involves two steps: (1) the attachment of an individual enzyme into an open half-cage and (2) the assembly of two half-cages into a full (closed) nanocage. DNA half-cages were constructed by folding a full-length M13 viral DNA[29] into the indicated shape based on a honeycomb lattice using the DNA origami technique[14]; a shape with two open sides was chosen to improve accessibility of the internal cavity to large proteins. Two half-cages were then linked into a full-cage by adding 24 short-bridge DNA strands that hybridize with the complementary ssDNA sequences extending from the edges of either half-cage. The DNA full-cage is ~ 54 nm $\times$ 27 nm $\times$ 26 nm with designed inner cavity dimensions of 20 nm $\times$ 20 nm $\times$ 17 nm. By design, 42 small nanopores (each ~ 2.5 nm in diameter) were introduced on each of the top and bottom surfaces of the DNA nanocage to permit the diffusion of small molecules (for example, enzyme substrates) across the DNA walls (Supplementary Fig. 1).

The formation of half and full DNA nanocages was first characterized using transmission electron microscopy (TEM) (Supplementary Figs 2 and 3) and gel electrophoresis (Supplementary Fig. 4), which indicate a nearly 100% yield for half-cages and a more than 90% yield for full-cages. To capture target enzymes into a half-cage, a previously reported succinimidyl 3-(2-pyridyldithio) propionate (SPDP) chemistry was used to crosslink a lysine residue on the protein surface to a thiol-modified oligonucleotide[19,30,31]. Two anchor probes of complementary sequence were displayed on the bottom of the half-cage cavity to capture a DNA-modified enzyme via sequence-specific DNA hybridization. As a demonstration of an enzyme cascade, a glucose oxidase (GOx)-attached half-cage was incubated with a horseradish peroxidase (HRP)-attached half-cage at a stoichiometric ratio of ~ 1:1, followed by the addition of bridge strands into solution to assemble a full-DNA nanocage containing a GOx/HRP pair. The inner cavity of a full nanocage is of sufficient size to encapsulate this enzyme pair (GOx is ~ 10 nm (ref. 32) and HRP ~ 5 nm in diameter (ref. 33)). Unencapsulated enzyme and excess short DNA strands were removed using agarose gel electrophoresis[29]. Details of the enzyme–DNA conjugation and optimization of the assembly are discussed in Supplementary Figs 5–10, Supplementary Table 1 and Supplementary Note 1.

Characterization of enzyme encapsulation. To verify the presence of both enzymes within a DNA nanocage, the co-localization of a Cy3-labelled GOx (green emission) and a Cy5-labelled HRP (red emission) was quantified by dual-colour fluorescence gel electrophoresis, where a gel band with over-lapped green and red colours was identified (see Supplementary Fig. 11). By comparison, the GOx-containing half-cage (Half[GOx]) shows the presence of only Cy3 (green), whereas a HRP-half-cage (Half[HRP]) shows the presence of only Cy5 (red). In addition, negatively stained TEM images were used to visualize DNA cages upon stoichiometrically controlled encapsulation of a single GOx (Fig. 1b) or a single GOx/HRP pair (Fig. 1c), where GOx and HRP were visible as brighter spots within the cage. To quantitatively analyse the yield of DNA nanocage encapsulation, two-colour total internal reflection fluorescence (TIRF) microscopy[34] (Fig. 2a) was used to characterize the fluorescence co-localization of a Cy3-labelled enzyme and a Cy5-labelled nanocage (Fig. 2b). Six different enzymes were tested and characterized for encapsulation, ranging from the smallest HRP (~ 44 kDa)[35], (malic dehydrogenase (MDH, ~ 70 kDa)[36], glucose-6-phosphate dehydrogenase (G6PDH, ~ 100 kDa)[37], lactic dehydrogenase (LDH, ~ 140 kDa)[38] and GOx (~ 160 kDa)[39] to the largest β-galactosidase (β-Gal, ~ 450 kDa)[40]. All six enzymes were successfully encapsulated within full DNA nanocages with high yields, ranging from 64 to 98% (Fig. 2c and Supplementary Table 2). The relatively low yield of β-Gal (64%) may be because of its large size (~ 16 nm in diameter), which is comparable to the inner diameter of the nanocage (~ 20 nm), likely resulting in steric hindrance for encapsulation. To evaluate how many copies of the same enzyme were encapsulated per DNA nanocage, the stepwise single-molecule fluorescence photobleaching was used to count the number of Cy3 fluorophores per cage (Fig. 2d). The number of copies of each enzyme per cage was estimated by normalizing the number of Cy3 fluorophores per DNA nanocage with the average number of Cy3 labels per free enzyme (see Supplementary Note 2). A majority of nanocage-encapsulated enzymes showed only one- or two-step photobleaching of Cy3, similar to the photobleaching of single free

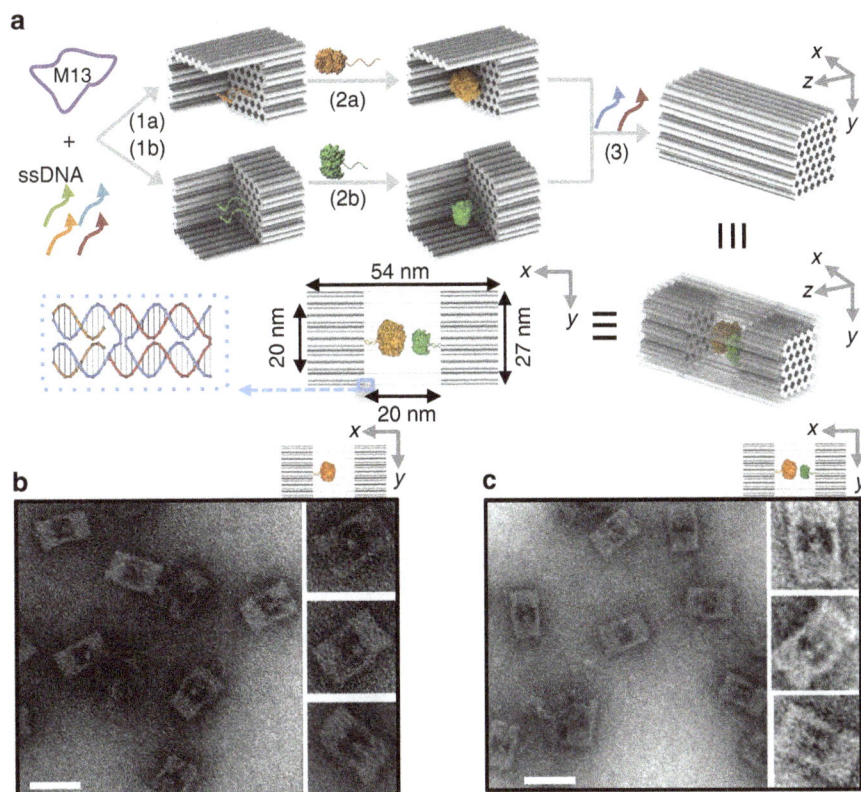

Figure 1 | Design and characterization of DNA nanocage-encapsulated enzymes with controlled stoichiometry. (**a**) Schematic representations of the assembly of a DNA nanocage encapsulating a pair of GOx (orange) and HRP (green) enzymes. Individual enzymes were first attached to half-cages, followed by the addition of linker strands (red) to combine the two halves into a full-cage. Small pores of honeycomb shape (~2.5 nm d.i.) were designed on the bottom of cages to facilitate the diffusion of substrate molecules in an out of the cage. (**b**) Negatively stained TEM images of DNA cages containing a single GOx (shown as less stained dots) and (**c**) a pair of GOx and HRP (shown as less stained dots). Scale bar, 50 nm.

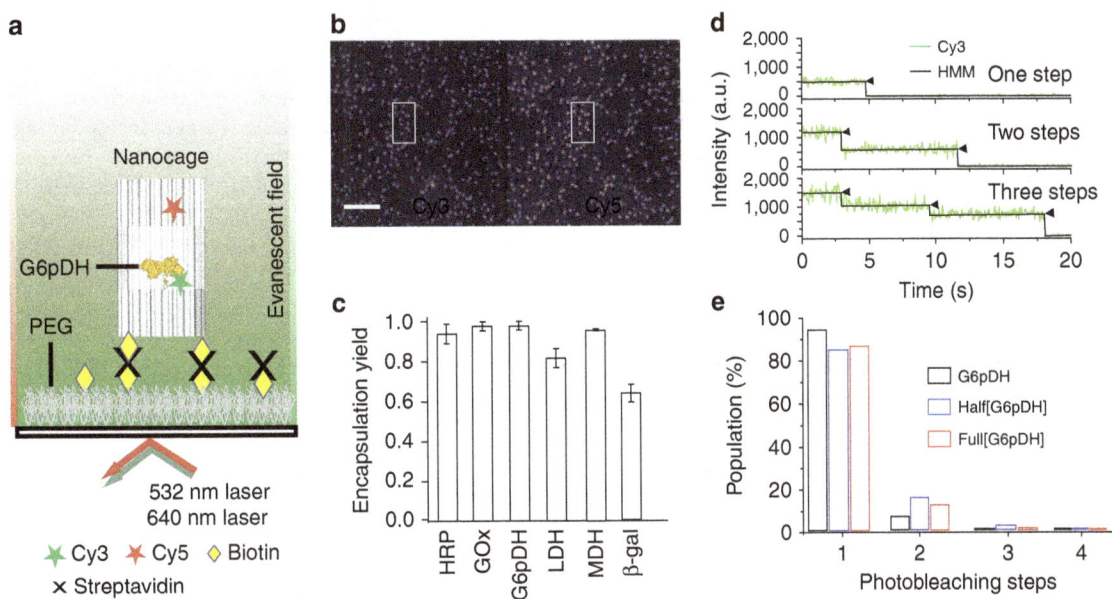

Figure 2 | Single-molecule fluorescence characterization of enzyme encapsulation. (**a**) Schematic illustration of single-molecule fluorescence co-localization of Cy3-labelled protein with Cy5-labelled cage using TIRF microscopy. DNA cages were captured on the surface by biotin-streptavidin interaction. (**b**) Representative field of view of enzyme-encapsulating cages under TIRF microscope. Examples of Cy3-Cy5 co-localization are highlighted using a pair of rectangles. Scale bar, 10 μm. (**c**) Quantified encapsulation yield for six different enzymes. The total number of molecules analysed for each protein is shown in Supplementary Table 2. The error bars represent the standard deviation obtained from the analysis of two to four movies of the sample from the same batch. (**d**) Fluorophore photobleaching trajectories with one, two, and three photobleaching steps. Photobleaching steps were quantitatively analysed by fitting the trajectories by HMM in QUB program. (**e**) Photobleaching statistics for Cy3-labelled proteins encapsulated within half-cages (Half[G6pDH]) or full-cages (Full[G6pDH]), as well as for an unencapsulated protein control (G6pDH). HMM, hidden-Markov modelling.

enzymes (Fig. 2e). These results suggest that most nanocages (~90%) contain exactly one enzyme per cage, as expected (Fig. 2e and Supplementary Tables 3 and 4).

Activity characterization of nanocaged enzymes. To evaluate the effect of DNA nanocages on enzyme activity, we first tested an encapsulated GOx/HRP pair (Fig. 3a). This pair of enzymes catalyses a reaction cascade beginning with the oxidation of glucose by GOx to generate hydrogen peroxide (H_2O_2). H_2O_2 is subsequently used by HRP to oxidize ABTS, producing a strong colorimetric signal. As shown in Fig. 3b, the overall activity of a co-assembled GOx/HRP cage (Full[GOx/HRP]) is ~8-fold higher than that of a control enzyme pair incubated with the same cage but without encapsulation. We hypothesized that two

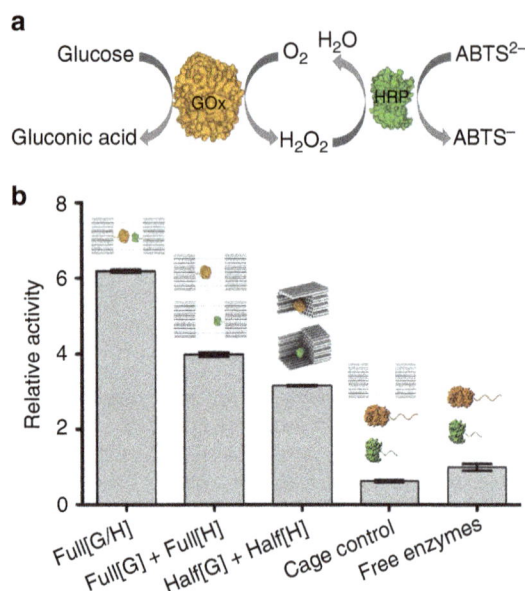

a

Glucose O₂ H₂O ABTS²⁻

Gluconic acid H₂O₂ ABTS⁻

b

Figure 3 | Activity characterization of encapsulated GOx/HRP pairs.
(**a**) Schematic representation of the GOx/HRP cascade. (**b**) Normalized cascade activities for a GOx/HRP pair encapsulated within a full-cage (Full[GOx/HRP]), two individual full-cages (Full[GOx] + Full[HRP]) and two individual half-cages (Half[GOx] + Half[HRP]), as well as unencapsulaed enzyme pairs with and without the presence of DNA cages. Assay conditions: 1 nM enzyme or 1 nM enzyme-DNA cage, 1 mM glucose and 2 mM ABTS in TBS buffer (pH 7.5), and monitoring absorbance at 410 nm. Error bars were generated as the s.d. of at least three replicates.

plausible effects could contribute to such a significant activity enhancement: (1) The proximity effect that brings the two enzymes close together and facilitates their substrate transfer, as described previously[19,21,41]; and/or (2) the unique environment provided by the high charge density of DNA helices within a nanocage. To separate the proximity effect from the charge density effect, we designed control experiments of DNA nanocages encapsulating only a single GOx or HRP enzyme, which clearly does not allow for substrate channelling between two proximal enzymes. For example, an equimolar mixture of two separate nanocages encapsulating either a single GOx or a single HRP (Full[GOx] + Full[HRP]) exhibited an ~4-fold increase in overall activity compared with the unencapsulated control enzymes. Similarly, an equimolar mixture of two half-cages encapsulating either a single GOx or a single HRP already showed an increase in overall activity by ~3-fold. Since there was no proximity effect in the case of two enzymes encapsulated into two different nanocages, the local environment modified by a DNA nanocage appears to be more important for the observed activity enhancement. Similarly, a half-cage was almost as effective in activity enhancement (3-fold) as a full-cage, suggesting that enzyme access to substrate does not play a role in this enhancement. Interestingly, a similar enhancement was reported previously upon conjugation of enzymes to a giant multi-branched DNA scaffold, without further explanation[42].

To test the generality of our observations, we evaluated the activity of six different enzymes upon encapsulation within DNA nanocages. As shown in Table 1, five of them (GOx, HRP, G6PDH, MDH and LDH) exhibited higher activity in nanocages than the free enzyme, with enhancements ranging from 3- to 10-fold. Detailed kinetic analyses show that the K_m (the Michaelis–Menten constant) varies little between encapsulated and free enzyme for most substrates (ranging from 0.5 to 2.4-fold of the free enzyme), suggesting that the porous DNA cages do not substantially hinder diffusion of small-molecule substrates. In contrast, a large increase in turnover number (k_{cat}) was observed for these five enzymes (ranging from 3.5- to 9.6-fold of the free enzyme), suggesting an inherently higher catalytic activity of the proteins. For all the raw kinetics data and TEM images of the assembled structures, please see Supplementary Figs 12–46. Strikingly, an inverse correlation was observed between enhanced turnover and size of the encapsulated enzyme (Fig. 4a). That is, the smaller HRP (44 kDa) and MDH (70 kDa) exhibited relatively large increases in turnover number of 9.6 ± 0.4- and 9.0 ± 0.7-fold, respectively, whereas the larger enzymes G6PDH, LDH and GOx exhibited smaller enhancements of 4.7 ± 0.1-,

Table 1 | Enzyme kinetic data for each individual enzyme encapsulated inside a DNA full-cage in comparison with the values for the free enzymes in solution.

Enzyme	pI	Molecular weight	Substrate	Free enzymes		Encapsulated enzymes	
				K_m (µM)	k_{cat} (s⁻¹)	K_m (µM)	k_{cat} (s⁻¹)
GOx	4.2	160 kDa	Glucose	6,200 ± 900	240 ± 10	3,000 ± 600	1,300 ± 50
HRP	8.8	44 kDa	H_2O_2	2.3 ± 0.5	32 ± 1	4.3 ± 0.6	290 ± 5
			ABTS	2,600 ± 400	59 ± 5	2,500 ± 200	560 ± 20
G6pDH	4.3	100 kDa	Glucose-6-phosphate	220 ± 20	130 ± 3	310 ± 30	460 ± 10
			NAD⁺	510 ± 50	100 ± 3	590 ± 40	480 ± 10
MDH	10.0	70 kDa	NADH	180 ± 50	51 ± 5	270 ± 50	460 ± 30
LDH	5.0	140 kDa	NADH	7.2 ± 1.3	46 ± 2	17.0 ± 1.5	190 ± 5
β-Gal	4.1	465 kDa	RBG	58.7 ± 16.0	8.5 ± 0.6*	95.5 ± 18.9	1.6 ± 0.1*

ABTS, 2,2'-azino-bis(3-ethylbenzothiazoline-6-sulphonic acid); GOx, glucose oxidase; G6pDH, glucose-6-phosphate dehydrogenase; HRP, horseradish peroxidase; LDH, lactic dehydrogenase; MDH, malic dehydrogenase; pI, isoelectric point.
The Michaelis–Menten plots of each enzyme and the conditions of the enzyme activity measurements can be found in the Supplementary Figs. The pI values of the enzymes were obtained from brenda-enzymes.org (refs 65–68).
*k_{cat} values for β-Gal groups were not calibrated.

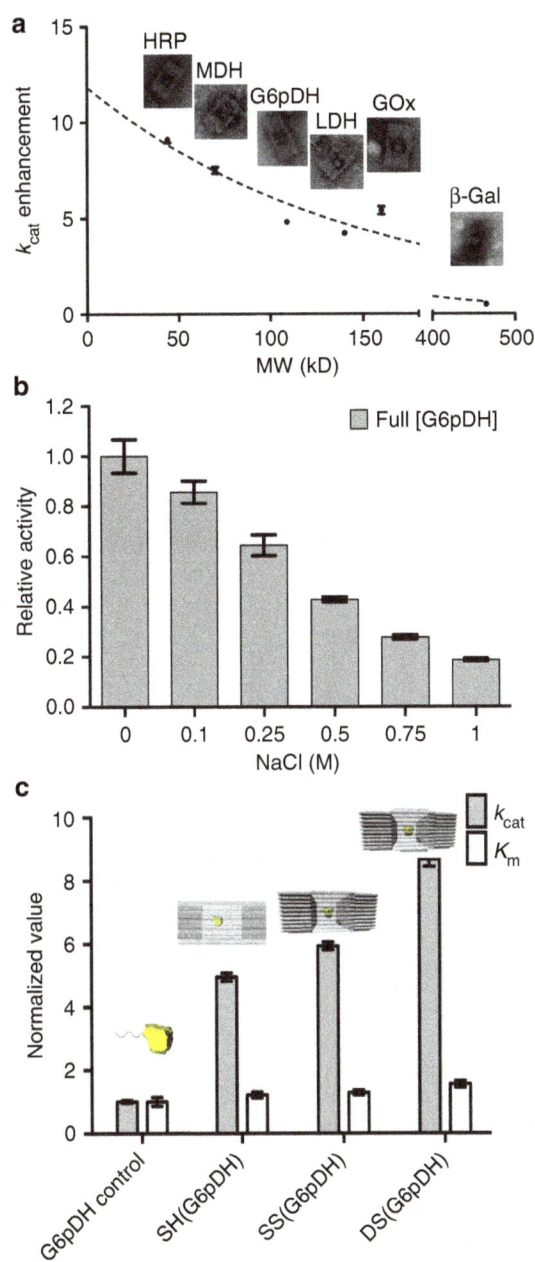

Figure 4 | Mechanistic study of the activity enhancement of DNA nanocage-encapsulated enzymes. (**a**) Relationship between turnover rate enhancement factor after encapsulation against enzyme molecular weight (fitted using one-phase decay function). (**b**) Nanocage-encapsulated G6pDH activity change after incubation with different amount of NaCl. Assay conditions: 0.5 nM enzyme-DNA cage, incubation with 1 mM glucose-6-phosphate and 1 mM NAD$^+$ in TBS buffer (pH 7.5), and monitoring absorbance at 340 nm. (**c**) Normalized k_{cat} and K_m values of free G6PDH and G6PDH that is encapsulated within different DNA cage: SH(G6pDH), a honeycomb lattice origami with a single layer; SS(G6pDH), a square-lattice origami with a single layer; and DS(G6pDH), a square-lattice origami with two layers. k_{cat} and K_m values of caged enzymes are normalized to that of free enzymes. Error bars were generated as the standard deviation of at least three replicates.

($\sim$16 nm in diameter, Supplementary Fig. 47) that is comparable to the inner cavity diameter ($\sim$20 nm) of the DNA nanocage. Alternatively, the β-Gal orientation may be unfavourable and block binding of substrate to the active site. Notably, in a control experiment polyphosphate inhibited the activity of β-Gal (Supplementary Fig. 48), suggesting that the local high density of backbone phosphates of the DNA nanocage might be responsible for the decrease in activity of β-Gal. The DNA cages retained their structural integrity during the enzymatic reactions (Supplementary Fig. 49).

To gain more detailed mechanistic insight into the enhancement of catalytic turnover, we applied a novel single-molecule fluorescence assay to characterize the activity of individual enzymes with and without encapsulation (Fig. 5). As shown in Fig. 5a,b, we used TIRF microscopy to record the repetitive turnover of substrates by individual G6pDH enzymes over time; coupling with a PMS/resazurin reaction[43] (see single-molecule enzyme activity assay in Supplementary Note 3) allowed us to detect stochastic fluctuations of enzyme turnover rates via transient spikes in intensity from the generation of the fluorescent product resorufin (Fig. 5c,d and Supplementary Figs 50–52). Such fluctuations have been observed for various enzymes before[43–45] and are thought to be induced by the conformational switching between more and less active sub-states[45–47]. Compared with a control without substrate, more frequent fluorescent spikes were observed with the addition of glucose-6-phosphate substrate (Fig. 5c,d). The average spike frequency was increased from $0.016 \pm 0.001\,\mathrm{s}^{-1}$ for unencapsulated enzymes, to $0.019 \pm 0.001\,\mathrm{s}^{-1}$ for the half-cage and $0.026 \pm 0.002\,\mathrm{s}^{-1}$ for the full-cage (Fig. 5e). Further analysis suggested that the fraction of active enzyme molecules was increased from $\sim$20% for unencapsulated enzymes to $\sim$27% for the half-cage and $\sim$31% for the full-cage (Fig. 5f). Taken together, the 1.6-fold higher spike frequency and the 1.5-fold increase in the fraction of active enzymes yield an $\sim$2.5-fold increase in G6pDH activity for the encapsulated compared with the unencapsulated enzyme (Fig. 5g), comparable to the $\sim$4-fold enhancement observed in the bulk assay. Conversely, a similar analysis of β-Gal activity showed an $\sim$3-fold lower activity of the full-cage enzyme (2.3 ± 0.5-fold lower in spike frequency compared with free enzyme whereas the fractions of active enzymes ($\sim$65%) were similar) compared with unencapsulated enzyme (Supplementary Fig. 51), also consistent with the bulk measurement.

The activity enhancement for DNA cage-encapsulated enzymes is consistent with recent reports of enhanced enzyme activity upon attachment to a long double-stranded DNA molecule (λDNA)[42], a two-dimensional rectangular DNA origami[48], or a DNA scaffold that bound to enzyme substrates[49,50], and further suggests that it may be a widespread effect of enzyme–DNA interactions. Several mechanisms have been previously proposed to explain these observed enhancements, including micro-environment composed of giant and ordered DNA molecules, molecular crowding and the substrates affinity to DNA scaffolds. We further suggested that the negatively charged phosphate backbones of DNA might also contribute to the activity enhancement. DNA is a negatively charged biopolymer because of its closely spaced backbone phosphates (leading to a linear negative charge density of $\sim$0.6 e Å^{-1}). Thus, upon encapsulation within a DNA nanocage, an enzyme is exposed to an environment full of negative charges that may resemble the relative abundance of polyanionic molecules and surfaces (including RNA and phospholipid membranes) within the cell. Phosphate is a known kosmotropic anion that increases the extent of hydrogen-bonded water structures (termed high-density or structured water)[51–53]. A DNA nanocage is thus expected to attract a strongly bound hydration layer of hydrogen-bonded

4.1 ± 0.1- and 5.4 ± 0.2-fold, respectively. No correlation was observed between enhancement and isoelectric point (pI), despite the wide range of pI values for these enzymes (ranging from 4.2 to 10.0). In contrast to these five enzymes, β-Gal is strongly inhibited upon encapsulation, possibly because of its large size

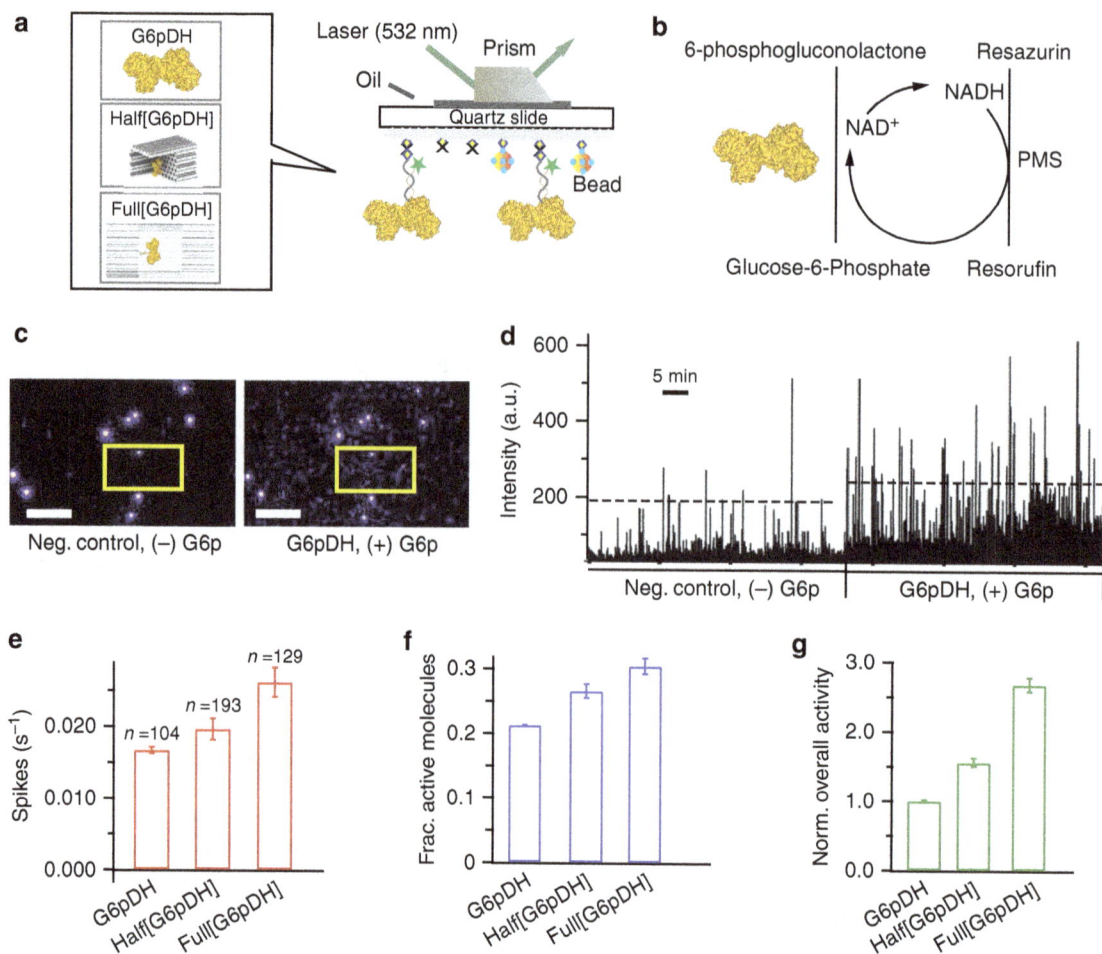

Figure 5 | Single-molecule kinetics of nanocage-encapsulated enzymes. (**a**) Schematic of the experimental TIRF set up for characterizing G6pDH encapsulated within a full-cage (Full[G6pDH]) and a half-cage (Half[G6pDH]), as well as an unencapsulated control. (**b**) A PMS/resazurin-coupled fluorescence assay used to characterize the activity of G6pDH. NAD^+ is first reduced to NADH by G6pDH, followed by PMS- catalyzed electron transfer from NADH to resazurin, producing a strongly fluorescent resorufin, which has an excitation/emission maximum at 544/590 nm. (**c**) TIRFM snapshots captured before and after the injection of substrate G6p. In presence of G6p, the field of view showed increased fluorescence due to the formation of resorufin (compare the boxed regions). Fluorescent beads (very bright spots present in both + G6p and − G6p images) were used as reference markers to correct for the drift. Scale bars, 10 μm. (**d**) Real-time traces of fluorescence spikes (resorufin production) for enzymes without and with the addition of G6p substrate. Ten single-molecule traces for each condition were concatenated. (**e–g**) Statistics of spike frequency, fraction of active molecules, and overall observed enzyme activity for G6pDH. The number of active molecules analysed is denoted by 'n' in **e**. The standard deviations for the spike frequency were calculated after randomly assigning the active molecules into three groups; those for the fractions of active molecules were calculated from three to four independent movies, and those for the normalized overall activity were estimated from the propagation of errors. All experiments were carried out at room temperature in 1× TBS buffer, pH 7.5, in the presence of 1 mM Mg^{2+} and 10% (w/v) PEG 8000.

water molecules inside its cavity[54–55]. Multiple studies[56–58] have described that proteins are more stable and active in a highly ordered, hydrogen-bonded water environment, possibly due to stabilization of the hydrophobic interactions of a folded protein through an increase in the solvent entropy penalty upon unfolding. Consistent with this model, polyphosphate has been shown to act as a generic chaperone stabilizing a variety of enzymes[59]. To further test whether this mechanism is at work in our nanocages, we titrated the concentration of NaCl (known to consist of chaotropic ions)[60,61] for the purpose of interrupting hydrogen-bonded water molecules. Consistent with our hypothesis, the activity of encapsulated enzymes significantly decreased with increasing NaCl concentration (reduced to ∼25% activity with 1 M NaCl as shown in Fig. 4b. A high concentration of Na^+ can shield the negative charge on the DNA surface, thus disrupting the surface-bound hydration layer. As a control, we observed that the bulky kosmotropic cation, triethylammonium, had a much less pronounced effect on enzymatic activity

(Supplementary Fig. 53). This model also allowed us to rationalize why we observed smaller enzymes to be more activated than larger enzymes, because their higher surface-to-volume ratio predicts a stronger impact of the hydration layer.

To further test this model, we investigated the effect of DNA helix density on the encapsulated enzyme activity. As shown in Fig. 4c, we designed three nanocages with walls that systematically increase the density of DNA helices, including: (1) a single-layer honeycomb pattern (SH) with ∼2–3 nm pores between helices; (2) a single-layer square pattern (SS) with smaller ∼0.5–1 nm pores between helices; and (3) a double-layer square pattern (DS). The helix density at the top and bottom surfaces thus increased from 0.12 helices per nm^2 for SH to 0.16 helices per nm^2 for the SS and DS designs. The k_{cat} of G6pDH encapsulated in the SH-cage was ∼4.7-fold higher than that of the free enzyme. As the density of DNA helices was increased, the k_{cat} of encapsulated G6pDH raised to ∼6-fold for the SS-cage and 8-fold for the DS-cage compared with the free enzyme

a

DNA layers

b

☐ Before digestion
☐ After digestion

Full [G/H]   Free GOx/HRP

c

◆ Free G6pDH
■ Full [G6pDH]

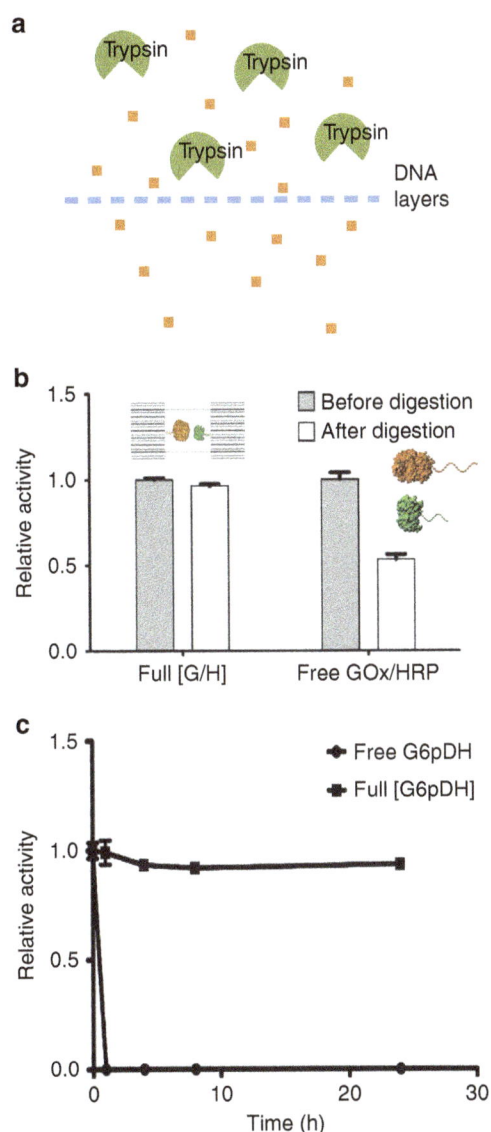

Figure 6 | Protection of nanocaged enzymes against protease-mediated degradation and aggregation. (**a**) Schematic representation illustrating how a DNA cage may block access of big proteins such as a protease to the interior of the cage, but still allow the penetration of small molecules. (**b**) Relative enzyme activity of encapsulated GOx/HRP pairs (Full[GOx/HRP]) and free GOx/HRP pairs (free GOx/HRP) before and after the addition of trypsin. Trypsin digestion conditions: enzyme or enzyme-DNA cage was incubated with 1,000 times excess trypsin for 24 h at 37 °C. Assay conditions: 0.5 nM enzyme or 0.5 nM enzyme-DNA cage, incubation with 1 mM glucose and 2 mM ABTS in 1 × TBS buffer (pH 7.5), and monitoring absorbance at 410 nm. (**c**) Relative activity data for free G6pDH and Full[G6pDH] (0.5 nM) with trypsin digestion for 0, 1, 4, 8 and 24 h. Digestion by incubation sample with 1,000 times amount of trypsin at 37 °C in 1 × TBS buffer (pH 7.5). Error bars were generated as the s.d. of at least three replicates.

control. A slight increase in K_m values was also observed from the SH-cage to the SS- and DS-cages, possibly due to a decrease in substrate diffusion through the DNA walls of these more tightly packed structures. For example, the K_m value of G6PDH increased from ∼410 µM in the SH-cage to ∼440 µM in the SS-cage and ∼530 µM in the DS-cage (Fig. 4c, Supplementary Fig. 54). Additional studies showed that activities of attached enzymes were enhanced by increasing the helix packing density for various one-, two- and three-dimensional DNA scaffolds

(Supplementary Fig. 55). These observations suggest that encapsulated enzymes exhibit higher activity within densely packed DNA cages, consistent with our model that the highly ordered, hydrogen-bonded water environment near closely spaced phosphate groups are responsible for this effect.

Nanocaged enzymes are protected from proteolysis. Self-assembled DNA nanostructures previously were found to be more resistant against nuclease degradation than single- or double-stranded DNA molecules[62,63]. Similarly, DNA nanocages should protect encapsulated enzymes from deactivation and aggregation under challenging biological conditions. As shown in Fig. 6, encapsulated GOx/HRP was highly resistant to digestion by trypsin (Fig. 6b), and retained more than 95% of its initial activity after incubation with trypsin for 24 h (Fig. 6b). A time-course experiment was also performed to demonstrate the stability of caged enzymes against trypsin digestion (Fig. 6c, Supplementary Figs 56–59). In contrast, free GOx/HRP only retained ∼50% of its initial activity after a similar incubation with trypsin. This result demonstrated the potential utility of DNA nanocages for protecting encapsulated proteins from biological degradation.

Discussion

In summary, we have developed a method for using a DNA nanocage to efficiently encapsulate enzymes with high yield. Using single-molecule characterization, we were able to quantify the copies of encapsulated enzymes per cage with demonstrated one enzyme per cage. Upon encapsulation, five of six tested metabolic enzymes exhibit turnover numbers 4–10-fold higher than that of the free enzyme. Conversely, the K_m values remain similar between encapsulated enzymes and free enzymes, indicating an uninterrupted diffusion of small-molecule substrates and products through the nanopores in the DNA cage. Application of a novel single-molecule enzyme assay showed that both the fraction of active enzyme molecules and their individual turnover numbers increase as a consequence of encapsulation. We therefore propose that the unique local environment created within a DNA nanocage, particularly the high density of negatively charged phosphate groups, enhances the activity of encapsulated enzymes, where the tightly bound, highly structured water layers on DNA surface may stabilize the active enzyme conformations. This effect appears consistent with recent independent evidence that many conserved metabolic enzymes are stabilized by polyphosphate and associate non-specifically with nucleic acids through cryptic binding sites[27,28,64], thus taking advantage of the high polyanionic DNA and RNA contents of the cell. DNA nanocages therefore may serve as a molecular tool to precisely sculpt the properties of the local environment of enzymes in smart-material and biotechnological application. DNA nanocages also demonstrated their value in protecting encapsulated enzymes from biological degradation through proteases. In the future, it may be feasible to construct precisely controlled, programmable DNA nanocages for theranostic nanodevices as therapeutic agents.

Methods
The design and characterization of DNA half-cages and full-cages. DNA origami half-cage and structures were designed with caDNAno (www.cadnano. org), each used one M13mp18 ssDNA as the scaffold. Detailed design schemes and DNA sequences are shown in the Supplementary Figs 1, 60 and 61. One or both of the half-cages contained single-stranded probe strands (four in each half-cage) extended towards the inside of the cage for binding with the DNA conjugated enzymes. Two of the half-cages can be linked together to form a fully enclosed full-cage though 24 linker strands. To form each of the half-cages, the M13mp18 ssDNA was mixed with the corresponding staples at a 1:10 molar ratio in 1 × TAE-Mg²⁺ buffer (40 mM Tris, 20 mM acetic acid, 2 mM EDTA and 12.5 mM

magnesium acetate, pH 8.0) and annealed from 80 to 4 °C for 37 h. The excess staple strands were removed by the filtration of the DNA cages solution using 100 kDa Amicon filter with $1 \times$ TAE-Mg^{2+} buffer for three times. To form a full-cage, 24 single-stranded DNA linkers were incubated with the two purified half-cages at a molar ratio of 5:1 for three hours at room temperature, in order to connect the two half-cages together.

Enzyme-DNA cage assembly. A 15-fold molar excess of oligonucleotide-conjugated enzyme was incubated with the DNA half-cage containing capture strands. Protein assembly was performed using an annealing protocol, in which the temperature was gradually decreased from 37 to 4 °C over 2 h and then held constant at 4 °C using an established procedure[30,31]. Two Enzyme-attached half-cages were then assembled into a full-cage by adding DNA linkers as described above. The DNA caged-enzymes were further purified by agarose gel electrophoresis to remove excess free enzymes (please see Supplementary Notes 1 and 4 for detailed information).

Single-molecule fluorescence microscopy. All single-molecule measurements were performed at room temperature using a TIRF microscope on PEGylated fused silica microscope slides. To passivate the microscope slides and functionalize the surface with biotin for selective immobilization of nanocages, a biotin- and PEG-coated surface was prepared by silylation with APTES, followed by incubation with a 1:10 mixture of biotin-PEG-SVA 5k:mPEG-SVA 5k as described previously[31]. A flow channel was constructed as described elsewhere[31]. To prepare the surface for enzyme or nanocage binding, a solution of 0.2 mg ml^{-1} streptavidin in T50 buffer (50 mM Tris-HCl, pH 8.0, 50 mM NaCl and 1 mM EDTA) was injected in to the flow channel, incubated for 10 min, and the excess streptavidin was flushed out thoroughly first with T50, then with $1 \times$ TAE-Mg^{2+}. For more detailed information, please see the Supplementary Notes 2 and 3.

Bulk solution enzyme assay. A 96-well plate reader was used to monitor enzyme activity through absorbance changes of the samples. The enzyme samples and substrates were loaded in the wells of the 96-well plate with a final concentration of caged enzymes ~ 0.5 nM in $1 \times$ TBS (Tris buffered saline with 1 mM $MgCl_2$, pH 7.5) for most assays. Detailed assay conditions are described in the Supplementary Methods.

References

1. Chen, A. H. & Silver, P. A. Designing biological compartmentalization. *Trends. Cell. Biol.* **12**, 662–670 (2012).
2. Hurtley, S. Location, location, location. *Science* **326**, 1205 (2009).
3. Kertelf, C. A., Heinhorst, S. & Cannon, G. C. Bacterial microcompartments. *Annu. Rev. Microbiol.* **64**, 391–408 (2010).
4. Kerfeld, C. A. et al. Protein structures forming the shell of primitive bacterial organelles. *Science* **309**, 936–938 (2005).
5. Graff, A., Winterhalter, M. & Meier, W. Nanoreactors from polymer-stabilized liposomes. *Langmuir.* **17**, 919–923 (2001).
6. Hartl, F. U. Molecular chaperones in cellular protein folding. *Nature* **381**, 571–580 (1996).
7. Comellas-Aragones, M. et al. A virus-based single-enzyme nanoreactor. *Nat. Nanotechnol.* **2**, 635–639 (2007).
8. Liu, Y. et al. Biomimetic enzyme nanocomplexes and their use as antidotes and preventive measures for alcohol intoxication. *Nat. Nanotechnol.* **8**, 187–192 (2013).
9. Sang, L. & Coppens, M. Effects of surface curvature and surface chemistry on the structure and activity of protein adsorbed in nanopores. *Phys. Chem. Chem. Phys.* **13**, 6689–6698 (2011).
10. Vriezema, D. M. et al. Self-assembled nanoreactors. *Chem. Rev.* **105**, 1445–1490 (2005).
11. Bruns, N. & Tiller, J. C. Amphiphilic network as nanoreactor for enzymes in organic solvents. *Nano Lett.* **5**, 45–48 (2005).
12. Betancor, L. & Luckarift, H. R. Bioinspired enzyme encapsulation for biocatalysis. *Trends. Biotechnol.* **26**, 566–572 (2008).
13. Fiedler, J. D., Brown, S. D., Lau, J. & Finn, M. G. RNA-directed packaging of enzymes within virus-like particles. *Angew. Chem. Int. Ed. Engl.* **49**, 9648–9651 (2010).
14. Douglas, S. M. et al. Self-assembly of DNA into nanoscale three-dimensional shapes. *Nature* **459**, 414–418 (2009).
15. Han, D. et al. DNA origami with complex curvatures in three-dimensional space. *Science* **332**, 342–346 (2011).
16. Ke, Y., Ong, L. L., Shih, W. M. & Yin, P. Three-dimensional structures self-assembled from DNA bricks. *Science* **338**, 1177–1183 (2012).
17. Kuzyk, A. et al. DNA-based self-assembly of chiral plasmonic nanostructures with tailored optical response. *Nature* **483**, 311–314 (2012).
18. Langecker, M. et al. Synthetic lipid membrane channels by designed DNA nanostructures. *Science* **338**, 932–936 (2012).
19. Fu, J., Liu, M., Liu, Y., Woodbury, N. W. & Yan, H. Interenzyme substrate diffusion for an enzyme cascade organized on spatially addressable DNA nanostructure. *J. Am. Chem. Soc.* **134**, 5516–5519 (2012).
20. Fu, J., Liu, M., Liu, Y. & Yan, H. Spatially-interactive biomolecular networks organized by nucleic acid nanostructures. *Acc. Chem. Res.* **45**, 1215–1226 (2012).
21. Wilner, O. I. et al. Enzyme cascades activated on topologically programmed DNA scaffolds. *Nat. Nanotechnol.* **4**, 249–254 (2009).
22. Andersen, E. S. et al. Self-assembly of a nanoscale DNA box with a controllable lid. *Nature* **459**, 73–76 (2009).
23. Douglas, S. M., Bachelet, I. & Church, G. M. A logic-gated nanorobot for targeted transport of molecular payloads. *Science* **335**, 831–834 (2012).
24. Juul, S. et al. Temperature-controlled encapsulation and release of an active enzyme in the cavity of a self-assembled DNA nanocage. *ACS Nano* **7**, 9724–9734 (2013).
25. Fu, Y. et al. Single-step rapid assembly of DNA origami nanostructures for addressable nanoscale bioreactors. *J. Am. Chem. Soc.* **135**, 696–702 (2013).
26. Linko, V., Eerikainen, M. & Kostiainen, M. A modular DNA origami-based enzyme cascade nanoreactor. *Chem. Commun.* **51**, 5351–5354 (2015).
27. Gray, M. J. et al. Polyphosphate is a primordial chaperone. *Mol. Cell.* **53**, 689–699 (2014).
28. Cieśla, J. Metabolic enzymes that bind RNA: yet another level of cellular regulatory network? *Acta Biochim. Pol.* **53**, 11–32 (2006).
29. Bellot, G., McClintock, M. A., Lin, C. X. & Shih, W. M. Recovery of intact DNA nanostructures after agarose gel-based separation. *Nat. Methods.* **8**, 192–194 (2011).
30. Liu, M. et al. A DNA tweezer-actuated enzyme nanoreactors. *Nat. Commun.* **6**, 712–719 (2013).
31. Fu, J. et al. Multi-enzyme complexes on DNA scaffolds capable of substrate channeling with an artificial swinging arm. *Nat. Nanotechnol.* **9**, 531–536 (2014).
32. Hecht, H. J., Kalisz, K., Hendle, J., Schmid, R. D. & Schomburg, D. Crystal structure of glucose oxidase from *Aspergillus niger* refined at 2-3 Å resolution. *J. Mol. Biol.* **229**, 153–172 (1993).
33. Henriksen, A., Schuller, D. J. & Gajhede, M. Structural interactions between horseradish peroxidase C and the substrate benzhydroxamic acid determined by X-ray crystallography. *Biochemistry* **37**, 8054–8060 (1998).
34. Widom, J. R., Dhakal, S., Heinicke, L. A. & Walter, N. G. Single-molecule tools for enzymology, structural biology, systems biology and nanotechnology: an update. *Arch. Toxicol.* **88**, 1965–1985 (2014).
35. Veitch, N. C. Horseradish peroxidase: a modern view of a classic enzyme. *Phytochemistry* **65**, 249–259 (2004).
36. Chapman, A. D., Cortés, A., Dafforn, T. R., Clarke, A. R. & Brady, R. L. Structural basis of substrate specificity in malate dehydrogenases: crystal structure of a ternary complex of porcine cytoplasmic malate dehydrogenase, alpha-ketomalonate and tetrahydoNAD. *J. Mol. Biol.* **285**, 703–712 (1999).
37. Rowland, P., Basak, A. K., Gover, S., Levy, H. R. & Adams, M. J. The three-dimensional structure of glucose 6-phosphate dehydrogenase from *Leuconostoc mesenteroides* refined at 2.0A resolution. *Structure.* **15**, 1073–1087 (1994).
38. Lovell, S. L. & Winzor, D. J. Effects of phosphate on the dissociation and enzymic stability of rabbit muscle lactate dehydrogenase. *Biochemistry* **13**, 3527–3531 (1974).
39. Hecht, H. J., Kalisz, H. M., Hendle, J., Schmid, R. D. & Schomburg, D. Crystal structure of glucose oxidase from *Aspergillus niger* refined at 2.3A resolution. *J. Mol. Biol.* **229**, 153–172 (1993).
40. Jacobson, R. H., Zhang, X. J., DuBose, R. F. & Matthews, B. W. Three-dimensional structure of beta-galactosidase from *E. coli. Nature* **369**, 761–766 (1994).
41. Erkelenz, M., Kuo, C. H. & Niemeyer, C. M. DNA-mediated assembly of cytochrome P450 BM3 subdomains. *J. Am. Chem. Soc.* **133**, 16111–16118 (2011).
42. Rudiuk, S., Venancio-Marques, A. & Baigl, D. Enhancement and modulation of enzymatic activity through higher-order structural changes of giant DNA-protein multibranch conjugates. *Angew. Chem. Int. Ed. Engl.* **51**, 12694–12698 (2012).
43. English, B. P. et al. Ever-fluctuating single enzyme molecules: Michaelis-Menten equation revisited. *Nat. Chem. Biol.* **2**, 87–94 (2006).
44. Liu, B., Baskin, R. J. & Kowalczykowski, S. C. DNA unwinding heterogeneity by RecBCD results from static molecules able to equilibrate. *Nature* **500**, 482–485 (2013).
45. Ramanathan, A., Savol, A., Burger, V., Chennubhotla, C. S. & Agarwal, P. K. Protein conformational populations and functionally relevant substates. *Acc. Chem. Res.* **47**, 149–156 (2014).
46. Hammes, G. G., Benkovic, S. J. & Hammes-Schiffer, S. Flexibility, diversity, and cooperativity: pillars of enzyme catalysis. *Biochemistry* **50**, 10422–10430 (2011).
47. Ramanathan, A. & Agarwal, P. K. Evolutionarily conserved linkage between enzyme fold, flexibility, and catalysis. *PLoS Biol.* **9**, 1–17 (2011).

48. Timm, C. & Niemeyer, C. M. Assembly and purification of enzyme-functionalized DNA origami structures. *Angew. Chem. Int. Ed. Engl.* **54,** 6745–6750 (2015).

49. Lin, J. & Wheeldon, I. Kinetic enhancements in DNA-enzyme nanostructures mimic the Sabatier principle. *ACS Catal.* **3,** 560–564 (2013).

50. Gao, Y. *et al.* Tuning enzyme kinetics through designed intermolecular interactions far from the active site. *ACS Catal.* **5,** 2149–2153 (2015).

51. Zhao, H. Effects of ions and other compatible solutes on enzyme activity, and its implication for biocatalysis using ionic liquids. *J. Mol. Catal. B Enzym.* **37,** 16–25 (2005).

52. Moelberta, S., Normandb, B. & Rios, P. D. L. Kosmotropes and chaotropes: modelling preferential exclusion, binding and aggregate stability. *Biophys. Chem.* **112,** 45–57 (2004).

53. Leberman, R. & Soper, A. K. Effect of high salt concentrations on water structure. *Nature* **378,** 364–366 (1995).

54. Jana, B. *et al.* Entropy of water in the hydration layer of major and minor grooves of DNA. *J. Phys. Chem. B* **110,** 19611–19618 (2006).

55. Chuprina, V. P. *et al.* Molecular dynamics simulation of the hydration shell of a B-DNA decamer reveals two main types of minor-groove hydration depending on groove width. *Proc. Natl Acad. Sci. USA* **88,** 593–597 (1991).

56. Zhao, H. *et al.* Effect of kosmotropicity of ionic liquids on the enzyme stability in aqueous solutions. *Bioorg. Chem.* **34,** 15–25 (2006).

57. Timasheff, S. N. Protein-solvent preferential interactions, protein hydration, and the modulation of biochemical reactions by solvent components. *Proc. Natl Acad. Sci. USA* **99,** 9721–9726 (2002).

58. Levy, Y. & Onuchic, J. N. Water and proteins: a love-hate relationship. *Proc. Natl Acad. Sci. USA* **101,** 3325–3326 (2004).

59. Grey, M. J. *et al.* Polyphosphate is a primordial chaperone. *Mol. Cell.* **53,** 689–699 (2014).

60. Marcus, Y. Effects of ions on the structure of water: structure making and breaking. *Chem. Rev.* **109,** 1346–1370 (2009).

61. Mei, Q. *et al.* Stability of DNA origami nanoarrays in cell lysate. *Nano Lett.* **11,** 1477–1482 (2011).

62. Jiang, Q. *et al.* DNA origami as a carrier for circumvention of drug resistance. *J. Am. Chem. Soc.* **134,** 13396–13403 (2012).

63. Castello, A. *et al.* Insights into RNA biology from an atlas of mammalian mRNA-binding proteins. *Cell* **149,** 1393–1406 (2012).

64. Wong, C. M., Wong, K. H. & Chen, X. D. Glucose oxidase: natural occurrence, function, properties and industrial application. *Appl. Microbiol. Biotechnol.* **78,** 927–938 (2008).

65. Guo, S. *et al.* One of the possible mechanisms for the inhibition effect of Tb(III) on peroxidase activity in horseradish (*Armoracia rusticana*) treated with Tb(III). *J. Biol. Inorg. Chem.* **13,** 587–597 (2008).

66. Sung, J. Y. & Lee, Y. N. Isoforms of glucose 6-phosphate dehydrogenase in *Deinococcus radiophilus*. *J. Microbiol.* **45,** 318–325 (2007).

67. Horikiri, S. *et al.* Electron acquisition system constructed from an NAD-independent D-lactate dehydrogenase and cytochrome c2 in *Rhodopseudomonas palustris* No. 7. *Biosci. Biotechnol. Biochem.* **68,** 516–522 (2004).

68. Eanes, R. Z. & Kun, E. Separation and characterization of aconitate hydratase isoenzymes from pig tissues. *Biochim. Biophys. Acta* **227,** 204–210 (1971).

Acknowledgements

This work is supported by an Army Research Office grant W911NF-11-1-0137 to H.Y. and Y.L., an Army Research Office MURI award W911NF-12-1-0420 to H.Y., N.G.W. and N.W.W. and a National Science Foundation grant 1033222 to N.W.W. and H.Y. H.Y. is also supported by the NIH Transformative award R01GM104960, National Science Foundation of China award 21329501 and the Presidential Strategic Initiative Fund from Arizona State University. J.F. is supported by an Army Research Office YIP award W911NF-14-1-0434, and start-up funds from Rutgers University.

Author contributions

Z.Z., J.F. and H.Y. conceived the concepts; Z.Z. and J.F. designed DNA nanocages, performed the enzyme-DNA structure assembly and activity assay and analysed data; J.F., M.L. and T.Z. performed enzyme-DNA conjugation and purification; S.D., A.J.-B. and N.G.W. designed and performed single-molecule fluorescence experiments and analysed data; Z.Z., J.F. and S.D. wrote the manuscript; all the authors discussed the results and commented on the manuscript.

Additional information

Ferroelastic switching in a layered-perovskite thin film

Chuanshou Wang[1,*], Xiaoxing Ke[2,3,*], Jianjun Wang[4,*], Renrong Liang[5], Zhenlin Luo[6], Yu Tian[1], Di Yi[7], Qintong Zhang[8], Jing Wang[1], Xiu-Feng Han[8], Gustaaf Van Tendeloo[2], Long-Qing Chen[4,9], Ce-Wen Nan[4], Ramamoorthy Ramesh[7] & Jinxing Zhang[1]

A controllable ferroelastic switching in ferroelectric/multiferroic oxides is highly desirable due to the non-volatile strain and possible coupling between lattice and other order parameter in heterostructures. However, a substrate clamping usually inhibits their elastic deformation in thin films without micro/nano-patterned structure so that the integration of the non-volatile strain with thin film devices is challenging. Here, we report that reversible in-plane elastic switching with a non-volatile strain of approximately 0.4% can be achieved in layered-perovskite Bi_2WO_6 thin films, where the ferroelectric polarization rotates by 90° within four in-plane preferred orientations. Phase-field simulation indicates that the energy barrier of ferroelastic switching in orthorhombic Bi_2WO_6 film is ten times lower than the one in $PbTiO_3$ films, revealing the origin of the switching with negligible substrate constraint. The reversible control of the in-plane strain in this layered-perovskite thin film demonstrates a new pathway to integrate mechanical deformation with nanoscale electronic and/or magnetoelectronic applications.

[1] Department of Physics, Beijing Normal University, 100875 Beijing, China. [2] EMAT (Electron Microscopy for Materials Science), University of Antwerp, Groenenborgerlaan 171, B-2020 Antwerpen, Belgium. [3] Institute of Microstructures and Properties of Advanced Materials, Beijing University of Technology, 100124 Beijing, China. [4] State Key Laboratory of New Ceramics and Fine Processing, School of Materials Science and Engineering, Tsinghua University, 100084 Beijing, China. [5] Tsinghua National Laboratory for Information Science and Technology, Institute of Microelectronics, Tsinghua University, 100084 Beijing, China. [6] National Synchrotron Radiation Laboratory and CAS Key Laboratory of Materials for Energy Conversion, University of Science and Technology of China, 230026 Hefei, China. [7] Department of Materials Science and Engineering, University of California, 94720 Berkeley, California, USA. [8] Beijing National Laboratory of Condensed Matter Physics, Institute of Physics, Chinese Academy of Science, 100190 Beijing, China. [9] Department of Materials Science and Engineering, The Pennsylvania State University, University Park, Pennsylvania, 16802 Pennsylvania, USA. * These authors contributed equally to this work. Correspondence and requests for materials should be addressed to C.-W.N. (email: cwnan@tsinghua.edu.cn) or to R.R. (email: rramesh@berkeley.edu) or to J.Z. (email: jxzhang@bnu.edu.cn).

In ferroic (ferroelectric[1,2], multiferroic[3,4]) materials and their heterostructures, a large reversible electric-field-driven elastic deformation can provide an effective pathway to achieve the coupling between lattice degree of freedom and other order such as spontaneous polarization, spin, orbital and so on[5–10]. This is crucial for designing sensing, data-storage and magnetoelectric devices with ultralow energy consumption[11,12]. In order to integrate these functionalities into applications, the pursuit of such an electric-field-induced elastic deformation in a thin film heterostructure, particularly a large ferroelastic strain controlled within a mono-domain structure, is highly desirable[13–15]. However, the ferroelastic switching (non-180° polar rotation) in traditional epitaxial ferroelectric or relaxor thin films suffers significant constraint from the substrate[16,17], inhibiting the intrinsic non-volatile in-plane elastic strain and the mechanical coupling at hetero-interfaces[18]. The main origins of clamping are the elastic constraint of the vertical deformation in a thin film on a substrate geometry as well as the domain-wall pinning around the film/substrate interface[19–25]. One effective way to reduce the substrate clamping is micro-patterning of those oxide thin films using a method such as focused ion beam (FIB) etching to optimize the aspect ratio of the thin films[24–27]. This, however, usually requires a complicated process in nanoscale thin film devices[15]. Very recently, chemical doping or epitaxial misfit strain[28–30] in tetragonal or rhombohedral ferroelectric thin films seems to be a possible solution to facilitate an in-plane ferroelastic switching. However, the polarization switching only exists around an array of twin-like domain or multi-domain structures, indicating that the elastic strain is still clamped and a controllable switching of a large-area mono-domain with pure elastic switching for practical applications is difficult[31].

Under these circumstances, there is a strong impetus to explore new ferroelectric material systems, where the ferroelastic switching and non-volatile in-plane strain in a thin film heterostructure may not be restricted by the substrate. The elastic switching and strong substrate clamping in ferroelectric thin films with a projection of out-of-plane polarization is dependent on the crystal symmetry and geometry of the nanostructures with specific domains[16,24]. They can be described as:

$$d_{33} = \left(\frac{\Delta S_3}{\Delta E_3} \right)_T \qquad (1)$$

$$d_{33}^{\mathrm{measured}} = d_{33} - \frac{2S_{13}^E}{E_3} Y_f^0 \chi_0 d_{31} E \qquad (2)$$

where S_3 (S_{13}^E) is the out-of-plane strain caused by the applied electric field E_3 (the subscript 3 is defined as out-of-plane direction and the subscript 1 is defined as the in-plane direction.); d_{33} and $d_{33}^{\mathrm{measured}}$ are the intrinsic piezoelectric coefficient and the effective piezoelectric coefficient, respectively; Y_f^0 is the generalized Young's modulus of the film and χ_0 is the distribution function of the stress[24]. It indicates that the out-of-plane and/or in-plane crystal deformation in a thin film structure is mostly restricted by the intrinsic d_{31} of the material. Considering a pure in-plane polarized thin film without any possible out-of-plane projection and switching (the intrinsic d_{31} is negligible), it will not give rise to out-of-plane deformation before and after the application of the out-of-plane or in-plane electric field. Polarization will only switch among the in-plane equivalent states so that the switching energy barrier or penalty from the change of volume will be quite small compared with the case in ferroelectric materials with the possibility of out-of-plane switching. Previous theoretical work has predicted that a pure in-plane ferroelectric polarization could be achieved in a ferroelectric thin film with an orthorhombic symmetry[32,33]. For traditional ferroelectric oxides (for example, $BiFeO_3$ (BFO), $PbTiO_3$ (PTO), $BaTiO_3$ (BTO) and so on), the stabilization of the orthorhombic phase with a robust ferroelastic switching and control of the ferroelectric domains at room temperature is very challenging[28–30,32,33]. Alternatively, Aurivillius-layered perovskites are ferroelectric with an orthorhombic ground state[34–37] and a high Curie temperature (ranging from 650 to 1,200 K). In these compounds, the primary ferroelectric polarization is along the [100] crystallographic direction due to the Bi displacement in the Bi_2O_2 fluorite-like sheet with respect to their octahedral[38,39], supporting a 90° elastic domain wall in their (001) plane of those single crystals[40,41]. Therefore, the full in-plane polarization of 001-oriented epitaxial thin film heterostructures of Aurivillius oxides could provide a new framework for the in-plane ferroelastic switching.

In this work, we take ferroelectric Bi_2WO_6 (BWO), the simplest structure in the Aurivillius family[38,42,43], as a model system to demonstrate an approach to obtain a large in-plane ferroelastic strain in 001-oriented epitaxial thin films. Scanning probe microscopy studies reveal that the ferroelectric polarization is fully constrained within the film plane without any out-of-plane projection, forming elastic 90° domain walls. A full in-plane polarization switching using planar large-area electrodes produces a reversible in-plane strain of approximately 0.4%, which has been confirmed by high-resolution (scanning) transmission electron microscopy (STEM) and X-ray diffraction. In combination with phase-field simulation, thickness-dependent and probe-based switching experiments provide further insights into the microscopic origin of the in-plane ferroelastic polar rotation with respect to fully clamped ferroelectric $PbTiO_3$ thin films.

Results

Epitaxial growth of Bi_2WO_6 thin film. Epitaxial BWO thin films (approximately 200 nm) have been grown using laser molecular beam epitaxy on perovskite substrates such as $SrTiO_3$ (STO) or $LaAlO_3$ (LAO). The detailed growth conditions can be found in the Methods section or elsewhere[44]. Our θ-2θ X-ray diffraction analysis of the film on a STO substrate reveals a high-quality epitaxial structure with a set of 001-oriented diffraction peaks as shown in Fig. 1a. A low-magnification TEM study and energy dispersive X-ray spectroscopy analysis confirm that the chemical composition of the as-grown epitaxial BWO film (Supplementary Fig. 1) is very close to that of the desired BWO phase. High-resolution atomic force microscopy (AFM; Fig. 1b) shows the step-flow growth of the layered-perovskite with a step height of ~ 0.81 nm (half a BWO unit cell, as seen in the inset), while in the bulk the lattice constants are a = 5.437 Å, b = 5.458 Å and c = 16.430 Å, respectively[42,43]. According to the *in situ* reflection high-energy electron diffraction (RHEED) pattern in Fig. 1c, the distance of characteristic diffraction spots of the film is $\sqrt{2}$ times smaller than the one of the STO substrate, indicating that the in-plane lattice constant of the film is approximately $\sqrt{2}$ times larger than the one of the STO substrate (a = 3.935 Å at growth temperature), which is ~ 5.564 Å. This reveals a 45° in-plane rotation of the crystal orientation around the c-axis between film and substrate. This epitaxial relationship between BWO and the STO substrate is schematically illustrated in Fig. 1d. The detailed measurements for θ-2θ X-ray diffraction, AFM and STEM are given in the Methods section. The epitaxial growth of high-quality BWO films provides us with a fundamental model system of Aurivillius-layered compounds and allows us to explore their domain structures and ferroelastic switching mechanism, which is the central theme of this work.

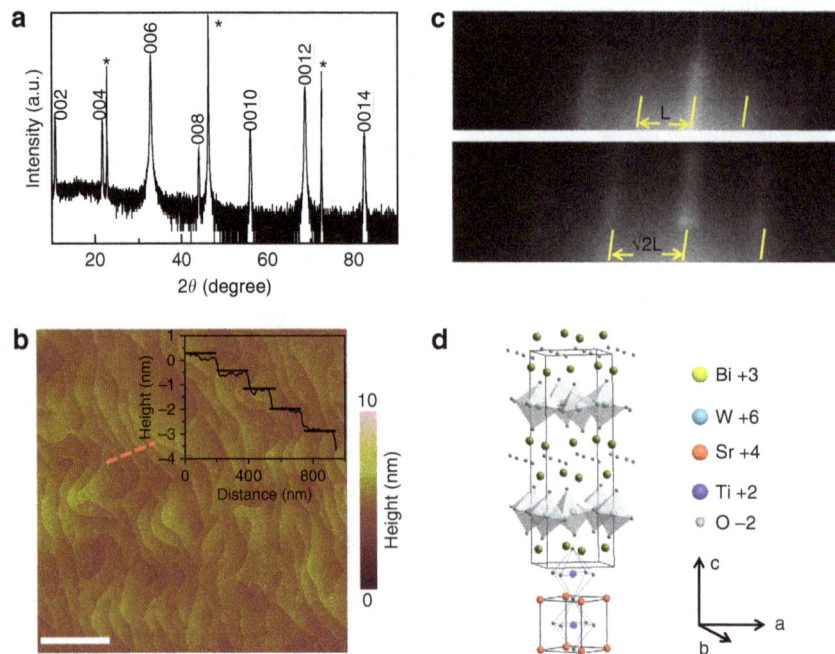

Figure 1 | Crystallographic structure of the epitaxial BWO films. (**a**) The θ-2θ X-ray diffraction pattern indicates a typical 001-oriented epitaxial Aurivillius oxide thin film (*indicates that the peaks belong to the STO substrate). (**b**) Topography of the high-quality epitaxial BWO thin film. The inset shows an average step height with about 0.81 nm derived from the area indicated as the red dashed line. Scale bar, 1 μm. (**c**) RHEED patterns (indicated by the yellow lines) before and after the epitaxial growth of ferroelectric BWO (100; top panel) on STO (100; bottom panel). The distance between them is indicated as the yellow double-sided arrows. (**d**) Schematic diagram of the epitaxial relationship between the film and the substrate.

Structure of ferroelectric/ferroelastic domain. The ferroelectric domain configuration of BWO thin films (approximately 200 nm) on STO substrates has been acquired using a scanning-probe-based technique as shown in the piezoresponse force microscopy (PFM) images in Fig. 2a,b. A schematic geometry of the PFM measurement on BWO film grown on STO substrate is illustrated in Fig. 2a. The polarization vectors are indicated by blue, red and cyan arrows in the in-plane domain image of Fig. 2b. A re-construction of the pure in-plane ferroelectric domain and polar vectors can be seen in Supplementary Fig. 2. More detailed PFM measurements can be seen in the Methods section. The study of the ferroelectric domain structure confirms that the ferroelectric polar directions of the BWO film is along the $<110>$ direction of the perovskite substrate. In this 001-oriented orthorhombic ferroelectric oxide, there are four preferred in-plane polarization directions, indicating the presence of 90° ferroelastic domain walls in BWO thin films grown on various perovskite substrates such as LAO and STO, which can be seen in the in-plane PFM image in Supplementary Fig. 3. The BWO films grown on STO suffer in-plane tensile strain (see X-ray diffraction analysis in Supplementary Fig. 3). An epitaxial tensile strain (approximately 1.7% lattice mismatch between BWO and STO) can further stabilize the in-plane ferroelectric polarization[32,33]. Therefore, the microscopic structure of the ferroelastic domain walls of the epitaxial BWO/STO heterostructure has been analysed.

A cross-sectional high-angle annular dark-field STEM (HAADF-STEM) image in Fig. 2c shows a 90° domain wall of the epitaxial BWO film at atomic resolution. The domain wall is ill-defined, but clearly visible from the diffraction patterns on both sides, as demonstrated in the insets. Across the ferroelastic domain wall (indicated by two arrows), ferroelectric a and b domains (viewed along [010] and [100] crystallographic directions, respectively) of BWO are revealed at each side. Different lattice constants (5.43 ± 0.01 Å versus 5.46 ± 0.01 Å) and

electron diffraction patterns are obtained along both zone axes as shown in the inset, where extra reflections of $(2n + 1, 0, m)$ viewed along [010] are indicated by solid green arrows, exampled by (101). On the other hand, these extra reflections are absent due to higher structural symmetry when viewed along the zone axes of [100], as indicated by hollow green arrows. The distinct differences in lattice constant between both domains are clearly evidenced by taking line scans over 25 unit cells from the [100] and [010] domains of BWO thin film (Fig. 2d) calibrated using internal reference of the STO substrate, indicating an approximately 0.4% elastic strain across the ferroelastic domain wall. Such an in-plane strain over a large area of the film has been further confirmed in reciprocal space mapping (RSM) as seen in Supplementary Fig. 4. The detailed measurements for the high-resolution STEM and RSM can be seen in the Methods section.

Controllable ferroelastic switching. In order to demonstrate an in-plane ferroelastic switching with 90° polarization rotation in the layered-perovskite BWO thin films with a negligible constraint from the substrate (illustrated schematically in Fig. 3a), Au-electrode patterns were fabricated on top of the BWO/STO heterostructure (film thickness approximately 200 nm) as seen in the AFM image in Fig. 3b. *In situ* in-plane PFM studies (Fig. 3c–f) before and after the application of the electric field (about 50–200 kV cm^{-1}) provide direct evidence that the ferroelastic domains are switchable without in-plane constraints. Figure 3c shows the as-grown multi-domain structure with four preferred polarizations before the application of the electric field. When an electrical bias was applied on the top-viewed electrode, the as-grown multiple-domain structures started to switch and were eventually erased (up to 200 kV cm^{-1}) within the electrodes. A ferroelectric mono-domain appeared with its polarization along the direction of the applied electric field (Fig. 3d). The emerging

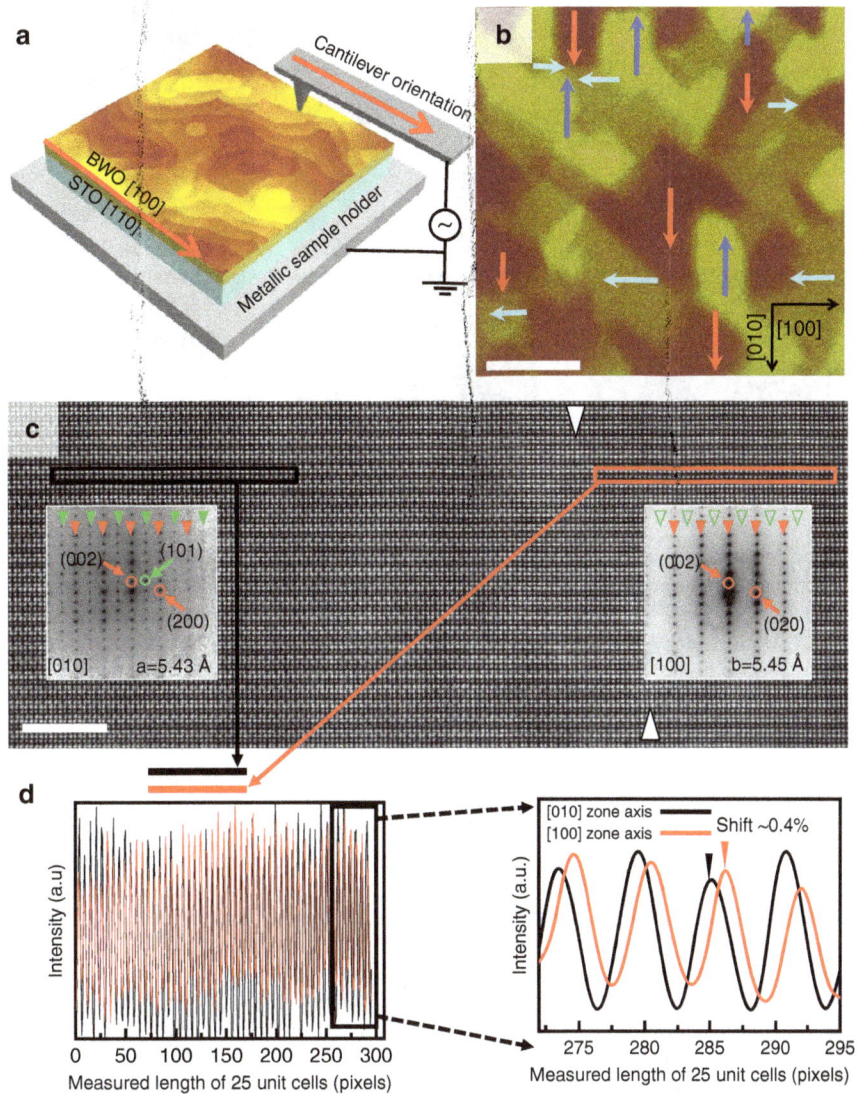

Figure 2 | Observation of ferroelectric-ferroelastic domains. (**a**) A Schematic geometry of the PFM measurement on BWO film grown on STO substrate (an AFM image was inserted). (**b**) The corresponding in-plane PFM image shows the in-plane ferroelectric domain structure with four preferred in-plane polarization directions labelled by blue, red and cyan arrows, respectively (the crystallographic directions of BWO film are labelled by black arrows). Scale bar, 500 nm. (**c**) Atomic-resolution HAADF-STEM image reveals a ferroelastic domain boundary (indicated by two white arrows) between two ferroelectric domains (*a* and *b* domains viewed along [010] and [100] zone axis, respectively); the inserted electron diffraction patterns confirm the change in orientation. Scale bar, 5 nm. (**d**) Line scans of 25 unit cells measured at both sides of the ferroelastic domain wall showing a spontaneous strain. The two periods of the intensity modulation (indicated by the red and black colour corresponding to the frame colours in **c**, respectively) are aligned to the starting peaks of unit cells, and show a distinctive shift with regard to each other after 25 unit cells (zoomed-in shown in the inset), corresponding to a measured strain of approximately 0.4%.

large-area mono-domain can be sequentially switched by 90° as a function of the electric field as shown in Fig. 3e,f. The microstructure and the correlated elastic strain before and after the ferroelastic domain switching have been investigated using dark-field TEM and HAADF-STEM. The switched mono-domain structure remains stable after at least 2 weeks across the orthorhombic crystal as shown in Supplementary Fig. 5, indicating that an electric-field-driven in-plane non-volatile elastic strain can be achieved in the whole region beneath the surface of the ferroelectric thin film.

The controllable ferroelastic strain of approximately 0.4% has been further demonstrated on the electrically switched *a* and *b* domains of BWO film using (S)TEM, as shown in Supplementary Fig. 6. Supplementary Figure 6a and e are the PFM images of the ferroelectric domains after the electrical switching. Both of two lamellae are then vertically extracted

(along and perpendicular to the directions of the electric field in Supplementary Fig. 6a,e) from the switched regions for (S)TEM investigations, respectively. Dark-field TEM (Supplementary Fig. 6b,f) reveals that a single region of one large domain is successfully patterned as characterized by the uniform contrast in-between the gold contacts. Diffraction patterns from each lamella are shown in Supplementary Fig. 6c,g, indicating the viewing directions of [100] and [010], that is, the electrically polarized directions of [010] (*b* domain) and [100] (*a* domain), respectively. Supplementary Figure 6d and h are the corresponding high-resolution HAADF-STEM images of two switched *b* and *a* domains. Line scans of 25 unit cells from each domain (calibrated using internal reference of STO substrate) show a controllable strain of approximately 0.4% between switched pure *a* and *b* domains (Supplementary Fig. 6i). It is noted that the strain value is not obtained directly

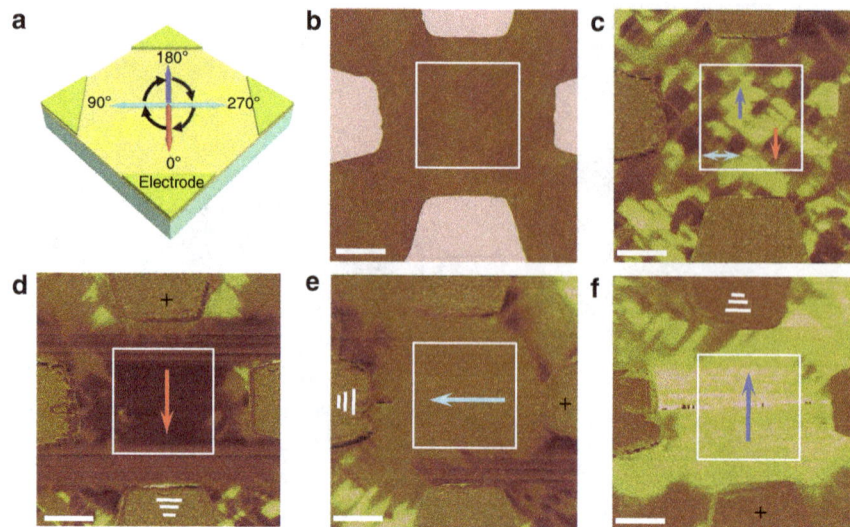

Figure 3 | Deterministic control of ferroelastic mono-domain switching via planar electrodes. (**a**) Schematic presentation of domain ferroelastic switching with four in-plane electrodes (red, blue and cyan arrows indicate four polarization directions and the black arrows indicate the switching with 90° step). (**b**) AFM image of the BWO thin film directly grown on STO substrate with four in-plane electrodes. (**c**) The PFM image shows the as-grown ferroelectric domain structure before the application of the in-plane electric field. (**d**) The PFM image shows a mono-domain structure when an in-plane electric field is applied on BWO. (**e,f**) Ferroelectric mono-domains with a polarization sequentially rotated by 90° (ferroelastic switching) when the in-plane electric field is further applied. Scale bar, 1 μm (**b–f**).

from the same region before and after electrical switching, while the diffraction patterns and (S)TEM results along [100] and [010] directions on any switched region calibrated using the STO substrate help provide the circumstantial evidence of this electrically controllable strain of ~0.4% on polarized domains. The detailed methods of the fabrication of the micro-sized electrodes, electric field control process and the *ex situ* TEM sample preparation can be found in Methods section.

Origin of the elastic switching with negligible constraint. To further elucidate the mechanism of the ferroelastic switching and in-plane non-volatile strain in BWO thin films, we studied the thickness-dependence of the switching field and used phase-field modelling to understand the electric-field-induced elastic mono-domain switching. First, the magnitude of substrate clamping and the consequent ferroelastic switching usually depends on the thickness of epitaxial oxide thin films[45,46]. For the BWO film with pure in-plane polarization, we studied the switching areal ratio as a function of electric field using planar electrodes in BWO heterostructures with thicknesses of 30, 120 and 200 nm, where we can see that domain size is essentially constant as a function of thickness (Supplementary Fig. 7). In Fig. 4a, we clearly observed that there is similar full switching field (approximately 50 kV cm^{-1}) for all these thicknesses, which means that the in-plane elastic switching is independent of the epitaxial strain release due to the increase of thickness. Unlike the domain switching in out-of-plane switching ferroelectrics, this result demonstrates that the in-plane polar rotation and free elastic switching is attributed to the intrinsic orthorhombic symmetry in this layered-perovskite.

We calculated the potential energy barriers (Fig. 4c) resulting from the ferroelastic switching in PTO epitaxial thin films on STO, PTO single crystals, BWO epitaxial thin films on STO and BWO single crystals, respectively. The detailed method for simulation is given in the Supplementary Note 1. As shown in Fig. 4b, there are four preferred ferroelectric polarizations along the <100> directions within the (001) plane of BWO, consistent with our experimental observations. In a fully clamped 001-oriented tetragonal PTO thin film, a 90° ferroelectric domain

switching (transition between *a* and *c* domain) driven by an electric field is energetically unfavourable (without an energy minimum around 90° in the blue curve) and only 180° switching is possible from a thermodynamic point of view[23]. In a PTO single crystal without substrate clamping, we observe two energy minima during the ferroelastic switching (green curve), indicating two elastic domain states with a decreased energy potential of about 27 MJ m^{-3} (corresponding to an electric field of approximately 380 kV cm^{-1} for the elastic domain switching). In an orthorhombic BWO, double-well energy potentials are revealed, where the energy potential required for 90° switching within the (001) plane is about 5 MJ m^{-3} for both the epitaxial film and single crystal (approximately 15 times lower than the one of a fully clamped PTO thin film) as indicated by the red and pink curves. This energy corresponds to an electric field of approximately 50 kV cm^{-1}, which is consistent with our experimental observations in the electric-field-driven elastic switching of the BWO thin film.

In order to compare the in-plane elastic switching in BWO and a fully clamped PTO thin film using planar electrodes, we used a PTO film on KaTiO$_3$ (KTO) substrate[47,48] to analyse the switching field and the switching energy when the polarization rotation occurs from twin-like to pure *a* or pure *b* domain (two variants of the in-plane polarization), where ferroelectric polarization is simulated to be fully in the film plane due to the epitaxial strain in PTO/KTO heterostructure. The detailed method and results for simulation are given in the Supplementary Note 1 and Supplementary Fig. 8. As shown in Supplementary Fig. 8a,b, the electric field for in-plane elastic switching from multi-domain to mono-domain in fully clamped PTO is above 4,000 kV cm^{-1}, which is over one order of magnitude larger than the one of BWO film. Such a large in-plane electric field for the elastic switching in a PTO film due to the substrate clamping is technically impractical[13–15]. The phase-field simulation further indicates that the substrate constraint on the ferroelastic switching is negligible in this intrinsic orthorhombic BWO ferroelectric thin film with a pure in-plane polarization in contrast to other ferroelectric thin films with tetragonal or rhombohedral ground states.

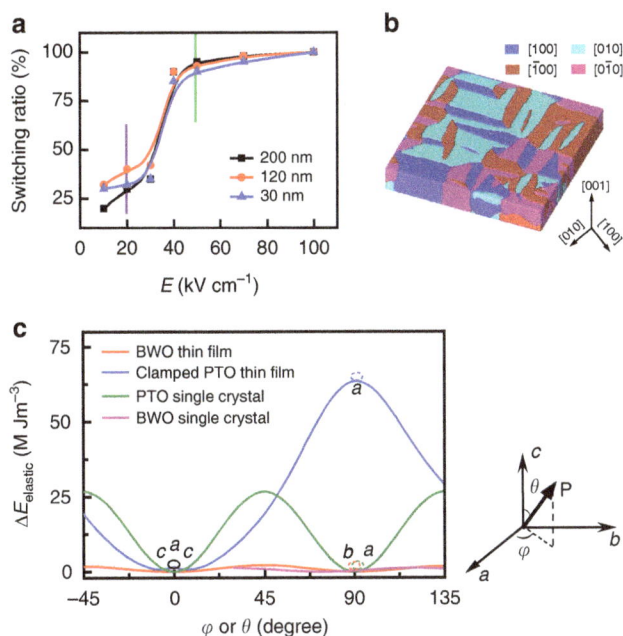

Figure 4 | Ferroelastic switching mechanism. (**a**) Ferroelastic switching areal ratio as a function of planar electric fields in BWO thin films with thicknesses of 30, 120 and 200 nm, respectively. It shows that the electric fields for elastic switching maintain in the BWO thin films with various thicknesses which have a similar threshold switching field about 20 kV cm^{-1} (indicated as purple line) and a full switching field about 50 kV cm^{-1} (indicated as green line). (**b**) Simulated domain structure of the 001-oriented BWO crystal with four preferred polarization directions indicated by blue, red, cyan and pink colour, respectively. The crystallographic directions of BWO are indicated as the black arrows, respectively. (**c**) Phase-field studies of the energy potential curves between PTO and BWO during the electric-field-driven ferroelastic domain switching in bulk and thin film, respectively; the energy potential for a ferroelastic switching in the BWO thin film on STO is approximately 15 times lower than the one for a ferroelastic switching in a fully clamped PTO thin film on STO, indicating that the substrate constraint on ferroelastic switching is negligible in BWO epitaxial film.

Nanoscale control of domain features and elastic switching. With the full understanding of the in-plane clamping-free elastic switching in BWO thin films, nanoscale manipulation of this elastic strain and mono-domain structure help further reveal the mechanism of the ferroelastic switching dynamics without the assistance of twin-like nanodomains. Scanning-probe-based technique is a powerful tool to generate a local electric field and control the ferroelectric polarization within a size of approximately 20 nm (refs 49,50). The nucleated mono-domain may depend on the tip-scanning size and direction, which is not restricted by the large voltage and geometry in planar electrodes. As illustrated schematically in Fig. 5a, a conductive epitaxial SrRuO₃ (SRO) was used as a bottom contact in the BWO/SRO/STO heterostructure during the nanoscale probe bias control, where a simulated distribution of the in-plane electric field beneath the nanoscale metallic AFM tip is also shown. For all the probe bias poling in the BWO/SRO heterostructures, out-of-plane domain switching is absent as confirmed by the out-of-plane domain image before and after the application of the electric field (Fig. 5b), indicating the polarization is fully in the film plane. This behaviour can be also confirmed by the dielectric measurement as a function of electric field (Supplementary Fig. 9g). In an approximately 200-nm thick BWO thin film, the in-plane ferroelectric polarization was totally switched under the application of

a probe bias of approximately $+10$ V as shown in Fig. 5c. The switched in-plane polarization can be determined by the slow-scanning direction of the AFM cantilever and the probe bias voltage. As seen in Fig. 5c and Supplementary Fig. 10, the final in-plane polar vector (blue arrow) of BWO is always anti-parallel to the slow-scanning direction (dark dash arrow) when a positive bias is applied on the AFM tip. However, when a negative bias is applied on the AFM tip, the local field with opposite directions is reversed, resulting in a final polarization parallel to the slow-scanning direction (the red arrow in Fig. 5d). The detailed control of domain switching with scanning-probe-bias is given in the Supplementary Note 2.

The dynamics of this probe-bias-induced pure in-plane switching are shown in Fig. 6. During the probe switching, the scanning-direction-dependent anisotropic in-plane component of a positive local bias applied on the AFM tip interacts with the in-plane polarization of BWO. Therefore, the in-plane polar vectors of BWO will be switched row-by-row by the local bias with opposite directions (dark and white arrows around the tip). In this way, polarization switching from a multi-domain state to a mono-domain state occurs, accompanying with the final polar vectors anti-parallel to the slow-scanning direction as seen in Fig. 6a. Consequently, an alternative positive and negative probe bias on the AFM tip will give rise to an opposite polarization (blue and red arrows) in the pure in-plane domain structure as shown in Fig. 6b. The ferroelastic switching of the BWO films with different thickness from 30 to 200 nm are also shown in Supplementary Fig. 11. Different with the strain-driven twin-like domain switching in tetragonal or rhombohedral ferroelectric thin films, the nanoscale manipulation of the in-plane ferroelastic switching dynamics in this intrinsic orthorhombic system is independent with the epitaxial strain. BWO films with -1.3% compressive strain show similar domain formation and switching behaviour as the one on STO substrate (Supplementary Fig. 12). This demonstrates that the switched mono-domain is totally decoupled with each other, which provides us a crucial pathway for the future exploration of domain-wall engineering and various topological structures[51–53] in ferroelectric materials.

Furthermore, a complete 90°ferroelastic switching sequence can be also reversibly controlled using the scanning-probe-based technique as seen in Fig. 7. This full in-plane rotation of polarization and a controllable ferroelastic mono-domain switching in thin film heterostructures without the assistance of twin-like nanodomains demonstrate a key step towards the integration of the ferroelasticity with a thin film geometry.

Discussion
To conclude, we use epitaxial layered-perovskite BWO to demonstrate a ferroelastic switching with a reversible non-volatile control of mono-domain in an orthorhombic heterostructure, which is absent in ferroelectrics with rhombohedral or tetragonal structure due to their strong substrate clamping and pinning of domain-wall motion. In combination with phase-field simulation, manipulation of the elastic mono-domain switching can be achieved by a planar electrode and a scanning-probe-based technique, indicating that this material can be a good model system for ferroelastic domain engineering. The reversible control of nanoscale ferroelectric/elastic switching at room temperature with a significant in-plane non-volatile strain in a thin film heterostructure also provides us with a promising framework to study the coupling of the lattice degree of freedom with other order in future micro/nano applications.

Methods
Thin film growth. BWO and BWO/SRO heterostructures were fabricated using laser molecular beam epitaxy (248 nm excimer) with *in situ* monitoring of RHEED

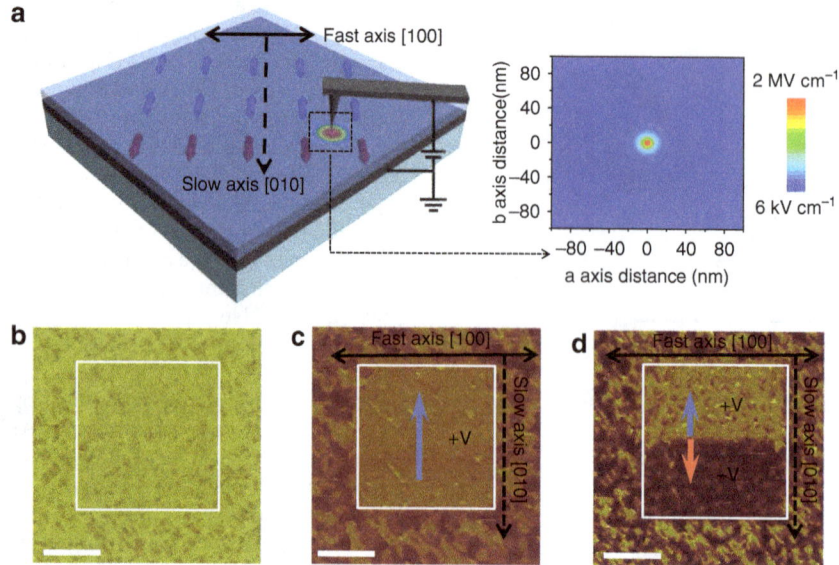

Figure 5 | Nanoscale control of multiple ferroelectric/ferroelastic domain features. (**a**) Schematic geometries of the in-plane ferroelastic domain structure of BWO film grown on STO substrate with SRO bottom electrode controlled by an out-of-plane electric field applied on the AFM tip and the distribution of the in-plane electric field underneath the tip. The blue and red arrows label two polar vectors perpendicular to the fast-scanning axis of the BWO [100] crystallographic direction (indicated by the black line with double-sided arrows). (**b**) The out-of-plane PFM image shows no out-of-plane projections of the ferroelectric polarization before and after the application of the probe bias. (**c**) The in-plane PFM image acquired after the probe-bias switching with a +10 V bias applied on the scanning tip, where the polarization direction of the switched domain is anti-parallel to the slow-scanning direction as illustrated by the blue arrow. (**d**) In-plane PFM image acquired after the probe-bias switching with a +10 and −10 V bias applied alternatively on the scanning tip, where the polarization directions of the switched domains are anti-parallel and parallel to the slow-scanning direction as illustrated by the red and blue arrows. Scale bar, 1 μm (**b–d**).

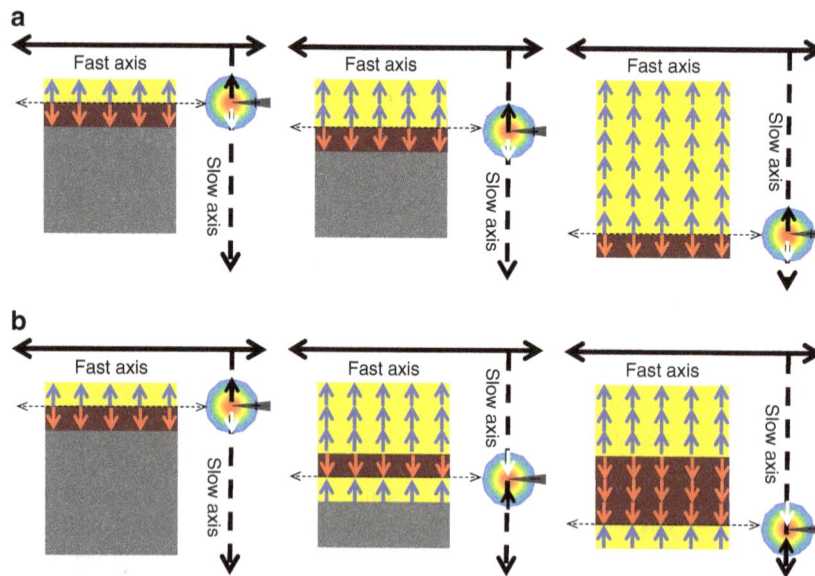

Figure 6 | Scanning-probe switching dynamics at nanoscale. (**a**) Schematics of the in-plane polarization switching process under the application of a scanning-direction-dependent anisotropic electric field. (**b**) Schematics of the in-plane polarization switching using an anisotropic electric field with positive and negative bias. The [100] and [010] crystallographic directions of BWO are indicated by the black line and dashed line with double-sided arrows, respectively.

during the growth. Atomically smooth substrates (001-oriented STO and LAO) were prepared by a combined HF-etching/anneal treatment. Stoichiometric SRO and BWO (5% excess of Bi_2O_3 in order to compensate the volatile Bi at high growth temperature) targets were ablated at a laser energy density of approximately $1 J cm^{-2}$ and a repetition rate of 1 Hz for the growth of SRO (thickness approximately 10 nm) and BWO (thickness up to 200 nm), respectively. During growth of SRO, the substrate temperature was maintained at 700 °C at an oxygen environment of 100 mtorr. For the growth of BWO, the substrate temperature was

720 °C at an oxygen pressure of 100 mtorr. Afterwards, the films were cooled to room temperature at 0.1 atm of oxygen with a cooling rate of 5 °C min^{-1}. The Au electrodes (approximately 50-nm thick) with micro-patterns were fabricated by magnetron sputtering assisted by electron beam lithography (JEOL JBX6300FS).

Characterizations of crystal and domain structures.
X-ray diffraction. X-ray θ-2θ scans were obtained by high-resolution X-ray diffraction (Lab XRD-6000, SHIMADZU). The detailed crystal structures was

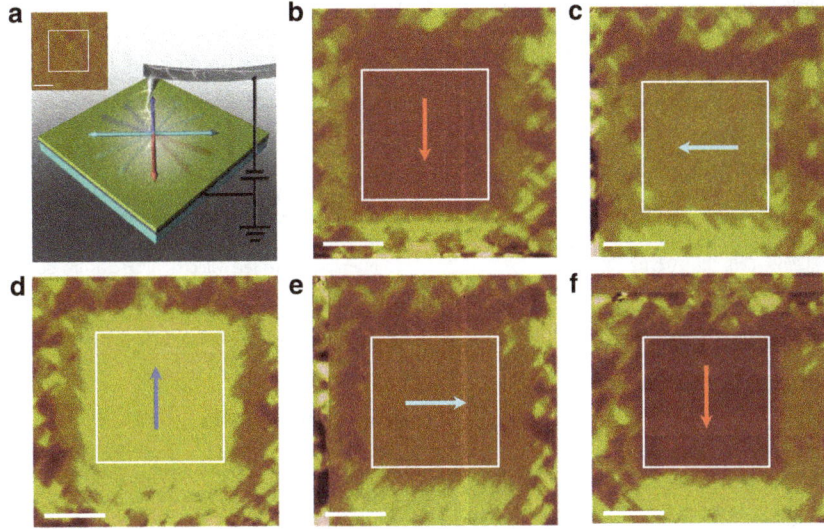

Figure 7 | A complete 90° ferroelastic switching sequence. (**a**) The schematic of the ferroelastic switching with scanning-probe-bias of −15 V in the BWO thin film (approximately 120 nm) with SRO bottom electrode grown on STO substrate. The inset is the topography of the BWO thin film (the region in the white box is the switched area). Scale bar, 1 μm. (**b**) The PFM image of the downward-polarized mono-domain configuration as indicated by the red arrow. The controllable 90° ferroelastic switching can be achieved from a downward polarization to left polarization (**c**) as indicated by the cyan arrow, from left polarization to upward polarization (**d**) as indicated by the blue arrow, from upward polarization to right polarization (**e**) as indicated by the cyan arrow and finally, the mono-domain configuration can be switched back to the initial downward state for a complete sequence (**f**) as indicated by the red arrow. Scale bar, 1 μm (**b**–**f**).

analysed by a four-circle film diffractometer with a Ge (220) × 2 incident-beam monochromator and stripe detector (Rigaku SmartLab Film Version, Cu-*Ka* radiation). For the strain values extracted from the rocking curve, the intensity distribution of diffraction spots could be described base on the kinetics of X-ray diffraction as:

$$I_c(S) = |f|^2 E_e^2 \frac{\sin^2 \frac{N_1}{2} \vec{s} \cdot \vec{a}}{\sin^2 \frac{1}{2} \vec{s} \cdot \vec{a}} \cdot \frac{\sin^2 \frac{N_2}{2} \vec{s} \cdot \vec{b}}{\sin^2 \frac{1}{2} \vec{s} \cdot \vec{b}} \cdot \frac{\sin^2 \frac{N_3}{2} \vec{s} \cdot \vec{c}}{\sin^2 \frac{1}{2} \vec{s} \cdot \vec{c}} \qquad (3)$$

The profile of diffraction spots of small crystallite depends mainly on the geometry size of the crystallite. From this expression, it could be deduced that all the Bragg spots from one crystallite have almost the same rocking curve profiles in the reciprocal space, including the same FWHM value. In our sample, there are two kinds of crystalline domains: *a* and *b*. It is assumed that these two kinds of domains have the same average domain size along [100] direction. Thus the diffraction spots from *a* or *b* domain in the reciprocal space, such as (0 0 18), (2 0 18), (− 2 0 18), should possess the same FWHM value in [100] direction. On the other hand, since the in-plane lattice parameters of BWO unit cell a and b are not the same, the diffraction spots at h0l(h≠0) from *a* and *b* domains should split along [100] direction, and the splitting interval stands for difference between lattice parameter a and b. In the case of BWO (2 0 18) or BWO (− 2 0 18) RSM, due to the overlap of diffraction spots from *a* and *b* domains, the splitting interval is obtained indirectly by subtraction the size-caused width from the FWHM. The final value is about 0.0037 rlu of STO. The BWO (2 0 18) or BWO (− 2 0 18) diffractions spots located around (1 1 4.285) in the reciprocal lattice units of STO. Therefore, the difference between lattice parameters a and b of BWO can be extracted by this method.

FIB milling of the TEM samples. The TEM samples (before and after the application of electric field) were prepared using FIB (FEI Helios NanoLab DualBeam system). Lamellae were prepared following the <110> direction of the STO substrate. A final cleaning of the sample surface was performed using 2 keV Ga⁺ ions with a small beam current to reduce amorphous layer on the lamellae. The thickness of the as-prepared lamellae is approximately 50–80 nm.

TEM imaging conditions. The high-resolution HAADF-STEM image was acquired using a FEI Titan Cube 60–300 microscope fitted with an aberration corrector for the probe-forming lens, operated at 300 kV. The STEM convergence semi-angle was approximately 21.4 mrad, providing a probe size of approximately 1.0 Å at 300 kV. Diffraction patterns and dark-field TEM images were acquired on a FEI Tecnai G2 microscope operated at 200 kV. Measured lattice constants of a and b of the as-grown thin film from HAADF-STEM images are a = 5.43 ± 0.01 Å and b = 5.46 ± 0.01 Å (error bar represents the s.d., which is calculated from 10 measurements from different areas using 2 samples).

AFM and PFM. The AFM measurements were carried out on tapping mode with ultra-sharp Si tips. The PFM measurements were carried out on a Bruker Multimode 8 AFM with commercially available TiPt-coated Si tips (Mikro Masch) with a tip curvature radius of <30 nm. The typical tip-scanning velocity was 2 μm ms⁻¹. The amplitude and frequency of the AC input were 1.5 V_pp and 22 kHz,

respectively. High-resolution PFM images were acquired on a wide array of samples. The polarization vectors have been re-constructed based on the domain images obtained from the cantilever scanning parallel and perpendicular to the BWO [100] axis, respectively.

References

1. Park, S.-E. & Shrout, T. R. Ultrahigh strain and piezoelectric behavior in relaxor based ferroelectric single crystals. *J. Appl. Phys.* **82**, 1804–1811 (1997).
2. Ren, X. B. Large electric-field-induced strain in ferroelectric crystals by point-defect-mediated reversible domain switching. *Nat. Mater.* **3**, 91–94 (2004).
3. Eerenstein, W., Mathur, N. D. & Scott, J. F. Multiferroic and magnetoelectric materials. *Nature* **442**, 759–765 (2006).
4. Ramesh, R. & Spaldin, N. A. Multiferroics: progress and prospects in thin films. *Nature* **6**, 21–29 (2007).
5. Zheng, H. *et al.* Multiferroic BaTiO₃-CoFe₂O₄ nanostructures. *Science* **30**, 661–663 (2004).
6. Eerenstein, W., Wiora, M., Prieto, J. L., Scott, J. F. & Mathur, N. D. Giant sharp and persistent converse magnetoelectric effects in multiferroic epitaxial heterostructures. *Nat. Mater.* **6**, 348–351 (2007).
7. Zhang, S. *et al.* Electric-field control of nonvolatile magnetization in Co₄₀Fe₄₀B₂₀/ Pb(Mg₁/₃Nb₂/₃)₀.₇Ti₀.₃O₃ structure at room temperature. *Phys. Rev. Lett.* **108**, 137203 (2012).
8. Yang, S. W. *et al.* Non-volatile 180° magnetization reversal by an electric field in multiferroic heterostructures. *Adv. Mater.* **26**, 7091–7095 (2014).
9. Franke, K. J. A. *et al.* Size dependence of domain pattern transfer in multiferroic heterostructures. *Phys. Rev. Lett.* **112**, 017201 (2014).
10. Lee, Y. *et al.* Large resistivity modulation in mixed-phase metallic systems. *Nat. Commun.* **6**, 5959 (2015).
11. Hu, J. M., Li, Z., Chen, L. Q. & Nan, C. W. High-density magnetoresistive random access memory operating at ultralow voltage at room temperature. *Nat. Commun.* **2**, 553 (2011).
12. Heron, J. T. *et al.* Deterministic switching of ferromagnetism at room temperature using an electric field. *Nature* **516**, 370–373 (2014).
13. Dawber, M., Rabe, K. M. & Scott, J. F. Physics of thin-film ferroelectric oxides. *Rev. Mod. Phys.* **77**, 1083–1124 (2005).
14. Setter, N. *et al.* Ferroelectric thin films: review of materials, properties, and applications. *J Appl. Phys.* **100**, 051606 (2006).
15. Scott, J. F. Applications of modern ferroelectrics. *Science* **315**, 954–959 (2007).
16. Lefki, K. & Dormans, G. J. M. Measurement of piezoelectric coefficients of ferroelectric thin films. *J. Appl. Phys.* **76**, 1764–1767 (1994).
17. Torah, R. N., Beeby, S. P. & White, N. M. Experimental investigation into the effect of substrate clamping on the piezoelectric behaviour of thick-film PZT elements. *J. Phys. D Appl. Phys.* **37**, 1074–1078 (2004).

18. Damjanovic, D. Ferroelectric, dielectric and piezoelectric properties of ferroelectric thin films and ceramics. *Rep. Prog. Phys.* **61**, 1267–1324 (1998).

19. Cruz, M. P. *et al.* Strain control of domain-wall stability in epitaxial BiFeO₃ (110) films. *Phys. Rev. Lett.* **99**, 217601 (2007).

20. Nelson, C. T. *et al.* Domain dynamics during ferroelectric switching. *Science* **334**, 968–971 (2011).

21. Chu, M. W., Szafraniak, I., Hesse, D., Alexe, M. & Gösele, U. Elastic coupling between 90° twin walls and interfacial dislocations in epitaxial ferroelectric perovskites: a quantitative high-resolution transmission electron microscopy study. *Phys. Rev. B* **72**, 174112 (2005).

22. Su, D. *et al.* Origin of 90° domain wall pinning in Pb(Zr₀.₂Ti₀.₈)O₃ heteroepitaxial thin films. *Appl. Phys. Lett.* **99**, 102902 (2011).

23. Gao, P. *et al.* Ferroelastic domain switching dynamics under electrical and mechanical excitations. *Nat. Commun.* **5**, 3801 (2014).

24. Nagarajan, V. *et al.* Dynamics of ferroelastic domains in ferroelectric thin films. *Nat. Mater.* **2**, 43–47 (2002).

25. Baek, S. H. *et al.* Ferroelastic switching for nanoscale non-volatile magnetoelectric devices. *Nat. Mater.* **9**, 309–314 (2010).

26. Baek, S. H. *et al.* Giant piezoelectricity on Si for hyperactive MEMS. *Science* **334**, 958–961 (2011).

27. Zhang, J. X. *et al.* A nanoscale shape memory oxide. *Nat. Commun.* **4**, 2768 (2013).

28. Yang, J. C. *et al.* Orthorhombic BiFeO₃. *Phys. Rev. Lett.* **109**, 247606 (2012).

29. Matzen, S. *et al.* Super switching and control of in-plane ferroelectric nanodomains in strained thin films. *Nat. Commun.* **5**, 4415 (2014).

30. Xu, R. J. *et al.* Ferroelectric polarization reversal via successive ferroelastic transitions. *Nat. Mater.* **14**, 79–86 (2015).

31. Mathur, N. A desirable wind up. *Nature* **454**, 591–592 (2008).

32. Pertsev, N. A., Zembilgotov, A. G. & Tagantsev, A. K. Effect of mechanical boundary conditions on phase diagrams of epitaxial ferroelectric thin films. *Phys. Rev. Lett.* **80**, 1988 (1998).

33. Schlom, D. G. *et al.* Strain tuning of ferroelectric thin films. *Ann. Rev. Mater. Res.* **37**, 589–626 (2007).

34. A-Paz de Araujo, C., Cuchiaro, J. D., McMillan, L. D., Scott, M. C. & Scott, J. F. Fatigue-free ferroelectric capacitors with platinum electrodes. *Nature* **374**, 627–629 (1995).

35. Park, B. H. *et al.* Lanthanum-substituted bismuth titanate for use in non-volatile memories. *Nature* **401**, 682–684 (1999).

36. Lee, H. N., Hesse, D., Zakharov, N. & Gösele, U. Ferroelectric Bi₃.₂₅La₀.₇₅Ti₃O₁₂ films of uniform a-axis orientation on silicon substrates. *Science* **296**, 2006–2009 (2002).

37. Chon, U., Jang, H. M., Kim, M. G. & Chang, C. H. Layered perovskites with giant spontaneous polarizations for nonvolatile memories. *Phys. Rev. Lett.* **89**, 087601 (2002).

38. Shimakawa, Y. *et al.* Crystal structure and ferroelectric properties of ABi₂Ta₂O₉ (A = Ca, Sr, and Ba). *Phys. Rev. B* **61**, 6559 (2000).

39. Machado, R., Stachiotti, M. G., Migoni, R. L. & Tera, A. H. First-principles determination of ferroelectric instabilities in Aurivillius compounds. *Phys. Rev. B* **70**, 214112 (2004).

40. Watanabe, T. & Funakubo, H. Controlled crystal growth of layered-perovskite thin films as an approach to study their basic properties. *J. Appl. Phys.* **100**, 051602 (2006).

41. Kitanaka, Y., Noguchi, Y. & Miyayama, M. Oxygen-vacancy-induced 90°-domain clamping in ferroelectric Bi₄Ti₃O₁₂ single crystals. *Phys. Rev. B* **81**, 094114 (2010).

42. Djani, H., Bousquet, E., Kellou, A. & Ghosez, P. First-principles study of the ferroelectric Aurivillius phase Bi₂WO₆. *Phys. Rev. B* **86**, 054107 (2012).

43. Djani, H., Hermet, P. & Ghosez, P. First-principles characterization of the *P2₁ab* ferroelectric phase of Aurivillius Bi₂WO₆. *J. Phys. Chem. C* **118**, 13514–13524 (2014).

44. Hamada, M., Tabata, H. & Kawai, T. Microstructure and dielectric properties of epitaxial Bi₂WO₆ deposited by pulsed laser ablation. *Thin Solid Films* **306**, 6–9 (1997).

45. Nagarajan, V. *et al.* Thickness dependence of structural and electrical properties in epitaxial lead zirconate titanate films. *J. Appl. Phys.* **86**, 595–602 (1999).

46. Pertsev, N. A. *et al.* Coercive field of ultrathin Pb(Zr₀.₅₂Ti₀.₄₈)O₃ epitaxial films. *Appl. Phys. Lett.* **83**, 3356–3358 (2003).

47. Kwak, B. S. *et al.* Strain relaxation by domain formation in epitaxial ferroelectric thin films. *Phys. Rev. Lett.* **68**, 3733–3736 (1992).

48. Lee, K. S., Choi, J. H., Lee, J. Y. & Baik, S. Domain formation in epitaxial Pb(Zr, Ti)O₃ thin films. *J. Appl. Phys.* **90**, 4095–4102 (2001).

49. Balke, N. *et al.* Deterministic control of ferroelastic switching in multiferroic materials. *Nat. Nanotechnol.* **4**, 868–875 (2009).

50. Balke, N. *et al.* Enhanced electric conductivity at ferroelectric vortex cores in BiFeO₃. *Nat. Phys.* **8**, 81–88 (2011).

51. Vasudevan, R. K. *et al.* Exploring topological defects in epitaxial BiFeO₃ thin films. *ACS Nano* **5**, 879–887 (2011).

52. McQuaid, R. G. P., Gruverman, A., Scott, J. F. & Gregg, J. M. Exploring vertex interactions in ferroelectric flux-closure domains. *Nano Lett.* **14**, 4230–4237 (2014).

53. Oh, Y. S. *et al.* Experimental demonstration of hybrid improper ferroelectricity and the presence of abundant charged walls in (Ca,Sr)₃Ti₂O₇ crystals. *Nat. Mater.* **14**, 407–413 (2015).

Acknowledgements

The work in Beijing Normal University is supported by the NSFC under contract numbers 51322207, 51332001 and 11274045. J.Z. also acknowledges the support from National Basic Research Program of China, under contract No. 2014CB920902. G.V.T. acknowledges the funding from the European Research Council under the Seventh Framework Program (FP7), ERC Advanced Grant No. 246791-COUNTATOMS. X.K. acknowledges the funding from NSFC (Grant No.11404016) and Beijing University of Technology (2015-RD-QB-19). J.W. acknowledges the funding from NSFC (Grant number 51472140). L.-Q.C. acknowledges the supporting by the U.S. Department of Energy, Office of Basic Energy Sciences, Division of Materials Sciences and Engineering under Award FG02-07ER46417. R.L. acknowledges Tsinghua National Laboratory for Information Science and Technology (TNList) Cross-discipline Foundation. Z.L. acknowledges the support from the NSFC (No.11374010 and No.11434009). Q.Z. and X.-F.H. acknowledge the funding support from NSFC (Grant No. 11434014). R.R. acknowledges support from the National Science Foundation (Nanosystems Engineering Research Center for Translational Applications of Nanoscale Multiferroic Systems) under grant number EEC-1160504.

Author contributions

J.Z. and C.W. conceived the experiments and prepared the manuscript. C.W. prepared the samples and performed the planar electric field switching. G.V.T. and X.K. designed and performed the (S)TEM measurements. J.W. performed the phase-field simulations. Y.T. and J.W. performed the PFM measurements and the probe switching. R.L. and Q.Z. helped the electrode fabrication and dielectric measurement. Z.L. performed the x-ray diffraction measurements. D.Y. provided the RHEED data. X.-F.H., G.V.T., L.-Q.C., C.-W.N. and R.R. were involved in the preparation and revision of the manuscript. All authors were involved in the analysis of the experimental and theoretical results.

Additional information

Competing financial interests: The authors declare no competing financial interests.

6

4Pi-RESOLFT nanoscopy

Ulrike Böhm[1], Stefan W. Hell[1] & Roman Schmidt[1]

By enlarging the aperture along the optic axis, the coherent utilization of opposing objective lenses (4Pi arrangement) has the potential to offer the sharpest and most light-efficient point-spread-functions in three-dimensional (3D) far-field fluorescence nanoscopy. However, to obtain unambiguous images, the signal has to be discriminated against contributions from lobes above and below the focal plane, which has tentatively limited 4Pi arrangements to imaging samples with controllable optical conditions. Here we apply the 4Pi scheme to RESOLFT nanoscopy using two-photon absorption for the on-switching of fluorescent proteins. We show that in this combination, the lobes are so low that low-light level, 3D nanoscale imaging of living cells becomes possible. Our method thus offers robust access to densely packed, axially extended cellular regions that have been notoriously difficult to super-resolve. Our approach also entails a fluorescence read-out scheme that translates molecular sensitivity to local off-switching rates into improved signal-to-noise ratio and resolution.

[1] Department of NanoBiophotonics, Max Planck Institute for Biophysical Chemistry, Am Fassberg 11, Göttingen 37077, Germany. Correspondence and requests for materials should be addressed to S.W.H. (email: stefan.hell@mpibpc.mpg.de) or to R.S. (email: roman.schmidt@mpibpc.mpg.de).

The three to seven fold improved axial resolution provided by 4Pi microscopy[1-3] in the 1990s marked a first step in the quest for radically improving the resolution in far-field fluorescence microscopy. Yet the resolution provided by 4Pi microscopy remained diffraction-limited, because by jointly using two opposing lenses for focusing the excitation and/or the fluorescence light, this method just optimized the focusing conditions for feature separation. Modern far-field fluorescence nanoscopy[4], or superresolution microscopy, such as the methods called stimulated emission depletion (STED), reversible fluorescent saturable optical transition (RESOLFT) and later also photoactivated localization microscopy (PALM)/stochastical optical reconstruction microscopy (STORM) fundamentally departed from such early superresolution concepts by discerning features through a molecular state transition. The use of a state transition for feature separation, typically a transition between a fluorescent (ON) and a non-fluorescent (OFF) state, opened the road to lens-based fluorescence microscopy with resolution that is conceptually not limited by diffraction.

Yet diffraction plays a role in these 'diffraction-unlimited' techniques because the resolution of these 'nanoscopy' methods still benefits strongly from focusing the light as sharply as possible. While in STED and RESOLFT, it is the focusing of the illumination light in sample space that matters, in PALM/STORM it is the focusing of the emitted light at the detector. Therefore, the optimization of focusing remains very timely. 4Pi arrangements can also facilitate the doubling of the detected fluorescence without compromising the resolution in the focal plane (x,y), and offer significantly sharper axial (z) intensity gradients than single lenses for both the illumination and the detected light. Consequently, the combination of 4Pi with STED, RESOLFT and PALM/STORM approaches currently offers the most powerful optical setting for three-dimensional (3D) fluorescence nanoscopy[5-7].

Yet 4Pi-type super-resolution arrangements are scarcely reported for STED and PALM and entirely unexplored for RESOLFT, a STED-derivative that typically uses reversibly switchable fluorescent proteins (RSFP) for providing the mandatory ON and OFF states. RSFP-based RESOLFT is particularly attractive because it operates with low light levels, making it gentle to living cells[8].

The difficulties of realizing a 4Pi setup are commonly attributed to the counter-alignment of the two high numerical aperture (NA) lenses. In practice, however, the alignment can be controlled and stabilized over many hours. Instead, a far more general problem that is inherent to all fluorescent imaging modalities comes to the fore. The fluorescence signal (that emanates from each sub-diffraction pixel volume under investigation) needs to be discriminated against 'background' signal from outside of this volume. This 'background' largely stems from optical aberrations that preclude precise spatial control of the illumination or fluorescence beam positions and, in case of STED, RESOLFT or PALM/STORM, from imperfections of the ON/OFF-state transfer (switching) process. Discrimination against this 'background' signal is most challenging along the optical (z) axis, especially when the probed volume is of sub-diffraction dimensions. Lack of sufficient discrimination along the z-axis (optical sectioning) manifests itself as artifacts in the image, particularly as 'ghost features' above and below the real features.

When describing the imaging process in the spatial frequency domain, the appearance of axial lobes corresponds to local depressions in the amplitude of the optical transfer function, that is, the modulation transfer function (MTF) of the microscope. Structural information of the sample can only be retrieved in those spatial frequency bands where the MTF is strong enough to convey a signal that sufficiently exceeds the local noise level.

In a 4Pi microscope, MTF depressions are typically restricted to sharp local minima at the so-called critical frequencies[9]. As their amplitude strongly depends on the aperture angle α of the objective lenses used[9], combinations of 4Pi with diffraction-unlimited super-resolution/nanoscopy methods such as isoSTED[5,10,11] and iPALM[6,12], have unfortunately been limited to imaging fixed samples that are more easily accessible with high angle lenses ($\alpha \geq 74°$, as for oil immersion lenses with NA ≥ 1.46). Furthermore, the imaged objects were rather thin and labelled very sparsely, as both properties alleviate the requirements on optical sectioning, that is, on suppressing ('background') signal from above and below the focal plane. Fortunately, in a coordinate-targeted nanoscopy method such as RESOLFT, the signal received from the targeted nanosized pixel volume scales with the average number of molecules located within this volume, allowing for tailoring of the pixel volume[11], and hence the resolution and the signal, to the actual imaging conditions to render the 'background' (mathematically) treatable.

Here we report the realization of 4Pi-RESOLFT nanoscopy, that is, of a conceptually diffraction-unlimited resolving method which, by virtue of 4Pi microscopy, provides spatially uniform 3D resolution for imaging (living) cells at the nanometre scale, offers strong optical sectioning due to multiple background suppressing mechanisms and operates at low light levels in 3D.

Results

On-switching order and optical sectioning. The effective point-spread-function (PSF) $h_{ef}(\mathbf{r})$ of a coordinate-targeted super-resolution imaging modality ultimately quantifies the 3D-coordinate range where the fluorescent molecules are allowed to yield measurable signal. If the fluorophores from a certain range are imaged onto a (confocal point) detector, $h_{ef}(\mathbf{r})$ is given by the normalized distribution $S^{ON}(\mathbf{r})$ showing where a molecule is allowed to be in the ON-state at the time of read-out, multiplied by a normalized function $h^{det}(\mathbf{r})$ that describes the detection probability:

$$h_{ef}(\mathbf{r}) = S^{ON}(\mathbf{r}) \cdot h^{det}(\mathbf{r}) \tag{1}$$

$S_{ON}(\mathbf{r})$ is proportional to a product of normalized terms h^{on} and $\widetilde{h}^{off}$ that describe the generation of the ON-state by the use of on- and off-switching processes, respectively. h^{on} describes the spatial probability to assume the ON-state in the absence of off-switching light. It can typically be written as a product of terms h_i^{on} that each express the relative probability for absorption of a single photon that drives a transition to a (virtual) state, and therefore scales with the intensity of the light patterns used (for example, $h_{exc}^{on} \sim I_{exc}^{on}$ for single-photon excitation with intensity I_{exc}^{on}; in case of two-photon excitation: $h_{2phexc}^{on} = h_{exc}^{on} \cdot h_{exc}^{on}$). $\widetilde{h}^{off}$ describes the effect of the off-switching light on a potential ON-state distribution; $\widetilde{h}^{off} = 1$ where molecules are always allowed to assume the ON-state, and $\widetilde{h}^{off} = 0$ where they are forced to stay in an OFF-state. Due to the forced assumption of an OFF-state by 'saturating' off-switching, $\widetilde{h}^{off}$ is usually much sharper than the off-switching light intensity patterns.

We formally define h^{det} as the first on-switching term $h_1^{on} \equiv h^{det}$ (because of its similar effect on h_{ef}), drop any h_i^{on} that does not significantly sharpen h_{ef} (for example, widefield-detection or sample-wide switching) and obtain:

$$h_{ef}(\mathbf{r}) = \prod_{i=1}^{O^{on}} h_i^{on}(\mathbf{r}) \cdot \widetilde{h}^{off}(\mathbf{r}) \tag{2}$$

Here the number of on-factors O^{on} denotes the on-(switching) order of h_{ef}, for example, $O^{on} = 2$ (excitation by single-photon absorption and confocal detection) for a typical confocal (STED) microscope. Optical sectioning can generally be improved by engineering $\tilde{h}^{\mathrm{off}}$ such that molecules in out-of-focus areas are switched off more effectively (Supplementary Fig. 1), or by requiring the absorption of multiple photons for the occupation of the ON-state, which increases O^{on}. The latter can be realized directly through standard two-photon absorption[13–17], or by requiring the sequential occupation of multiple real states to reach the ON-state[18]. Such sequential state occupation is easily realized using the switching steps offered by RSFP (that are central to the RSFP-based RESOLFT concept[4,19]).

The 4Pi-RESOLFT modality. In this study, we devised a coordinate-targeted 4Pi-RESOLFT modality that utilizes negative-switching RSFP (that is, those that are switched off at a wavelength that is also used for generating the fluorescent signal, Fig. 1a) and that resorts to all the processes mentioned above for strong optical sectioning. Concretely, we opted for the RSFP Dronpa-M159T[20–22], which stands out by relatively fast switching kinetics with comparatively low background. At each scanning position, the local RSFP molecules were cycled through their ON- and OFF-states by consecutive light pulses that defined our RESOLFT imaging sequence (Fig. 1a). In the initial step ('activation' pulse), we applied a μs-long train of 170–fs pulses at 90 MHz/780 nm in a focal pattern h_{ac} to (partially) transfer ('activate') local RSFP to their meta-stable 'active'-state by two-photon absorption, as described by the activation distribution $S^B(\mathbf{r})$. Subsequently, we applied a μs–ms-long 'deactivation' pulse of continuous-wave (CW) irradiation at 488 or 491 nm, focused to form a hollow deactivation pattern (for example, a 'z-donut' h_{zd}, Fig. 1a, Supplementary Methods). This drove active RSFP outside the targeted pixel volume

(for example, above and below the focal plane) back to their inactive state, which effectively denied them a further excitation to the fluorescent ON-state and thus improved the spatial ON/OFF-contrast during the following 'read-out' pulse. We finally probed the remaining active RSFP by a second μs–ms-long CW pulse at 491 nm with a focal pattern h_{ro}, which transferred them to their fluorescent (ON) state, and detected the fluorescence through a confocal pinhole.

Our scheme thus entails a number of advantages for live-cell 3D-imaging. First, RSFP are inherently live-cell compatible protein markers, and selection of sufficiently bright and stable RSFP is readily available[8,20,23]. Second, optical sectioning benefits from the additional switching step (activation) involved in the RSFP switching cycle with respect to modalities that do not make use of a meta-stable state. This additional switching becomes especially powerful if it is mediated by two-photon absorption in a 4Pi configuration, as O^{on} rises to 4 and the activation and read-out patterns (h_{ac}, h_{ro}) can be setup to a limited zone of overlap in the focal region (Fig. 1a). While overlapping several pattern h_i^{on} also forms the basis of 4Pi microscopy of type C using two-photon excitation[24], here we do not require coherent double-lens (4Pi) detection of the emitted fluorescence, and therefore do not need broad-band intra-cavity dispersion compensation. The scheme presented here thus acts to the same effect with much less technical complexity. Finally, activation by two-photon absorption entails much less photo damage than two-photon excitation, as it takes place at a time during the switching cycle when virtually no markers can assume their excited fluorescent state.

Under ideal conditions, the effective PSF h_{ef} of such a system is virtually free of axial lobes (Fig. 1a) even without a deactivation pulse. In practice, incomplete deactivation and optical aberrations

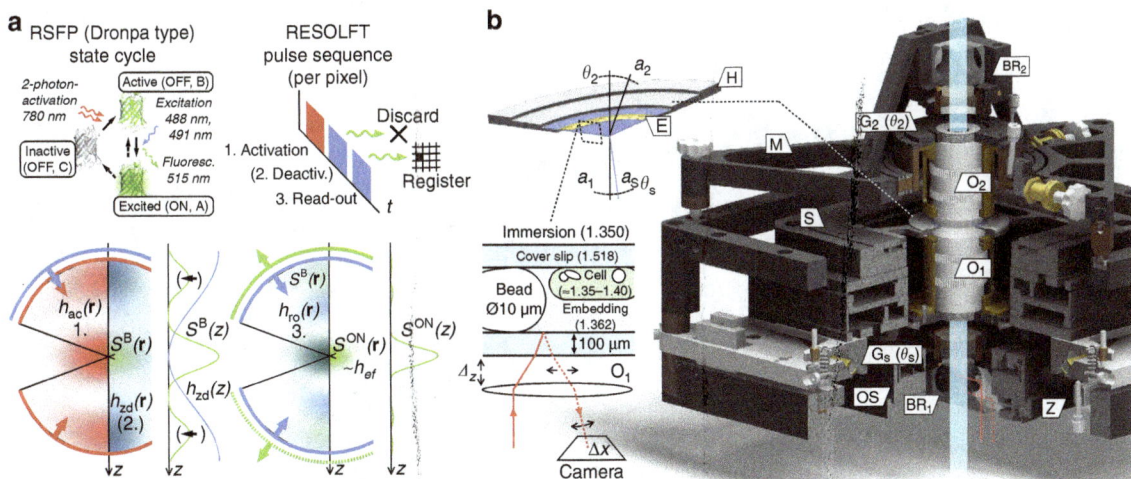

Figure 1 | 4Pi-RESOLFT principle and sample optics. (a) Coherent double-lens illumination cycles RSFP markers between dark (OFF) and bright (ON) states to generate spatial ON/OFF-contrast. For each pixel, an activation light pulse (focal pattern h_{ac}) induces two-photon activation of RSFP (state transition C->B) in a pattern $S^B(\mathbf{r})$ with axial side-maxima (lobes) that are optionally suppressed by a subsequently applied deactivation pulse (h_{zd}, B<->A->C). Fluorescence generated by the ON-state A is detected during read-out (B<->A->C) by a pattern h_{ro}. Its mutual overlap with $S^B(\mathbf{r})$ is constrained to the focal centre, resulting in an effective PSF $h_{\mathrm{ef}} \sim S^{\mathrm{ON}}$ that exhibits ≈100 nm axial FWHM and exceptionally low side-maxima. Profiles show on-axis values. **(b)** The upright 4Pi unit of the microscope. Cells are mounted on a ring-shaped sample holder (H), between two cover glasses fixed at 10 μm distance by spacer beads and epoxy resin (E). The set of refractive indices (in brackets) of the immersion and embedding medium, cover slip thickness and correction collar settings of the objective lenses (O_1, O_2) diminishes aberrations from the sample. The sample stage (S) is mounted on a vertically movable (Z) goniometer (G_S), accepts the sample holder (H) and provides five degrees-of-freedom for coarse xyz-positioning and z-scanning of the sample, as well as tip-/tilt-alignment (θ_S) of the cover slip normal (a_S) to the optic axis of O_1 (a_1). O_1 itself is mounted on a xyz-piezo stage (OS) that provides online fine control over the displacement of both foci. A triangular mount (M) allows for tip/tilt-(θ_2) and coarse xyz-alignment of O_2 (axis a_2) with respect to O_1, and can conveniently be detached to change the sample. Two polarizing beam splitter/quarter-wave retarder pairs ($BR_{1,2}$) clean up and tune the polarization of the incident beam pairs to opposing circular states. One beam splitter furthermore serves as a port for an alignment laser beam that provides optical feedback for online-stabilization of the axial sample position (Δz); the beam traverses the respective objective lens off-axis (solid red path), gets reflected at the embedding medium interface and is imaged onto a camera (dotted red path).

may give rise to lobe amplitudes that are still relevant. To counteract these effects, we applied dedicated lobe deactivation by h_{zd} and developed low-aberration[25], live-cell 4Pi optics (Fig. 1b, Methods). These measures enabled volume scans of over 5-μm-thick mammalian cells without noticeable bleaching at an axial (z) resolution in the 100 nm range and axial lobes of only $\sim$15% of the main peak of the z-response, that is, measured on laterally (xy) integrated data (Fig. 2). The base acquisition time of 7–21 s μm^{-3} (depending on the brightness of the labelled structure) was short enough to capture the subtly moving cytoskeleton of a living cell (Fig. 2b,c, total acquisition times incl. drift correction overhead b: 115 min per 703 μm^3, c: 160 min per 400 μm^3). For highly mobile organelles, such as mitochondria (Fig. 2a), fixation of the sample by paraformaldehyde incubation (Supplementary Methods) offered a means to prevent motion blur. While this treatment irreversibly arrests the cell, its potential to introduce structural artifacts is very low with respect to staining/embedding protocols that involve membrane permeabilization.

To resolve smaller features, we implemented a second switching pattern for deactivation of RSFP around the focal centre: A hollow '3D-donut' h_{3d} (Fig. 3a), created by a single focused 4Pi beam pair[26] (Methods), allowed us to tune h_{ef} to a near-isotropic resolution below the diffraction limit (Fig. 3a). Calculations using a vectorial diffraction theory[27] predicted on-axis MTF values of over 40% of the MTF maximum within the MTF bandwidth up to a resolution of 30 nm. This feature keeps the signal well over the noise level in most applications and exemplifies the improvement brought about by higher order on-switching in comparison to modalities of second order such as those reported in isoSTED

microscopy[5] (Fig. 3b). Furthermore, the confinement of the fluorescent on-state, that is, of h_{ef}, to sub-diffraction 3D volumes means that fewer fluorophores are interrogated at any point in time. This reduction in number of interrogated molecules (that are inherently co-localized) greatly facilitates the quantitative assessment of the properties of the fluorescent labels as they vary in the sample. We found that in time-resolved recordings, the on/off separation contrast decayed over time, hinting to the contribution of multiple deactivation rates. Thus, we introduced a 'rate-gated' RESOLFT detection scheme that improved both the signal-to-noise ratio in the image and the resolution by discriminating individual signal components (Fig. 3c,d, Methods).

Following this approach, we recorded images of Lifeact-Dronpa-M159T-expressing cells and adjusted rate-gating and the RESOLFT pulse sequence for target resolutions of 30-50 nm; the parameters were established by a PSF simulation using measured rate kinetics. Optical xz-sections taken perpendicular to the run of solitary actin fibre bundles confirmed the effectiveness of rate-gating (Fig. 3c,d) and the overall shape of the effective PSF (Fig. 3e). Illumination with the z-donut-shaped (h_{zd}) focus for 1 ms at an average light power of 1.8 μW (488 nm, CW) was sufficient to virtually eliminate lobe background from the image (Fig. 3e, $+h_{zd}$), while the low gradients around the central zero of h_{zd} with respect to h_{3d} facilitated the mutual alignment of these patterns (Supplementary Fig. 1). Turning to the finer structured actin network inside the cell body, we measured apparent feature sizes well below 40 nm (Fig. 4a–c). At a relaxed target resolution of 50 nm and an acquisition time of 3.3 min μm^{-2}, we observed the time evolution of the actin scaffold at a vertical contact region of two neighbouring cells (Fig. 4e–g).

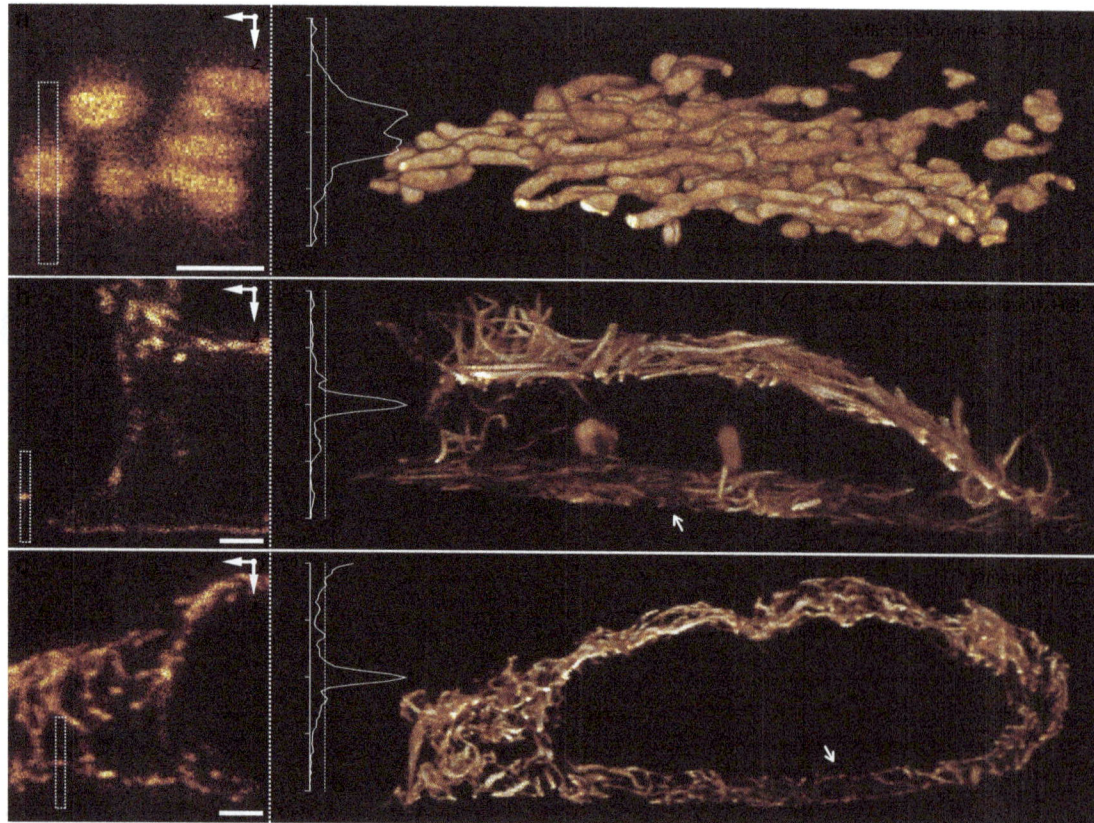

Figure 2 | 4Pi-RESOLFT imaging exhibits only minor axial lobes. 4Pi-RESOLFT raw data (left) and volume renderings (right) of Dronpa-M159T targeted to (**a**) the lumen of mitochondria, (**b**) actin microfilaments and (**c**) intermediate filaments of the cytoskeleton. The sample in **a** was subject to PFA fixation to freeze the motion of mitochondria; the filament networks in **b,c** were recorded from living cells, and exhibit regions of reduced density adjacent to the cover slip (arrows). Estimates of the z-response (insets), measured as box profiles over extended structures, exhibit only minor axial lobes in the 15 % range. Fast-to-slow order of scan axes, xzy. Pulse parameters, E_{ac}, E_{zd}, $E_{ro} = 1.6$ mW $\cdot$ 50 μs, 18 μW $\cdot$ 50 μs, 3.1 μW $\cdot$ 50 μs. Scale bars, 1μm.

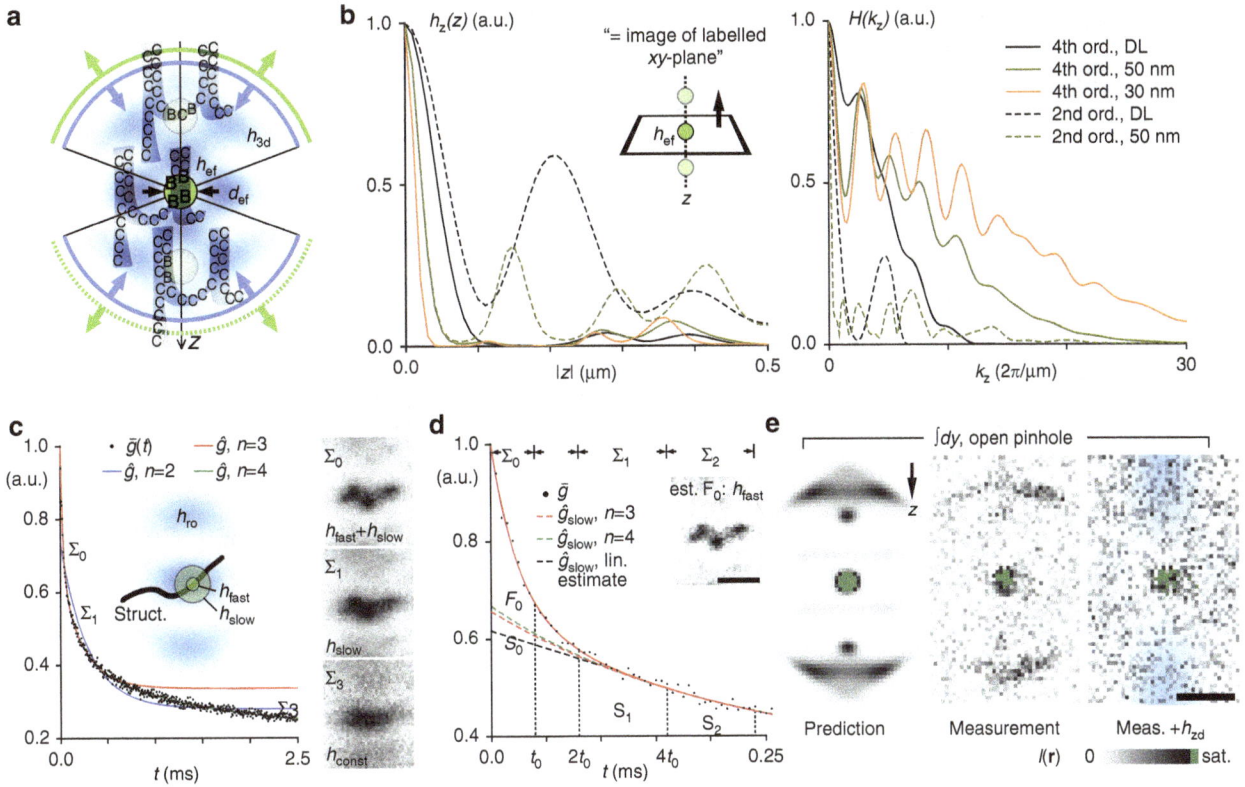

Figure 3 | 4Pi-RESOLFT image formation with <100 nm isotropic resolution. (**a**) A hollow switching pattern h_{3d} confines the central effective PSF to a spot with diameter d_{ef} by switching activated markers (B) back to their inactive state (C). Side-lobes due to inefficient switching at low off-centre h_{3d} amplitudes rise in relative strength as d_{ef} is reduced. μ, labelled structure. (**b**) Simulated z-response $h_z(z)$ (laterally integrated h_{ef}) and axial MTF profile $H(k_z)$ of the 4Pi-RESOLFT microscope (fourth on-order switching, solid lines) at different target resolutions d_{ef}. DL, diffraction limit. Graphs for an isoSTED microscope under similar conditions are included for reference (second on-order, dotted lines). (**c**) Normalized time-resolved, mean fluorescence signal $\bar{g}(t)$ collected from an xz-section through an actin fibre bundle (struct.) in a cell expressing Lifeact-Dronpa-M159T. Target resolution 50 nm, read-out pattern h_{ro} with a total power of $P_{ro} = 3.1\,\mu W$ incident on the sample. An n-component multi-exponential fit to the data corresponds to n apparent switching speeds $\hat{\lambda}_i$. A minimum of $n = 3$ is required to adequately represent the data from the beginning of the read-out pulse $t = 0$ up to 0.5 ms, $\hat{\lambda}_i = 40.5, 4.3, 0.0\,ms^{-1}$ (for up to 2.5 ms: $n = 4$, $\hat{\lambda}_i = 41, 5.9, 0.85, 0.0\,ms^{-1}$). Images $\Sigma_{0,1,3}$ integrated over time regimes that are dominated by fast (h_{fast}), slow (h_{slow}) and about constant PSF components (h_{const}) exhibit a declining resolution. (**d**) Rate-gated 4Pi-RESOLFT. Extrapolation of the initial contribution of h_{slow} ($= S_0$), based on integrated images Σ_1 ($\approx S_1$) and Σ_2 ($\approx S_2$), $t_0 = 40\,\mu s$, provides an estimate of the partial image generated by h_{fast} ($F_0 \approx \Sigma_0 - S_0$, inset), improving resolution and image fidelity over Σ_0. Details are provided in Methods. (**e**) Rate-gated xz-sections through actin fibres, recorded with open pinhole to boost out-of-focus signal. The measured (y-integrated) side-lobe structure closely resembles the numerical prediction and can be further suppressed (right) by an additional z-donut h_{zd} (overlay, $E_{zd} = 1.8\,\mu W \cdot 1.0\,ms$). Simulation parameters, numerical aperture 1.20, refractive index 1.362, pinhole diameter 0.5 airy units (**e**: open pinhole). Pulse parameters, E_{ac}, E_{3d}, $E_{ro} = 1.6\,mW \cdot 0.2\,ms$, $1.3\,\mu W \cdot 1.6\,ms$, $3.1\,\mu W \cdot 2.5\,ms$ (**e**: 0.5 ms). Scale bars, 250 nm.

Discussion

Using the current RSFP Dronpa-M159T, rate-gating allowed us to obtain images based on-switching speeds (switch-off half-time $T^{1/2} = 10$–$17\,\mu s$ at $11.5\,kW\,cm^{-2}$ illumination intensity, Fig. 3c, Supplementary Fig. 2, Supplementary Table 1) that were over an order of magnitude faster than the previously reported corresponding values for rapid switching RSFP ($T^{1/2}$, rsEGFP2: 250 μs (ref. 23), Dronpa-M159T: 230 μs (ref. 28)). Still the recording speed of our 4Pi-RESOLFT nanoscopy scheme can be made substantially faster by parallelization using a multi-spot scanning arrangement. In this case, the recording time of a certain sample area or volume would be cut down by the number of individual recording channels, that is, by the degree of parallelization.

In this study, we opted for cellular structures that are more demanding for 3D-superresolution imaging due to their high spatial density and wide axial extension. Under conditions exacerbated by the optical inhomogeneity of living cells, the signal from a (sub-diffractive) ensemble is easily buried in background (lobe) fluorescence beyond recovery. Nevertheless, owing to the consistently robust MTF of our 4Pi-RESOLFT

scheme (Fig. 3b), we obtained raw (Fig. 2, insets) and rate-gated image data (Fig. 4e–g) that were conclusive without the mathematical post-processing (that is, deconvolution) dedicated to lobe-removal that is usually applied in 4Pi-based methods. The actin network seen in the exemplary time-lapse recording (Fig. 4e–g) appeared particularly crowded and extended over 8 μm along the optic axis, which forced the light to pass through several micrometres of cellular material from all angles. Still, the rearrangement of the entwined actin fibres could be traced in great detail, which was possible because the obtained images were practically devoid of axial lobes.

Notably, our scheme of reducing the global refractive index (RI) variance (Fig. 1b) turned out to sufficiently mitigate sample-induced aberrations without adding the complexity associated with adaptive optical elements. The most prominent residual aberration effect was a position-dependent 4Pi phase offset that stemmed from the uneven thickness of the cell layer; it has been accounted for during our recordings by the automated correction mechanism that also counteracted thermal drift (Supplementary Methods).

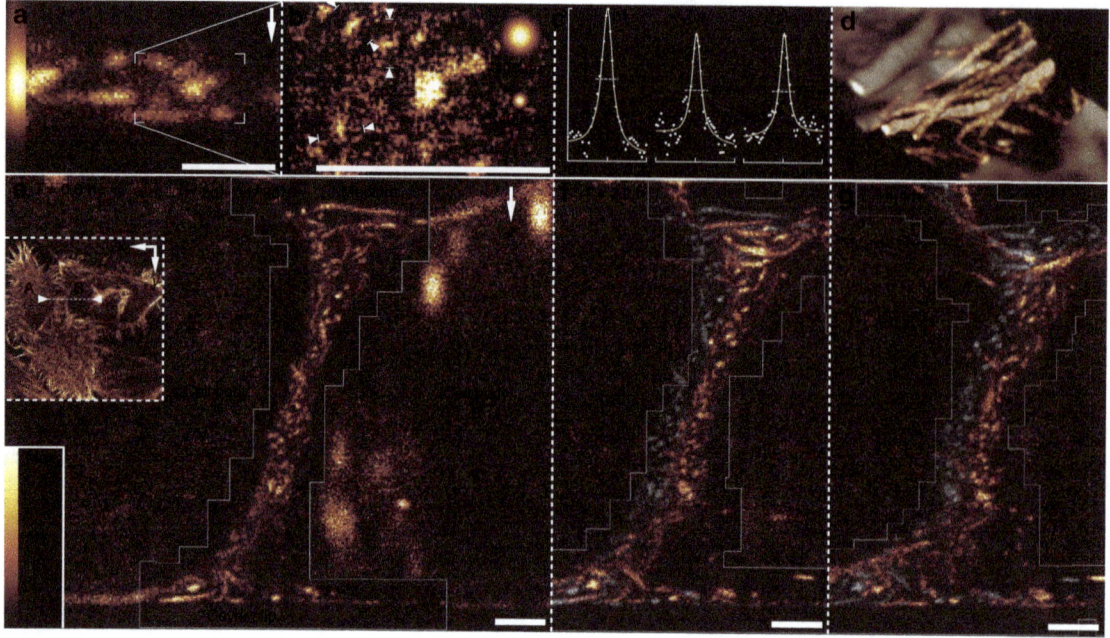

Figure 4 | 3D nanoscopy with strong optical sectioning. xz-sections of live HeLa cells expressing Lifeact-Dronpa-M159T. (**a**) Overview (optical xz-section) of actin fibre bundles at an axial base resolution in the 100 nm range. Left inset, confocal reference. (**b**) Addition of a 3D deactivation donut ($+h_{3d}$, $E_{3d} = 2.6\,\mu\text{W} \cdot 3.2\,\text{ms}$) to the RESOLFT pulse sequence reveals Dronpa patterns with apparent feature sizes well below 40 nm (inset, Gaussian reference spheres); (**c**) Lorentzian fits, plus a linear local background, to box profiles p1–3 over marked features in **b** along different directions. Numbers indicate full widths at half maximum (FWHM) over background. (**d**) Rendering of the volume surrounding **a**. (**e-g**) Time (T) evolution of an 8-μm-thick, densely labelled, vertical contact region between two adjacent cells (xz-section as marked in the xy-overview). Grayscale overlays of the preceding time step (**f,g**) aid in the tracking of individual features. A narrowed region of interest was generated online from initial overview scans ($-h_{3d}$, grey outline) at each time frame and imaged at 50 nm target resolution ($+h_{3d}$, grey outline, E_{3d}, $E_{zd} = 1.3\,\mu\text{W} \cdot 1.6\,\text{ms}$, $1.8\,\mu\text{W} \cdot 0.5\,\text{ms}$). Despite the challenging imaging conditions, stacked actin structures are unambiguously resolved across the full axial extent of the cell layer. xz-panels depict rate-gated 4Pi-RESOLFT raw data, solely subjected to noise reduction. Fluorescence intensities $I(r)$. Common pulse parameters (**b,e-g**), E_{ac}, $E_{ro} = 1.6\,\text{mW} \cdot 0.2\,\text{ms}$, $3.1\,\mu\text{W} \cdot 0.5\,\text{ms}$. Scale bars, 1 μm.

In conclusion, by realizing 4Pi-RESOLFT nanoscopy based on RSFPs, we have demonstrated exceptional optical sectioning in coordinate-targeted far-field fluorescence nanoscopy, which greatly facilitates nanometre scale 3D fluorescence imaging in living cells. Many accepted constraints to the sample can be lifted, which opens up an imaging regime that has so far been systematically avoided.

Methods

4Pi sample optics for live-cell imaging. In a 4Pi arrangement, the RI (n) difference between the material forming a living mammalian cell ($n \approx 1.35$–1.40) and the surrounding culture medium (typically $n \approx 1.33$) is a major source of optical aberrations. Aberrations generally reduce the attainable signal-to-noise ratio (S/N) by blurring the light intensity distributions in the focal region and by raising the intensity of the central minimum of the off-switching light patterns (Figs 1a and 3a). We therefore devised a mounting procedure that minimizes optical aberrations by raising the RI of a standard cell culture medium to $n = 1.362$ (Supplementary Fig. 3, Supplementary Methods) and by designing the optical setup accordingly (Fig. 1b, Supplementary Figs 4 and 5): The 4Pi foci are jointly created by two 1.20 NA water immersion objectives that are outfitted with individual tip/tilt-correction to prevent aberrations that arise from lens-coverslip misalignment[29]. The refractive indices of the embedding ($n = 1.362$) and immersion media ($n = 1.350$), the correction collar settings and the cover slip thickness were chosen to minimize spherical aberrations[25] over at least 10 μm of sample depth. The changes in the optical path lengths of the two 4Pi-interferometer arms due to z-scanning of the sample were compensated[30] by synchronous position adjustment of the main beam splitter cube (Supplementary Fig. 4).

A single-focus 3D light pattern for deactivation. To resolve features smaller than 100 nm, we added a RSFP deactivation beam to the microscope. It was imprinted with a circular phase ramp that was subsequently imaged into the back pupil planes of both objective lenses. In contrast to the configuration of a single-lens RESOLFT setup, the direction of rotation of the phase ramp was oriented in countersense with respect to the circular beam polarization at the back pupil planes, which produced a 4Pi off-switching pattern h_{3d} that completely surrounded a central zero[26] (Fig. 3a). Applying h_{3d} during the deactivation phase of

the RESOLFT switching cycle squeezed the central full width at half maximum (FWHM) d_{ef} of h_{ef} and allowed us to tune h_{ef} to a resolution below the diffraction limit.

RESOLFT imaging with rate-gated detection. In the present case of RESOLFT imaging of negative-switching RSFP, deactivation of fluorophores during read-out gives rise to a time-dependent signal and hence a time-dependent effective PSF $h_{ef}(\mathbf{r},t)$. The time-resolved image $g(\mathbf{r},t)$ obtained by imaging a structure $s(\mathbf{r})$ is thus given by

$$g(\mathbf{r}, t) = s(\mathbf{r}) \otimes h_{ef}(\mathbf{r}, t). \tag{3}$$

We simplistically assume a deactivation rate $\lambda(\mathbf{r})$ that only depends on the read-out intensity $I_{ro}(\mathbf{r})$ and therefore obtain

$$h_{ef}(\mathbf{r}, t) = h_{ef}(\mathbf{r}) \cdot e^{-\lambda(\mathbf{r}) \cdot t} = h_{ef}(\mathbf{r}) \cdot e^{-\lambda \cdot I_{ro}(\mathbf{r}) \cdot t} \tag{4}$$

with t denoting time relative to the start of the read-out pulse. The deactivation pattern $h_{3d}(\mathbf{r})$ typically confines the effective volume from which fluorescence is collected to a region of FWHM d_{ef} around the primary zero of the deactivation pattern. This region is much narrower than the FWHM d_{ro} of the diffraction pattern used for read-out:

$$d_{ef} \ll d_{ro} \tag{5}$$

The read-out intensity can then be considered constant, and $h_{ef}(\mathbf{r},t)$ follows a mono-exponential decay at a rate that only depends on the peak read-out intensity $I_{ro}^0 := I_{ro}(0)$. Thus, $g(\mathbf{r},t)$ becomes separable and transforms into:

$$g(\mathbf{r}, t) = g(\mathbf{r}) \cdot g(t) := (s(\mathbf{r}) \otimes h_{ef}(\mathbf{r})) \cdot e^{-\lambda \cdot I_{ro}^0 \cdot t} \tag{6}$$

Data from test structures, however, exhibit a distinct multi-exponential behaviour that requires additional components for a proper fit (Fig. 3c):

$$\widehat{g}(\mathbf{r}, t) = \sum_{i=0}^{n-1} \widehat{c}_i(\mathbf{r}) \cdot e^{-\widehat{\lambda}_i(\mathbf{r}) \cdot t} \tag{7}$$

where n components with ordered switching rates $\widehat{\lambda}_i$, $\widehat{\lambda}_i > \widehat{\lambda}_{i+1}$ and coefficients $\widehat{c}_i$ are fitted to the data ('$\wedge$' marks fit results). While we attribute the fastest rate $\widehat{\lambda}_0$ to signal from unimpaired RSFP at the focal centre, the presence of additional rates suggests the co-existence of RSFP species that exhibit significantly slower switching kinetics. A slowed-down switching observed after fixation supports this notion.

Unintended processes during image recording also potentially contribute to the observed signal behaviour, for example, the re-activation of RSFP by the read-out light which generates a constant background.

Without loss of generality, we assume a position independent mixture of n species with discrete switching rates λ_j. Since the deactivation pattern h_{3d} shrinks the effective PSF by a λ-dependent factor, $h_{ef}(\mathbf{r},t)$ has to be generalized to a superposition of n individual h_{ef}^j that each correspond to a λ_j. Furthermore, we assume our experimental parameters are chosen such that the in-focus part h_{fast} of h_{ef}^0 obeys the analogue to equation (5), and pool the remaining contributions in h_{slow}:

$$h_{ef}(\mathbf{r},t) = h_{fast}(\mathbf{r},t) + h_{slow}(\mathbf{r},t) := h_{fast}(\mathbf{r}) \cdot e^{-\lambda_0 \cdot I_{ro}^0 \cdot t} + \sum_{j=1}^{n-1} h_{slow}^j(\mathbf{r}) \cdot e^{-\lambda_j \cdot I_{ro}(\mathbf{r}) \cdot t} \quad (8)$$

Consequently, the apparent resolution of the acquired image is less than the potential resolution provided by h_{fast} and declines over time, as faster components vanish first (Fig. 3c).

To access the full image information that is mediated by h_{fast}, we implemented an unmixing scheme that isolates the fastest switching signal component (rate λ_0), and that we hence termed 'rate-gating': according to equation (3), the image generated by a PSF equation (8) takes on the form

$$g(\mathbf{r},t) = g_{fast}(\mathbf{r},t) + g_{slow}(\mathbf{r},t) := g_0(\mathbf{r}) \cdot e^{-\lambda_0 \cdot I_{ro}^0 \cdot t} + \int_0^{\lambda_1} c(\mathbf{r},\Lambda) \cdot e^{-\Lambda t} d\Lambda \quad (9)$$

whereby g_{slow} is represented by a continuum of exponentials with switching speeds $\Lambda \in [0,\lambda_1]$ and coefficients $c(\mathbf{r},\Lambda)$ as the result of $s \otimes h_{slow}$.

Hence, a fit $\widehat{g}(\mathbf{r},t)$ to an imaged structure according to equation (7), in principle, provides a position invariant estimate for $\lambda_0 = \widehat{\lambda}_0/I_{ro}^0$ by $\widehat{g}_{fast}$, on top of a local approximation of $g_{slow}(\mathbf{r},t)$. In practice however, an insufficient photon count often prohibits local fitting of equation (9). We therefore implemented a robust approximation scheme for $g_{fast}(\mathbf{r})$ that only relies on parameters that can be extracted from a fit $\widehat{g}(t)$ to the global (that is, from a region much larger than the corresponding diffraction limit) spatial average $\overline{g}(t)$ of the measured data:

First, we estimated the time t_0 at which the integrated signal exhibits the maximum S/N with respect to g_{fast}:

$$t_0 := \operatorname{argmax}\left(\frac{\int_0^T \overline{g}_{fast}(t) dt}{\sqrt{\int_0^T \overline{g}_{fast}(t) + \overline{g}_{slow}(t) dt}} \right) \quad (10)$$

Locally calculated values for t_0 would slightly differ, but as this only affects the statistical error of the result, equation (10) is usually sufficiently precise. Second, we determined a cut-off time t_1 such that

$$g_{fast}(t) \ll g_{slow}(t) \quad (t > t_1) \quad (11)$$

which is usually the case for $t_1 = 2t_0$. By further choosing t_2 and t_3 such that $t_i > t_{i+1}(i = 0..3)$, we partition the measured signal into time bins $\sum_{0,1,2}$ (Fig. 3d),

$$\sum_{0,1,2}(\mathbf{r}) := \int_{0,t_1,t_2}^{t_0,t_2,t_3} g(\mathbf{r},t) dt \approx F(\mathbf{r}) + S_0(\mathbf{r}), S_1(\mathbf{r}), S_2(\mathbf{r}) \quad (12)$$

with $F(\mathbf{r})$ and $S(\mathbf{r})$ denoting the time integrals over the fast and slow components:

$$F(\mathbf{r}) := \int_0^{t_0} g_{fast}(\mathbf{r},t) dt, \quad S_{0,1,2} := \int_{0,t_1,t_2}^{t_0,t_2,t_3} g_{slow}(\mathbf{r},t) dt \quad (13)$$

Finally, we estimated $F(\mathbf{r})$, and thereby $g_0(\mathbf{r})$, by linear extrapolation in either zeroth or first order:

$$F(\mathbf{r})^{0th} := \sum_0(\mathbf{r}) - u \cdot \sum_1(\mathbf{r}) \quad (14)$$

$$F(\mathbf{r})^{1st} := F(\mathbf{r})^{0th} - v \cdot \left(\sum_1(\mathbf{r}) - \sum_2(\mathbf{r}) \right) \quad (15)$$

with $u,v = u,v(t_{0..3})$ denoting geometrical factors that account for the particular choice of the $t_{0..3}$. Narrowing the integration intervals defined by $t_{0..3}$ and moving them closer to $t=0$, just as the inclusion of the first extrapolation order, reduces the systematic error, but also raises the statistic error due to a reduced photon count. To mitigate this effect, and owing to equation (11), we substituted $\sum_{1,2}$ with their respective resolution-neutral local averages, for example, by applying a Gaussian filter with a FWHM sufficiently far below the FWHM of $h_{slow}^{1,2}$.

References

1. Hell, S. W. Double-confocal scanning microscope. European Patent 0491289 (1992).
2. Hell, S. W. & Stelzer, E. H. K. Properties of a 4pi confocal fluorescence microscope. J. Opt. Soc. Am. A Opt. Image Sci. Vis. 9, 2159–2166 (1992).
3. Gustafsson, M. G. L., Agard, D. A. & Sedat, J. W. Sevenfold improvement of axial resolution in 3D widefield microscopy using two objective lenses. Proc. Soc. Photo-Opt. Instrum. Eng. 2412, 147–156 (1995).
4. Eggeling, C., Willig, K. I., Sahl, S. J. & Hell, S. W. Lens-based fluorescence nanoscopy. Q. Rev. Biophys. 48, 178–243 (2015).
5. Schmidt, R. et al. Spherical nanosized focal spot unravels the interior of cells. Nat. Methods 5, 539–544 (2008).
6. Shtengel, G. et al. Interferometric fluorescent super-resolution microscopy resolves 3D cellular ultrastructure. Proc. Natl Acad. Sci. USA 106, 3125–3130 (2009).
7. Hell, S. W., Schmidt, R. & Egner, A. Diffraction-unlimited three-dimensional optical nanoscopy with opposing lenses. Nat. Photon. 3, 381–387 (2009).
8. Grotjohann, T. et al. Diffraction-unlimited all-optical imaging and writing with a photochromic GFP. Nature 478, 204–208 (2011).
9. Lang, M. C., Engelhardt, J. & Hell, S. W. 4Pi microscopy with linear fluorescence excitation. Opt. Lett. 32, 259–261 (2007).
10. Schmidt, R. et al. Mitochondrial cristae revealed with focused light. Nano Lett. 9, 2508–2510 (2009).
11. Ullal, C. K., Schmidt, R., Hell, S. W. & Egner, A. Block copolymer nanostructures mapped by far-field optics. Nano Lett. 9, 2497–2500 (2009).
12. Kanchanawong, P. et al. Nanoscale architecture of integrin-based cell adhesions. Nature 468, 580–584 (2010).
13. Denk, W., Strickler, J. H. & Webb, W. W. 2-photon laser scanning fluorescence microscopy. Science 248, 73–76 (1990).
14. Schneider, M., Barozzi, S., Testa, I., Faretta, M. & Diaspro, A. Two-photon activation and excitation properties of PA-GFP in the 720-920-nm region. Biophys. J. 89, 1346–1352 (2005).
15. Glaschick, S. et al. Axial resolution enhancement by 4Pi confocal fluorescence microscopy with two-photon excitation. J. Biol. Phys. 33, 433–443 (2007).
16. Moneron, G. & Hell, S. W. Two-photon excitation STED microscopy. Opt. Express 17, 14567–14573 (2009).
17. York, A. G., Ghitani, A., Vaziri, A., Davidson, M. W. & Shroff, H. Confined activation and subdiffractive localization enables whole-cell PALM with genetically expressed probes. Nat. Methods 8, 327–333 (2011).
18. Hell, S. W. Improvement of lateral resolution in far-field light microscopy using two-photon excitation with offset beams. Opt. Commun. 106, 19–24 (1994).
19. Hell, S. W. Toward fluorescence nanoscopy. Nat. Biotechnol. 21, 1347–1355 (2003).
20. Stiel, A. C. et al. 1.8 angstrom bright-state structure of the reversibly switchable fluorescent protein dronpa guides the generation of fast switching variants. Biochem. J. 402, 35–42 (2007).
21. Willig, K. I., Stiel, A. C., Brakemann, T., Jakobs, S. & Hell, S. W. Dual-label STED nanoscopy of living cells using photochromism. Nano Lett. 11, 3970–3973 (2011).
22. Testa, I. et al. Nanoscopy of living brain slices with low light levels. Neuron 75, 992–1000 (2012).
23. Grotjohann, T. et al. rsEGFP2 enables fast RESOLFT nanoscopy of living cells. eLife 1, e00248 00241-00214 (2012).
24. Gugel, H. et al. Cooperative 4pi excitation and detection yields sevenfold sharper optical sections in live-cell microscopy. Biophys. J. 87, 4146–4152 (2004).
25. Wan, D. S., Rajadhyaksha, M. & Webb, R. H. Analysis of spherical aberration of a water immersion objective: application to specimens with refractive indices 1.33-1.40. J. Microsc. 197, 274–284 (2000).
26. Schmidt, R. 3D Fluorescence Microscopy with Isotropic Resolution on the Nanoscale (PhD thesis, Univ. Heidelberg, 2008).
27. Richards, B. & Wolf, E. Electromagnetic diffraction in optical systems. II. structure of the image field in an aplanatic system. Proc. R. Soc. Lond. A 253, 358–379 (1959).
28. Testa, I., D'Este, E., Urban, N. T., Balzarotti, F. & Hell, S. W. Dual channel RESOLFT nanoscopy by using fluorescent state kinetics. Nano Lett. 15, 103–106 (2015).
29. Arimoto, R. & Murray, J. M. A common aberration with water-immersion objective lenses. J. Microsc. 216, 49–51 (2004).
30. Hell, S. W., Schrader, M. & VanderVoort, H. T. M. Far-field fluorescence microscopy with three-dimensional resolution in the 100-nm range. J. Microsc. 187, 1–7 (1997).

Acknowledgements

We thank C. Gregor for providing the plasmid Lifeact-Dronpa-M159T, T. Gilat and E. Rothermel for cloning support and assistance with cell culture preparation, B. Thiel for Inspector software support and A. Pucher-Diehl for precision mechanic support for custom microscope parts. S.W.H. and R.S. acknowledge a grant by the Deutsche Forschungsgemeinschaft within SFB 775.

Author contributions

U.B. prepared the samples and carried out the measurements. U.B. and R.S. designed and built the RESOLFT microscope and evaluated the data. R.S. designed and implemented the 4Pi unit and the data acquisition and rate-gating algorithms. The project was defined and supervised by R.S. and S.W.H. All authors discussed the project as it evolved and edited the final version of the paper. The paper was written by R.S. and S.W.H. with contributions by U.B.

Additional information

Competing financial interests: The authors declare no competing financial interests.

7

Structural semiconductor-to-semimetal phase transition in two-dimensional materials induced by electrostatic gating

Yao Li[1], Karel-Alexander N. Duerloo[2], Kerry Wauson[3] & Evan J. Reed[2]

Dynamic control of conductivity and optical properties via atomic structure changes is of technological importance in information storage. Energy consumption considerations provide a driving force towards employing thin materials in devices. Monolayer transition metal dichalcogenides are nearly atomically thin materials that can exist in multiple crystal structures, each with distinct electrical properties. By developing new density functional-based methods, we discover that electrostatic gating device configurations have the potential to drive structural semiconductor-to-semimetal phase transitions in some monolayer transition metal dichalcogenides. Here we show that the semiconductor-to-semimetal phase transition in monolayer $MoTe_2$ can be driven by a gate voltage of several volts with appropriate choice of dielectric. We find that the transition gate voltage can be reduced arbitrarily by alloying, for example, for $Mo_xW_{1-x}Te_2$ monolayers. Our findings identify a new physical mechanism, not existing in bulk materials, to dynamically control structural phase transitions in two-dimensional materials, enabling potential applications in phase-change electronic devices.

[1] Department of Applied Physics, Stanford University, Stanford, California 94305, USA. [2] Department of Material Science and Engineering, Stanford University, Stanford, California 94305, USA. [3] Klipsch School of Electrical and Computer Engineering, New Mexico State University, Las Cruces, New Mexico 88003, USA. Correspondence and requests for materials should be addressed to E.J.R. (email: evanreed@stanford.edu).

Structural phase transitions yielding a change of electrical conductivity are a topic of long-standing interest and importance[1,2]. Two of the most studied phase-change material classes for electronic and optical applications are metal oxide materials[3,4] and GeSbTe alloys[5], both having a large electrical contrast. For example, the metal oxide material vanadium dioxide (VO_2) is reported to exhibit a structural metal–insulator transition near room temperature at ultrafast timescales, which can be triggered by various stimuli including heating[6], optical[7] excitations and strain[8]. GeSbTe alloys can undergo reversible switching between amorphous and crystalline states with different electrical resistivity and optical properties. This is usually achieved by Joule heating employed in phase-change memory applications[9,10]. These materials are distinguished from the myriad materials that exhibit atomic structural changes by the proximity of a phase boundary to ambient conditions.

Another group of materials that can undergo phase transitions are layered transition metal dichalcogenides (TMDs), which have received recent attention as single- and few-layer materials, although research on bulk TMDs dates back decades[11,12]. Early attention has been focused primarily on electronic transitions between incommensurate and commensurate charge density wave[13,14] phases and superconducting phases[15]. Some TMDs have been found to exist in multiple crystal structures[16], and transitions between them have been demonstrated in group V TMDs ($TaSe_2$ and TaS_2) utilizing an scanning tunnelling microscope (STM) tip[17,18]. These reported transitions in $TaSe_2$ and TaS_2 are between two metallic phases. Recently, group VI TMDs have attracted increasing attention because they can exist in a semiconducting phase[19]. Recent computational work indicates that structural transitions between phases of large electrical contrast in some exfoliated two-dimensional (2D) group VI TMDs can be driven by mechanical strain[20]. Excess charges transferred from chemical surroundings are also reported to induce structural phase transitions in 2D group VI TMDs[21-24]. One would like to know the threshold charge density required to induce these transitions and whether these transitions could be dynamically controlled by electrostatic gating, utilizing standard electronic devices.

Here we show the potential of phase control in some monolayer TMDs using electrostatic gating device configurations. In this work, we use density functional theory (DFT) to determine the phase boundaries of single-layer MoS_2, $MoTe_2$, $TaSe_2$ and the alloy $Mo_xW_{1-x}Te_2$. We consider MoS_2 because it has received considerable attention as an exceptionally stable semiconductor, and $MoTe_2$ because DFT calculations indicate that its energy difference between semiconducting and semimetallic phases is exceptionally small among Mo- and W-TMDs[20]. We calculate the phase boundaries at conditions of constant charge and constant voltage, the electrical analogues to mechanical conditions of constant volume and constant pressure, respectively. We find that a surface charge density of less than $-0.04e$ or greater than $0.09e$ per formula unit is required to observe the semiconductor-to-semimetal phase transition in undoped monolayer $MoTe_2$ under constant-stress conditions (e is the elementary electric charge) and a much larger value of approximately $-0.29e$ or $0.35e$ per formula unit is required in the undoped monolayer MoS_2 case. The charge densities discussed in this work refer to excess charge density and should not be misinterpreted as the electron or hole density in a charge-neutral material that one might obtain from chemical doping. We also study the potential of phase control in monolayer $MoTe_2$ and $TaSe_2$ through electrostatic gating using a capacitor structure. We discover that a gate voltage as small as a few volts for some choices of gate dielectric can be applied to drive the phase transition in

monolayer $MoTe_2$ using a capacitor structure. While the required field magnitudes are large and may be challenging to achieve, we find that the transition gate voltage may be reduced to 0.3–1 V and potentially lower by substituting a specific fraction of W atoms within $MoTe_2$ monolayers to yield the alloy $Mo_xW_{1-x}Te_2$. To accomplish these calculations, we have developed a DFT-based model of the electrostatically gated structure (Supplementary Figs 1 and 2; Supplementary Note 1). This approach is validated by comparing to direct DFT simulations in Supplementary Fig. 3 and Supplementary Note 2.

Results

Crystal structures. TMDs are a class of layered materials with the formula MX_2, where M is a transition metal atom and X is a chalcogen atom. Each monolayer is composed of a metal layer sandwiched between two chalcogenide layers, forming a X–M–X structure[16] that is three atoms thick. The weak interlayer attraction of TMDs allows exfoliation of these stable three-atom-thick layers. Given the crystal structures reported in the bulk, we expect that exfoliated monolayer TMDs have the potential to exist in the crystal structures shown in Fig. 1. Figure 1 shows the X atoms with trigonal prismatic coordination, octahedral coordination or a distorted octahedral coordination around the M atoms[16,20,25,26]. We will refer to these three structures of the monolayer as the 2H phase, 1T phase and 1T' phase, respectively. Symmetry breaking in the 1T' leads to a rectangular primitive unit cell.

Among these 2D TMDs, the Mo- and W-based materials have attracted the most attention because their 2H crystal structures are semiconductors with photon absorption gaps in the 1–2 eV (ref. 27) range, showing potential for applications in ultrathin flexible and nearly transparent 2D electronics. Radisavljevic et al.[28] fabricated single-layer MoS_2 transistors of high mobility, large current on/off ratios and low standby power dissipation. Unlike group IV and group V TMDs (for example, $TaSe_2$ and TaS_2), which have been observed in the metallic 1T crystal structure[16], DFT calculations on the group VI TMDs (Mo and W based) freestanding monolayers indicate that the 1T structure is unstable in the absence of external stabilizing influences[20]. However, group VI TMDs do have a stable octahedrally coordinated structure of large electrical conductivity, which is a distorted version of the 1T phase and referred to as 1T' structure (Fig. 1). On the basis of DFT calculation results, Kohn–Sham

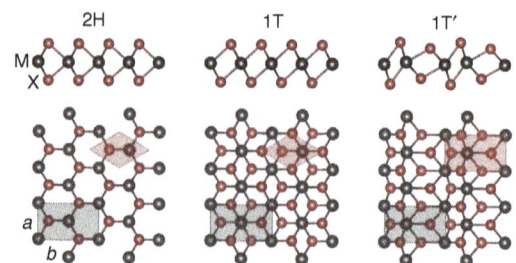

Figure 1 | Three crystal structures of monolayer TMDs. The top schematics show cross-sectional views and the bottom schematics show basal plane views. The grey atoms are transition metal atoms and the red atoms are chalcogen atoms; in all three phases, a layer of transition metal atoms (M) is sandwiched between two chalcogenide layers (X). The semiconducting 2H phase has trigonal prismatic structure, and the metallic 1T and semimetallic 1T' phases have octahedral and distorted octahedral structures, respectively. The grey shadow represents a rectangular computational cell with dimensions $a \times b$, and the red shadow represents the primitive cell.

states of this 1T′ crystal structure have metallic or semimetallic characteristics, consistent with previous experiments[16]. This octahedral-like 1T′ crystal structure has been observed in WTe$_2$ under ambient conditions[16,29], in MoTe$_2$ at high temperature[29] and in lithium-intercalated MoS$_2$ (ref. 25). There is recent experimental evidence that few layer films of the T′ phase of MoTe$_2$ exhibit a bandgap that varies from 60 meV to zero with variations in number of layers[30].

The relative energies of Mo- and W-based TMDs monolayer crystals shown in Fig. 1 have been calculated using semilocal DFT with spin–orbit coupling, shown in Supplementary Fig. 4. These results are consistent with experimental evidence that the bulk form of WTe$_2$ is stable in the metallic 1T′ phase, while other Mo- and W-dichalcogenides are stable in the semiconducting 2H phase[16]. These calculations indicate that the switch from semiconducting 2H phase to semimetallic 1T′ phase in monolayer MoTe$_2$ requires the least energy (31 meV per formula unit), suggesting the potential for a transition that is exceptionally close to ambient conditions. Therefore, we choose to focus on determining the phase boundary of monolayer MoTe$_2$. While the computed energy difference between 2H and 1T′ is considerably larger for MoS$_2$ (548 meV per formula unit), we also compute phase boundaries for this monolayer at constant charge because it has received more attention in the laboratory to date. Among 2D group VI TMDs, monolayer MoS$_2$ has attracted the most experimental attention for its stability and relative ease of exfoliation and synthesis. Monolayer MoTe$_2$ has also been exfoliated[31,32] and its synthesis is a fast-developing field.

Energy calculations for charged monolayers. We examine two distinct thermodynamic constraints for a system containing a charged monolayer. In one scenario, the monolayer is constrained to be at constant excess charge, as shown in Fig. 2a; in the other, the monolayer is constrained to be at constant voltage, as shown in Fig. 2b. These are the electrical analogues to mechanical conditions of constant volume and constant pressure, respectively. The electrical contact depictions in Fig. 2b and subsequent figures are schematic and could be accomplished in other manners, for

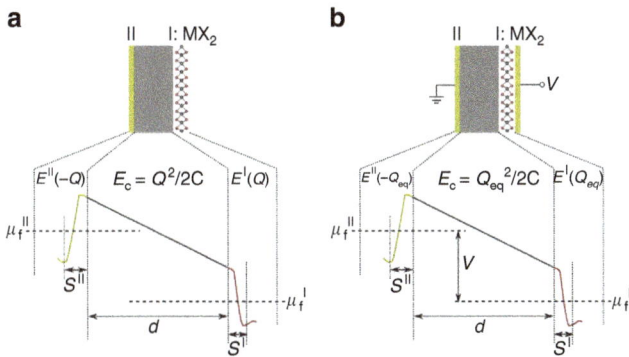

Figure 2 | Energy calculations for systems containing a charged monolayer. Layer I is a monolayer TMDs with a Fermi level μ_f^I, and plate II has a Fermi level of μ_f^{II}. Charge Q on the monolayer TMDs is fixed in **a**, whereas the voltage V is fixed in **b** giving rise to an equilibrium charge Q_{eq}. A dielectric medium of thickness d and capacitance C, which can be vacuum, is sandwiched between the monolayer and the plate. Distance s^I is the separation between the centre of the monolayer and the right surface of the dielectric medium, and s^{II} is the separation between the surface atoms of plate II and left surface of the dielectric medium. The total energy in the fixed charge case is the sum of three parts: energy stored in the dielectric medium E_c, energy of the plate II E^{II}, and energy of the charged monolayer TMDs E^I.

example, side contacts. In Fig. 2b, the charge is assumed to be stored in the monolayer rather than the metal contact. Layer I is a monolayer TMDs with a Fermi level μ_f^I, and plate II has a Fermi level of μ_f^{II}. A dielectric medium of thickness d and capacitance C is sandwiched between monolayer TMDs and plate II. This dielectric medium can be vacuum. Distance s^I is the separation between the centre of monolayer TMDs I and the right surface of the dielectric medium, while s^{II} is the separation between the surface atoms of plate II and the left surface of the dielectric medium. (See Supplementary Figs 5–7, Supplementary Table 1 and Supplementary Note 3 for more details about distance parameters.)

When the charge Q on the monolayer is fixed, the total energy of the system $E(Q)$ is the sum of three parts: energy stored in the dielectric medium (E_c), energy of the plate II (E^{II}) and energy of the charged monolayer TMDs (E^I), as shown in Fig. 2.

$$E(Q) = E^I(Q, s^I) + E^{II}(-Q, s^{II}) + E_c$$
$$= E^I(Q, s^I) + E^{II}(-Q, s^{II}) + \frac{Q^2}{2C}, \quad (1)$$

where C is the capacitance of the dielectric medium. $E^I(Q = 0, s^I)$ is the ground-state energy of the electrically neutral monolayer TMDs and $E^I(Q, s^I) - E^I(Q = 0, s^I)$ is the energy required to move electrons Q from the Fermi level of the monolayer TMDs to the dielectric surface. $E^{II}(-Q, s^{II})$ is defined analogously. We take the monolayers to be undoped in this work.

The first term in equation (1), $E^I(Q, s^I)$, is calculated using DFT for each phase of the monolayer TMDs to yield a $E(Q)$ for each monolayer phase (see Supplementary Figs 1 and 2, and Supplementary Note 1 for calculation details). The phase change does not enter into the third term in equation (1) or change the capacitance of the dielectric medium C.

We take plate II to be a bulk metal with a work function W so that the second term in equation (1) can be approximately written as:

$$E^{II}(-Q, s^{II}) = -QW. \quad (2)$$

When the voltage is fixed rather than the charge, the grand potential $\Phi_G(Q, V)$ becomes the relevant thermodynamic energy defined as:

$$\Phi_G(Q, V) = E(Q) - QV, \quad (3)$$

where $E(Q)$ is computed using equation (1). The QV term in this expression represents external energy supplied to the system when the charge Q flows through an externally applied voltage V. The equilibrium charge Q_{eq} can be calculated through minimization of the grand potential at a given gate voltage V.

$$\left.\frac{\partial \Phi_G(Q, V)}{\partial Q}\right|_{Q = Q_{eq}} = 0 \quad (4)$$

Applying the computed $Q_{eq}(V)$ to equation (3), we can obtain the equilibrium grand potential as a function of gate voltage $\Phi_G^{eq}(V)$.

$$\Phi_G^{eq}(V) = \Phi_G(Q_{eq}(V), V) = E(Q_{eq}(V)) - Q_{eq}(V)V \quad (5)$$

Hereafter, we omit the superscript 'eq' for the equilibrium grand potential $\Phi_G^{eq}(V)$.

In addition to the electrical constraint, the nature of the mechanical constraint on the monolayer is also expected to play a role in the phase boundary, discussed in Supplementary Note 4.

Phase boundary at constant charge. The distinction between the constant charge and voltage cases is most important when a phase transformation occurs. We discover that the transition between semiconducting 2H-TMDs and semimetallic 1T′-TMDs can be driven by excess electric charge (positive or negative) in the monolayer. A constant charge condition exists when the charge on the monolayer remains constant during the phase transition as if it is electrically isolated. An approximate condition of constant charge could exist when adsorbed atoms or molecules donate charge to the monolayer.

Figure 3 presents the energy difference between the 2H and 1T′ phases as a function of the charge density in the monolayer. Figure 3a is a schematic of the system. The monolayer TMD is a distance d away from the electron reservoir (metal electrode). Because the dielectric medium is vacuum in this schematic, E^I and E_c in equation (1) can be combined, which can be understood from Fig. 2a, and equation (1) can be rewritten as:

$$E(Q) = E^I(Q, s^I) + E^{II}(-Q, s^{II}) + E_c$$
$$= E^I(Q, s^I + d) + E^{II}(-Q, s^{II}) \qquad (6)$$

When computing the energy difference between a system where the monolayer is in the 1T′ phase and another system where the monolayer is in the 2H phase $E_{T'}(Q) - E_H(Q)$, the terms $E^{II}(-Q, s^{II})$ in equation (6) cancel, leading to,

$$E_{T'}(Q) - E_H(Q) = \left[E^I(Q, s^I + d) + E^{II}(-Q, s^{II}) \right]_{T'}$$
$$- \left[E^I(Q, s^I + d) + E^{II}(-Q, s^{II}) \right]_H \qquad (7)$$
$$= E_{T'}^I(Q, s^I + d) - E_H^I(Q, s^I + d)$$

Equation (7) shows that the energy difference depends on $s^I + d$ rather than s^I and d independently. Variation of the results of Fig. 3b,c with the separation $s^I + d$ (chosen to be 15 Å in Fig. 3) is weak or none as shown in Supplementary Fig. 7.

The blue lines are constant-stress (stress-free) cases, in which both phases exhibit minimum energy lattice constants and atomic positions. This condition is expected to hold when the monolayer is freely suspended or is not constrained by friction on a substrate. The red lines represent constant-area cases, where the monolayer is clamped to its 2H lattice constants. This condition

might be expected to hold when there is a strong frictional interaction between the monolayer and substrate preventing the monolayer from relaxing freely.

Figure 3b shows that semiconducting 2H-MoTe$_2$ has lower free energy and is the equilibrium state when the monolayer is electrically neutral or minimally charged. For the stress-free case (blue line), when the charge density is between $-0.04\,e$ and $0.09\,e$ per formula unit, 2H-MoTe$_2$ is the thermodynamically stable phase. These charge densities correspond to -3.7×10^{13} and $8.2 \times 10^{13}\,e\,cm^{-2}$, respectively. Outside this range, semimetallic 1T′-MoTe$_2$ will become the equilibrium phase and a transition from the semiconducting 2H phase to the semimetallic 1T′ phase will occur.

In the constant-area case (red line) in Fig. 3b, a considerably larger charge density is required to drive the phase transition. This suggests that the precise transition point may be sensitive to the presence of a substrate and that the detailed nature of the mechanical constraint of the monolayer may play a substantive role in the magnitude of the phase boundaries. The higher transition charge in this case can be understood by considering that the energy of the strained T′ phase is higher than that of the zero stress T′ phase, pushing the phase boundary to larger charge states.

Figure 3c shows that the transition in monolayer MoS$_2$ requires much larger charge density than the MoTe$_2$ case. If the negative charge density is $>0.29\,e$ per MoS$_2$ formula unit, semimetallic 1T′-MoS$_2$ will have lower free energy and be more stable. For negative charge densities $<0.29\,e$ per formula unit, semiconducting 2H-MoS$_2$ will be energetically favourable. This is consistent with previous experimental reports that adsorbed species donating negative charge to monolayer MoS$_2$ can trigger a trigonal prismatic to octahedral structure transformation[24,33]. MoS$_2$ single layer is reported to adopt a distorted octahedral structure when bulk MoS$_2$ is first intercalated with lithium to form Li$_x$MoS$_2$ with $x \approx 1.0$ and then exfoliated by immersion in distilled water[25]. This is qualitatively consistent with our

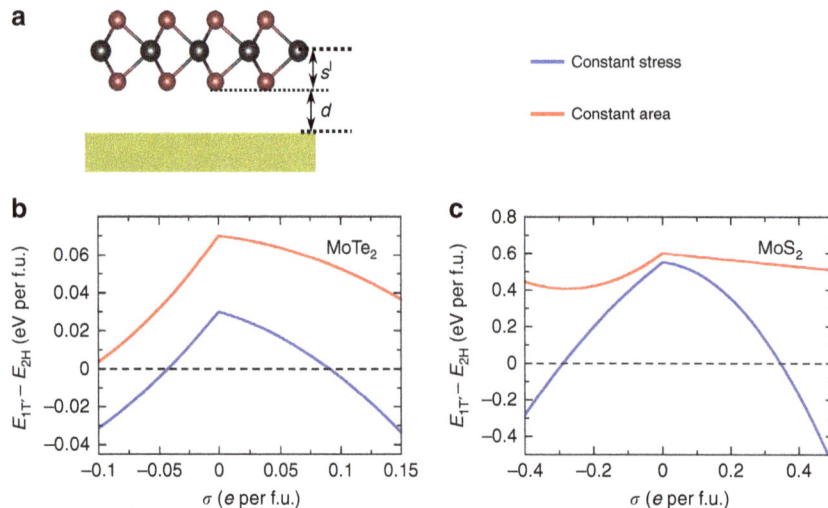

Figure 3 | Phase boundary at constant charge in monolayer MoTe$_2$ and MoS$_2$. (**a**) Schematic representation of a monolayer TMDs separated by vacuum from an electron reservoir, for example, the surface of a metal. The internal energy difference between 2H and 1T′ phases $E_{1T'} - E_{2H}$ changes with respect to the charge density σ as shown in **b** and **c**. The units are per formula unit (f.u.). The blue line represents constant-stress (stress-free) case, in which both 2H and 1T′ are structure relaxed. The red line represents the constant-area case, in which the monolayer is clamped to its 2H lattice constants. (**b**) Semiconducting 2H-MoTe$_2$ is a stable phase and semimetallic 1T′-MoTe$_2$ is metastable when the monolayer is charge neutral. However, 1T′-MoTe$_2$ is more thermodynamically favourable when the monolayer is charged beyond the positive or negative threshold values. The charge thresholds exhibit a significant dependence on the relaxation of lattice constants, indicating that the precise transition point may be sensitive to the presence of a substrate. (**c**) MoS$_2$ is stable in the 2H structure when charge neutral. The magnitude of charge required for the transition to 1T′ is larger than for MoTe$_2$. In both cases, transition at constant stress is more easily induced than the transition at constant area.

prediction that a negative charge density $> 0.29\,e$ per MoS$_2$ may trigger the phase transition from 2H phase to 1T′ phase MoS$_2$. See Supplementary Fig. 8 and Supplementary Note 5 for an intuitive discussion of the mechanism for the charge-induced structural phase transition.

Phase boundary at constant voltage. Another relevant type of electrical constraint is fixed voltage or electron chemical potential. This constraint is most applicable when the monolayer is in an electrostatic gating structure similar to field-effect transistors made using monolayers. Such a device structure enables a dynamical approach to achieve semiconductor/semimetal phase control in monolayer TMDs, suggesting intriguing applications for ultrathin flexible 2D electronic devices including phase-change memory.

Many distinct electrostatic gating device structures can be utilized to realize this dynamic control through a change in carrier density or electron chemical potential of the monolayer. Here we consider a capacitor structure shown in Fig. 4a. A monolayer of MoTe$_2$ is deposited on top of a dielectric layer of thickness d, which we take to be HfO$_2$ with a large dielectric constant of 25 (ref. 34). Monolayer and dielectric are sandwiched between two metal plates between which a voltage V is applied. High-dielectric constant material HfO$_2$ is chosen to increase the capacitance and hence increase the charge density in the monolayer. The metal plate is chosen to be aluminum with a work function of 4.08 eV. The curves in Fig. 4 assume the monolayer to be at a state of constant stress, with both 2H and 1T′ phases structurally relaxed. We compute the total energy and equilibrium grand potential of this system using equations (1–5).

Plotted in Fig. 4b is the total energy (equation 1) of the capacitor shown in Fig. 4a as a function of charge density in monolayer MoTe$_2$. Two black dashed lines depict common tangents between 2H and 1T′ energy surfaces, the slopes of which are defined by the set of equations,

$$\left(\frac{\partial E_H}{\partial Q}\right)_{Q_H} = \left(\frac{\partial E_{T'}}{\partial Q}\right)_{Q_{T'}} = \frac{E_H(Q_H) - E_{T'}(Q_{T'})}{Q_H - Q_{T'}} \qquad (8)$$

where $V_t = \left(\frac{\partial E_H}{\partial Q}\right)_{Q_H} = \left(\frac{\partial E_{T'}}{\partial Q}\right)_{Q_{T'}}$ is the transition gate voltage.

Plotted in Fig. 4c is the equilibrium grand potential (equation 5) as a function of the gate voltage. Two transition voltages are labelled also using black dashed lines. Figure 4b,c

shows that a transition gate voltage of -1.6 V or 4.4 V can be applied to drive the phase transition in monolayer MoTe$_2$ using the capacitor in Fig. 4a. The experimental breakdown voltage for a 4.5-nm-thick HfO$_2$ is reported to be as large as 3.825 V (ref. 35), which is larger than twice the magnitude of the negative transition voltage. This breakdown field in HfO$_2$ is larger than some other reports and may depend on the details of growth[36,37]. Therefore, employing an appropriate dielectric is likely to be critical here in observing the phase change. Ionic liquids may be employed to help address the challenge of achieving large voltages. Ionic gating has been applied to a variety of TMDs to investigate superconductivity[38] by measuring I–V curves. However, when a large voltage is applied, it may be challenging to probe structural phase transitions from I–V curves alone due to a large density of charge in the TMDs. Structural characterization approaches, such as Raman spectroscopy, may provide a more direct probe of electrically induced structure phase transitions in monolayer TMDs.

While the curves in Fig. 4 assume that the monolayer is at a state of zero stress across the transition, Fig. 5 presents calculations for MoTe$_2$ at constant stress (Fig. 5a) and constant area (Fig. 5b) utilizing the capacitor structure shown in Fig. 4a. These phase diagrams predict the thermodynamically favoured phase as a function of voltage V and thickness d of the HfO$_2$ dielectric medium. In each phase diagram, there exist two phase boundaries, the positions of which vary with the work function W of the capacitor plate. The 2H semiconducting phase of MoTe$_2$ is stable between the two phase boundaries, and metallic 1T′-MoTe$_2$ is stabilized by application of sufficiently positive or negative gate voltages. The transition voltages increase with the thickness of the dielectric layer. For a capacitor containing a HfO$_2$ dielectric layer of thickness < 5 nm, a negative gate voltage of approximately -2 V may be applied to drive the semiconductor-to-semimetal phase transition at constant stress (Fig. 5a) but the required voltage increases to approximately -4 V at constant area in Fig. 5b. In analogue with the changes in charge density phase boundaries shown in Fig. 3b, the voltage magnitudes for the transition are larger in constant-area conditions (Fig. 5b) than at constant stress (Fig. 5a). If the substrate constrains the area of the monolayer across the transition through friction, the voltages in Fig. 5b are expected to be applicable. The figure also shows a reported experimental breakdown voltage of a 4.5-nm-thick HfO$_2$ film[35].

Field-effect transistors based on few-layered MoTe$_2$ have been reported in ref. 39 using a 270-nm-thick SiO$_2$ gate dielectric layer

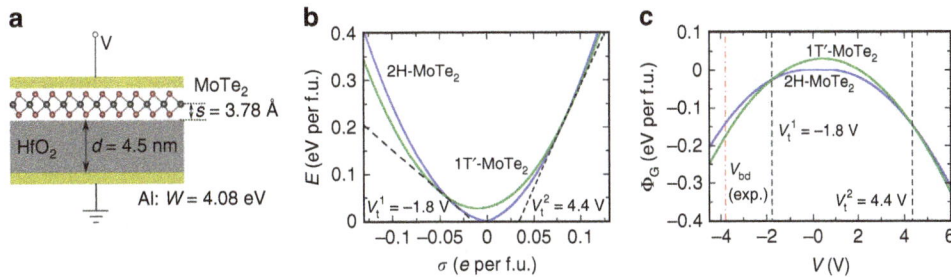

Figure 4 | Phase boundary at constant voltage and stress. (**a**) Monolayer MoTe$_2$ deposited on top of a HfO$_2$ layer of thickness $d = 4.5$ nm, which is on top of an aluminum plate of work function $W = 4.08$ eV. Voltage V is applied between the monolayer and the aluminum plate. (**b**) Plotted is the total energy E of the capacitor shown in **a** as a function of the charge density σ on monolayer MoTe$_2$. (**c**) Plotted is the grand potential Φ_G as a function of the gate voltage V. The blue line represents a capacitor containing 2H-MoTe$_2$, whereas the green line represents a capacitor containing 1T′-MoTe$_2$. The two black dashed lines in **b** depict common tangents between the 2H and 1T′ energy surfaces, and in **c** represent intersections of the 2H and 1T′ grand potentials indicating two transition voltages V_t^1 and V_t^2. Between the two transition voltages, semiconducting 2H-MoTe$_2$ has a lower grand potential and is thermodynamically stable. Outside this range, 1T′ will be more stable. The red dashed line in **c** represents a breakdown voltage[35] obtained experimentally for a HfO$_2$ film of thickness 4.5 nm. The separation between MoTe$_2$ centre and the surface of HfO$_2$ is assumed to be $s = 3.78$ Å, and both 2H and 1T′ are structurally relaxed (constant stress) in **b** and **c**.

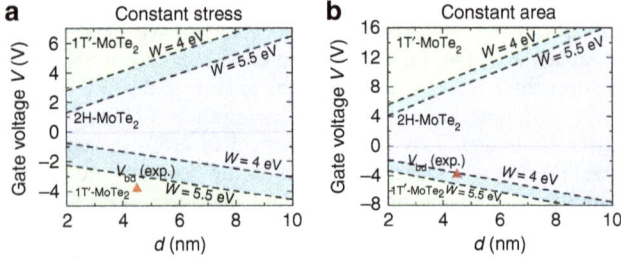

Figure 5 | Phase control of MoTe$_2$ through gating at constant stress and constant area. (a,b) Plotted are phase stabilities of monolayer MoTe$_2$ with respect to gate voltage V and dielectric thickness d using the capacitor structure shown in Fig. 4a. In each phase diagram, there exist two phase boundaries that vary with the work function W of the capacitor plate. Between the two phase boundaries, semiconducting 2H is more stable, and outside 1T' is the stable structure. The required transition gate voltage is smaller in the constant-stress case (**a**) than in the constant-area case (**b**). For a constant-stress scenario, a negative gate voltage as small as -1 to -2 V can trigger the semiconducting-to-semimetallic phase transition in monolayer MoTe$_2$. The red triangle represents the breakdown voltage of a 4.5-nm-thick HfO$_2$ film obtained experimentally[35].

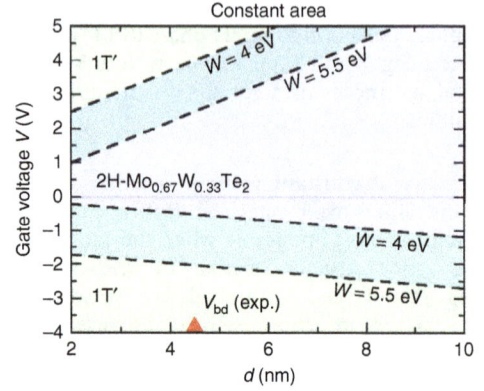

Figure 6 | Reducing transition gate voltages with the alloy Mo$_x$W$_{1-x}$Te$_2$. Plotted is the phase stability of a representative alloyed monolayer Mo$_{0.67}$W$_{0.33}$Te$_2$ with respect to the gate voltage V and the dielectric thickness d using the capacitor structure shown in Fig. 4a. The transition is assumed to occur at constant monolayer area. The magnitudes of the transition gate voltages in this alloy are smaller than those of pure MoTe$_2$ monolayer indicating the potential for phase boundary engineering.

(3.9 dielectric constant) with gate voltages as large as -50 V. For monolayer MoTe$_2$ (rather than few layers), our model predicts that a gate voltage >200 V is required to drive the phase transition for this device configuration. Both the increase of dielectric thickness (from 5 to 270 nm) and the decrease of dielectric constant (from 25 for HfO$_2$ to 3.9 for SiO$_2$ (ref. 35)) will result in larger transition gate voltages than shown in Fig. 5. To observe the 2H-1T' phase transition in a device, choosing a dielectric medium of large dielectric constant and dielectric performance will be critical.

Reducing transition gate voltages with the alloy Mo$_x$W$_{1-x}$Te$_2$. Monolayer alloys present the possibility for reducing the required gate voltage by varying the chemical composition. Recently, monolayer alloys of Mo- and W-dichalcogenides have attracted increasing attention for their tunable properties[40–44]. We hypothesize that the 2H-1T' transition gate voltage can be tuned to lower values by alloying MoTe$_2$-WTe$_2$ monolayers. This is because in monolayer MoTe$_2$, the 2H phase is energetically favourable by 31 meV per formula unit relative to the 1T' phase, whereas in monolayer WTe$_2$, the energy of the 1T' phase is 123 meV per formula lower than the 2H phase, as shown in Supplementary Fig. 4. Therefore, one might expect the energy difference between the two charge-neutral phases to be tunable through zero with alloy composition. The smaller the energy difference is, the closer the phase boundary is to ambient condition and the smaller the external force required to drive the phase transition in monolayer TMDs[20]. Therefore, controlling alloy composition is likely to enable tuning of the transition gate voltage.

Earlier experimental reports of the synthesis of the bulk alloy Mo$_x$W$_{1-x}$Te$_2$ (ref. 45) and detailed calculations on monolayers[44] indicate that the phase changes from 2H to 1T' with increase in W fraction $1-x$. This indicates that the free energy difference between the 2H and 1T' phases can be made arbitrarily small by varying x, enabling an arbitrary reduction of the gate voltage. However, the precise value of x required to achieve a particular transition voltage is likely to depend on a number of factors including synthesis conditions and mechanical constraints[44,45]. Here we study an approximate representative atomic configuration for this alloy for $x = 0.67$, displayed in Supplementary Fig. 9, with the knowledge that some variation of

computed phase diagram can occur with the choice of configuration. Detailed cluster expansion calculations for these monolayer alloys are presented in ref. 44.

For the alloy configuration we employed in this work (assumed at constant area), the 2H phase is a semiconductor with a semilocal quasiparticle Kohn–Sham bandgap of ~ 0.9 eV, and its free energy is 15 meV lower than the 1T' phase at constant monolayer area, which is metallic or semimetallic. Figure 6 shows that the 2H-1T' phase transition in this alloy can be driven by negative gating of a smaller gate voltage than pure MoTe$_2$ monolayer. For example, assuming HfO$_2$ medium of 4.5-nm thickness and capacitor plate of 4.0 eV work function, the magnitude of negative transition gate voltage can be reduced from 3.6 V (MoTe$_2$, constant-area case) to 0.4 V in the constant-area case of Mo$_{0.67}$W$_{0.33}$Te$_2$ monolayer.

One might expect that the transition gate voltage in monolayers can be tuned and reduced potentially arbitrarily by controlling the chemical composition of this and other potentially alloys. To enable a structural phase transition driven by a small gate voltage, elements should be selected for alloying so that the energy difference between charge-neutral 2H and 1T' phases can be tuned through zero with alloy composition. Alternative mechanical constraints placed on the alloy monolayer (for example, constant stress) can also be expected to shift the phase boundary and transition gate voltage.

Phase transition in Ta-based TMDs. Electrically induced structural phase changes between 2H and 1T phases in the Ta-based TMDs, TaSe$_2$ and TaS$_2$, have been reported in experiments using a STM tip[17,18], although the mechanism for this reported effect may differ from the charge-induced effect reported in the present work. As a supplement to the previous calculations on group VI TMDs, we have computed the constant-stress phase diagram of monolayer TaSe$_2$ in the capacitor gating structure shown in Fig. 7a. Figure 7b shows that the phase diagram of TaSe$_2$ has a phase boundary at only positive gate voltage, qualitatively different from MoTe$_2$ and MoS$_2$. We find that this difference results from the metallic nature of both 2H- and 1T-TaSe$_2$, further discussed in Supplementary Note 5.

A qualitative difference between these calculations and the STM experiment[17] is the observation of the transition at both signs of STM bias, suggesting that other effects could be at play in the experiment. Further quantitative comparison with the

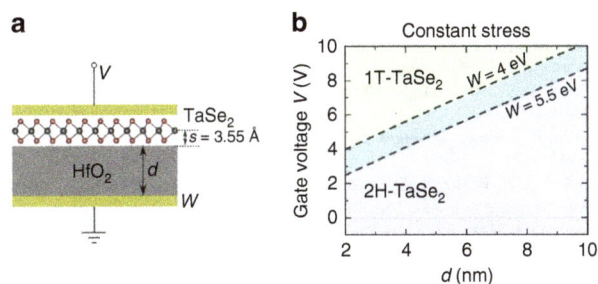

Figure 7 | Phase control of monolayer TaSe$_2$ through gating at constant stress. (b) Plotted is the computed phase diagram of monolayer TaSe$_2$ in the capacitor gating structure as shown in **a**. Unlike MoTe$_2$, the phase diagram of TaSe$_2$ only has one phase boundary, which corresponds to a positive gate voltage. Below the phase boundary, 2H-TaSe$_2$ is more stable; above the boundary, 1T has lower energy and is more stable. Intuition for this qualitative difference from MoTe$_2$ is provided in Supplementary Note 5.

experiment is made challenging by the small separation between the monolayer and the STM tip (Supplementary Fig. 3; Supplementary Note 2).

Discussion

The electrical dynamical control of structural phase in monolayer TMDs has exciting potential applications in ultrathin flexible 2D electronic devices. If the kinetics of the transformation are suitable, nonvolatile phase-change memory[9] may be an application. One might expect 2D materials to have energy consumption advantages over bulk materials due to their small thickness. If the kinetics is sufficiently fast, another potential application may be subthreshold swing reduction in field-effect transistors to overcome the scaling limit of conventional transistors[4]. In addition, the change in the transmittance of light due to the phase transition of monolayer TMDs may be employed in infrared optical switching devices, such as infrared optical shutters and modulators for cameras, window coating and infrared antennas with tunable resonance.

To summarize, we have identified a new mechanism, electrostatic gating, to induce a structural semiconductor-to-semimetal phase transition in monolayer TMDs. We have computed phase boundaries for monolayer MoTe$_2$, MoS$_2$ and TaSe$_2$. We discover that changing carrier density or electron chemical potential in the monolayer can induce a semiconductor-to-semimetal phase transition in monolayer TMDs. We find that a surface charge density less than $-0.04\,e$ or greater than $0.09\,e$ per formula unit is required to observe the semiconductor-to-semimetal phase transition in monolayer MoTe$_2$ under constant-stress conditions, and a significantly larger value of approximately $-0.29\,e$ or $0.35\,e$ per formula unit is required in the monolayer MoS$_2$ case. A capacitor structure can be employed to dynamically control the semiconductor-to-semimetal phase transition in monolayer MoTe$_2$ with a gate voltage ~ 2–4 V for MoTe$_2$. These transition charges and voltages are expected to vary considerably with the nature of the mechanical constraint of the monolayer and also potentially the presence of dopants or Fermi level pinning. While the gate voltages required to observe the transition in MoTe$_2$ are likely near breakdown and could be challenging to realize in the lab, we find that the voltage magnitudes can be reduced arbitrarily by alloying Mo atoms with substitutional W atoms to create the alloy Mo$_x$W$_{1-x}$Te$_2$.

Methods

Electronic structure calculations. All periodic DFT calculations were performed within the Vienna Ab Initio Simulation Package[46], version 5.3.3, using the

projector augmented-wave[47] method and the plane-wave basis set with a kinetic energy cutoff of 350 eV. Electron exchange and correlation effects were treated using the Generalized Gradient Approximation (GGA) functional of Perdew, Burke and Ernzerhof[48]. An $18 \times 18 \times 1$ Monkhorst-Pack[49] k-point mesh was utilized to sample the Brillouin zone. The convergence thresholds for electronic and ionic relaxations were chosen to be 0.5×10^{-8} eV per MX$_2$ formula unit and 0.5×10^{-7} eV per MX$_2$ formula unit, respectively. A Gaussian smearing of 50 meV was used. The computational cell length is 36 Å along the c axis. Spin–orbit coupling is employed in all DFT calculations. The ionic relaxations were performed using conjugate gradient algorithm.

All calculations in this work were performed at zero ionic temperature, omitting the vibrational component of the free energy. Reference 20 has shown that inclusion of vibrational free energy and temperature would shift the phase boundaries closer to ambient conditions and lower the energy required to switch the phases. Therefore, one would expect inclusion of these effects to decrease the magnitude of the transition charge density and gate voltage calculated in this work. Also, the change of bandgap width is expected to affect 2H-1T' phase boundary, as further discussed in Supplementary Fig. 10 and Supplementary Note 6. See Supplementary Note 7 for a discussion on vacuum electronic states.

References

1. Lencer, D. *et al.* A map for phase-change materials. *Nat. Mater.* **7**, 972–977 (2008).
2. Wong, H.-S. P. *et al.* Phase change memory. *Proc. IEEE* **98**, 2201–2227 (2010).
3. Wong, H.-S. P. *et al.* Metal-oxide RRAM. *Proc. IEEE* **100**, 1951–1970 (2012).
4. Zhou, Y. & Ramanathan, S. Correlated electron materials and field effect transistors for logic: a review. *Crit. Rev. Solid State Mater. Sci.* **38**, 286–317 (2013).
5. Shu, M. J. *et al.* Ultrafast terahertz-induced response of GeSbTe phase-change materials. *Appl. Phys. Lett.* **104**, 251907 (2014).
6. Aetukuri, N. B. *et al.* Control of the metal-insulator transition in vanadium dioxide by modifying orbital occupancy. *Nat. Phys.* **9**, 661–666 (2013).
7. Cavalleri, A. *et al.* Femtosecond structural dynamics in VO$_2$ during an ultrafast solid-solid phase transition. *Phys. Rev. Lett.* **87**, 237401 (2001).
8. Kikuzuki, T. & Lippmaa, M. Characterizing a strain-driven phase transition in VO2. *Appl. Phys. Lett.* **96**, 132107 (2010).
9. Wuttig, M. & Yamada, N. Phase-change materials for rewriteable data storage. *Nat. Mater.* **6**, 824–832 (2007).
10. Lee, S.-H., Jung, Y. & Agarwal, R. Highly scalable non-volatile and ultra-low-power phase-change nanowire memory. *Nat. Nanotechnol.* **2**, 626–630 (2007).
11. Bulaevskii, L. N. Structural transitions with formation of charge-density waves in layer compounds. *Sov. Phys. Uspekhi* **19**, 836–843 (1976).
12. McMillan, W. L. Theory of discommensurations and the commensurate-incommensurate charge-density-wave phase transition. *Phys. Rev. B* **14**, 1496–1502 (1976).
13. Thomson, R. E., Burk, B., Zettl, A. & Clarke, J. Scanning tunneling microscopy of the charge-density-wave structure in 1T-TaS$_2$. *Phys. Rev. B* **49**, 16899–16916 (1994).
14. Scruby, C. B., Williams, P. M. & Parry, G. S. The role of charge density waves in structural transformations of 1T TaS$_2$. *Philos. Mag.* **31**, 255–274 (1975).
15. Castro Neto, A. H. Charge density wave, superconductivity, and anomalous metallic behavior in 2D transition metal dichalcogenides. *Phys. Rev. Lett.* **86**, 4382–4385 (2001).
16. Wilson, J. A. & Yoffe, A. D. The transition metal dichalcogenides discussion and interpretation of the observed optical, electrical and structural properties. *Adv. Phys.* **18**, 193–335 (1969).
17. Zhang, J., Liu, J., Huang, J. L., Kim, P. & Lieber, C. M. Creation of nanocrystals through a solid-solid phase transition induced by an STM tip. *Science* **274**, 757–760 (1996).
18. Kim, J.-J. *et al.* Observation of a phase transition from the T phase to the H phase induced by a STM tip in 1T-TaS$_2$. *Phys. Rev. B* **56**, R15573–R15576 (1997).
19. Mak, K. F., Lee, C., Hone, J., Shan, J. & Heinz, T. F. Atomically thin MoS$_2$: a new direct-gap semiconductor. *Phys. Rev. Lett.* **105**, 136805 (2010).
20. Duerloo, K.-A. N., Li, Y. & Reed, E. J. Structural phase transitions in two-dimensional Mo- and W-dichalcogenide monolayers. *Nat. Commun.* **5**, 4214 (2014).
21. Eda, G. *et al.* Coherent atomic and electronic heterostructures of single-layer MoS$_2$. *ACS Nano* **6**, 7311–7317 (2012).
22. Gao, G. *et al.* Charge mediated semiconducting-to-metallic phase transition in molybdenum disulfide monolayer and hydrogen evolution reaction in new 1T' phase. *J. Phys. Chem. C* **119**, 13124–13128 (2015).
23. Kan, M. *et al.* Structures and phase transition of a MoS$_2$ monolayer. *J. Phys. Chem. C* **118**, 1515–1522 (2014).
24. Kang, Y. *et al.* Plasmonic hot electron induced structural phase transition in a MoS$_2$ monolayer. *Adv. Mater.* **26**, 6467–6471 (2014).

25. Jiménez Sandoval, S., Yang, D., Frindt, R. F. & Irwin, J. C. Raman study and lattice dynamics of single molecular layers of MoS_2. *Phys. Rev. B* **44**, 3955–3962 (1991).

26. Kan, M., Nam, H. G., Lee, Y. H. & Sun, Q. Phase stability and Raman vibration of the molybdenum ditelluride ($MoTe_2$) monolayer. *Phys. Chem. Chem. Phys.* **17**, 14866–14871 (2015).

27. Wang, Q. H., Kalantar-Zadeh, K., Kis, A., Coleman, J. N. & Strano, M. S. Electronics and optoelectronics of two-dimensional transition metal dichalcogenides. *Nat. Nanotechnol.* **7**, 699–712 (2012).

28. Radisavljevic, B., Radenovic, A., Brivio, J., Giacometti, V. & Kis, A. Single-layer MoS_2 transistors. *Nat. Nanotechnol.* **6**, 147–150 (2011).

29. Brown, B. E. The crystal structures of WTe_2 and high-temperature $MoTe_2$. *Acta Crystallogr.* **20**, 268–274 (1966).

30. Keum, D. H. *et al.* Bandgap opening in few-layered monoclinic $MoTe_2$. *Nat. Phys.* **11**, 482–486 (2015).

31. Ruppert, C., Aslan, O. B. & Heinz, T. F. Optical properties and band gap of single- and few-layer $MoTe_2$ crystals. *Nano Lett.* **14**, 6231–6236 (2014).

32. Lezama, I. G. *et al.* Indirect-to-direct band gap crossover in few-layer $MoTe_2$. *Nano Lett.* **15**, 2336–2342 (2015).

33. Lin, Y.-C., Dumcenco, D. O., Huang, Y.-S. & Suenaga, K. Atomic mechanism of the semiconducting-to-metallic phase transition in single-layered MoS_2. *Nat. Nanotechnol.* **9**, 391–396 (2014).

34. Ray, S. K., Mahapatra, R. & Maikap, S. High-k gate oxide for silicon heterostructure MOSFET devices. *J. Mater. Sci. Mater. Electron* **17**, 689–710 (2006).

35. Kang, L. *et al.* Electrical characteristics of highly reliable ultrathin hafnium oxide gate dielectric. *IEEE Electron Device Lett.* **21**, 181–183 (2000).

36. Lee, J.-H. *et al.* Characteristics of ultrathin $HfO2$ gate dielectrics on strained-Si0.74Ge0.26 layers. *Appl. Phys. Lett.* **83**, 779–781 (2003).

37. Wolborski, M., Rooth, M., Bakowski, M. & Hallén, A. Characterization of $HfO2$ films deposited on 4H-SiC by atomic layer deposition. *J. Appl. Phys.* **101**, 124105 (2007).

38. Shi, W. *et al.* Superconductivity series in transition metal dichalcogenides by ionic gating. *Sci. Rep.* **5**, 12534 (2015).

39. Pradhan, N. R. *et al.* Field-effect transistors based on few-layered α-$MoTe_2$. *ACS Nano* **8**, 5911–5920 (2014).

40. Chen, Y. *et al.* Tunable band gap photoluminescence from atomically thin transition-metal dichalcogenide alloys. *ACS Nano* **7**, 4610–4616 (2013).

41. Zhang, M. *et al.* Two-dimensional molybdenum tungsten diselenide alloys: photoluminescence, Raman scattering, and electrical transport. *ACS Nano* **8**, 7130–7137 (2014).

42. Kutana, A., Penev, E. S. & Yakobson, B. I. Engineering electronic properties of layered transition-metal dichalcogenide compounds through alloying. *Nanoscale* **6**, 5820–5825 (2014).

43. Tongay, S. *et al.* Two-dimensional semiconductor alloys: Monolayer Mo1-xWxSe2. *Appl. Phys. Lett.* **104**, 012101 (2014).

44. Duerloo, K.-A. N. & Reed, E. J. Structural phase transitions by design in monolayer alloys. *ACS Nano.* doi: 10.1021/acsnano.5b04359 (2015).

45. Champion, J. A. Some properties of (Mo, W) (Se, Te) 2. *Br. J. Appl. Phys.* **16**, 1035 (1965).

46. Kresse, G. & Furthmüller, J. Efficient iterative schemes for ab initio total-energy calculations using a plane-wave basis set. *Phys. Rev. B* **54**, 11169–11186 (1996).

47. Blöchl, P. E. Projector augmented-wave method. *Phys. Rev. B* **50**, 17953–17979 (1994).

48. Perdew, J. P., Burke, K. & Ernzerhof, M. Generalized gradient approximation made simple. *Phys. Rev. Lett.* **77**, 3865–3868 (1996).

49. Monkhorst, H. J. & Pack, J. D. Special points for Brillouin-zone integrations. *Phys. Rev. B* **13**, 5188–5192 (1976).

Acknowledgements

Our work was supported in part by the US Army Research Laboratory, through the Army High Performance Computing Research Center, Cooperative Agreement W911NF-07-0027. This work was also partially supported by NSF grants EECS-1436626 and DMR-1455050, Army Research Office grant W911NF-15-1-0570, Office of Naval Research grant N00014-15-1-2697 and a seed grant from Stanford System X Alliance. We thank Philip Kim for discussions.

Author contributions

Y.L., K.-A.N.D. and E.J.R. designed the simulations and the framework for thermodynamic analysis; Y.L. performed the simulations and subsequent numerical data analysis; Y.L. and K.W. performed the preliminary simulations. Y.L. and E.J.R. interpreted the data and wrote the paper.

Additional information

Visualizing the orientational dependence of an intermolecular potential

Adam Sweetman[1], Mohammad A. Rashid[1], Samuel P. Jarvis[1], Janette L. Dunn[1], Philipp Rahe[1] & Philip Moriarty[1]

Scanning probe microscopy can now be used to map the properties of single molecules with intramolecular precision by functionalization of the apex of the scanning probe tip with a single atom or molecule. Here we report on the mapping of the three-dimensional potential between fullerene (C_{60}) molecules in different relative orientations, with sub-Angstrom resolution, using dynamic force microscopy (DFM). We introduce a visualization method which is capable of directly imaging the variation in equilibrium binding energy of different molecular orientations. We model the interaction using both a simple approach based around analytical Lennard–Jones potentials, and with dispersion-force-corrected density functional theory (DFT), and show that the positional variation in the binding energy between the molecules is dominated by the onset of repulsive interactions. Our modelling suggests that variations in the dispersion interaction are masked by repulsive interactions even at displacements significantly larger than the equilibrium intermolecular separation.

[1] School of Physics and Astronomy, University of Nottingham, Nottingham NG7 2RD, UK. Correspondence and requests for materials should be addressed to A.S. (email: adam.sweetman@nottingham.ac.uk).

The nature of intermolecular interactions underpins a vast array of physical and chemical phenomena, and is a scientific theme that straddles the disciplines of physics, chemistry and biology. Particular impetus has been given to the study of intermolecular forces at the single-molecule level due to the stunning advances in ultrahigh resolution scanning probe imaging pioneered by Gross et al.[1]. Three-dimensional (3D) force maps were acquired over planar organic molecules that bore a striking resemblance to the classic textbook 'ball-and-stick' models. These advances were first realized via the controllable functionalization of the scanning probe tip with a single pre-selected atom or molecule, which provides a unique level of control with which to investigate the atomic and molecular scale properties of matter, and also helps to eliminate the most troublesome aspect of scanning probe experiments, that is, the uncertainty surrounding the tip structure.

Although this tip functionalization strategy is now commonly applied to single CO molecules to allow intramolecular imaging[1,2], the technique has application well beyond imaging, and similar protocols have also been used to study the electronic[3,4] and mechanical[5] properties of single molecules trapped in the tip-sample junction, and to quantitatively measure intermolecular interactions[6-8]. There has also been considerable interest centred around the possibility of using this technique to directly visualize intermolecular interactions[9], although considerable debate surrounds the interpretation of these results[2,10-12].

In this paper, we discuss the results of a series of experiments—and their interpretation on the basis of both simple analytical potentials and DFT—that map the orientational dependence of the 3D potential between two-complex molecules. By measuring the full 3D potential we are able to apply a novel visualization method that directly shows the variation in the equilibrium binding energy for the molecular system for different relative orientations of the molecules. We also discuss the feasibility of detecting the variation in dispersion forces due to molecular rotation via DFM.

Results

Experimental results. Figure 1 shows representative constant height Δf images, taken from a 3D grid, acquired at decreasing tip-sample separation over three surface-adsorbed C_{60} molecules in different orientations, using a C_{60}-terminated tip.

At larger separations a featureless circular attractive interaction is observed (Fig. 1b), but on closer approach intricate intramolecular features are resolved (Fig. 1c,d), followed by their intense 'sharpening' (Fig. 1e,f). This evolution in contrast is similar to the onset of sub-molecular features during imaging of planar molecules with flexible tips[1,12]. However, because in this experiment both molecules have a complex structure, the intramolecular features in these images cannot be easily assigned to the molecular structure of the surface molecule as is the case for images taken with simple (that is, atomic point-like) tip terminations.

Converting the acquired Δf grid into a map of potential allows us to create similar constant height images of the tip-sample force and potential (Fig. 2a,b and Supplementary Fig. 1). Although constant height slices of force and energy provide the closest visual analogue to how the data are collected, these images necessarily conflate the value of the tip-sample energy and the topographic height of the molecule at a given position. Consequently, topographically higher features dominate the constant height image due to their being effectively shifted in z, even if the range of energies at these positions is identical to other locations over the molecule.

Representative single $U(z)$ curves may be extracted (Fig. 2c) and allow a selection of the energy minimum values at different positions to be observed, but this is an indirect, and not necessarily intuitive, method of analysing the variation in intermolecular potential across the molecules.

In Fig. 2d, we instead show an image constructed by searching each vertical column in the 3D data set (that is, each $U(z)$ curve) for the value of the potential energy minimum, and then projecting this minimum value over the xy plane of the grid, which we hereafter refer to as a 'U_{min}' image. This provides an

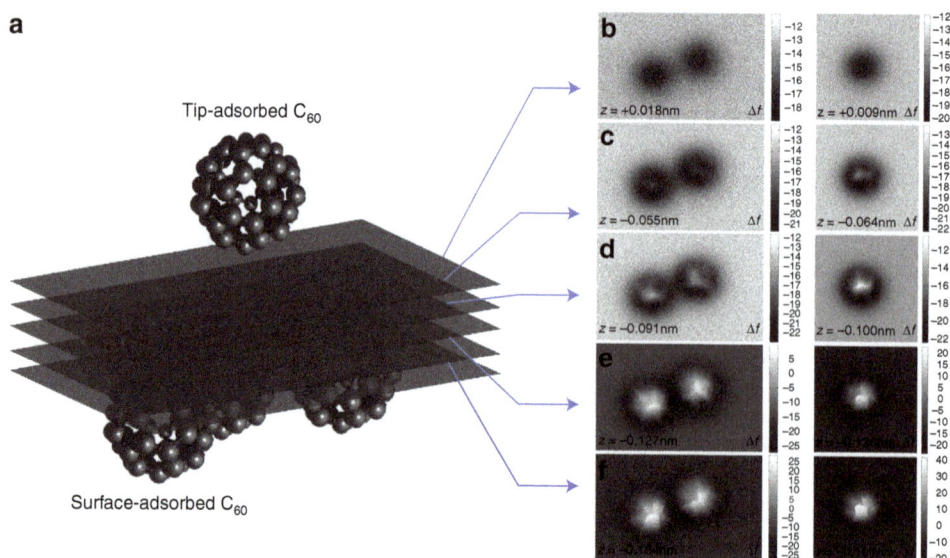

Figure 1 | Experimental data acquisition protocol. (**a**) cartoon showing the method of data acquisition for 3D potential mapping—a single C_{60} molecule is attached to the tip of the scanning probe microscope and brought close to a group of surface-adsorbed molecules. Constant height scans are acquired at decreasing tip-sample separation, with active drift compensation between each scan, and the variation in the frequency shift Δf measured. (**b-f**) representative Δf images (in Hz) at decreasing tip-sample height. Tip-sample heights shown for each image are given relative to the Δf set point used for atom-tracking over the molecule. The slightly different z heights for the two data sets result from the slightly different tracking heights used in each case. Image sizes: $3.5 \times 2.2\,nm^2$ and $2.5 \times 2.5\,nm^2$.

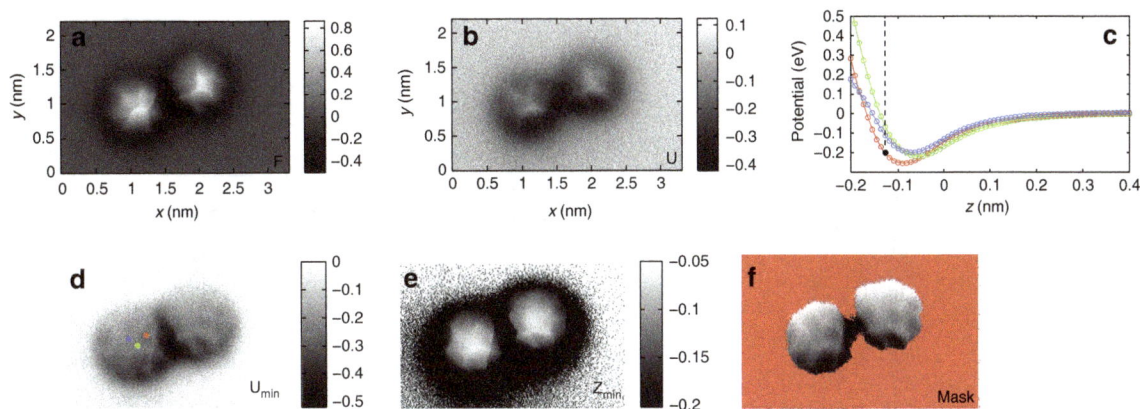

Figure 2 | Experimental measurement of variation in potential between C_{60} molecules in different orientations. Constant height images of (**a**) force (in nN) and (**b**) energy (in eV). (**c**) Representative $U(z)$ curves taken at different positions across the left hand C_{60} molecule, dotted line shows the height of the force and energy slices shown in **a,b**. (**d**) Image showing the variation in the value of the energy at the minimum in the $U(z)$ curve (in eV) at each position in the grid. The positions of the curves shown in **c** are marked. (**e**) As for **d** but showing instead the z height at which the minimum occurs, note that the black regions indicate parts of the grid where the minimum is found at the lowest tip-sample separation (that is, no turnaround detected). (**f**) Variation in energy minimum masked using the minimum in z position. Red shading indicates locations where the minimum in the intermolecular potential is not present in the $U(z)$ curve.

immediate and intuitive way of visualizing the strength of the equilibrium interaction as the relative position of the tip- and sample-adsorbed molecules is varied. We note that, because of the near-unique high-rotational symmetry of the C_{60} molecules, displacements in the xy plane should be equivalent to changes in rotational orientation.

We highlight here that some care must be taken in the interpretation of these images, as the value of the minimum in the potential energy curve only has a directly interpretable physical meaning when the actual minimum of the potential is present in a given $U(z)$ curve (that is, the turn-around in the $U(z)$ curve is present in the data set). If the minimum is not reached, then the closest point of approach will usually be identified as the minimum value. We therefore also map the height of the potential energy minimum in terms of z, which yields a complementary map of z_{min}. By masking the U_{min} map with the z_{min} map we can exclude those curves which do not contain the $U(z)$ turn-around, and visualize only the region of the image, which can be interpreted directly as representing the intermolecular interaction minimum (Fig. 2f). Application of this visualization technique also reveals a gradient in the value of the minimum in the potential across the molecule, most likely related to an asymmetric mounting of the molecule on the tip. Since this gradient directly affects the spread in the energy values we therefore only discuss the variation observed in the region located over the centre of the molecule, where the variation due to the gradient is small compared to the variation produced by the changes in molecular orientation (Supplementary Figs 16–18).

The same technique may also be applied to the 3D force field and raw 3D Δf measurements (see Supplementary Fig. 3). Although these maps do not have such a direct physical interpretation as for the minimum in the potential, they still provide an extremely powerful technique for visualizing the relative interaction over the molecule. Interestingly, we note a strong qualitative similarity in the appearance of these images and recent data acquired using a profile-corrected constant height technique by Moreno et al.[13]. We also note that Mohn et al.[14] recovered a pssuedo-topographic Δf image from a 3D data set, and experimentally it has been shown how to operate in the $\Delta f = 0$ regime[15], which might, in principle, produce similar imaging if applied to intermolecular measurements. Critically,

however, none of these earlier works directly measured and visualised the physical quantity of interest here: the variation in the value of the minimum in the intermolecular potential.

Our data demonstrate that as the relative orientations of the tip and surface C_{60} molecules are varied the potential minimum between the two molecules varies of the order 60 meV. A key question is therefore—what is the origin of this variation? A common approach to evaluating the C_{60}–C_{60} intermolecular interaction is to model molecular energy variation using the Girifalco potential[7,16], but this simplified model assumes a uniform spherical interaction, and does not give any information about sub-molecular variation in the potential. In particular, given the extended 3D nature of the molecule, it is not immediately clear how the attractive and repulsive components of the intermolecular potential contribute to the variation in the magnitude, and position, of the energy minimum. Following recent studies investigating the variation in dispersion force as a function of molecular size[6], there is also an open question as to whether the difference in the dispersion interaction can be observed for changes in the orientation of extended molecules. C_{60}, with its near-spherical symmetry represents a particularly important test bed for this hypothesis.

Computational results. To interpret our results, we modelled our experimental system with two different approaches (as described in the Methods section). First, we used a simple Lennard–Jones (L–J) potential for two C_{60} molecules, coupled with a modified version of the flexible tip model introduced by Hapala et al.[12], to simulate the C_{60} interaction. The simple nature of the model means that it is computationally inexpensive and thus can be exploited to generate high-resolution 3D data sets of comparable data density to those we obtain experimentally (Fig. 3d–g). Second, to test the validity of our empirical model, we compare the results of the L–J calculations to simulations of the same C_{60}–C_{60} interaction performed using the *ab initio* CP2K DFT code. The significant computational cost of the *ab initio* simulations precludes the calculation of a full 3D grid as for the L–J simulations, and we therefore instead compare 2D xz slices taken across the centre of the molecule–molecule interaction (Fig. 3b,c). In this comparison, we modelled a prototypical high symmetry orientation (hexagon face on hexagon face, hereafter referred to as

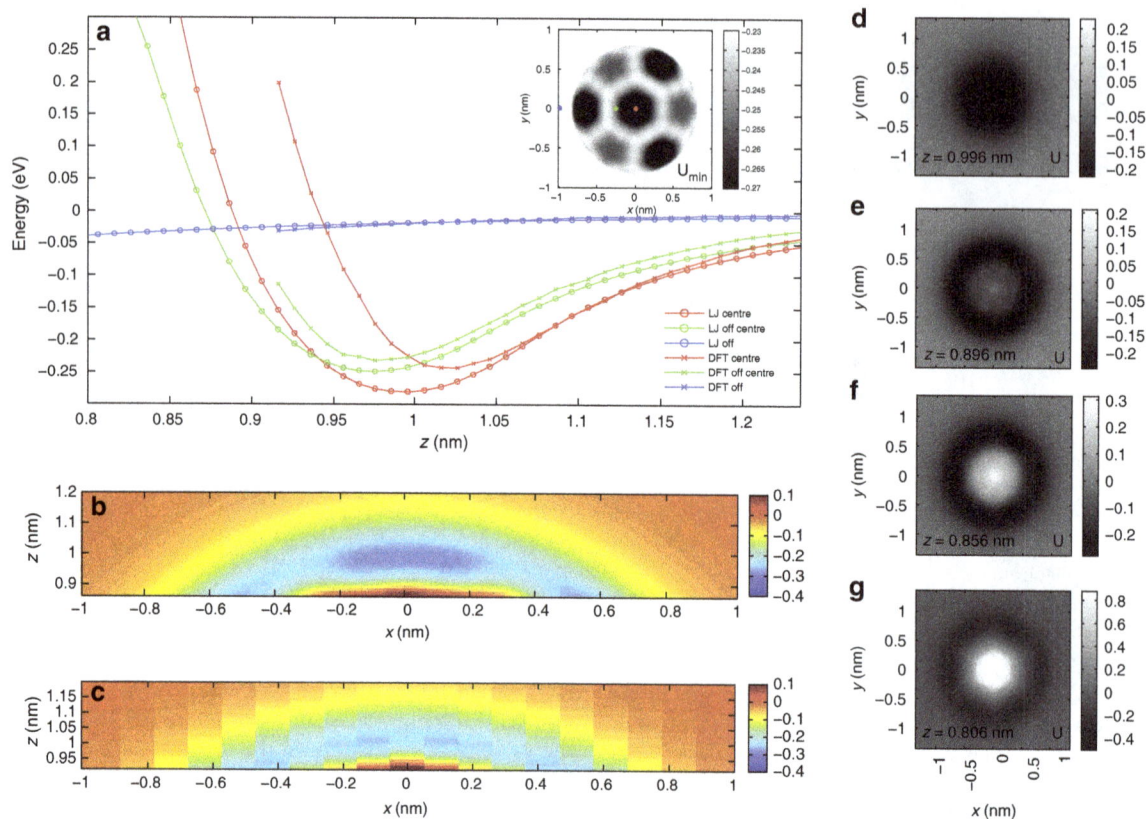

Figure 3 | Modelling the C$_{60}$–C$_{60}$ interaction. (**a**) Comparison of potentials calculated by L-J modelling and DFT. z heights are defined as the vertical separation of the two molecules measured from the molecular centres. Curves are shown for the same initial xy coordinates in both simulations. Inset: Complete xy U_{min} (in eV) image for L-J simulation. (**b,c**) xz plot of calculated energies for L-J and DFT simulations, respectively. (**d-g**) Constant height xy energy images (in eV) from L-J simulations, relative heights labelled on each image.

Hex–Hex) for the two molecules using both simulations methods. In general, we find good agreement between the two techniques, noting in particular that the potential gradients in both the attractive and repulsive branches of the potential curve are very similar. We do, however, observe some quantitative differences between the empirical L-J simulations and *ab initio* DFT. Specifically, for the L-J parameters chosen ($\epsilon_a = 2.5$ meV and $r_a = 1.966$ Å (ref. 16)), the maximum depth and width of the well is slightly larger that for the DFT simulations, as is the variation in the range of U$_{min}$ (ΔU_{min} for L-J xz plot ~ 50 meV compared with ~ 20 meV for the DFT xz plot over the same range (Supplementary Figs 20–22)). We note, however, that variation in the positions of the minimum is almost identical (ΔZ_{min} for L-J xz plot ~ 0.09 nm, compared with $\Delta Z_{min} \sim 0.11$ nm for the DFT xz plot). Furthermore, it is clear that tuning the choice of L-J parameters based on the DFT results could improve the quantitative agreement between the two simulation methods, but here we prefer to use those L-J parameters derived from previous experimental work and which are also consistent with our earlier publications, rather than arbitrarily adjusting the L-J parameters. These results imply that while the L-J model is a simplification of the complex intermolecular interaction, it nonetheless appears to be sufficient to model much of the essential physics underpinning the variation in intermolecular potential.

Although we stress that the high-symmetry Hex–Hex configuration used in the simulations is not the configuration of the C$_{60}$ molecules in the experimental data set shown in Fig. 2, we nonetheless observe a number of qualitatively similar features in both the simulations and experiment. In particular, the simulations reproduce the 'sharpening' of the features observed in

the constant-height experimental images, in line with the sharpening reported for CO-terminated tips. In addition, the appearance of the simulated U$_{min}$ image is qualitatively similar to that acquired in the experiment, which reveals the complex variation in potential minimum as the molecular positions are varied. Interestingly, the L-J simulations overestimate the depth of the potential relative to the DFT calculations, but better reproduce the variation in U$_{min}$ observed experimentally, with a variation of ~ 50 to 60 meV in the U$_{min}$ image depending on molecular orientation (Supplementary Figs 16–18). We also note that simulations performed with other tip-sample molecular configurations, such as those found for C$_{60}$ adsorption on the Si(111)-7 $\times$ 7 substrate, produce much more complex patterns in the constant height, and U$_{min}$, images (see Supplementary Figs 7–13), qualitatively similar to those observed experimentally.

Discussion

Because of the simple additive, and analytical, nature of the L-J model, it is possible to decompose the interaction into its attractive and repulsive components, and ascertain if we might in principle be able to observe rotational variation in the dispersion interaction between the C$_{60}$ molecules. To assess the relative influence of the repulsive and attractive elements of the potential on the value of the potential minimum we investigated the change in the potential for several orientations of the tip and sample C$_{60}$ (Fig. 4a). We then plot the modulus (i.e. the absolute value) of the differences in the total energies, and the separate energies from the r^6 and r^{12} terms, between these orientations and the high symmetry 'Hex–Hex' configuration (Fig. 4b), and then extract the differences in these ΔU_{r^6} and $\Delta U_{r^{12}}$ terms (Fig. 4c). Specifically,

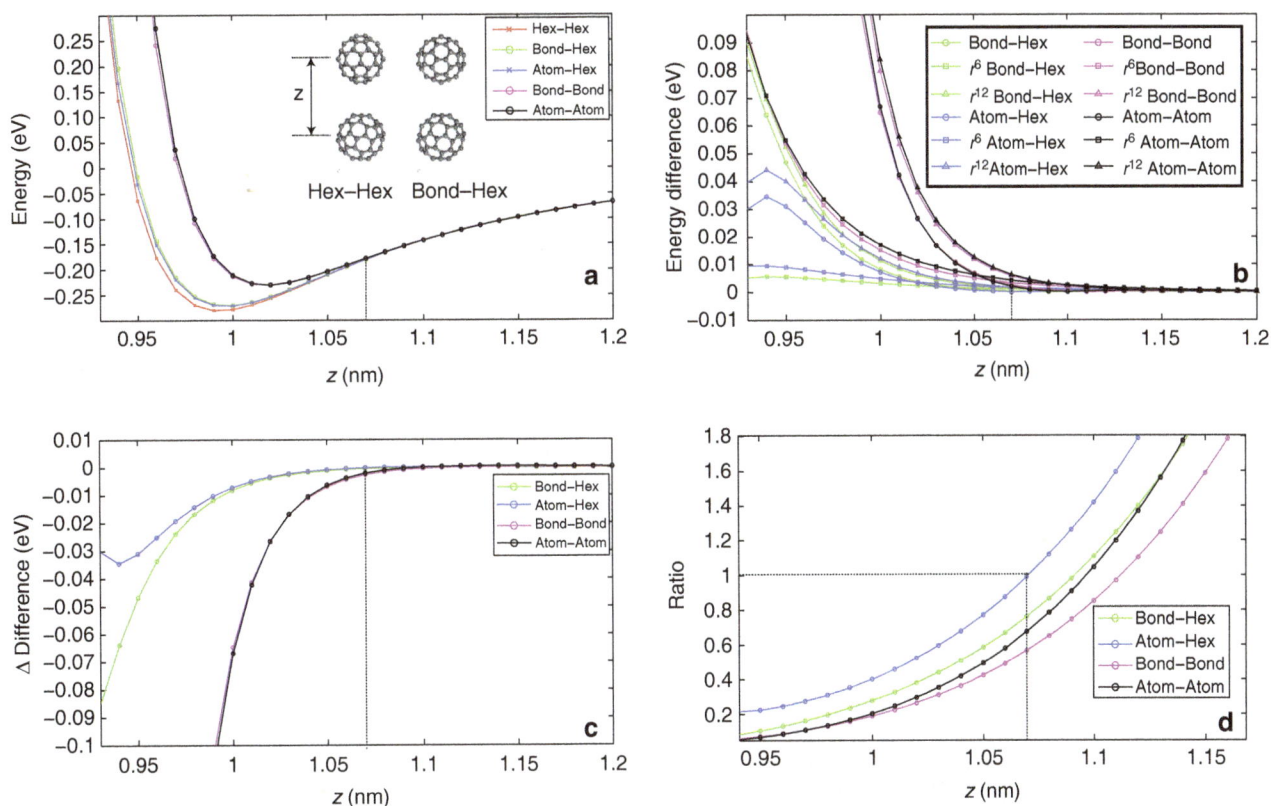

Figure 4 | Investigating the origin of the variation in potential energy minimum. (**a**) Variation in potential energy between two C_{60} molecules due to variation in rotational orientation, the terms in the legend refer to the facing part of the molecule. Inset, representative ball-and-stick models showing two of the simulated orientations. (**b**) Modulus of the difference in U_{total}, U_{r^6}, and $U_{r^{12}}$ between stated orientation and the Hex–Hex orientation. (**c**) Difference of variation in U_{r^6} and $U_{r^{12}}$ terms (that is, $|\Delta U_{r^6}| - |\Delta U_{r^{12}}|$). Negative values indicate the $U_{r^{12}}$ term dominates the difference in energy between the two configurations. (**d**) Ratio of variation in U_{r^6} and $U_{r^{12}}$ terms. Values above 1 indicate the U_{r^6} term has a larger contribution. Dotted lines indicate the separation at which the U_{r^6} term first becomes dominant.

we define $|\Delta U_{r^6}| - |\Delta U_{r^{12}}|$ as the modulus of the difference of the ΔU_{r^6} term (between the stated orientation and the Hex–Hex orientation), minus the modulus of the difference in the $\Delta U_{r^{12}}$ terms (between the stated orientation and the Hex–Hex orientation).

This quantity gives the relative influence of the two terms in defining the difference in the energy curves between the two orientations. If the difference in r^6 terms at a given separation is greater (that is, dispersion forces vary significantly between different orientations), then this quantity will be positive. If the difference in r^{12} terms is large (that is, Pauli forces vary significantly between different orientations), then it will be negative. If both quantities contribute equally to the difference in energy, then the term will be approximately zero. Surprisingly, we observe that the difference is negative and, consequently, the differences observed in the total energies, even in the part of the well where the potential gradient is positive, are dominated by repulsive interactions. Here we wish to make it explicitly clear that, for intermolecular separations greater than the equilibrium value (that is, before the energy 'turn-around'), the magnitude of the r^6 term is indeed larger in all cases, and dominates the r^{12} term, but that contribution of the r^6 term is very similar for all the orientations.

It must be noted, however, that the interplay between the two terms is somewhat subtle. If we examine the ratio of the differences (Fig. 4d), then it is clear that the r^6 term does begin to dominate the difference in the energies at around 1.06–1.1 nm separation. However, by reference to (Fig. 4c) it becomes clear that at this separation the difference in the potentials is <5 meV, that is, below our experimental sensitivity. Therefore, our

modelling suggests that at the point at which the potential curves for different orientations become experimentally distinguishable, the difference between them is dominated by repulsive, rather than dispersive, interactions. As such it seems likely that although the magnitude of the variation in energy due to the variation in dispersion interaction under rotation of the molecule might in principle be within the noise limit of current DFM techniques, its direct measurement will always be hindered by the intrinsic convolution of the variation in energy due to repulsive forces, even at intermolecular separations significantly greater than the equilibrium value, where the gradient in the potential is positive.

We have presented 3D mapping of the variation in intermolecular interaction under changes in rotational orientation of a complex molecule with sub-Angstrom resolution via the functionalization of a scanning probe tip. Using a novel visualization method we can directly observe the variation in the value and position of the minimum of the potential energy as the orientation of the molecules is varied. By comparison of our results to both simple analytical and *ab initio* simulations, we are able to show that the variation in binding energy across the molecule is dominated by the onset of repulsive interactions between the front-most parts of the molecules. Surprisingly, we also find that variation in the net attractive part of the potential due to rotation of the molecules is still dominated by the repulsive forces, and the majority of the molecule only adds a uniform background to the potential. We anticipate that similar experimental techniques to those described here could be utilized to intuitively visualise the reactivity across complex interatomic and intermolecular potentials, including molecules with polarized or hydrogen bonding end groups.

Methods

Experimental methods. Clean Si(111)-7 × 7 surfaces were prepared by flash annealing a silicon wafer to 1,200 °C, rapid cooling to 900 °C and then slow cooling to room temperature. A low coverage of C_{60} was prepared by depositing the molecules from a home made tantalum pocket deposition source onto the room temperature substrate. Post-deposition, the sample was transferred into the scan head of an Omicron Nanotechnology LT DFM operating in UHV at cryogenic temperatures, and left to cool to 5 K before imaging.

Commercial qPlus sensors (Omicron Nanotechnology GmbH) with electrochemically etched tungsten wire tips were introduced into the scan head without any further preparation. The sensors were first prepared on clean Si(111)-7 × 7 surfaces by standard STM techniques until good STM/DFM resolution was achieved. Single C_{60} molecules were transferred to the tip by close approach to surface-adsorbed molecules, and the functionalization of the tip was checked by inverse imaging of the tip adsorbed molecule on the surface adatoms (Supplementary Figs 2,4 and 5)[7]. In all experiments an oscillation amplitude (A_0) of 110 pm was used, and the tip-sample bias was set to 0 V. Three-dimensional Δf volumes over the molecules were collected via the 'slice' method[17] and site specific (short-range) Δf values were extracted using the 'on-off' method[18,19] then converted to potentials using the Sader–Jarvis algorithm[20]. Due to the long acquisition times required, residual thermal drift and piezoelectric creep were corrected using a custom atom-tracking and scripting setup[21,22]. Further details on the experimental setup, data processing steps and additional experimental data sets may be found in the Supplementary Methods.

Flexible tip model and simulated spectroscopy procedure. To simulate DFM images, we adapted the method proposed by Hapala et al.[12] to model the interaction between a sample and a CO-functionalized DFM tip. In our simulation the functionalized tip is assumed to consist of a tip base (outermost atom of the tip) and a probe. The probe is the flexible end of the model tip, and is allowed to move around the tip base. In our simulation, the probe is a C_{60} molecule consisting of 60 carbon atoms acting as a single effective probe particle attached to the tip base (Supplementary Fig. 6). Each atom in the probe experiences three forces; (i) a Lennard–Jones (L–J) force due to the tip base, (ii) a sum of all pairwise L–J forces due to interactions with atoms in the sample and (iii) a lateral harmonic force from the tip base. The net force on the probe is calculated by summing up all the forces experienced by each atom on the probe. The L–J interactions between atoms α and β are written as

$$\mathbf{F}_{\alpha\beta}(\mathbf{R}) = 12\epsilon_{\alpha\beta}\mathbf{R}\left(\frac{r_{\alpha\beta}^{12}}{r^{14}} - \frac{r_{\alpha\beta}^{6}}{r^{8}}\right) \quad (1)$$

$$U_{\alpha\beta}(r) = \epsilon_{\alpha\beta}\left(\frac{r_{\alpha\beta}^{12}}{r^{12}} - \frac{2r_{\alpha\beta}^{6}}{r^{6}}\right), \quad (2)$$

where $r = |\mathbf{R}|$ is the distance between atoms α and β, $\epsilon_{\alpha\beta} = \sqrt{\epsilon_\alpha\epsilon_\beta}$ is the pair-binding energy and $r_{\alpha\beta} = r_\alpha + r_\beta$ is the equilibrium separation of the two atoms with ϵ_α and r_α being the atomic parameters. In our calculations the L–J parameters for the carbon atoms were set to $\epsilon_\alpha = 2.5$ meV and $r_\alpha = 1.966$ Å to ensure consistency with the work of Girifalco et al.[16] and our own earlier work[7]. For the tip base a value of $r_\alpha = 5.0$ Å was chosen, in order to take into account the larger size of the C_{60} molecule. The probe lateral stiffness and apex L–J parameter were set to $k_{xy} = 0.5$ N m^{-1} and $\epsilon_\alpha = 1,000$ meV, respectively (Supplementary Figs 14–15). We acquired the simulation data by scanning the sample laterally with a step of Δx, $\Delta y = 0.1$ Å. At each lateral position we placed the tip base at an initial separation $z_0 = 22$ Å from the surface molecule and approached the sample (in our simulations another C_{60} molecule) in steps of $\Delta z = 0.1$ Å until $z = 17.5$ Å allowing the probe position to be relaxed at each step due to the combined force of the sample and tip base. Note, however, that for ease of comparison to the DFT simulations all molecular separations discussed in the paper are given relative to the initial vertical core-core separation of the probe C_{60} from the surface C_{60}.

Density functional theory. DFT calculations were performed using the same initial high symmetry geometry (as described in main text) as the L–J simulations using the open source CP2K/Quickstep code[23,24] utilising a hybrid Gaussian and plane-wave method[25]. Goedecker, Teter and Hutter pseudopotentials[26] and the Perdew Burke Ernzerhof generalized gradient approximation method[27] were used with a 300 Ry plane-wave energy cutoff. To account for dispersion interactions we employed the Grimme DFT-D3 method[28], which well reproduced the C_{60}–C_{60} pair potential (Supplementary Fig. 19) A double-zeta Gaussian basis set plus polarization (DZVP-MOLOPT)[29] was used with a force convergence criterion for geometry relaxation of 0.05 eV Å^{-1}. Geometry relaxation was carried out by allowing all atoms to relax other than those hexagonal faces of each molecule furthest apart from one another (to simulate attachment to the surface/tip).

References

1. Gross, L., Mohn, F., Moll, N., Liljeroth, P. & Meyer, G. The chemical structure of a molecule resolved by atomic force microscopy. *Science* **325**, 1110–1114 (2009).
2. Jarvis, S. P. Resolving intra-and inter-molecular structure with non-contact atomic force microscopy. *Int. J. Mol. Sci.* **16**, 19936 (2015).
3. Schull, G., Frederiksen, T., Arnau, A., Sanchez-Portal, D. & Berndt, R. Atomic-scale engineering of electrodes for single-molecule contacts. *Nat. Chem.* **6**, 23–27 (2010).
4. Schull, G., Frederiksen, T., Brandbyge, M. & Berndt, R. Passing current through touching molecules. *Phys. Rev. Lett.* **103**, 11–14 (2009; 0910.1281).
5. Hauptmann, N. et al. Force and conductance during contact formation to a C_{60} molecule. *New. J. Phys.* **14**, 073032 (2012).
6. Wagner, C. et al. Non-additivity of molecule-surface van der Waals potentials from force measurements. *Nat. Commun.* **5**, 5568 (2014).
7. Chiutu, C. et al. Precise orientation of a single C_{60} molecule on the tip of a scanning probe microscope. *Phys. Rev. Lett.* **108**, 268302 (2012).
8. Sun, Z., Boneschanscher, M. P., Swart, I., Vanmaekelbergh, D. & Liljeroth, P. Quantitative atomic force microscopy with carbon monoxide terminated tips. *Phys. Rev. Lett.* **106**, 46104 (2011).
9. Zhang, J. et al. Real-space identification of intermolecular bonding with atomic force microscopy. *Science* **342**, 611–614 (2013).
10. Sweetman, A. M. et al. Mapping the force field of a hydrogen-bonded assembly. *Nat. Commun.* **5**, 3931 (2014).
11. Hämäläinen, S. K. et al. Intermolecular contrast in atomic force microscopy images without intermolecular bonds. *Phys. Rev. Lett.* **113**, 186102 (2014).
12. Hapala, P. et al. Mechanism of high-resolution STM/AFM imaging with functionalized tips. *Phys. Rev. B* **90**, 085421 (2014).
13. Moreno, C., Stetsovych, O., Shimizu, T. K. & Custance, O. Imaging three-dimensional surface objects with submolecular resolution by atomic force microscopy. *Nano Lett.* **15**, 2257 (2015).
14. Mohn, F., Schuler, B., Gross, L. & Meyer, G. Different tips for high-resolution atomic force microscopy and scanning tunneling microscopy of single molecules. *Appl. Phys. Lett.* **102**, 073109 (2013).
15. Rode, S., Schreiber, M., Kühnle, A. & Rahe, P. Frequency-modulated atomic force microscopy operation by imaging at the frequency shift minimum: The dip-df mode. *Rev. Sci. Instrum.* **85**, 043707 (2014).
16. Girifalco, L. a. Interaction potential for carbon (C_{60}) molecules. *J. Phys. Chem.* **95**, 5370–5371 (1991).
17. Neu, M. et al. Image correction for atomic force microscopy images with functionalized tips. *Phys. Rev. B* **89**, 205407 (2014).
18. Lantz, M. A. et al. Quantitative measurement of short-range chemical bonding forces. *Science* **291**, 2580–2583 (2001).
19. Sweetman, A. & Stannard, A. Uncertainties in forces extracted from non-contact atomic force microscopy measurements by fitting of long-range background forces. *Beilstein J. Nanotechnol.* **5**, 386–393 (2014).
20. Sader, J. E. & Jarvis, S. P. Accurate formulas for interaction force and energy in frequency modulation force spectroscopy. *Appl. Phys. Lett.* **84**, 1801 (2004).
21. Abe, M. et al. Room-temperature reproducible spatial force spectroscopy using atom-tracking technique. *Appl. Phys. Lett.* **87**, 173503 (2005).
22. Rahe, P. et al. Flexible drift-compensation system for precise 3D force mapping in severe drift environments. *Rev. Sci. Instrum.* **82**, 063704 (2011).
23. Hutter, J, Iannuzzi, M., Schiffmann, F. & VandeVondele, J. CP2K: atomistic simulations of condensed matter systems. *WIREs Comput. Mol. Sci.* **4**, 1525 (2013).
24. VandeVondele, J. et al. Quickstep: Fast and accurate density functional calculations using a mixed gaussian and plane waves approach. *Comput. Phys. Commun.* **167**, 103–128 (2005).
25. Lippert, G., Jurg, H. & Parinello, M. A hybrid gaussian and plane wave density functional scheme. *Mol. Phys.* **92**, 477–488 (1997).
26. Goedecker, S., Teter, M. & Hutter, J. Separable dual-space gaussian pseudopotentials. *Phys. Rev. B* **54**, 1703–1710 (1996).
27. Perdew, J. P., Burke, K. & Ernzerhof, M. Generalized gradient approximation made simple. *Phys. Rev. Lett.* **77**, 3865–3868 (1996).
28. Grimme, S., Antony, J., Ehrlich, S. & Krieg, H. A consistent and accurate ab initio parametrization of density functional dispersion correction (dft-d) for the 94 elements h-pu. *J. Chem. Phys.* **132**, 154104 (2010).
29. VandeVondele, J. & Hutter, J. Gaussian basis sets for accurate calculations on molecular systems in gas and condensed phases. *J. Chem. Phys.* **127**, 114105 (2007).

Acknowledgements

We gratefully acknowledge valuable discussions with P. Hapla and S. Hämäläinen regarding implementation of the L–J model, and important comments on the manuscript by J. Leaf. A.S. gratefully acknowledges the support of the Leverhulme Trust via fellowship ECF-2013-525. S.P.J. thanks the Engineering and Physical Sciences Research Council (EPSRC) for the award of fellowship EP/J500483/1, and The Leverhulme Trust for the award of fellowship ECF-2015-005, respectively. P.J.M. thanks EPSRC and the Leverhulme Trust, respectively, for Grants No. EP/G007837/1 and F00/114 BI. M.R.A. acknowledges funding from the University of Nottingham via the Vice-Chancellor's Scholarship for Research. P.R. received funding from the People Programme (Marie Curie Actions) of the European Union's Seventh Framework Programme (FP7/2007–2013) under REA grant agreement no. [628439]. We also acknowledge the support of the University of Nottingham High Performance Computing Facility (in particular, Dr Colin Bannister).

Author contributions

A.S. and P.M. conceived and designed the experiments. A.S. carried out the experiments and analysed the experimental data. A.S., M.A.R., J.L.D and S.P.J. designed and ran the L–J calculations. S.P.J. performed and analysed the DFT calculations. P.R. designed the atom tracking equipment and scripting. A.S. and P.M. wrote the paper. All authors read and commented on the final manuscript.

Additional information

Competing financial interests: The authors declare no competing financial interests.

Pure and stable metallic phase molybdenum disulfide nanosheets for hydrogen evolution reaction

Xiumei Geng[1], Weiwei Sun[1], Wei Wu[1], Benjamin Chen[2], Alaa Al-Hilo[1], Mourad Benamara[3], Hongli Zhu[4], Fumiya Watanabe[1], Jingbiao Cui[5] & Tar-pin Chen[1]

Metallic-phase MoS_2 (M-MoS_2) is metastable and does not exist in nature. Pure and stable M-MoS_2 has not been previously prepared by chemical synthesis, to the best of our knowledge. Here we report a hydrothermal process for synthesizing stable two-dimensional M-MoS_2 nanosheets in water. The metal-metal Raman stretching mode at $146\,cm^{-1}$ in the M-MoS_2 structure, as predicted by theoretical calculations, is experimentally observed. The stability of the M-MoS_2 is associated with the adsorption of a monolayer of water molecules on both sides of the nanosheets, which reduce restacking and prevent aggregation in water. The obtained M-MoS_2 exhibits excellent stability in water and superior activity for the hydrogen evolution reaction, with a current density of $10\,mA\,cm^{-2}$ at a low potential of $-175\,mV$ and a Tafel slope of $41\,mV$ per decade.

[1]Department of Physics and Astronomy, University of Arkansas at Little Rock, 2801 South University Avenue, Little Rock, Arkansas 72204, USA. [2]Department of Physics, University at Buffalo, Buffalo, New York 14260, USA. [3]Institute for Nanoscale Materials Science and Engineering, University of Arkansas, Fayetteville, Arkansas 72701, USA. [4]Department of Mechanical and Industrial Engineering, Northeastern University, Boston, Massachusetts 02115, USA. [5]Department of Physics and Materials Science, University of Memphis, Memphis, Tennessee 38152, USA. Correspondence and requests for materials should be addressed to X.G. (email: gengxiu0758@gmail.com) or to J.C. (email: jcui@memphis.edu) or to T.-p.C. (email: txchen@ualr.edu).

The metallic-phase MoS_2 (M-MoS_2) is a single-layer S-Mo-S structure in which each molybdenum atom is surrounded by six sulfur atoms in an octahedral lattice. With dense active sites and an electronic conductivity that is six orders of magnitude higher than that of the semiconductor phase of MoS_2 (S-MoS_2) (ref. 1), M-MoS_2 has emerged as a promising candidate for a broad range of applications and is expected to exhibit better performance than its semiconducting counterpart, in particular in the hydrogen evolution reaction (HER) and as a photocatalyst and supercapacitor[2–5]. These attractive properties of M-MoS_2 are offset by the fact that a pure phase of M-MoS_2 is very challenging to prepare, because it is highly unstable. Over the past few decades, many efforts to develop a process to prepare stable and highly pure M-MoS_2 have largely proved futile. A complicated method for the preparation of metastable M-MoS_2 using lithium intercalation of S-MoS_2 has been reported[5–8]. However, the reported M-MoS_2 suffered from coexistence with S-MoS_2 in a relative proportion of 50–80% (refs 3,9). In addition, the intermediary Li_xMoS_2 and intercalator of n-butyllithium are both dangerous materials that may self-heat[8] and are highly pyrophoric in air[5]. The preparation procedures are also tedious and can take as long as 3 days[10,11]. The previously mentioned M-MoS_2 easily transformed to S-MoS_2 due to S–S van der Waals interactions[12–14]. Therefore, an efficient synthesis method for the production of a large quantity of stable and pure M-MoS_2 is still highly desirable.

M-MoS_2 could be an intermediary state during the synthesis of S-MoS_2, which is a stable form in many chemical reactions[15,16]. This approach provides a possible method for capturing the metastable M-MoS_2 even though it is thermodynamically unstable and may rapidly transform to S-MoS_2 at elevated temperatures. In fact, S-MoS_2 can transform into an octahedral structure of M-MoS_2 under a high pressure of 35 GPa (refs 17–19), which implies that the metallic phase of MoS_2 may be achievable at a relatively low temperature, a specific pressure and a proper chemical environment. Therefore, we employ octahedral MoO_3 as the starting material in a pressurized hydrothermal process, to directly grow highly pure M-MoS_2 in water. The as-prepared M-MoS_2 is highly stable in water due to the reduced stacking of the layered materials, which are separated by water molecules. Urea, which is a weak reducing agent, plays a key role in the formation of M-MoS_2. The structure of octahedral MoO_3 can be maintained under acidic conditions. To satisfy the expected growth conditions, thioacetamide is chosen as the sulfur source and the reaction is maintained at a low pH value (pH ~4). The synthesis of M-MoS_2 is performed at 200 °C by mixing MoO_3, thioacetamide and urea in an autoclave. The as-prepared M-MoS_2 is pure and stable with high activity for the HER.

Results

Highly pure metallic phase MoS_2. Figure 1a shows a typical scanning electron microscopic (SEM) image of the as-prepared M-MoS_2 sample. The MoS_2 nanosheets are 100 nm in size and a few nanometres in thickness. The morphology of S-MoS_2 is similar to that of M-MoS_2 on the nanoscale (Supplementary Fig. 1). The results from energy dispersive X-ray spectroscopy indicated that the S to Mo atomic ratio of the as-prepared M-MoS_2 is ~2.05 (Supplementary Fig. 2). The most notable difference between M-MoS_2 and S-MoS_2 is the symmetry of the sulfur in their structures. The structural variations result in significant differences in their characteristic Raman features. Figure 1b shows the Raman spectra of M-MoS_2 and S-MoS_2 synthesized at 200 °C and 240 °C, respectively. A strong Raman band is observed at $146\ cm^{-1}$ and attributed to Mo–Mo stretching vibrations in M-MoS_2. Calculations have predicted this band at the K points of the Brillouin zone for octahedrally

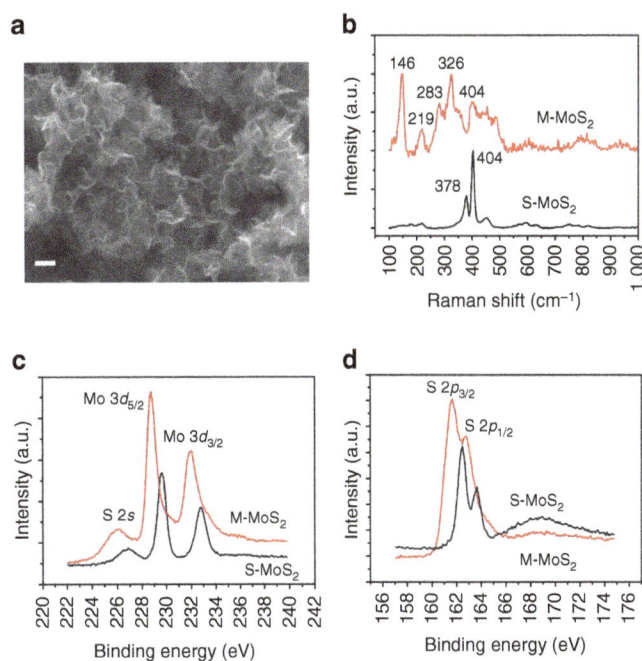

Figure 1 | Morphology of M-MoS_2 and phase identification of M-MoS_2 and S-MoS_2. (**a**) SEM image of M-MoS_2 showing the nanosheet structures. Scale bar, 100 nm. (**b**) Raman shift of M-MoS_2 and S-MoS_2. (**c**,**d**) High-resolution X-ray photoelectron spectroscopy (XPS) spectra of Mo $3d$ (**c**) and S $2p$ (**d**) for M-MoS_2 and S-MoS_2.

coordinated distorted MoS_2 (ref. 12). Our results confirm the existence of this band in M-MoS_2. The Raman shifts (that is, ~219, 283 and $326\ cm^{-1}$) are also associated with the phonon modes in M-MoS_2. S-MoS_2 exhibits typical Raman shifts of ~378 and $404\ cm^{-1}$ for E^1_{2g} and A_{1g}, respectively, which are substantially different from those of M-MoS_2.

The phase identification of the synthesized M-MoS_2 and S-MoS_2 was further studied using X-ray photoelectron spectroscopy and the results are shown in Fig. 1c,d. The Mo $3d$ spectra consist of peaks located at approximately 228.7 and 231.8 eV, which correspond to the $3d_{5/2}$ and $3d_{3/2}$ components, respectively, of Mo^{4+} in M-MoS_2. However, the Mo $3d$ peaks of the M-MoS_2 samples also exhibit weak shoulders at the positions as those in S-MoS_2 due to the transformation of M-MoS_2 in vacuum and X-ray illumination. The details for the peak deconvolution are provided in Supplementary Fig. 3 and Supplementary Note 1. The Mo $3d$ peaks of M-MoS_2 shifted to lower binding energies by ~1 eV with respect to the corresponding peaks in S-MoS_2. This result is consistent with the previously obtained relaxation energy of 1 eV for M-MoS_2 derived from S-MoS_2 (refs 20,21). Similarly, the S $2p$ peaks of M-MoS_2 were located at ~161.6 and 162.7 eV, and associated with S $2p_{3/2}$ and $2p_{1/2}$, respectively, which are also ~1 eV lower than the corresponding peaks in S-MoS_2. It is important to note that the prominent S peak located at ~168 eV was absent, indicating that the sulfur atoms remain unoxidized in M-MoS_2. However, this peak was observed in S-MoS_2 after annealing at 240 °C.

Morphologies and optical properties of MoS_2. Figure 2a shows optical images of the two types of MoS_2 dispersed in water. M-MoS_2 does not appear to exhibit bulk aggregation, whereas S-MoS_2 aggregates at the bottom of the solution. This result may be because M-MoS_2 possesses hydrophilic surfaces, whereas S-MoS_2 possesses relatively hydrophobic surfaces[5,22]. After sonication, both M-MoS_2 and S-MoS_2 can uniformly

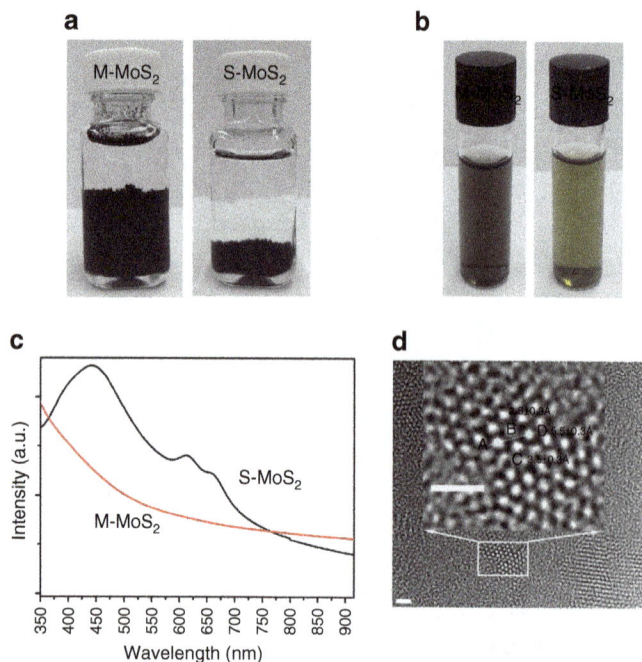

Figure 2 | Morphology and property of M-MoS$_2$ and S-MoS$_2$.
(**a**) As-prepared M-MoS$_2$ and S-MoS$_2$ in water. (**b**) Suspension of M-MoS$_2$ and S-MoS$_2$ in water. (**c**) Ultraviolet–visible absorption of M-MoS$_2$ and S-MoS$_2$. (**d**) HRTEM image of M-MoS$_2$. The region indicated by the squares is enlarged to show the atomic structure of M-MoS$_2$. Scale bar, 1 nm.

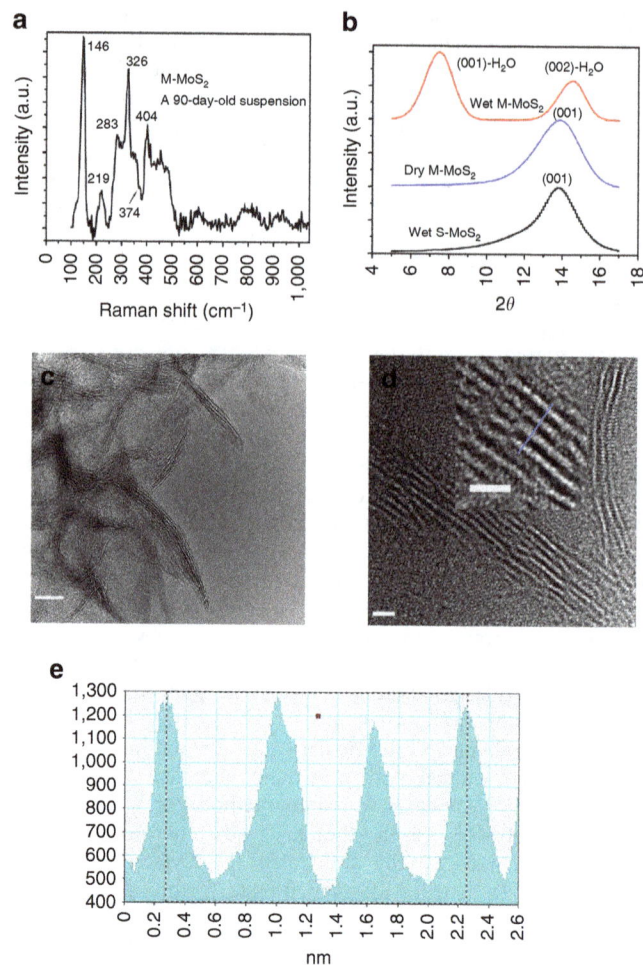

Figure 3 | Raman and structural characterization to investigate the stability of M-MoS$_2$. (**a**) Raman shift of M-MoS$_2$ stored for 90 days at room temperature. (**b**) XRD patterns of wet M-MoS$_2$, dry M-MoS$_2$ and wet S-MoS$_2$. (**c**) Low-resolution TEM image of M-MoS$_2$. Scale bar, 10 nm. (**d**) HRTEM image of M-MoS$_2$. The region indicated by the square is enlarged to show the layered structure of M-MoS$_2$. Scale bar, 2 nm. (**e**) Line scan of the HRTEM image indicated by the blue line in **d**, indicating a layer-to-layer spacing of 0.65 nm.

disperse in water (Fig. 2b). However, M-MoS$_2$ was grey colour, whereas S-MoS$_2$ was green colour. These colours were associated with their optical properties, which were confirmed by optical absorption spectroscopy. Figure 2c shows the ultraviolet–visible absorption spectra of M-MoS$_2$ and S-MoS$_2$ in water. Two typical absorption peaks located at 613 and 660 nm were observed for S-MoS$_2$ and these peaks are associated with the energy split from the valence band spin–orbital coupling in S-MoS$_2$ with large lateral dimensions[23]. In addition, another optical absorption band located at ~442 nm was observed. This band most probably results from the quantum effect of smaller lateral-sized S-MoS$_2$ nanosheets. The absorption spectrum of M-MoS$_2$ has no salient absorption bands but a monotonic change that is indicative of its metallic property.

The atomic structure of M-MoS$_2$ was investigated using a high-resolution scanning transmission electron microscope (HRTEM) at different magnifications (Supplementary Fig. 4) and the high-magnification image is shown in Fig. 2d. The hexagonal atomic arrangement of Mo indicates that each individual nanosheet consists of single crystal domains. Mo atoms on the basal plane have a Mo–Mo distance from A to D of 5.5 ± 0.3 Å, A to B of 2.8 ± 0.3 Å and A to C of 3.5 ± 0.3 Å. However, the crystal lattice may convert into an amorphous structure under the high-energy electron beam of TEM as observed by Lukowski et al.[2].

Stability of the M-MoS$_2$. The as-prepared M-MoS$_2$ nanosheets in water exhibited good stability over time. We believe that the high purity of M-MoS$_2$ plays an important role in its stability. 1T-MoS$_2$ (metallic phase), which was derived from 2H-MoS$_2$ (semiconductor phase), consists of electronic heterostructures of 2H-MoS$_2$ and 1T-MoS$_2$ in individual layers of MoS$_2$ (ref. 9). The coexistence of semiconductor MoS$_2$ with a metallic phase in a single layer resulted in an expedited transformation of 1T back to 2H MoS$_2$. Therefore, the structure was reported to be stable for 12 days with the observation of the 2H structure in water for

chemically exfoliated MoS$_2$ (ref. 12). The samples obtained in this study were stored in water for more than 90 days at room temperature and no obvious change in the Raman shift was observed (Fig. 3a). The characteristic peaks of the stored M-MoS$_2$ were similar to those observed for the freshly prepared samples. This observation indicates that the as-prepared M-MoS$_2$ was highly stable in water. It is important to note that a traceable transformation of M-MoS$_2$ to S-MoS$_2$ may occur, as suggested by the very weak shoulder at 374 cm^{-1}. The pure metallic phase in the individual layers contributes to the stability of the material. A comparison of the Raman spectra of M-MoS$_2$ stored under different conditions and S-MoS$_2$ is shown in Supplementary Fig. 5.

Another possible stability mechanism may arise from the monolayer of water molecules that was adsorbed on both sides of the M-MoS$_2$ nanosheets. This layer prevents aggregation and stacking, and maintains the octahedral structure in M-MoS$_2$. The existence of the water molecule layer was confirmed by XRD measurements (Fig. 3b). The peak at 7.5°, which corresponds to a spacing of 11.8 Å, was identified as (001)-H$_2$O for M-MoS$_2$-H$_2$O, which contains a bilayer of water molecules between the adjacent layers of MoS$_2$ nanosheets[24]. As the layer thickness of MoS$_2$ was

$\sim$6.2 Å and the thickness of a monolayer of water was 2.8 Å, the total thickness of a MoS_2 layer surrounded by a water bilayer would be 11.8 Å. This result was previously confirmed by XRD^{25}. The peak located at 14.6° was assigned to the second-order reflection of $M-MoS_2-H_2O$.

XRD was also performed on the $M-MoS_2$ nanosheets after they were dried for 24 h in a vacuum chamber at a pressure of 6 mTorr. The XRD pattern of the dried $M-MoS_2$ is similar to that of $S-MoS_2$ with a peak at 13.88° (Fig. 3b), which corresponds to the stacked $M-MoS_2$ multilayers. This result suggests that the spacing between the Mo planes for $M-MoS_2$ was 6.4 ± 0.1 Å. This spacing is larger than that of bulk MoS_2 (that is, 6.15 Å). The diffraction peak at 7.5° disappeared due to desorption of the water molecules on the nanosheet surfaces. The TEM and HRTEM images indicate that $M-MoS_2$ was composed of single-layer or several-layer structures (Fig. 3c,d). The lattice fringe spacing of $M-MoS_2$ was measured to be 1.97 nm for a four-layer nanosheet, which produced a Mo–Mo spacing of 6.5 ± 0.1 Å (Fig. 3e). This result is consistent with the value obtained from the XRD measurements. The increased interlayer spacing in $M-MoS_2$ may contribute to its stability due to the weakened S–S van der Waals interactions and reduced restacking probability.

$M-MoS_2$ contained adsorbed water molecules, which was confirmed by XRD measurements. However, this phenomenon was not observed in $S-MoS_2$ nanostructures. This difference between the two phases of MoS_2 reflects the different surface properties (that is, $M-MoS_2$ is hydrophilic, whereas $S-MoS_2$ is hydrophobic). These properties are determined by the atomic structures of $M-MoS_2$ and $S-MoS_2$. The Mo atom in $M-MoS_2$ is located in an octahedral S coordination, whereas the Mo atom in $S-MoS_2$ coordinated by six S atoms in a trigonal prismatic arrangement with S acting as a hydrophobic site and Mo serving as a hydrophilic site[26]. These different surface properties were also observed regardless of whether they were dispersed in water or deposited as thin films on substrates (Supplementary Fig. 6). Contact angle measurements were conducted on both the $M-MoS_2$ and $S-MoS_2$ films. A contact angle of 25° was observed for $M-MoS_2$ and an angle of 118° was determined for $S-MoS_2$, confirming the highly hydrophilic surface of $M-MoS_2$. The details are provided in Supplementary Fig. 7.

Metallic phase MoS_2 for HER. The different surface properties, atomic structure and morphology of $M-MoS_2$ and $S-MoS_2$ on substrates led to significant differences in the HER. Hydrogen production measurements were conducted for the $M-MoS_2$ and $S-MoS_2$ nanosheets deposited on glassy carbon electrodes using a three-electrode cell in a 0.5-M sulfuric acid electrolyte. Figure 4a,b show the polarization curves and corresponding Tafel plots of the $M-MoS_2$ and $S-MoS_2$ nanosheets compared with those of Pt. It is important to note that the Tafel plot employs a logarithmic current density axis. A low potential of − 175 mV at a current density of $10 \, mA \, cm^{-2}$ and a Tafel slope of 41 mV per decade were obtained for $M-MoS_2$. This observed potential for hydrogen production represents the lowest value among the reported experimental data for the use of only MoS_2 materials. The long-term stability of the $M-MoS_2$ electrode for the HER was investigated. After 1,000 cycles of continuous operation, the exfoliated MoS_2 nanosheets exhibited <12% decay in the electrocatalytic current density (Fig. 4c). The advantage of using $M-MoS_2$ nanosheets becomes more apparent through electrochemical impedance spectroscopy analysis. The electrochemical impedance spectroscopy was performed under the HER conditions. The Nyquist plots are shown in Fig. 4d and revealed a dramatically decreased charge transfer resistance for the $M-MoS_2$ nanosheets ($\sim$1 Ω) relative to the $S-MoS_2$ nanosheets ($\sim$50 Ω). Furthermore, we measured the resistivity of the $M-MoS_2$ (300 nm

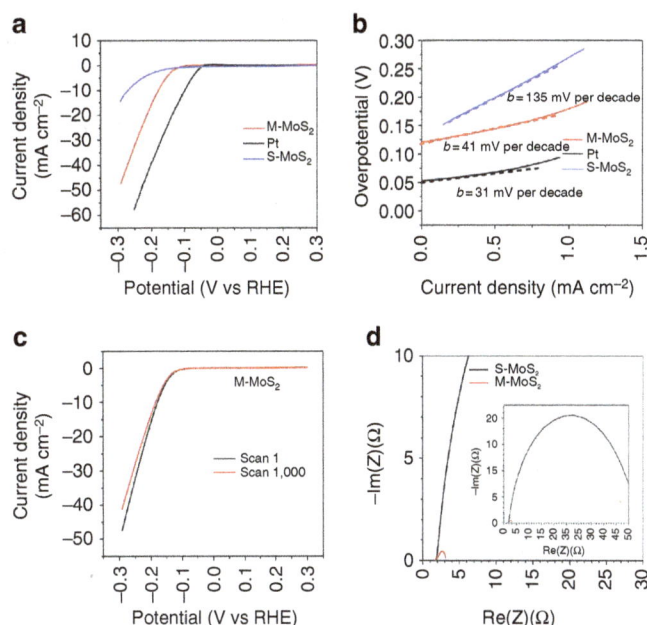

Figure 4 | HER activity of the synthesized MoS_2 nanosheets. (a) Polarization curves of the $M-MoS_2$ and $S-MoS_2$ nanosheets. **(b)** Corresponding Tafel plots obtained from the polarization curves. The Tafel slopes were $\sim$41 and $\sim$135 mV per decade for $M-MoS_2$ and $S-MoS_2$, respectively. **(c)** Polarization curves of the $M-MoS_2$ nanosheets after 1 and 1,000 cycles of continuous operation. **(d)** Nyquist plots of $M-MoS_2$ and $S-MoS_2$.

thick) and $S-MoS_2$ (340 nm thick) films using a four-probe method under ambient conditions. The sheet resistances are $1.61 \times 10^4 \, \Omega/\square$ for $M-MoS_2$ and $2.26 \times 10^9 \, \Omega/\square$ for $S-MoS_2$, which correspond to a resistivity of $0.482 \, \Omega \cdot cm$ for $M-MoS_2$ and $7.68 \times 10^4 \, \Omega \cdot cm$ for $S-MoS_2$. This result indicates that $M-MoS_2$ has a conductivity that is five orders of magnitude higher than that of $S-MoS_2$.

It is important to note that the growth of $M-MoS_2$ and the HER measurements are highly reproducible. The growth process of $M-MoS_2$ failed only once in 96 trials. The failure was caused by washing the autoclave with HNO_3 and we believe that a trace amount of residual HNO_3 may have oxidized the products via different chemical reactions. The autoclave was typically cleaned by sonication in water. We have performed the HER 62 times using our synthesized $M-MoS_2$ and all of the trials yielded results similar to those shown in Fig. 4a,b. The exceptional performance of $M-MoS_2$ for hydrogen production may be due to the high purity, dense active sites, good conductivity and excellent hydrophilic property of $M-MoS_2$. By contrast, $S-MoS_2$ exhibited little HER activity with a potential of − 274 mV at $10 \, mA \, cm^{-2}$ and a Tafel slope of 135 mV per decade. The poor performance of $S-MoS_2$ in the HER was due to few active sites, aggregation and its hydrophobic surface. The exchange current density of $M-MoS_2$, $S-MoS_2$ and Pt is discussed in Supplementary Discussion. Although $M-MoS_2$ still lags behind Pt, improvement in the HER activity can be accomplished by enhancing the conductivity of the nanoscale $M-MoS_2$. Therefore, metallic MoS_2 exhibits great potential for use as a replacement for Pt in the HER as a low-cost and highly active catalyst for practical applications.

Discussion

The mechanism underlying the formation of the stable metallic phase of MoS_2 is directly associated with the weak reducer urea, which can precisely and effectively reduce MoO_3 to $M-MoS_2$ as shown in equation (1). The absence of a urea reducer while maintaining other growth conditions would result in a product dominated by amorphous MoS_3, as shown in equation (2) (see Supplementary Methods and Supplementary Fig. 8). Other reducers, such as hydrazine, reduce MoO_3 to $S-MoS_2$ only due to the relatively strong electron donating ability of hydrazine.

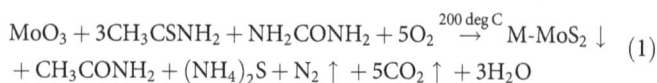

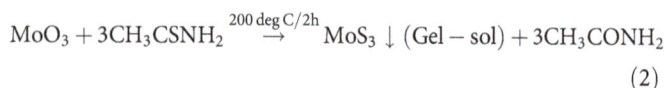

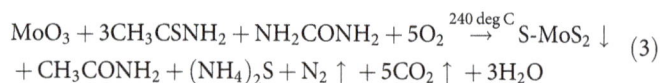

$$MoO_3 + 3CH_3CSNH_2 + NH_2CONH_2 + 5O_2 \overset{200\,\text{deg C}}{\rightarrow} M\text{-}MoS_2 \downarrow \quad (1)$$
$$+ CH_3CONH_2 + (NH_4)_2S + N_2 \uparrow + 5CO_2 \uparrow + 3H_2O$$

$$MoO_3 + 3CH_3CSNH_2 \overset{200\,\text{deg C}/2h}{\rightarrow} MoS_3 \downarrow (Gel-sol) + 3CH_3CONH_2$$
$$(2)$$

$$MoO_3 + 3CH_3CSNH_2 + NH_2CONH_2 + 5O_2 \overset{240\,\text{deg C}}{\rightarrow} S\text{-}MoS_2 \downarrow \quad (3)$$
$$+ CH_3CONH_2 + (NH_4)_2S + N_2 \uparrow + 5CO_2 \uparrow + 3H_2O$$

During the synthesis of $M-MoS_2$, the three most important factors are solvent, pH and temperature. First, water serves as a solvent and is adsorbed on the surface of $M-MoS_2$ to form a water molecular layer that prevents layer stacking. When an organic solvent, such as ethanol or dimethylformamide, was added to the reaction, no desirable product was produced (see the Supplementary Methods and Supplementary Fig. 9). No solid product was found in the autoclave when a 1:1 ratio of water to ethanol was used. Second, the pH controls the steric structure of the starting materials and the formation of $M-MoS_2$. When the pH value was ≤ 4, the octahedral coordination of MoO_3 can be maintained, which may lead to the formation of $M-MoS_2$. At higher pH values (pH 8) achieved through the addition of ammonia, the MoO_3 structure converts into MoO_4^{2-} tetrahedral centres. This change causes the reaction to yield no desirable products. Third, the temperature also plays a key role in the formation of $M-MoS_2$. In this study, pure $M-MoS_2$ was obtained at 200 °C and $S-MoS_2$ was formed at 240 °C in the reaction listed in equation (3).

Based on these discussions, the metastable $M-MoS_2$ intermediate can be formed and stabilized in a pressurized hydrothermal process with a suitable reducer and temperature. The proposed mechanism involves the formation of a crystal phase of $M-MoS_2$ from amorphous clustered MoS_3 with S_2^- edge bonds reduced by urea. In fact, many investigations have revealed the structural evolution from MoS_3 to $S-MoS_2$ without the MoS_2 metallic phase during the evolution process[27-29]. We believe that the high temperature and strong reducer used in the previous studies may have only resulted in the formation of semiconductor MoS_2. This result was also confirmed in this study, because only $S-MoS_2$ was obtained when the temperature was increased to 240 °C. The evolution process for the formation of $M-MoS_2$ is shown in Fig. 5, which illustrates the growth of MoS_3 from one cluster to two and finally to a nanosheet. The central part possesses the structure of $M-MoS_2$ and the S_2^- bond on the edge can be removed by urea to continue the growth to form a single layer of $M-MoS_2$ (ref. 30). A control experiment was carried out by stopping the growth after 90 min, which resulted in a mixture of MoS_3 and $M-MoS_2$ in the solution. The existence of MoS_3 and its corresponding peak at $120\,cm^{-1}$ peak was confirmed by Raman spectroscopy[31]. Furthermore, a Raman band located at $548\,cm^{-1}$, which was due to the S_2^- stretching vibration in MoS_3, was observed. Energy dispersive X-ray spectroscopy measurements indicate a ratio of 1:3 for Mo and S (See Supplementary Fig. 10b). The Mo–Mo vibration mode at $146\,cm^{-1}$ starts to appear and become more prominent as growth time increased, indicating the transformation of MoS_3 to $M-MoS_2$ (see Supplementary Fig. 10c). After continuing the reaction for 6 h, only $M-MoS_2$ was observed in the product. At this point, the Raman peak located at $120\,cm^{-1}$ disappeared and

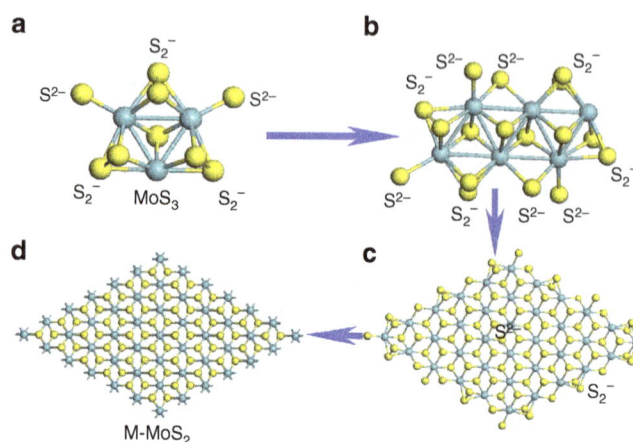

Figure 5 | Schematic illustrations of the evolution process of M-MoS₂ from MoS₃. (a) One MoS_3 cluster with S^{2-} and S_2^-. **(b)** Combination of two MoS_3 clusters with the breaking of S_2^-. **(c)** Several MoS_3 cluster forming MoS_2 with S^{2-} in the centre and MoS_3 with S_2^- at the edge. **(d)** Stable $M-MoS_2$ crystal structure.

the intensity of the $146\,cm^{-1}$ peak reached its maximum. This result indicates the complete conversion of MoS_3 to $M-MoS_2$.

A hydrothermal process has been used to grow $S-MoS_2$. To the best of our knowledge, the preparation of $M-MoS_2$ using a hydrothermal method has not been previously reported. We believe that the growth conditions used in this study play a key role in the formation and stabilization of $M-MoS_2$. $M-MoS_2$ is an unstable phase that can easily transform to $S-MoS_2$. Therefore, the octahedral coordination of $M-MoS_2$ must be maintained during deposition. For this reason, we used MoO_3 as the starting material, because it possesses the same octahedral structure as $M-MoS_2$. The metallic phase can be obtained when the crystal structure can be maintained during the conversion process from MoO_3 to MoS_2. Urea, which is a weak reducer, is able to maintain the crystal structure at a low pH value and optimum temperature (200 °C) for hydrothermal growth. As mentioned in the previous sections, only $S-MoS_2$ was obtained at a higher growth temperature of 240 °C, because the octahedral structure of $M-MoS_2$ cannot be maintained under the growth conditions used. Once the $M-MoS_2$ nanosheets were formed at 200 °C, the adsorption of water molecules on their surfaces due to the hydrophilic nature of the surface provides an environment to maintain the octahedral structure and preventing restacking of the layers. Therefore, $M-MoS_2$ can be stored in water for several months without significant transition to $S-MoS_2$. However, once $M-MoS_2$ is removed from water and dried in a vacuum, it will slowly convert to $S-MoS_2$. This observation indicates that the stability of $M-MoS_2$ is dependent on the environment. Although this study was focused on the influence of water, the environmental stability of $M-MoS_2$ under other conditions, such as different gases and solutions other than water, should be considered in future investigations.

In conclusion, highly pure and stable $M-MoS_2$ nanosheets were synthesized in solution using a rational and controllable approach. A unique Raman peak located at $146\,cm^{-1}$ due to $M-MoS_2$ was experimentally observed, which confirms the theoretical prediction. A growth mechanism was proposed and is supported by the experimental data. Excellent activity of $M-MoS_2$ for the HER was achieved and the material has the potential to replace Pt in practical applications. These results provide opportunities for studying the metallic phase of MoS_2, which has potential technological applications in renewable energy, energy storage and heterogeneous catalysis.

Methods

Preparation of M-MoS$_2$ and S-MoS$_2$. MoO$_3$ powder (CAS number is 1311-27-5) and urea (CAS number is 57-13-6) were purchased from the Sigma-Aldrich Company. Thioacetamide (CAS number is 62-55-5) was purchased from the Fisher Scientific Company. The autoclave (model number 4749) was ordered from the Parr Instrument Company.

Twelve milligrams of MoO$_3$, 14 mg of thioacetamide and 0.12 g of urea were dissolved in 10 ml of deionized water and stirred for 2 h. Then, the solution was placed in an autoclave and loaded into a furnace (from MTI), which has been heated to 200 °C. The temperature of the oven was maintained at 200 °C for 12 h. Next, the reaction was terminated by rapidly cooling the solution to room temperature by removing the autoclave from the oven. The prepared M-MoS$_2$ was collected and washed with deionized water several times, followed by storage in deionized water before use. The same procedure was used to synthesize S-MoS$_2$ at 240 °C. Two-dimensional nanosheets were observed and no significant difference was observed between M-Mo$_2$ and S-MoS$_2$ based on their nanoscale morphology.

Ultraviolet–visible–infrared spectroscopy. Ultraviolet–visible–infrared absorption spectroscopy of the M-MoS$_2$ and S-MoS$_2$ dispersions in water were recorded using a Perkin Elmer Lambda 25 spectrophotometer.

TEM observations. TEM and HRTEM images and electron diffraction were performed on a FEI Tecnai G2 F20 S-Twin microscope operated at an accelerating voltage of up to 200 kV. The TEM samples were prepared by sonication at 500 W for ~5 min and 25 µl of the supernatant were dropped onto holey carbon grids.

SEM observations. The SEM images were obtained using a field-emission gun SEM (Quanta 400 FEG FEI). The samples were dispersed in water and then dropped onto the Si/SiO$_2$ substrates.

X-ray photoelectron spectroscopy. X-ray photoelectron spectroscopy was conducted using an Axis Ultra DLD (Kratos) system. The vacuum of the chamber was 1×10^{-9} Torr. A monochromatic aluminum K_α source with a source power of 150 W (15 kV × 10 mA) was used. The pass energy was 160 eV for wide scans and 40 eV for narrow scans.

Raman spectroscopy. Raman spectroscopy was performed using a LabRam HR800 UV NIR and 532-nm laser excitation with working distances on a × 50 lens. The Raman spectra of M-MoS$_2$ and S-MoS$_2$ were recorded by depositing the samples on silicon substrates.

XRD patterns. The XRD patterns of M-MoS$_2$ and S-MoS$_2$ were recorded for two theta values ranging from 4° to 18°, to characterize the interlayer spacing. The characterization was performed on a Bruker AXS-D8 Advance powder X-ray diffractometer using Cu/Ka radiation ($\lambda = 1.5406$ Å) with a step size of 0.02° and a dwell time of 3.0 s.

Electrocatalytic hydrogen evolution. The electrochemical measurements were performed using a three-electrode electrochemical station (Gamry Instruments). All of the measurements were performed in a solution consisting of 50 ml of a 0.5-M H$_2$SO$_4$ electrolyte prepared using 18 M deionized water purged with H$_2$ gas (99.999%). The MoS$_2$ suspension in water was dropped onto glassy carbon as the working electrode. All of the working electrodes were prepared using the MoS$_2$ solutions. The solution concentrations of M-MoS$_2$ and S-MoS$_2$ were ~0.3 mg ml^{-1} and ~0.4 mg ml^{-1}, respectively, in water. The diameter of the glass carbon electrode was 0.3 cm and the area was ~0.07 cm^2. The amount of deposited catalyst on the electrode was ~ 43 µg cm^{-2} for M-MoS$_2$ and 114 µg cm^{-2} for S-MoS$_2$. Platinum foil was used as the counter electrode and a saturated calomel was used as the reference electrode. The reversible hydrogen electrode was calibrated using platinum as both the working and counter electrodes to + 0.3 V versus the saturated calomel reference electrode. The performance of the hydrogen evolution catalyst was measured using linear sweep voltammetry from + 0.3 to − 0.3 V versus reversible hydrogen electrode with a scan rate of 5 mV s^{-1}.

References

1. Eda, G. *et al.* Photoluminescence from chemically exfoliated MoS$_2$. *Nano Lett.* **11**, 5111–5116 (2011).
2. Lukowski, M. A. *et al.* Enhanced hydrogen evolution catalysis from chemically exfoliated metallic MoS$_2$ nanosheets. *J. Am. Chem. Soc.* **135**, 10274–10277 (2013).
3. Voiry, D. *et al.* Conducting MoS$_2$ nanosheets as catalysts for hydrogen evolution reaction. *Nano Lett.* **13**, 6222–6227 (2013).
4. Bai, S., Wang, L., Chen, X., Du, J. & Xiong, Y. Chemically exfoliated metallic MoS$_2$ nanosheets: a promising supporting co-catalyst for enhancing the photocatalytic performance of TiO$_2$ nanocrystals. *Nano Res.* **8**, 175–183 (2015).
5. Acerce, M., Voiry, D. & Chhowalla, M. Metallic 1T phase MoS$_2$ nanosheets as supercapacitor electrode materials. *Nat. Nanotechnol.* **10**, 313–318 (2015).
6. Py, M. A. & Haering, R. R. Structural destabilization induced by lithium intercalation in MoS$_2$ and related compounds. *Can. J. Phys.* **61**, 76–84 (1983).
7. Mattheiss, L. F. Band structures of transition-metal–dichalcogenide layer compounds. *Phys. Rev. B* **8**, 3719–3740 (1973).
8. Zheng, J. *et al.* High yield exfoliation of two-dimensional chalcogenides using sodium naphthalenide. *Nat. Commun.* **5**, 2995 (2014).
9. Eda, G. *et al.* Coherent atomic and electronic heterostructures of single-layer MoS$_2$. *ACS Nano* **6**, 7311–7317 (2012).
10. Heising, J. & Kanatzidis, M. G. Structure of restacked MoS$_2$ and WS$_2$ elucidated by electron crystallography. *J. Am. Chem. Soc.* **121**, 638–643 (1999).
11. Chou, S. S. *et al.* Controlling the metal to semiconductor transition of MoS$_2$ and WS$_2$ in solution. *J. Am. Chem. Soc.* **137**, 1742–1745 (2015).
12. Jiménez Sandoval, S., Yang, D., Frindt, R. F. & Irwin, J. C. Raman study and lattice dynamics of single molecular layers of MoS$_2$. *Phys. Rev. B* **44**, 3955–3962 (1991).
13. Calandra, M. Chemically exfoliated single-layer MoS$_2$: stability, lattice dynamics, and catalytic adsorption from first principles. *Phys. Rev. B* **88**, 245428 (2013).
14. Güller, F., Llois, A. M., Goniakowski, J. & Noguera, C. Prediction of structural and metal-to-semiconductor phase transitions in nanoscale MoS$_2$, WS$_2$, and other transition metal dichalcogenide zigzag ribbons. *Phys. Rev. B* **91**, 075407 (2015).
15. Liang, Y. *et al.* Rechargeable Mg batteries with graphene-like MoS$_2$ cathode and ultrasmall Mg nanoparticle anode. *Adv. Mater.* **23**, 640–643 (2011).
16. Feldman, Y., Wasserman, E., Srolovitz, D. J. & Tenne, R. High-rate, gas-phase growth of MoS$_2$ nested inorganic fullerenes and nanotubes. *Science* **267**, 222–225 (1995).
17. Enyashin, A. N. *et al.* New route for stabilization of 1T-WS$_2$ and MoS$_2$ phases. *J. Phys. Chem. C.* **115**, 24586–24591 (2011).
18. Lin, Y. C., Dumcenco, D. O., Huang, Y. S. & Suenaga, K. Atomic mechanism of the semiconducting-to-metallic phase transition in single-layered MoS$_2$. *Nat. Nanotechnol.* **9**, 391–396 (2014).
19. Nayak, A. P. *et al.* Pressure-induced semiconducting to metallic transition in multilayered molybdenum disulphide. *Nat. Commun.* **5**, 3731 (2014).
20. Kappera, R. *et al.* Phase-engineered low-resistance contacts for ultrathin MoS$_2$ transistors. *Nat. Mater.* **13**, 1128–1134 (2014).
21. Wang, H. *et al.* Electrochemical tuning of vertically aligned MoS$_2$ nanofilms and its application in improving hydrogen evolution reaction. *Proc. Natl Acad. Sci. USA* **110**, 19701–19706 (2013).
22. Lee, J. *et al.* Two-dimensional layered MoS$_2$ biosensors enable highly sensitive detection of biomolecules. *Sci. Rep.* **4**, 7352 (2014).
23. Mak, K. F., Lee, C., Hone, J., Shan, J. & Heinz, T. F. Atomically thin MoS$_2$: a new direct-gap semiconductor. *Phys. Rev. Lett.* **105**, 136805 (2010).
24. Joensen, P., Crozier, E. D., Alberding, N. & Frindt, R. F. A study of single-layer and restacked MoS$_2$ by x-ray diffraction and x-ray absorption spectroscopy. *J. Phys. C: Solid State Phys.* **20**, 4043–4053 (1987).
25. Qin, X. R. *Scanning Tunneling Microcopy of Layered Materials* (PhD Thesis. Simon Fraser Univ., 1992).
26. Farimani, A. B., Min, K. & Aluru, N. R. DNA base detection using a single-layer MoS$_2$. *ACS Nano* **8**, 7914–7922 (2014).
27. Liu, K.K. *et al.* Growth of large-area and highly crystalline MoS$_2$ thin layers on insulating substrates. *Nano Lett.* **12**, 1538–1544 (2012).
28. Koroteev, V. O., Bulusheva, L. G., Okotrub, A. V., Yudanov, N. F. & Vyalikh, D. V. Formation of MoS$_2$ nanoparticles on the surface of reduced graphite oxide. *Phys. Stat. Solid. B* **248**, 2740–2743 (2011).
29. Liao, L. *et al.* MoS$_2$ formed on mesoporous graphene as a highly active catalyst for hydrogen evolution. *Adv. Funct. Mater.* **23**, 5326–5333 (2013).
30. Diemann, E., Weber, T. & Muller, A. Modeling the thiophene HDS reaction on a molecular level. *J. Catal.* **148**, 288–303 (1994).
31. Chang, C. H. & Chan, S. S. Infrared and raman studies of amorphous MoS$_3$ and poorly crystalline MoS$_2$. *J. Catal.* **72**, 139–148 (1981).

Acknowledgements

This work was supported by the National Science Foundation of US (Award Number EPS-1003970) and the NASA RID programme. We thank Dr Allan Thomas for discussions on the electrochemical impedance spectroscopy measurements. We thank the Center for Integrative Nanotechnology Sciences at the UALR for experimental assistance.

Author contributions

X.G., T.C. and J.C. conceived and designed the experiments. X.G. and J.C wrote the paper with input from all of the authors. B.C. edited the manuscript. All of the experimental work was conducted by X.G with W.W., W.S., A.A., M.B., H.Z., F.W. and J.C. The experimental data were analysed by X.G. with J.C. and T.C., and the results were discussed by all of the authors.

Additional information

Competing financial interests: The authors declare no competing financial interests.

Exceptional damage-tolerance of a medium-entropy alloy CrCoNi at cryogenic temperatures

Bernd Gludovatz[1], Anton Hohenwarter[2], Keli V.S. Thurston[1,3], Hongbin Bei[4], Zhenggang Wu[5], Easo P. George[4,5,†] & Robert O. Ritchie[1,3]

High-entropy alloys are an intriguing new class of metallic materials that derive their properties from being multi-element systems that can crystallize as a single phase, despite containing high concentrations of five or more elements with different crystal structures. Here we examine an equiatomic medium-entropy alloy containing only three elements, CrCoNi, as a single-phase face-centred cubic solid solution, which displays strength-toughness properties that exceed those of all high-entropy alloys and most multi-phase alloys. At room temperature, the alloy shows tensile strengths of almost 1 GPa, failure strains of $\sim 70\%$ and K_{Jic} fracture-toughness values above 200 MPa m$^{1/2}$; at cryogenic temperatures strength, ductility and toughness of the CrCoNi alloy improve to strength levels above 1.3 GPa, failure strains up to 90% and K_{Jic} values of 275 MPa m$^{1/2}$. Such properties appear to result from continuous steady strain hardening, which acts to suppress plastic instability, resulting from pronounced dislocation activity and deformation-induced nano-twinning.

[1] Materials Sciences Division, Lawrence Berkeley National Laboratory, Berkeley, California 94720, USA. [2] Department of Materials Physics, Montanuniversität Leoben and Erich Schmid Institute of Materials Science, Austrian Academy of Sciences, Leoben 8700, Austria. [3] Department of Materials Science and Engineering, University of California, Berkeley, California 94720, USA. [4] Materials Sciences and Technology Division, Oak Ridge National Laboratory, Oak Ridge, Tennessee 37831, USA. [5] Department of Materials Sciences and Engineering, University of Tennessee, Knoxville, Tennessee 37996, USA. † Present address: Institute for Materials, Ruhr University, 44801 Bochum, Germany. Correspondence and requests for materials should be addressed to E.P.G. (email: easo.george@rub.de) or to R.O.R. (email: roritchie@lbl.gov).

Equiatomic multi-component metallic materials, referred to variously as high-entropy alloys (HEAs), multi-component alloys or compositionally complex alloys, have generated considerable excitement in the materials science community of late as a new class of materials that derive their properties not from a single dominant constituent, such as iron in steels, but rather from multiple principal elements with the potential for unique combinations of mechanical properties compared with conventional alloys[1-19]. Much of the interest is predicated on the belief that many new alloys with useful properties are likely to be discovered near the centres (as opposed to the corners) of phase diagrams in compositionally complex systems[17].

One of the extensively investigated high-entropy alloys, an equiatomic, face-centred cubic (fcc) metallic alloy comprising five transition elements, Cr, Mn Fe, Co and Ni, was introduced in 2004 (ref. 1), although it was not for a decade that its mechanical properties were first systematically characterized[6-8,13-15]. CrMnFeCoNi (often termed the Cantor alloy[1]) displays strongly temperature-dependent strength and ductility with only a small strain-rate dependence[6,7]. Furthermore, between room temperature and 77 K, the alloy displays fracture toughness, K_{IIc} values at crack initiation that remain well above 200 MPa m$^{1/2}$ associated with an increase in tensile strength ($763 \rightarrow 1,280$ MPa) and ductility ($0.5 \rightarrow 0.7$), making it not simply an ideal material for cryogenic applications but putting it among the most damage-tolerant materials in that temperature range[8].

Although this excellent combination of properties can be related to progressively increasing strain hardening with hardening exponents above 0.4 (refs 7,8), it remains unclear why this particular combination of elements with very different crystal structures produces a single-phase microstructure[7,10-12,20,21], whereas many others with comparable configurational entropies do not[5]. In fact, a relatively small number of the reported multi-element high-entropy alloys are simple solid solutions[22]. Cantor et al.[1] produced an alloy with 20 elements in equal atomic ratios that crystallized as a very brittle multi-phase microstructure indicating that high configurational entropy by itself is unable to suppress the formation of intermetallic phases comprising the constituent elements. As pointed out recently[16], while the equiatomic composition maximizes configurational entropy, it does not necessarily minimize the total Gibbs free energy of a multi-component solid solution, and increasing the number of constituent elements could actually lead to the formation of undesirable intermetallic phases. Clearly, it is the nature of the alloying elements and not just their sheer number that is relevant.

However, there is also a question of the role of high configurational entropy in these materials with respect to properties, particularly how such high-entropy alloys compare with other equiatomic multi-element systems. Here we examine a variant of the single-phase CrMnFeCoNi high-entropy alloy in which two of the elements have been removed. The resulting CrCoNi alloy, an equiatomic 'medium-entropy alloy' (MEA), has a single-phase, fcc crystal structure[23], whose uniaxial tensile properties have recently been reported[24]. The experimental results (X-ray diffraction and backscattered electron, BSE, images)[23] are consistent with the CrCoNi ternary phase diagrams[25], which indicate that the equiatomic composition is a single-phase solid solution at elevated temperatures. (XRD and BSE analyses of the five-component CrMnFeCoNi alloy after casting and homogenization[5] showed that it too is single-phase fcc and remains so after recrystallization when examined by transmission electron microscopy[7]. In addition, three-dimensional atom probe tomography on the five-component CrMnFeCoNi alloy in the cast/homogenized state[21] and after severe plastic deformation[20] have shown that it retains its true single-phase character at the much finer atomic scale.

Significantly, we find here that the fracture toughness properties of the three-component CrCoNi MEA are even better than those of the five-component CrMnFeCoNi HEA, and are further enhanced with decrease in temperature between 293 and 77 K, making it one of the toughest metallic materials reported to date.

Results

Microstructure. The CrCoNi MEA was produced from high-purity elements ($>99.9\%$ pure) which were arc-melted under argon atmosphere and drop-cast into rectangular cross-section copper moulds followed by cold forging and cross rolling at room temperature into sheets of roughly 10 mm thickness (Fig. 1a). Following recrystallization, optical microscopy (Fig. 1b), scanning electron microscopy (SEM) (Fig. 1c) and electron back-scattered diffraction (EBSD; Fig. 1d) images taken from the cross-section of the sheets revealed an equiaxed grain structure with a variable grain size of 5–50 μm and numerous recrystallization twins (inset of Fig. 1c); the equiatomic elemental distribution of the alloy can be seen from energy-dispersive X-ray (EDX) spectroscopy in Fig. 1e. Uniaxial tensile specimens and compact-tension C(T) fracture-toughness specimens were cut from the sheets using electrical discharge machining; the C(T) samples were fatigue precracked and subsequently side-grooved, in general accordance with ASTM standard E1820 (ref. 26).

Strength and ductility. Using uniaxial, dog-bone-shaped tensile specimen, we measured stress–strain curves at room temperature (293 K), in a mixture of dry ice and ethanol (198 K), and in liquid nitrogen (77 K). Results in Fig. 2a show a ~50% increase in both yield strength, σ_y, and ultimate tensile strength, σ_{UTS}, with decreasing temperature to values of $\sigma_y = 657$ MPa and $\sigma_{UTS} = 1,311$ MPa at 77 K. The tensile ductility (strain to failure, ε_f) similarly increased by ~25% to ~0.9, leading to an increase in fracture energy of more than 80%, associated with a high strain-hardening exponent, n, of 0.4. (Note that compared with pure Ni, this material displays both higher strain hardening and higher elongation to failure[24], consistent with the widely accepted Considère's criterion that higher work-hardening ability promotes ductility by postponing plastic (geometric) instability.)

The yield strength of this alloy is not particularly high, although it does significantly strain harden to give low-temperature tensile strengths above 1 GPa. However, as discussed below, its outstanding characteristic is a combination of high strength, ductility and especially fracture toughness which is enhanced significantly at cryogenic temperatures. This refers to its damage tolerance which is invariably the most important property for the application of a structural material.

Fracture toughness. To assess the fracture toughness of the CrCoNi alloy and account for both the elastic and extensive plastic contributions involved in the deformation process and during crack growth, we applied nonlinear-elastic fracture mechanics analysis to determine J-based crack-resistance curves, that is, J_R as a function of crack extension Δa, as shown in Fig. 2b. At room temperature, our C(T) specimens show fracture toughness, J, values in excess of 200 kJ m^{-2} at crack initiation, which increased to above 400 kJ m^{-2} with crack extensions of slightly more than 2 mm, the maximum extent of cracking permitted for this geometry by ASTM standards[26]. Despite the much higher strength at lower temperatures, at 77 K the critical J increased even further to above 350 kJ m^{-2} at crack initiation and to almost 950 kJ m^{-2} at full extension of the crack. Given that the requirements for J-dominant conditions, that is, b, $B >> 10$ (J per σ_{flow}), where b is the uncracked ligament width (sample width, $W - a$), B the sample thickness and σ_{flow} the flow

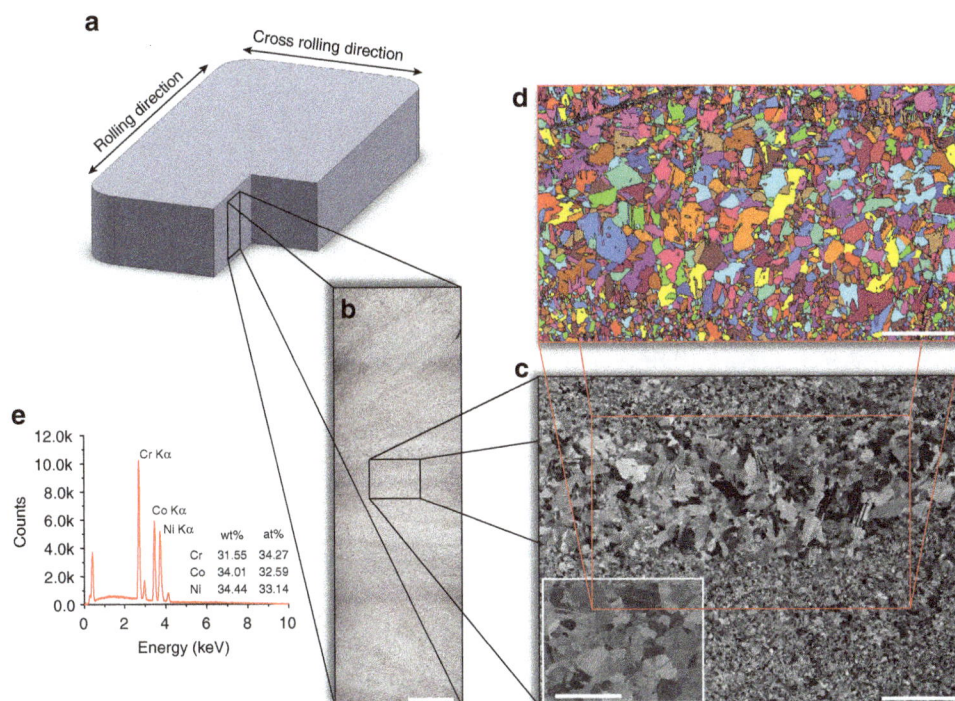

Figure 1 | Processing and microstructure of the medium-entropy alloy CrCoNi. (a) The material was processed by arc melting, drop casting, forging and rolling into sheets of roughly 10 mm thickness from which samples for cross-sectional analysis, tensile tests and fracture toughness tests were machined. **(b)** Optical microscopy image shows the varying degree of deformation through the thickness of the sheets. **(c)** Scanning electron microscopy images reveal the non-uniform grain size of the material resulting from the deformation gradients, equiaxed grains and numerous annealing twins after recrystallization (inset). **(d)** Grain maps from electron back-scatter diffraction scans confirm the varying grain size and show the fully recrystallized microstructure. **(e)** Energy-dispersive X-ray spectroscopy verifies the equiatomic character of the alloy. The scale bars in **b,c** and the inset of **c** and **d** are 1 mm, 200 μm, 20 μm and 150 μm, respectively.

stress $(\sigma_{flow} = (\sigma_y + \sigma_{UTS})/2)$, were met, the standard J–K equivalence (mode I) relationship, $K_J = (J\ E')^{1/2}$, was used to determine stress-intensity K values corresponding to these measured J toughnesses. (Here, $E' = E$, the Young's modulus in plane stress and $E/(1-v^2)$ in plane strain, v is the Poisson's ratio, where values of E and v were determined at each temperature using resonance ultrasound spectroscopy methods described elsewhere[27].) Fracture toughnesses for the CrCoNi alloy, defined at crack initiation, were strictly valid by ASTM Standard E1820 (ref. 26), with measured K_{JIc}, values of 208 MPa m$^{1/2}$ ($J_{Ic} = 212$ kJ m^{-2}) at 293 K increasing to 273 MPa m$^{1/2}$ ($J_{Ic} = 363$ kJ m^{-2}) at 77 K. ASTM valid crack-growth toughnesses, defined at $\Delta a \sim 2$ mm, were significantly higher with critical stress-intensity values above ~ 290 MPa m$^{1/2}$ ($J \sim 400$ kJ m^{-2}) at 293 K, rising up to ~ 430 MPa m$^{1/2}$ ($J \sim 900$ kJ m^{-2}) at 77 K.

Deformation and failure mechanisms. The extremely high fracture toughness values of the CrCoNi alloy (Fig. 2) are associated with fully ductile fracture, with a pronounced stretch-zone at crack initiation (Fig. 2c) and failure by microvoid coalescence (Fig. 2d). The volume fraction of the void-initiating inclusions was lower than in the five-component CrMnFeCoNi alloy[8], which is partly an effect of removing Mn that is known to increase the number of inclusions[6]. The particles here were analysed by EDX spectroscopy and found to be Cr-rich (insets of Fig. 2d), whereas in the five-component HEA, both Cr and Mn-rich particles were found[8]. In both alloys, we believe that these particles are oxide inclusions that typically form when alloys containing reactive elements are melted. Consistent with this, a recent study identified $MnCr_2O_4$ oxide particles in an induction

melted CrMnFeCoNi HEA by EDX analysis[28]. To quantify their effects on ductility and fracture, further studies are needed in the future that would accurately determine the relative volume fractions of the void-initiating inclusions in the MEA and HEA and identify their chemistry and crystal structure by TEM after extraction from the voids.

While the yield strength and K_{JIc} fracture toughness of the medium-entropy CrCoNi and high-entropy CrMnFeCoNi alloys are comparable, the tensile strengths, tensile ductility and work of fracture of the CrCoNi alloy are significantly higher, by respectively ~ 15, ~ 30 and $\sim 50\%$, at room temperature. At cryogenic temperatures, the strengths of the two alloys are comparable ($\sigma_{UTS} \sim 1,300$ MPa at 77 K), but the K_{JIc} fracture toughness, tensile ductility and work of fracture are again markedly higher in the CrCoNi alloy, by ~ 25, ~ 27 and $\sim 31\%$, respectively. The yield strength, σ_y, at 77 K is slightly below that of the CrMnFeCoNi alloy, which we believe is due to the non-uniform grain size of our present material (Fig. 1b–d). Consistent with this notion, in a previous study where the grain size of CrCoNi was uniform and comparable to that of the CrMnFeCoNi alloy, the tensile properties of the three-component alloy were found to exceed those of the five-component alloy at all temperatures[24].

Discussion

To seek the origins of such strength, ductility and fracture resistance between 293 and 77 K, we conducted detailed SEM analysis of the vicinity of the propagated crack; this was performed on samples sliced in two through the thickness to ensure that deformation conditions had been in fully plane strain (Fig. 3a). The EBSD scans taken in the wake of the propagated

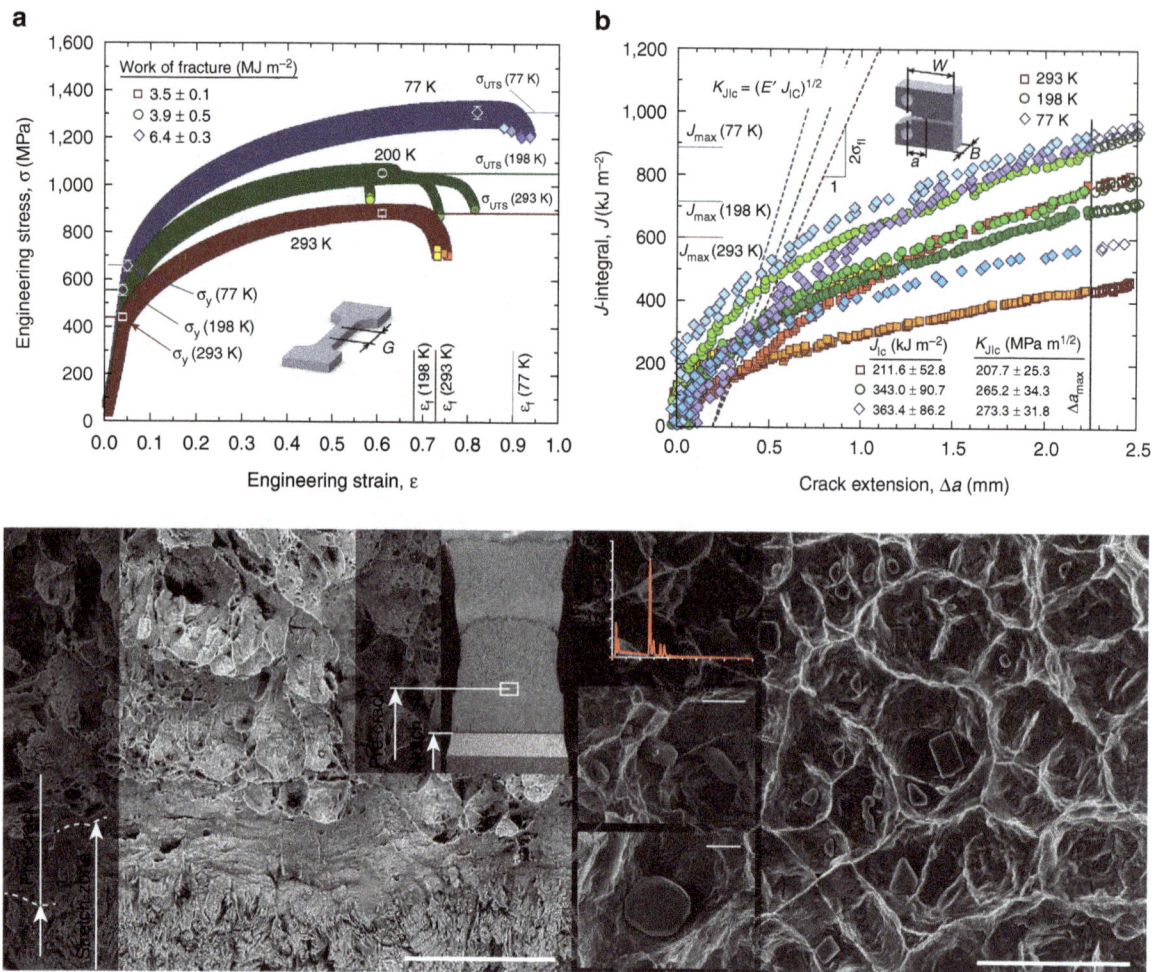

Figure 2 | Mechanical properties and failure characteristics of the CrCoNi medium-entropy alloy. (**a**) Tensile tests show a significant increase in yield strength, σ_y, ultimate tensile strength, σ_{UTS} and strain to failure, ε_f, with decreasing temperature from room temperature, 293 K, to cryogenic temperatures, 198 and 77 K. In the same temperature range, the work of fracture increases from 3.5 MJ m^{-2} to 6.4 MJ m^{-2}. (**b**) Fracture toughness tests on compact-tension, C(T), specimens show an increasing fracture resistance with crack extension and crack initiation, K_{JIc}, values of 208, 265 and 273 MPa m$^{1/2}$ at 293, 198 and 77 K, respectively. (**c**) Stereo microscopy and scanning electron microscopy images show a clear transition from the notch to the pre-crack and a pronounced stretch-zone between the pre-crack and the fully ductile fracture region of a sample that was tested at 198 K. (**d**) The fracture surface shows ductile dimpled fracture and Cr-rich particles that act as void initiation sides. (Data points shown are mean ± s.d.; see Supplementary Table 1 for exact values.) The scale bars in **c** and **d**, and the insets of **d** are 75, 5 and 2 μm, respectively.

crack of a sample tested at 293 K (Fig. 3b), ahead of the crack tip of a sample tested at 198 K (Fig. 3c) and at a crack flank of a sample tested at 77 K (Fig. 3d) show grain misorientations as gradual changes in colour within individual grains indicative of significant amounts of dislocation plasticity. Similarly, back-scattered electron (BSE) scans taken on specimens fractured at room (Fig. 3b) and cryogenic temperatures (Fig. 3d) show the formation of pronounced dislocation cell structures akin to the five-component CrMnFeCoNi HEA where dislocation motion is associated with glide of ½<110> dislocations on {111} planes[7], a typical deformation mechanism for fcc materials, which we presume also occurs in our three-component CrCoNi MEA. In addition, the EBSD scans show a few recrystallization twins in all samples (approximately one or two per grain) as well as the presence of deformation-induced nano-twins at 77 K (Fig. 3d). The BSE images, however, clearly reveal that deformation-induced nano-twinning is a dominant deformation mechanism occurring initially at room temperature but with increasing intensity at 198 and 77 K. From the images in Fig. 3, the nano-twins in the EBSD scans become very clear by overlaying the scan on an image quality, IQ, map of the same

data set, which permits the measurement of the typical misorientation angle of 60° for twinning (Fig. 3e). We conclude from these results that between room and cryogenic temperatures where the strength, ductility and toughness of the medium-entropy CrCoNi are all simultaneously enhanced, nano-twinning contributes an important additional deformation mode that helps alleviate the deleterious effects of high strength that would normally be expected to result in lower toughness[29].

We did not observe deformation nano-twinning at room temperature in the five-component alloy, where deformation at 293 K is solely carried by dislocation slip[7,8], specifically, involving the rapid movement of partial dislocations and the much slower planar slip of undissociated dislocations[30], although as with the present three-component alloy, twinning became a major deformation mode at 77 K. We believe that the earlier onset of deformation nano-twinning is key to the exceptional damage-tolerance of this medium-entropy alloy. Although in most materials the achievement of strength and toughness is invariably a compromise[29]—high strength is often associated with lower toughness and vice versa—it has become increasingly apparent that the presence of twinning as the dominant

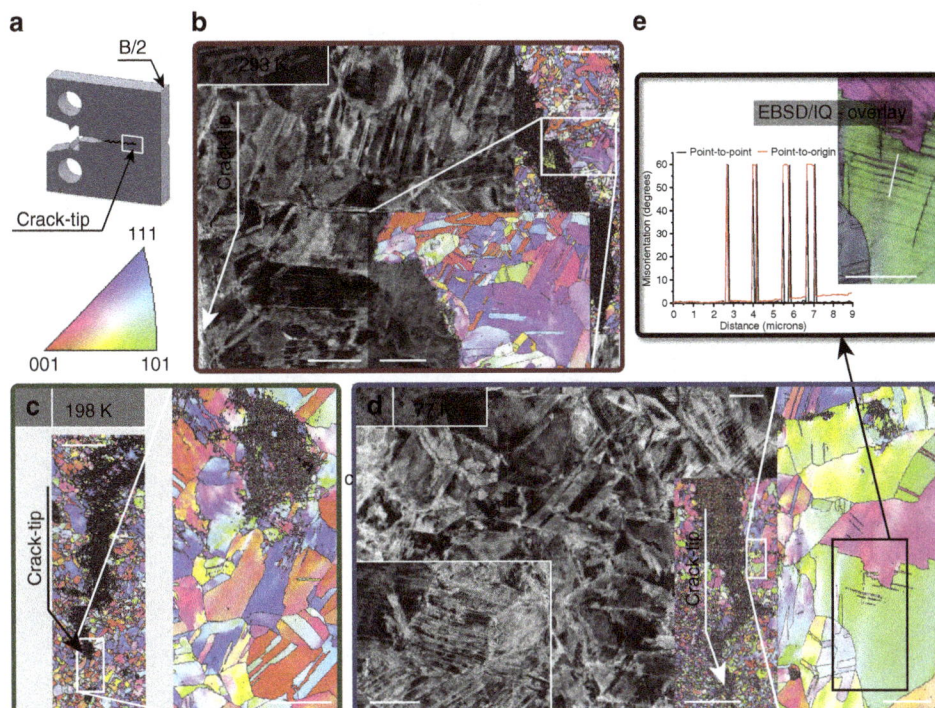

Figure 3 | Deformation mechanisms in CrCoNi between 293 and 77 K. (**a**) After testing, some samples were sliced in two along the half-thickness mid-plane, and the crack-tip regions in the centre of the samples (plane strain) were investigated in the scanning electron microscope using back-scattered electrons (BSE) and electron-backscatter diffraction (EBSD). (**b**) EBSD scans in the wake of the propagated crack of a sample tested at room temperature show a few recrystallization twins and grain misorientations indicative of dislocation plasticity whereas BSE scans reveal cell formation and nano-twinning as additional deformation mechanism. (**c**) Similar to room temperature behaviour, EBSD scans of samples tested at 198 K show recrystallization twins and misorientations indicative of dislocation plasticity ahead of the propagated crack-tip. (**d**) Samples tested at 77 K show pronounced nano-twinning and the formation of dislocation cells (BSE), whereas EBSD scans reveal dislocation plasticity in the form of grain misorientations, some recrystallization twins and deformation induced nano-twins. (**e**) An arbitrarily chosen path on an EBSD image overlaid on an image quality (IQ) map shows 60° misorientations typical for the character of such deformation twins. (The IQ map measures the quality of the collected EBSD patterns and is often used to visualize microstructural features.) The scale bars of the BSE image, the EBSD image and the inset of the EBSD image in **b** are 5, 75 and 25 µm, respectively; the ones of the EBSD image and its inset in **c** are 50 and 10 µm, respectively. The BSE image and its corresponding inset, and the EBSD image and its inset have scale bars of 10, 5, 200 and 15 µm, respectively. The scale bar in **e** is 15 µm.

deformation mechanism serves to 'defeat this conflict', specifically by providing a steady source of strain hardening, which promotes ductility by delaying the onset of plastic instability by necking, and an additional deformation mode besides dislocation plasticity to accommodate the imposed strain. In addition to the high- and medium-entropy alloys, there are now several other materials known to benefit from twinning, including copper thin films[31–34] and 11–15 wt.% (Hadfield) Mn-steels (used in the mining industry for rock crushers because of their hardness and fracture resistance) and their modern variant known as twinning-induced plasticity steels[35–42], which have application in the automobile industry.

The current medium-entropy CrCoNi alloy, however, appears to optimize these features to achieve literally unparalleled mechanical performance at low temperatures. Although solid-solution hardening provides the ideal hardening mechanism for cryogenic use, the increasing role of nano-twinning with decreasing temperature, as is evident from the apparently denser network of nano-twins at 77 K (inset in Fig. 3d) compared with room temperature (Fig. 2b), acts to progressively further enhance damage-tolerance (strength, ductility and toughness) with decreasing temperature, to achieve extremely high strain-hardening exponents on the order of 0.4.

Such damage-tolerant properties of the CrCoNi medium-entropy alloy are literally unprecedented for mechanical behaviour at cryogenic temperatures. For a material with a tensile strength of 1.3 GPa to display ductilities (failure strains) of 90%, and 'valid' crack-growth fracture toughnesses that exceeds 430 MPa m$^{1/2}$, all at liquid-nitrogen temperatures, is exceptional and clearly exceed the excellent cryogenic properties of our previously reported CrMnFeCoNi high-entropy alloy[8]. Its ductility compares favourably to high-Mn twinning-induced plasticity steels[35–42] and strength and toughness are comparable to the very best cryogenic steels, for example, certain austenitic stainless steels[43–47] and high-Ni steels[48–51]; in addition, the strength, ductility and toughness of the CrCoNi alloy exceed the properties of all medium- and high-entropy alloys reported to date (Fig. 4). Moreover, with a uniformly fine grain size, it is eminently feasible that the strength, ductility and toughness properties of this CrCoNi alloy may further improve.

With respect to high-entropy alloys in general, by comparing the CrCoNi and CrMnFeCoNi alloys, the current work does lend credence to our belief that it is the nature of elements in complex solid solutions that is more important than their mere number. Indeed, in terms of (valid) crack-initiation and crack-growth toughnesses, the CrCoNi medium-entropy alloy represents one of the toughest materials in any materials class ever reported.

Methods

Materials processing and microstructural characterization. The CrCoNi MEA was produced from high-purity elements (>99.9% pure), which were arc-melted under argon atmosphere and drop-cast into rectangular cross-section copper moulds measuring $25.4 \times 19.1 \times 127$ mm. The ingots were homogenized at 1,200 °C for 24 h in vacuum, cut in half length-wise and then cold-forged and

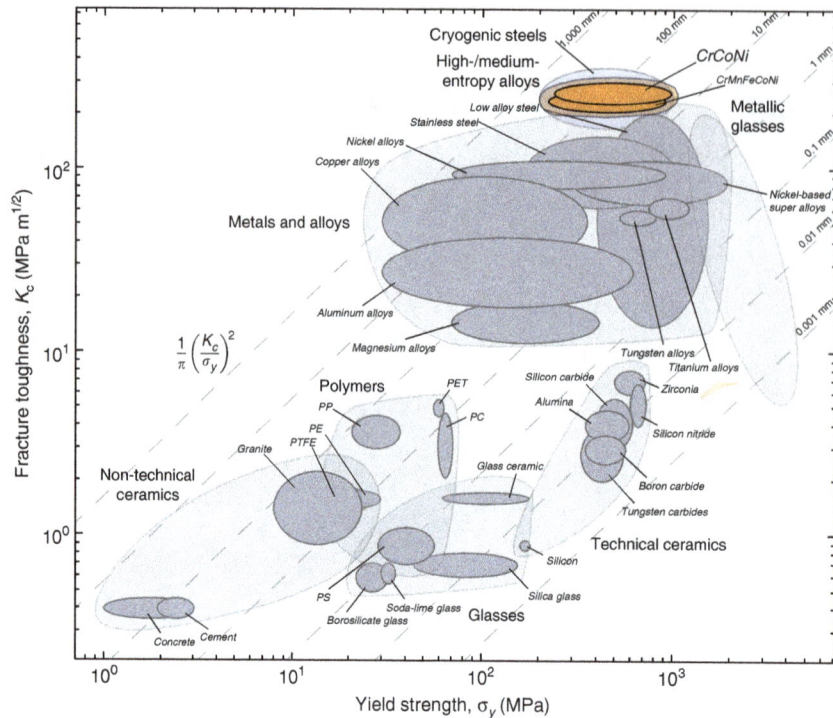

Figure 4 | Ashby map of fracture toughness versus yield strength for various classes of materials. The investigated medium-entropy alloy CrCoNi compares favourably with materials classes like metals and alloys and metallic glasses. Its combination of strength and toughness (that is damage tolerance) is comparable to cryogenic steels, for example, certain austenitic stainless steels[43–47] and high-Ni steels[48–51], and exceeds all high- and medium-entropy alloys reported to date.

cross-rolled at room temperature along the side that is 25.4 mm to a final thickness of ~10 mm, as shown in Fig. 1a (total reduction in thickness of ~60%). Each piece was subsequently annealed at 800 °C for 1 h in air leading to sheets with a fully recrystallized microstructure consisting of equiaxed grains ~5–50 μm in size.

To analyse the microstructure of the material after processing, two pieces were cut from the recrystallized sheets perpendicular to the rolling direction, embedded in conductive resin and metallographically polished in stages to a final surface finish of 0.04 μm using colloidal silica. For optical microscopy analysis, one polished surface was chemically etched using a standard solution for austenitic steels (10 ml H_2O, 1 ml HNO_3, 5 ml HCl and 1 g $FeCl_3$); the other was analysed as is in an LEO (Zeiss) 1525 FE-SEM (Carl Zeiss, Oberkochen, Germany) scanning electron microscope (SEM) operated at 20 kV in the back-scattered electron mode.

Mechanical characterization. Rectangular dog-bone-shaped tensile specimens with a gauge length of 12.7 mm were machined from the recrystallized sheets by electrical discharge machining. Both sides of the specimen were ground using SiC paper resulting in a final thickness of ~1.5 mm and a gauge width of ~3.0 mm. The gauge length was marked with Vickers microhardness indents (300 g load) to enable elongations to be measured after fracture using a Nikon travelling microscope. Tensile tests were performed at an engineering strain rate of 10^{-3} s^{-1} in a screw-driven Instron 4,204 load frame. Groups of four samples were tested at three different temperatures ($N = 12$); at room temperature (293 K), in a bath of dry ice and ethanol (198 K), and in a bath of liquid nitrogen (77 K).

The elongation of the gauge length of each sample was measured after testing, and engineering stress–strain curves were calculated from the load-displacement data. Yield strength, σ_y, ultimate tensile strength, σ_u, and elongation to failure, ε_f, were determined from the uniaxial tensile stress–strain curves and are shown in Supplementary Table 1 as mean ± s.d. for each set of tests at the individual temperatures. True stress–strain curves were calculated from the engineering stress–strain curves and strain-hardening exponents, n, were determined for each temperature based on the constitutive law $\sigma = \kappa\varepsilon^n$, where σ and ε are, respectively, the true stress and plastic strain, k is a scaling constant and n the strain-hardening exponent; n values are also listed in Supplementary Table 1.

Nine ($N = 9$) compact-tension C(T) specimens, of nominal width $W = 18$ mm and thickness $B = 9$ mm, were prepared in strict accordance with ASTM standard E1820 (ref. 26) using electrical discharge machining (EDM). Notches, 6.6 mm in length with notch root radii of ~100 μm, were cut using EDM; before pre-cracking, the faces of all samples were metallographically ground and polished in stages to a final 1 μm surface finish to allow accurate crack-length measurements using optical microscopy. All the samples were fatigue pre-cracked and tested using an electroservo-hydraulic MTS 810 load frame (MTS Corporation, Eden Prairie, MN, USA) controlled by an Instron 8800 digital controller (Instron Corporation,

Norwood, MA, USA). Fatigue pre-cracks were created under load control (tension–tension loading) at a stress intensity range of $\Delta K = K_{\max} - K_{\min}$ of 15 MPa m$^{1/2}$ and a constant frequency of 10 Hz (sine wave) with a load ratio $R = 0.1$, where R is the ratio of minimum to maximum applied load. During pre-cracking, the crack length was optically checked from both sides of the sample to ensure a straight crack front with crack extension monitored using an Epsilon clip gauge of 3 mm ($-1/+2.5$ mm) gauge length (Epsilon Technology, Jackson, WY, USA) mounted at the load-line of the sample; final crack lengths, a were in the range of 8.1–12.6 mm ($a/W \sim 0.45$–0.7) and thus were well above the ASTM standard's minimum length requirement for a pre-crack of 1.3 mm. To improve the constraint conditions at the crack tip during testing, all the samples were side-grooved using EDM to depths of ~1 mm, which resulted in a net sample thickness of $B_N \sim 7$ mm; this reduction in thickness did not exceed 20–25%, as mandated by ASTM Standard E1820 (ref. 26).

Nonlinear-elastic fracture mechanics methodologies were used to incorporate both the elastic and inelastic contributions to the measurement of the fracture toughness; specifically, the change in crack resistance with crack extension, that is, crack-resistance curve (R-curve) behaviour, was characterized in terms of the J-integral as a function of crack growth at three different temperatures: 293, 198 and 77 K. The samples were tested under displacement control at a constant displacement rate of 2 mm min^{-1}. The onset of cracking as well as subsequent subcritical crack growth were determined by periodically unloading the sample (~20% of the peak-load) to record the elastic unloading compliance using an Epsilon clip gauge of 3 mm ($-1/+7$ mm) gauge length (Epsilon Technology, Jackson, WY, USA) mounted in the load-line of the sample. Crack lengths, a_i were calculated from the compliance data obtained during the test using the compliance expression of a C(T) sample at the load-line[26]:

$$a_i/W = 1.000196 - 4.06319u + 11.242u^2 - 106.043u^3 + 464.335u^4 - 650.677u^5, \quad (1)$$

where

$$u = \frac{1}{\left[B_e E C_{c(i)} \right]^{1/2} + 1}. \quad (2)$$

$C_{c(i)}$ is the rotation-corrected, elastic unloading compliance and B_e the effective sample thickness of a side-grooved sample calculated as $B_e = B - (B - B_N)^2/B$. (Initial and final crack lengths were additionally verified by post-test optical measurements.) For each crack length data point, a_i, the corresponding J-integral was computed as the sum of elastic, $J_{el\,(i)}$, and plastic components, $J_{pl\,(i)}$, such that the J-integral can be written as follows:

$$J_i = K_I^2/E' + J_{pl(i)}, \quad (3)$$

where $E' = E$, the Young's modulus, in plane stress and $E/(1 - \nu^2)$ in plane strain; ν

is Poisson's ratio. K_i, the linear elastic stress intensity corresponding to each data point on the load-displacement curve, was calculated for the C(T) geometry from:

$$K_i = \frac{P_i}{(BB_N W)^{1/2}} f(a_i/W),\qquad(4)$$

where P_i is the applied load at each individual data point and $f(a_i/W)$ is a geometry-dependent function of the ratio of crack length, a_i, to width, W, as listed in the ASTM standard. The plastic component of J_i can be calculated from the following equation:

$$J_{pl(i)} = \left[J_{pl(i-1)} + \left(\frac{\eta_{pl(i-1)}}{b_{(i-1)}} \right) \frac{A_{pl(i)} - A_{pl(i-1)}}{B_N} \right] \left[1 - \gamma_{(i-1)} \left(\frac{a_{(i)} - a_{(i-1)}}{b_{(i-1)}} \right) \right],\qquad(5)$$

where $\eta_{pl\,(i-1)} = 2 + 0.522\, b_{(i-1)}/W$ and $\gamma_{pl\,(i-1)} = 1 + 0.76\, b_{(i-1)}/W$. $A_{pl\,(i)} - A_{pl\,(i-1)}$ is the increment of plastic area underneath the load-displacement curve, and b_i is the uncracked ligament width (that is, $b_i = W - a_i$). Using this formulation, the value of J_i can be determined at any point along the load-displacement curve and together with the corresponding crack lengths, the $J - \Delta a$ resistance curve created. (Here, Δa is the difference of the individual crack lengths, a_i, during testing and the initial crack length, a, after pre-cracking.)

The intersection of the resistance curve with the 0.2 mm offset/blunting line ($J = 2\,\sigma_0\Delta a$; where σ_0 is the flow stress) defines a provisional toughness J_Q, which can be considered as a size-independent (valid) fracture toughness, J_{Ic}, provided the validity requirements for J-field dominance and plane-strain conditions prevail, that is , that B, $b_0 > 10\, J_Q/\sigma_0$, where b_0 is the initial ligament length. The fracture toughness expressed in terms of the stress intensity was then computed using the standard $J - K$ equivalence (mode I) relationship $K_{JIc} = (E' J_{Ic})^{1/2}$. Values for E and ν at the individual temperatures were determined by resonance ultrasound spectroscopy using the procedure described in Haglund et al.[27]; at 293, 198 and 77 K, Young's moduli, E of 229, 235 and 241 GPa and Poisson's ratios, ν of 0.31, 0.30 and 0.30 were used, respectively.

To discern the mechanisms underlying the measured fracture toughness values and investigate the microstructure in the vicinity of the crack tip and wake in the plane-strain region in the interior of the sample after testing, one sample from each of the tested temperatures was sliced in two, each with a thickness of $\sim B/2$. For each sample, one half was embedded in conductive resin, progressively polished to a 0.04 μm surface finish using colloidal silica, and analysed in the SEM in back-scattered electron mode as well as by electron back-scatter diffraction, EBSD using a TEAM EDAX analysis system (Ametek EDAX, Mahwah, NJ, USA).

The remaining ligament of all other samples was cycled to failure at a ΔK of ~ 30 MPa m$^{1/2}$, a frequency of 100 Hz (sine wave) and a load ratio $R = 0.5$ so that both the initial and the final crack lengths could be optically determined with precision from the change in fracture mode. In addition, the mating fracture surfaces of each sample were examined in the SEM at an accelerating voltage of 20 kV in the secondary electron mode. Particles inside the microvoids of samples tested at 293 and 77 K were analysed using an Energy Dispersive Spectroscopy (EDS) system from Oxford Instruments (Model 7426, Oxford, England). EDS analyses were performed on five randomly chosen particles from samples tested at both room and liquid nitrogen temperature to determine their chemical composition.

References

1. Cantor, B., Chang, I. T. H., Knight, P. & Vincent, A. J. B. Microstructural development in equiatomic multicomponent alloys. *Mater. Sci. Eng. A* **375–377**, 213–218 (2004).
2. Yeh, J.-W. et al. Nanostructured high-entropy alloys with multiple principal elements: novel alloy design concepts and outcomes. *Adv. Eng. Mater.* **6**, 299–303 (2004).
3. Hsu, C.-Y., Yeh, J.-W., Chen, S.-K. & Shun, T.-T. Wear resistance and high-temperature compression strength of Fcc CuCoNiCrAl0.5Fe alloy with boron addition. *Metall. Mater. Trans. A* **35**, 1465–1469 (2004).
4. Senkov, O. N., Wilks, G. B., Scott, J. M. & Miracle, D. B. Mechanical properties of Nb25Mo25Ta25W25 and V20Nb20Mo20Ta20W20 refractory high entropy alloys. *Intermetallics* **19**, 698–706 (2011).
5. Otto, F., Yang, Y., Bei, H. & George, E. P. Relative effects of enthalpy and entropy on the phase stability of equiatomic high-entropy alloys. *Acta Mater.* **61**, 2628–2638 (2013).
6. Gali, A. & George, E. P. Tensile properties of high- and medium-entropy alloys. *Intermetallics* **39**, 74–78 (2013).
7. Otto, F. et al. The influences of temperature and microstructure on the tensile properties of a CoCrFeMnNi high-entropy alloy. *Acta Mater.* **61**, 5743–5755 (2013).
8. Gludovatz, B. et al. A fracture-resistant high-entropy alloy for cryogenic applications. *Science* **345**, 1153–1158 (2014).
9. Manzoni, A., Daoud, H., Völkl, R., Glatzel, U. & Wanderka, N. Phase separation in equiatomic AlCoCrFeNi high-entropy alloy. *Ultramicroscopy* **132**, 212–215 (2013).
10. Yao, M. J., Pradeep, K. G., Tasan, C. C. & Raabe, D. A novel, single phase, non-equiatomic FeMnNiCoCr high-entropy alloy with exceptional phase stability and tensile ductility. *Scripta Mater.* **72–73**, 5–8 (2014).
11. Tasan, C. C. et al. Composition dependence of phase stability, deformation mechanisms, and mechanical properties of the CoCrFeMnNi high-entropy alloy system. *JOM* **66**, 1993–2001 (2014).
12. Deng, Y. et al. Design of a twinning-induced plasticity high entropy alloy. *Acta Mater.* **94**, 124–133 (2015).
13. He, J. Y. et al. Effects of Al addition on structural evolution and tensile properties of the FeCoNiCrMn high-entropy alloy system. *Acta Mater.* **62**, 105–113 (2014).
14. He, J. Y. et al. Steady state flow of the FeCoNiCrMn high entropy alloy at elevated temperatures. *Intermetallics* **55**, 9–14 (2014).
15. Stepanov, N. et al. Effect of cryo-deformation on structure and properties of CoCrFeNiMn high-entropy alloy. *Intermetallics* **59**, 8–17 (2015).
16. Zhang, F. et al. An understanding of high entropy alloys from phase diagram calculations. *CALPHAD* **45**, 1–10 (2014).
17. Cantor, B. Multicomponent and high entropy alloys. *Entropy* **16**, 4749–4768 (2014).
18. Senkov, O. N., Senkova, S. V., Woodward, C. & Miracle, D. B. Low-density, refractory multi-principal element alloys of the Cr–Nb–Ti–V–Zr system: microstructure and phase analysis. *Acta Mater.* **61**, 1545–1557 (2013).
19. Miracle, D. B. et al. Exploration and development of high entropy alloys for structural applications. *Entropy* **16**, 494–525 (2013).
20. Schuh, B. et al. Mechanical properties, microstructure and thermal stability of a nanocrystalline CoCrFeMnNi high-entropy alloy after severe plastic deformation. *Acta Mater.* **96**, 258–268 (2015).
21. Laurent-Brocq, M. et al. Insights into the phase diagram of the CrMnFeCoNi high entropy alloy. *Acta Mater.* **88**, 355–365 (2015).
22. Kozak, R., Sologubenko, A. & Steurer, W. Single-phase high-entropy alloys—an overview. *Z. Kristallogr.* **230**, 55–68 (2015).
23. Wu, Z., Bei, H., Otto, F., Pharr, G. M. & George, E. P. Recovery, recrystallization, grain growth and phase stability of a family of FCC-structured multi-component equiatomic solid solution alloys. *Intermetallics* **46**, 131–140 (2014).
24. Wu, Z., Bei, H., Pharr, G. M. & George, E. P. Temperature dependence of the mechanical properties of equiatomic solid solution alloys with face-centered cubic crystal structures. *Acta Mater.* **81**, 428–441 (2014).
25. Kaufman, L. & Nesor, H. *Co-Cr-Ni Phase Diagram, ASM Alloy Phase Diagrams Database.* (eds Villars, P., Okamoto, H. & Cenzual, K.) http://www1.asminternational.org/AsmEnterprise/APD (ASM International, 2006).
26. E08 Committee. *E1820-13 Standard Test Method for Measurement of Fracture Toughness* (ASTM International, 2013).
27. Haglund, A., Koehler, M., Catoor, D., George, E. P. & Keppens, V. Polycrystalline elastic moduli of a high-entropy alloy at cryogenic temperatures. *Intermetallics* **58**, 62–64 (2015).
28. Laplanche, G., Horst, O., Otto, F., Eggeler, G. & George, E. P. Microstructural evolution of a CoCrFeMnNi high-entropy alloy after swaging and annealing. *J. Alloys Compd.* **647**, 548–557 (2015).
29. Ritchie, R. O. The conflicts between strength and toughness. *Nat. Mater.* **10**, 817–822 (2011).
30. Zhang, Z.-J. et al. Nanoscale origins of the damage tolerance of the high-entropy alloy CrMnFeCoNi. *Nat. Commun.* **6**, 10143 (2015).
31. Dao, M., Lu, L., Shen, Y. F. & Suresh, S. Strength, strain-rate sensitivity and ductility of copper with nanoscale twins. *Acta Mater.* **54**, 5421–5432 (2006).
32. Lu, L., Chen, X., Huang, X. & Lu, K. Revealing the maximum strength in nanotwinned copper. *Science* **323**, 607–610 (2009).
33. Lu, K., Lu, L. & Suresh, S. Strengthening materials by engineering coherent internal boundaries at the nanoscale. *Science* **324**, 349–352 (2009).
34. Singh, A., Tang, L., Dao, M., Lu, L. & Suresh, S. Fracture toughness and fatigue crack growth characteristics of nanotwinned copper. *Acta Mater.* **59**, 2437–2446 (2011).
35. Hadfield, R. A. Hadfield's manganese steel. *Science* **12**, 284–286 (1888).
36. Schumann, V. H. Martensitische Umwandlung in austenitischen Mangan-Kohlenstoff-Stählen. *Neue Hütte* **17**, 605–609 (1972).
37. Remy, L. & Pineau, A. Twinning and strain-induced F.C.C.→H.C.P. transformation in the Fe-Mn-Cr-C system. *Mater. Sci. Eng.* **28**, 99–107 (1977).
38. Kim, T. W. & Kim, Y. G. Properties of austenitic Fe-25Mn-1Al-0.3C alloy for automotive structural applications. *Mater. Sci. Eng. A* **160**, 13–15 (1993).
39. Grässel, O., Frommeyer, G., Derder, C. & Hofmann, H. Phase transformations and mechanical properties of Fe-Mn-Si-Al TRIP-steels. *J. Phys. IV* **07**, C5-383–C5-388 (1997).
40. Grässel, O., Krüger, L., Frommeyer, G. & Meyer, L. W. High strength Fe-Mn-(Al, Si) TRIP/TWIP steels development—properties—application. *Int. J. Plast.* **16**, 1391–1409 (2000).

41. Frommeyer, G., Brüx, U. & Neumann, P. Supra-ductile and high-strength manganese-TRIP/TWIP steels for high energy absorption purposes. *ISIJ Int.* **43**, 438–446 (2003).

42. Chen, L., Zhao, Y. & Qin, X. Some aspects of high manganese twinning-induced plasticity (TWIP) steel, a review. *Acta Metall. Sin. Engl. Lett.* **26**, 1–15 (2013).

43. Read, D. T. & Reed, R. P. Fracture and strength properties of selected austenitic stainless steels at cryogenic temperatures. *Cryogenics* **21**, 415–417 (1981).

44. Mills, W. J. Fracture toughness of type 304 and 316 stainless steels and their welds. *Int. Mater. Rev.* **42**, 45–82 (1997).

45. Sokolov, M. *et al. Effects of Radiation on Materials: 20th International Symposium* (eds Rosinski, S., Grossbeck, M., Allen, T. & Kumar, A.) 125–147 (ASTM International, 2001).

46. Shindo, Y. & Horiguchi, K. Cryogenic fracture and adiabatic heating of austenitic stainless steels for superconducting fusion magnets. *Sci. Technol. Adv. Mater.* **4**, 319 (2003).

47. Sa, J. W. *et al. Twenty-First IEEE/NPS Symposium on Fusion Engineering 2005* 1–4 (IEEE, 2005).

48. Strife, J. R. & Passoja, D. E. The effect of heat treatment on microstructure and cryogenic fracture properties in 5Ni and 9Ni steel. *Metall. Trans. A* **11**, 1341–1350 (1980).

49. Syn, C. K., Morris, J. W. & Jin, S. Cryogenic fracture toughness of 9Ni steel enhanced through grain refinement. *Metall. Trans. A* **7**, 1827–1832 (1976).

50. Pense, A. W. & Stout, R. D. Fracture toughness and related characteristics of the cryogenic nickel steels. *Weld. Res. Counc. Bull* **205**, 1–43 (1975).

51. Stout, R. D. & Wiersma, S. J. *Advances in Cryogenic Engineering Materials.* (eds Reed, R. P. & Clark, A. F.) 389–395 (Springer, 1986).

Acknowledgements

This research was sponsored by the U.S. Department of Energy, Office of Science, Office of Basic Energy Sciences, Materials Sciences and Engineering Division, through the Materials Science and Technology Division at the Oak Ridge National Laboratory (for H.B, Z.W. and E.P.G.) and the Mechanical Behavior of Materials Program (KC13) at the Lawrence Berkeley National Laboratory (for B.G., K.V.S.T. and R.O.R.).

Author contributions

B.G., E.P.G. and R.O.R. designed the research; H.B. and E.P.G. made the alloy; B.G., A.H., K.V.S.T., H.B. and Z.W. mechanically characterized the alloy; B.G., A.H., K.V.S.T., H.B., E.P.G. and R.O.R. analysed and interpreted the data; B.G., E.P.G. and R.O.R. wrote the manuscript.

Additional information

Temperature-feedback upconversion nanocomposite for accurate photothermal therapy at facile temperature

Xingjun Zhu[1,*], Wei Feng[2,*], Jian Chang[3], Yan-Wen Tan[3], Jiachang Li[2], Min Chen[2], Yun Sun[2] & Fuyou Li[1,2,4]

Photothermal therapy (PTT) at present, following the temperature definition for conventional thermal therapy, usually keeps the temperature of lesions at 42–45 °C or even higher. Such high temperature kills cancer cells but also increases the damage of normal tissues near lesions through heat conduction and thus brings about more side effects and inhibits therapeutic accuracy. Here we use temperature-feedback upconversion nanoparticle combined with photothermal material for real-time monitoring of microscopic temperature in PTT. We observe that microscopic temperature of photothermal material upon illumination is high enough to kill cancer cells when the temperature of lesions is still low enough to prevent damage to normal tissue. On the basis of the above phenomenon, we further realize high spatial resolution photothermal ablation of labelled tumour with minimal damage to normal tissues in vivo. Our work points to a method for investigating photothermal properties at nanoscale, and for the development of new generation of PTT strategy.

[1] Institutes of Biomedical Sciences, Fudan University, 220 Handan Road, Shanghai 200433, China. [2] Department of Chemistry, State Key Laboratory of Molecular Engineering of Polymers, Fudan University, 220 Handan Road, Shanghai 200433, China. [3] Department of Physics, Fudan University, 220 Handan Road, Shanghai 200433, China. [4] Collaborative Innovation Center of Chemistry for Energy Materials, Fudan University, 220 Handan Road, Shanghai 200433, China. * These authors contributed equally to this work. Correspondence and requests for materials should be addressed to F.L. (email: fyli@fudan.edu.cn).

The past few decades have witnessed significant efforts in the treatment of cancer[1,2]. Among the existing treatment methods, thermal therapy has become an important treatment modality[3]. Conventional approaches including radiofrequency or microwave ablation have been widely used in clinic[4,5]. However, these approaches relying on macroscopic heat sources have a relatively large destruction range causing normal tissue damages and even some serious systemic side effects[6,7]. Photothermal therapy (PTT), using photoabsorbing molecules or nanoparticles as microscopic heat sources, is expected to improve the therapeutic accuracy and reduce injury to normal tissues[8–13]. However, the current method to monitor PTT regards the entire lesion containing PTT agents as a macroscopic heat source and keep the overall temperature of the lesion (here defined as apparent temperature) at high level in line with the temperature definition for conventional thermal therapy (usually 42–45 °C)[14,15]. In some cases, the temperature is even higher[16,17]. Such a high apparent temperature can damage normal tissues adjacent to the lesions due to massive heat transfer, therefore, leading to more side effect and inhibiting the therapeutic accuracy of PTT. In our view, distinguished from traditional thermal therapy methods using macroscopic heat source and requiring apparent temperature as reference, PTT uses heat source at nanoscale, so correspondingly the temperature of those nanoparticles (here defined as the eigen temperature) during photothermal process should be the prerequisite to determine the temperature threshold for effective and minimally harmful PTT. Although some previous studies have referred to the measurement of the eigen temperature of gold nanostructures[18,19], to date, no data were reported to utilize the eigen temperature during PTT in a real biosystem to achieve therapeutic effect with high spatial resolution under facile apparent temperature. It still remains unsolved to seek for an adequate thermal-sensitive system that is stable and not affected by the complex biological condition to report the eigen temperature of photothermal agents, thus determining the temperature threshold for accurate and facile therapy. If there exists a suitable way to monitor the eigen temperature during PTT, then it will not only open a window to learn the temperature of PTT agents at microscopic level, but also innovate the PTT strategy for better therapeutic efficacy.

To monitor the eigen temperature of PTT agents in biological systems, temperature-sensing luminescent materials are appropriate options as the optical signals provide high resolution and sensitivity. A series of luminescent temperature-sensing probes have been developed including organic dyes, polymers, QDs and lanthanide-based upconversion nanophosphors (Ln-UCNPs)[20–25]. Among those, Ln-UCNPs which allow the conversion of lower-energy light in the near-infrared (NIR) region into higher energy emissions have many advantages to be served as imaging or therapy agents such as superior photostability, non-blinking, absence of autofluorescence of biological tissue and low-energy NIR radiation[26–43]. On the basis of the above merits, Ln-UCNPs are ideal probes to real-time sensing the eigen temperature of PTT agents in biological system. Here we build a carbon-coated core-shell upconversion nanocomposite NaLuF$_4$:Yb,Er@NaLuF$_4$@Carbon (csUCNP@C) to investigate the possible difference between the apparent and eigen temperature and evaluate the therapeutic effect of implementing photothermal therapy at low apparent temperature, but a much higher eigen temperature in real biosystems. Moreover, this kind of nanocomposite can also be served as theranostic agents as the upconversion core and carbon shell endorse it with a good imaging quality and PTT efficacy.

Results

Characterization of temperature-feedback csUCNP@C. The synthetic procedure of csUCNP@C is shown in Supplementary Fig. 1. The carbon shell generated heat under 730-nm irradiation and the core of NaLuF$_4$:Yb,Er provided thermal-sensitive upconversion luminescence (UCL) emission under 980-nm excitation (Fig. 1). As shown by transmission electron microscopy (TEM), the NaLuF$_4$:Yb,Er nanoparticles (UCNPs, the core) were uniform in morphology with a diameter of ~ 25 nm (Fig. 2b). After coating with a non-doping NaLuF$_4$ layer, the shape of the formed nanoparticles was changed to rod-like and the size was increased to ~ 50 nm in length and ~ 40 nm in width (Fig. 2b), indicating the formation of the core-shell nanoparticles NaLuF$_4$:Yb,Er@NaLuF$_4$ (csUCNPs). X-ray powder diffraction (Supplementary Fig. 2c) patterns of UCNPs and csUCNPs were indexed as the hexagonal phase of NaLuF$_4$. The as-prepared oleate-capped csUCNPs were transformed into the aqueous phase using an acid-based ligand removal method reported by Capobianco *et al.*[44] To coat the shell layer of carbon, a glucose solution (0.4 mmol ml^{-1}) containing hydrophilic csUCNPs

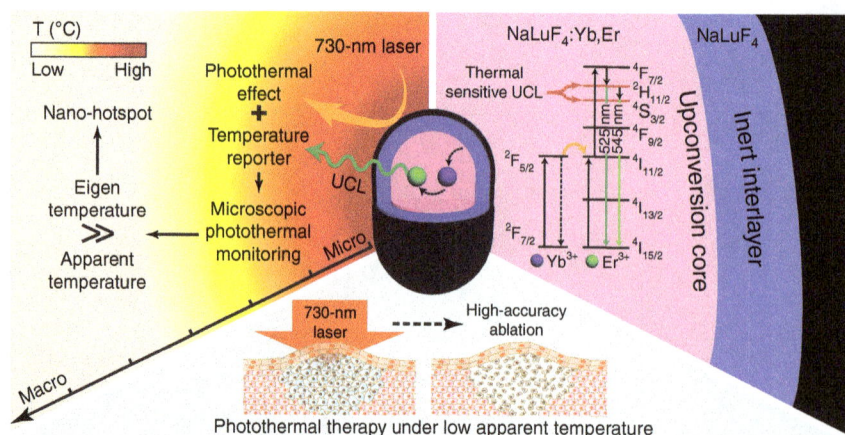

Figure 1 | Schematic of csUCNP@C for accurate PTT at facile temperature. The csUCNP@C exhibit both UCL emission and photothermal effect. With temperature-sensitive UCL emission, csUCNP@C was used to monitor the change in microscopic temperature of the photoabsorber (carbon shell) under 730-nm irradiation. The eigen temperature of csUCNP@C was much higher than the apparent temperature observed macroscopically, indicating that csUCNP@C acted as a nano-hotspot at the microscopic level. By utilizing the high eigen temperature during photothermal process, accurate PTT, which prevent the damage to normal tissues can be realized.

$(2 \, \text{mg ml}^{-1})$ was hydrothermally treated at $160\,°C$ for 2 h. Fourier-transform infrared spectroscopy indicated that the oleate species on csUCNPs were totally removed after acid treatment (Supplementary Fig. 4a) and a carbon layer was successfully coated on csUCNPs after the hydrothermal treatment, as the stretching bands referred to as $C-H$ and $C=C$ bonds appeared. A new coating layer was also visible in the TEM images (Fig. 2b; Supplementary Fig. 2b) and the monodispersity of the nanoparticles kept good (Supplementary Fig. 3). In particular, Raman spectra confirmed the dominance of an ordered conjugated π-bond structure in the outside shell layer, as the intensity ratio of the peaks of graphitic carbon and amorphous carbon reached 4.3 (Supplementary Fig. 4c). The highly graphitic components in the carbon shell, inducing effective π-plasmon, accounted for the broad absorption from visible to NIR region of csUCNP@C (Fig. 2d; Supplementary Fig. 5). Thus, the colour of the nanocomposites turned from

white of csUCNP to deep brown (Fig. 2d inset). After carbon coating, the average hydrodynamic diameter was increased from 56.5 to 77.0 nm (Supplementary Fig. 6b), and the weight percentage (wt%) of the organic components was enhanced from ~ 0.5 to ~ 3.7wt% (Supplementary Fig. 4d). These factors indicated that a carbon layer was successfully coated on the surface of csUCNPs by hydrothermal treatment. The as synthesized csUCNP@C can be easily dispersed in water and other buffers because polymerization of glucose occurred before carbonization and the hydrophilic polymer chains located at the outer layer of csUCNP@C, which make the nanoparticles very stable in aqueous phase[45]. Dynamic light scattering (DLS) data showed that the colloidal suspensions can be preserved for 2 weeks without any aggregation (Supplementary Fig. 7).

Importantly, the insertion of a non-doping NaLuF$_4$ interlayer (~ 7.5-nm thick) between the carbon shell and upconversion core of NaLuF$_4$:Yb,Er can both enhance the UCL emission and prevent luminescent quenching by the carbon shell. When the concentration of the luminescence centre of Er^{3+} was consistent, csUCNPs emitted stronger UCL at 540 nm (increased 4.9-fold) compared with UCNPs. The control NaLuF$_4$:Yb,Er@Carbon (UCNP@C) without a NaLuF$_4$ protective layer displayed an 83% decrease in UCL intensity, whereas csUCNP@C with a protective layer maintained 82% of the UCL intensity (Fig. 2c). Thus, the introduction of NaLuF$_4$ is necessary for fabricating the temperature-feedback PTT agent.

Photothermal properties of csUCNP@C were also evaluated by measuring the photothermal conversion efficiency under 730-nm laser irradiation $(1 \, \text{W cm}^{-2})$. Detailed data are shown in Supplementary Information (Supplementary Figs 8 and 9). The final heat-generation efficiency (η) is 38.1 % that is higher than those widely studied PTT agents such as gold nanorods (21%) and nanoshells (13%), $Cu_{2-x}Se$ (22%) and Cu_9S_5 (25.7%)[46,47]. The high heat-generation efficiency indicated that the carbon shell of csUCNP@C is a kind of excellent PTT agent to realize the therapeutic temperature at even lower laser-irradiation dosage. It should be noted that the 2-in-1 design that combined temperature-sensitive upconversion luminescence with photothermal carbon shell can realize both good temperature-sensing property and photothermal effect. Although other carbon materials such as carbon dots also exhibit special luminescent property that is marked by tunable emission wavelength and broad excitation spectra[48,49], their

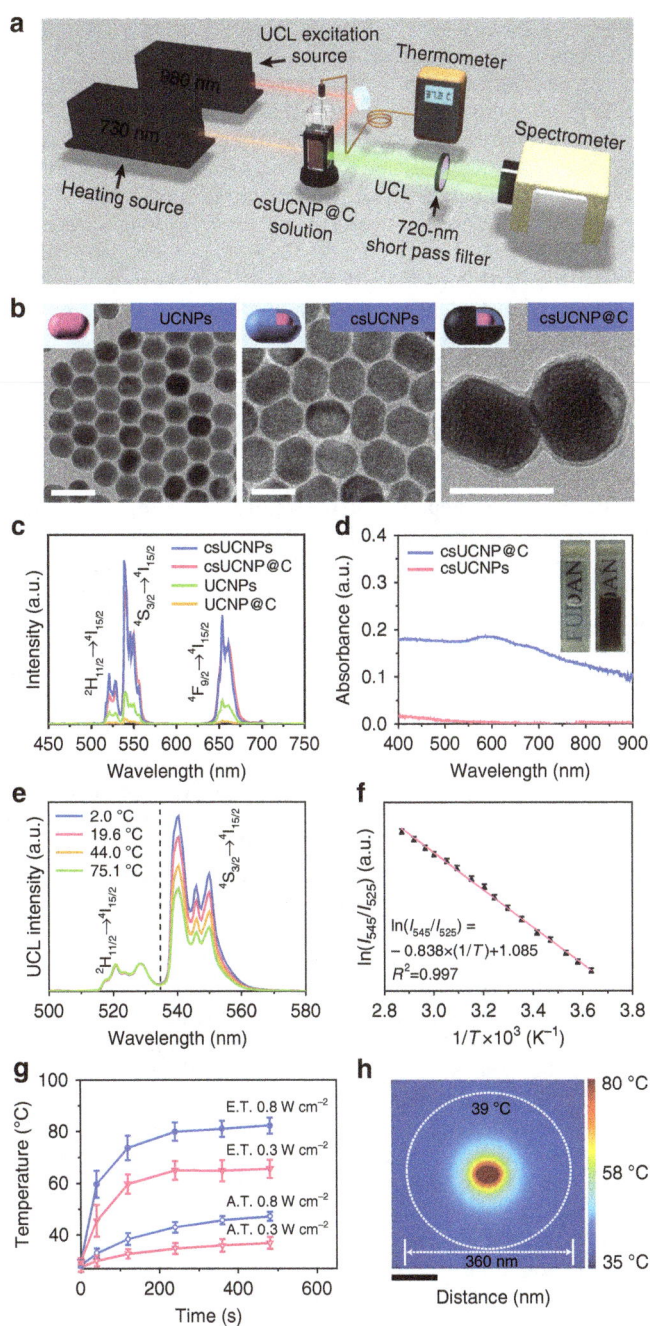

Figure 2 | Characterization and temperature-sensing properties of csUCNP@C. (**a**) Schematic diagram of the detection of the eigen temperature of csUCNP@C. (**b**) TEM images of UCNPs (left), csUCNPs (middle) and csUCNP@C (right). Scale bar, 50 nm. (**c**) UCL emission spectra of UCNPs, UCNPs@C, csUCNPs and csUCNP@C in the aqueous dispersion with the same concentration of luminescence centre Er^{3+}. (**d**) Powder absorption spectra of csUCNPs and csUCNP@C. (**e**) UCL emission spectra of Er^{3+}-doping csUCNP@C at different temperatures by external heating. The peaks were normalized at 528.5 nm. (**f**) A plot of $\ln(I_{525}/I_{545})$ versus $1/T$ to calibrate the thermometric scale for csUCNP@C. I_{525} and I_{545} indicate the UCL emission of the $^2H_{11/2} \rightarrow {}^4I_{15/2}$ and $^4S_{3/2} \rightarrow {}^4I_{15/2}$ transitions, respectively. Average values of I_{525}/I_{545} under different temperature were given to fit the calibration curve based on three times measurements of UCL spectrum. Error bars were defined as s.d. (**g**) Elevation of apparent temperature (A.T.) and eigen temperature (E.T.) of csUCNP@C $(1 \, \text{mg ml}^{-1})$ in aqueous dispersion under irradiation with a 730-nm laser at 0.8 and $0.3 \, \text{W cm}^{-2}$. Average value of A.T. and E.T. under different time points were given based on three times measurements. Error bars were defined as s.d. (**h**) Finite element method (FEM) simulation of the heat conduction of a single csUCNP@C nanoparticle induced by the photothermal process. Scale bar, 100 nm.

absorption spectrum usually located in ultraviolet to blue region and their photothermal effect under NIR laser irradiation has not been proved yet. Also, the temperature-sensing properties of carbon dots have not been clarified.

Observation of the eigen temperature of csUCNP@C. In the Er^{3+}-doped upconversion system, $^2H_{11/2} \to {}^4I_{15/2}$ (UCL emission centred at 525 nm) and $^2S_{3/2} \to {}^4I_{15/2}$ (centred at 545 nm) transitions were in close proximity to a thermal equilibrium ruled by the Boltzmann factor (equation 1)[24]:

$$(I_{525})/(I_{545}) = C \exp(-\Delta E/kT), \quad (1)$$

where I_{525} and I_{545} are the UCL emission of the $^2H_{11/2} \to {}^4I_{15/2}$ and $^4S_{3/2} \to {}^4I_{15/2}$ transitions, respectively; C is a constant determined by the degeneracy, spontaneous emission rate and photon energies of the emitting states in the host materials; ΔE is the energy gap separating the two excited states; k is the Boltzmann constant; T is temperature using the Kelvin scale. The changes in the UCL intensities as a function of temperature make it possible to quantitatively monitor the eigen temperature fluctuation of csUCNP@C when irradiated.

To simultaneously measure the apparent temperature and upconversion emission spectrum of the aqueous solution containing csUCNP@C, we designed and set-up a system by introducing a thermometer and upconversion emission spectroscope, as shown in Fig. 2a. First, we obtained a calibration curve to determine the relationship between UCL intensity and temperature. By heating the solution of csUCNP@C with a temperature controller (Supplementary Fig. 10), the UCL emission at 525 nm was correspondingly enhanced (Fig. 2e). The dependence of $\ln(I_{525}/I_{545})$ on the inverse temperature ($1/T$), which showed a linear behaviour (Fig. 2f), was well fitted as $\ln(I_{525}/I_{545}) = 1.085 - 0.838 \times (1/T)$ (T given in K). Judging from the signal changes of UCL intensity changes, csUCNP@C is a good optical nanothermometer with a relatively high sensitivity of 1% signal change K^{-1} around 35–40 °C and a high temperature resolution of about 0.5 K. Although some previous works have reported other methods for intracellular thermometry with excellent temperature resolution and sensitivity[50–52], those thermometry methods rely on single band emission without an internal ratiometric. Hence, the accuracy of temperature sensing will be strongly affected by other environmental factors, which can change the emission intensity such as absorption, scattering, tissue motions, autofluorescence and quenching centres. Moreover, their combination with photothermal therapy have not been exploited and the advantages of marked local temperature changes of PTT agents have not been fully recognized to improve PTT. The optical temperature sensor in csUCNP@C, $NaLuF_4$:Yb,Er, has a couple of temperature-sensitive emission bands (centred at 525 and 545 nm) for ratiometric thermometry, which is insusceptible to disturbance of environmental factors. The closely combined structure of temperature sensor and photothermal component in csUCNP@C is indispensable to monitor local temperature elevation of PTT agents during photothermal process and thus give us the chance to explore a new strategy of photothermal therapy with much less normal tissue damage.

To observe the difference of eigen and apparent temperature, csUCNP@C solution under 730-nm irradiation at various time points were recorded using the thermometer (Fig. 2a) to compare the eigen temperature of csUCNP@C and the apparent temperature. As shown in Fig. 2g, the eigen temperature of csUCNP@C was much higher than the apparent temperature at each time point, with a significant difference of $\sim$34 K at equilibrium with a power density of 0.8 W cm^{-2} at 730-nm irradiation. Under low-power illumination at 0.3 W cm^{-2} for

8 min, the apparent temperature of the solution was 36.6 °C, whereas the eigen temperature of csUCNP@C soared to 65.5 °C by determining the ratio of UCL emission at 525 to 545 nm. Even when irradiated for 2 min, the eigen temperature of csUCNP@C reached 59.7 °C, whereas the apparent temperature of the solution was 31.6 °C. To the best of our knowledge, this is the first time such a significant difference between eigen temperature of the PTT agent and apparent temperature of its surrounding system has been reported. It should be noted that $NaLuF_4$:Yb,Er as temperature sensor has a temperature resolution of 0.5 °C and sensitivity of 1% signal change per degree, which can be perfectly qualified for differentiating the remarkable temperature elevation (for example, eigen temperature elevation is 30.0 °C with 730-nm irradiation at 0.3 W cm^{-2} for 2 min) of PTT agents in microscopic state.

To shed light on the temperature distribution at the microscopic level, we used a single-particle model to simulate the possible temperature distribution around the nanoparticles (heating centre) in solution. As shown in Fig. 2h, the temperature of the nanoparticles was kept at 80 °C (in accordance with the highest eigen temperature in our experiment). Water was chosen as the ambient condition for the nanoparticles and the external boundary was kept at 35 °C. The model simulated a steady state with heat conduction mode (see Methods). The simulation showed that the temperature of water declined with increasing distance from the nanoparticles, below the lethal temperature to cancer cells (39 °C) at $\sim$360 nm in diameter (Fig. 2h). Therefore, it is reasoned that under 730-nm laser irradiation, csUCNP@C in the aqueous phase become localized hot spots whose eigen temperature is much higher than the ambient temperature, and thus their effective range for killing cancer cells by PTT is in the nanoscale. Therefore, it was theoretically proved that PTT can be performed at low apparent temperature and high spatial accuracy.

Facile and high-accuracy PTT *in vitro*. To confirm that PTT is effective at low apparent temperature, experiments at the macroscopic level were carried out. First, non-labelled HeLa cells were irradiated by a 730-nm laser (0.3 W cm^{-2}) to figure out the laser-induced heating effect and the final apparent temperature elevation in the laser spot, T_0 (35.6 °C, Supplementary Fig. 13b), was set as a benchmark. ΔT shown in Fig. 3b was the temperature increment subtracting T_0 after photothermal process or external heating. Calcein acetoxymethyl ester (Calcein AM) and propidium iodide (PI) double staining were used to confirm the state (live or dead) of cells. When csUCNP@C-incubated HeLa cells were irradiated under 0.3 W cm^{-2} for less than 3 min, Calcein AM/PI staining indicated that csUCNP@C-incubated cells were dead when ΔT was 1.4 °C. Non-labelled cells with external heating holder indicated that cells were alive when ΔT was 1.4 °C and were dead only when ΔT reached 3.6 °C. These results suggest that the photothermal effect of csUCNP@C can effectively kill the cells without heating the surrounding solution to a high apparent temperature.

As proof of concept experiments to investigate the accuracy of PTT, we mixed csUCNP@C-labelled HeLa cells with non-labelled cells together and let them adhere to the culture dish. In our designed PTT system (Fig. 3a), two lasers, one at 980 nm and the other at 730 nm, were used to generate upconversion emission and the photothermal effect of csUCNP@C, respectively. Before 730-nm irradiation, csUCNP@C-treated cells (incubating dosage, 200 µg ml^{-1}), which have a relatively wide distribution of csUCNP@C in cytoplasm with a uptake of 4.4 pg csUCNP@C per cell, could be stained with Calcein AM (Supplementary Fig. 11a–d) showing they were all alive. Methyl thiazolyl tetrazolium (MTT) assays also confirmed that csUCNP@C have

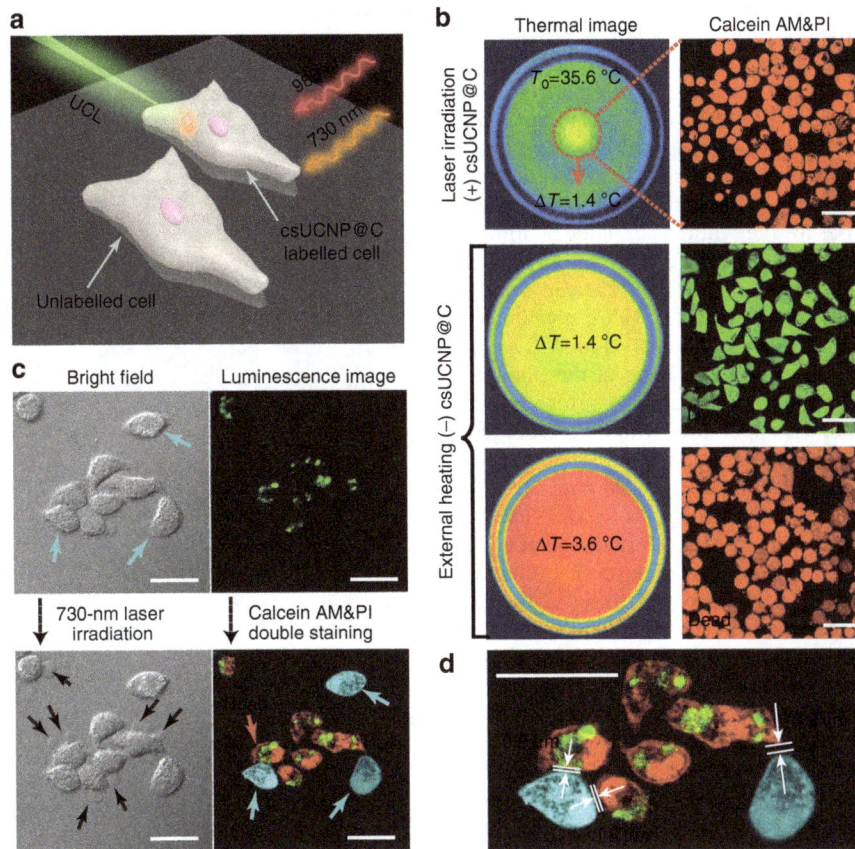

Figure 3 | csUCNP@C for high-accuracy PTT at cell level. (a) Schematic diagram of PTT in cells. **(b)** Thermal images and Calcein AM and PI double-stained images of HeLa cells treated with photothermal ablation or external heating. Non-labelled cells were irradiated by 730-nm laser (0.3 W cm^{-2}) and the final apparent temperature elevation in the laser spot ($T_0 = 35.6$ °C) was set as a benchmark. ΔT was the difference between apparent temperature and T_0. In external heating, cells were alive when $\Delta T = 1.4$ °C and dead when $\Delta T = 3.6$ °C. With 730-nm laser irradiation, csUCNP@C labelled cells were dead when $\Delta T = 1.4$ °C indicating that the eigen temperature of csUCNP@C had reached to a lethal temperature to the cells even though the apparent temperature was still safe. Scale bar, 50 μm. **(c)** Photothermal therapy of HeLa cells under 730-nm laser irradiation at 0.3 W cm^{-2} for 5 min. Cells labelled with csUCNP@C showed a strong UCL signal in the cytoplasm (green). The signal is collected in the wavelength region of 520–550 nm. After 730-nm irradiation, dead cells showed conspicuous cytoplasm leakage which labelled with black arrows. Calcein AM (cyan) and PI (red) double-staining showed that only the cells labelled with csUCNP@C were dead. Scale bar, 30 μm. **(d)** Amplified image of the luminescent cell images in **c**. The distance between the adjacent live and dead cells was measured. The minimum distance was ~0.9 μm. Scale bar, 30 μm.

no toxic effect on cells under this incubating dosage (Supplementary Fig. 11e,f). A low-power density (0.3 W cm^{-2}) of 730-nm laser was adopted for photothermal cell ablation. Determined by UCL intensity changes on the csUCNP@C-labelled HeLa cells (Supplementary Fig. 14), a huge eigen temperature elevation was observed after 730-nm laser irradiation from 37 °C to nearly 60 °C (Supplementary Fig. 15). Such a high eigen temperature reflects enormous localized heat which is very critical in killing the cancer cells. As shown in Fig. 3c, only the cells with upconversion emission displayed red fluorescence from PI ($\lambda_{ex} = 633$ nm), whereas the others without upconversion signals showed the fluorescence signal from Calcein AM (cyan, $\lambda_{ex} = 488$ nm). The results indicated that, after 730-nm irradiation, the csUCNP@C-labelled cells were dead and the free cells were still alive. Bright-field image of the csUCNP@C-labelled cells showed membrane damage and leakage of the cytoplasm. Moreover, laser irradiation of csUCNP@C-labelled HeLa cells and non-labelled C2C12 cells indicated that photothermal effect can only occurred in the cancer cells without damaging normal cells (Supplementary Fig. 12). MTT assays also confirmed these results by a remarkable decrease in cell viability after 730-nm irradiation (Supplementary Fig. 10f). By further decreasing the power density of the 730-nm laser to 0.2 W cm^{-2}, early stage cell apoptosis was observed only in csUCNP@C-labelled cells

(Supplementary Fig. 13a), using the Annexin V-FITC/PI double staining method. On the basis of the cell selective ablation experiment (Fig. 3b,c; Supplementary Fig. 15), we can conclude that high temperature of a limited space is enough to kill cells, while the overall temperature changes little. In other words, with the microscopic temperature monitoring technology, there is no longer a need to use overall temperature to monitor PTT. The usage of overall temperature of cells for PTT, which only involves the average value of high-temperature region containing csUCNP@C and low-temperature region without csUCNP@C, will neglect the significance of localized high temperature in killing the cancer cells and obviously restrict the therapeutic accuracy. It is worth noting that the distance between PTT-affected cells and unaffected cells in the confocal image (Fig. 3c,d) was at the micrometre level and the minimum distance was only 0.9 μm, that is, the photothermal ablation under a low apparent temperature was proved to have very high spatial resolution at the microscopic level.

Facile and high-accuracy PTT *in vivo*. To further confirm the feasibility of photothermal therapy at low apparent temperature and investigate the heat-conduction process in the living body, it is crucial to know about the eigen temperature fluctuations of csUCNP@C under laser irradiation in biological tissue. As it is difficult to obtain the temperature calibration curve and to investigate the heat conduction process in living animals due to some limiting factors in instruments, tissue phantoms were used to simulate the temperature elevation of csUCNP@C in biological tissue (Fig. 4a). The tissue phantom is consisted of gelatin and a certain amount of haemoglobin and intralipid to simulate the absorption and scattering properties of real tissue (see Supplementary Methods for the synthesis of tissue phantom). The temperature calibration curve (from 0 to 100 °C) obtained in a single-layer phantom containing csUCNP@C showed a linear behaviour [$\ln(I_{525}/I_{545}) = 0.924 - 0.687 \times (1/T)$] ($T$ given in K, Fig. 4b), which was similar to that detected in solution (Fig. 2f). A double-layer phantom where the lower layer of the phantom containing csUCNP@C (Layer 1, ∼10-mm thick) simulating the tumour area for therapy and the upper layer without csUCNP@C (Layer 2, ∼4-mm thick) simulating normal tissue (Fig. 4a) was used to investigate the heat conduction in facile and excessive photothermal process. The photothermal effect of csUCNP@C in the phantom tissues was triggered by 730-nm irradiation at 0.3 or $0.8 \, W \, cm^{-2}$. The apparent temperatures of layer 1 ('tumour') and layer 2 ('normal tissue') were recorded using a thermal camera, and the eigen temperature of csUCNP@C in layer 1 was calculated by the upconversion emission spectrum (Fig. 4a). Following $0.8 \, W \, cm^{-2}$ irradiation for 8 min, layer 1 was heated to 44.5 °C (apparent temperature) and layer 2 also underwent an obvious temperature elevation to 43.3 °C due to massive heat conduction, that is, the overheating photothermal effect. However, following mild laser irradiation at $0.3 \, W \, cm^{-2}$, the apparent temperature of layer 1 was moderate at 37.3 °C and layer 2 only showed slight heat conduction (35.2 °C), as shown in Fig. 4c. Meanwhile, the eigen temperature of csUCNP@C within the phantom obtained by the UCL spectrum was 56.7 °C, which is sufficient to kill cancer cells (Fig. 4d). Therefore, it can be safely concluded that csUCNP@C as a PTT agent can take effect under mild apparent temperature elevation, while excessive laser power aggravates injury of normal tissue. It should be noted that in previous publications on PTT, an overheating effect existed during the therapy process due to the absence of a temperature-feedback unit in the photothermal agent.

The above-mentioned results encouraged us to further assess the therapeutic effect of csUCNP@C in small living animals. The csUCNP@C-incubated (4.4 pg per cell) HeLa cells (1×10^7 cells) were subcutaneously transplanted into nude mice for tumour growth. Fourteen days after transplantation, a tumour with a diameter of ∼0.6 cm was observed and displayed strong UCL signals collected by a 720-nm short pass filter (Fig. 4f; Supplementary Fig. 19a). The apparent temperature in the tumour area was recorded by an infrared thermal imaging device. The eigen temperature in the tumour area was calculated by upconversion luminescence of csUCNP@C. On the basis of the therapeutic data and the heat conduction behaviour under two sets of 730-nm laser power density (0.3 and $0.8 \, W \, cm^{-2}$) in Fig. 4c, we have summarized and proposed a model of temperature-feedback photothermal treatment system and make a demonstration of this feedback treatment with the strategies given in this work (Supplementary Fig. 18). In this model, controller (in this work, controller is experimenter) will make decision on the treatment strategy. Strategy box stores a series of photothermal therapy strategies for selection (in this work, strategies are 1# and 2# with 730-nm laser at 0.3 and $0.8 \, W \, cm^{-2}$, respectively.). csUCNP@C receives the strategy output and is served as photothermal agent and eigen temperature reporter for feedback signal input. Eigen temperature of csUCNP@C with different power density of 730-nm laser irradiation reported by UCL spectra (Supplementary Fig. 17)

showed that both sets of power density can result in an enough high eigen temperature to ablate cancer cells (61.5 °C with $0.3 \, W \, cm^{-2}$ and 73.1 °C with $0.8 \, W \, cm^{-2}$). The temperature elevation through heat conduction in layer 2 ('normal tissue') is

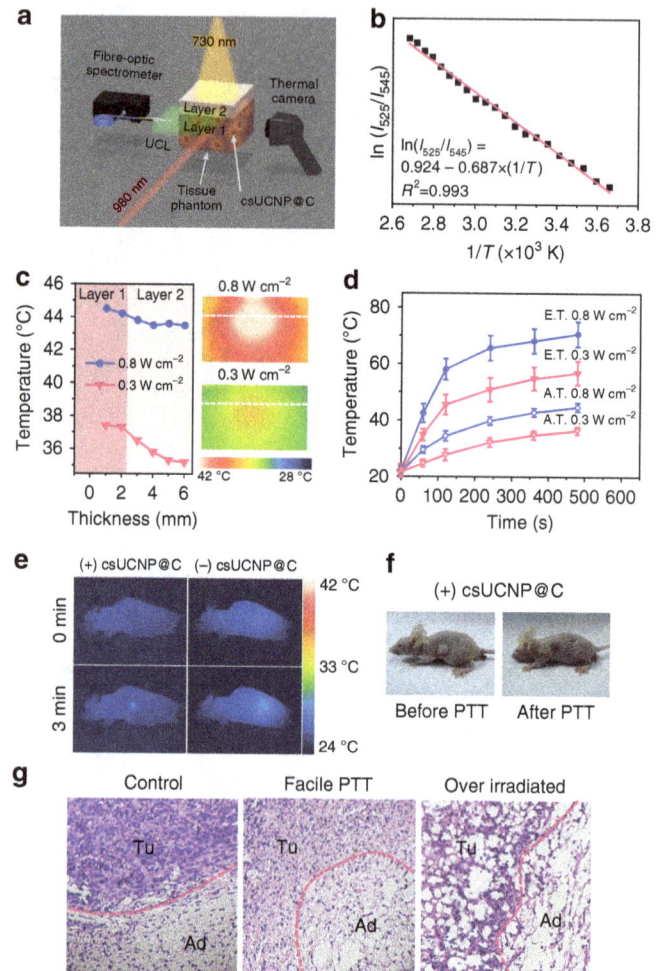

Figure 4 | csUCNP@C for high-accuracy PTT *in vivo*. (a) Schematic diagram of the feasibility of PTT *in vivo* using a tissue phantom. (b) A plot of $\ln(I_{525}/I_{545})$ versus $1/T$ to calibrate the thermometric scale for csUCNP@C in the tissue phantom. (c) Apparent temperature versus the thickness of the tissue phantom (from layer 1 to layer 2) under 730-nm irradiation at two different power densities (left panel). Thermal images of longitudinal sections of the phantom within two irradiation power densities to show the heat conduction process. The white dashed line separates the simulated 'tumour' and 'normal tissue' (right panel). (d) Elevation of apparent temperature (A.T.) and eigen temperature (E.T.) of csUCNP@C in the tissue phantom under irradiation with 730-nm laser at 0.8 and $0.3 \, W \, cm^{-2}$. Average values of A.T. and E.T. under different time points were given based on three times measurement. Error bars were defined as s.d. (e) Thermal images of nude mice with (left panel) and without (right panel) csUCNP@C-labelled HeLa cell tumours under 730-nm irradiation ($0.3 \, W \, cm^{-2}$). (f) Representative photos of nude mice transplanted with csUCNP@C-labelled HeLa cells under 730-nm irradiation ($0.3 \, W \, cm^{-2}$). (g) H&E histologic section of the border of tumour and normal fat tissue. The tumour region (Tu) and the adipocytes (Ad) in normal fat tissue of the mice without 730-nm irradiation (left, control) is compact and the tumour cells are stretched. Following photothermal treatment (middle, Facile PTT), the tumour region became loose and fragile and the tumour cells are atrophic. The adipocytes (Ad) in normal fat tissue are intact with minimal damage. However, following high-power irradiation with the 730-nm laser ($0.8 \, W \, cm^{-2}$), both Tu and Ad suffered extreme damage (right, Over irradiated).

confined within 2 °C under 0.3 W cm^{-2} irradiation. Considering that the body temperature is 37 °C and 2 °C elevation will not exceed 40 °C that cause protein denaturation[53], treatment strategy 1# with a power density of 730-nm laser at 0.3 W cm^{-2} was chosen for therapy. Tumour-bearing mice with csUCNP@C-labelled were exposed to a 730-nm laser (0.3 W cm^{-2}) for 3 min and the difference between final apparent temperature (at 3 min) and the initial temperature (at 0 min) was controlled at ~1.5 K (Fig. 4e). The tumour-bearing mice without csUCNP@C treatment (control) under 0.3 W cm^{-2} 730-nm laser irradiation showed a slight increase in apparent temperature (Fig. 4e). The tumour in each mouse in the treatment group was exposed to the 730-nm laser every day. Five days later, the tumours in the treatment group shrank and were finally eliminated without any regrowth (Supplementary Figs 19b and 21; Supplementary Table 1). In contrast, neither csUCNP@C-treatment nor laser irradiation affected tumour growth and the tumour size increased rapidly. Reference groups (including untreated mice, mice exposed to the 730-nm laser only, and csUCNP@C-treated mice without 730-nm irradiation) showed an average lifespan for tumour-bearing mice of ~22 days, while the mice in the PTT experimental group survived for over 40 days without mortality (Supplementary Fig. 19b,c). Hematoxylin and eosin (H&E) histopathological analysis indicated that malignant cells at the tumour site (Tu) in the facile PTT treatment group (730-nm laser, 0.3 W cm^{-2}) showed obvious shrinkage and fragmented nuclei (Fig. 4g), whereas the control group showed no conspicuous necrosis (Fig. 4g). Furthermore, the adjacent subcutaneous adipocytes (Ad) showed intact morphology in the control group and in the facile PTT treatment group (0.3 W cm^{-2}), while both Tu and Ad sites in the 730-nm over-irradiated group (0.8 W cm^{-2}) that had a high apparent temperature (Supplementary Fig. 16) were severely damaged (Fig. 4g). Hence, by using csUCNP@C as a photothermal agent, PTT at a mild apparent temperature is successfully achieved in living animals without damaging normal tissue.

Furthermore, using folic-acid-modified csUCNP@C (FA-csUCNP@C) as the photothermal agent, targeting PTT of the HeLa tumour-bearing balb/c mice under mild apparent temperature was investigated. After intravenous injection of FA-csUCNP@C (2 mg ml^{-1}, 200 µl) for 2 h, strong UCL signals were detected in tumour region (Supplementary Fig. 22a). Ex vivo UCL imaging indicated the biodistribution of csUCNP@C in other organs (Heart, Liver, Spleen, Lung, Kidney and Tumour) (Supplementary Fig. 22b). Histological and serum biochemistry assays suggested no evident toxic effects in vivo within one week of FA-csUCNP@C administration when compared with the untreated group (Supplementary Fig. 20; Supplementary Methods for detailed experimental procedures). After 730-nm irradiation (0.3 W cm^{-2}) for 3 min three times per day for 6 days, the tumours shrank and were finally eliminated (Supplementary Figs 22c and 23; Supplementary Table 2), and the PTT-treated mice survived for over 2 months without mortality (Supplementary Fig. 22d). Thus, we have successfully proved that FA-csUCNP@C can be used as a theranostic agent in vivo.

Discussion

We demonstrated carbon-coated core-shell upconversion nanocomposite NaLuF$_4$:Yb,Er@NaLuF$_4$@Carbon (csUCNP@C) for monitoring of microscopic temperature in photothermal process. Under laser irradiation at 730 nm, the carbon shell serves as an excellent photothermal agent for cancer therapy and simultaneously heats up the nanocomposite. By analysing upconversion luminescence emitted from the NaLuF$_4$:Yb,Er core of csUCNP@C during the photothermal process, microscopic temperature of photothermal agent was detected and found to be much greater than the temperature at macroscopic level. By utilizing this phenomenon, we selectively ablated csUCNP@C-labelled cancer cells under mild apparent temperature without harming adjacent non-labelled cells. The minimum separation between ablated cell and ambient preserved cell is 0.9 µm. High spatial resolution photothermal ablation in vivo of tumour with minimal damage to normal tissues was also realized at low apparent temperature. In stark contrast to the existing principle of PTT focusing on elevating the apparent temperature to overheating level, which can cause severe adverse effects in normal tissues near the tumour area, our approach relying on csUCNP@C as a temperature-feedback photothermal agent indicated that an effective photothermal treatment with high accuracy can be realized under moderate conditions. This point proves that the strategy to ensure enough heating of the photothermal agent to ablate the labelled cancer cell and to simultaneously circumvent heat conduction to non-labelled normal tissues is possible. The indispensable advantage of this work is to use a microscopic temperature-feedback system to point out an optimized irradiation dose for facile photothermal therapy, which is different from the common understanding and cannot be achieved by previous macroscopic temperature measuring method. If automatic controlling device with automatic spectrum analysing and laser power controlling abilities is integrated, then the strategies of temperature-feedback photothermal therapy will be more diversified and can be conducted more conveniently. It is reasonable to presume that PTT possessing high therapeutic accuracy and mild treatment conditions can do more sophisticated operations such as precise lymphadenectomy, embolization of tumour microvessels and low-injury intervention treatment around vital organs. Our work presented a powerful tool to give an insight into the photothermal process at microscopic level. By utilizing the merits of the presented temperature-feedback upconversion nanocomposite, the mode and concept of PTT will be changed profoundly.

Methods

Synthesis of oleate-coating NaLuF$_4$:Yb,Er/Tm nanoparticles. Spherical-like oleate-coating NaLuF$_4$:Yb,Er nanoparticles (OA-UCNPs) were synthesized via a modified solvothermal method[54]. In a typical procedure, 1 mmol lanthanide chloride (78% mol Lu, 20% mol Yb and 2% mol Er) were mixed with 6 ml oleic acid (OA) and 17 ml 1-octadecene in a 100-ml three-necked flask. The resulting mixture was degassed at 90 °C for 20 min, and then heated to 160 °C for 1 h to form a transparent solution. After that, the solution was cooled to room temperature. Then 2.5 mmol NaOH and 4 mmol NH$_4$F dissolved in 5 ml methanol were added and the mixture was degassed for another 30 min at 90 °C. Thereafter, the solution was heated to 300 °C as quickly as possible and the temperature was maintained for 1 h under an argon atmosphere. When the reaction was complete, an excess amount of ethanol was poured into the solution at room temperature. Nanoparticles were collected by centrifugation and washed three times with ethanol/cyclohexane (1:1 v/v). The as-obtained nanoparticles OA-UCNPs were dispersed in 5 ml cyclohexane for the following synthesis.

Synthesis of NaLuF$_4$:Yb,Er@NaLuF$_4$ nanoparticles. Rod-like oleate-coating NaLuF$_4$:Yb,Er@NaLuF$_4$ nanoparticles (OA-csUCNPs) were prepared by epitaxial growth on OA-UCNPs via a similar solvothermal method. One millimole of LuCl$_3$ was mixed with 12 ml oleic acid and 15 ml 1-octadecene in a 100-ml three-necked flask. Analogous to the procedure in synthesizing OA-UCNPs, the resulting mixture was degassed at 90 °C for 20 min, and then heated to 160 °C for 1 h to form a transparent solution and then cooled to room temperature. After that, 5 ml cyclohexane solution containing OA-UCNPs was added into the system dropwise. The system was kept at 80 °C for 30 min to evaporate cyclohexane. Then 2.5 mmol NaOH and 4 mmol NH$_4$F dissolved in 5 ml methanol were added into the mixture, degassed at 90 °C for 30 min, and finally maintained at 300 °C for 1 h. OA-csUCNPs were collected by centrifugation, washed three times with ethanol/cyclohexane (1:1, v/v), subsequently washed with acetone and isolated by centrifugation. The products were dried and stored for further use.

Synthesis of carbon-coated csUCNPs. The OA-csUCNPs underwent a ligand-exchange process to make them water-soluble according to a reported method[44]. Briefly, OA-csUCNPs were dispersed in 10 ml aqueous solution (pH = 4) by adding 0.1 mol l[−1] HCl and stirring for 2 h. Diethyl ether was used to extract oleic acid yielded from the protonated oleate ligand. After extraction was carried out three times, the products in the water layer were collected by centrifugation and washed three times with deionized water. The resulting ligand-free csUCNPs were easily dispersed in water. Then a certain amount of ligand-free csUCNPs were added into 0.4 M glucose aqueous solution and were redispersed by sonication. The concentration of nanoparticles was kept at 2 mg ml[−1]. The solution was transferred to a 50-ml autoclave, sealed and hydrothermally treated at 160 °C for 2 h. Finally, the system was allowed to cool to room temperature and carbon-coated csUCNPs (csUCNP@C) were collected by centrifugation and washed three times with deionized water.

Synthesis of folic acid-conjugated csUCNP@C. To synthesize FA functionalized csUCNP@C, 10 mg FA was mixed with 10 mg 4-dimethylaminopyridine and 20 mg csUCNP@C in 2 ml anhydrous dimethylformamide under a N₂ atmosphere. sixteen milligrams of N,N′-dicyclohexylcarbodiimide and 20 µl triethylamine was then added into the mixture and stirred for 24 h. An excess of diethyl ether was added to precipitate FA-conjugated csUCNP@C (FA-csUCNP@C). The as obtained FA-csUCNP@C were washed with methanol once and washed with ethanol twice and finally redispersed in deionized water.

Simulation of temperature distribution in one-nanoparticle system. Here we used a finite element method to simulate the heat conduction process of a single nanoparticle in the aqueous phase. An ellipse with a long axis of 50 nm and short axis of 40 nm was set as the nanoparticle. A circle with a diameter of 1 µm was set as the water surroundings. The temperature of the nanoparticle was set at 80 °C and the external boundary of water was set at 35 °C. The model only considered the heat-transfer process in the static condition. The temperature distribution presented is the final equilibrium state. The initial temperature was set at 35 °C.

Calculation of the photothermal conversion efficiency. According to the method described in the literature, the total energy conservation for the system can be expressed by equation 2.

$$\sum_i m_i C_{p,i} \frac{dT}{dt} = Q_{cs} + Q_B - Q_{sur}, \tag{2}$$

where m and C_p are the mass and heat capacity of water, respectively, T is the solution temperature, Q_{cs} is the energy induced by carbon shell of csUCNP@C, Q_B is the baseline energy induced by the sample cell, and Q_{sur} is heat conduction away from the surface by air.

Q_{cs} is caused by the π-plasmon of the carbon shell under irradiation of 730-nm laser:

$$Q_{cs} = I(1 - 10^{-A_{730}})\eta, \tag{3}$$

where I is the laser power, η is the conversion efficiency from incident laser energy to thermal energy, and A_{730} is the absorbance of carbon shell of csUCNP@C at wavelength of 730 nm. On the other hand, Q_B, expressing heat dissipated from light absorbed by the sample cell, was measured independently to be 28.4 mW using a quartz cuvette containing pure water without csUCNP@C. Moreover, Q_{sur} is in proportion to temperature for the outgoing thermal energy, as given by equation 4:

$$Q_{sur} = hS(T - T_{amb}), \tag{4}$$

where h is heat transfer coefficient, S is the surface area of the container, and T_{amb} is ambient temperature of the surroundings.

According to equation 4, when the system temperature will reach a maximum, the heat input is equal to heat output:

$$Q_{CS} + Q_B = hS(T_{max} - T_{amb}), \tag{5}$$

where T_{max} is the equilibrium temperature. The 730-nm laser heat-conversion efficiency (η) can be determined by substituting equation 3 for Q_{cs} into equation 5 and rearranging to get

$$\eta = \frac{hS(T_{max} - T_{amb}) - Q_B}{I(1 - 10^{-A_{730}})}, \tag{6}$$

where Q_B was measured independently to be 28.4 mW, the ($T_{max} - T_{amb}$) was 22.3 °C according to Supplementary Fig. 8a, I is 1 W cm[−2], A_{730} is the absorbance (1.254) of csUCNP@C at 730 nm (Supplementary Fig. 9). Here hS is calculated by introducing θ, is defined as the expression below:

$$\theta = \frac{T - T_{amb}}{T_{max} - T_{amb}}, \tag{7}$$

and a sample system time constant τ_s

$$\tau_s = \frac{\sum_i m_i C_{p,i}}{hS}, \tag{8}$$

which is substituted into equation 5 and rearranged to yield

$$\frac{d\theta}{dt} = \frac{1}{\tau_s}\left[\frac{Q_{cs} + Q_B}{hS(T_{max} - T_{amb})} - \theta\right], \tag{9}$$

At the cooling stage of the aqueous dispersion of the csUCNP@C, the light source was shut off, the $Q_{cs} + Q_B = 0$, reducing the equation 10:

$$dt = -\tau_s \frac{d\theta}{\theta}, \tag{10}$$

and integrating, giving the expression:

$$t = -\tau_s \ln\theta, \tag{11}$$

Therefore, time constant for heat transfer from the system is determined to be $\tau_s = 120.5$ s by applying the linear time data from the cooling period (after 400 s) versus negative natural logarithm of θ (Supplementary Fig. 8b). In addition, the m is 0.5 g and the C is 4.2 J g[−1]. Thus, according to equation 11, the hS is deduced to be 17.4 mW °C[−1]. Substituting 17.4 mW °C[−1] into the hS into equation 6, the 730-nm laser heat conversion efficiency (η) of csUCNP@C can be calculated to be 38.1%.

Eigen temperature measurement of csUCNP@C in solution. To obtain temperature-upconversion luminescence calibration curve in solution (Apparatus diagram is shown in Supplementary Fig. 10.), a quartz cuvette containing csUCNP@C aqueous dispersion (2 ml, 0.5 mg ml[−1]) was placed in Edinburgh FLS-920 fluorescence spectrometer with an external temperature controller. Aqueous solution was heated from temperature from 5 to 100 °C and the corresponding UCL from 500 to 580 nm was recorded with excitation of a continuous wave (CW) 980-nm laser (50 mW cm[−2]). The evolutions of the ratio of UCL emission peaks centred at 525 and 545 nm as function of temperature are used as calibration curve for eigen temperature monitoring. To evaluate the photothermal process of csUCNP@C (Apparatus diagram is shown in Fig. 2a.), a quartz cuvette containing an aqueous dispersion (2 ml) of csUCNP@C (0.5 mg ml[−1]) was irradiated with an optical fibre coupled 730-nm diode-laser (Weining Technology Development Co., Ltd. China) for 8 min at a laser power of 0.3 and 0.8 W cm[−2]. A CW 980-nm laser was used to generate the upconversion luminescence. UCL spectra, from 500 to 580 nm, were collected by Edinburgh FLS-920 fluorescence spectrometer at different time intervals (0, 40, 120, 240, 360 and 480 s). The eigen temperature of csUCNP@C was determined from the ratio of the luminescence peaks which centred at 525 nm and 545 nm. Apparent temperature changes of the solution were recorded by a thermocouple thermometer.

Cell culture and confocal UCL imaging in vitro. HeLa cells and C2C12 myoblasts were provided by the Institute of Biochemistry and Cell Biology, SIBS, CAS (China). The cell lines used in this work (HeLa cells and C2C12 myoblasts) do not appear in the list of mis-identified cell lines made by International Cell Line Authentication Committee (ICLAC). Cells were grown in RPMI 1640 (Roswell Park Memorial Institute medium) supplemented with 10% fetal bovine serum at 37 °C and 5% CO₂. Cells (5×10^8 l[−1]) were plated on 14-mm glass coverslips under 100% humidity and allowed to adhere for 24 h. After washing with PBS, the cells were incubated in a serum-free medium containing 200 µg ml[−1] csUCNP@C at 37 °C for 2 h under 5% CO₂, and then washed with PBS three times to get rid of excess nanoparticles. Confocal UCL imaging was performed on our designed laser scanning UCL microscope with an Olympus FV1000 scanning unit. The set-ups of the confocal UCL microscopy and upconversion luminescence in vivo imaging system are detailed in ref. 55,56, respectively. The cells were excited by a CW laser operating at 980 nm (Connet Fiber Optics, China) with the focused power of ~19 mW. A 60 × oil-immersion objective lens was used and luminescence signals were detected in the wavelength region of 500–580 nm. To qualitatively assess the photothermal effect in vitro, cells were irradiated with a 730-nm laser at a power density of 0.3 W cm[−2] for 5 min, and then stained with PI (propidium iodide) and calcein AM. PI signals were collected at 600-680 nm excited with a CW 543 nm laser. Calcein AM signals were collected at 500–580 nm excited with a CW 488-nm laser. To assess the apoptosis promoting effect of csUCNP@C in vitro, cells were irradiated with a 730-nm laser at a low-power density of 0.2 W cm[−2] for 5 min, and then stained with PI and Annexin V-FITC. Annexin V-FITC signals were collected at 500–550 nm excited with a CW 488-nm laser.

Temperature mapping of csUCNP@C labelled cells. For eigen temperature monitoring in cell, confocal UCL imaging was performed on our designed laser scanning UCL microscope with an Olympus FV1000 scanning unit. The HeLa cells were incubated in a serum-free medium containing 200 µg ml[−1] csUCNP@C at 37 °C for 2 h under 5% CO₂, and then washed with PBS three times to get rid of excess nanoparticles. Then csUCNP@C labelled HeLa cells were excited by a CW laser operating at 980 nm (Connet Fiber Optics, China) with the focused power of ~19 mW. A 60 × oil-immersion objective lens was used and luminescence signals were detected in the wavelength region of 540–570 nm (I_{545}) and 515–535 nm (I_{525}), respectively. Eigen temperature mapping of csUCNP@C labelled cells before and after 730-nm irradiation (0.3 W cm[−2] for 5 min) was achieved by determining the ratio of I_{545} and I_{525} in confocal images based on calibration formula $((I_{545})/(I_{525}) = C \exp(-\Delta E/kT))$.

In vitro photothermal cytotoxicity of csUCNP@C.
In vitro quantitative photothermal cytotoxicity of csUCNP@C was measured by performing MTT assays on HeLa cells. Cells were seeded into a 96-well cell culture plate at 5×10^4 per well, under 100% humidity, and were cultured at $37\,^\circ C$ and 5% CO_2 for 24 h; different concentrations of csUCNP@C (0, 50, 100, 150, 200, 300 and 400 $\mu g\,ml^{-1}$, diluted in RPMI 1640) were then added to the wells. The cells were subsequently incubated for 3 h at $37\,^\circ C$ under 5% CO_2. Thereafter, the cells were exposed to an NIR laser (730 nm, 0.3 $W\,cm^{-2}$) for 0 and 5 min, respectively, and then incubated for another 24 h. After that, MTT (5 μl; 5 $mg\,ml^{-1}$) was added to each well and the plate was incubated for an additional 4 h at $37\,^\circ C$ under 5% CO_2. Following the addition of 10% SDS (50 μl per well), the assay plate was allowed to stand at room temperature for 12 h. The optical density OD_{570} value (*Abs.*) of each well, with background subtraction at 690 nm, was measured by means of a Tecan Infinite M200 monochromator-based multifunction microplate reader. The following formula was used to calculate the inhibition of cell growth:

Cell viability(%) = (mean of *Abs.* value of treatment group/mean *Abs.* value of control)100%

Temperature monitoring in tissue phantom.
Apparatus configuration and the measurement of UCL-temperature calibration curve are similar to that in aqueous solution, but the aqueous solution of csUCNP@C was replaced by tissue phantom (see Supplementary Methods for the synthesis of tissue phantom). To evaluate the photothermal process of csUCNP@C in tissue phantom, a double-layer phantom where the lower layer of the phantom containing csUCNP@C (Layer 1, $\sim$10-mm thick) simulating the tumour area for therapy and the upper layer without csUCNP@C (Layer 2, $\sim$4-mm thick) simulating normal tissue was prepared. Schematic diagram of temperature monitoring in tissue phantom is shown in Fig. 4a. Tissue phantom was irradiated with an optical fibre-coupled 730-nm diode-laser (Weining Technology Development Co., Ltd. China) for 8 min at a laser power of 0.3 and 0.8 $W\,cm^{-2}$. A CW 980-nm laser was used to generate the UCL. UCL spectra, from 500 to 580 nm, were collected by fibre-optic spectrometer (PG2000 Pro, Ideaoptics, China) at different time intervals (0, 40, 120, 240, 360 and 480 s) for the calculation of eigen temperature. Thermal camera (FLIR E40) was used to record the apparent temperature. Investigation of heat conduction from layer 1 to layer 2 was also carried out by reading the temperature value in every other millimetre around the borderline of layer 1 and layer 2 from the thermal images.

Tumour xenografts.
Animal procedures were in agreement with the guidelines of the Institutional Animal Care and Use Committee, School of Pharmacy, Fudan University. For *in vivo* photothermal therapy with csUCNP@C pre-labelled tumour, HeLa cells were incubated in a serum-free medium containing 200 $\mu g\,ml^{-1}$ csUCNP@C at $37\,^\circ C$ for 2 h under 5% CO_2, and then washed with PBS three times to get rid of excess nanoparticles. After that, HeLa cells were collected by incubation with 0.05% trypsin-EDTA. Cells were collected by centrifugation and resuspended in sterile PBS. Cells (10^7 cells per site) were subcutaneously implanted into 4-week-old male athymic nude mice. Photothermal therapy was performed when the tumours reached an average diameter of 0.6 cm. For *in vivo* photothermal therapy, HeLa cells were collected by incubation with 0.05% trypsin-EDTA. Cells were collected by centrifugation and resuspended in sterile PBS. Cells (10^8 cells per site) were subcutaneously implanted into 4-week-old female athymic Balb/c mice. Photothermal therapy was performed when the tumours reached an average diameter of 0.6 cm.

UCL bioimaging in vivo.
UCL imaging *in vivo* was performed with an *in vivo* imaging system designed by our group, using two external 0 ~ 5 W adjustable CW 980-nm lasers (Connet Fiber Optics Co., China) as the excited source and an Andor DU897 EMCCD as the signal collector. Excitation was provided by the CW laser at 980 nm and UCL signals were collected using a 720-nm short-pass filter. In the case of csUCNP@C-labelled HeLa cells transplantation, UCL imaging was performed when the tumour reached an average diameter of 0.6 cm. In the case of *in vivo* targeting imaging, FA-csUCNP@C were intravenously injected into HeLa cell tumour-bearing athymic Balb/c mice. Whole-body imaging of the nude mice was performed 1 h after the injection.

Eigen temperature monitoring in vivo.
Eigen temperature monitoring *in vivo* was conducted in csUCNP@C labelled HeLa tumour-bearing nude mice. 0–5 W adjustable CW 980-nm lasers was chosen as the excitation source of UCL and an optical fibre-coupled 730-nm diode-laser was used for photothermal excitation. Fibre-optic spectrometer was used to collect the UCL signals and a 720-nm short-pass filter was installed in front of the probe of the spectrometer. Under 730-nm laser irradiation at 0, 0.3 and 0.8 $W\,cm^{-2}$ for 3 min, upconversion emission spectra were collected by fibre-optic spectrometer and the eigen temperature under different laser-power density were calculated by the luminescence spectra with the temperature calibration curve got from the tissue phantom.

Photothermal therapy in vivo.
An optical fibre-coupled 730-nm diode-laser (Weining Technology Development Co., Ltd. China) was used to irradiate tumours during the experiments. For photothermal treatment, the 730-nm laser beam with a diameter of $\sim$10 mm was focused on the tumour area at the power density of 0.3 $W\,cm^{-2}$ for 3 min. Infrared thermal images were taken by an FLIR E40 thermal imaging camera. Tumour sizes of treatment group (csUCNP@C-labelled tumour with 730-nm irradiation in csUCNP@C-labelled HeLa cells transplantation and FA-csUCNP@C targeted tumour with 730-nm irradiation in targeting photothermal therapy *in vivo*) and reference groups (untreated mice, mice exposed to the 730-nm laser only, and csUCNP@C-treated mice without 730-nm irradiation) were measured every day after treatment. Each group contained five mice for relatively rational evaluation. The tumour sizes were measured using a caliper and calculated as volume = (tumour length) $\times$ (tumour width)^2/2. Relative tumour volumes were normalized and were calculated as V/V_0 (V_0 is the tumour volume when the treatment was initiated). According to the guidelines of Institutional Animal Care and Use Committee, School of Pharmacy, Fudan University, the maximum permitted tumour size is 20 mm in an average diameter for mice. The tumours' size in this work is confined within this criterion.

References

1. Neidle, S. & Thurston, D. E. Chemical approaches to the discovery and development of cancer therapies. *Nat. Rev. Cancer* **5**, 285–296 (2005).
2. Davis, M. E., Chen, Z. & Shin, D. M. Nanoparticle therapeutics: An emerging treatment modality for cancer. *Nat. Rev. Drug Discov.* **7**, 771–782 (2008).
3. Chu, K. F. & Dupuy, D. E. Thermal ablation of tumours: biological mechanisms and advances in therapy. *Nat. Rev. Cancer* **14**, 199–208 (2014).
4. Gillams, A. R. The use of radiofrequency in cancer. *Br. J. Cancer* **92**, 1825–1829 (2005).
5. Simon, C. J., Dupuy, D. E. & Mayo-Smith, W. W. Microwave ablation: Principles and applications. *Radiographics* **25**, S69–S83 (2005).
6. Shinohara, K. Thermal ablation of prostate diseases: Advantages and limitations. *Int. J. Hyperther.* **20**, 679–697 (2004).
7. Jansen, M. C. *et al.* Adverse effects of radiofrequency ablation of liver tumours in the netherlands. *Br. J. Surg.* **92**, 1248–1254 (2005).
8. Hirsch, L. R. *et al.* Nanoshell-mediated near-infrared thermal therapy of tumors under magnetic resonance guidance. *Proc. Natl Acad. Sci. USA* **100**, 13549–13554 (2003).
9. Huang, X., El-Sayed, I. H., Qian, W. & El-Sayed, M. A. Cancer cell imaging and photothermal therapy in the near-infrared region by using gold nanorods. *J. Am. Chem. Soc.* **128**, 2115–2120 (2006).
10. Kam, N. W. S., O'Connell, M., Wisdom, J. A. & Dai, H. Carbon nanotubes as multifunctional biological transporters and near-infrared agents for selective cancer cell destruction. *Proc. Natl Acad. Sci. USA* **102**, 11600–11605 (2005).
11. Seo, W. S. *et al.* FeCo/graphitic-shell nanocrystals as advanced magnetic-resonance-imaging and near-infrared agents. *Nat. Mater.* **5**, 971–976 (2006).
12. Yang, K. *et al.* Graphene in mice: Ultrahigh *in vivo* tumor uptake and efficient photothermal therapy. *Nano Lett.* **10**, 3318–3323 (2010).
13. Robinson, J. T. *et al.* Ultrasmall reduced graphene oxide with high near-infrared absorbance for photothermal therapy. *J. Am. Chem. Soc.* **133**, 6825–6831 (2011).
14. Huang, X. *et al.* Freestanding palladium nanosheets with plasmonic and catalytic properties. *Nat. Nano* **6**, 28–32 (2011).
15. Tian, Q. *et al.* Sub-10 nm Fe$_3$O$_4$@Cu$_{2-x}$S core–shell nanoparticles for dual-modal imaging and photothermal therapy. *J. Am. Chem. Soc.* **135**, 8571–8577 (2013).
16. Lovell, J. F. *et al.* Porphysome nanovesicles generated by porphyrin bilayers for use as multimodal biophotonic contrast agents. *Nat. Mater.* **10**, 324–332 (2011).
17. Cheng, L. *et al.* PEGylated WS$_2$ nanosheets as a multifunctional theranostic agent for *in vivo* dual-modal CT/photoacoustic imaging guided photothermal therapy. *Adv. Mater.* **26**, 1886–1893 (2014).
18. Kyrsting, A., Bendix, P. M., Stamou, D. G. & Oddershede, L. B. Heat profiling of three-dimensionally optically trapped gold nanoparticles using vesicle cargo release. *Nano Lett.* **11**, 888–892 (2010).
19. Freddi, S. *et al.* A molecular thermometer for nanoparticles for optical hyperthermia. *Nano Lett.* **13**, 2004–2010 (2013).
20. Wang, X.-D., Wolfbeis, O. S. & Meier, R. J. Luminescent probes and sensors for temperature. *Chem. Soc. Rev.* **42**, 7834–7869 (2013).
21. Löw, P., Kim, B., Takama, N. & Bergaud, C. High-spatial-resolution surface-temperature mapping using fluorescent thermometry. *Small* **4**, 908–914 (2008).
22. Okabe, K. *et al.* Intracellular temperature mapping with a fluorescent polymeric thermometer and fluorescence lifetime imaging microscopy. *Nat. Commun.* **3**, 705 (2012).
23. Vlaskin, V. A., Janssen, N., van Rijssel, J., Beaulac, R. M. & Gamelin, D. R. Tunable dual emission in doped semiconductor nanocrystals. *Nano Lett.* **10**, 3670–3674 (2010).
24. Vetrone, F. *et al.* Temperature sensing using fluorescent nanothermometers. *ACS Nano* **4**, 3254–3258 (2010).

25. Sedlmeier, A., Achatz, D. E., Fischer, L. H., Gorris, H. H. & Wolfbeis, O. S. Photon upconverting nanoparticles for luminescent sensing of temperature. *Nanoscale* **4**, 7090–7096 (2012).

26. Wang, J. *et al.* Enhancing multiphoton upconversion through energy clustering at sublattice level. *Nat. Mater.* **13**, 157–162 (2014).

27. Bünzli, J.-C. G. Lanthanide luminescence for biomedical analyses and imaging. *Chem. Rev.* **110**, 2729–2755 (2010).

28. Idris, N. M. *et al. In vivo* photodynamic therapy using upconversion nanoparticles as remote-controlled nanotransducers. *Nat. Med.* **18**, 1580–1585 (2012).

29. Jayakumar, M. K. G., Idris, N. M. & Zhang, Y. Remote activation of biomolecules in deep tissues using near-infrared-to-UV upconversion nanotransducers. *Proc. Natl Acad. Sci. USA* **109**, 8483–8488 (2012).

30. Park, Y. I. *et al.* Theranostic probe based on lanthanide-doped nanoparticles for simultaneous *in vivo* dual-modal imaging and photodynamic therapy. *Adv. Mater.* **24**, 5755–5761 (2012).

31. Wang, Y.-F. *et al.* Nd^{3+}-sensitized upconversion nanophosphors: Efficient *in vivo* bioimaging probes with minimized heating effect. *ACS Nano* **7**, 7200–7206 (2013).

32. Huang, P. *et al.* Lanthanide-doped $LiLuF_4$ upconversion nanoprobes for the detection of disease biomarkers. *Angew. Chem. Int. Ed.* **53**, 1252–1257 (2014).

33. Tang, Q. *et al.* Color tuning and white light emission via in situ doping of luminescent lanthanide metal–organic frameworks. *Inorg. Chem.* **53**, 289–293 (2013).

34. Lu, Y. *et al.* Tunable lifetime multiplexing using luminescent nanocrystals. *Nat. Photon.* **8**, 32–36 (2014).

35. Johnson, N. J. J., Korinek, A., Dong, C. & van Veggel, F. C. J. M. Self-focusing by ostwald ripening: A strategy for layer-by-layer epitaxial growth on upconverting nanocrystals. *J. Am. Chem. Soc.* **134**, 11068–11071 (2012).

36. Yi, G. S. & Chow, G. M. Synthesis of hexagonal-phase $NaYF_4$:Yb,Er and $NaYF_4$:Yb,Tm nanocrystals with efficient up-conversion fluorescence. *Adv. Funct. Mater.* **16**, 2324–2329 (2006).

37. Fan, W. *et al.* Rattle-structured multifunctional nanotheranostics for synergetic chemo-/radiotherapy and simultaneous magnetic/luminescent dual-mode imaging. *J. Am. Chem. Soc.* **135**, 6494–6503 (2013).

38. Ma, P. *et al.* Rational design of multifunctional upconversion nanocrystals/polymer nanocomposites for cisplatin (IV) delivery and biomedical imaging. *Adv. Mater.* **25**, 4898–4905 (2013).

39. Shen, J., Zhao, L. & Han, G. Lanthanide-doped upconverting luminescent nanoparticle platforms for optical imaging-guided drug delivery and therapy. *Adv. Drug Deliver. Rev.* **65**, 744–755 (2013).

40. Chen, G., Qiu, H., Prasad, P. N. & Chen, X. Upconversion nanoparticles: design, nanochemistry, and applications in theranostics. *Chem. Rev.* **114**, 5161–5214 (2014).

41. Shan, G. B., Weissleder, R. & Hilderbrand, S. A. Upconverting organic dye doped core-shell nano-composites for dual-modality NIR imaging and photothermal therapy. *Theranostics* **3**, 267–274 (2013).

42. Dong, B. *et al.* Multifunctional $NaYF_4 : Yb^{3+}, Er^{3+}$@Ag core/shell nanocomposites: Integration of upconversion imaging and photothermal therapy. *J. Mater. Chem.* **21**, 6193–6200 (2011).

43. Zhou, J., Liu, Z. & Li, F. Upconversion nanophosphors for small-animal imaging. *Chem. Soc. Rev.* **41**, 1323–1349 (2012).

44. Bogdan, N., Vetrone, F., Ozin, G. A. & Capobianco, J. A. Synthesis of ligand-free colloidally stable water dispersible brightly luminescent lanthanide-doped upconverting nanoparticles. *Nano Lett.* **11**, 835–840 (2011).

45. Sun, X. & Li, Y. Colloidal carbon spheres and their core/shell structures with noble-metal nanoparticles. *Angew. Chem. Int. Ed.* **43**, 597–601 (2004).

46. Hessel, C. M. *et al.* Copper selenide nanocrystals for photothermal therapy. *Nano Lett.* **11**, 2560–2566 (2011).

47. Tian, Q. *et al.* Hydrophilic Cu_9S_5 nanocrystals: A photothermal agent with a 25.7% heat conversion efficiency for photothermal ablation of cancer cells *in vivo. ACS Nano* **5**, 9761–9771 (2011).

48. Liu, C. *et al.* One-step synthesis of surface passivated carbon nanodots by microwave assisted pyrolysis for enhanced multicolor photoluminescence and bioimaging. *J. Mater. Chem.* **21**, 13163–13167 (2011).

49. Liu, C. *et al.* Nano-carrier for gene delivery and bioimaging based on carbon dots with PEI-passivation enhanced fluorescence. *Biomaterials* **33**, 3604–3613 (2012).

50. Kucsko, G. *et al.* Nanometre-scale thermometry in a living cell. *Nature* **500**, 54–58 (2013).

51. Shang, L., Stockmar, F., Azadfar, N. & Nienhaus, G. U. Intracellular thermometry by using fluorescent gold nanoclusters. *Angew. Chem. Int. Ed.* **52**, 11154–11157 (2013).

52. Arai, S., Lee, S.-C., Zhai, D., Suzuki, M. & Chang, Y. T. A molecular fluorescent probe for targeted visualization of temperature at the endoplasmic reticulum. *Sci. Rep.* **4**, 6701 (2014).

53. Dewey, W. C. Arrhenius relationships from the molecule and cell to the clinic. *Int. J. Hyperther.* **25**, 3–20 (2009).

54. Li, Z. & Zhang, Y. An efficient and user-friendly method for the synthesis of hexagonal-phase $NaYF_4$:Yb,Er/Tm nanocrystals with controllable shape and upconversion fluorescence. *Nanotechnology* **19**, 345606 (2008).

55. Yu, M. *et al.* Laser scanning up-conversion luminescence microscopy for imaging cells labeled with rare-earth nanophosphors. *Anal. Chem.* **81**, 930–935 (2009).

56. Xiong, L. *et al.* High contrast upconversion luminescence targeted imaging *in vivo* using peptide-labeled nanophosphors. *Anal. Chem.* **81**, 8687–8694 (2009).

Acknowledgements

We thank National Basic Research Program of China (2015CB931800, 2013CB733700), National Natural Science Foundation of China (21527801, 21231004) and the CAS/SAFEA International Partnership Program for Creative Research Teams for financial support.

Author contributions

The manuscript was written by X.Z, W.F. and F.L. The experiment and analysis were carried out by X.Z., W.F., J.C., J.L., M.C. and Y.S. The experimental work and the manuscript were supervised by Y.-W.T., W.F. and F.L.

Additional information

Frequency comb transferred by surface plasmon resonance

Xiao Tao Geng[1,2], Byung Jae Chun[3], Ji Hoon Seo[4], Kwanyong Seo[4], Hana Yoon[5], Dong-Eon Kim[1,2], Young-Jin Kim[3] & Seungchul Kim[1,2]

Frequency combs, millions of narrow-linewidth optical modes referenced to an atomic clock, have shown remarkable potential in time/frequency metrology, atomic/molecular spectroscopy and precision LIDARs. Applications have extended to coherent nonlinear Raman spectroscopy of molecules and quantum metrology for entangled atomic qubits. Frequency combs will create novel possibilities in nano-photonics and plasmonics; however, its interrelation with surface plasmons is unexplored despite the important role that plasmonics plays in nonlinear spectroscopy and quantum optics through the manipulation of light on a subwavelength scale. Here, we demonstrate that a frequency comb can be transformed to a plasmonic comb in plasmonic nanostructures and reverted to the original frequency comb without noticeable degradation of $<6.51 \times 10^{-19}$ in absolute position, 2.92×10^{-19} in stability and 1 Hz in linewidth. The results indicate that the superior performance of a well-defined frequency comb can be applied to nanoplasmonic spectroscopy, quantum metrology and subwavelength photonic circuits.

[1] Max Planck Center for Attosecond Science, Max Planck POSTECH/KOREA Res. Initiative, Pohang, Gyeongbuk 376-73, South Korea. [2] Department of Physics, Center for Attosecond Science and Technology (CASTECH), POSTECH, Pohang, Gyeongbuk 376-73, South Korea. [3] School of Mechanical and Aerospace Engineering, Nanyang Technological University (NTU), 50 Nanyang Avenue, Singapore 639798, Singapore. [4] Department of Energy Engineering, Ulsan National Institute of Science and Technology (UNIST), Ulsan 689-798, South Korea. [5] Energy Storage Department, Korea Institute of Energy Research (KIER), Daejeon 305-343, South Korea. Correspondence and requests for materials should be addressed to Y.-J.K. (email: yj.kim@ntu.edu.sg) or to S.K. (email: inter99@postech.ac.kr).

The frequency comb of mode-locked femtosecond lasers has led to remarkable advances in high-resolution spectroscopy[1,2], broadband calibration of astronomical spectrographs[3,4], time/frequency transfer over long distances[5,6], absolute laser ranging[7–10] and inter-comparison of atomic clocks[11,12]. It provides millions of well-defined optical modes over a broad spectral bandwidth with high-level phase coherence referenced to an atomic clock. Recently, the potential of frequency comb has expanded to microscopic applications; high inter-mode coherence within a short pulse duration enabled manipulating atomic qubits[13], operating quantum logic gates and performing high-speed molecular detection by coherent Raman spectroscopy through harnessing inter-mode beat frequencies between two frequency combs at different repetition rates[14].

Coupling surface plasmons (SPs)[15,16], collective charge oscillations produced by the resonant interaction of light and free electrons on the interface of metallic and dielectric materials, to frequency comb creates numerous advantages. First, SP can allow for the frequency comb to access nanoscopic volumes that surpass the diffraction limit[17]. Second, the field enhancement by localized SP enables the highly sensitive detection of weak signals, even from a single molecule (for example, surface-enhanced Raman scattering)[18]. Third, next-generation photonic devices and circuits can be implemented within a small subwavelength volume by all-optical control of light properties (amplitude, phase and polarization state) in plasmonic nanostructures within ultrafast time scales[19–22]. However, the superior performance of the frequency comb, such as absolute frequency uncertainty, high-frequency stability and narrow linewidth, could deteriorate during the photon-plasmon conversion process. For exploring novel combination of frequency comb and SP resonance, it is prerequisite to verify that frequency comb maintains its performance under plasmonic resonance; however, there have been no studies to date.

In the following, we report that frequency comb successfully maintains core performances in photon-plasmon conversion by exploiting plasmonic extraordinary transmission through a subwavelength plasmonic hole array. This implies that the original frequency comb can be transformed into a form of plasmonic comb on metallic nanostructures and reverted to an original frequency comb without noticeable degradation in absolute frequency position, stability and linewidth. The superior performance of well-defined frequency combs can therefore be applied to various nanoplasmonic spectroscopy, coherent quantum metrology and subwavelength photonic circuits.

Results

Frequency comb transferred by SP resonance.

Figure 1 shows the experimental apparatus to characterize the conservation of frequency comb for the conversion from photon to SP. The frequency comb is split into reference and measurement beams; one part of the beam transmits through an acousto-optic modulator (AOM) for a frequency shift of 40 MHz to construct a reference frequency comb and the other part of the beam passes through the plasmonic sample. The frequency comb structure in SP resonance was generated by the exploitation of a metallic nanohole array used for extraordinary optical transmission (EOT) that converted photon into SP. The small diameter of each hole prevents light passing through the sample based on classical optics. However, the SP-mediated tunnelling effect of nanohole array drastically enhances optical transmittance[23]. These intriguing optical phenomena have been studied widely for high-resolution chemical sensing, ultrafast optical modulation, wavelength-tunable optical filtering and subwavelength lithography[24,25]. The physical origin of EOT has been attributed to resonant SP

polaritons (SPPs)[26]. The appropriate geometrical and material parameters of nanohole array excite the SPP mode that allows the transmission of light that contains plasmonic information inside an EOT sample. The resonant nature of the SP changes the transmitted spectral distribution, depending on sample design, input polarization and incident angle. Plasmonic EOT can also induce wavelength-dependent changes in optical frequency and phase in addition to wavelength-dependent transmittance. The optical frequency of a single frequency comb mode transmitted through the plasmonic sample via SP resonance (f_{MEA}) can be expressed as

$$f_{MEA} = nf_r + f_{ceo} + \Delta f_{sp} \tag{1}$$

where f_r is the pulse repetition frequency, f_{ceo} the carrier-envelope offset frequency, and Δf_{sp} the frequency and phase change generated by SP resonance. Meanwhile, the optical frequency of the single mode passing through the reference path (f_{REF}) can be expressed as

$$f_{REF} = nf_r + f_{ceo} + f_{AOM} \tag{2}$$

where f_{AOM} denotes the intentional frequency shift by AOM. The detection of the heterodyne beat-frequency generated by the interference between the reference and measurement beams enables the measurement of optical frequency difference, ($f_{REF} - f_{MEA}$) at a radio-frequency (RF) regime using a fast avalanche photodiode. This resultant frequency difference can be simplified to $f_{AOM} - \Delta f_{sp}$, where f_{AOM} works as the high-frequency carrier to isolate Δf_{sp} from the relatively strong low frequency noise components.

Plasmonic extraordinary transmission.

For transmitting frequency combs through the subwavelength holes by SP resonance, there are three important geometric parameters: hole diameter (d), hole pitch (l) and Au film thickness (t; Fig. 2a). For maximum optical transmission at a wavelength of 840 nm, three parameters were optimized by solving Maxwell's equations using finite-difference time-domain (FDTD) method. Figure 2b,c show the calculated plasmonic field distribution through the optimized sample. The electric field around the hole was significantly enhanced by SP in the periodic apertures, delivering the optical energy through the hole. Figure 2a shows the scanning electron microscope image of the fabricated nanohole array; all dimensions were matched with optimized design parameters within a geometric error of <5%. Figure 2d shows that the transmitted optical spectrum coincided with the numerical FDTD results and validated the numerical analysis. Minor deviations between the two spectrums are expected by focusing geometry onto the plasmonic sample. The plasmonic resonance conditions are dissimilar in given transverse electric-transverse magnetic polarization if the angle of incidence is not surface normal. As a result, plasmonic sample shows different transmission spectra for transverse electric-transverse magnetic polarization at the incidence angle of 45° (Fig. 2e); therefore, optical transmission of our sample is dominated by the plasmonic EOT, not classical diffraction theory.

Frequency comb structure after plasmonic transmission.

The transmitted frequency combs through the plasmonic sample results in an interference with the reference frequency comb to verify the frequency comb structure after the photon-plasmon mode conversion by the EOT (Fig. 3a). For comparative analysis, interference signals were obtained at three different wavelength regimes with optical band-pass filters, representing on-resonance (840 nm) and off-resonance (800 and 900 nm) positions.

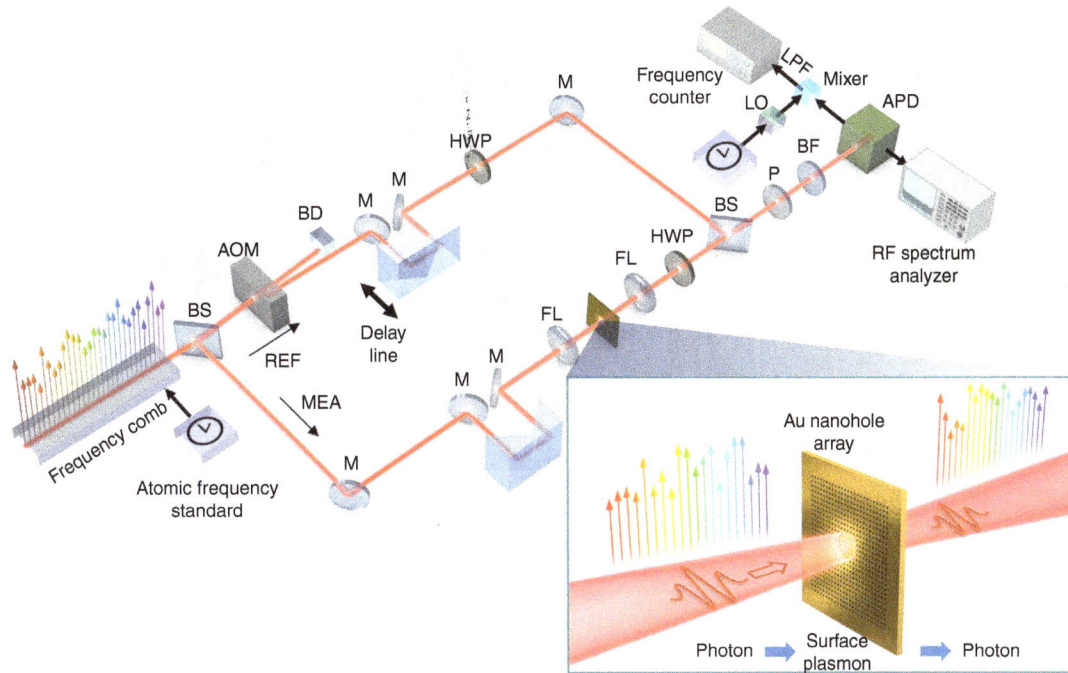

Figure 1 | Generation and characterization of plasmonic frequency comb. Part of the frequency comb experiences plasmonic mode conversion by passing through the plasmonic sample. The sample consists of a subwavelength nanohole array on an Au thin-film, enabling the conversion from photon to SP (from SP to photon). The other part of the frequency comb is used as a reference beam to compare with the frequency comb passed through the plasmonic sample. The frequency combs at two different paths are combined and monitored by APD. The characteristics of the frequency comb at measurement path are analysed by an RF spectrum analyser and a frequency counter. APD, avalanche photo-detector; BD, beam dumper; BF, band-pass filter; BS, beam splitter; FL, focusing lens; HWP, half-wave plate; LO, local oscillator; LPF, low-pass filter; M, mirror; MEA, measurement beam path; P, polarizer; REF, reference beam path.

The coherence of a large number of frequency comb modes can be deteriorated by temporal and spectral plasmonic dispersion, phase noise and frequency noise during the propagation through the plasmonic EOT sample. The frequency comb fundamentally suffers from phase and frequency noises when passing through the optical medium (for example, ambient air and optical fibre) exposed to environmental variations, such as vibration, temperature variation and humidity change. Therefore, it has been an important task to monitor and compensate the temporal and spectral dispersion, phase noise and frequency noise generated in the medium, as reported through long optical-fibre[6] and through ambient air[27]. SPs also suffer from the dispersion and phase change by the medium and environmental disturbances, which have not been investigated with the frequency comb for their quantitative or qualitative analysis. Propagating SPs through the EOT sample experience phase delay depending on their wavelengths and spatial locations before and after tunnelling through each subwavelength hole; this phase delay can be additionally induced by the plasmonic dynamic damping, imperfect sample geometry, surface roughness of the metal film or air refractive index change around the sample. Therefore, the total summation of the electromagnetic waves at the output side of each hole may contain temporal and spectral dispersion, phase distortion and frequency change.

Most noise sources of the frequency comb can be categorized into intra-cavity and extra-cavity sources; intra-cavity noise sources (including cavity length change, cavity loss fluctuations and pump noise) cause frequency noise whereas extra-cavity noise sources (induced by path-length fluctuation, shot noise from the limited power or noise generated during super-continuum generation) result in time-varying phase noise floor[5]. In this investigation, plasmonic mode conversion by the EOT was

considered as an extra-cavity noise source that provided wavelength-dependent power attenuation, phase shift and frequency noise, similar to the supercontinuum generation process. Noise contributions should be observed at $f_{AOM} - \Delta f_{SP}$ in the form of linewidth broadening, frequency shift, signal-to-noise (S/N) ratio reduction, increased phase noise or a higher Allan deviation if the plasmonic frequency comb suffers from phase or frequency noise during the plasmonic mode conversion.

Linewidth broadening and S/N ratio reduction in plasmonic mode conversion process was initially evaluated by measuring RF beat linewidth of $f_{AOM} - \Delta f_{SP}$ at three different wavelength regimes (Fig. 3b). With different resolution bandwidths (RBWs), there was no substantial degradation in the linewidth at 840 nm before and after the installation of the plasmonic sample in the beam path. The high-level S/N ratio of ~60 dB beat signal indicates that the plasmonic EOT provide no significant phase noise to the frequency comb.

Phase noise and frequency stability was measured for the quantitative analysis of frequency-dependent noise contributions. Figure 4a shows the phase noise spectrum obtained by monitoring one of high harmonics of the beat frequencies at ~1.2 GHz with and without the plasmonic sample; this confirms that there was no noticeable frequency noise inclusion. For high-precision frequency position measurement, the beat frequency between reference and measurement frequency comb was measured by a frequency counter for 3,000 s, resulting in 0.24 mHz frequency difference with a s.d. of 61 mHz (Fig. 4b). This corresponds to 6.51×10^{-19}, which proves that plasmonic mode conversion provides no substantial degradation in the frequency accuracy of the frequency comb. The stability of the beat signal was measured to be 4.08×10^{-18} without the plasmonic sample, 4.37×10^{-18} with the plasmonic sample at

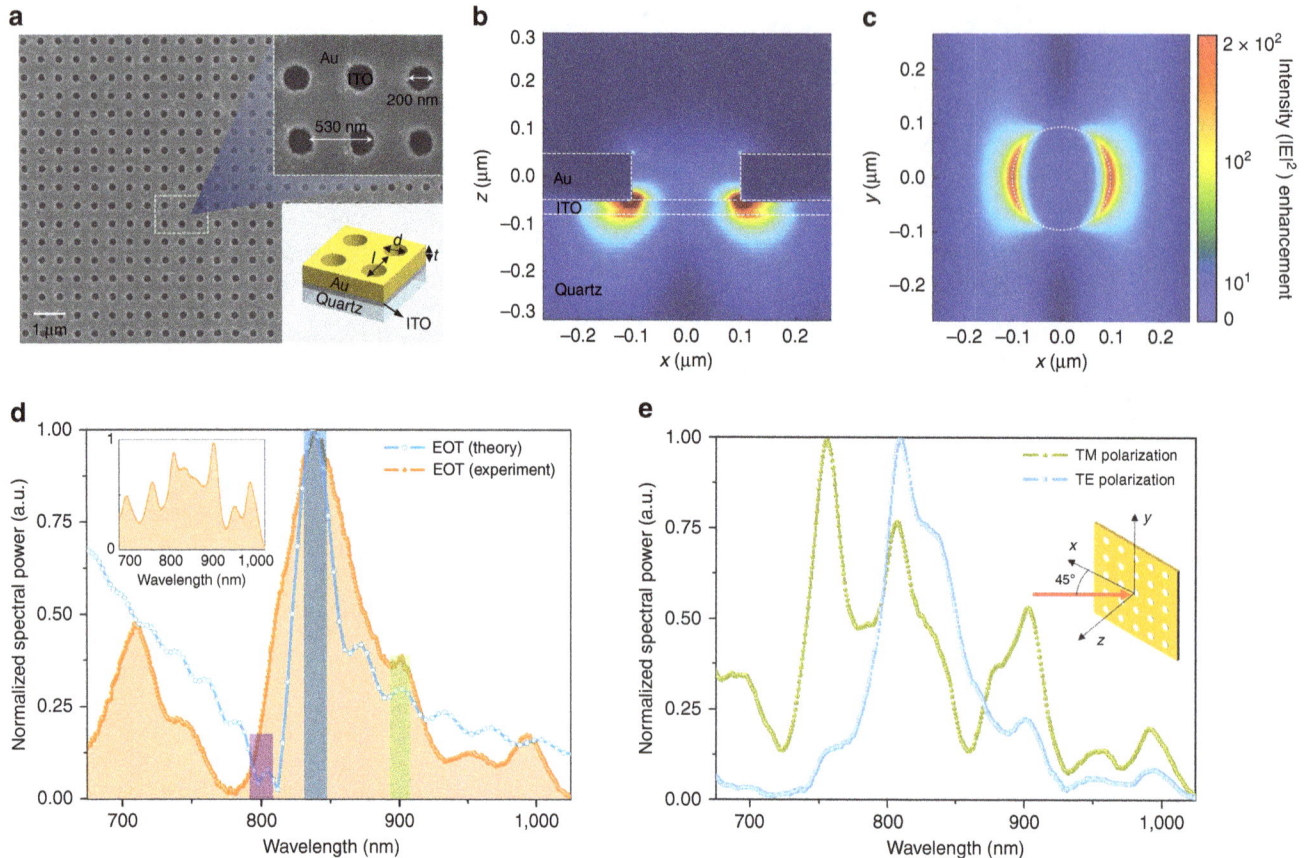

Figure 2 | Numerical simulation and characterization of fabricated plasmonic sample. (a) Scanning electron microscope image of the fabricated subwavelength nanohole array for plasmonic EOT. The fabricated nanohole has the diameter (d) of 200 nm, pitch (l) of 530 nm and thickness (t) of 100 nm on 25-nm-thick ITO-coated quartz substrate. **(b)** Calculated intensity distribution of an plasmonic sample taken from the side. **(c)** Calculated intensity distribution at the interface between Au and ITO layer. **(d)** Theoretical (blue line) and experimental (orange line) spectrum of transmitted frequency combs through the plasmonic sample. Purple, blue and green bars represent the selected spectral components (800, 840 and 900 nm) to characterize frequency comb, respectively. Inset (top left) shows the original spectrum of the frequency comb. **(e)** Polarization dependent transmission spectrum through the plasmonic sample at an incident angle of 45°.

resonance wavelength of 840 nm for an averaging time of 100 s, respectively (Fig. 4c). At the off-resonance wavelength, the stability of beat signal was 4.59×10^{-18}, signifying almost no difference between on- and off-plasmonic resonance stabilities. All the experiments pointed that plasmonic mode conversion causes no substantial degradation to the frequency comb in terms of linewidth, frequency position, S/N ratio and frequency stability.

Discussions

All hundreds of thousands optical modes in the frequency comb were firstly converted from photonic to plasmonic mode at the input side of the plasmonic EOT sample and then reverted to photonic mode at the other output side of the sample. It is known to be practically difficult to directly measure the optical frequency of the plasmonic mode so the characteristics of the plasmonic comb were measured here in the far field. Because the plasmonic and photonic modes are assumed to be mutually coherent, if there is any change in the frequency comb characteristics during the plasmonic propagation (in plasmonic mode) through the sample, it should be monitored at the output side in the far field (in photonic mode). Therefore, the beat-frequency detection using the transmitted photonic mode in the far-field regime enabled us to compare the qualities of the plasmonic comb with the original frequency comb, which cannot be implemented in the

near-field regime. As the result of the comparison, there were no noticeable degradation in linewidth, frequency shift, S/N ratio, phase noise and Allan deviation. This implies that SP, the collective electrons, can be regarded as information carrier as precise as the optical frequency comb.

The frequency comb passing through the plasmonic EOT sample experiences the different physical process with the light reflection at a metallic mirror. Although both of the SP resonance and the surface reflection are governed by free-electron oscillation in conduction band of metals, the SP resonance additionally requires the specific momentum matching between incident photon and SP, whose relationship is determined by the plasmonic dispersion relation. Therefore, it is natural to maintain the coherence during the light reflection at metal surface (governed by frequency conservation), which is not the case in plasmonic structures (governed by frequency and momentum matching). Once the incident photon (in photonic mode) is converted into SP, it will propagate through the metal as the form of SPPs (in plasmonic mode). This plasmonic propagation causes temporal and spectral dispersions, phase variations and frequency changes, which may degrade the inherently high coherence of the optical frequency comb.

Plasmonic EOT is governed by not only the hole geometry[28] but also hole pitch. Therefore, the incidence angle tuning of the input beam can provide the change in plasmonic coupling mode without dimensional changes, which can possibly cause some

a

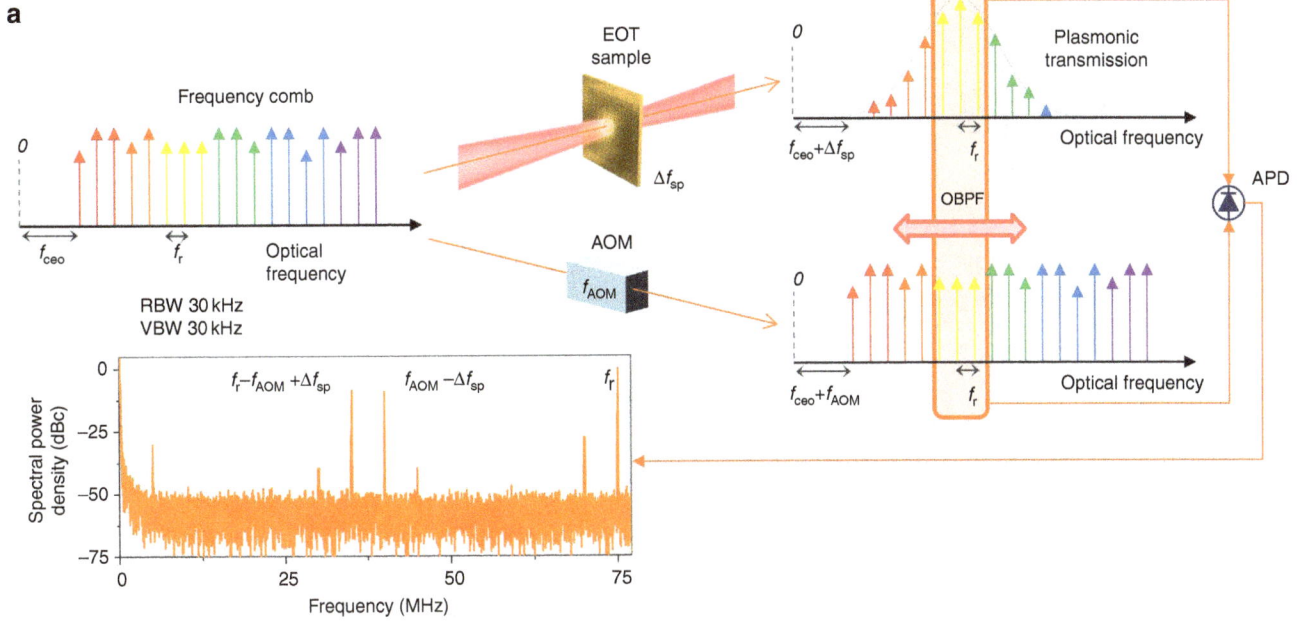

b

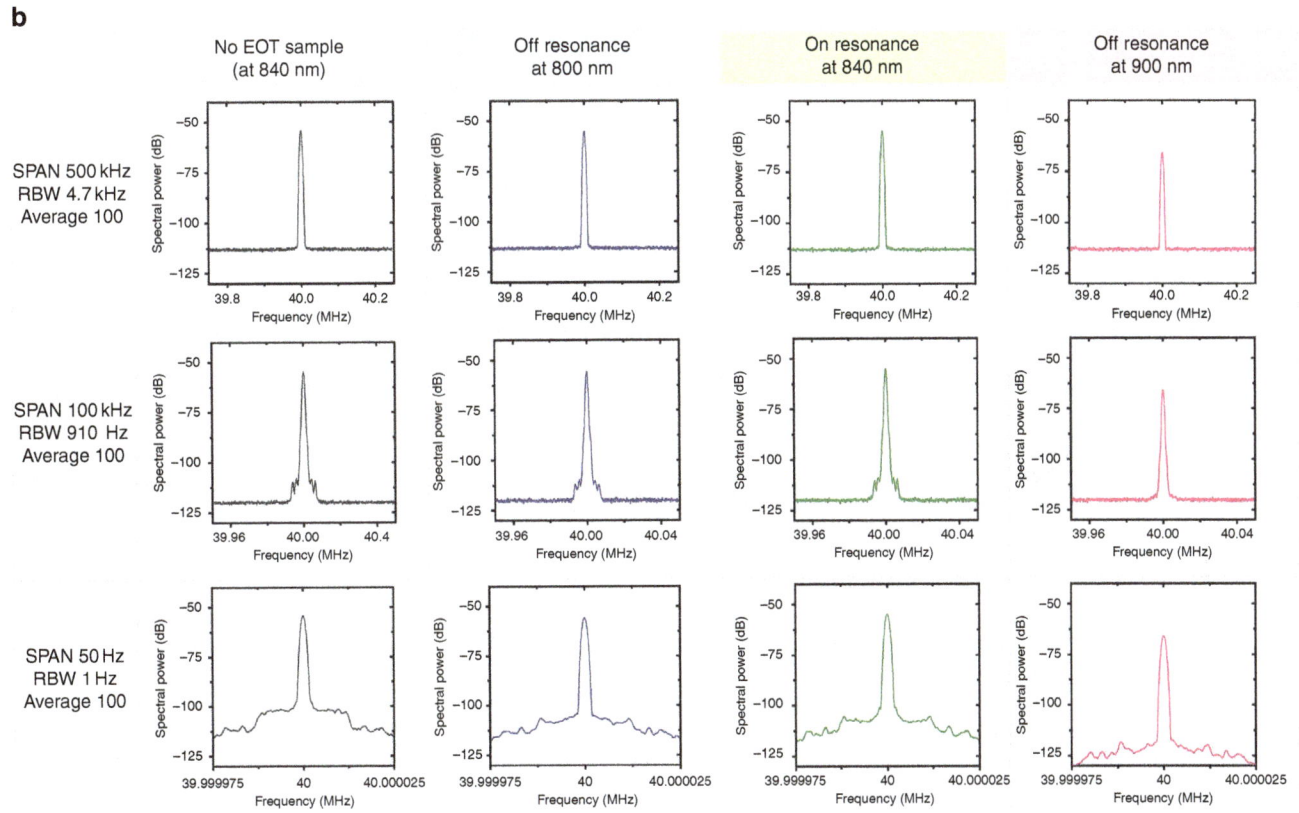

Figure 3 | Evaluation of the plasmonic frequency comb by EOT. (**a**) Generation of RF beats by the interference between frequency-shifted (40 MHz) reference combs and plasmonic EOT combs. The beat spectra of plasmonically transmitted frequency comb and the reference comb are measured at three wavelengths: one at a strong plasmonic resonance position (a 840-nm centre wavelength with a 10-nm bandwidth), two at off-resonance positions (a 800-nm centre wavelength with a 40-nm bandwidth and 900 nm with a 10-nm bandwidth) using three optical band-pass filters. These are compared with a beat spectrum at a 840-nm wavelength, acquired without the plasmonic sample. (**b**) Linewidth measurement of RF beats with different span, RBWs and VBWs. There was no noticeable linewidth degradation by the plasmonic transduction (<1 Hz, limited by RBW of the instrument). APD, avalanche photo-detector; OBPF, optical band-pass filter; VBWs, video bandwidths.

degradation in the frequency characteristics of the frequency comb by providing different plasmonic field distribution and enhancement. To test this, the beat spectrum was monitored

while the sample was rotated by up to 45° (for transverse magnetic wave) as shown in Figure 2e. For the given condition, all frequency characteristics were maintained in the same level with

a

b

c

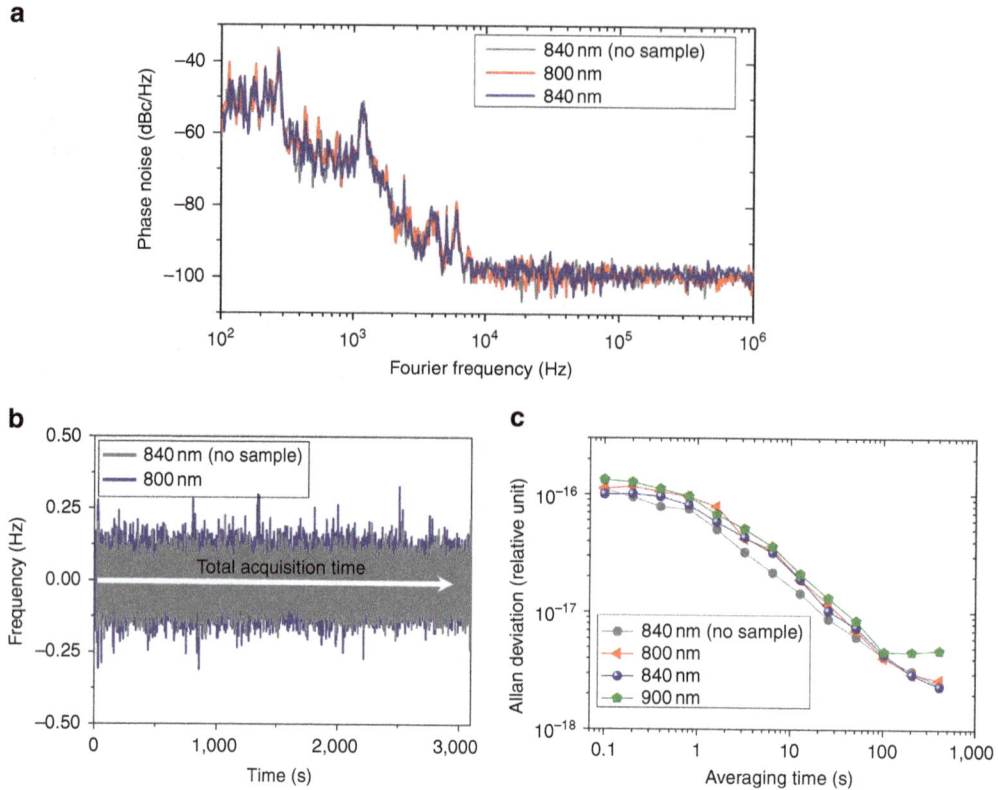

Figure 4 | Plasmonic frequency comb: phase noise and frequency stability. (**a**) Phase noise spectra at a 1.2-GHz RF carrier at on- and off-resonance wavelengths. (**b**) Time trace of the beat frequency with and without the plasmonic sample over 3,000 s. (**c**) Allan deviations of frequency stability with varying average time at the positions of on-resonance and off-resonance wavelengths.

normal incidence case, which shows that no performance degradation exist depending on plasmonic coupling or geometrical parameters of the sample.

The linewidth broadening by plasmonic EOT was evaluated to be <1 Hz, which is limited by RBW of the instrument in use (Fig. 3b). A single RF beat-frequency corresponds to the superposition of small RF beat contributions of $>10^4$ frequency comb modes, which proves that there is no significant wavelength-dependent frequency or phase noise during the plasmonic EOT. There was minor increase in spectral power in the pedestal peaks at 12, 17 and 21 Hz when the frequency comb passed through the plasmonic sample; this is expected to be caused by the vibrational and thermal noises at the plasmonic sample. The beat frequency, $f_{AOM} - \Delta f_{SP}$, was found to be exactly the same as the driving frequency of the AOM in all measured spectra shown in Fig. 3b, which implies that the absolute frequency position is well maintained in the plasmonic mode. The ambient temperature and vibration on EOT sample were not intentionally controlled so as to evaluate the performance in normal laboratory environment conditions. Our results show that the frequency comb structures are well maintained under environmental disturbances, for example, temperature variation, mechanical vibration and air fluctuation. This will enable us to develop high-sensitivity frequency-comb-referenced SP sensors working in harsh environments. The phase noise spectra in Fig. 4a also shows a number of minor peaks at 0.2, 1.5, 300 and 600 kHz other than the low-frequency spectral peaks at 12, 17 and 21 Hz observed in Fig. 3b. At higher frequency than 10 kHz, there is a flat noise floor without other spectral peaks or broad pedestals. In S/N ratio measurement, the S/N ratio theoretically could reach $68 \sim 75$ dB in a 100 kHz RBW because there are $10^4 \sim 10^5$ frequency comb modes in the pass-band of the optical

filter transmittance. The experimental S/N ratio with the plasmonic sample on-resonance position was ~ 60 dB; this minor deviation could come from imperfect intensity balancing, polarization matching and spatial beam mode-matching. The S/N ratio at 900 nm was 54 dB, relatively lower than that at 800 nm because the quantum efficiency of the avalanche photodetector at 900 nm is $\sim 20\%$ lower than that at 800 nm and the filter bandwidth at 900 nm is 25% of that at 800 nm.

In this article, we have studied SP resonance effects on frequency comb structure in the plasmonic EOT of light through a subwavelength metallic nanohole array. The frequency comb was transduced to plasmonic mode in the sample and reverted to photonic mode without significant changes in linewidth, frequency shift, S/N ratio, phase noise and Allan deviation. The linewidth broadening was <1 Hz (instrument limited), frequency inaccuracy was 6.51×10^{-19}, S/N ratio was higher than 60 dB, Allan deviation increased by 2.92×10^{-19} at 100 s averaging time. This outstanding frequency comb performance in plasmonic nanostructures enables a highly sensitive, high accurate and broadband measurement with direct traceability to standards. This inclusion of frequency comb has the potential to accelerate progresses in various plasmonic applications such as bio-chemical spectroscopy or sensing, quantum optics and sub-diffraction-limit biomedical-imaging. With the aid of SP, frequency-comb-referenced high-speed coherent anti-stokes Raman spectroscopy[14] can be implemented in much smaller nanoscopic volume being requested for single-molecule detection, for example, surface-enhanced coherent anti-stokes Raman spectroscopy[29]. A large number of optical modes in a frequency comb as the time and frequency standard can be coupled at the same time with SP for broadband quantum metrology for entangled atomic qubits or information carrier in subwavelength scale[13,30].

Localized field enhancement of SP will enable highly efficient nonlinear optics[31] coupled with high precision of frequency comb, which is prerequisite for novel sub-diffraction-limit nonlinear biomedical imaging and spectroscopy.

Methods

Frequency comb. A Ti:sapphire femtosecond laser delivers 4.8 fs pulses at a repetition rate of 75 MHz over a broad spectral bandwidth from 1.03 to 2.06 eV (Venteon UB, Venteon). For establishing a frequency comb, the pulse repetition frequency (f_r) and carrier-envelope offset frequency (f_{ceo}) were precisely locked to a reference Rb atomic clock (FS725, Stanford Research Systems) with the aid of a f–$2f$ interferometer and phase-locked control loops (AVR32, TEM-Messtechnik & XPS800-E, Menlosystems). One part of the beam was diverted to and transmitted through an AOM for the frequency shift of 40 MHz to construct a reference frequency comb. If the plasmonic frequency comb suffers from the phase or frequency noise during the plasmonic mode conversion, the noise contributions should be observed at $f_{AOM} - \Delta f_{SP}$ in the forms of linewidth broadening, frequency shift, S/N ratio reduction, increased phase noise or higher Allan deviation. The frequency comb excited the plasmonic sample with the whole-broadband spectrum in a loose focusing geometry with an aspheric lens of 100 mm focal length. The focused peak intensity at the plasmonic sample was set to be < 0.1 MW cm^{-2} not to exceed the thermal damage threshold (~ 1 TW cm^{-2} for Au). Input polarization state was set linear and its direction is parallel to the x axis of periodic holes on plasmonic sample as denoted in Fig. 2e.

Plasmonic EOT sample: design and development. We exploited FDTD solution (XFDTD8.3, Lumerical) to solve Maxwell's equation for plasmonic near-field distribution and transmitted spectrum. Through a series of iterative computations, the optimal geometric parameters were determined as $d = 200$ nm, $l = 530$ nm and $t = 100$ nm. The thickness, t, was designed to be much thicker than the Au skin depth (~ 20 nm) here to block the direct transmission through the Au film. The designed nanohole array was fabricated using electron-beam lithography (Raith 150) onto 25-nm-thick ITO-coated quartz substrate.

Evaluation of plasmonic frequency comb. For comparative analysis, interference signals were obtained at three different wavelength regimes – one at plasmonic on-resonance (840 nm) and the others at off-resonance positions (800 and 900 nm) – using optical band-pass filters. The resulting interference beat signal was obtained by high-speed avalanche photodiode and analysed using a high-resolution RF spectrum analyzer (N9020A, Agilent) and a RF frequency counter (53230A, Keysight Technologies). An exemplary RF spectrum is shown in Fig. 3b; the repetition rate (f_r) is located at 75 MHz, the beat frequency ($f_{AOM} - \Delta f_{sp}$) between the frequency-shifted reference frequency comb and the plasmonic frequency comb is at ~ 40 MHz, and the beat frequency ($f_r - f_{AOM} + \Delta f_{sp}$) between the other nearby reference frequency comb modes and the plasmonic frequency comb is at ~ 35 MHz. Other minor spurious peaks are due to the imperfect sinusoidal modulation of AOM; their positions match with beat frequencies between the f_{AOM}-harmonics and the reference frequency comb of $2f_{AOM} - f_r$, $2f_r - 3f_{AOM}$, $3f_{AOM} - f_r$ and $2(f_r - f_{AOM})$.

References

1. Hänsch, T. W. Nobel Lecture: passion for precision. *Rev. Mod. Phys.* **78**, 1297–1309 (2006).
2. Hall, J. L. Nobel Lecture: defining and measuring optical frequencies. *Rev. Mod. Phys.* **78**, 1279–1295 (2006).
3. Steinmetz, T. *et al.* Laser frequency combs for astronomical observations. *Science* **321**, 1335–1337 (2008).
4. Wilken, T. *et al.* A spectrograph for exoplanet observations at the centimetre-per-second level. *Nature* **485**, 611–614 (2012).
5. Predehl, K. *et al.* A 920-kilometer optical fiber link for frequency metrology at the 19th decimal place. *Science* **336**, 441–444 (2012).
6. Giorgetta, F. R. *et al.* Optical two-way time and frequency transfer over free space. *Nat. Photon.* **7**, 434–438 (2013).
7. Newbury, N. R. Searching for applications with a fine-tooth comb. *Nat. Photon.* **5**, 186–188 (2011).
8. Minoshima, K. & Matsumoto, H. High-accuracy measurement of 240-m distance in an optical tunnel by use of a compact femtosecond laser. *Appl. Opt.* **39**, 5512–5517 (2000).
9. Coddington, I., Swann, W. C., Nenadovic, L. & Newbury, N. R. Rapid and precise absolute distance measurements at long range. *Nat. Photon.* **3**, 351–356 (2009).
10. Lee, J., Kim, Y.-J., Lee, K., Lee, S. & Kim, S.-W. Time-of-flight measurement with femtosecond light pulses. *Nat. Photon.* **4**, 716–720 (2010).
11. Rosenband, T. *et al.* Frequency ratio of Al + and Hg + single-ion optical clocks; metrology at the 17th decimal place. *Science* **319**, 1808–1812 (2008).
12. Reinhardt, S. *et al.* Test of relativistic time dilation with fast optical atomic clocks at different velocities. *Nat. Phys.* **3**, 861–864 (2007).
13. Hayes, D. *et al.* Entanglement of atomic qubits using an optical frequency comb. *Phys. Rev. Lett.* **104**, 140501 (2010).
14. Ideguchi, T. *et al.* Coherent Raman spectro-imaging with laser frequency combs. *Nature* **502**, 355–358 (2013).
15. Stockman, M. Nanoplasmonics: past, present, and glimpse into future. *Opt. Express* **19**, 22029–22106 (2011).
16. Kauranen, M. & Zayats, A. V. Nonlinear plasmonics. *Nat. Photon.* **6**, 737–748 (2012).
17. Willets, K. A. & Van Duyne, R. P. Localized surface plasmon resonance spectroscopy and sensing. *Annu. Rev. Phys. Chem.* **58**, 267–297 (2007).
18. Zhang, Y. *et al.* Coherent anti-Stokes Raman scattering with single-molecule sensitivity using a plasmonic Fano resonance. *Nat. Commun.* **5**, 4424 (2014).
19. Tame, M. S. *et al.* Quantum plasmonics. *Nat. Phys.* **9**, 329–340 (2013).
20. Melikyan, A. *et al.* High-speed plasmonic phase modulators. *Nat. Photon.* **8**, 229–233 (2014).
21. Haffner, C. *et al.* All-plasmonic Mach–Zehnder modulator enabling optical high-speed communication at the microscale. *Nat. Photon.* **9**, 525–528 (2015).
22. Dennis, B. S. *et al.* Compact nanomechanical plasmonic phase modulators. *Nat. Photon.* **9**, 267–273 (2015).
23. Ebbesen, T. W., Lezec, H. J., Chaemi, H. F., Thio, T. & Wolff, P. A. Extraordinary optical transmission through sub-wavelength hole arrays. *Nature* **391**, 667–669 (1998).
24. Barnes, W. L., Dereux, A. & Ebbesen, T. W. Surface plasmon subwavelength optics. *Nature* **424**, 824–830 (2003).
25. Wurtz, G. A. & Zayats, A. V. Nonlinear surface plasmon polaritonic crystals. *Laser Photon. Rev.* **2**, 125–135 (2008).
26. Garcia-Vidal, F. J., Martin-Moreno, L., Ebbesen, T. W. & Kuipers, L. Light passing through subwavelength apertures. *Rev. Mod. Phys.* **82**, 729–787 (2010).
27. Newbury, N. R. & Swann, W. C. Low-noise fiber-laser frequency combs. *J. Opt. Soc. Am. B* **24**, 1756–1770 (2007).
28. Yue, W. *et al.* Enhanced extraordinary optical transmission (EOT) through arrays of bridged nanohole pairs and their sensing applications. *Nanoscale* **6**, 7917–7923 (2014).
29. Steuwe, C., Kaminski, C. F., Baumberg, J. J. & Mahajan, S. Surface enhanced coherent anti-Stokes Raman scattering on nanostructured gold surfaces. *Nano Lett.* **11**, 5339–5343 (2011).
30. Altewischer, E., Van Exter, M. P. & Woerdman, J. P. Plasmon-assisted transmission of entangled photons. *Nature* **418**, 304–306 (2002).
31. Almeida, E. & Prior, Y. Rational design of metallic nanocavities for resonantly enhanced four-wave mixing. *Sci. Rep.* **5**, 10033 (2015).

Acknowledgements

This work was supported by the Basic Science Research Program (NRF-2013R1A1A2004932; NRF-2014R1A1A1004885), by Global Research Laboratory Program (Grant No. 2009-00439), by the Leading Foreign Research Institute Recruitment Program (Grant No. 2010-00471), by the Max Planck POSTECH/KOREA Research Initiative Program (Grant No. 2011-0031558), by the NRF Grant (No. 2010-0021735), by the Leading Foreign Research Institute Recruitment Program (Grant No. 2012K1A4A3053565) through the NRF funded by the MEST. This work was also supported by a Grant (14CTAP-C077584-01) from Infrastructure and Transportation Technology Promotion Research Program funded by Ministry of Land, Infrastructure and Transport of Korean government. This work was also supported by Singapore National Research Foundation (NRF-NRFF2015-02) and Singapore Ministry of Education under its Tier 1 Grant (RG85/15).

Author contributions

The project was planned and overseen by Y.-J.K., D.E.K. and S.K. Plasmonic sample was prepared and characterized by J.H.S., H.Y. and K.S. Frequency comb experiments were performed by X.T.G., B.J.C., Y.-J.K. and S.K. All authors contributed to the manuscript preparation.

Additional information

Arbitrary cross-section SEM-cathodoluminescence imaging of growth sectors and local carrier concentrations within micro-sampled semiconductor nanorods

Kentaro Watanabe[1,2], Takahiro Nagata[1], Seungjun Oh[1], Yutaka Wakayama[1], Takashi Sekiguchi[1], János Volk[3] & Yoshiaki Nakamura[2]

Future one-dimensional electronics require single-crystalline semiconductor free-standing nanorods grown with uniform electrical properties. However, this is currently unrealistic as each crystallographic plane of a nanorod grows at unique incorporation rates of environmental dopants, which forms axial and lateral growth sectors with different carrier concentrations. Here we propose a series of techniques that micro-sample a free-standing nanorod of interest, fabricate its arbitrary cross-sections by controlling focused ion beam incidence orientation, and visualize its internal carrier concentration map. ZnO nanorods are grown by selective area homoepitaxy in precursor aqueous solution, each of which has a (0001): +c top-plane and six {1-100}:m side-planes. Near-band-edge cathodoluminescence nanospectroscopy evaluates carrier concentration map within a nanorod at high spatial resolution (60 nm) and high sensitivity. It also visualizes +c and m growth sectors at arbitrary nanorod cross-section and history of local transient growth events within each growth sector. Our technique paves the way for well-defined bottom-up nanoelectronics.

[1] International Center for Materials Nanoarchitectonics, National Institute for Materials Science, 1-1 Namiki, Ibaraki 305-0044, Japan. [2] Graduate School of Engineering Science, Osaka University, 1-3 Machikaneyama-cho, Osaka 560-8531, Japan. [3] MTA EK Institute of Technical Physics and Materials Science, Konkoly Thege M. ut 29-33, Budapest 1121, Hungary. Correspondence and requests for materials should be addressed to K.W. (email: watanabe@ee.es.osaka-u.ac.jp).

Bottom-up growths (for example, wet chemical growth[1,2] and molecular beam epitaxy[3,4]) of semiconductor nanocrystals are paid attention because of the limitation of semiconductor microfabrication. Especially, a free-standing nanowire or nanorod has several exotic properties applicable to unique one-dimensional electronics; high flexibility (large fracture strain) for strain-engineered ultrafast nanowire field-effect transistors[5] and for efficient nanopiezotronic devices[6], large surface-to-volume ratio for sensitive solar cells[7] and gas sensors[8], and wave-guides for Fabry-Perot nanolasers[9-11]. A nanowire with uniform electric properties besides uniform diameter is essential for one-dimensional electronics with high device designability and reproducibility, which should be driven by well-defined current density.

Up to now, electrical properties of a bottom-up nanocrystal and its device properties are reported assuming electrical uniformities. However, this is idealistic because each crystallographic plane surface has an atomic arrangement with unique chemical activity (for example, etching rate[12,13], host crystal growth rate[14,15] and incorporation rates of point-defects), as is reported on plane-dependent donor concentration in ZnO bulk crystal[13,16] and amphoteric Si-doping in GaAs substrates[17,18]. Thus, a nanocrystal grown in multiple crystallographic orientations has corresponding growth sectors with different electric properties, which is critical for above electronic applications. Further, growth sectors within a nanocrystal are not fully described macroscopically by pure crystallographic planes at constant growth rates[19-23]. (For example, a free-standing nanowire with uniform diameter is also idealistic[24].) In reality, a growth form of a nanocrystal may evolve with growth duration depending on the local and temporal growth environment on each crystallographic plane, where spontaneous surface roughening or new plane formation may take place. Thus, it is difficult to foresee the internal spatial distributions of growth sectors. We demand a novel experimental technique which reveals local electrical properties within a nanocrystal, especially those due to the growth sectors, at high spatial resolution and high sensitivity.

A recent report on nanowire Hall effect measurement reveals carrier transport properties of an entire free-standing nanowire between micro-fabricated electrical contacts[25]. However, there is no adequate technique that can probe such electrical non-uniformities within a free-standing nanocrystal at sufficient sensitivity. Atom probe tomography[26,27] and cross-sectional scanning transmission electron microscopy[28] may visualize dopant species at atomic resolutions. However, none of them has sufficient sensitivities to probe local concentration and investigate electrical activity of point-defects within nominally undoped semiconductors. Cross-sectional scanning tunnelling microscopy[29] may visualize electrical activities of dopants at sub-surfaces at atomic resolution. However, atomically smooth specimen cleavage is required to obtain noticeable contrasts of impurity atoms or vacancies, which is not available for a free-standing nanocrystals of interest up to now.

Contrary, cathodoluminescence (CL) nanospectroscopy is promising for its high sensitivity ($\sim 10^{15}\,\mathrm{cm}^{-3}$) of shallow level point-defects and nanometric resolution achievable by low-energy electron beam (e-beam) probe[30-34]. Cross-sectional CL imaging of growth sector interface is reported on hydrothermally grown ZnO bulk crystals with $(0001){:}{+}c$, $\{1{-}100\}{:}m$ and $\{1{-}101\}{:}{+}p$ planes[35] and on hydride vapour phase epitaxy grown bulk GaN $+c$ films containing hexagonal micro-pits with $+p$ planes[36,37]. Plane-dependent CL behaviours of ZnO nanorod are reported recently[38]. However, cross-sectional CL technique is not applied to any semiconductor free-standing nanocrystal up to now.

ZnO single-crystalline free-standing nanorod array is chosen to demonstrate our technique, as it is suitable for low-cost optoelectronics and piezoelectronics. Wurzite ZnO is a piezoelectric wide-gap semiconductor with direct bandgap ($E_g = 3.37\,\mathrm{eV}$), a large exciton binding energy ($E_b = 60\,\mathrm{meV}$)[39] and a large piezoelectric coefficient along its polar c-axis ($d_{cc} = 12.4\,\mathrm{pm\,V}^{-1}$)[40], which can be synthesized at low temperature and at atmospheric pressure in precursor aqueous solution: by simple chemical synthesis[41-54] or by electrodeposition[55-58].

Here we demonstrate a series of techniques to micro-sample a free-standing nanocrystal of interest and to visualize its internal local carrier concentration, especially those due to growth sectors, from arbitrary cross-sectional orientations. Circular growth windows are fabricated by e-beam lithography in a polymethyl methacrylate (PMMA) e-beam-resist film spin-coated on a single-crystalline ZnO (0001) substrate. ZnO single-crystalline free-standing nanorod arrays are then homoepitaxially grown at a time in precursor aqueous solution from circular growth window arrays with different diameters (D_w; Supplementary Fig. 1). Scanning electron microscopy (SEM) observation shows that any nanorod grows from each circular growth window to have a $+c$ top-plane and six m side-planes (Supplementary Figs 2 and 3). A nanorod is micro-sampled and its cross-section is fabricated by controlling crystallographic orientation of focused ion beam (FIB) incidence. Cross-section CL imaging of the nanorod visualizes $+c$ and m growth sectors within an entire nanorod at high spatial resolution ($<60\,\mathrm{nm}$). CL nanospectroscopy on $+c$ and m planes reveals their local carrier concentration differences and its quantitative accuracy is confirmed by 'differential' I–V measurements of an individual ZnO free-standing nanorod (Supplementary Fig. 4).

Results

Room temperature CL nanospectroscopy of ZnO nanorods. Local CL properties are studied for ZnO free-standing nanorods grown homoepitaxially at arrayed circular windows. ZnO nanorod arrays of $D_w = 100$ and $500\,\mathrm{nm}$ are investigated by SEM observation and near-band-edge (NBE)/visible (Visible) CL imaging at 3.0 keV from bird's-eye view angles (Fig. 1a,b). Further, local CL spectroscopy is performed at 4.5 keV on a nanorod of $D_w = 500\,\mathrm{nm}$ at spot-CL mode. Eight spot-CL spectra at corresponding e-beam spot positions 1–8 are shown in Fig. 1c, which are correlated by the same colour: position 1 at top $+c$ plane, positions 2–7 at side m plane from top to the bottom and position 8 on the ZnO $+c$ substrate covered with transparent PMMA film. Spectral band of NBE and Visible CL imaging are also indicated by the rectangular grey shadows. Visible CL emission band at 2.1 eV exhibits much more uniform spatial distribution and spectral shape within the nanorod (Supplementary Note 2). Contrary, NBE CL images show that $+c$ top-planes are darkest and m side-planes gradually become brighter towards the nanorod bottom by one order of magnitude, regardless of D_w. Spot-CL spectroscopy also shows that each NBE CL peak consists of intrinsic emission (3.28 eV) and red-shifted emission (3.19 eV). Intrinsic NBE CL emission is dominant at position 1, however, red-shifted NBE CL emission intensity increases to be dominant and then saturated as the spot position goes from 2 to 7. Considering that SEM e-beam probes CL from ZnO surface-to-depth of Kanaya–Okayama electron range (192 nm for 4.5 keV electron)[59], red-shifted emission is unique to m plane and m polar growth thickness seems to be increasing from 0 at the nanorod top-plane to $>200\,\mathrm{nm}$ at the nanorod bottom. Spatial distribution of NBE CL emission efficiency within a nanorod is illustrated in Fig. 1d, which is further investigated in

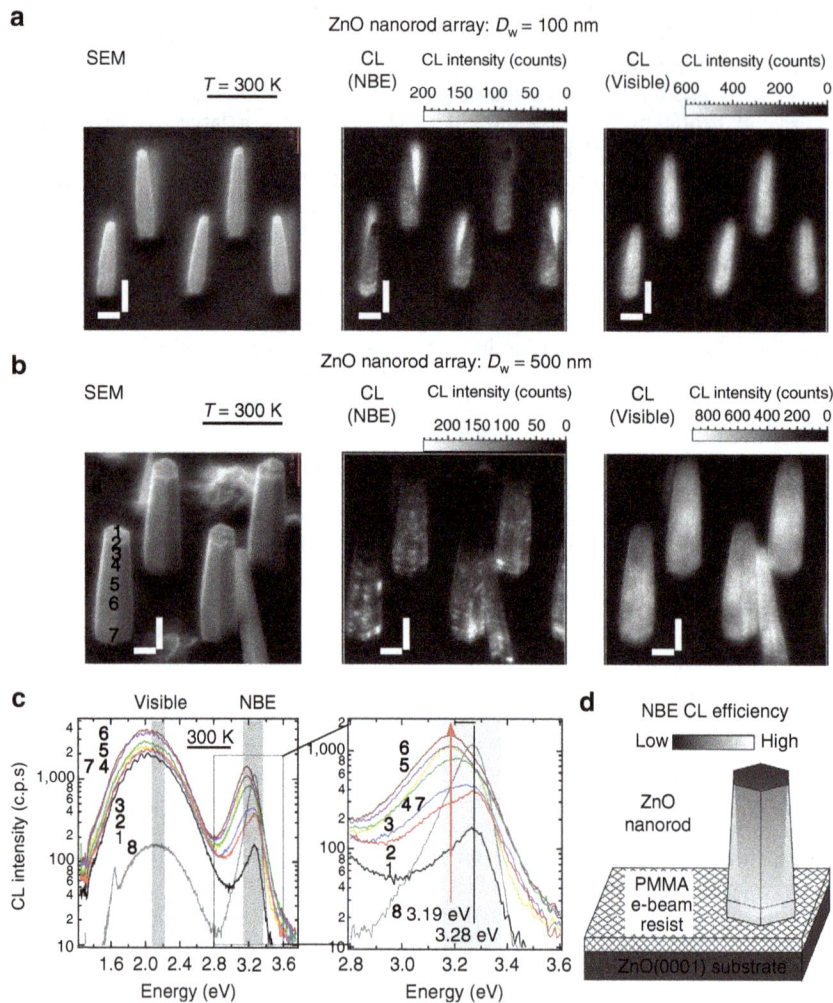

Figure 1 | Bird's-eye view NBE/Visible CL imaging of ZnO free-standing nanorods. SEM and monochromatic (NBE/Visible) CL images of ZnO free-standing nanorod arrays, probed by 3.0 keV e-beam. Each nanorod is homoepitaxially grown from individual growth window with a variety of diameter: (**a**) small diameter growth windows ($D_w = 100$ nm) and (**b**) large diameter growth windows ($D_w = 500$ nm). Nanorods misorientation in $D_w = 100$ nm array takes place during the SEM observation due to electrical charge-up. A bright shadow behind each nanorod array in NBE CL image are signal from ZnO substrate, which is excited due to PMMA thinning by the concentrated e-beam dose. (**c**) Spot-CL spectroscopy at positions 1–8 in **b**, probed by 4.5 keV e-beam. Spectral bands for NBE/Visible CL imaging are highlighted with transparent grey zones. (**d**) A schematic of ZnO free-standing nanorod. Local NBE CL emission efficiency within the nanorod is illustrated by grey scale. The specimen is tilted by 45 degree from ZnO (0001) plane normal. All horizontal and vertical scale bars indicate lengths of 0.5 μm and 1.0 μm, respectively.

Figs 3 and 4. Note that each nanorod exhibits some locally bright NBE CL emission spots (CL spots) on m side-planes. As spectrum 6 on a CL spot shows the same spectral shape as those at other spot positions on m plane, the CL spots have the same properties of m plane.

Low-temperature CL spectroscopy at nanorod top/side plane. To reveal the origin of red-shifted NBE CL emission of m plane, temperature-dependent CL spectroscopy is performed on $+c$ top-plane and m side-plane of a nanorod ($D_w = 500$ nm). A bird's-eye view SEM image of a ZnO free-standing nanorod indicates each e-beam scan area by corresponding solid box superimposed (Fig. 2a). Temperature-dependent 3.0 keV NBE CL spectra are obtained in Area-CL mode at higher wavelength resolution: at 300 K (top), at 80 K (middle) and at 10 K (bottom; Fig. 2b). Each CL peak at 10 K is assigned as follows: 3.37 eV to radiative recombination of free excitons (FX), 3.361 eV to neutral donor-bound excitons (D^0X) observed in naturally n-type ZnO, 3.324 eV to its two electron satellite (TES-D^0X) and two lower energy peaks with 0.072 eV interspacing to nth LO phonon

replica of D^0X peak (D^0X-nLO; $n = 1$ and 2). The energy difference between D^0X (1 s state) and TES-D^0X (2s and 2p states) peaks of 37 meV equals the three-fourth of donor-binding energy $E_D = 49$ meV in the hydrogenic effective mass approach[60]. At elevated temperatures, D^0X peaks quench and FX peaks emerge at 80 K and FX peak locates at 3.280 eV on $+c$ top-plane and at 3.200 eV on m side-plane at 300 K (Fig. 2b). The detailed temperature dependence between 80 and 300 K also reveals the emergence of redshifted FX emission peak above 220 K (Fig. 2c). Such a 80-meV redshift larger than E_D is not attributed to the donor-binding energy itself. Rather, it is attributed to the band-gap shrinkage or band-tailing of a heavily donor-doped semiconductor observable at elevated temperature, both of which originates in donor ionization and its local perturbation of conduction and valence band-edges.

Local carrier concentration evaluated by CL nanospectroscopy. Here, room temperature FX peak energy in Fig. 2b is attributed to the local residual carrier concentration n (cm^{-3}). Room temperature band-edge emission energy of a heavily donor-doped

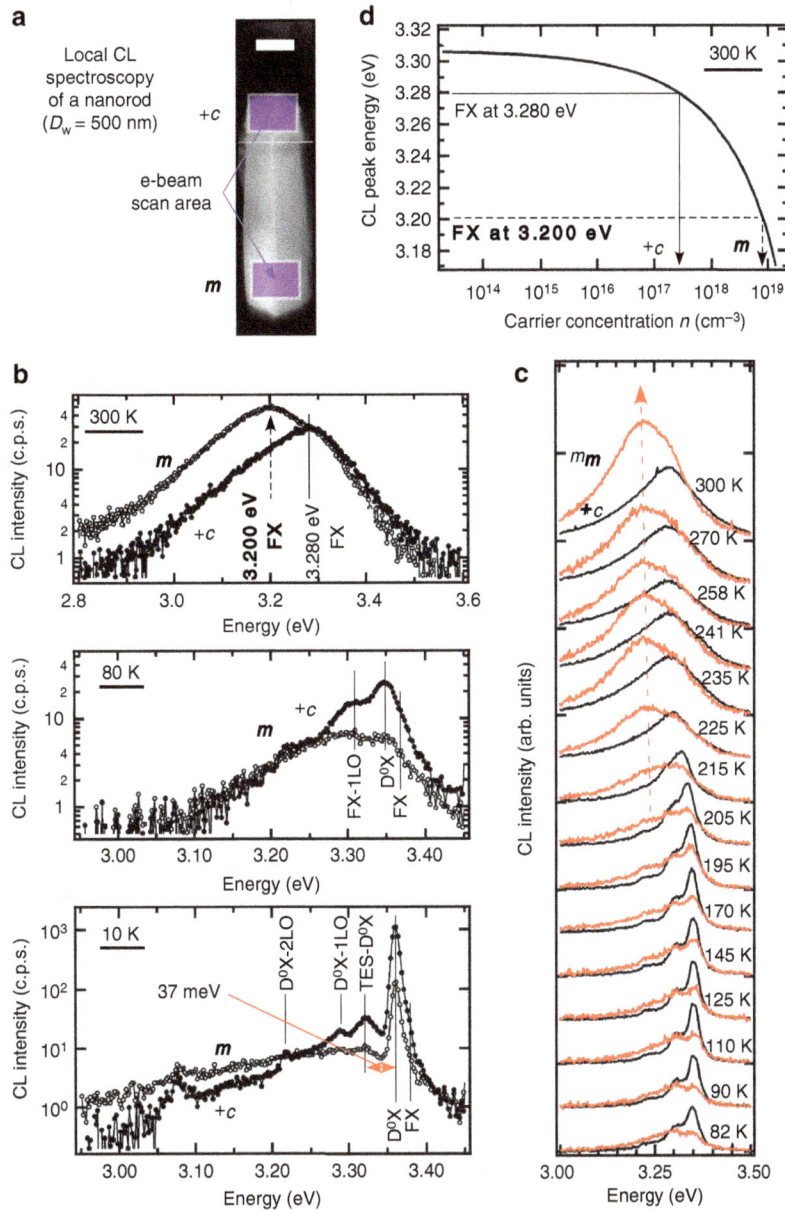

Figure 2 | Temperature-dependent high-resolution CL spectroscopy of a nanorod $+c$ top-plane and m side-plane. (**a**) Bird's-eye view SEM image of a ZnO nanorod investigated by Area-CL spectroscopy. E-beam scan areas on $+c$ top-plane and m side-plane are indicated by transparent purple boxes. A horizontal scale bar indicates length of 0.5 μm. (**b**) Area-CL spectra on $+c$ and m planes at 300, 80 and 10 K, probed by 3.0 keV e-beam. CL peaks are assigned to free exciton (FX), neutral donor-bound exciton (D^0X), its nth LO phonon replica (D^0X-nLO) and its two electron satellite (TES-D^0X). (**c**) Detailed temperature dependence of Area-CL spectra between 80 and 300 K. Emergence of the redshifted FX peak is clearly observed. (**d**) Correlation between FX peak energy E_{FX} (eV) and carrier electron concentration n (cm^{-3}). Residual carrier concentrations n of $+c$ and m growth sectors are evaluated using their correlation: $E_{FX}(n) = 3.307 - 8.39 \times 10^{-15} n^{2/3} - 3.64 \times 10^{-8} n^{1/3}$ (ref. 61).

semiconductor redshifts due to its band-gap shrinkage or band-tailing. Giles *et al.* investigated *n*-type ZnO bulk crystals with different carrier concentrations n (cm^{-3}) by photoluminescence spectroscopy and reported their FX emission energy: $E_{FX}(n)$ (eV) $= 3.307 - 8.39 \times 10^{-15} \cdot n^{2/3} - 3.64 \times 10^{-8} \cdot n^{1/3}$ (solid curve in Fig. 2d)[61]. Based on this work, we obtain $n_{+c} = 2.8 \times 10^{17}$ cm^{-3} at $+c$ top-plane and $n_m = 8.2 \times 10^{18}$ cm^{-3} at m side-planes. Carrier concentration gap at 300 K is evidenced qualitatively by FX emission intensity gap between dark $+c$ top-plane and bright m side-plane, which is intensified by residual carrier concentration.

Also, room temperature NBE CL emission spectroscopy of a 'ZnO nanocolumn', the nanorod at earlier growth stage, is performed at Spot-CL mode (Supplementary Fig. 3). This

nanocolumn also exhibits 30 meV redshift of FX emission at its m side-plane. NBE CL emission energy of ZnO nanocolumn is also attributed to the local carrier concentration: 3.28 eV emission to $n_{+c} = 2 \times 10^{17}$ cm^{-3} in the axial $+c$ growth sector and 3.25 eV emission to $n_m = 2 \times 10^{18}$ cm^{-3} in the lateral m growth sector. The growth duration-dependent energy redshift also supports our idea that the energy redshift of NBE CL emission is originated from the difference of donor incorporation rates, rather than a certain donor-binding energy.

A quantitative accuracy of n values given by CL nanospectroscopy is evaluated in comparison with a net carrier concentration estimated from 'differential' *I–V* measurements of a ZnO free-standing nanorod (Supplementary Fig. 4). Here, σ_e is net electrical conductivity in the nanorod. Electron mobility

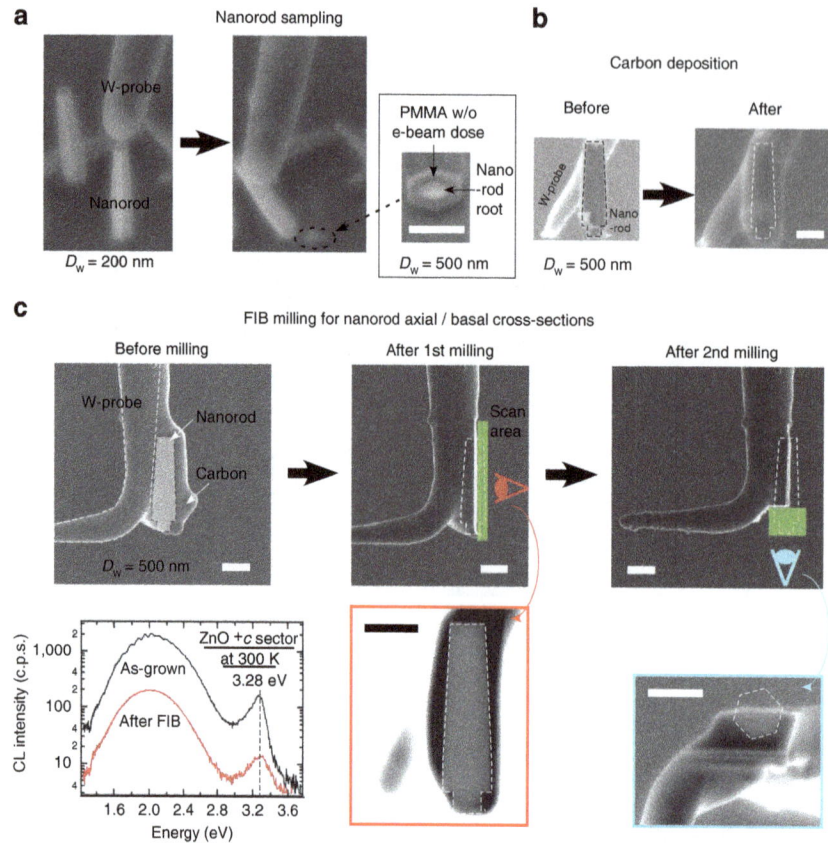

Figure 3 | Nanorod micro-sampling and cross-section fabrication by controlling FIB incidence orientation. (a) A nanorod ($D_w = 200$ nm) micro-sampling observed by SEM. The inset SEM image shows the circular nanorod root left after the removal of a nanorod ($D_w = 500$ nm). **(b)** SEM images of a micro-sampled nanorod before and after the amorphous carbon deposition. Nanorod position is indicated by dashed lines. **(c)** A series of FIB milling process observed by scanning ion microscope imaging. Nanorod positions and FIB milling scan areas at each step are indicated by dashed line and half-transparent box, respectively. Axial and basal cross-sections fabricated by first and second FIB milling, respectively, are observed consecutively by SEM. The inset at the bottom left shows Spot-CL spectra of the nanorod $+c$ top-plane before and after the FIB milling, probed by 4.5 keV e-beam. All scale bars indicate lengths of 1 μm.

$\mu_e = 1 \times 10^1$ cm^2 V^{-1} s^{-1} in the nanorod is assumed as aqueous solution synthesized ZnO films with Hall mobility of $\mu_e = 12.5$ cm^2 V^{-1} s^{-1} at $n = 6.66 \times 10^{17}$ cm^{-3} are reported[62]. Considering slightly tapered nanorod shape, 'differential' I–V measurements of an individual ZnO free-standing nanorod yields net electrical conductivity, $\sigma_e = 0.7$ Ω^{-1} cm^{-1}. Calculated net residual carrier concentration $n = \sigma_e/(e\mu_e) = 4 \times 10^{17}$ cm^{-3} falls between n values of $+c$ and m growth sectors, which suggests that n evaluation by CL nanospectroscopy is quantitative.

Arbitral cross-section of micro-sampled nanorod made by FIB. Next we show a sequential process to fabricate cross-sections of a micro-sampled free-standing nanorod in arbitral crystallographic orientation. First, a nanorod is micro-sampled using a tungsten (W) probe attached to a nanomanipulator (Fig. 3a). Note that inset SEM image shows a circular root of a nanorod remained attached to the substrate. This root surrounded by PMMA film with a hexagonal white area without e-beam dose, which is originally covered by the nanorod. This micro-sampling reveals circular cylindrical root of the hexagonal nanorod, which suggests that the nanorod growth is limited by PMMA film and that m plane growth thickness is evaluated from the diameter gap at the nanorod bottom. The micro-sampled nanorod on the W-probe is then embedded in a carbon film by carbon deposition from different orientations (Fig. 3b).

The W-probe with the nanorod is then transferred into FIB system to fabricate nanorod cross-sections at arbitrary

crystallographic orientations by two-step FIB milling (Fig. 3c). Sequential scanning ion microscope images show first FIB milling in axial direction and second FIB milling in basal direction. SEM image of each cross-section observed from eye mark direction is also displayed, which highlights each nanorod cross-section with a bright contrast. Note that the SEM image of second basal cross-section shows that first axial cross-section is fabricated exactly on the nanorod axis, which evidences the high position controllability of this FIB milling technique.

Impact of Ga ion beam milling on surface CL properties is studied by 4.5 keV CL spectra at the nanorod top-plane before and after FIB milling (Fig. 3c inset). After the two-step FIB milling, the CL intensity has dropped to one-tenth due to the residual damaged layer of cross-section surface. Nevertheless, CL spectral shape is retained to investigate local CL properties of the nanorod.

Growth sectors visualized at arbitrary nanorod cross-section. Here we show cross-sectional CL observation of a micro-sampled nanorod to visualize its internal $+c/m$ growth sectors. Figure 4a,b shows SEM, NBE CL and Visible CL images of first axial cross-section and second basal cross-section at 3.0 kV, respectively. P1–P3 in Fig. 4c are NBE CL line profiles along Z axis ($// c$) in Fig. 4a. CL probing regions of P1–P3 are indicated by red areas at nanorod core (centre), nanorod shell (side-plane intersection) and nanorod shell (side-plane centre), respectively, in SEM image in Fig. 4b. Fig. 4d is Panchromatic, NBE and

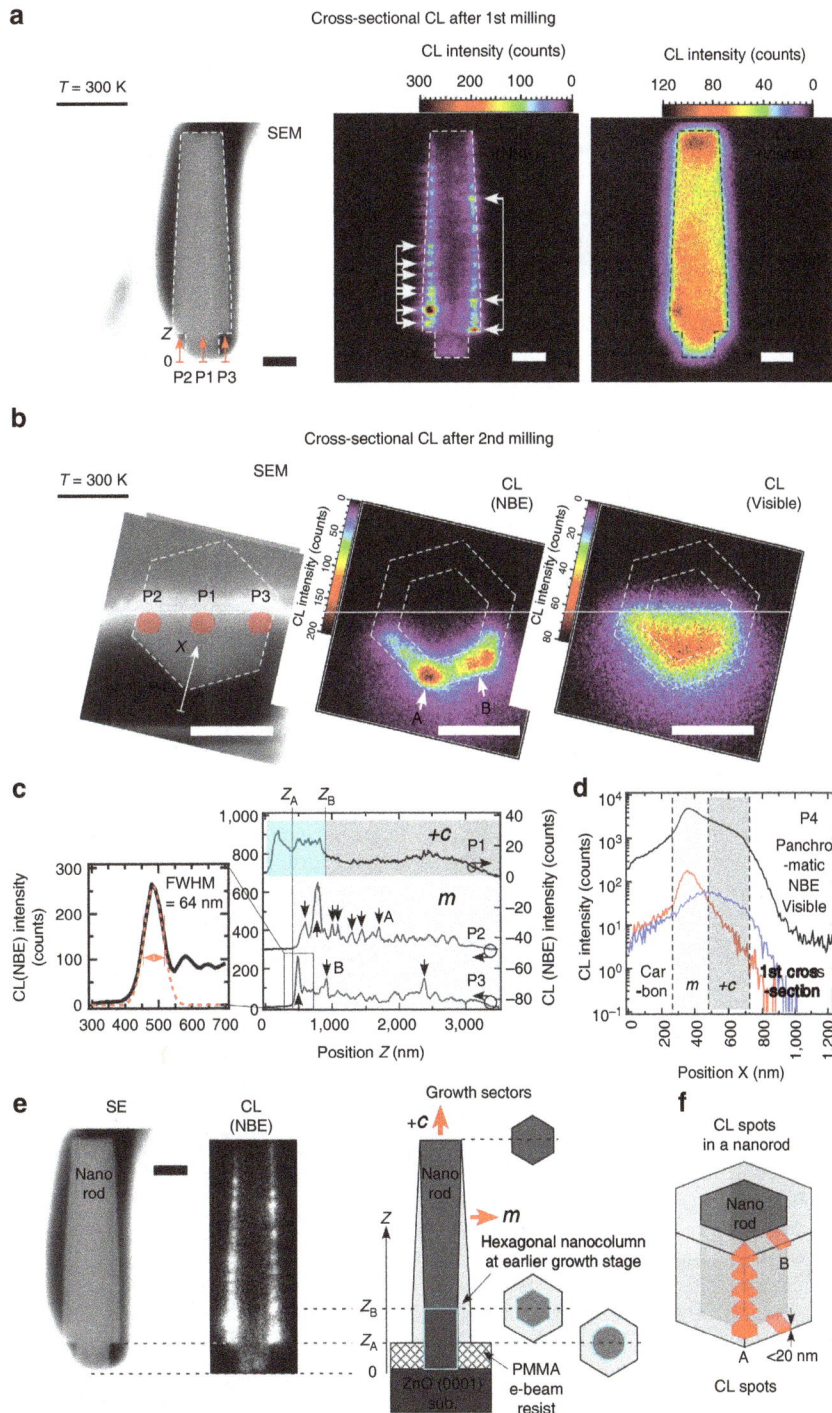

Figure 4 | High-resolution cross-sectional CL imaging of a micro-sampled nanorod. SEM and CL (NBE/Visible) images of first axial cross-section (**a**) and those of second basal cross-section (**b**), both of which are probed by 3.0 keV e-beam. Original nanorod volume is indicated by dashed line for eye-guide. (**c**) NBE CL profiles P1–P3 along coordinates Z at P1–P3 in **a**. CL spatial resolution is evaluated from FWHM of profile P3 around the sharpest CL spot. (**d**) CL (Panchromatic/NBE/Visible) profiles along coordinate X. (**e**) Schematic representation of $+c$ and m growth sectors within the nanorod cross-section, revealed from comparison between SEM and NBE CL images. Nanorod bottom ($Z = Z_A$) and slight NBE CL intensity drop in $+c$ growth sector ($Z = Z_B$) are also indicated in **c,e**. CL profiles P1–P4 and a hexagonal nanocolumn at earlier growth stage are indicated by aqua zones and schematically by an aqua box in a $+c$ growth sector. All scale bars indicate lengths of 0.5 μm. (**f**) Schematic illustration of spatial distribution and shape of CL spots. Observed CL spots in **a–c** are indicated by solid arrows and categorized into either spot A or spot B by their shape and location.

Visible CL profiles along X axis in Fig. 4b. Figure 4b,d shows CL quenches near the first axial cross-section because of the damaged surfaces. Two NBE CL images show the two regions with distinct intensity difference: dark hexagonal column core and bright hexagonal shell, whereas two Visible CL images also show them

with opposite and weaker contrasts. Note that the bright shell region in NBE CL image corresponds to the diameter gap at the nanorod bottom in SEM image. The diameter gap is formed at an earlier growth stage. A single hexagonal nanocolumn with its diameter equivalent to $D_w = 0.5$ μm is formed by the coalescence of multiple nuclei within each growth window. Subsequently, the lateral growth of the nanocolumn is allowed above PMMA film surface to form a diameter gap (Supplementary Fig. 2b). Thus, we assign the dark core and bright shell in NBE CL images to $+c$ and m growth sectors of the nanorod, respectively, where each sector has different donor concentration: $n_{+c} = 2.8 \times 10^{17}\,\text{cm}^{-3}$ in axial $+c$ growth sector and $n_m = 8.2 \times 10^{18}\,\text{cm}^{-3}$ in lateral m growth sectors, as investigated in Fig. 2. Figure 4e summarizes schematic spatial distribution of $+c$ and m growth sectors together with corresponding SEM and highlighted NBE CL images.

Also, CL spots are observed in NBE CL images, as indicated by solid arrows and labelled as spot A and B in Fig. 4a,b and as their locations and shapes are illustrated in Fig. 4f. They extend laterally across m growth sector and CL spots locate more at the m region intersection (spot A) than at their centres (spot B), comparing P2 with P3. A bright CL spot at nanorod bottom is chosen to evaluate the spatial resolution of CL imaging. P3 is locally fitted well with Gaussian curve $[I(Z) = I_0 \cdot \exp\{-(Z-Z_0)^2/2\sigma^2\}]$ of full-width at half-maximum (FWHM) $= 2\sqrt{(2\ln 2)} \cdot \sigma = 64\,\text{nm}$. As this FWHM is comparable to Kanaya–Okayama range[59] of 3.0 keV electron (97 nm), the FWHM is attributed to the lateral resolution of CL imaging.

Note that cross-sectional NBE CL image exhibits two growth turning points, $(Z_A, Z_B) = (402\,\text{nm}, 871\,\text{nm})$, where the axial $+c$ growth domain contains a slightly brighter region ($0 < Z < Z_B$; Fig. 4c,e). As Z_A matches with PMMA film thickness (0.3 μm), the turning point at $Z = Z_A$ is attributed to PMMA film surface. Sequential SEM observation of ZnO nanorods at earlier growth stages reveals that the Z_B–Z_A matches with the typical height (0.5 μm) of each hexagonal nanocolumn above PMMA film (Supplementary Fig. 2b). The turning point at $Z = Z_B$ is then attributed to the formation of single hexagonal nanocolumn (nanorod at earlier stage). CL profiles P1–P4 and a hexagonal nanocolumn at earlier growth stage are indicated by aqua zones and schematically by an aqua box in a $+c$ growth sector (Fig. 4c–e). Spot-CL spectroscopy of a nanocolumn reveals NBE CL emission redshift on its side-plane, which evidences that the top-plane and side-planes of the nanocolumn already have distinct difference in their donor incorporation rates (Supplementary Fig. 3b). As nuclei within a growth window also experience axial and lateral growths, the slightly bright region within this nominal $+c$ growth sector in Fig. 4e is accountable by the lateral m growth sectors of nuclei exposed to the cross-section. Thus, cross-sectional CL imaging of an individual nanorod reveals its growth history from minor contrasts within each growth sector as well as its local carrier concentrations from major contrasts between growth sectors.

Discussion

The origin of unintentional donors is remained to be discussed. Here we tentatively attribute the residual carriers to interstitial hydrogen donors incorporated from growth environment for the following five reasons: (i) wide-scan X-ray photoemission spectroscopy of reference ZnO homoepitaxial thin film (Supplementary Fig. 5e) does not detect any other possible ZnO donor element than hydrogen, such as Al or Ga, and thus its concentration is typically below 0.1 atomic %; (ii) narrow-scan X-ray photoemission spectroscopy (Supplementary Fig. 5f,g) and Raman spectroscopy (Supplementary Fig. 5h) of the ZnO thin film evidence the presence of hydrogen atoms at bond-centred interstitial sites in the form of [-OH] groups[63–67]; (iii) Raman peak observed at $330\,\text{cm}^{-1}$ (Supplementary Fig. 5h) is energetically equivalent to the 37 meV gap in 10 K CL peaks in Fig. 2b and thus can be attributed to $1s \rightarrow 2p$ transitions of

interstitial hydrogen donor[66–68], although it can also be ascribed to second-order vibration mode (E_2^{high}–E_2^{low}) of intrinsic ZnO at $333\,\text{cm}^{-1}$ (ref. 69); (iv) D^0X peak in Fig. 2b matches I_4 peak (3.363 eV) associated with H donor[60]; (v) the $E_D = 49$ meV calculated from CL spectrum at 10 K agrees with $E_D = 46.1$ meV of hydrogen donor reported[60]. Interstitial hydrogen donor in our ZnO nanorod may be originated from $Zn(OH)_2$ precursors and incorporated as [-OH] group at oxygen site (Supplementary Note 1). This model is supported by the report on enhanced hydrogen donor incorporation into ZnO (0001) films grown in aqueous solution at higher $[OH^-]$ concentration ($n \sim 10^{17}\,\text{cm}^{-3}$ at pH = 8 and $n = 1.79 \times 10^{19}\,\text{cm}^{-3}$ at pH = 10.9)[62].

Also, we discuss origins of CL spots appeared in m growth sectors. The minimum thickness of CL spot origin is at least half an order of magnitude smaller (<20 nm). CL spots extends from the interface of growth sectors along the nanorod basal plane. NBE CL images also show that each CL spot extends laterally, populate preferentially around nanorod corners, and populate more on nanorods of $D_w = 500$ nm than those of $D_w = 100$ nm. Inspired by the above indirect observations, we tentatively consider that CL spots origin might be related with the growth mode of the nanorod side-planes. However, further studies are required to demonstrate this idea.

In summary, we demonstrate that cross-sectional CL technique evaluates local carrier concentration quantitatively at high spatial resolution and at high sensitivity. It also visualizes internal growth sectors of an entire semiconductor nanorod from arbitrary crystallographic orientations, and even reveals nanorod growth history. Our model also gives suggestions how to improve nanorods: (i) nanorod side-plane growth should be minimized as it is the origin of non-uniform diameter and local high carrier concentration; (ii) for aqueous solution growth, amine additives might be promising to enhance uniformities of nanorod diameter and its electrical properties, as they adsorb selectively on nanorod non-polar side-planes and suppress nanorod lateral growth[41,42]. Above findings are quite general and are valid for various luminescent semiconductor nanocrystals, regardless of semiconductor species and growth methods.

Methods

Selective area homoepitaxy in precursor aqueous solution. In the precursor aqueous solution, ZnO hexagonal nanorods are grown at a time from size-controlled circular holes patterned in PMMA film on an atomically flat ZnO $+c$ substrate (selective area homoepitaxy)[53,54]. The c-axis polarities of these nanorods are expected to be matched considering the recent report on aqueous solution grown ZnO nanowire[70].

First, Zn-terminated ZnO $+c$ substrate was ultrasonically rinsed in acetone, ethanol and deionized water and then annealed at 1,000 °C for 8 h in oxygen atmosphere at 1 atm to make atomically flat surfaces. Then, 300-nm-thick PMMA photoresist is formed on the substrate by spin-coating (Supplementary Fig. 1a). Two hundred circular growth windows per each diameters ($D_w = 100, 150, 200, 300, 400, 500$ nm) are opened in the PMMA film by e-beam lithography to form a 2×100 trigonal lattice with $2.0\,\mu\text{m} + D_w$ interspacing (Supplementary Fig. 1b). Buffered aqueous solution (0.2 l) of equimolar (8×10^{-3} M) zinc nitrate hexahydrate ($>99.9\%$, Wako) and hexamethylenetetramine ($>99.0\%$, Wako) without any chemical additive is prepared at room temperature: $T_{aq} = 25$ °C. The specimen substrate is mounted upside-down in the solution and sealed in the polytetrafluoroethylene (PTFE) container. Then, the container is introduced into the multi-purpose oven set at $T_{set} = 85.0$ °C and heated for 3.5 h. Therein, ZnO deposition starts at $T_{aq} > 60$ °C and ZnO nanorods array grow stationary at $T_{aq} = 79$ °C (Supplementary Note 1). After the heating, the container is cooled naturally down to room temperature and the specimen substrate is removed from the container, rinsed by the deionized water and finally dried by nitrogen gas blow. Contrary to our previous publications[30,53,54], PMMA film is not removed from ZnO substrate after the growth.

SEM-CL observation of a ZnO nanorod. SEM-CL observation is conducted in Nanoprobe-CL system[30], which is based on a Schottky SEM (Hitachi High-Technologies SU6600). This system equips a SEM specimen cooling stage (Thermal Block Company), which is modified to mount a triaxial piezoelectric nanomanipulator (Kleindiek Nanotechnik MM3A-EM) and triaxial coaxial cables for nano-manipulation and electrical nano-probing. Low-temperature measurements down to 10 K are available by flowing He or N_2 cooling gas beneath the specimen. SEM-CL nanospectroscopy is performed using e-beam of 3.0 keV and 2.35 nA or that of 4.5 keV and 2.75 nA. E-beam excites CL from ZnO surface to the depth of Kanaya–Okayama range, 97 nm at 3.0 keV and 192 nm at 4.5 keV, respectively[59]. CL is collected by ellipsoidal mirror, dispersed in the spectrometer (Horiba iHR320) and detected by multi-channel charge-coupled device detector (Andor Tech. DU420A-BU2) for CL spectroscopy or by photomultiplier tube (Hamamatsu R943-02) for monochromatic CL imaging. Panchromatic CL analysis is also available using the optical path bypassing the spectrometer. The 300-nm-thick PMMA film on the ZnO substrate is transparent for ZnO CL emission range but plays as an e-beam stopping layer to suppress strong CL from high-quality ZnO substrate. The 3.0-keV CL imaging were performed as short as possible, as e-beam with energy higher than 3.0 keV penetrate PMMA film to excite NBE emission of ZnO substrate and the e-beam dose decompose the PMMA film gradually. CL spectroscopy is performed either in Spot-CL mode at moderate energy resolution (32 meV) and in Area-CL mode at high energy resolution (4.3 meV), where focused e-beam is either spotted or scanned on a nanorod of interest.

Nanorod micro-sampling and cross-section fabrication by FIB. ZnO nanorods are micro-sampled and their cross-sections are made at desired crystallographic orientations by two-step FIB technique. First, specimens were installed in Nanoprobe-CL system[30]. Single ZnO free-standing nanorod is picked up by electrochemically etched W-probe attached to the nanomanipulator. The probe is then transferred to the carbon coater where carbon film is deposited from various directions to embed the nanorod in a thick amorphous carbon layer. Here, the carbon medium is expected to play several roles: (i) surface protective layer and heat sink from Ga ion beam bombardment; (ii) non-luminescent embedding medium for CL imaging; (iii) embedding medium with appropriate hardness for smooth cross-sections. Each nanorod cross-section is fabricated by two-step FIB milling, coarse milling at 30 keV followed by fine milling at 10 keV, to minimize thickness of damaged layer of the cross-section surface and to improve the S/N ratio of CL imaging.

References

1. Peng, X.-G., Wickham, J. & Alivisatos, A. P. Kinetics of II-VI and III-V colloidal semiconductor nanocrystal growth: 'Focusing' of size distributions. *J. Am. Chem. Soc.* **120**, 5343–5344 (1998).
2. Peng, X.-G. *et al.* Shape control of CdSe nanocrystals. *Nature* **404**, 59–61 (2000).
3. Nakamura, Y., Watanabe, K., Fukuzawa, Y. & Ichikawa, M. Observation of the quantum-confinement effect in individual Ge nanocrystals on oxidized Si substrates using scanning tunneling spectroscopy. *Appl. Phys. Lett.* **87**, 133119 (2005).
4. Nakamura, Y., Amari, S., Naruse, N., Mera, Y. & Ichikawa, M. Self-assembled epitaxial growth of high density beta-FeSi2 nanodots on Si (001) and their spatially resolved optical absorption properties. *Cryst. Growth Des.* **8**, 3019–3023 (2008).
5. Li, Y. *et al.* Dopant-free GaN/AlN/AlGaN radial Nanowire. *Nano Lett.* **6**, 1468–1473 (2006).
6. Wang, Z.-L. Nanopiezotronics. *Adv. Mater.* **19**, 889–892 (2007).
7. Tian, B. *et al.* Coaxial silicon nanowires as solar cells and nanoelectronic power sources. *Nature* **449**, 885–890 (2007).
8. Cui, Y., Wei, Q.-Q., Park, H.-K. & Lieber, C. M. Nanowire nanosensors for highly sensitive and selective detection of biological and chemical species. *Science* **293**, 1289–1292 (2001).
9. Huang, M. H. *et al.* Room-temperature ultraviolet Nanowire nanolasers. *Science* **292**, 1897–1899 (2001).
10. Duan, X., Huang, Y., Agarwal, R. & Lieber, C. M. Single-nanowire electrically driven lasers. *Nature* **421**, 241–245 (2003).
11. Chu, S. *et al.* Electrically pumped waveguide lasing from ZnO nanowires. *Nat. Nanotechnol* **6**, 506–510 (2011).
12. Mariano, A. N. & Hanneman, R. E. Crystallographic polarity of ZnO crystals. *J. Appl. Phys.* **34**, 384–388 (1963).
13. Heiland, G. & Kunstmann, P. Polar surfaces of zinc oxide crystals. *Surf. Sci* **13**, 72–84 (1969).
14. Sekiguchi, T., Miyashita, S., Obara, K., Shishido, T. & Sakagami, N. Hydrothermal growth of ZnO single crystals and their optical characterization. *J. Cryst. Growth* **214/215**, 72–76 (2000).
15. He, Y. *et al.* Crystal-plane dependence of critical concentration for nucleation on hydrothermal ZnO nanowires. *J. Phys. Chem. C* **117**, 1197–1203 (2013).
16. Sakagami, N. *et al.* Variation of electrical properties on growth sectors of ZnO single crystals. *J. Cryst. Growth* **229**, 98–103 (2001).
17. Ballingall, J. M. & Wood, C. E. C. Crystal orientation dependence of silicon auto compensation in molecular beam epitaxial gallium arsenide. *Appl. Phys. Lett.* **41**, 947–949 (1982).
18. Wang, W. I., Mendez, E. E., Kuan, T. S. & Esaki, L. Crystal orientation dependence of silicon doping in molecular beam epitaxial AlGaAs/GaAs heterostructures. *Appl. Phys. Lett.* **47**, 826–828 (1985).
19. Wulff, G. Zur frage der geschwindigkeit des wachstums und der auflosung der krystallflachen. *Z. Kristallogr.* **34**, 449–530 (1901).
20. Chernov, A. A. The kinetics of the growth forms of crystals. *Sov. Phys. Crystallogr* **7**, 728–730 (1963).
21. Borgström, L. H. Die geometrische bedingung fur die entstehung von kombinationen. *Z. Kristallogr.* **62**, 1 (1925).
22. Alexandru, H. V. A macroscopic model for the habit of crystals grown from solutions. *J. Cryst. Growth* **5**, 115–124 (1969).
23. Singh, M., Verma, P., Tung, H.-H., Bordawekar, S. & Ramkrishna, D. Screening crystal morphologies from crystal structure. *Cryst. Growth Des.* **13**, 1390–1396 (2013).
24. Dubrovskii, V. G., Timofeeva, M. A., Tchernycheva, M. & Bolshakov, A. D. Lateral growth and shape of semiconductor nanowires. *Semiconductors* **47**, 50–57 (2013).
25. Storm, K. *et al.* Spatially resolved Hall effect measurement in a single semiconductor nanowire. *Nat. Nanotechnol* **7**, 718–722 (2012).
26. Chen, W. *et al.* Boron distribution in the core of Si nanowire grown by chemical vapor deposition. *J. Appl. Phys.* **111**, 094909 (2012).
27. Chen, W. *et al.* Incorporation and redistribution of impurities into silicon nanowires during metal-particle-assisted growth. *Nat. Commun* **5**, 4134 (2014).
28. van Benthem, K. *et al.* Three-dimensional imaging of individual hafnium atoms inside a semiconductor device. *Appl. Phys. Lett.* **87**, 034104 (2005).
29. Çelebi, C. C. *et al.* Surface induced asymmetry of acceptor wave functions. *Phys. Rev. Lett.* **104**, 086404 (2010).
30. Watanabe, K. *et al.* Band-gap deformation potential and elasticity limit of semiconductor free-standing nanorods characterized *in situ* by scanning electron microscope-cathodoluminescence nanospectroscopy. *ACS Nano* **9**, 2989–3001 (2015).
31. Watanabe, K., Nakamura, Y. & Ichikawa, M. Conductive optical-fibre STM probe for local excitation and collection of cathodoluminescence at semiconductor surfaces. *Opt. Exp* **21**, 19261–19268 (2013).
32. Watanabe, K. *et al.* Scanning tunneling microscope-cathodoluminescence measurement of the GaAs/AlGaAs heterostructure. *J. Vac. Sci. Technol. B* **27**, 1874–1880 (2009).
33. Watanabe, K., Nakamura, Y. & Ichikawa, M. Spatial resolution of imaging contaminations on the GaAs surface by scanning tunneling microscope-cathodoluminescence spectroscopy. *Appl. Surf. Sci.* **254**, 7737–7741 (2008).
34. Watanabe, K., Nakamura, Y. & Ichikawa, M. Measurements of local optical properties of Si-doped GaAs (110) surfaces using modulation scanning tunneling microscope cathodoluminescence spectroscopy. *J. Vac. Sci. Technol. B* **26**, 195–200 (2008).
35. Mass, J. *et al.* Cathodoluminescence characterization of hydrothermal ZnO crystals. *Superlattices Microstruct.* **38**, 223–230 (2005).
36. Lee, W. *et al.* Cathodoluminescence study of nonuniformity in hydride vapor phase epitaxy-grown thick GaN films. *J. Electron Microscopy* **61**, 25–30 (2012).
37. Lee, W. *et al.* Cross sectional CL study of the growth and annihilation of pit type defects in HVPE grown (0001) thick GaN. *J. Cryst. Growth* **351**, 83–87 (2012).
38. Lee, W.-W., Kim, S.-B., Yi, J., Nichols, W. T. & Park, W.-I. Surface polarity-dependent cathodoluminescence in hydrothermally grown ZnO hexagonal rods. *J. Phys. Chem. C* **116**, 456–460 (2012).
39. *Numerical Data and Fundamental Relationships in Science and Technology* Vol. 17 of Landolt-Bornstein new series (Springer, 1982).
40. Christman, J. A., Woolcott, Jr. R. R., Kingon, A. I. & Nemanich, R. J. Piezoelectric measurements with atomic force microscopy. *Appl. Phys. Lett.* **73**, 3851–3853 (1998).
41. Law, M., Greene, L. E., Johnson, J. C., Saykally, R. & Yang, P. Nanowire dye-sensitized solar cells. *Nat. Mater.* **4**, 455–459 (2005).
42. Zhou, Y., Wu, W., Hu, G., Wu, H. & Cui, S. Hydrothermal synthesis of ZnO nanorods arrays with the addition of polyethyleneimine. *Mater. Res. Bull.* **43**, 2113–2118 (2008).
43. Vayssieres, L., Keis, K., Lindquist, S.-E. & Hagfeldt, A. Purpose-built anisotropic metal oxide materials: 3D high oriented microrod array of ZnO. *J. Phys. Chem B* **105**, 3350–3352 (2001).
44. Vayssieres, L. Growth of arrayed nanorods and nanowires of ZnO from aqueous solutions. *Adv. Mater.* **15**, 464–466 (2001).
45. Greece, L. E. *et al.* Low-temperature wafer-scale production of ZnO Nanowire arrays. *Angew. Chem. Int. Ed.* **42**, 3031–3034 (2003).
46. Tian, Z. R., Voigt, J. A., Liu, J., Mckenzie, B. & Mcdermott, M. J. Biomimetic arrays of oriented helical ZnO nanorods and columns. *J. Am. Chem. Soc.* **124**, 12954–12955 (2002).

47. Govender, K., Boyle, D. S., Kenway, P. B. & O'Brien, P. Understanding the factors that govern the deposition and morphology of thin films of ZnO from aqueous solution. *J. Mater. Chem.* **14,** 2575–2591 (2004).

48. Sun, Y., Riley, D. J. & Ashfold, M. N. R. Mechanism of ZnO nanotube growth by hydrothermal methods on ZnO film-coated Si substrates. *J. Phys. Chem. B* **110,** 15186–15192 (2006).

49. Ashfold, M. N. R., Doherty, R. P., Ndifor-Angwafor, N. G., Riley, D. J. & Sun, Y. The kinetics of the hydrothermal growth of ZnO nanostructures. *Thin Solid Films* **515,** 8679–8683 (2007).

50. Xu, S., Lao, C., Weintraub, B. & Wang, Z.-L. Density-controlled growth of aligned ZnO nanowire arrays by seedless chemical approach on smooth surfaces. *J. Mater. Res.* **23,** 2072–2077 (2008).

51. Xu, S. *et al.* Optimizing and improving the growth quality of ZnO nanowire arrays guided by statistical design of experiments. *ACS Nano* **3,** 1803–1812 (2009).

52. Xu, S. & Wang, Z.-L. One-dimensional ZnO nanostructures: solution growth and functional properties. *Nano Res* **4,** 1013–1098 (2011).

53. Volk, J. *et al.* Highly uniform epitaxial ZnO nanorods arrays for nanopiezotronics. *Nanoscale Res. Lett.* **4,** 699–704 (2009).

54. Erdérlyi, R. *et al.* Investigations into the impact of the template layer on ZnO Nanowire arrays made using low temperature wet chemical growth. *Cryst. Growth Des.* **11,** 2515–2519 (2011).

55. Pauporté, T. h., Lincot, D., Viana, B. & Pellé, F. Toward laser emission of epitaxial nanorods arrays of ZnO grown by electrodeposition. *Appl. Phys. Lett.* **89,** 233112 (2006).

56. Belghiti, H. E., Pauporté, T. & Lincot, D. Mechanistic study of ZnO nanorod array electrodeposition. *Phys. Stat. Solid.* **205,** 2360–2364 (2008).

57. Könenkamp, R. *et al.* Thin film semiconductor deposition on free-standing ZnO columns. *Appl.Phys. Lett.* **77,** 2575–2577 (2000).

58. Weintraub, B., Deng, Y. & Wang, Z.-L. Position-controlled Seedless growth of ZnO nanorods arrays on a polymer substrate via wet chemical synthesis. *J. Phys. Chem. C* **111,** 10162–10165 (2007).

59. Kanaya, K. & Okayama, S. Penetration and energy-loss theory of electrons in solid targets. *J. Phys. D* **5,** 43–58 (1972).

60. Meyer, B. K. *et al.* Bound exciton and donor-acceptor pair recombinations in ZnO. *Phys. Stat. Sol. (b)* **241,** 231–260 (2004).

61. Giles, N. C. *et al.* Effects of phonon coupling and free carriers on band-edge emission at room temperature in n-type ZnO crystals. *Appl. Phys. Lett.* **89,** 251906 (2006).

62. Zhang, Y. B., Goh, G. K., Ooi, K. F. & Tripathy, S. Hydrogen-related n-type conductivity in hydrothermally grown epitaxial ZnO films. *J. Appl. Phys.* **108,** 083716 (2010).

63. Reynolds, J. G. *et al.* Shallow acceptor complex in p-type ZnO. *Appl. Phys. Lett.* **102,** 152114 (2013).

64. Lavrov, E. V., Weber, J., Börrnert, F., Van der Walle, C. G. & Helbig, R. Hydrogen-related defects in ZnO studied by infrared absorption spectroscopy. *Phys. Rev. B* **66,** 165205 (2002).

65. Lavrov, E. V., Börrnert, F. & Weber, J. Dominant hydrogen-oxygen complex in hydrothermally grown ZnO. *Phys. Rev. B* **71,** 035205 (2005).

66. Lavrov, E. V., Herklotz, F. & Weber, J. Identification of two hydrogen donors in ZnO. *Phys. Rev. B* **79,** 165210 (2009).

67. Koch, S. G., Lavrov, E. V. & Weber, J. Interplay between interstitial and substitutional hydrogen donors in ZnO. *Phys. Rev. B* **89,** 235203 (2014).

68. Janotti, A. & Van der Walle, C. G. Fundamentals of zinc oxide as a semiconductor. *Rep. Prog. Phys* **72,** 126501 (2009).

69. Cuscó, R. *et al.* Temperature dependence of Raman scattering in ZnO. *Phys. Rev. B* **75,** 165202 (2007).

70. Consonni, V. *et al.* Selective area growth of well-ordered ZnO nanowire arrays with controllable polarity. *ACS Nano* **8,** 4761–4770 (2014).

Acknowledgements

This work was supported by JSPS KAKENHI (grant numbers 23760022 and 26790046) and JSPS-HAS Bilateral Joint Research Projects.

Author contributions

K.W. conceived and designed this work. K.W. fabricated the specimen with the assistance of J.V., S.O., T.N. and Y.W. K.W. developed Nanoprobe-CL system and performed micro-sampling, FIB milling, cross-sectional CL nanospectroscopy, and 'differential' *I–V* measurements of a single nanorod. K.W., T.N., and Y.N. performed XPS, XRD, and Raman spectroscopy of referential ZnO thin-films, respectively. T.S. joined some discussions. K.W. wrote the manuscript with the assistance of J.V., T.N. and Y.N. All authors have given approval to the manuscript.

Additional information

Realization of mid-infrared graphene hyperbolic metamaterials

You-Chia Chang[1,2], Che-Hung Liu[1,3], Chang-Hua Liu[3], Siyuan Zhang[4], Seth R. Marder[4], Evgenii E. Narimanov[5], Zhaohui Zhong[1,3] & Theodore B. Norris[1,3]

While metal is the most common conducting constituent element in the fabrication of metamaterials, graphene provides another useful building block, that is, a truly two-dimensional conducting sheet whose conductivity can be controlled by doping. Here we report the experimental realization of a multilayer structure of alternating graphene and Al_2O_3 layers, a structure similar to the metal-dielectric multilayers commonly used in creating visible wavelength hyperbolic metamaterials. Chemical vapour deposited graphene rather than exfoliated or epitaxial graphene is used, because layer transfer methods are easily applied in fabrication. We employ a method of doping to increase the layer conductivity, and our analysis shows that the doped chemical vapour deposited graphene has good optical properties in the mid-infrared range. We therefore design the metamaterial for mid-infrared operation; our characterization with an infrared ellipsometer demonstrates that the metamaterial experiences an optical topological transition from elliptic to hyperbolic dispersion at a wavelength of 4.5 μm.

[1] Center for Photonics and Multiscale Nanomaterials, University of Michigan, 2200 Bonisteel Blvd., Ann Arbor, Michigan 48109, USA. [2] Department of Physics, University of Michigan, 450 Church St, Ann Arbor, Michigan 48109, USA. [3] Department of Electrical Engineering and Computer Science, University of Michigan, 1301 Beal Avenue, Ann Arbor, Michigan 48109, USA. [4] School of Chemistry and Biochemistry, Georgia Institute of Technology, 901 Atlantic Drive, Atlanta, Geogia 30332, USA. [5] School of Electrical and Computer Engineering and Birck Nanotechnology Center, Purdue University, 1205 West State Street, West Lafayette, Indiana 47907, USA. Correspondence and requests for materials should be addressed to T.N. (email: tnorris@umich.edu).

Hyperbolic metamaterials (HMMs) are artificially structured materials designed to attain an extremely anisotropic optical response, in which the permittivities associated with different polarization directions exhibit opposite signs[1-3]. Such anisotropic behaviour results in an isofrequency surface in the shape of a hyperboloid, which supports propagating high k-modes and exhibits an enhanced photonic density of states. Many interesting applications have been enabled by HMMs. For example, the spontaneous emission rate of quantum emitters can be modified if they are brought close to a HMM[4], and similarly, the scattering cross-section of small scatterers near a HMM is enhanced[5]. The near-field radiative heat transfer associated with HMMs becomes super-Planckian[6]. Also, the propagating high k-modes supported by HMM are exploited to achieve sub-diffraction-limited images using a hyperlens[7]. Some natural materials such as bismuth, graphite and hexagonal boron nitride exhibit hyperbolic dispersion in specific spectral ranges[8-10], while artificial HMMs are most commonly realized with two categories of structures such as metal-dielectric multilayers[4,7] and metallic nanorod arrays[11]. The former structure can be fabricated layer by layer using vapour deposition, and the latter is often obtained by electrochemical deposition of a metal on porous anodic aluminium oxide. In both cases, metal is the essential element to provide the conducting electrons that make the extreme anisotropicity possible. Metals can also be replaced by doped semiconductors for realizing HMMs in the infrared range[12].

In this paper, we explore the realization of a particular HMM, in which the role of the metal in providing a conducting layer is taken over by graphene[13-21]. Graphene is a two-dimensional (2D) semi-metal with a thickness of only one atom[22,23]. It has been shown that doped graphene is a good infrared plasmonic material in terms of material loss[24]. As a truly 2D material that only conducts in the plane, graphene by nature has the anisotropicity required for HMMs. As the thinnest material imaginable, graphene also makes an ideal building block for multilayer structures, as it enables the minimum possible period and therefore the highest possible cutoff for the high k-modes[14,25], which has been limited in metal and semiconductor-based HMMs by the non-negligible thickness of those materials. The conductivity of graphene, unlike that of metals, can be effectively modulated by electrical gating (see Supplementary Fig. 1 and Supplementary Note 1) or optical pumping[26,27]. This unique advantage has been demonstrated in other graphene-based metamaterials[28], and can potentially be exploited to realize a tunable HMM, in which the photonic density of states can be controlled electronically on demand. In addition, graphene shows much richer optoelectronic behaviour than metals, and the massless Dirac quasi-particles in graphene also give rise to very different carrier dynamics compared with other semiconductors. Various photodetection mechanisms, such as thermoelectric, bolometric, photovoltaic, photo-gating and photo-Dember effects, have been demonstrated with graphene[29-32]. Graphene multilayer structures can therefore serve as a unique platform in optoelectronics, incorporating the unusual photonic behaviour of HMMs into graphene detectors or other optoelectronic devices. For example, an ultrathin super-absorber enabled by HMM could be incorporated into graphene detectors to enhance the light absorption[18]. A brief summary of this report is as follows. The design criterions and material choices for realizing the graphene HMM are discussed. Chemical vapour deposited (CVD) graphene is identified as a good practical choice in the mid-infrared range when it is heavily doped. A chemical doping method is developed to obtain the desired high carrier density and ellipsometry is used to characterize the optical conductivity of monolayer graphene. The metamaterial with

multilayer structure is fabricated by repetitive graphene transfer and dielectric deposition. We characterize the effective permittivities of the fabricated metamaterial with ellipsometry to demonstrate the hyperbolic dispersion in the mid-infrared range.

Results

Design of graphene HMM. Figure 1 shows the structure of the graphene-based HMM, which consists of alternating dielectric and graphene layers. Similar graphene-dielectric multilayer structures have been proposed and analysed theoretically by different groups and shown to function as a HMM operating at terahertz (THz) and mid-infrared frequencies[13-21]. Various applications have also been discussed. For example, in our previous work we have calculated theoretically the Purcell factor of a graphene-based HMM with a finite number of layers[17], and we have simulated numerically the light coupling from free space into a graphene-based HMM slab with a metallic grating[18]. In spite of the large body of theoretical work on graphene-based HMM, no experimental demonstrations have yet been reported, the primary reason being the challenge in obtaining a sufficiently high level of doping in the graphene layers in the required multilayer structure.

The graphene-dielectric multilayer structure can be homogenized and viewed as a metamaterial using the effective medium approximation (EMA). The effective out-of-plane and in-plane permittivities of this metamaterial can be derived by taking the long-wavelength limit of the Bloch theory[13-16]:

$$\varepsilon_{eff,\perp} = \varepsilon_d,$$
$$\varepsilon_{eff,\parallel} = \varepsilon_d + i\frac{\sigma Z_0}{2\pi}\left(\frac{\lambda}{d}\right). \tag{1}$$

Here ε_d is the permittivity of the dielectric layer, d is the dielectric thickness and σ is the optical conductivity of graphene. Z_0 is the vacuum impedance. Here graphene, as a 2D material, is treated as an infinitely thin layer described by its in-plane sheet conductivity. As indicated by equation 1, the graphene-dielectric multilayer system forms a uniaxial anisotropic metamaterial. $\varepsilon_{eff,\perp}$ is the same as the constituent dielectric and is always positive. On the other hand, the real part of $\varepsilon_{eff,\parallel}$ becomes negative if

$$\text{Im } \sigma > 2\pi(d/\lambda)(\varepsilon_d/Z_0). \tag{2}$$

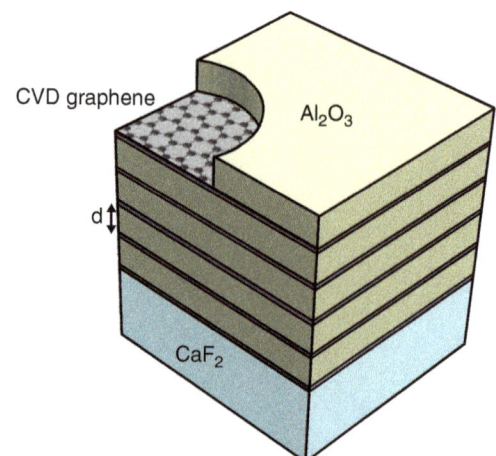

Figure 1 | The schematic representation of the graphene-dielectric multilayer structure that turns into a HMM at mid-infrared frequencies. It consists of five periods of alternating CVD graphene sheets and Al_2O_3 layers on a CaF_2 substrate. The thickness d of the Al_2O_3 layer is ~10 nm.

When this criterion is satisfied, the isofrequency surface becomes a hyperboloid and we obtain HMM. Such an isofrequency surface allows the existence of propagating high k-modes, which can be traced back to the coupled plasmon modes in the graphene-dielectric multilayer structure[17]. The criterion described by equation 2 determines the wavelength at which the optical topological transition between elliptical and hyperbolic dispersions occurs[4].

While most previous theoretical work has concentrated on using high-mobility graphene that may be obtained from mechanically exfoliated or epitaxially grown samples, we use CVD graphene because it is the most realistic choice for practical fabrication of a multilayer structure[33]. Growth of large-area CVD graphene is well established, and it can be transferred onto arbitrary surfaces using poly(methyl methacrylate) (PMMA) as the carrier material. In spite of its advantage in fabrication, CVD graphene often has a higher degree of disorder, which is typically manifested by a reduced mobility (usually on the order of thousands $cm^2 V^{-1} s^{-1}$). As a result of the lower crystal quality, the stronger carrier scattering in typical polycrystalline CVD graphene enhances the free-carrier absorption at THz frequencies, which can be understood from the theoretical optical conductivity of graphene[34–36]

$$\sigma(\omega) = \frac{\sigma_0}{2}\left(\tanh\frac{\hbar\omega + 2E_F}{4k_BT} + \tanh\frac{\hbar\omega - 2E_F}{4k_BT}\right)$$
$$- i\frac{\sigma_0}{2\pi}\log\left[\frac{(\hbar\omega + 2E_F)^2}{(\hbar\omega - 2E_F)^2 + (2k_BT)^2}\right] + i\frac{4\sigma_0}{\pi}\frac{E_F}{\hbar\omega + i\hbar\gamma},$$

(3)

where σ_0 equals to $e^2/(4\hbar)$, E_F is the Fermi energy relative to the Dirac point and γ is the intraband scattering rate. In this expression, the first two terms correspond to interband transitions, while the third term is the Drude-like intraband conductivity. Figure 2 shows a plot of the theoretical optical conductivity given by equation 3 with parameters typical for doped polycrystalline CVD graphene. To realize a good HMM, we need graphene with a large positive imaginary conductivity to

interact with light, but with a small real conductivity to minimize the material loss. As indicated by Fig. 2, graphene is lossy at high frequencies when $\hbar\omega > 2E_F$ because of interband transitions. On the other hand, at low frequencies when $\hbar\omega \lesssim \hbar\gamma$, graphene also exhibits a large loss because of the intraband free-carrier absorption enabled by scattering. Because CVD graphene typically has a $\hbar\gamma$ of tens of meV, it is a lossy material at THz frequencies[37]. As shown by Fig. 2, however, there is a spectral range between the two lossy regions, such that the imaginary part of the conductivity exceeds the real part. As this spectral range lies in the mid-infrared part of the spectrum, CVD graphene-based HMM operates better in the mid-infrared than the THz region. Also, Fig. 2 indicates that doping can improve the properties of graphene for realizing a HMM. A large E_F can turn off the interband absorption by the Pauli blocking and increase the Imσ required for achieving negative $\varepsilon_{eff,\parallel}$. Furthermore, doping can also suppress the intraband scattering by screening charged impurities[37,38].

Characterization of the optical conductivity of grapheme. Because graphene is the key building block of the metamaterial, it is important to have an accurate measurement on the optical conductivity of the actual CVD graphene layers used to fabricate the sample. Although the theoretical optical conductivity given by equation 3 provides a good guideline for designing the graphene HMM, real CVD graphene layers can have imperfections or extrinsic properties that are not taken into account by equation 3. We therefore need to characterize actual graphene samples and examine the scope of validity of equation 3.

In our previous work, we have developed a technique based on ellipsometry to measure the optical conductivity of truly 2D materials[39]. In this technique, the analysis used in conventional ellipsometry is modified to handle the infinitely thin 2D material whose properties are fully described by the 2D optical conductivity. To characterize actual CVD graphene samples with this technique, we have prepared two kinds of samples, unintentionally doped and the chemically doped CVD graphene, on CaF₂ substrates by the standard PMMA transfer method. Even without chemical treatment, unintentionally doped CVD graphene is p-type because of adsorbed gas molecules and residual ammonium persulfate from the transfer process[40,41]. The chemically doped CVD graphene is prepared by a solution process that leaves a sub-monolayer of Tris (4-bromophenyl)ammoniumyl hexachloroantimonate (also known as 'magic blue'), a somewhat air-stable p-type dopant, on the surface (see Methods section, Supplementary Fig. 2 and Supplementary Note 2)[42,43]. Figure 3a shows the optical conductivities of both samples measured with ellipsometry. The optical conductivities shown here are mathematically described by cubic splines without assuming an *a priori* theoretical expression like equation 3. Consistent with Fig. 2, in the mid-infrared range the chemically doped graphene has a larger imaginary conductivity, which is necessary for creating the extreme anisotropicity in the metamaterial.

Although the spline-fitted conductivity of actual CVD graphene sample shown in Fig. 3a is useful in many applications, a conductivity model based on a theoretical expression such as equation 3 provides more physical insight and requires fewer unknown parameters to perform the fit. The latter is important when we want to parameterize the homogenized metamaterial, which will be discussed in next section. In Fig. 3b, we examine how well equation 3 works for our chemically doped CVD graphene samples. In fitting the ellipsometer data, we express the optical conductivity $\sigma(\omega)$ by the model given by equation 3 with E_F and γ being the only two unknown fitting parameters. We also

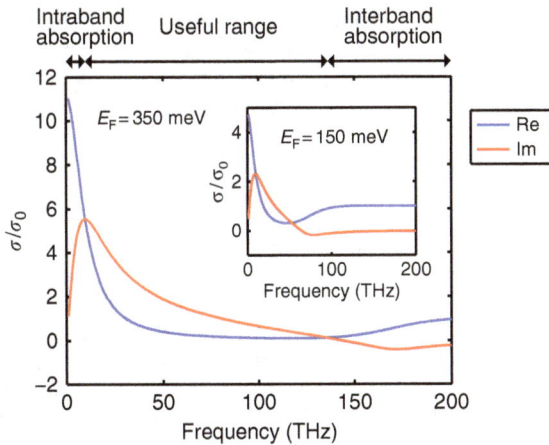

Figure 2 | The theoretical optical conductivity of graphene. It is plotted with $E_F = 350$ meV and $\hbar\gamma = 40$ meV. These numbers correspond to heavily doped CVD graphene. At the high-frequency end of the spectrum, graphene is lossy because of the interband absorption. At the low-frequency end, graphene is again lossy because of the intraband free-carrier absorption. There is a useful spectral range in between, where the imaginary part of the optical conductivity exceeds the real part. In this particular example, the useful wavelengths range from 2 to 30 µm in the mid-infrared range. The inset shows another example of lightly doped CVD graphene with $E_F = 150$ meV and $\hbar\gamma = 40$ meV. The useful wavelength range is smaller when the doping is lower.

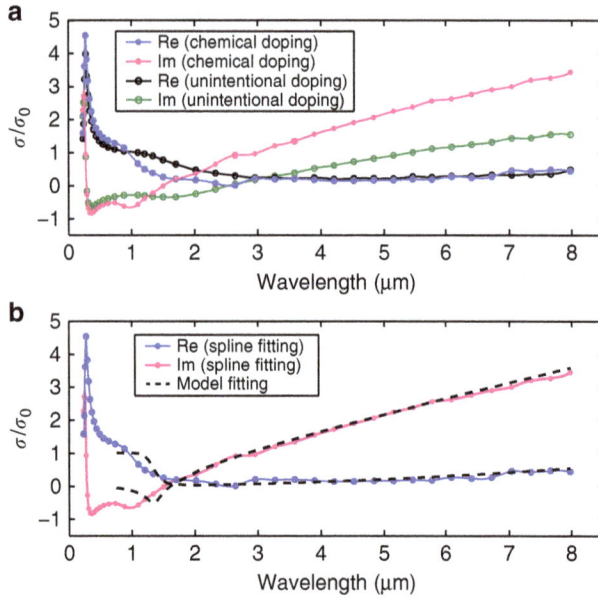

Figure 3 | The optical conductivity of CVD graphene measured by ellipsometry. (**a**) The real and imaginary part of the optical conductivity of the chemically doped CVD graphene (blue and magenta curves) and the unintentionally doped CVD graphene (black and green curves). These curves are mathematically expressed by cubic splines, and the markers denote the control points of the splines. The chemically doped CVD graphene has a larger imaginary conductivity in the mid-infrared range. (**b**) The real and imaginary part of the optical conductivity of the chemically doped CVD graphene. The blue and magenta curves are obtained by fitting with cubic splines, and the black dash lines are obtained by using the model given by equation 3. The model fitting is consistent with the spline fitting in the mid-infrared range. The extracted E_F and $\hbar\gamma$ from the model fitting are 460 and 23 meV, respectively, which corresponds to a mobility of $\sim 2{,}000\,cm^2\,V^{-1}\,s^{-1}$.

show in the same figure the spline-fitted conductivity obtained from the same set of data. It is apparent that the resulting conductivity based on equation 3 overlaps very well with the spline-fitted conductivity throughout the mid-infrared range, assuring the validity of using equation 3 for the mid-infrared metamaterial. We extract from the fit that $E_F = 460\,meV$ and $\hbar\gamma = 23\,meV$. A mobility of $\sim 2{,}000\,cm^2\,V^{-1}\,s^{-1}$ can be calculated from these numbers using the relationship $\mu = e\pi\hbar V_F^2/(\hbar\gamma E_F)$, where μ is the mobility and V_F is the Fermi velocity.

In the mid-infrared range, the optical conductivity is mostly determined by intraband transitions, which are described by the Drude-like term in equation 3. Our result is consistent with ref. 37, which shows that the Drude model can successfully fit the measured absorption spectrum of CVD graphene over a broad range of infrared wavelengths. We do not apply equation 3 in the ultraviolet to visible wavelength range because the many-body correction has been shown to be important[44,45]. There is some discrepancy between the model and spline fits in the near-infrared ($\sim 1.5\,\mu m$, that is, near the wavelength corresponding to interband transitions close to the Fermi level). The origin of this discrepancy is not quantitatively understood, but may be related to spatial inhomogeneity in the Fermi energy or other disorder effects. Since the optical topological transition wavelength of our HMM is very far from this spectral region, and the fit is excellent over the entire mid-infrared range, the failure of the simple model in the near-infrared region does not affect the behaviour of the material in the mid-infrared, which is the region of concern in

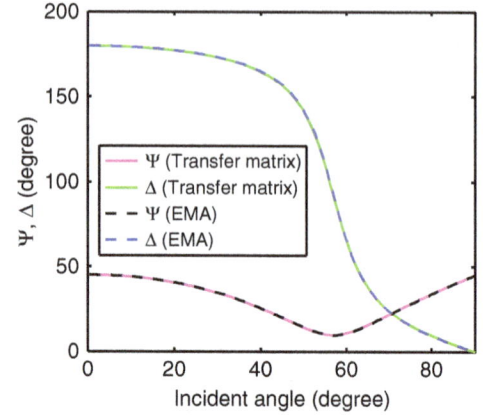

Figure 4 | Calculation of ellipsometric angles with exact transfer-matrix method and EMA. Ellipsometric angles Ψ and Δ are defined by $r_p/r_s = (\tan\Psi)e^{i\Delta}$, where r_p and r_s are the reflection coefficients for p and s light, respectively. They are the quantities an ellipsometer measures. The transfer-matrix method calculates the response of five periods of graphene-dielectric multilayer structure, while the EMA simulates a homogenized anisotropic layer with the permittivities given by equation 1. This calculation shows that the EMA is an accurate approximation for the structure. The wavelength used in this simulation is 6 μm. The material properties are $\varepsilon_d = 2.1$ and $\sigma = (0.43 + 1.98i)\,\sigma_0$. Thickness $d = 10\,nm$. The substrate has a refractive index of 1.39.

this work. Equation 3 thus provides an excellent description for the mid-infrared conductivity. Other imperfections that are typically present in transferred CVD graphene samples, such as the existence of small multilayer graphene patches and holes (see Supplementary Fig. 3 and Supplementary Note 3), can also contribute to the deviations observed in Fig. 3b (ref. 46).

Measurement of the effective permittivity of graphene HMM. We have fabricated the multilayer structure shown in Fig. 1, which consists five periods of alternating CVD graphene and Al_2O_3. The CVD graphene is transferred by the PMMA method and doped with Tris (4-romophenyl) ammoniumyl hexachloroantimonate ('magic blue'). The Al_2O_3 dielectric layer is grown by atomic layer deposition (ALD). We choose Al_2O_3 as the dielectric material, because it has negligible loss at the mid-infrared wavelengths up to 8 μm. The dielectric thickness is chosen to be $\sim 10\,nm$ to create an optical topological transition in the mid-infrared range.

To characterize the metamaterial, we use infrared ellipsometry, which is appropriate to probe the effective permittivity of a metamaterial, since it measures the sample with free-space plane waves and the transverse wave vector ($k_0\sin\theta$) associated with the free-space plane waves is very small ($k_0\sin\theta d \ll 1$, where θ is the angle of incidence). We are therefore probing the low k-modes of the metamaterial, ensuring the validity of the long-wavelength approximation. Although the long-wavelength approximation is evidently satisfied for our metamaterial ($d/\lambda < 1/300$ in our case), we still need to confirm the validity of the EMA with a rigorous transfer-matrix calculation, since the EMA is derived for an infinite periodic system, while our metamaterial has only five periods. In Fig. 4 we show the transfer-matrix calculation of five periods of graphene-dielectric multilayer structure and the EMA calculation with the structure homogenized into an anisotropic layer, with the permittivities of the homogenized anisotropic layer given by equation 1. Here we calculate the ellipsometric angles Ψ and Δ, the quantities an ellipsometer acquires directly, at different incident angles. Ψ and Δ are defined by $r_p/r_s = (\tan\Psi)e^{i\Delta}$, where

r_p and r_s are the reflection coefficients for p and s light, respectively. Numbers used in the simulation are chosen according to measured material properties of the individual layers. As demonstrated by Fig. 4, the two methods give very close results, confirming that the five-period graphene-dielectric structure, in the low k-regime probed by ellipsometry, can be accurately treated as a metamaterial with the effective permittivities given by equation 1. In fact, in the low k-regime, even one period of the graphene-dielectric unit cell can be homogenized by the same EMA formula given by equation 1 and still reproduce the optical properties accurately (see Supplementary Figs 4 and 5 and Supplementary Note 4). However, the high k-regime is where the real interest of HMM lies, and as discussed in Supplementary Note 4, the high k optical properties depend on the number of unit cells in the metamaterial. The five-period structure in our experimental realization of graphene HMM is chosen to create desirable high k optical properties.

The results of infrared ellipsometry, ellipsometric angles Ψ and Δ for our HMM sample, are shown in Fig. 5a,b, from which we extract the effective permittivities by fitting the acquired data. A robust and physical fitting in ellipsometry requires correct prior knowledge about the sample parameters, which allows us to use a minimal number of unknowns. Since our simulation in Fig. 4 demonstrates that the EMA is an accurate description for the multilayer structure, we can apply equation 1 in fitting the data. More precisely, we fit the experimental data to a layer of an anisotropic material on a CaF$_2$ substrate with the permittivities of the anisotropic material given by equation 1. In equation 1, we know everything except the optical conductivity of graphene σ, as we have measured the thickness d independently after depositing each Al$_2$O$_3$ layer, and we have measured the refractive index of the ALD-grown Al$_2$O$_3$ in the relevant spectral range independently on a reference sample (see Methods section). Furthermore, as shown by Fig. 3b, considering the mid-infrared range with only the intraband response, the expression of equation 3 is a good description for the optical conductivity of the actual CVD graphene layers. Therefore, we can apply equation 3 and parameterize the optical conductivity with only E_F and γ. As a result of this independent knowledge of the sample, only two unknowns, E_F and γ, are sufficient to fit the experimental data of the multilayer metamaterial.

The fitted results of the ellipsometric angles Ψ and Δ are plotted as the blue dash lines in Fig. 5a,b. We restrict the wavelengths range of the fitting to 3.5–8 µm, where the lower bound is limited by the requirement of intraband-only response in the application of equation 3, and the upper bound is because of the limited transparent spectral range of Al$_2$O$_3$. As shown by Fig. 5, we are able to reproduce all six Ψ and Δ curves acquired at different incident angles with only two free parameters in the fitting. The extracted E_F is 365 meV, and the extracted $\hbar\gamma$ is 41 meV. The extracted E_F is lower than the value we typically obtain from chemically doped monolayer CVD graphene, because some dopants are lost in the ALD process because of the vacuum environment and the elevated temperature. The obtained

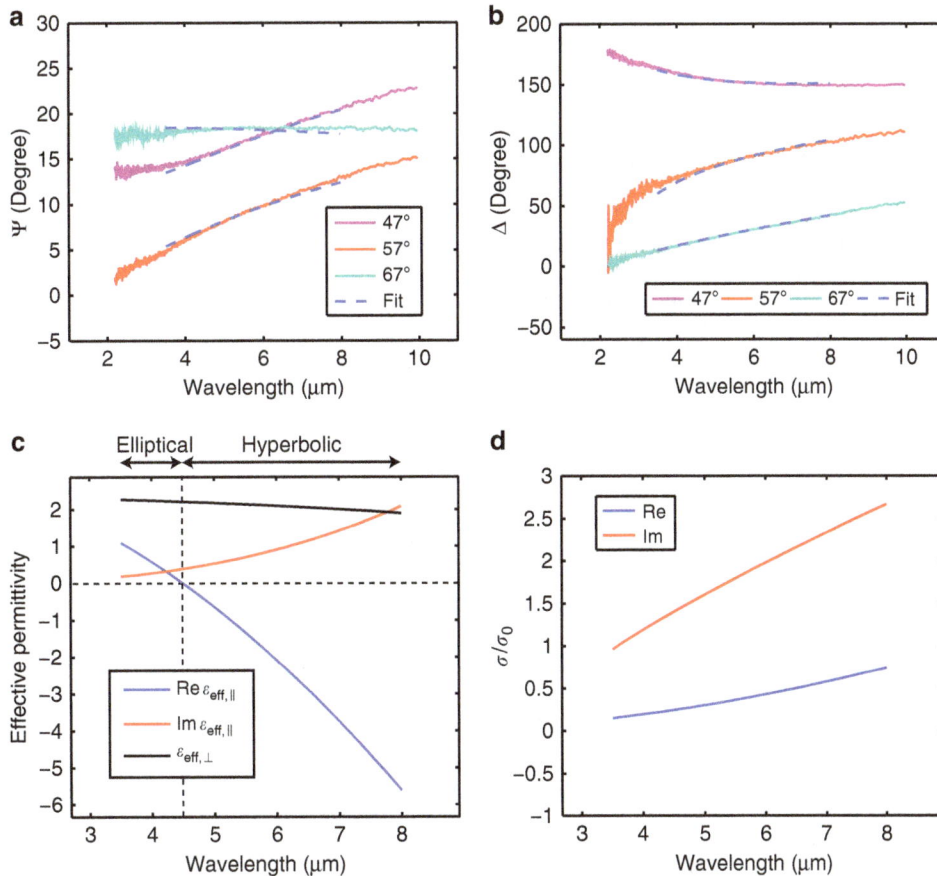

Figure 5 | Extraction of the effective permittivity of the graphene HMM. (a,b) The ellipsometric angles Ψ and Δ acquired from the graphene-dielectric multilayer structure. The measurement is performed at incident angles of 47°, 57° and 67°. The blue dash lines show the fitting by homogenizing the multilayer structure into a metamaterial with the effective permittivities given by equation 1. We extract from the fitting that $E_F = 365$ meV and $\hbar\gamma = 41$ meV. **(c)** The extracted effective permittivity of the metamaterial, which exhibits an optical topological transition from elliptical to hyperbolic dispersion at 4.5 µm. When the wavelength is at 6 µm, $\varepsilon_{\text{eff},\parallel}$ equals $2.1 + 0.9i$ and $\varepsilon_{\text{eff},\perp}$ equals 2.1. **(d)** The extracted optical conductivity of the constituent CVD graphene in the metamaterial.

scattering rate $\hbar\gamma$ is higher than the value of graphene on CaF_2 substrate shown in Fig. 3. This can be explained by the fact that the carrier scattering in graphene depends on the surrounding environment, from which we conclude that sandwiching graphene between Al_2O_3 increases the carrier scattering.

Figure 5c shows the effective permittivity of the graphene metamaterial given by the extracted values of E_F and γ. $\varepsilon_{eff,\perp}$ is always positive because it equals the permittivity of Al_2O_3. On the other hand, the real part of $\varepsilon_{eff,\parallel}$ changes from positive to negative at 4.5 µm indicating an optical topological transition from an elliptical metamaterial to a HMM. This graphene metamaterial is therefore a transverse epsilon-near-zero metamaterial at the wavelength of 4.5 µm (ref. 14). The imaginary part of $\varepsilon_{eff,\parallel}$ is several times smaller than the real part in most of the spectral range with hyperbolic dispersion, indicating that the loss of this HMM is reasonably low. In Fig. 5d, we plot the optical conductivity of the constituent graphene sheet of the metamaterial using the extracted E_F and γ.

Discussion

Our characterization by the infrared ellipsometry demonstrates that the graphene-dielectric multilayer structure indeed experiences an optical topological transition from an elliptical to a hyperbolic dispersion in the mid-infrared range, confirming the theoretical predictions in previous works[13–21]. Our metamaterial sample has an optical topological transition at a wavelength of 4.5 µm, and maintains good hyperbolic properties up to 8 µm. The upper bound of the wavelength range is limited by the absorption in Al_2O_3 and CVD graphene. While the absorption in the dielectric layer can be overcome by replacing Al_2O_3 with other infrared transparent materials such as ZnSe, the absorption in CVD graphene is limited by the quality of graphene. Recently, there have been reports of the growth of large-area CVD graphene with the quality of a single crystal[47], and new transfer process for CVD graphene without degrading the mobility[48]. With higher quality CVD graphene, the intraband absorption resulted from scattering could potentially be suppressed. The transition wavelength, as determined by equation 2, can be shifted by choosing the dielectric thickness or controlling the doping of graphene. The latter is especially useful if it can be done by the electrical gating. Shifting the transition wavelength farther into the infrared can be done by using lightly doped graphene or thicker dielectric. We have also realized a graphene HMM with the same structure except that the CVD graphene layers were not chemically doped (see Supplementary Fig. 6 and Supplementary Note 5), resulting in a transition wavelength red-shifted to 7.2 µm. On the other hand, blue shifting the transition wavelength is limited by the highest doping and the thinnest dielectric layers achievable in practice. While the structure reported in this work has only five periods, the procedure developed here can be repeated to scale up the graphene HMM. Some applications of HMMs do not require a large number of periods; for example, only a few periods is sufficient to produce a Purcell factor close to a semi-infinite structure, according to the theoretical calculations in ref. 17.

Methods

Sample fabrication. The graphene-dielectric multilayer structure with five periods is fabricated on a CaF_2 wedge. The CVD graphene is grown on copper foil (Graphenea Inc) and transferred to the substrate using the standard PMMA transfer technique[33,46]. The copper foil is etched using an ammonium persulfate solution. The size of the CVD graphene we transfer is ~10 mm by 10 mm. After transferring each graphene layer, we dope the graphene by soaking the sample in a 0.25 mM solution of Tris (4-bromophenyl)ammoniumyl hexachloroantimonate 'magic blue' in dichloromethane for 10 min, and then rinse the sample with dichloromethane (see Supplementary Fig. 2 and Supplementary Note 2). The Al_2O_3 dielectric layer is deposited by the ALD at 150 °C using trimethylaluminium as the

Al precursor and H_2O as the oxygen precursor. The number of cycles used in the ALD process is calibrated to grow ~10 nm of Al_2O_3 on graphene, with the thickness characterized by an ellipsometer (Woollam M-2000). The procedure is repeated to fabricate five periods of the graphene-Al_2O_3 unit cell. We have also confirmed that the chemical doping with Tris (4-bromophenyl) ammoniumyl hexachloroantimonate does not affect the Al_2O_3 layer and the substrate. We have found that although nitric acid can also p-dope graphene effectively[37,39], it is not a good dopant for making the multilayer structure because of damage to the thin Al_2O_3 layer. The substrate is wedged to avoid back side reflection in the ellipsometry measurement. We also characterize the graphene-dielectric multilayer structure with the Woollam M-2000 ellipsometer after depositing each Al_2O_3 layer and after transferring each graphene layer. With the acquired ellipsometry data, we extract an average Al_2O_3 thickness of 10.4 nm.

Ellipsometry characterization. The optical conductivity of monolayer graphene is measured by the ellipsometric analysis method described in ref. 39. Two ellipsometers designed for different spectral ranges, Woollam M-2000 and Woollam IR-VASE, are used for the wavelengths from 230 nm to 1.64 µm and the wavelengths above 2 µm, respectively. The data are acquired at three angles of incidence: 47°, 57° and 67°. The spot sizes of M-2000 and IR-VASE are 3 mm by 5.5 mm and 8 mm by 20 mm, respectively when the incident angle is 57°. We mask the samples for the IR-VASE measurement because the spot size is larger than the graphene area. To obtain the refractive index of Al_2O_3, we have prepared a sample with ALD-grown Al_2O_3 film on a CaF_2 wedge. We measure the sample with both ellipsometers, and fit the refractive index of Al_2O_3 with the Sellmeier equation. The measurement of the effective permittivities of the graphene-dielectric multilayer structure is performed by the IR-VASE ellipsometer with the same settings described above.

References

1. Smith, D. R. & Schurig, D. Electromagnetic wave propagation in media with indefinite permittivity and permeability tensors. *Phys. Rev. Lett.* **90,** 077405 (2003).
2. Poddubny, A., Iorsh, I., Belov, P. & Kivshar, Y. Hyperbolic metamaterials. *Nat. Photon.* **7,** 948–957 (2013).
3. Jacob, Z. *et al.* Engineering photonic density of states using metamaterials. *Appl. Phys. B* **100,** 215–218 (2010).
4. Krishnamoorthy, H. N., Jacob, Z., Narimanov, E., Kretzschmar, I. & Menon, V. M. Topological transitions in metamaterials. *Science* **336,** 205–209 (2012).
5. Guclu, C., Campione, S. & Capolino, F. Hyperbolic metamaterial as super absorber for scattered fields generated at its surface. *Phys. Rev. B* **86,** 205130 (2012).
6. Biehs, S. A., Tschikin, M., Messina, R. & Ben-Abdallah, P. Super-Planckian near-field thermal emission with phonon-polaritonic hyperbolic metamaterials. *Appl. Phys. Lett.* **102,** 131106 (2013).
7. Liu, Z., Lee, H., Xiong, Y., Sun, C. & Zhang, X. Far-field optical hyperlens magnifying sub-diffraction-limited objects. *Science* **315,** 1686–1686 (2007).
8. Narimanov, E. E. & Kildishev, A. V. Naturally hyperbolic. *Nat. Photon.* **9,** 214–216 (2015).
9. Dai, S. *et al.* Tunable phonon polaritons in atomically thin van der Waals crystals of boron nitride. *Science* **343,** 1125–1129 (2014).
10. Caldwell, J. D *et al.* Sub-diffractional volume-confined polaritons in the natural hyperbolic material hexagonal boron nitride. *Nat. Commun.* **5,** 5221 (2014).
11. Noginov, M. A. *et al.* Bulk photonic metamaterial with hyperbolic dispersion. *Appl. Phys. Lett.* **94,** 151105 (2009).
12. Hoffman, A. J. *et al.* Negative refraction in semiconductor metamaterials. *Nat. Mater.* **6,** 946–950 (2007).
13. Iorsh, I. V., Mukhin, I. S., Shadrivov, I. V., Belov, P. A. & Kivshar, Y. S. Hyperbolic metamaterials based on multilayer graphene structures. *Phys. Rev. B* **87,** 075416 (2013).
14. Othman, M. A. K., Guclu, C. & Capolino, F. Graphene–dielectric composite metamaterials: evolution from elliptic to hyperbolic wavevector dispersion and the transverse epsilon-near-zero condition. *J. Nanophoton.* **7,** 073089–073089 (2013).
15. Othman, M. A. K., Guclu, C. & Capolino, F. Graphene-based tunable hyperbolic metamaterials and enhanced near-field absorption. *Opt. Express* **21,** 7614–7632 (2013).
16. Wang, B., Zhang, X., García-Vidal, F. J., Yuan, X. & Teng, J. Strong coupling of surface plasmon polaritons in monolayer graphene sheet arrays. *Phys. Rev. Lett.* **109,** 073901 (2012).
17. DaSilva, A. M., Chang, Y. C., Norris, T. B. & MacDonald, A. H. Enhancement of photonic density of states in finite graphene multilayers. *Phys. Rev. B* **88,** 195411 (2013).
18. Chang, Y. C. *et al.* Mid-infrared hyperbolic metamaterial based on graphene-dielectric multilayers. *Proc. SPIE* **9544,** 954417 (2015).
19. Nefedov, I. S., Valagiannopoulos, C. A. & Melnikov, L. A. Perfect absorption in graphene multilayers. *J. Opt.* **15,** 114003 (2013).

20. Sreekanth, K. V., De Luca, A. & Strangi, G. Negative refraction in graphene-based hyperbolic metamaterials. *Appl. Phys. Lett.* **103**, 023107 (2013).

21. Andryieuski, A., Lavrinenko, A. V. & Chigrin, D. N. Graphene hyperlens for terahertz radiation. *Phys. Rev. B* **86**, 121108 (2012).

22. Novoselov, K. S. *et al.* Electric field effect in atomically thin carbon films. *Science* **306**, 666–669 (2004).

23. Novoselov, K. S. A. *et al.* Two-dimensional gas of massless Dirac fermions in graphene. *Nature* **438**, 197–200 (2005).

24. Jablan, M., Buljan, H. & Soljacic, M. Plasmonics in graphene at infrared frequencies. *Phys. Rev. B* **80**, 245435 (2009).

25. Kidwai, O., Zhukovsky, S. V. & Sipe, J. E. Effective-medium approach to planar multilayer hyperbolic metamaterials: strengths and limitations. *Phys. Rev. A* **85**, 053842 (2012).

26. Liu, M. *et al.* A graphene-based broadband optical modulator. *Nature* **474**, 64–67 (2011).

27. Bao, Q. *et al.* Atomic-layer graphene as a saturable absorber for ultrafast pulsed lasers. *Adv. Funct. Mater.* **19**, 3077–3083 (2009).

28. Lee, S. H., Choi, J., Kim, H. D., Choi, H. & Min, B. Ultrafast refractive index control of a terahertz graphene metamaterial. *Sci. Rep.* **3**, 2135 (2013).

29. Gabor, N. M. *et al.* Hot carrier–assisted intrinsic photoresponse in graphene. *Science* **334**, 648–652 (2011).

30. Yan, J. *et al.* Dual-gated bilayer graphene hot-electron bolometer. *Nat. Nanotechnol.* **7**, 472–478 (2012).

31. Liu, C. H., Chang, Y. C., Norris, T. B. & Zhong, Z. Graphene photodetectors with ultra-broadband and high responsivity at room temperature. *Nat. Nanotechnol.* **9**, 273–278 (2014).

32. Liu, C. H. *et al.* Ultrafast lateral photo-Dember effect in graphene induced by nonequilibrium hot carrier dynamics. *Nano Lett.* **15**, 4234–4239 (2015).

33. Li, X. *et al.* Large-area synthesis of high-quality and uniform graphene films on copper foils. *Science* **324**, 1312–1314 (2009).

34. Falkovsky, L. A. & Pershoguba, S. S. Optical far-infrared properties of a graphene monolayer and multilayer. *Phys. Rev. B* **76**, 153410 (2007).

35. Falkovsky, L. A. & Varlamov, A. A. Space-time dispersion of graphene conductivity. *Eur. Phys. J. B* **56**, 281–284 (2007).

36. Stauber, T., Peres, N. M. R. & Geim, A. K. Optical conductivity of graphene in the visible region of the spectrum. *Phys. Rev. B* **78**, 085432 (2008).

37. Yan, H. *et al.* Tunable infrared plasmonic devices using graphene/insulator stacks. *Nat. Nanotechnol.* **7**, 330–334 (2012).

38. Hwang, E. H., Adam, S. & Sarma, S. D. Carrier transport in two-dimensional graphene layers. *Phys. Rev. Lett.* **98**, 186806 (2007).

39. Chang, Y. C., Liu, C. H., Liu, C. H., Zhong, Z. & Norris, T. B. Extracting the complex optical conductivity of mono- and bilayer graphene by Ellipsometry. *Appl. Phys. Lett.* **104**, 261909 (2014).

40. Schedin, F. *et al.* Detection of individual gas molecules adsorbed on graphene. *Nat. Mater.* **6**, 652–655 (2007).

41. Bae, S. *et al.* Roll-to-roll production of 30-inch graphene films for transparent electrodes. *Nat. Nanotechnol.* **5**, 574–578 (2010).

42. Tarasov, A. *et al.* Controlled doping of large-area trilayer MoS_2 with molecular reductants and oxidants. *Adv. Mater.* **27**, 1175–1181 (2015).

43. Paniagua, S. A. *et al.* Production of heavily n-and p-doped CVD graphene with solution-processed redox-active metal–organic species. *Mater. Horiz.* **1**, 111–115 (2014).

44. Mak, K. F., Shan, J. & Heinz, T. F. Seeing many-body effects in single-and few-layer graphene: observation of two-dimensional saddle-point excitons. *Phys. Rev. Lett.* **106**, 046401 (2011).

45. Yang, L., Deslippe, J., Park, C. H., Cohen, M. L. & Louie, S. G. Excitonic effects on the optical response of graphene and bilayer graphene. *Phys. Rev. Lett* **103**, 186802 (2009).

46. Liang, X. *et al.* Toward clean and crackless transfer of graphene. *ACS Nano* **5**, 9144–9153 (2011).

47. Hao, Y. *et al.* The role of surface oxygen in the growth of large single-crystal graphene on copper. *Science* **342**, 720–723 (2013).

48. Banszerus, L. *et al.* Ultrahigh-mobility graphene devices from chemical vapor deposition on reusable copper. *Sci. Adv.* **1**, e1500222, 2015).

Acknowledgements

This work was supported by the National Science Foundation (NSF) Center for Photonic and Multiscale Nanomaterials (DMR 1120923). This work was performed in part at the Lurie Nanofabrication Facility, a member of the National Nanotechnology Infrastructure Network, which is supported in part by the National Science Foundation. This research was also funded by the National Science MRSEC Program, DMR-0820382. Z.Z. thanks the support from NSF CAREER Award (ECCS-1254468).

Author contributions

Y.-C.C. and T.B.N. conceived the experiments. Y.-C.C., Che.-H.L. and Cha.-H.L. fabricated the samples. S.Z and S.R.M identified suitable dopants and suggested procedures for doping studies. Y.-C.C. performed the measurements. All authors discussed the results. Y.-C.C. and T.B.N. co-wrote the manuscript and all authors provided comments.

Additional information

15

Seamless growth of a supramolecular carpet

Ju-Hyung Kim[1,2,3,4], Jean-Charles Ribierre[3], Yu Seok Yang[3], Chihaya Adachi[3], Maki Kawai[4], Jaehoon Jung[1,5], Takanori Fukushima[6] & Yousoo Kim[1]

Organic/metal interfaces play crucial roles in the formation of intermolecular networks on metal surfaces and the performance of organic devices. Although their purity and uniformity have profound effects on the operation of organic devices, the formation of organic thin films with high interfacial uniformity on metal surfaces has suffered from the intrinsic limitation of molecular ordering imposed by irregular surface structures. Here we demonstrate a supramolecular carpet with widely uniform interfacial structure and high adaptability on a metal surface via a one-step process. The high uniformity is achieved with well-balanced interfacial interactions and site-specific molecular rearrangements, even on a pre-annealed amorphous gold surface. Co-existing electronic structures show selective availability corresponding to the energy region and the local position of the system. These findings provide not only a deeper insight into organic thin films with high structural integrity, but also a new way to tailor interfacial geometric and electronic structures.

[1] Surface and Interface Science Laboratory, RIKEN, 2-1 Hirosawa, Wako, Saitama 351-0198, Japan. [2] Department of Chemical Engineering, Pukyong National University, 365 Sinseon-ro, Nam-gu, Busan 608-739, Republic of Korea. [3] Center for Organic Photonics and Electronics Research (OPERA), Kyushu University, 744 Motooka, Nishi, Fukuoka 819-0395, Japan. [4] Department of Advanced Materials Science, The University of Tokyo, 5-1-5 Kashiwanoha, Kashiwa, Chiba 277-8561, Japan. [5] Department of Chemistry, University of Ulsan, 93 Daehak-ro, Nam-gu, Ulsan 44610, Republic of Korea. [6] Chemical Resources Laboratory, Tokyo Institute of Technology, 4259 Nagatsuta, Midori-ku, Yokohama 226-8503, Japan. Correspondence and requests for materials should be addressed to J.J. (email: jjung2015@ulsan.ac.kr) or to T.F. (email: fukushima@res.titech.ac.jp) or to Y.K. (email: ykim@riken.jp).

Organic thin films (OTFs) have received much attention for their potential in various electronic and optoelectronic device applications, since they have outstanding advantages (that is, low cost, low weight and mechanical flexibility) in comparison with standard inorganic technologies[1-6]. In particular, it has been recognized in recent years that organic/metal (O/M) interfaces at which charge carriers are injected into OTFs play a crucial role in the operation and performance of organic devices[7-14]. Various interactions at the O/M interface, such as surface–molecule and intermolecular interactions, are of great importance in the formation of intermolecular networks and organic epitaxy, and have a strong correlation with their electronic structures[7,9,13,15-19]. Although the purity and uniformity of OTFs at the O/M interfaces have profound effects on the operating characteristics of organic devices[7], the formation of OTFs with high interfacial uniformity on metal surfaces has suffered from the intrinsic limitation of molecular ordering imposed by surface step edges, due to the relatively weak intermolecular interactions between organic molecules. Extensive research efforts in the development of OTFs with high interfacial uniformity on metal surfaces have mostly focused on examining the formation of covalent bonding and supramolecular complexes between organic precursors via laborious post-treatments, such as epitaxial surface polymerization and metal–organic coordination[20-25]. Recently, one-dimensional (1D) molecular assembly across step edges was achieved on a specific metal surface via hydrogen bonding[26,27], and at the solid–liquid interface, the formation of supramolecular networks with high uniformity on rough graphene has been studied in depth[28]. However, the creation of widely uniform OTFs on metal surfaces that maintain structural integrity based on non-bonding intermolecular interactions remains a challenge for further development and enhancement of organic devices fabricated by one-step deposition of organic molecules.

In this article, we successfully demonstrate the formation of a supramolecular carpet (SMC), with a widely uniform interfacial structure and high adaptability, on a metal surface, via a one-step deposition process. The geometric and electronic structures of the SMC were investigated by means of scanning tunnelling microscopy/spectroscopy (STM/STS) and density functional theory (DFT) calculations. For this work, bis[1,2,5]thiadiazolotetracyanoquinodimethane (BTDA-TCNQ)[29] was used as the key building block with which to realize the SMCs, not only on a single crystal gold (Au) surface but also on a pre-annealed amorphous Au surface. Tetracyanoquinodimethane (TCNQ) is one of the strongest organic electron acceptors to have found wide use in organic devices, and forms a strongly bonded donor–acceptor complex with tetrathiafulvalene (TTF)[17,19,30-32]. Thus, we used a model system of BTDA-TCNQ, which integrates the structural properties of both TCNQ and TTF in a single molecule, for realizing self-organization into an ordered domain. BTDA-TCNQ is a TCNQ derivative fused with 1,2,5-thiadiazole rings, exerting strong intermolecular interactions in a planar fashion to form network structures in the single crystal[29] and even in its charge-transfer crystals[33,34]. The rhombic structure of BTDA-TCNQ exhibits twofold symmetry with two mirror planes, allowing perpendicular alignment of two electrostatically opposite symmetry axes consisting of electronegative and electropositive end groups, respectively. It, therefore, enables equivalent intermolecular interactions along the four sides of the molecule. Such rhombic structure also facilitates access of neighbouring molecules to each other via high electrostatic interactions in the four directions, which leads to topographically favourable intermolecular interactions. The strong non-bonding intermolecular interactions, balanced with surface–molecule interfacial

interactions and site-specific rearrangements of the BTDA-TCNQ molecules near surface step edges, enable the step-flow growth mode for the SMC formation[35,36]. This can lead to an extension of the SMC over step edges of the Au surface without loss of its structural integrity, and results in a covering with high interfacial uniformity over multiple surface steps and terraces, even on the pre-annealed amorphous Au surface prepared on a glass substrate. In addition, different types and dimensionalities of the interfacial electronic structures are distinctively observed in the SMC of BTDA-TCNQ on the Au(111) surface, corresponding to the energy region and the local position of the system, implying that various types of electronic structures projected onto the SMC can be selectively accessible. These results suggest that the SMC has great potential for applications in organic electronics, and also provide important guidelines to develop novel materials forming seamless OTFs on various surfaces.

Results

Intermolecular network of BTDA-TCNQ/Au(111). On deposition of BTDA-TCNQ on the Au(111) surface, decoration at the step edge occurs first, and then growth of a highly ordered SMC proceeds from the step edge across the lower terrace without further nucleation occurring on the terrace (see Fig. 1a,b). This step-flow growth mode of the SMC exhibits a quasi-epitaxial nature with an in-plane azimuthal orientation of BTDA-TCNQ (see Supplementary Figs 1 and 2). The reconstructed herringbone structures of Au(111) beneath the SMC are readily observed in

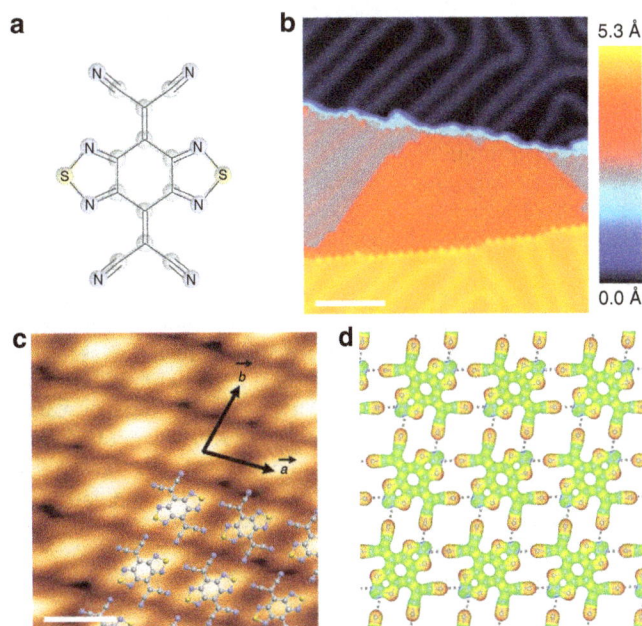

Figure 1 | Network structure of the BTDA-TCNQ molecules.
(**a**) The chemical structure of BTDA-TCNQ. (**b**) STM image of BTDA-TCNQ/Au(111), which shows that the molecular network of BTDA-TCNQ grows from the step edge at the lower terrace (sample bias voltage (V_s) = 1,000 mV, tunnelling current (I_t) = 1.00 nA, scale bar (S), 10.0 nm). The reconstructed herringbone structures of Au(111) beneath the BTDA-TCNQ molecules are clearly evident. Height scale is indicated in colour, and the step height of Au(111) is found to be around 2.3 Å.
(**c**) Close-up STM image of the network structure formed by BTDA-TCNQ (V_s = 2,000 mV, I_t = 1.0 nA, S = 1.0 nm). Molecular models reveal that the unit cell is noticeably similar to that of the coplanar molecular network in the crystal structure of BTDA-TCNQ. (**d**) The electrostatic potential map of the BTDA-TCNQ network. Blue to red corresponds to positive to negative charges. The distance for the S–N contacts (indicated by grey dotted lines) is 2.80 Å, and the calculated binding energy is ~0.70 eV per molecule.

the STM images, which reveals that the adsorption of BTDA-TCNQ on the Au(111) terraces does not involve quenching of the Au(111) Shockley surface state, as will be discussed below in relation to the STS spectra. These experimental observations indicate that the surface–molecule interfacial interactions of BTDA-TCNQ/Au(111) are sufficiently weak so that lateral diffusion of the molecules across the Au(111) terraces is possible[35,36]. Contrary to the weak adsorption characteristics of BTDA-TCNQ/Au(111), the partial positive and negative charges of the S and N atoms of BTDA-TCNQ, respectively, introduce effective attractive interactions (that is, the S–N intermolecular interactions) between the neighbouring molecules, which determine the array of the molecules in the SMC, as shown in Fig. 1c,d. Detailed analysis of the STM images (see Fig. 1c) gave lattice constants for the SMC unit cell of $a = b = 10.0 \pm 1.0$ Å and $\gamma = 80 \pm 5°$, which are noticeably similar to those achieved by the S–N intermolecular interactions of the coplanar molecular network in the single crystal structure of BTDA-TCNQ[33,34].

DFT calculations were also performed to examine the formation of SMC and to confirm its stability in terms of intermolecular interactions. On the basis of the experimental observations of weak surface–molecule interactions, the substrate was not considered in the calculations. For comparisons among a variety of initial geometries, various molecular orientations rotated by 10° increments were considered with the experimentally observed periodicity of the intermolecular network structure. On the basis of our computational approach, three nonequivalent local potential minima were found corresponding to the varied molecular orientations (see Supplementary Fig. 3). The experimentally observed azimuthal orientation is most favourable with the highest binding energy per molecule of ~0.70 eV (see Fig. 1d), which is more stable than the other two local potential minima by >0.30 eV. Note that the calculated binding energy of ~0.70 eV per molecule is highly competitive with the dipole polarization energy arising from donor–acceptor complexation in the charge-transfer crystal[37,38]. The higher stability of the geometric structure of the SMC, in comparison with the followed second and third local potential minima, mainly originates from the relative orientations of electronegative cyano (–CN) groups between the neighbouring molecules. The electrostatic repulsions between the –CN groups are easily expected in both second and third local potential minima rather than the S–N intermolecular interactions (see Supplementary Fig. 3), which implies that the S–N intermolecular interactions are an impetus for the formation of SMC. The calculated lattice constants of the SMC unit cell ($a = b = 9.9$ Å and $\gamma = 81.5°$) are almost identical to our STM observations, and the calculated distance for the S–N contacts (~2.80 Å) is also well matched with that of the crystal structure of BTDA-TCNQ (~3.04 Å; ref. 29).

Seamless growth of SMC with high structural integrity. In the high-coverage regime of BTDA-TCNQ shown in Fig. 2a, the SMC grows very compactly, covering the whole Au(111) surface, and is extended over multiple surface steps and terraces in a downhill direction, without losing its structural integrity. It is worth noting that single molecular domains of the SMC with a few hundred nanometres in length and width could be achieved on the Au(111) surface, of which sizes are meaningfully large when recent lithography techniques are considered for further applications[39,40]. Interestingly, the outstanding growth characteristics of BTDA-TCNQ are maintained on the pre-annealed amorphous Au surface, as shown in Fig. 2b. The amorphous Au surface was prepared on the cleaned glass substrate by thermal evaporation in a vacuum chamber, and followed by a thermal annealing process without any other preliminary treatments.

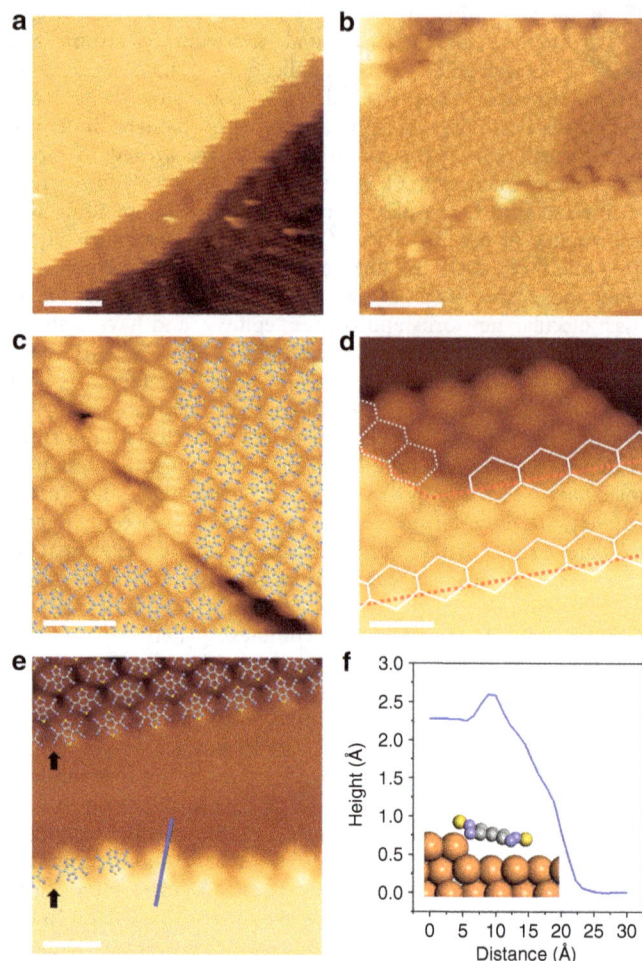

Figure 2 | Highly ordered SMC of BTDA-TCNQ on the Au surfaces. (**a**) STM image of the high-coverage regime of BTDA-TCNQ on Au(111) ($V_s = 2,000$ mV, $I_t = 0.5$ nA, $S = 12.0$ nm), showing that the highly ordered SMC of BTDA-TCNQ is extended over multiple surface steps and terraces without losing its structural integrity. (**b**) STM image of an SMC of BTDA-TCNQ on the pre-annealed amorphous Au surface ($V_s = 2,000$ mV, $I_t = 0.3$ nA, $S = 5.0$ nm). (**c**) STM image of a structural defect in an SMC of BTDA-TCNQ between two misaligned domains on Au(111) ($V_s = 1,000$ mV, $I_t = 0.1$ nA, $S = 2.0$ nm). (**d**) STM image of an SMC of BTDA-TCNQ of which the structural integrity is not interrupted by the surface step edge of Au(111) ($V_s = 300$ mV, $I_t = 0.5$ nA, $S = 2.0$ nm). The BTDA-TCNQ molecules with the slanting adsorption structure (enclosed with white solid lines) and the rearranged flat-lying adsorption structure (enclosed with white dotted lines) are clearly shown near the step edges of Au(111) (indicated by red dotted lines). (**e**) STM image of the BTDA-TCNQ molecules which lean down from the upper terrace to the lower terrace at the step edge ($V_s = 500$ mV, $I_t = 0.5$ nA, $S = 2.0$ nm). The step edges of Au(111) are indicated by black arrows. (**f**) The height profile along the blue line in **e**. corresponding to the slanting adsorption structure of BTDA-TCNQ at the step edge is presented here. The slanting adsorption structure of BTDA-TCNQ is schematically illustrated in the inset of **f**.

Note that the amorphous Au surface tends toward the close-packed (111) facet with irregular steps and terraces after annealing, because Au(111) is the most thermodynamically stable facet of Au (see Supplementary Fig. 4). Although structural defects of the SMC, such as domain boundaries and dislocations, are occasionally observed between misaligned domains on Au(111), the structural integrity is not essentially interrupted by the step edges (see Fig. 2c,d). These results imply that the S–N intermolecular interactions in the molecular

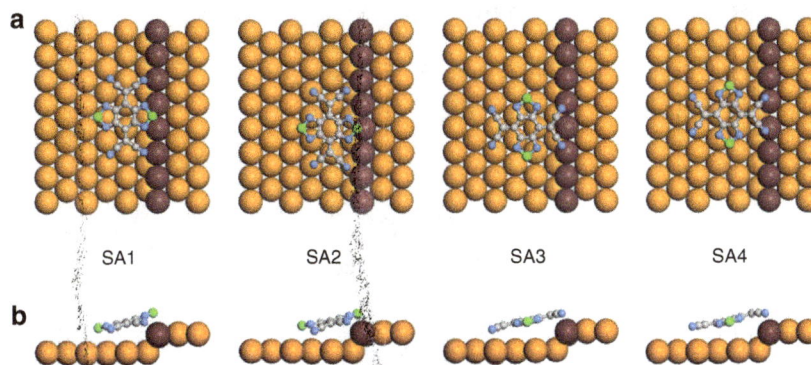

Figure 3 | DFT calculations for the SA structure of BTDA-TCNQ at the step edge of Au(111). (a,b) Optimized geometries of four different hypothetical model structures of BTDA-TCNQ adsorbed at the step edge of Au(111) (designated 'SA1', 'SA2', 'SA3' and 'SA4', respectively) by DFT calculations. The S–S molecular axes of the molecules are perpendicular to the step line in 'SA1' and 'SA2', and parallel in 'SA3' and 'SA4'. Side views of **a** are shown in **b**.

ordering of BTDA-TCNQ remain intact across the surface steps. This feature distinguishes the behaviour of the BTDA-TCNQ system from that of most other supramolecular assemblies on metal surfaces involving the structural complexities near surface step edges. The adsorption structure of BTDA-TCNQ at the step edges plays a decisive role in maintaining the S–N intermolecular interactions between the two adjacent terraces, thus preserving the structural integrity of the SMC over multiple surface steps and terraces. When the BTDA-TCNQ molecules initially decorate the step edges, they lean down from the upper terrace to the lower terrace (that is, the slanting adsorption (SA) structure), as shown in Fig. 2e,f. The STM observations indicate that the S–S molecular axis (the molecular axis along the two S atoms) of BTDA-TCNQ in this structure is perpendicular to the step line, and thus the two S atoms are located on the upper and lower terraces, respectively (see Fig. 2e,f). In consideration of the symmetry of BTDA-TCNQ, the SA structure enables symmetrical intermolecular interactions on the upper and lower terraces, leading to the intact intermolecular interactions to be valid across the surface steps. Differently from the weak adsorption nature of BTDA-TCNQ on the Au(111) terraces, the surface–molecule interfacial interactions localized near the step edges are the primary factors that determine the SA structure, as will be discussed in relation to the DFT calculations and the STS spectra.

For the BTDA-TCNQ molecules that are aligned along parallel straight lines of the step edges, the SA structure at the step edges is preserved even in the high-coverage regime (see Fig. 2d), thus acting as a bridge for maintaining the S–N intermolecular interactions across the adjacent terraces. However, the BTDA-TCNQ molecules that are adsorbed out of the parallel straight lines at the step edges are displaced from step edges and rearranged onto the lower terraces (that is, the rearranged flat-lying adsorption structure), as shown in Fig. 2d. These phenomena imply that the S–N intermolecular interactions in supramolecular assembly dominate the surface–molecule interactions even at the relatively reactive step edges, and thus result in the SMC formation with structural integrity maintained over wide areas. In addition, the potential energy barrier for the diffusion of molecule crossing the upper step edge is significantly higher than the lower step edge, because the step dipole is mostly formed below the upper step edge[41–43]. The SMC, therefore, grows across the surface steps only in the downhill direction.

Adsorption structure of BTDA-TCNQ at the step edges. To gain deeper insight into the SA structure of BTDA-TCNQ at the step edges, DFT calculations were performed in consideration of four different adsorption models (designated 'SA1', 'SA2', 'SA3' and 'SA4', respectively) as shown in Fig. 3 and Table 1. Among

Table 1 | Relative energies, adsorption energies and optimized nearest atomic distances of S···Au and –CN···Au of four adsorption models, 'SA1', 'SA2', 'SA3' and 'SA4'.

	SA1	SA2	SA3	SA4
E_{rel} (eV)	0.00	0.17	0.52	0.53
E_{ads} (eV)	2.64	2.47	2.12	2.11
Step edge				
S···Au (Å)	3.12 (×2)	2.54		
–CN···Au (Å)	2.18	2.56	2.31	2.34
Terrace				
S···Au (Å)	3.34	3.33	3.61	3.69
–CN···Au (Å)	3.17	2.36	2.68	3.14

The optimized distances at the step edge and on the terrace are presented, respectively. In 'SA1', a bidentate fashion of S···Au at the step edge is denoted with '× 2'.

these structure models, the most stable adsorption structure of the BTDA-TCNQ molecule at the step edge was found to be 'SA1' with the adsorption energy of 2.64 eV. In 'SA1', the S–S molecular axis of BTDA-TCNQ is perpendicular to the step line, and the two S atoms are located on the upper and lower terraces, respectively. Compared with 'SA3' that is the most stable adsorption structure with the S–S molecular axis parallel to the step line, 'SA1' shows higher adsorption energy by 0.52 eV, which is consistent with the experimental observations of the SA structure at the step edges.

The detailed geometric interpretation of the computational results indicates that the –CN groups of the BTDA-TCNQ molecule significantly contribute to determination of the adsorption structure at the step edges. Side views of the molecular adsorption structures clearly reveal that the –CN groups are displaced from the molecular plane and become closer to the surface, compared with the S atoms (see Fig. 3b). In the most stable 'SA1', the nearest atomic distance between the molecule and the reactive Au atoms of the step edge (–CN···Au$_{edge}$) is 2.18 Å, and is much shorter than the S···Au$_{edge}$ distance of 3.12 Å even though the S atom of BTDA-TCNQ is interacting with the two Au$_{edge}$ atoms. The higher stability of 'SA1' than 'SA3' also can be deduced from the shorter –CN···Au$_{edge}$ distance in 'SA1' than that of 'SA3' by 0.13 Å. Combined with the –CN···Au$_{edge}$ interactions, the relatively weak S···Au$_{edge}$ interactions in 'SA1' also possibly participate in determining the adsorption structure when no existing S···Au$_{edge}$ interaction in 'SA3' is considered. In addition, the charge density difference map of the most stable 'SA1' structure clearly reveals the partial charge transfer from the

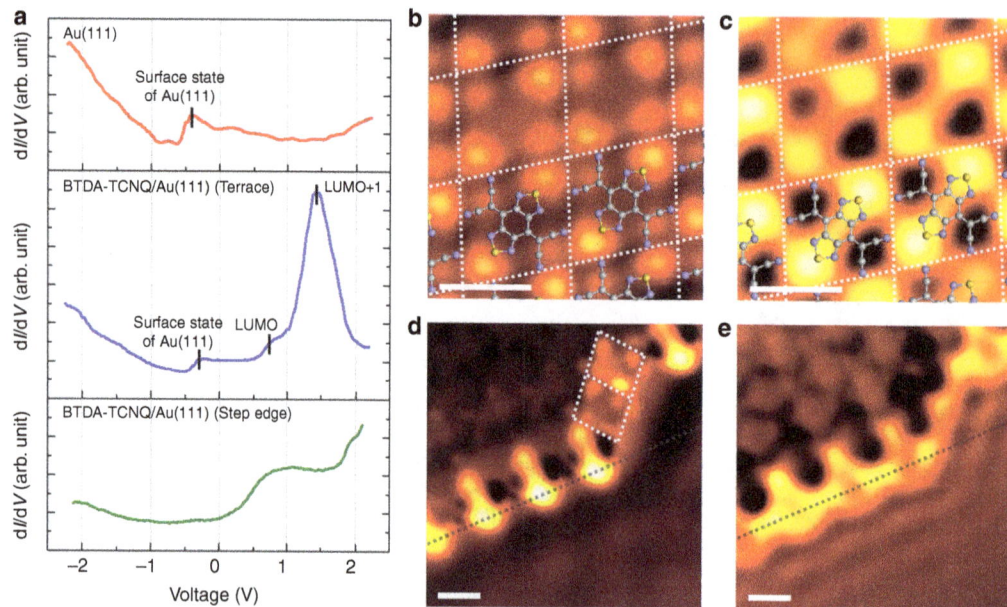

Figure 4 | STS spectra and STS mapping images of BTDA-TCNQ/Au(111). (**a**) STS spectra of (top) the bare Au(111) surface, (middle) the BTDA-TCNQ molecule on the Au(111) terrace and (bottom) the BTDA-TCNQ molecule at the Au(111) step edge. (**b,c**) STS mapping images of BTDA-TCNQ on the Au(111) terrace. The electron probability distributions are clearly observed for the LUMO state of BTDA-TCNQ in **b** ($V_s = 800$ mV, $I_t = 0.8$ nA, $S = 1.0$ nm) and the LUMO + 1 state of BTDA-TCNQ in **c** ($V_s = 1,500$ mV, $I_t = 0.8$ nA, $S = 1.0$ nm). Each BTDA-TCNQ molecule is enclosed with white dotted lines in **b** and **c**. (**d,e**) STS mapping images of BTDA-TCNQ at the Au(111) step edge. Parallel straight lines of the step edge are indicated by grey dotted lines in **d** and **e**, and the BTDA-TCNQ molecules that are rearranged onto the lower terrace from the step edge are enclosed with white dotted lines in **d**. The STS mapping image in **d** ($V_s = 500$ mV, $I_t = 1.0$ nA, $S = 1.0$ nm) reveals the LUMO state for BTDA-TCNQ with the rearranged flat-lying adsorption structure and the LUMO + 1 state for BTDA-TCNQ with the slanting adsorption structure at the lower sample bias than that in **b** and **c**. The STS mapping image in **e** ($V_s = 2,000$ mV, $I_t = 1.0$ nA, $S = 1.0$ nm) also indicates that the unoccupied MO states of BTDA-TCNQ are highly hybridized with the Au surface at the step edge, forming quasi-1D electron dispersion along the step-edge line.

surface to the molecule through the –CN groups at the step edge (see Supplementary Fig. 5), and the corresponding amount of the partial charge transfer evaluated by Bader population analysis is 0.68e. The computationally estimated feature of the partial charger transfer strongly supports the surface–molecule interfacial interactions localized at the step edges, resulting from the step dipole as will be discussed in relation to the STS spectra.

Electronic structures of BTDA-TCNQ on the Au surface. To elucidate the electronic structures of BTDA-TCNQ on the Au surface in more detail, intensive STS and STS mapping measurements were performed, as shown in Fig. 4. The Au(111) Shockley surface state, the band edge of which is ~500 meV below the Fermi level (E_F), remains clearly evident in the STS spectrum of the BTDA-TCNQ molecule adsorbed on the Au(111) terrace, indicating the weak adsorption character of BTDA-TCNQ/Au(111) (refs 44,45). The lowest unoccupied molecular orbital (LUMO) and the LUMO + 1 states of BTDA-TCNQ are distinctly observed in the STS spectrum (at ~ + 800 mV and ~ + 1,500 mV, respectively). STS mapping reveals the electron probability distributions for these unoccupied molecular orbital (MO) states (see Fig. 4b,c), demonstrating that the LUMO state is locally distributed around the –CN groups and the S atoms, forming a rhombic structure, whereas the LUMO + 1 state shows a more conjugated electronic structure along the S–S molecular axis. In general, the unoccupied MO states of BTDA-TCNQ are largely distributed around the S atoms, even though their electronic structures are distinct from each other. The conjugated character of the unoccupied MO states of BTDA-TCNQ results in n-type semiconducting behaviour, which is clarified in a field-effect transistor (FET) configuration (see Supplementary Fig. 6).

Note that the distinct electronic structures of the unoccupied MO states lead to strong bias dependence in STM imaging in positive sample bias region (see Supplementary Fig. 1), because STM imaging strongly reflects the integration of the electronic states that can contribute to electron tunnelling[46].

At the step edge of Au(111), however, the STS spectrum indicates a rather strong interaction between BTDA-TCNQ and the Au surface. The unoccupied feature of BTDA-TCNQ at the step edge shows a steep initial rise near the E_F, and then maintains a nearly constant level before another steep rise at ~ + 1,700 mV. Such substantial broadening of the unoccupied states is mainly caused by hybridization with the electronic states of the metal surface, and thus the feature for the step-edge state of Au(111) is not distinguishable in the STS spectrum of BTDA-TCNQ adsorbed at the step edge[42]. STS mapping reveals that the unoccupied MO states of the BTDA-TCNQ molecules show a significant downward shift toward the E_F at the step edge. The LUMO + 1 state for the SA structure and the LUMO state for the rearranged flat-lying adsorption structure of BTDA-TCNQ at the step edge are observed at ~ + 500 mV, as shown in Fig. 4d. These phenomena mainly originate from the step dipole described as the Smoluchowski effect[41–43]. The step dipole, with an opposite direction to that of the surface dipole, induces a strong localized electric field at the step edge, resulting in Stark shifts of the adsorbed molecules near the step edge[41]. Since the SA structure of BTDA-TCNQ covers the step edge completely, the Stark shift in the SA structure is larger than that of the rearranged flat-lying adsorption structure. On the basis of a detailed analysis of the STS mapping data (see Supplementary Fig. 7), the Stark shifts of the unoccupied MO states of BTDA-TCNQ near the step edge are ~ − 900 meV and ~ − 300 meV for the SA structure and the rearranged flat-lying adsorption structure, respectively.

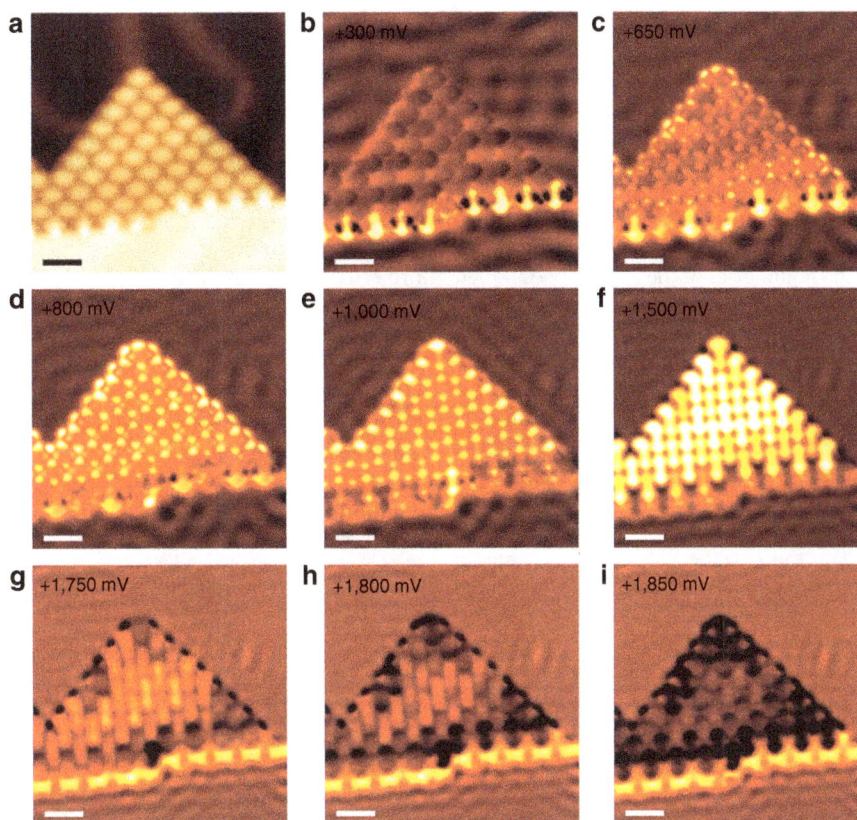

Figure 5 | STS mapping of an SMC of BTDA-TCNQ on the Au(111) surface. (**a**) STM image of an SMC of BTDA-TCNQ on the Au(111) surface ($V_s = 500$ mV, $I_t = 1.0$ nA, $S = 2.0$ nm). (**b-i**) STS mapping images of repeating measurements over the same surface area as that in **a**, with varying sample bias ($I_t = 1.0$ nA, $S = 2.0$ nm). The Au(111) surface states are transparently observed through the SMC of BTDA-TCNQ in **b** ($V_s = 300$ mV) and **c** ($V_s = 650$ mV). The electron probability distributions of the MO states of BTDA-TCNQ are the most prominent features in **d** ($V_s = 800$ mV), **e** ($V_s = 1,000$ mV) and **f** ($V_s = 1,500$ mV). The SMC of BTDA-TCNQ exhibits a highly delocalized electron distribution which is confined within the boundaries of the SMC in **g** ($V_s = 1,750$ mV), **h** ($V_s = 1,800$ mV) and **i** ($V_s = 1,850$ mV). The quasi-1D electron dispersion along the step edge resulting from the hybridization of BTDA-TCNQ and Au at the step is also evident in **g-i**.

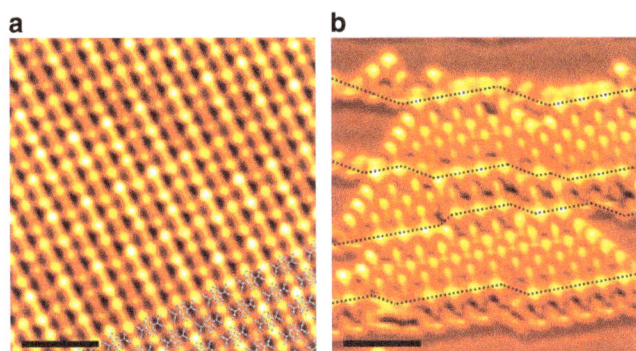

Figure 6 | Widely uniform anisotropic MO states of BTDA-TCNQ/ Au(111). (**a,b**) STS mapping images of an SMC of BTDA-TCNQ on the Au(111) surface in the energy range corresponding to the LUMO+1 state ($V_s = 1,500$ mV, $I_t = 1.0$ nA). Widely uniform anisotropic MO states of the SMC are clearly observed in **a** ($S = 3.0$ nm), which can be extended over multiple surface steps and terraces as in **b** ($S = 5.0$ nm). Step-edge lines are indicated with black dotted lines in **b**, revealing small sizes (~ 5 nm in width) of the (111) facets.

These results strongly suggest that the localized electric fields induced by the step dipoles exert influences on the partial charge transfers from the step edges to the unoccupied MO states of BTDA-TCNQ. Considering the LUMO state of BTDA-TCNQ on the Au(111) terrace (at $\sim +800$ meV relative to the E_F), the Stark shift of ~ -900 meV possibly induces the partial charge

transfer to the LUMO state for the SA structure from the step edge, and the LUMO+1 state consequentially represents the nearest unoccupied MO state with respect to the E_F at the step edge (see Supplementary Fig. 7). The computationally estimated charge density difference map also strongly supports the partial charge transfer from the step edge to the molecule (see Supplementary Fig. 5). In addition, the STS spectrum and mapping images show that the unoccupied MO states for the SA structure of BTDA-TCNQ are highly hybridized with the Au surface at the step edge above $\sim +1,800$ meV with respect to the E_F, forming quasi-1D electron dispersion along the step-edge line (see Fig. 4e).

Electronic structures between the SMC and the Au surface. The interfacial electronic structures between the SMC of BTDA-TCNQ and the Au(111) surface were also investigated as a function of energy and of local position, by repeating STS mapping over the same area of surface with the sample bias voltage varied between $+300$ mV and $+2,000$ mV (see Fig. 5). The interfacial electronic structures of the SMC on the Au(111) terrace are characterized into three main categories according to energy ranges. First, since the MO states of BTDA-TCNQ do not appear near the E_F as indicated in the STS spectrum (see Fig. 4a), the Au(111) surface states are transparently observed through the SMC in the lower energy region near $+300$ meV with respect to the E_F (see Fig. 5b,c). In this energy range, the STS mapping images exhibit strong standing-wave patterns of the surface-state electrons, which are continuous between two media (that is, the

SMC and the Au(111) terrace) without scattering at the boundaries of the SMC[47]. Second, in the energy range corresponding to the MO states of BTDA-TCNQ, the electronic structures of the SMC can be attributed to the electron probability distributions of the MO states (see Fig. 5d–f). Thus, the STS mapping data reveal the localized MO states of the SMC independently of the Au(111) surface states, which also substantiates the weak adsorption nature of BTDA-TCNQ/Au(111) (ref. 48). Strongly scattered standing-wave patterns, which are caused by the reflection of the surface-state electrons at the boundaries of the SMC, are observed on the Au(111) terrace in this energy region. Compared with the lower energy region (near +300 meV), the observation of the scattering of surface-state electrons indicates that it is the localized MO states of the SMC that serve as potential walls for reflecting the surface-state electrons rather than the geometric footprint of the molecule[49,50]. Finally, above the highly scattering region of the surface-state electrons, the contribution of the MO states of BTDA-TCNQ decreases, as indicated in the STS spectrum (see Fig. 4a). In this higher energy region (near +1,900 meV with respect to the E_F), the SMC exhibits a highly delocalized electron distribution, which is confined within the boundaries of the SMC (see Fig. 5g–i). The delocalized electron distribution shows a pattern of two-dimensional dispersive electronic states, and faintly scattered standing-wave patterns appear on the Au(111) terrace near the boundaries of the SMC. Considering the weak adsorption character of BTDA-TCNQ on the Au(111) terraces and the dissipation of the MO states in this higher energy region, it is the electronic states of Au(111) underneath the SMC that essentially contribute to the dispersive electronic states observed in the SMC[16,17]. On the other hand, the quasi-1D electron dispersion along the step edge resulting from the hybridization of BTDA-TCNQ and Au at the step is observed in this higher energy region separately from the terrace (see Fig. 5g–i), which implies that the surface steps modulate the dimensionality of the electron dispersion. Interestingly, these different types and dimensionalities of the interfacial electronic structures co-exist in the SMC of BTDA-TCNQ on the Au(111) surface. These results suggest that various types of electronic structures projected onto the SMC can be selectively available corresponding to the energy region and the local position of the system. Note that the SMC of BTDA-TCNQ on the Au(111) surface shows widely uniform anisotropic MO states in the energy region corresponding to the LUMO + 1 state (see Fig. 6). Such anisotropic characteristics originate predominantly from the distribution of the LUMO + 1 states along the S atoms and the uniaxial orientation of BTDA-TCNQ via the S–N intermolecular interactions, which are also maintained over multiple surface steps and terraces as shown in Fig. 6b. These results also imply that careful design of favourable intermolecular interactions facilitates the tailoring of direction and alignment of localized MO states in OTFs without the loss of structural integrity.

Discussion

We present a well-designed system, BTDA-TCNQ on the Au surface, to demonstrate the formation of SMC with widely uniform interfacial structure and high adaptability on the metal surface. The strong non-bonding intermolecular interactions induced by the partial positive charge of the S atoms and the partial negative charge of the N atoms of BTDA-TCNQ, and the surface–molecule interactions on both surface steps and terraces enable the step-flow growth mode for the self-organized SMC domain. Owing to the well-balanced interfacial interactions, the SMC of BTDA-TCNQ on the Au surface can extend over multiple surface steps and terraces without losing its structural integrity, accompanying the unidirectional alignment of the MO states. This results in a highly uniform interfacial structure, even on the pre-annealed amorphous Au surface prepared on the glass substrate. In consideration of the typical amorphous Au electrodes widely used in electronics, our findings on the pre-annealed amorphous Au surface may have an impact on the field of organic electronics in terms of molecular orientation, alignment and ordering on the electrodes, which have profound effects on the electrical and optical properties[51,52]. In addition, selective availability of co-existing interfacial electronic structures projected onto the SMC of BTDA-TCNQ is one of the remarkable features of the SMC on the Au surface. We anticipate that the findings of this work will provide not only a deeper insight into the formation of OTFs with high film integrity at the O/M interfaces, but also a new way to tailor the properties of interfacial geometric and electronic structures for future applications of organic devices.

Methods

Low-temperature STM/STS experiments. All the experiments were performed using a low-temperature STM (Omicron GmbH) with an electrochemically etched tungsten tip in an ultrahigh-vacuum (UHV) chamber in which the base pressure was maintained at below 8.0×10^{-11} Torr. The Au(111) and amorphous Au surfaces were cleaned by several cycles of sputtering with argon ions (Ar$^+$) and annealing at 800 K. Low-temperature STM surface scanning at atomic resolution was used to confirm the cleanliness of each Au surface. BTDA-TCNQ was synthesized and purified in accordance with a previous report[26,30]. Using a Knudsen cell, BTDA-TCNQ (in powdered form at room temperature) was thermally evaporated onto each Au surface under UHV conditions. Evaporation temperature was 473 K for BTDA-TCNQ, which was monitored using a K-type (alumel–chromel) thermocouple. The Au surface was maintained at room temperature during and after each thermal evaporation to confirm a thermodynamically stable phase (~1 h), then cooled to 4.7 K. Note that the domain size of the SMC in the step-flow regime is mainly determined by the amount of adsorbed molecules at sub-monolayer coverage, which can be simply controlled by the deposition rate and time. In the deposition and annealing processes, no intentional temperature control was performed for the Au surface, and the substrate was merely located in the room temperature chamber. When the sample was immediately cooled to 4.7 K (cooling time of ~1 h) even without annealing time after deposition, there were no significant changes observed in the formation of SMC as well following the step-flow growth scheme. The cooled sample was also not damaged if the sample temperature repeatedly increased up to room temperature in the UHV chamber.

All STM and STS measurements were performed at 4.7 K, and STS signals were measured with lock-in detection by applying a modulation of 50 mV (r.m.s.) to the sample bias voltage at 797 Hz. Note that all STS spectra were obtained at a fixed tip-sample separation (feed-back off), whereas STS mapping was performed in the constant current mode, with the sample bias voltage varied from +300 mV to +2,000 mV.

DFT calculations. We used vdW-TS method[53] implemented in the Vienna *Ab initio* Simulation Package (VASP) code[54,55], which can take account of van der Waals interactions, to examine the formation of SMC. The core electrons were replaced by projector-augmented wave pseudopotentials[56], expanded in a basis set of plane waves up to a cutoff energy of 400 eV. The substrate-free DFT calculations were performed, based on the experimental observations of weak surface–molecule interactions. The minimum-size supercell including one molecule was used according to the experimentally observed periodicity of the two-dimensional intermolecular network. For comparisons among a variety of initial geometries, various molecular orientations rotated by 10° increments were prepared with respect to the experimentally proposed molecular arrangement. The calculations using DFT-D2 method[57] were also performed to verify the computational reliability, leading to a good agreement with the results obtained with vdW-TS method. During ionic relaxation, the molecular geometry in a planar fashion was maintained by symmetry constraints, but the volume and shape of the slab were freely adjusted toward a local potential minimum. To ensure two-dimensional molecular architecture, we initially separated the periodically replicated slabs with a large vacuum region of 15 Å. The resultant vacuum regions of the optimized slabs were about 6.0 Å, in which the interactions between the molecular layers are expected to be negligible in obtaining proper geometries and energies of the intermolecular network structures of BTDA-TCNQ. Ionic (electronic) relaxations were performed until atomic forces (energies) were less than 0.05 eV Å^{-1} (10^{-7} eV). The k-point sampling of the Brillouin zone was performed with $8 \times 8 \times 1$ Γ-centred grids in the periodic DFT calculations.

As regards the adsorption structure of BTDA-TCNQ at the step edge of Au(111), we used a (6 × 1)-Au(455) supercell to investigate the adsorption of isolated BTDA-TCNQ molecule, which provides the nearest interatomic distance of longer than 8 Å along the step-edge line and the large distance from the molecule to the next step edge longer than 15 Å. The slab model was composed of six Au layers and vacuum region of ~20 Å, in which the two bottom layers were fixed in their bulk positions. Bader population analysis[58] was also performed to evaluate the partial charge transfer from the surface to the molecule at the step edge. Dipole correction was applied to avoid undesired interactions between periodic slab images. The k-point sampling of the Brillouin zone was performed with $2 \times 2 \times 1$ Γ-centred grids in the periodic DFT calculations.

FET fabrication and characterization. BTDA-TCNQ was incorporated into the FET configuration with a bottom-gate and bottom-contact structure. A highly doped n-type Si wafer with a thermally grown SiO_2 dielectric layer (of ~300 nm thickness) was used as a substrate. The Au source and drain electrodes (of ~40 nm thickness) were deposited onto the pre-cleaned substrate by thermal evaporation in a vacuum chamber, and then annealed at 300 °C for ~1 h in a N_2-filled glove-box. The channel length and width between the Au source and drain electrodes were 50 μm and 975 μm, respectively. After cooling the substrate to room temperature, BTDA-TCNQ (of ~50 nm thickness) was evaporated at the rate of 0.1 Å s^{-1}. The transfer and output characteristics (see Supplementary Fig. 6) were measured using a semiconductor parameter analyser (Agilent B1500A) in the N_2-filled glove-box at room temperature.

References

1. Park, S. H. et al. Bulk heterojunction solar cells with internal quantum efficiency approaching 100%. Nat. Photon. 3, 297–302 (2009).
2. Sekitani, T. et al. Stretchable active matrix organic light-emitting diode display using printable elastic conductors. Nat. Mater. 8, 494–499 (2009).
3. Sekitani, T., Zschieschang, U., Klauk, H. & Someya, T. Flexible organic transistors and circuits with extreme bending stability. Nat. Mater. 9, 1015–1022 (2010).
4. Zhang, W. Supramolecular linear heterojunction composed of graphite-like semiconducting nanotubular segments. Science 334, 340–343 (2011).
5. Uoyama, H., Goushi, K., Shizu, K., Nomura, H. & Adachi, C. Highly efficient organic light-emitting diodes from delayed fluorescence. Nature 492, 234–238 (2012).
6. Kim, R. H. et al. Non-volatile organic memory with sub-millimetre bending radius. Nat. Commun. 5, 3583 (2014).
7. Forrest, S. R. Ultrathin organic films grown by organic molecular beam deposition and related techniques. Chem. Rev. 97, 1793–1896 (1997).
8. Ishii, H., Sugiyama, K., Ito, E. & Seki, K. Energy level alignment and interfacial electronic structures at organic/metal and organic/organic interfaces. Adv. Mater. 11, 605–625 (1999).
9. Hooks, D. E., Fritz, T. & Ward, M. D. Epitaxy and molecular organization on solid substrates. Adv. Mater. 13, 227–241 (2001).
10. Zhu, X.-Y. Electronic structure and electron dynamics at molecule-metal interfaces: implications for molecule-based electronics. Surf. Sci. Rep. 56, 1–83 (2004).
11. Braun, S., Salaneck, W. R. & Fahlman, M. Energy-level alignment at organic/metal and organic/organic interfaces. Adv. Mater. 21, 1450–1472 (2009).
12. Kawano, K. & Adachi, C. Evaluating carrier accumulation in degraded bulk heterojunction organic solar cells by a thermally stimulated current technique. Adv. Funct. Mater. 19, 3934–3940 (2009).
13. Yamane, H., Kanai, K., Ouchi, Y., Ueno, N. & Seki, K. Impact of interface geometric structure on organic–metal interface energetics and subsequent films electronic structure. J. Electron Spectrosc. Relat. Phenom. 174, 28–34 (2009).
14. Ma, H., Yip, H.-L., Huang, F. & Jen, A. K.-Y. Interface engineering for organic electronics. Adv. Funct. Mater. 20, 1371–1388 (2010).
15. Barth, J. V., Costantini, G. & Kern, K. Engineering atomic and molecular nanostructures at surfaces. Nature 437, 671–679 (2005).
16. Temirov, R., Soubatch, S., Luican, A. & Tautz, F. S. Free-electron-like dispersion in an organic monolayer film on a metal substrate. Nature 444, 350–353 (2006).
17. Gonzalez-Lakunza, N. et al. Formation of dispersive hybrid bands at an organic-metal interface. Phys. Rev. Lett. 100, 156805 (2008).
18. Bartels, L. Tailoring molecular layers at metal surfaces. Nat. Chem. 2, 87–95 (2010).
19. Tseng, T.-C. et al. Charge-transfer-induced structural rearrangements at both sides of organic/metal interfaces. Nat. Chem. 2, 374–379 (2010).
20. Grill, L. et al. Nano-architectures by covalent assembly of molecular building blocks. Nat. Nanotechnol. 2, 687–691 (2007).
21. Zwaneveld, N. A. A. et al. Organized formation of 2D extended covalent organic frameworks at surfaces. J. Am. Chem. Soc. 130, 6678–6679 (2008).
22. Abdurakhmanova, N. et al. Stereoselectivity and electrostatics in charge-transfer Mn- and Cs-TCNQ$_4$ networks on Ag(100). Nat. Commun. 3, 940 (2012).
23. Kley, C. S. et al. Highly adaptable two-dimensional metal–organic coordination networks on metal surfaces. J. Am. Chem. Soc. 134, 6072–6075 (2012).
24. Li, Y. et al. Coordination and metalation bifunctionality of Cu with 5, 10, 15, 20-tetra (4-pyridyl) porphyrin: toward a mixed-valence two-dimensional coordination network. J. Am. Chem. Soc. 134, 6401–6408 (2012).
25. Floris, A., Comisso, A. & Vita, A. D. Fine-tuning the electrostatic properties of an alkali-linked organic adlayer on a metal substrate. ACS Nano 7, 8059–8065 (2013).
26. Schnadt, J. et al. Extended one-dimensional supramolecular assembly on a stepped surface. Phys. Rev. Lett. 100, 046103 (2008).
27. Schnadt, J. et al. Interplay of adsorbate-adsorbate and adsorbate-substrate interactions in self-assembled molecular surface nanostructures. Nano Res. 3, 459–471 (2010).
28. Li, B. et al. Self-assembled air-stable supramolecular porous networks on graphene. ACS Nano 7, 10764–10772 (2013).
29. Yamashita, Y., Suzuku, T., Mukai, T. & Saito, G. Preparation and properties of a tetracyanoquinodimethane fused with 1,2,5-thiadiazole units. J. Chem. Soc. Chem. Commun. 1985, 1044–1045 (1985).
30. Torrance, J. B. The difference between metallic and insulating salts of tetracyanoquinodimethone (TCNQ): how to design an organic metal. Acc. Chem. Res. 12, 79–86 (1979).
31. Alves, H., Molinari, A. S., Xie, H. & Morpurgo, A. F. Metallic conduction at organic charge-transfer interfaces. Nat. Mater. 7, 574–580 (2008).
32. Kirtley, J. R. & Mannhart, J. Organic electronics: when TTF met TCNQ. Nat. Mater. 7, 520–521 (2008).
33. Suzuki, T. et al. Clathrate formation and molecular recognition by novel chalcogen-cyano interactions in tetracyanoquinodimethanes fused with thiadiazole and selenadiazole rings. J. Am. Chem. Soc. 114, 3034–3043 (1992).
34. Suzuki, T., Fukushima, T., Yamashita, Y. & Miyashi, T. An absolute asymmetric synthesis of the [2 + 2] cycloadduct via single crystal-to-single crystal transformation by charge-transfer excitation of solid-state molecular complexes composed of arylolefins and bis[1,2,5]thiadiazolotetracyanoquinodimethane. J. Am. Chem. Soc. 116, 2793–2803 (1994).
35. Vlieg, E. Understanding crystal growth in vacuum and beyond. Surf. Sci. 500, 458–474 (2002).
36. Wagner, S. R., Lunt, R. R. & Zhang, P. Anisotropic crystalline organic step-flow growth on deactivated Si surfaces. Phys. Rev. Lett. 110, 086107 (2013).
37. Aragó, J., Sancho-García, J. C., Ortí, E. & Beljonne, D. Ab initio modeling of donor–acceptor interactions and charge-transfer excitations in molecular complexes: the case of terthiophene–tetracyanoquinodimethane. J. Chem. Theory Comput. 7, 2068–2077 (2011).
38. Sini, G., Sears, J. S. & Brédas, J.-L. Evaluating the performance of DFT functionals in assessing the interaction energy and ground-state charge transfer of donor/acceptor complexes: tetrathiafulvalene-tetracyanoquinodimethane (TTF-TCNQ) as a model case. J. Chem. Theory Comput. 7, 602–609 (2011).
39. Beesley, D. J. et al. Sub-15-nm patterning of asymmetric metal electrodes and devices by adhesion lithography. Nat. Commun. 5, 3933 (2014).
40. Cho, H. et al. Replication of flexible polymer membranes with geometry-controllable nano-apertures via a hierarchical mould-based dewetting. Nat. Commun. 5, 3137 (2014).
41. Wandelt, K. Properties and influence of surface defects. Surf. Sci. 251, 387–395 (1991).
42. Avouris, P., Lyo, I.-W. & Molinàs-Mata, P. STM studies of the interaction of surface state electrons on metals with steps and adsorbates. Chem. Phys. Lett. 240, 423–428 (1995).
43. Kamna, M. M., Stranick, S. J. & Weiss, P. S. Imaging substrate-mediated interactions. Isr. J. Chem. 36, 59–62 (1996).
44. Chen, W., Madhavan, V., Jamneala, T. & Crommie, M. F. Scanning tunneling microscopy observation of an electronic superlattice at the surface of clean gold. Phys. Rev. Lett. 80, 1469–1472 (1998).
45. Kim, J.-H. et al. Direct observation of adsorption geometry for the van der Waals adsorption of a single π-conjugated hydrocarbon molecule on Au (111). J. Chem. Phys. 140, 074709 (2014).
46. Wisendanger, R. Scanning Probe Microscopy and Spectroscopy: Methods and Applications (Cambridge Univ. Press, 1994).
47. Repp, J., Meyer, G. & Rieder, K.-H. Snell's law for surface electrons: refraction of an electron gas imaged in real space. Phys. Rev. Lett. 92, 036803 (2004).
48. Soe, W.-H., Manzano, C., De Sarkar, A., Chandrasekhar, N. & Joachim, C. Direct observation of molecular orbitals of pentacene physisorbed on Au (111) by scanning tunneling microscope. Phys. Rev. Lett. 102, 176102 (2009).
49. Gross, L. et al. Scattering of surface state electrons at large organic molecules. Phys. Rev. Lett. 93, 056103 (2004).
50. Seufert, K. et al. Controlled interaction of surface quantum-well electronic states. Nano Lett. 13, 6130–6135 (2013).
51. Dimitrakopoulos, C. D. & Malenfant, P. R. L. Organic thin film transistors for large area electronics. Adv. Mater. 14, 99–117 (2002).
52. Yokoyama, D. Molecular orientation in small-molecule organic light-emitting diodes. J. Mater. Chem. 21, 19187–19202 (2011).
53. Tkatchenko, A. & Scheffler, M. Accurate molecular van der Waals interactions from ground-state electron density and free-atom reference data. Phys. Rev. Lett. 102, 073005 (2009).
54. Kresse, G. & Hafner, J. Ab initio molecular dynamics for liquid metals. Phys. Rev. B 47, 558–561 (1993).

55. Kresse, G. & Furthmüller, J. Efficient iterative schemes for *ab initio* total-energy calculations using a plane-wave basis set. *Phys. Rev. B* **54**, 11169–11186 (1996).

56. Kresse, G. & Joubert, D. From ultrasoft pseudopotentials to the projector augmented-wave method. *Phys. Rev. B* **59**, 1758–1775 (1999).

57. Grimme, S. Semiempirical GGA-type density functional constructed with a long-range dispersion correction. *J. Comp. Chem.* **27**, 1787–1799 (2006).

58. Tang, W., Sanville, E. & Henkelman, G. A grid-based Bader analysis algorithm without lattice bias. *J. Phys. Condens. Matter* **21**, 084204 (2009).

Acknowledgements

We thank Dr Eisuke Ohta and Dr Yuki Suna for the synthesis of BTDA-TCNQ. We are grateful for access to the RIKEN Integrated Cluster of Clusters supercomputer system. T.F. acknowledges funding and support from a Grant-in-Aid for Scientific Research on Innovative Areas 'π-Figuration' (No. 26102001) from the Ministry of Education, Culture, Sports, Science and Technology (MEXT). J.J. acknowledges financial support from 2015 Research Fund of University of Ulsan. J.-H.K. also acknowledges support by Leading Foreign Research Institute Recruitment Program through the National Research Foundation of Korea (NRF-2010-00453).

Author contributions

J.-H.K., J.-C.R., C.A., M.K., J.J., T.F. and Y.K. conceived and designed the system, and Y.K. directed the project. J.-H.K. and Y.K. managed the overall experiment. T.F. prepared the BTDA-TCNQ molecules, and J.-H.K. performed the experiment. J.-H.K. acquired and analysed the STM/STS data. J.J. performed and analysed the DFT calculations. J.-H.K., J.-C.R. and Y.S.Y. fabricated the FET devices and analysed the device characteristics. All the authors discussed the results. J.-H.K. wrote the manuscript with contributions from all the other authors.

Additional information

16

A light-driven three-dimensional plasmonic nanosystem that translates molecular motion into reversible chiroptical function

Anton Kuzyk[1], Yangyang Yang[2,3,†], Xiaoyang Duan[1,4], Simon Stoll[1], Alexander O. Govorov[5], Hiroshi Sugiyama[2,3], Masayuki Endo[2] & Na Liu[1,4]

Nature has developed striking light-powered proteins such as bacteriorhodopsin, which can convert light energy into conformational changes for biological functions. Such natural machines are a great source of inspiration for creation of their synthetic analogues. However, synthetic molecular machines typically operate at the nanometre scale or below. Translating controlled operation of individual molecular machines to a larger dimension, for example, to 10–100 nm, which features many practical applications, is highly important but remains challenging. Here we demonstrate a light-driven plasmonic nanosystem that can amplify the molecular motion of azobenzene through the host nanostructure and consequently translate it into reversible chiroptical function with large amplitude modulation. Light is exploited as both energy source and information probe. Our plasmonic nanosystem bears unique features of optical addressability, reversibility and modulability, which are crucial for developing all-optical molecular devices with desired functionalities.

[1] Max Planck Institute for Intelligent Systems, Heisenbergstrasse 3, D-70569 Stuttgart, Germany. [2] Institute for Integrated Cell-Material Sciences (WPI-iCeMS), Kyoto University, Yoshida-ushinomiyacho, Sakyo-ku, Kyoto 606-8501, Japan. [3] Department of Chemistry, Graduate School of Science, Kyoto University, Kitashirakawa-oiwakecho, Sakyo-ku, Kyoto 606-8502, Japan. [4] Kirchhoff Institute for Physics, University of Heidelberg, Im Neuenheimer Feld 227, D-69120 Heidelberg, Germany. [5] Department of Physics and Astronomy, Ohio University, Athens, Ohio 45701, USA. † Present address: Shanghai Key Laboratory of Chemical Biology, School of Pharmacy, East China University of Science and Technology, 130 Meilong Road, Shanghai 200237, China. Correspondence and requests for materials should be addressed to A.K. (email: kuzyk@is.mpg.de) or to N.L. (email: laura.liu@is.mpg.de).

When designing active nanoscale devices, three prerequisites are of paramount importance. First, an efficient energy source for triggering conformation changes at the nanoscale is crucial[1]. Equally important is the reversible control over conformation of individual nanostructures. Last but not least is the ability to report such nanoscale conformation changes and translate them into tunable functionalities[2,3]. Among a variety of energy sources, light represents a unique stimulus to power an operation[4]. Different from chemical fuels that unavoidably introduce contaminants in a system, light is clean and waste-free. Also, in contrast to chemical fuels that crucially depend on diffusion kinetics, light offers high spatial and temporal resolution as it can be switched on and off rapidly. Most importantly, light can deliver noninvasive read-out of an optically active system, thus allowing for monitoring an operation in real time.

Here we demonstrate an all-optically controlled plasmonic nanosystem in the visible range using DNA nanotechnology. Our system can amplify the sub-nanometre conformation changes of azobenzene through the active host nanostructure[5,6] and consequently translate the light-induced molecular motion of azobenzene into reversible plasmonic chiroptical response, which can be in situ read out by optical spectroscopy. The plasmonic nanostructure comprises two gold nanorods (AuNRs) assembled on a reconfigurable DNA origami template[7–14]. A photo-responsive active site is introduced on the template with an azobenzene-modified DNA segment[15]. Light can cyclically 'write' and 'erase' the conformation states of the nanostructure through photoisomerization of azobenzene at a localized region. Different conformation states are read by probe light.

Results

Design of the photoresponsive nanostructures.
Photoisomerization of azobenzene[16] (Fig. 1) is widely used for the construction of light-driven artificial molecular machines. In particular, azobenzene can be incorporated into DNA strands for reversible control of DNA hybridization[15,17–19] (Fig. 1b). Our active nanostructure is based on a three-dimensional (3D) reconfigurable DNA origami template (Fig. 1c), which consists of two 14-helix bundles (80 nm × 16 nm × 8 nm), folded from a long single-stranded DNA scaffold, with the help of hundreds of staple strands (Supplementary Methods, Supplementary Tables 1 and 2 and Supplementary Figs 1–5). The two linked origami bundles form a chiral object with a tunable angle[14] (Supplementary Fig. 2). The active function of the structure is enabled by introducing an azobenzene-modified DNA segment on the template, which works as a recognition site to receive light stimuli. This photoresponsive segment comprises two DNA branches, which are extended from the two origami bundles, respectively. One branch possesses a double-stranded DNA (dsDNA) 20-base-pair part linked by disulfide bonds with azobenzene-modified oligonucleotides (Azo-ODN 1)[18,19]. The other branch contains Azo-ODN 2, which is pseudo-complementary to Azo-ODN 1. Azo-ODN 1 and Azo-ODN 2 contain three and four azobenzene modifications, respectively (Supplementary Methods and Supplementary Fig. 3). Multiple azobenzene modifications are essential for efficient photoregulation of DNA hybridization[17,20]. On ultraviolet light illumination, the azobenzene molecules in Azo-ODNs are converted to cis-form, resulting in dehybridization of the Azo-ODN duplex. The photoresponsive segment is opened and the conformation of the origami nanostructure is therefore 'relaxed'. In contrast, on visible light illumination, the azobenzene molecules are converted to trans-form and Azo-ODNs can be hybridized into the Azo-ODN duplex.

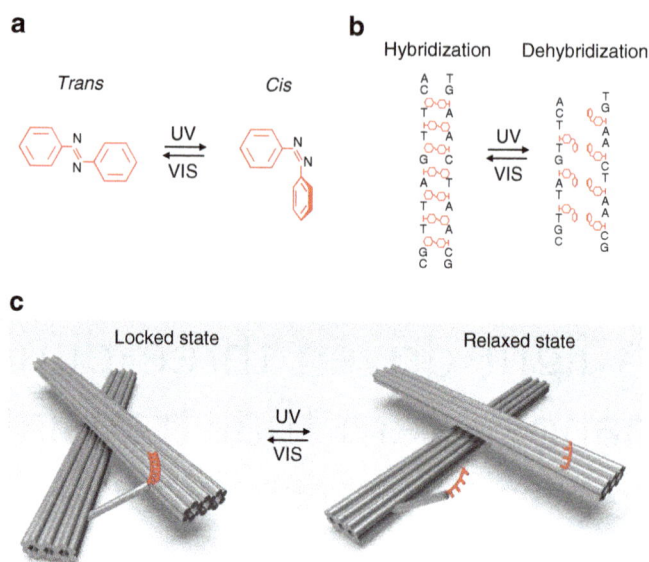

Figure 1 | Light-induced conformation changes of DNA origami nanostructures. (**a**) Trans–cis photoisomerization of an azobenzene molecule by ultraviolet (UV) and visible (VIS) light illumination. (**b**) Hybridization and dehybridization of azobenzene-modified DNA oligonucleotides controlled by trans–cis photoisomerization of azobenzene through UV and VIS light illumination. (**c**) Photoregulation of the DNA origami template between the locked and relaxed states by UV and VIS light illumination. The active function of the origami structure is enabled by introducing the azobenzene-modified DNA segment (red) in **b** on the template, which works as a recognition site to receive light stimuli for triggering light-induced motion.

Therefore, the photoresponsive segment is locked. The dsDNA part is employed here to define a rigid angle between the two origami bundles for a stable chiral conformation.

Photoregulation of the DNA origami templates.
It has been reported that the illumination time and temperature affect the hybridization and dehybridization kinetics of the Azo-ODN duplex[18,20]. To ensure good switching efficiency and simultaneously avoid origami damage, the sample was kept at a temperature of 40 °C during all switching experiments[18]. As a representative case, the locked state of the origami template was with a right-handed conformation, in which the angle between the two bundles was ∼50°. The sample was first illuminated by ultraviolet light (365 nm) for 15 min and then by visible light (450 nm) for 10 min. Transmission electron microscopy (TEM) images of the sample after ultraviolet and visible light illumination are shown in Fig. 2a,c, respectively (for additional TEM images see Supplementary Figs 5–11. Statistic histograms of the acute angle between two linked origami bundles based on an assessment of ∼400 origami structures after ultraviolet and visible light illumination are presented in Fig. 2b,d, respectively, and Supplementary Fig. 12. As shown in Fig. 2b, after ultraviolet light illumination, a broad distribution over angles is observed, with a maximum magnitude occurring around 90°. This reveals that the origami structures have been turned into the relaxed state by ultraviolet light. Here 90° is more favourable owing to the electrostatic repulsion between the two bundles within one origami structure. On the other hand, after visible light illumination, a maximum magnitude over angles occurs around 50°, which is in accordance with our structure design (Supplementary Fig. 2). This elucidates that a majority of the

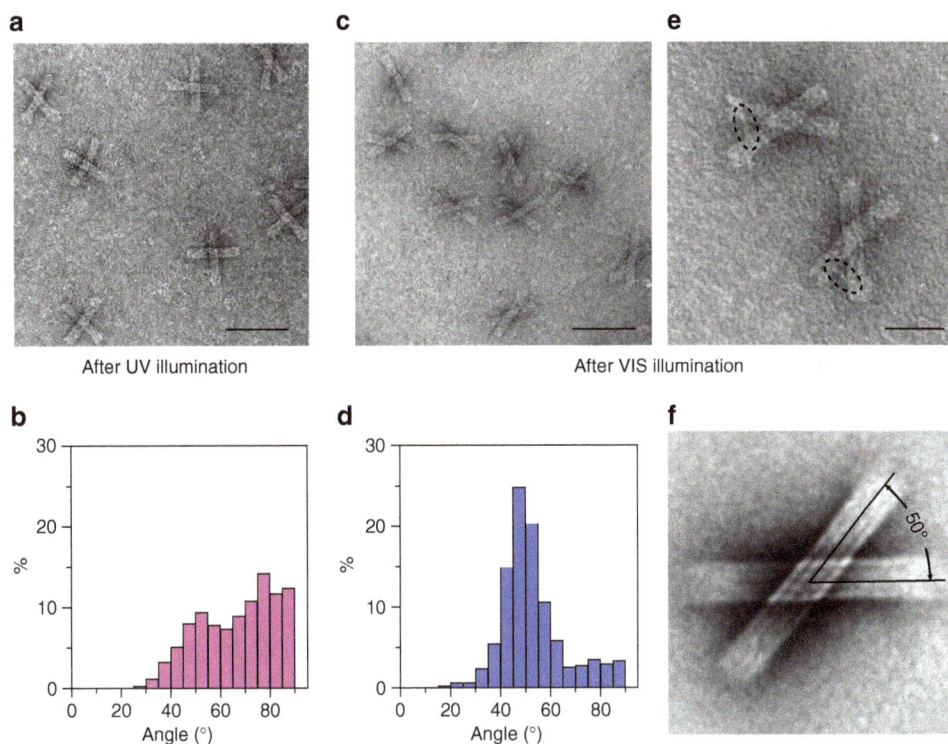

Figure 2 | Structural characterization of the DNA origami nanostructures. (**a**) TEM image of the DNA origami nanostructures after ultraviolet (UV) light illumination. (**b**) Statistic histogram of the acute angle between two linked origami bundles after UV light illumination. The number of the analysed structures: 463. A broad distribution over angles is observed. (**c**) TEM image of the DNA origami nanostructures after visible (VIS) light illumination. The locked state is designed to be right-handed. (**d**) Statistic histogram of the acute angle between two linked origami bundles after VIS light illumination. The number of the analysed structures: 541. A maximum magnitude over angles occurs around 50°, which is in accordance to our structure design. (**e**) Enlarged view of the origami structures in the locked state. The dsDNA branch, which links the two origami bundles to define the angle, is clearly visible. (**f**) Averaged TEM image reconstructed from locked origami structures. It evidently demonstrates the excellent structural homogeneity and high angle accuracy within the locked structures. Scale bars, 100 nm (**a,c**); 50 nm (**e**).

origami structures have been driven by visible light to the designated locked state. An enlarged view of the origami structures in the locked state is shown in Fig. 2e. The dsDNA branch, which links the two origami bundles to define the angle, is clearly visible in the individual structures in Fig. 2e. An averaged TEM image reconstructed from the perfectly locked origami structures (∼ 120) is presented in Fig. 2f. It demonstrates the excellent structural homogeneity and high angle accuracy within the locked structures. The TEM characterization reveals that *trans–cis* photoisomerization of azobenzene, which is associated with a molecular length change of ∼ 3.5 Å (ref. 21) can be efficiently amplified by the origami structures into their distinct conformation changes on the order of 30 nm (Supplementary Fig. 2). This corresponds to an amplification factor of ∼ 100. Certainly, this amplification factor can be further increased by designing larger origami frames or larger angle changes. Moreover, given the remarkable precision of addressability afforded by DNA, this translation can be well controlled in an individual nanostructure at a localized region, which serves as an active recognition site in response to light stimuli.

Light-driven 3D plasmonic nanosystem. Positioning of plasmonic nanoparticles with high precision offered by DNA[22–30] further endows our light-driven systems with unique optical functionalities. To this end, two AuNRs are assembled on one origami template to form a 3D plasmonic chiral nanostructure (Fig. 3a). Twelve binding sites on each origami bundle are

extended with capture strands for robust assembly of one AuNR (38 nm × 10 nm) functionalized with DNA complementary to the capture strands. The length of the binding site area is ∼ 36 nm (Supplementary Fig. 2). To ensure a high positioning accuracy of the AuNRs on origami, an additional thermal annealing procedure was carried out. Detailed information on the AuNR functionalization and assembly can be found in Supplementary Methods and Supplementary Fig. 13.

When light interacts with the 3D chiral nanostructure, plasmons are excited in the two AuNRs that are placed in close proximity. The excited plasmons are collectively coupled in the cross conformation, leading to plasmonic chiroptical response[31–35]. The resulting plasmonic circular dichroism (CD)[36] spectra are very sensitive on conformation changes, ideal for optically monitoring the conformation evolution in real time[14].

Figure 3b and Supplementary Fig. 14 show TEM images of the plasmonic nanostructures. A high assembly yield of the AuNR dimers on the origami templates has been achieved. Owing to a higher affinity of the AuNRs to the carbon film of the TEM grid compared with that of DNA, the AuNR pairs appear side by side in the TEM images. To *in situ* monitor the dynamic process associated with the conformation changes triggered by light, the CD response of a plasmonic sample was measured during visible (450 nm) and ultraviolet (365 nm) light illumination, respectively, using a J-815 CD spectrometer (Jasco). To be more specific, visible and ultraviolet light is used to 'write' and 'erase' the handed state of the plasmonic system, respectively, while circularly polarized light is used to 'read' the state changes. For a better elucidation, two representative CD spectra recorded after

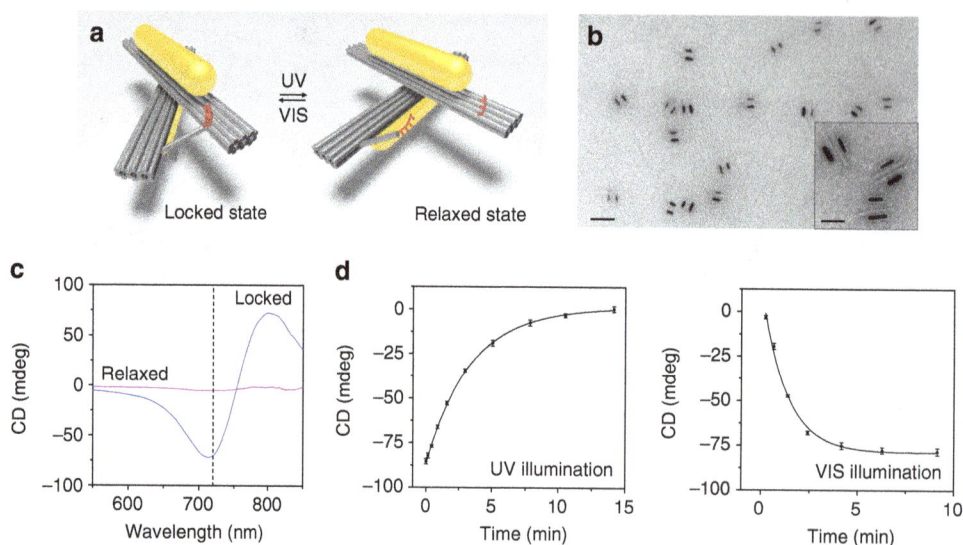

Figure 3 | Light-driven 3D plasmonic nanosystem. (**a**) Schematic of the 3D plasmonic nanosystem regulated by ultraviolet (UV) and visible (VIS) light illumination for switching between the locked right-handed and relaxed states. Two AuNRs are assembled on one origami template to form a 3D plasmonic chiral nanostructure. (**b**) TEM images of the plasmonic nanostructures in the locked right-handed state. Scale bars, 200 and 50 nm in the large image and in the inset image, respectively. (**c**) Measured CD spectra after UV (purple) and VIS (blue) illumination. (**d**) Kinetic characterization of the 3D plasmonic nanostructures switching from the locked right-handed state to the relaxed state and vice versa on UV and VIS illumination. The experimental data can be well fit by first-order reaction kinetics with rate constants of 5×10^{-3} and $1.3 \times 10^{-2} \, s^{-1}$ for UV and VIS illumination, respectively. The error bars represent one s.d. from the mean.

the system has achieved stable states for visible and ultraviolet light illumination are presented in Fig. 3c. The spectra were recorded within a wavelength range of 550–850 nm. The CD spectrum after visible light illumination is characterized by a bisignate dip-to-peak profile (in blue), which is typical for a right-handed system. This demonstrates that visible light has successfully driven the plasmonic system to the locked state, in which the conformation is 'written' as right-handed. On the other hand, the CD response after ultraviolet light illumination decreases significantly as shown by the purple curve. The plasmonic system has been converted into the relaxed state and the previous right-handed conformation is therefore 'erased'. A CD intensity modulation as high as 10 times between the two states has been achieved, demonstrating excellent photoresponsivity of the active plasmonic system. In this regard, molecular motion of azobenzene is spatially amplified by the host nanostructures and optically reflected through distinct chiroptical response changes. For the additional details of optical characterization, see Supplementary Methods and Supplementary Figs 15 and 16.

To provide deeper insight, theoretical calculations were performed using the commercial software COMSOL Multiphysics based on a finite element method (Supplementary Methods), and the results are shown in Supplementary Fig. 17. The CD spectra were calculated as the difference of extinction for the left- and right-handed circularly polarized light. The assembled nanostructures were randomly dispersed in solution, and therefore averaging over different orientations was carried out. To account for the inhomogeneous spectral broadening resulting from the polydispersity of the AuNRs, the dielectric function of Au was modified by including an extra damping coefficient. Overall, the agreement between the experiment and theory is good.

Next, the CD intensities at 720 nm, that is, approximately at the spectral dip position, are presented as a function of visible and ultraviolet light illumination time in Fig. 3d. The conversion from the right-handed state on ultraviolet light illumination took ~15 min to achieve the stable relaxed state, whereas the conversion from the relaxed state on visible light illumination

took ~10 min to reach the stable right-handed state. The data curves can be well fit by first-order reaction kinetics with rate constants of 5×10^{-3} and $1.3 \times 10^{-2} \, s^{-1}$ for ultraviolet and visible illumination, respectively. Also, the reversibility of the conversion is examined by alternative ultraviolet and visible light illumination in cycles for 15 and 10 min per exposure, respectively (Fig. 4a and Supplementary Fig. 16). The CD intensity was recorded at 720 nm. As shown in Fig. 4b, excellent reversibility of the chiroptical response is achieved between the two states with large signal modulations. In brief, 'writing', 'erasing' and 'reading' actions can be coordinated efficiently with such a bistable system, in which each state can be converted into the other by light and consequently be reported by light.

Discussion

The realization of light-driven plasmonic systems based on DNA nanotechnology offers many advantages for effectively manipulating materials and information at the nanoscale. From the material aspect, DNA as one of the most flexible materials in nanotechnology possesses unique biochemical specificity, remarkable spatial accuracy and ease of addressability[37–39]. Inclusion of chemical species that can execute reversible transformations by light such as azobenzene endows such hybrid systems with both spatial and temporal precision. From the information aspect, light as a stimulus renders 'writing' and 'erasing' of the conformation states in a reversible way possible without addition of any reagent. Meanwhile, light also serves as an information probe to read dynamic state changes in real time. Our plasmonic system may launch a new generation of sensing platforms[40,41], as light-induced structural changes could be optically tracked and successively retrieved. Finally, by harvesting light energy, individual nanostructures may generate collective actions and directly translate controlled molecular motion to a macroscopic level[42].

Methods

Materials. DNA scaffold strands (p7650) were purchased from Tilibit Nano-systems. Unmodified staple strands (purification: desalting) were purchased from

a

b

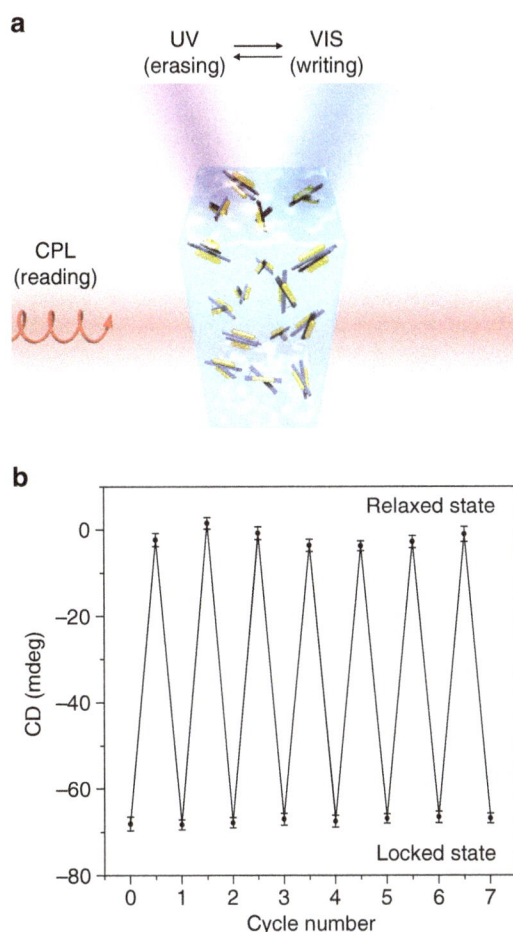

Figure 4 | 'Writing', 'erasing' and 'reading' of the 3D plasmonic nanostructures by light. (**a**) Reversible conversion of the plasmonic nanostructures between the relaxed and locked right-handed states by ultraviolet (UV) and visible (VIS) light illumination, which performs the 'erasing' and 'writing' behaviour, respectively. The resulting conformation states are probed by circularly polarized light (CPL) in real time, which performs the 'reading' behaviour. (**b**) CD intensity recorded at 720 nm (Fig. 3c) during alternative UV and VIS illumination in multiple cycles. Excellent reversibility of the chiroptical response is achieved between the two states with large signal modulations. The error bars represent one s.d. from the mean.

Eurofins MWG. Capture strands for the AuNRs (purification: desalting) were purchased from Sigma-Aldrich. Thiol-modified strands (purification: HPLC) were purchased from http://www.biomers.net/. Azobenzene-modified DNA strands were obtained following a previously published procedure[18]. Agarose for electrophoresis and SYBR Gold nucleic acid stain were purchased from Life Technologies. Uranyl formate for negative TEM staining was purchased from Polysciences, Inc. AuNRs were purchased from Sigma-Aldrich (catalogue no. 716812). Other chemicals were purchased either from Carl-Roth or from Sigma-Aldrich.

Design and preparation of the DNA origami templates. The design of the DNA origami structures was adopted from a previous study[14]. The strand routing diagram of the origami structures can be found in Supplementary Fig. 1. The sequences of the staple strands and modifications used for photoswitching are provided in Supplementary Table 1. The origami structures were prepared by thermal annealing (Supplementary Table 2) and purified by agarose gel electrophoresis (Supplementary Fig. 4) For details, see Supplementary Methods.

Light-driven conformational switching. For ultraviolet illumination, a 3-W light-emitting diode (Köhler Technologie-Systeme GmbH & Co. KG) with emission wavelength centred at 365 nm (± 5 nm) was used. For visible light illumination, a white light-emitting diode (M7RX, LED LENSER) and a bandpass optical filter centred at 450 nm with a 40 nm bandpass region (FB450-40, Thorlabs)

were used. For TEM characterizations, the DNA origami sample was incubated at 40 °C and pH 8. First, the origami sample was exposed to ultraviolet light illumination for 15 min, and part of the sample was used for TEM investigations. The rest of the sample was exposed to visible light illumination for 10 min and then used for subsequent TEM investigations.

TEM characterization. The DNA origami structures (with or without AuNRs) were imaged using a Philips CM 200 TEM operating at 200 kV. For imaging, the DNA origami structures (with or without AuNRs) were deposited on freshly glow-discharged carbon/formvar TEM grids. The TEM grids were treated with a uranyl formate solution (0.75%) for negative staining of the DNA structures. Angles between the origami bundles in individual structures were obtained by manual analysis of the TEM images. The acute angles were chosen for analysis. Class average images (Fig. 2f) were obtained using EMAN2 software[43].

Optical characterizations. CD and ultraviolet–visible measurements were performed with a J-815 Circular Dichroism Spectrometer (Jasco) using Quartz SUPRASIL cuvettes (105.203-QS, Hellma Analytics) with a path length of 10 mm. ultraviolet–visible measurements were also performed with a BioSpectrometer (Eppendorf). Technical details can be found in Supplementary Methods.

References

1. Ballardini, R., Balzani, V., Credi, A., Gandolfi, M. T. & Venturi, M. Artificial molecular-level machines: which energy to make them work? *Acc. Chem. Res.* **34,** 445–455 (2001).
2. Browne, W. R. & Feringa, B. L. Making molecular machines work. *Nat. Nanotechnol.* **1,** 25–35 (2006).
3. Coskun, A., Banaszak, M., Astumian, R. D., Stoddart, J. F. & Grzybowski, B. A. Great expectations: can artificial molecular machines deliver on their promise? *Chem. Soc. Rev.* **41,** 19–30 (2011).
4. Balzani, V., Credi, A. & Venturi, M. Light powered molecular machines. *Chem. Soc. Rev.* **38,** 1542–1550 (2009).
5. Muraoka, T., Kinbara, K. & Aida, T. Mechanical twisting of a guest by a photoresponsive host. *Nature* **440,** 512–515 (2006).
6. Hoersch, D., Roh, S.-H., Chiu, W. & Kortemme, T. Reprogramming an ATP-driven protein machine into a light-gated nanocage. *Nat. Nanotechnol.* **8,** 928–932 (2013).
7. Rothemund, P. W. K. Folding DNA to create nanoscale shapes and patterns. *Nature* **440,** 297–302 (2006).
8. Andersen, E. S. et al. Self-assembly of a nanoscale DNA box with a controllable lid. *Nature* **459,** 73–76 (2009).
9. Kuzuya, A., Sakai, Y., Yamazaki, T., Xu, Y. & Komiyama, M. Nanomechanical DNA origami 'single-molecule beacons' directly imaged by atomic force microscopy. *Nat. Commun.* **2,** 449 (2011).
10. Gerling, T., Wagenbauer, K. F., Neuner, A. M. & Dietz, H. Dynamic DNA devices and assemblies formed by shape-complementary, non-base pairing 3D components. *Science* **347,** 1446–1452 (2015).
11. Marras, A. E., Zhou, L., Su, H.-J. & Castro, C. E. Programmable motion of DNA origami mechanisms. *Proc. Natl Acad. Sci. USA* **112,** 713–718 (2015).
12. Han, D., Pal, S., Liu, Y. & Yan, H. Folding and cutting DNA into reconfigurable topological nanostructures. *Nat. Nanotechnol.* **5,** 712–717 (2010).
13. Douglas, S. M., Bachelet, I. & Church, G. M. A logic-gated nanorobot for targeted transport of molecular payloads. *Science* **335,** 831–834 (2012).
14. Kuzyk, A. et al. Reconfigurable 3D plasmonic metamolecules. *Nat. Mater.* **13,** 862–866 (2014).
15. Asanuma, H. et al. Synthesis of azobenzene-tethered DNA for reversible photo-regulation of DNA functions: hybridization and transcription. *Nat. Protoc.* **2,** 203–212 (2007).
16. Kumar, A. S. et al. Reversible photo-switching of single azobenzene molecules in controlled nanoscale environments. *Nano Lett.* **8,** 1644–1648 (2008).
17. Kamiya, Y. & Asanuma, H. Light-driven DNA nanomachine with a photoresponsive molecular engine. *Acc. Chem. Res.* **47,** 1663–1672 (2014).
18. Endo, M., Yang, Y., Suzuki, Y., Hidaka, K. & Sugiyama, H. Single-molecule visualization of the hybridization and dissociation of photoresponsive oligonucleotides and their reversible switching behavior in a DNA nanostructure. *Angew. Chem. Int. Ed.* **51,** 10518–10522 (2012).
19. Yang, Y., Endo, M., Hidaka, K. & Sugiyama, H. Photo-controllable DNA origami nanostructures assembling into predesigned multiorientational patterns. *J. Am. Chem. Soc.* **134,** 20645–20653 (2012).
20. Liang, X., Mochizuki, T. & Asanuma, H. A supra-photoswitch involving sandwiched DNA base pairs and azobenzenes for light-driven nanostructures and nanodevices. *Small* **5,** 1761–1768 (2009).
21. Kumar, G. S. & Neckers, D. C. Photochemistry of azobenzene-containing polymers. *Chem. Rev.* **89,** 1915–1925 (1989).
22. Tan, S. J., Campolongo, M. J., Luo, D. & Cheng, W. Building plasmonic nanostructures with DNA. *Nat. Nanotechnol.* **6,** 268–276 (2011).

23. Tørring, T., Voigt, N. V., Nangreave, J., Yan, H. & Gothelf, K. V. DNA origami: a quantum leap for self-assembly of complex structures. *Chem. Soc. Rev.* **40**, 5636–5646 (2011).

24. Kuzyk, A. *et al.* DNA-based self-assembly of chiral plasmonic nanostructures with tailored optical response. *Nature* **483**, 311–314 (2012).

25. Acuna, G. P. *et al.* Fluorescence enhancement at docking sites of DNA-directed self-assembled nanoantennas. *Science* **338**, 506–510 (2012).

26. Samanta, A., Banerjee, S. & Liu, Y. DNA nanotechnology for nanophotonic applications. *Nanoscale* **7**, 2210–2220 (2015).

27. Elbaz, J., Cecconello, A., Fan, Z., Govorov, A. O. & Willner, I. Powering the programmed nanostructure and function of gold nanoparticles with catenated DNA machines. *Nat. Commun.* **4**, 2000 (2013).

28. Li, Y., Liu, Z., Yu, G., Jiang, W. & Mao, C. Self-assembly of molecule-like nanoparticle clusters directed by DNA nanocages. *J. Am. Chem. Soc.* **137**, 4320–4323 (2015).

29. Tian, Y. *et al.* Prescribed nanoparticle cluster architectures and low-dimensional arrays built using octahedral DNA origami frames. *Nat. Nanotechnol.* **10**, 637–644 (2015).

30. Schreiber, R. *et al.* Hierarchical assembly of metal nanoparticles, quantum dots and organic dyes using DNA origami scaffolds. *Nat. Nanotechnol.* **9**, 74–78 (2014).

31. Fan, Z. & Govorov, A. O. Plasmonic circular dichroism of chiral metal nanoparticle assemblies. *Nano Lett.* **10**, 2580–2587 (2010).

32. Auguié, B., Alonso-Gómez, J. L., Guerrero-Martínez, A. & Liz-Marzán, L. M. Fingers crossed: optical activity of a chiral dimer of plasmonic nanorods. *J. Phys. Chem. Lett.* **2**, 846–851 (2011).

33. Zhang, S. *et al.* Photoinduced handedness switching in terahertz chiral metamolecules. *Nat. Commun.* **3**, 942 (2012).

34. Ma, W. *et al.* Chiral plasmonics of self-assembled nanorod dimers. *Sci. Rep.* **3**, 1934 (2013).

35. Shen, X. *et al.* 3D plasmonic chiral colloids. *Nanoscale* **6**, 2077–2081 (2014).

36. Guerrero-Martínez, A. *et al.* From individual to collective chirality in metal nanoparticles. *Nanotoday* **6**, 381–400 (2011).

37. Seeman, N. C. DNA in a material world. *Nature* **421**, 427–431 (2003).

38. Krishnan, Y. & Simmel, F. C. Nucleic acid based molecular devices. *Angew. Chem. Int. Ed.* **50**, 3124–3156 (2011).

39. Jones, M. R., Seeman, N. C. & Mirkin, C. A. Programmable materials and the nature of the DNA bond. *Science* **347**, 1260901 (2015).

40. Lal, S., Link, S. & Halas, N. J. Nano-optics from sensing to waveguiding. *Nat. Photon.* **1**, 641–648 (2007).

41. Joshi, G. K. *et al.* Ultrasensitive photoreversible molecular sensors of azobenzene-functionalized plasmonic nanoantennas. *Nano Lett.* **14**, 532–540 (2014).

42. Iamsaard, S. *et al.* Conversion of light into macroscopic helical motion. *Nat. Chem.* **6**, 229–235 (2014).

43. Tang, G. *et al.* EMAN2: an extensible image processing suite for electron microscopy. *J. Struct. Biol.* **157**, 38–46 (2007).

Acknowledgements

We thank A. Jeltsch and R. Jurkowska for assistance with CD spectrometry. We thank M. Kelsch for assistance with TEM. TEM images were collected at the Stuttgart Center for Electron Microscopy (StEM). N.L. was supported by the Sofja Kovalevskaja Award from the Alexander von Humboldt Foundation. A.K. was supported by a postdoctoral fellowship from the Alexander von Humboldt Foundation. A.K. and N.L. were supported by a Marie Curie CIG Fellowship. We also thank for the financial support from the European Research Council (ERC) Starting Grant 'Dynamic Nano'. A.O.G. was supported by the U.S. Army Research Office under grant number W911NF-12-1-0407 and by Volkswagen Foundation (Germany). M.E. was supported by JSPS KAKENHI (grant numbers 15H03837, 24104002 and 26620133).

Author contributions

A.K. and N.L. conceived the experiments; A.K. designed the DNA origami nanostructures; A.K., Y.Y., S.S., H.S. and M.E. prepared the nanostructures; A.K. and S.S. performed TEM and CD characterization; X.D. carried out the theoretical calculations; A.O.G. offered useful suggestions; A.K. and N.L. wrote the manuscript; all authors discussed the results, analysed the data and commented on the manuscript.

Additional information

Selectively enhanced photocurrent generation in twisted bilayer graphene with van Hove singularity

Jianbo Yin[1,*], Huan Wang[1,*], Han Peng[2,*], Zhenjun Tan[1,3], Lei Liao[1], Li Lin[1], Xiao Sun[1,3], Ai Leen Koh[4], Yulin Chen[2], Hailin Peng[1] & Zhongfan Liu[1]

Graphene with ultra-high carrier mobility and ultra-short photoresponse time has shown remarkable potential in ultrafast photodetection. However, the broad and weak optical absorption ($\sim$2.3%) of monolayer graphene hinders its practical application in photodetectors with high responsivity and selectivity. Here we demonstrate that twisted bilayer graphene, a stack of two graphene monolayers with an interlayer twist angle, exhibits a strong light–matter interaction and selectively enhanced photocurrent generation. Such enhancement is attributed to the emergence of unique twist-angle-dependent van Hove singularities, which are directly revealed by spatially resolved angle-resolved photoemission spectroscopy. When the energy interval between the van Hove singularities of the conduction and valance bands matches the energy of incident photons, the photocurrent generated can be significantly enhanced (up to $\sim$80 times with the integration of plasmonic structures in our devices). These results provide valuable insight for designing graphene photodetectors with enhanced sensitivity for variable wavelength.

[1] Center for Nanochemistry, Beijing Science and Engineering Center for Nanocarbons, Beijing National Laboratory for Molecular Sciences, College of Chemistry and Molecular Engineering, Peking University, 202 Chengfu Road, Haidian District, Beijing 100871, China. [2] Clarendon Laboratory, Department of Physics, University of Oxford, Parks Road, Oxford OX1 3PU, UK. [3] Academy for Advanced Interdisciplinary Studies, Peking University, Beijing 100871, China. [4] Stanford Nano Shared Facilities, Stanford University, Stanford, California 94305, USA. * These authors contributed equally to this work. Correspondence and requests for materials should be addressed to Z.L. (email: zfliu@pku.edu.cn) or to H.P. (email: hlpeng@pku.edu.cn) or to Y.C. (email: yulin.chen@physics.ox.ac.uk).

The unique Dirac-cone band structure makes graphene a promising material for photodetection. Its linearly dispersive band structure near Fermi level results in massless Dirac fermion type of carriers, large Fermi velocity ($\sim 1/300$ of the speed of light) and surprisingly high carrier mobility[1–4]. In graphene device, the photovoltage generation time is shorter than 50 fs, which is associated with the carrier heating time[5]. In addition, the rapid cooling process of photoexcited carriers ($\sim$ picoseconds) in the monolayer graphene results in a quick annihilation of photoelectrical signal in the electric circuit[5–12]. These advantages of the monolayer graphene facilitate its applications associated with ultrafast photodetection, such as high-speed optical communications[13–17] and terahertz oscillators[18]. However, it remains a great challenge to achieve high photoresponsivity and selectivity in the monolayer-graphene-based detectors due to the weak and broadband absorption (only 2.3%, from the ultraviolet to the infrared)[19] and the short photocarrier cooling time ($\sim$ picoseconds)[5–12].

On the other hand, twisted bilayer graphene (tBLG) is non-AB stacked bilayer graphene in which one graphene monolayer sheet rotates by a certain angle (θ) relative to the other (Fig. 1a). Recent theoretical studies of tBLG have shown that the Dirac band dispersions change dramatically and become strongly warped with small twist angles ($\theta \le 5°$)[20–23]. Even at relatively large twist angles, the electronic coupling between the two monolayers, albeit weak, can still introduce new band structures[20–27]. Unlike the parabolic band structure in AB-stacked bilayer graphene[28–34], the band structure of tBLG with large twist angle (typically larger than 5°)[20,35] maintains linear near the Dirac point and thus it inherits some unique properties of monolayer graphene[36]. Away from the Dirac point, Dirac cones of the two individual

monolayers intersect and form saddle points in reciprocal space of tBLG[24], leading to the formation of van Hove singularities (VHSs) in the density of state (DOS)[25,26,35,37,38], which then gives rise to some interesting phenomena such as enhanced optical absorption, Raman G-band resonance and enhanced chemical reactivity of tBLG[27,37,39–45].

In this study, to address the problem of low photoresponsivity and selectivity in the monolayer graphene photodetection, we explore the high-performance photodetector based on tBLG with VHSs. For the first time, we report that the VHSs in tBLG leads to a prominent photocurrent enhancement of tBLG photodetectors with a wavelength selectivity under incident light irradiation.

Results

Structure and Raman spectra. tBLG samples were grown on copper foil via chemical vapour deposition (CVD) method and then transferred to heavily doped Si substrate, which was capped with 90 nm SiO₂. As shown in typical optical image and scanning electron microscopy images (Fig. 1b,c), both the overlayer and underlayer in tBLG exhibit hexagonal shapes with sharp edges, which implies highly crystalline qualities of tBLG domains[46–48]. The interlayer twist angle can be measured from the relative misalignment of the straight edges, which is consistent with the observation by transmission electron microscopy (TEM) (Supplementary Fig. 1 and Supplementary Note 1). tBLG domains with different twist angles can be readily obtained in our samples (Fig. 1d), which provide a platform for the study of θ-dependent light–matter interactions. The highly crystalline quality and clean interface between two monolayers of our CVD sample are evidenced by the moiré pattern in high-resolution

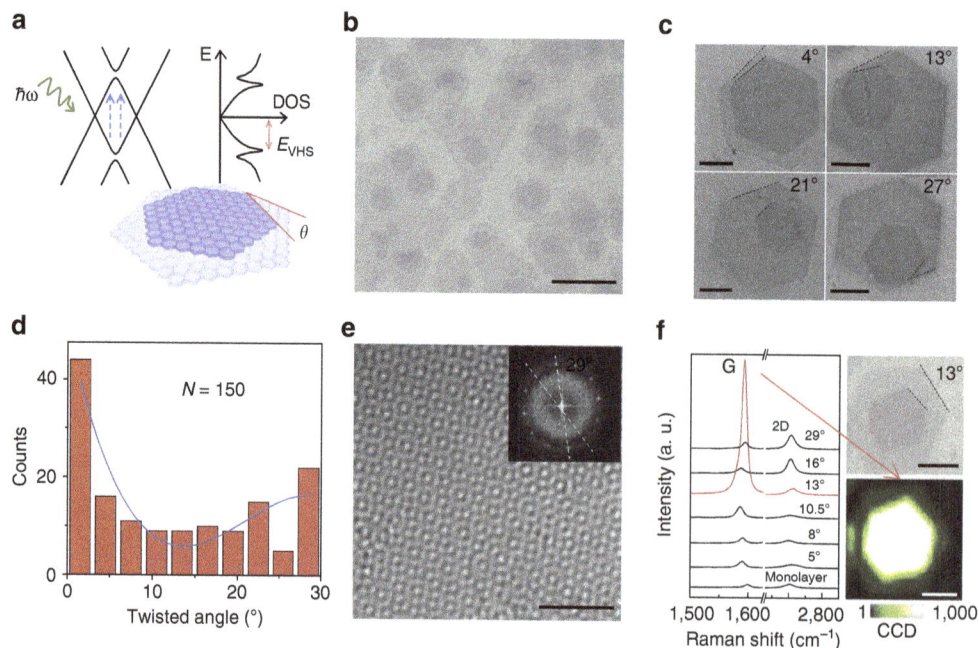

Figure 1 | Structures and Raman spectra of tBLG with different twist angles. (a) Schematics for band structure with minigaps (top left) and the corresponding DOS with VHSs (top right) in tBLG (bottom). Blue arrows describe the photoexcitation process as the energy interval of two VHSs ($2E_{VHS}$) matches the energy of incident photon. **(b)** The optical image of tBLG domains grown by CVD on Cu and then transferred onto SiO₂ (90 nm)/Si substrate. Scale bar, 30 µm. **(c)** Scanning electron microscopy (SEM) images of tBLG domains with different twist angles on SiO₂/Si. The twist angles are measured from the edges of over- and underlayer of tBLG domains. Scale bars, 5 µm. **(d)** Histogram of twist angles measured from tBLG domains in the CVD sample as shown in **b**. **(e)** Typical high-resolution TEM (HRTEM) image of tBLG. The periodicity of the moiré pattern is ~ 0.455 nm. The inset is the fast Fourier transform (FFT) of the image, showing that the twist angle is 29°. Scale bar, 2 µm. **(f)** Left column, Raman spectra of monolayer graphene and tBLG domains with twist angle of 5°, 8°, 10.5°, 13°, 16° and 29°, respectively. The incident laser wavelength is 532 nm (2.33 eV). Top right: the optical image of 13° tBLG domain on SiO₂/Si. Bottom right: G-band intensity mapping image of the 13° tBLG domain shows uniformity of the intensity enhancement of Raman G-band. Scale bars, 10 µm.

TEM image (Fig. 1e). This clean interface guarantees the interaction and coupling of electronic states from the over- and underlayer of tBLG. This interlayer electronic coupling is also proved by the enhanced G-band peak in Raman spectra (Fig. 1f). Taking 13° tBLG domain as an example, the Raman G-band intensity displays a tremendous enhancement of ~20 folds under 532 nm laser (2.33 eV), which is consistent with the previously reported results[27,28,39–43]. This Raman G-band enhancement implies that an interlayer coupling introduces new band structures in tBLG. In addition, the enhanced G-band intensity of 13° tBLG domain was found to be uniform across the whole domain as shown in the G-band mapping image (Fig. 1f), which further confirms the high quality of our CVD tBLG samples. The Raman G-band enhancement is believed to correlate with the formation of VHSs in tBLG[27,37,39–42,49].

Micro-ARPES spectra of tBLG. To unravel the nature of VHSs, we directly investigate the band structures of CVD-grown tBLG domains using spatially resolved angle-resolved photoemission spectroscopy with submicrometre spatial resolution (micro-ARPES). Owing to the twist angle (θ) between over- and underlayer of tBLG, the two sets of (six) Dirac points originated from each layer are rotated relatively by the angle θ as well (see Fig. 2a), which we mark as the **k** (left cones) and $\mathbf{k}_\theta$ (right cones) points, respectively. The band structures of a tBLG domain are shown in Fig. 2, where the constant energy contours (Fig. 2b), the band dispersions cutting across (Fig. 2c) and perpendicular (Fig. 2d) to the two adjacent Dirac points are presented, respectively. In Fig. 2b, the stacking plots of the band contours at different binding energies clearly depict the typical two Dirac-cone dispersions of tBLG and each preserves the linear dispersion of monolayer graphene. One of the Dirac cone exhibits a weaker intensity and higher electron doping level, indicating its origin from the underlayer graphene, as the photoelectrons from the bottom layer are screened by the top layer (thus leading to a weaker intensity), and being closer to the Cu substrate also increases its charge transfer[50,51].

By measuring the separation between the two Dirac points (Fig. 2b,c), we can determine the twist angle (θ) of this tBLG domain as 19.1° (Supplementary Fig. 2 and Supplementary Note 2). Without interlayer coupling, the two Dirac cones in Fig. 2 shall intersect and cross each other at higher binding energy. Instead, the band structure at Fig. 2b clearly shows fine structures at the intersection (indicated by red arrows in Fig. 2b) and the dispersion in Fig. 2c shows the opening of the gap at the crossing point of the dispersions from the two Dirac cones, which is indicated by the faint intensity in the spectra intensity map (left panel) and the dip in the DOS plot (right panel, indicated by red arrows). This gap opening in the band structure is a typical anticrossing behaviour introduced by interlayer electronic coupling[24], which leads to the formation of the VHS (Fig. 1a). In addition, from Fig. 2d, one can see that the anticrossing affects the hyperbolic curve as well and results in split and parallel dispersions.

With the same method, we further studied tBLG domains with various different twist angles and tracked the positions of VHSs with respect to the twist angles, as can be seen in Fig. 2e. At small angles, the value of E_{VHS} increases almost linearly with θ, in consistence with the theoretical prediction (Supplementary Fig. 2). This dependence also helps explain the Raman G-band enhancement at specific twist angle (Fig. 1f) for a given incident laser frequency. If the energy of incident photon matches the energy interval of the two VHSs of tBLG ($\hbar\omega \approx 2E_{\mathrm{VHS}}$, see Fig. 1a), the electrons are excited and transit between the fine band structure, causing the increase of the intensity of Raman G

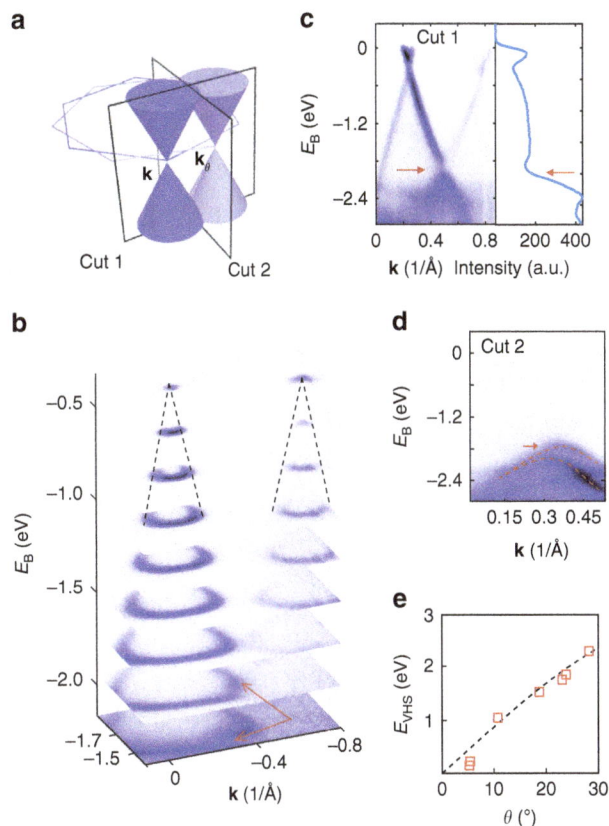

Figure 2 | Micro-ARPES spectra of tBLG. (**a**) Schematic illustration of the first primitive Brillouin zones (hexagons) and Dirac cones of over- and underlayer of tBLG. (**b**) Stacking plot of constant-energy contours at different binding energies (E_B) of tBLG. (**c**) ARPES spectra along Cut 1 as labelled in **a**. The right curve is energy spectrum density curve (EDC) integrated from the spectrum. (**d**) ARPES spectra along Cut 2 as labelled in **a**. Red arrows in **b**,**c** and **d** indicate the minigap band topology and the split parallel branches arising from interlayer coupling. (**e**) E_{VHS} versus the twist angle (θ) of tBLG domains. The E_{VHS}, measured from micro-ARPES data of tBLG, is the energy interval between the minigap (VHS) and Dirac point. The E_{VHS} varies almost linearly with twist angle. The black dashed line is a theoretical curve.

band peak (see Supplementary Fig. 3 and Supplementary Note 3 for details).

Selectively enhanced photocurrent generation of tBLG. The strong light–matter interaction of tBLG selectively enhanced by the VHSs can also enhance the generation of photocurrent under illumination. As an example, two adjacent tBLG domains with twist angle of 13° and 7° transferred onto SiO$_2$ (90 nm)/Si were etched into a strip and then embedded into two-terminal devices in parallel (Fig. 3a,b). Raman spectroscopy and two-dimensional maps of the two adjacent tBLG domains were first measured under the 532-nm laser (2.33 eV). As expected, the G-band intensity of whole 13° domain exhibits a uniformly 20-fold enhancement as compared with the 7° domain (see Fig. 3c), as the energy interval of the two VHSs in 13° domain matches the energy of incident photon ($\hbar\omega \approx 2E_{\mathrm{VHS}}$). To generate photocurrent selectively, interfacial junctions of tBLG-metal electrodes were used to separate the photoexcited electrons and holes under illumination[52,53]. As shown in current–bias voltage curves, both tBLG domains produce pronounced photocurrent

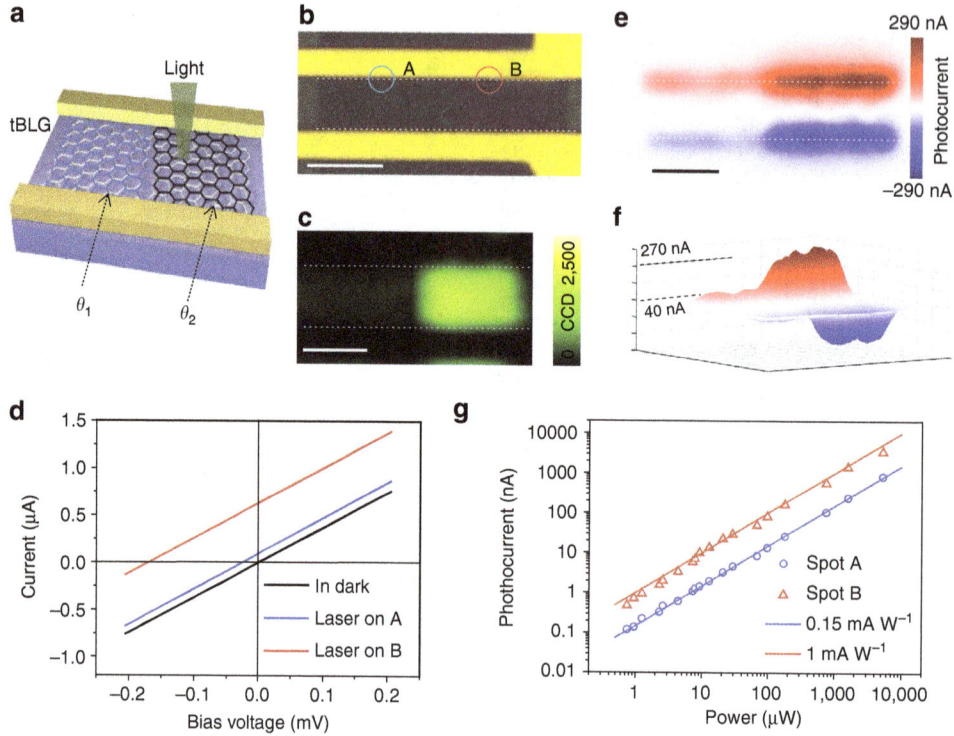

Figure 3 | Selectively enhanced photocurrent generation in tBLG photodetection devices. (**a**) Schematic illustration of a tBLG photodetection device. The channel comprises of two adjacent tBLG domains with different twist angles of θ_1 and θ_2, respectively. (**b**) Optical image of the tBLG photodetection device. The θ_1 and θ_2 are 7° and 13°, respectively. (**c**) Raman G-band intensity mapping image under 532 nm (2.33 eV) laser. 13° tBLG domain exhibits an enhanced G-band intensity. (**d**) Current versus source-drain bias (*I*–*V*) curve without laser on and with laser focusing on 7° (spot A) and 13° (spot B) tBLG domains, respectively. The intercepts at current axis represent the net photocurrents. (**e**) Scanning photocurrent images of the same tBLG device. A 532-nm laser with power of 200 μW is focused on the device, while the net photocurrent is amplified and then detected by a lock-in amplifier. All the photocurrents here are generated without source-drain bias and gate bias. (**f**) Three-dimensional view of the scanning photocurrent image of the same tBLG device. (**g**) Photocurrents generated from 7° (spot A) and 13° (spot B) tBLG domains as a function of incident power, respectively. The white dashed lines in **c** and **e** show the positions of graphene–metal electrode interfaces, respectively. Scale bars, 5 μm (all).

shifts (Fig. 3d). Remarkably, the 13° tBLG domain generates a much larger net photocurrent (0.63 μA) at zero bias than that of the 7° domain (0.097 μA), originated from selectively enhanced light–matter interaction of 13° tBLG domain with the 532-nm laser.

We further conducted net photocurrent mapping of the device by using scanning photocurrent microscopy, in which the net photocurrent was recorded while scanning a focused 532-nm laser spot with a diameter of ∼1 μm over the device (Fig. 3e,f). The photocurrent was observed to exhibit contrary directions at the two graphene–metal electrode interfaces in the device. Significantly, the intensity of photocurrent generated at 13° tBLG domain is ∼6.6 times stronger than that at the 7° tBLG domain. This twist angle-related photocurrent enhancement holds great promise in high-selectivity photodetection applications.

To further evaluate the photoresponsivity of tBLG, we performed photocurrent measurements of tBLG devices under different incident power of 532 nm laser illumination, respectively. As shown in Fig. 3g, the photocurrents from 7° and 13° tBLG domains both increase as the incident power rises from ∼1 μW to ∼5 mW. The photoresponsivity of 7° and 13° tBLG domain is measured as ∼0.15 and ∼1 mA W^{-1}, respectively, indicating a robust and strong enhancement in 13° tBLG domain under different incident power of 532 nm laser illumination.

From the unravelling of band structures, the energy interval of the two VHSs ($2E_{VHS}$) of 13° tBLG domain is ∼2.34 eV, which matches the energy of incident photon (2.33 eV, $\lambda = 532$ nm) and thus leads to a strong light–matter interaction. When we changed the wavelength of incident laser from 532 to 632.8 nm (1.96 eV),

the photocurrent was found to be selectively enhanced in a 10.5° tBLG domain device with $2E_{VHS}$ of ∼1.89 eV (Supplementary Fig. 4 and Supplementary Note 4). To further investigate the correlation of $2E_{VHS}$ with $\hbar\omega$ in photocurrent generation of 13° and 10.5° tBLG domains, $\hbar\omega$ was gradually changed from 1.77 to 2.48 eV (500 to 700 nm in wavelength), while the power of incident laser was kept unchanged. As shown in Fig. 4a, the photocurrents of 13° and 10.5° tBLG domains exhibit peaks at ∼2.30 and ∼1.94 eV, agreeing well with $2E_{VHS}$ values (∼2.34 and ∼1.89 eV), respectively.

Discussion
The origin of the photocurrent enhancement can be understood qualitatively when taking the unique electronic state of tBLG into account. In the photoexcitation process, the interband transition has to satisfy both momentum and energy conservation. For momentum conservation, the electrons are confined to transit between states with the same **k** value in reciprocal space, owing to the very small momentum of incident photons. As for energy conservation, the energy difference between these two states equals to $\hbar\omega$. When $\hbar\omega \approx 2E_{VHS}$, the initial and final states are both near VHSs (thus with enhanced DOS, see Fig. 1a). Specifically, the effect of VHSs on the photoexcitation process can be evaluated by joint DOS (JDOS), which is defined as:

$$JDOS(\omega) = \frac{1}{4\pi^3} \int \delta[E_c(\mathbf{k}) - E_v(\mathbf{k}) - \hbar\omega]d\mathbf{k} \qquad (1)$$

where E_C and E_V represent the energies of the conduction and valence bands, respectively. JDOS is associated with the process in which an electron absorbs a photon with energy $\hbar\omega = E_c - E_v$ and then transits from conduction to valence band. A calculated JDOS shows an abrupt increase associated with the VHSs when $\hbar\omega \approx 2E_{VHS}$ (ref. 40). This leads to an enhanced photoexcitation process, consistent with experimental observations in Raman (Fig. 1f) and absorption spectra[27,45]. As a result, the intensified photoexcitation process may result in the enhanced photocurrent generation.

Besides the efficient photoexcitation, the improvement of separation efficiency of excited carriers can facilitate the photocurrent generation in tBLG. A gate voltage applied on the tBLG photodetection device can manipulate the doping level of

graphene in channel and thus change the value of Seebeck coefficient, which may simultaneously lead to the photocurrent change[54–57]. As shown in Fig. 4b, the photocurrent of 13° tBLG domain has a 2.6-fold increase from ~25 to ~66 nA when the back-gate voltage decreases from 0 to −20 V. As the back-gate voltage increases from 0 to 20 V, the photocurrent first flips its polarity at 2.5 V and then reaches a value of about −82 nA. The inset in Fig. 4b shows the two band profiles of graphene–metal electrode junctions in the tBLG photodetection device under the applied back-gate voltage. From the transfer curve (Supplementary Fig. 5 and Supplementary Note 5), we believe that the positive gate voltage manipulates the graphene in channel from p- to n-type doping, which gives rise to the change of Seebeck coefficient. In contrast, owing to the Fermi-level pinning of graphene underneath the metal electrodes, its Seebeck coefficient keeps unchanged. Therefore, the difference of these two Seebeck coefficients could be tuned and flipped by gate voltage, which leads to the value and polarity change of photocurrent.

The responsivity of tBLG is measured as ~1 mA W^{-1} at the resonance frequency, which is about 20 times enhancement compared with that of mechanically exfoliated monolayer graphene (~0.05 mA W^{-1}) with similar device configuration (Supplementary Fig. 6). To further improve the responsivity, we have integrated tBLG with plasmonic electrode structures as shown in Fig. 5a. A tBLG domain and an adjacent monolayer domain were embedded into the same two-terminal electrodes. A finger-patterned plasmonic structure (Ti/Au, 5/45 nm in thickness) with 110 nm finger width and 300 nm pitch[58] were fabricated on the tBLG domain as shown in Fig. 5b,c. The Raman mapping image in Fig. 5d exhibits uniformly enhanced G-band intensity, which confirms that the interval of two VHSs ($2E_{VHS}$) of the tBLG domain matches the energy of incident photon (532 nm and 2.33 eV). The scanning photocurrent results of the device (Fig. 5e,f) show that the photocurrents of tBLG and the adjacent monolayer domain are measured as ~10 and ~0.95 nA,

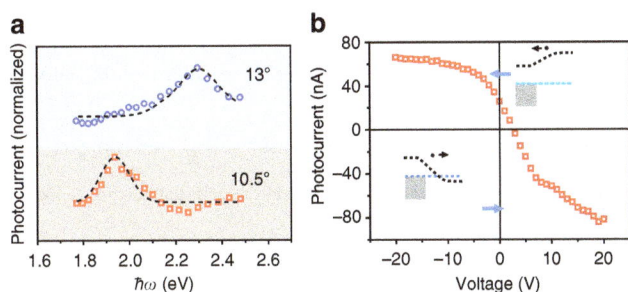

Figure 4 | The variation of photocurrent with photon energy and gate voltage. (**a**) Photocurrent versus energy of incident photon ($\hbar\omega$). tBLG domains with 10.5° and 13° twist angles show different peak positions. Incident photons with energy ω near $2E_{VHS}$ generate an enhanced photocurrent, while photons with energy lower or higher than $2E_{VHS}$ excite ordinary optoelectronic processes. Dotted lines were used to guide the eyes. The plots are normalized with that of AB-stacked bilayer graphene. (**b**) Plot of photocurrent as function of gate voltage. Insets are the corresponding band profiles, where the grey boxes, blue dotted lines and black dotted lines represent Ti electrodes, Fermi levels and positions of Dirac points of tBLG, respectively.

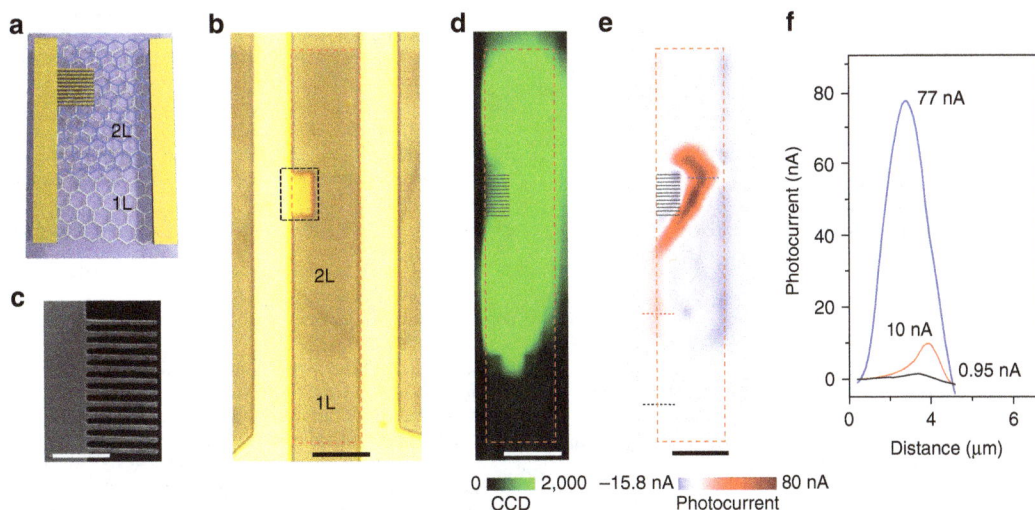

Figure 5 | tBLG photodetector integrated with plasmonic structure. (**a**) Schematic illustration of the detector. The channel comprises graphene monolayer domain and tBLG domain which are labelled by '1L' and '2L', respectively. The electrode is integrated with finger structure. (**b**) Optical image of the tBLG photodetector. (**c**) Scanning electron microscopy (SEM) image of the finger structure labelled by black dashed rectangle in **b**. (**d**) Raman G-band intensity mapping image of the device under 532 nm laser. The tBLG domain exhibits an enhanced G-band intensity. (**e**) Scanning photocurrent image of the photodetector. The exciting light is a focused 532 nm laser with power of 30 μW. (**f**) Line-scanning photocurrent of the photodetector. The blue, red and black curves correspond to photocurrent distributions along the tBLG near figure structure, tBLG and graphene monolayer, which are labelled by blue, red and black dashed lines in **e**. The maximum photocurrent at tBLG with the plasmonic structure is 77 nA, while the maximum photocurrents at the tBLG and monolayer graphene are 10 and 0.95 nA, respectively. All the photocurrents here are generated without source-drain bias and gate bias. The red dashed rectangles in **b,d** and **e** correspond to the channel of the photodetector. Scale bar, 1 μm (finger structure in **c**). Scale bars, 5 μm (others).

respectively. The enhancement of about 10.5 times is achieved. Remarkably, the photocurrent generated on tBLG near the finger structure is further enhanced to the value of ~ 77 nA, which is ~ 80-fold enhancement compared with the photocurrent of adjacent monolayer graphene channel. The mechanism of such strong photoresponsivity enhancements can be ascribed to the combination of the selective resonance enhancement of VHSs in tBLG and plasmonic enhancement of the finger structure.

In summary, we have experimentally demonstrated that the light–matter interaction in tBLG with VHSs is dependent on the interlayer twist angle (θ) and then can lead to a selective enhancement in photocurrent generation. Micro-ARPES was performed to unravel the band structure of CVD-grown tBLG and reveal the emergence of θ-dependent VHSs. The photocurrent of tBLG photodetectors exhibits a ~ 6.6-fold enhancement at a suitable θ, when the energy interval of the VHSs ($2E_{VHS}$) matches the energy of incident photon ($\hbar\omega$). By integrating plasmonic structures, the responsivity of tBLG photodetector can be significantly enhanced by ~ 80 folds compared with the monolayer graphene. Our results open a new route to graphene-based optoelectronic applications.

Methods

tBLG growth and characterization.
The tBLG was grown on copper foil in a home-made low-pressure CVD system. The growth was carried out under the flow of H_2 and CH_4 (600:1 in volume) with a pressure of 600 Pa at 1,030 °C for 40 min. The samples were characterized by Olympus BX51 microscope, scanning electron microscopy (Hitachi S-4800 operated at 2 kV), TEM (FEI Tecnai F30 operated at 300 kV for the diffraction image and FEI 80-300 Cs image-corrected Titan operated at 80 kV for the moiré pattern image) and Raman spectroscopy (Horiba HR800). The micro-ARPES measurements were carried out at the spectromicroscopy beamline at Elettra Synchrotron Radiation lab in Italy, with energy resolution of 50 meV, spatial resolution of 0.8 μm and angle resolution of 0.5°. Before the micro-ARPES measurements, in-situ annealing at 350 °C was carried out to clean the sample surface.

Device fabrication and measurement.
tBLG samples on copper were transferred to a highly doped Si substrate with 90 nm SiO_2 with the help of poly(methyl methacrylate)[59]. The Ti/Au (20/30 nm) electrodes and Ti/Au (5/45 nm) figure structures were fabricated by electron-beam lithography and the following electron-beam evaporation. The electrical measurements were performed by Keithley SCS-4200. The photoelectrical measurements were performed by a scanning photocurrent microscopy. In the set-up, 532 and 632.8 nm, and Supercontinuum Laser Sources (NKT Photonic) were used as laser sources. The chopper-modulated (~ 1 kHz) laser beams were focused to ~ 1 μm on the device using $\times 100$ objective and the short-circuit photocurrents were then measured by pre-amplifier and lock-in amplifier. When scanning the laser spot over the device, the induced photocurrents and beam positions were recorded and displayed simultaneously with the assistance of a computer, which communicated with lock-in amplifier and motorized stage (with device on it). A voltage source (Keithley 2400) was used to supply the gate voltage. All the electrical and photoelectrical measurements were performed in air at room temperature.

References

1. Novoselov, K. S. et al. Two-dimensional gas of massless Dirac fermions in graphene. Nature 438, 197–200 (2005).
2. Novoselov, K. S. et al. Electric field effect in atomically thin carbon films. Science 306, 666–669 (2004).
3. Zhang, Y. B. et al. Experimental observation of the quantum Hall effect and Berry's phase in graphene. Nature 438, 201–204 (2005).
4. Geim, A. K. & Novoselov, K. S. The rise of graphene. Nat. Mater. 6, 183–191 (2007).
5. Tielrooij, K. J. et al. Generation of photovoltage in graphene on a femtosecond timescale through efficient carrier heating. Nat. Nanotechnol. 10, 437–443 (2015).
6. Sun, D. et al. Ultrafast relaxation of excited Dirac fermions in epitaxial graphene using optical differential transmission spectroscopy. Phys. Rev. Lett. 101, 157402 (2008).
7. George, P. A. et al. Ultrafast optical-pump terahertz-probe spectroscopy of the carrier relaxation and recombination dynamics in epitaxial graphene. Nano Lett. 8, 4248–4251 (2008).
8. Tani, S. et al. Ultrafast carrier dynamics in graphene under a high electric field. Phys. Rev. Lett. 109, 166603 (2012).
9. Graham, M. W. et al. Photocurrent measurements of supercollision cooling in graphene. Nat. Phys. 9, 103–108 (2013).
10. Tielrooij, K. J. et al. Photoexcitation cascade and multiple hot-carrier generation in graphene. Nat. Phys. 9, 248–252 (2013).
11. Tse, W.-K. & Das Sarma, S. Energy relaxation of hot Dirac fermions in graphene. Phys. Rev. B 79, 235406 (2009).
12. Bistritzer, R. & MacDonald, A. H. Electronic cooling in graphene. Phys. Rev. Lett. 102, 206410 (2009).
13. Xia, F. N. et al. Ultrafast graphene photodetector. Nat. Nanotechnol. 4, 839–843 (2009).
14. Mueller, T. et al. Graphene photodetectors for high-speed optical communications. Nat. Photonics 4, 297–301 (2010).
15. Gan, X. et al. Chip-integrated ultrafast graphene photodetector with high responsivity. Nat. Photonics 7, 883–887 (2013).
16. Pospischil, A. et al. CMOS-compatible graphene photodetector covering all optical communication bands. Nat. Photonics 7, 892–896 (2013).
17. Schall, D. et al. 50 GBit/s photodetectors based on wafer-scale graphene for integrated silicon photonic communication systems. ACS Photonics 1, 781–784 (2014).
18. Boubanga-Tombet, S. et al. Ultrafast carrier dynamics and terahertz emission in optically pumped graphene at room temperature. Phys. Rev. B 85, 035443 (2012).
19. Nair, R. R. et al. Fine structure constant defines visual transparency of graphene. Science 320, 1308–1308 (2008).
20. Shallcross, S. et al. Quantum interference at the twist boundary in graphene. Phys. Rev. Lett. 101, 056803 (2008).
21. Shallcross, S. et al. Emergent momentum scale, localization, and van Hove singularities in the graphene twist bilayer. Phys. Rev. B 87, 245403 (2013).
22. Shallcross, S. et al. Electronic structure of turbostratic graphene. Phys. Rev. B 81, 1 (2010).
23. Landgraf, W. et al. Electronic structure of twisted graphene flakes. Phys. Rev. B 87, 075433 (2013).
24. Ohta, T. et al. Evidence for interlayer coupling and moire periodic potentials in twisted bilayer graphene. Phys. Rev. Lett. 109, 186807 (2012).
25. Li, G. H. et al. Observation of Van Hove singularities in twisted graphene layers. Nat. Phys. 6, 109–113 (2010).
26. Yan, W. et al. Angle-dependent van Hove singularities in a slightly twisted graphene bilayer. Phys. Rev. Lett. 109, 126801 (2012).
27. Havener, R. W. et al. Angle-resolved Raman imaging of inter layer rotations and interactions in twisted bilayer graphene. Nano Lett. 12, 3162–3167 (2012).
28. Ohta, T. et al. Controlling the electronic structure of bilayer graphene. Science 313, 951–954 (2006).
29. McCann, E. Asymmetry gap in the electronic band structure of bilayer graphene. Phys. Rev. B 74, 161403 (2006).
30. Castro, E. V. et al. Biased bilayer graphene: semiconductor with a gap tunable by the electric field effect. Phys. Rev. Lett. 99, 216802 (2007).
31. McCann, E. & Fal'ko, V. I. Landau-level degeneracy and quantum hall effect in a graphite bilayer. Phys. Rev. Lett. 96, 086805 (2006).
32. Guinea, F. et al. Electronic states and Landau levels in graphene stacks. Phys. Rev. B 73, 245426 (2006).
33. Ohta, T. et al. Interlayer interaction and electronic screening in multilayer graphene investigated with angle-resolved photoemission spectroscopy. Phys. Rev. Lett. 98, 206802 (2007).
34. Zhang, Y. B. et al. Direct observation of a widely tunable bandgap in bilayer graphene. Nature 459, 820–823 (2009).
35. Luican, A. et al. Single-layer behavior and its breakdown in twisted graphene layers. Phys. Rev. Lett. 106, 126802 (2011).
36. dos Santos, J. et al. Graphene bilayer with a twist: electronic structure. Phys. Rev. Lett. 99, 256802 (2007).
37. Coh, S. et al. Theory of the Raman spectrum of rotated double-layer graphene. Phys. Rev. B 88, 165431 (2013).
38. Brihuega, I. et al. Unraveling the intrinsic and robust nature of van Hove Singularities in twisted bilayer graphene by scanning tunneling microscopy and theoretical analysis. Phys. Rev. Lett. 109, 196802 (2012).
39. Kim, K. et al. Raman spectroscopy study of rotated double-layer graphene: misorientation-angle dependence of electronic structure. Phys. Rev. Lett. 108, 246103 (2012).
40. Sato, K. et al. Zone folding effect in Raman G-band intensity of twisted bilayer graphene. Phys. Rev. B 86, 125414 (2012).
41. Carozo, V. et al. Resonance effects on the Raman spectra of graphene superlattices. Phys. Rev. B 88, 085401 (2013).
42. He, R. et al. Observation of low energy Raman modes in twisted bilayer graphene. Nano Lett. 13, 3594–3601 (2013).
43. Ni, Z. et al. Reduction of Fermi velocity in folded graphene observed by resonance Raman spectroscopy. Phys. Rev. B 77, 235403 (2008).
44. Liao, L. et al. van Hove singularity enhanced photochemical reactivity of twisted bilayer graphene. Nano Lett. 15, 5585–5589 (2015).

45. Wang, Y. Y. *et al.* Stacking-dependent optical conductivity of bilayer graphene. *ACS Nano* **4**, 4074–4080 (2010).
46. Zhou, H. L. *et al.* Chemical vapour deposition growth of large single crystals of monolayer and bilayer graphene. *Nat. Commun.* **4**, 2096 (2013).
47. Yan, Z. *et al.* Toward the synthesis of wafer-scale single-crystal graphene on copper foils. *ACS Nano* **6**, 9110–9117 (2012).
48. Ma, T. *et al.* Repeated growth-etching-regrowth for large-area defect-free single-crystal graphene by chemical vapor deposition. *ACS Nano* **8**, 12806–12813 (2014).
49. Ni, Z. H. *et al.* G-band Raman double resonance in twisted bilayer graphene: evidence of band splitting and folding. *Phys. Rev. B* **80**, 125404 (2009).
50. Giovannetti, G. *et al.* Doping graphene with metal contacts. *Phys. Rev. Lett.* **101**, 026803 (2008).
51. Zhou, S. Y. *et al.* Substrate-induced bandgap opening in epitaxial graphene. *Nat. Mater.* **6**, 770–775 (2007).
52. Xia, F. N. *et al.* Photocurrent imaging and efficient photon detection in a graphene transistor. *Nano Lett.* **9**, 1039–1044 (2009).
53. Mueller, T. *et al.* Role of contacts in graphene transistors: a scanning photocurrent study. *Phys. Rev. B* **79**, 245430 (2009).
54. Song, J. C. W. *et al.* Hot carrier transport and photocurrent response in graphene. *Nano Lett.* **11**, 4688–4692 (2011).
55. Gabor, N. M. *et al.* Hot carrier-assisted intrinsic photoresponse in graphene. *Science* **334**, 648–652 (2011).
56. Xu, X. D. *et al.* Photo-thermoelectric effect at a graphene interface junction. *Nano Lett.* **10**, 562–566 (2010).
57. Sun, D. *et al.* Ultrafast hot-carrier-dominated photocurrent in graphene. *Nat. Nanotechnol.* **7**, 114–118 (2012).
58. Echtermeyer, T. J. *et al.* Strong plasmonic enhancement of photovoltage in graphene. *Nat. Commun.* **2**, 458 (2011).
59. Reina, A. *et al.* Transferring and identification of single- and few-layer graphene on arbitrary substrates. *J. Phys. Chem. C* **112**, 17741–17744 (2008).

Acknowledgements

We are grateful to Dr Yao Guo and Mr Chen Peng from Department of Electronics, Peking University, for their suggestions in device fabrication, and Mr Ziwei Li from School of Physics, Peking University, for the calculation regarding to plasmonic structures. We acknowledge financial support from the National Basic Research Program of China (numbers 2014CB932500, 2011CB921904 and 2013CB932603), the National Natural Science Foundation of China (numbers 21173004, 21222303, 51121091 and 51362029), the National Program for Support of Top-Notch Young Professionals and Beijing Municipal Science and Technology Commission (Z131100003213016). Part of this work was performed at the Stanford Nano Shared Facilities.

Author contributions

J.Y. and H.L.P. conceived and designed the experiments. J.Y and H.W. performed the synthesis and structural characterization. J.Y. made the devices and carried out optoelectronic measurements. Z.T., L. Liao, L. Lin and X.S. assisted in experimental work and contributed to the scientific discussions. H.P. and Y.L.C. preformed micro-ARPES. A.L.K. and H.L.P. conducted the TEM, high-resolution TEM and aberration-corrected high-resolution TEM experiments. J.Y., H.L.P. and H.P. wrote the paper. H.L.P., Z.L. and Y.L.C. supervised the project. All the authors discussed the results and commented on the manuscript.

Additional information

Creating single-atom Pt-ceria catalysts by surface step decoration

Filip Dvořák[1,*], Matteo Farnesi Camellone[2,*], Andrii Tovt[1], Nguyen-Dung Tran[2,3], Fabio R. Negreiros[2,†], Mykhailo Vorokhta[1], Tomáš Skála[1], Iva Matolínová[1], Josef Mysliveček[1], Vladimír Matolín[1] & Stefano Fabris[2,3]

Single-atom catalysts maximize the utilization of supported precious metals by exposing every single metal atom to reactants. To avoid sintering and deactivation at realistic reaction conditions, single metal atoms are stabilized by specific adsorption sites on catalyst substrates. Here we show by combining photoelectron spectroscopy, scanning tunnelling microscopy and density functional theory calculations that Pt single atoms on ceria are stabilized by the most ubiquitous defects on solid surfaces—monoatomic step edges. Pt segregation at steps leads to stable dispersions of single Pt^{2+} ions in planar PtO_4 moieties incorporating excess O atoms and contributing to oxygen storage capacity of ceria. We experimentally control the step density on our samples, to maximize the coverage of monodispersed Pt^{2+} and demonstrate that step engineering and step decoration represent effective strategies for understanding and design of new single-atom catalysts.

[1] Charles University in Prague, Faculty of Mathematics and Physics, V Holešovičkách 2, Prague 18000, Czech Republic. [2] CNR-IOM DEMOCRITOS, Istituto Officina dei Materiali, Consiglio Nazionale delle Ricerche, Via Bonomea 265, Trieste 34136, Italy. [3] SISSA, Scuola Internazionale Superiore di Studi Avanzati, Via Bonomea 265, Trieste 34136, Italy. * These authors contributed equally to this work. † Present address: Universidade Federal do ABC. Av. dos Estados, 5001 Bairro Bangu, Santo André SP CEP 09210-580, Brasil. Correspondence and requests for materials should be addressed to J.M. (email: josef.myslivecek@mff.cuni.cz) or to S.F. (email: fabris@democritos.it).

Single-atom catalysts represent the limiting realization of supported metal catalysts with metal load ultimately dispersed as single atoms[1,2]. This maximizes the utilization of supported metals and helps development of sustainable catalytic technologies for renewable energies and environmental applications with reduced precious metal contents[3,4]. A central prerequisite for understanding and knowledge-based design of single-atom catalysts is the identification of specific adsorption sites on catalyst supports that provide the stabilization of single metal atoms under reaction conditions at elevated temperatures and pressures. For oxide supports, understanding specific adsorption sites presently concentrates on low-index oxide facets[5-9]. Single-atom catalysts are, however, nanostructured large-area materials; thus, a question arises whether single supported atoms can be stabilized at defect sites of nanostructured oxide supports.

Highly dispersed platinum (Pt) ions on ceria qualify as single-atom catalysts[2] and hold a promise of radical reduction of Pt load in critical large-scale catalytic applications—hydrogen production[3], three-way catalytic converters[10] and fuel cells[11]. Ceria surfaces provide a limited amount of low coordinated surface sites where Pt^{2+} ions can adsorb and remain stable in real applications[2,3,10,11]. Recent studies on large-area ceria samples identify the necessity of nanostructuring the ceria substrates for obtaining supported Pt^{2+} ions[2,4] and propose a square-planar PtO_4 unit as a Pt^{2+}-containing surface moiety[4]. In the present model study on the single crystalline $CeO_2(111)$ surface, we demonstrate

that single-ion dispersions of Pt^{2+} are stabilized at monolayer (ML)-high ceria step edges. Pt^{2+} ions at step edges are located in PtO_4 units that can be considered the elementary building blocks of Pt^{2+}/ceria single-atom catalysts. The PtO_4 units incorporate excess O and can act as oxygen source for redox reactions. Besides clarifying the nature of Pt^{2+} stabilization on ceria, our study demonstrates the importance of step edges—the most common surface defects on oxide supports[12]—for single-atom catalyst stabilization. We experimentally adjust the step density on the ceria supports for maximizing the load of monodispersed Pt^{2+} ions. This identifies step engineering[13] and step decoration[14,15] as advanced techniques for designing new single-atom catalysts.

Results

Pt deposits on highly defined $CeO_2(111)$ surfaces. The experiments were performed on model $CeO_2(111)$ surfaces prepared as 20 to 40 Å thick ceria films on Cu(111) using procedures that allow adjusting the density of ML-high steps[16] and the density of surface oxygen vacancies on ceria surface[17]. On these highly defined surfaces, we deposit 0.06 ML of Pt, anneal at 700 K in ultra-high vacuum (UHV) and observe stabilization of Pt^{2+} species and/or nucleation of Pt clusters with scanning tunnelling microscopy (STM) and with photoelectron spectroscopy (PES). Deposition and annealing of Pt on $CeO_2(111)$ surfaces containing low concentrations of defects—ML-high steps and surface oxygen vacancies (Fig. 1a)—yield

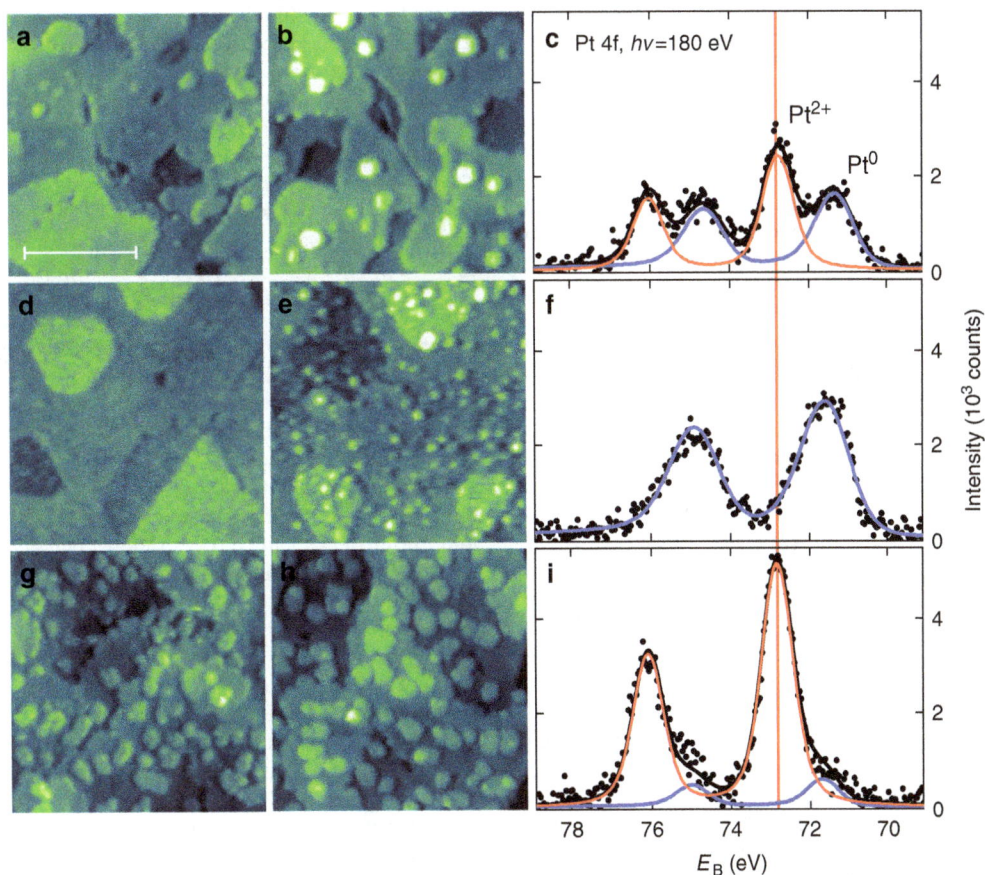

Figure 1 | Nucleation of Pt and stabilization of Pt^{2+} on ceria surfaces containing controlled amount of surface defects. (a–c) $CeO_2(111)$ surface with low density of surface oxygen vacancies and ML-high steps. (d–f) $CeO_{1.7}$ surface with increased density of surface oxygen vacancies. (g–i) $CeO_2(111)$ surface with increased density of ML-high steps. (a,d,g) STM images of clean surfaces before deposition of Pt. (b,e,h) STM images after deposition of 0.06 ML Pt and annealing at 700 K in UHV. All STM images 45 × 45 nm², tunnelling current 25–75 pA, sample bias voltage 2.5–3.5 V. Scale bar, 20 nm (a). (c,f,i) PES spectra of the Pt deposit after annealing. All PES spectra were acquired with photon energy $hv = 180$ eV (black points). Fits indicate metallic (Pt^0, blue line) and ionic (Pt^{2+}, red line) contributions to Pt 4f signal. E_B is the photoelectron binding energy.

metallic Pt^0 clusters (Fig. 1b,c) coexisting with ionic Pt^{2+} species (Fig. 1c). To determine whether the charge of the supported Pt species is selectively induced by a specific defect type, we repeat the experiment varying independently the amount of surface O vacancies—up to 0.16 ML, creating $CeO_{1.7}$ surface (Fig. 1d–f)—and the amount of ML-high steps on the $CeO_2(111)$ surface—up to 0.15 ML (Fig. 1g–i). We observe that surface oxygen vacancies do not promote the dispersion of Pt^{2+} species but lead to small metallic Pt^0 clusters (Fig. 1e,f)[18]. On the other hand, the increased step density leads to almost complete oxidation of the Pt deposit to Pt^{2+} (Fig. 1h,i) proving that step edges selectively promote the stabilization of Pt^{2+} species. Detailed STM images allow to exclude formation of three-dimensional and two-dimensional PtO_x clusters (Supplementary Fig. 1), and allow to conclude that Pt^{2+} species are incorporated in the ceria step edges. Nucleation of Pt^0 clusters and stabilization of Pt^{2+} species represent concurrent processes. Differently to Pt^0 clusters (Fig. 1b,e), Pt^{2+} species at the step edges are not discernible in empty states STM imaging (cf. Fig. 1g,h without and with Pt deposit) because of their electronic structure. STM imaging in occupied states on metal-supported ceria is unavailable[19].

The possibility to adjust the density of ML-high steps on the model $CeO_2(111)$ surfaces[16] allows us to obtain a quantitative correlation between the step density and the amount of Pt^{2+} species. We prepare $CeO_2(111)$ samples with step density between 0.06 and 0.20 ML[16] and deposit 0.06 or 0.18 ML Pt at 300 K. Parameters of the prepared samples are summarized in Supplementary Table 1. After annealing at 700 K the amount of Pt stabilized in the form of Pt^{2+} is determined by PES. For quantification, all relevant parameters—the density of ceria steps, deposited amount of Pt and amount of stabilized Pt^{2+}—are expressed in ML where 1 ML corresponds to the density of Ce atoms on the $CeO_2(111)$ surface, that is, $7.9 \times 10^{14}\,cm^{-2}$. The density of steps is defined as the density of Ce atoms located at the ceria step edges[20].

The amount of Pt^{2+} ions as a function of the step density is plotted in Fig. 2a. For the higher Pt coverage 0.18 ML, the analysis reveals a linear dependence between the amount of stabilized Pt^{2+} ions and the step density (Fig. 2a, blue symbols), confirming the activation of Pt oxidation to Pt^{2+} and the localization of Pt^{2+} at the surface steps. The highest step density 0.20 ML allows converting up to 80% of the Pt deposit to Pt^{2+}. The degree of oxidation of the Pt deposit increases with decreasing the amount of deposited Pt to 0.06 ML (Fig. 2a, black symbols). In this case, up to 90% of Pt converts to Pt^{2+}. The concentrations of ceria step edges and the amount of Pt^{2+} stabilized on the surface obey a classical supply-and-demand scenario characteristic for single-atom catalysts[1,4,10]: when sufficient step edges, the amount of oxidized Pt^{2+} is limited by the amount of deposited Pt. Otherwise, the amount of oxidized Pt^{2+} is limited by the step density regardless of the amount of deposited Pt. The Pt deposit exceeding the available step sites cannot be oxidized and nucleates as metallic Pt^0 clusters on the surface. Besides the high oxidative power of the step edges towards Pt, our quantitative analysis reveals also the capacity of ceria step edges to accommodate a high density of Pt^{2+} ions. Up to 0.16 ML of Pt^{2+} ions can be stabilized by the sample containing 0.18 ML of Pt deposit and 0.20 ML of steps (Fig. 2a, blue symbols). This corresponds to 80% of the step-edge sites being occupied by Pt^{2+}. In the whole range of the step densities between 0.06 and 0.20 ML, the occupation of the step-edge sites by Pt^{2+} varies between 50 and 80%.

Stability and charge state of $Pt^{2+}/CeO_2(111)$ samples. The necessity of annealing the Pt deposit on ceria in UHV at 700 K for obtaining Pt^{2+} stabilization in our experiment indicates the

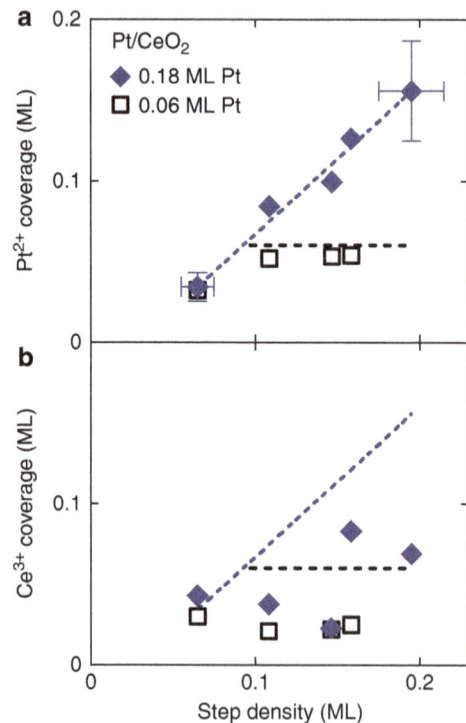

Figure 2 | Capacity of stepped $CeO_2(111)$ surface to accommodate Pt^{2+}. (a) Amount of Pt^{2+} stabilized on $CeO_2(111)$ substrates with different density of steps for 0.18 ML (blue symbols) and 0.06 ML (black symbols) of deposited platinum. Pt not stabilized in the form of Pt^{2+} remains metallic. Lines represent guides to the eyes. Blue line is a linear fit of 0.18 ML Pt data. Black line represents the maximum achievable amount of Pt^{2+} in the case of 100% oxidation of Pt for 0.06 ML of deposited Pt. (b) Reduction of the ceria surface accompanying the stabilization of Pt^{2+} ions determined by resonant PES expressed as a coverage of the surface by Ce^{3+} ions. Lines represent guides to the eyes from **a** and indicate the Pt^{2+} concentration. The Ce^{3+} concentration is lower or equal to the Pt^{2+} concentration on all samples.

activated nature of Pt segregation at the ceria steps and oxidation, and implies considerable thermal stability of Pt^{2+} ions on ceria. High-temperature annealing represents a prerequisite for obtaining Pt^{2+} ions also in the experiments on large-area nanostructured ceria samples[3,4]. Once created, Pt^{2+} ions remain stable on repeated annealing at 700 K in UHV. The Pt^{2+} ions in our experiment also remain stable on adsorption and thermal desorption of CO in UHV (Supplementary Fig. 2), or on exposure to air at ambient conditions (Supplementary Fig. 3).

Parallel to the charge state of the Pt deposit we determine the charge state of the CeO_2 support, in particular the concentration of surface Ce^{3+} ions that is indicative of reduction of the ceria surface. Contrary to the case of stabilizing Ni^{2+} ions on ceria[21], we observe that Pt oxidation during annealing is not accompanied by a corresponding reduction of $CeO_2(111)$ surface (Fig. 2b). This rules out the direct participation of ceria into the observed Pt oxidation at steps and indicates the involvement of other oxidizing agents in the Pt^{2+} stabilization, such as excess oxygen atoms. In the UHV environment of our experiments, the eligible source of excess oxygen can be water adsorbing in sub-ML amounts from background atmosphere (Supplementary Fig. 4) and undergoing dissociation on reduced ceria and Pt/ceria substrates[22,23]. In the large-area Pt^{2+}/CeO_2 catalysts displaying high concentration of Pt^{2+} ions and exceptional redox reactivity, excess O atoms may be incorporated during the synthesis that proceeds in air[3,10].

Segregation of Pt at CeO$_2$(111) step edges. *Ab initio* density functional theory (DFT) calculations allow to interpret the above experimental results. We calculate the segregation thermodynamics and the atomic and electronic structures of Pt atoms in representative adsorption sites on CeO$_2$(111) surfaces. The results

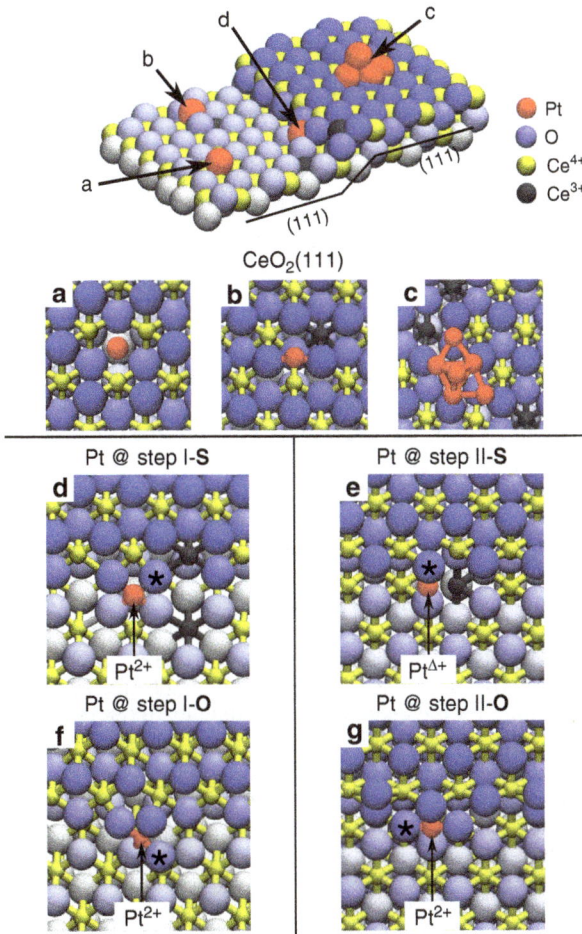

CeO$_2$(111)

Pt @ step I-S   Pt @ step II-S

Pt @ step I-O   Pt @ step II-O

Figure 3 | Pt adsorption sites on the CeO$_2$(111) surface obtained from DFT calculations. (**a**) Pt adatom in a surface O vacancy, (**b**) on the stoichiometric CeO$_2$(111) terrace and (**c**) supported Pt$_6$ cluster. (**d**) Pt adatom at the stoichiometric step I (step I-S) and (**e**) at the stoichiometric step II (step II-S). (**f**) Pt adatom at the step I with excess O (step I-O) and (**g**) at the step II with excess O (step II-O). Binding energies and Bader charges are summarized in Table 1. (**d–g**) The * symbol denotes the O atom removed to calculate the O vacancy formation energy reported in Table 2.

for the lowest-energy configurations are summarized in Fig. 3 and Table 1. The model adsorption sites include oxygen vacancies (Fig. 3a), regular sites (Fig. 3b) and Pt clusters[24] (Fig. 3c) on the CeO$_2$(111) terrace, as well as two low-energy ML-high steps, which we label following ref. 25 as step I (Fig. 3d) and step II (Fig. 3e). A detailed list of the systems considered in the DFT analysis is reported in the Supplementary Note 1. The steps I and II represent the preferred types of steps at the CeO$_2$(111) surfaces at temperatures < 1,000 K (ref. 25). On our experimental samples, the steps I and II appear in equal proportion as evidenced from the absence of triangularly shaped islands in Fig. 1a,g[25,26].

In agreement with the experiment, our calculations predict the preferential segregation of Pt adatoms at the steps I and II, independently on the local step geometry and stoichiometry. The binding energies of Pt at the steps are 1.6–3.4 eV higher than at stoichiometric or defective (111) terraces (Table 1). This driving force for Pt segregation at the steps is in qualitative agreement with recent calorimetric studies of other metal clusters on CeO$_2$(111)[27,28], see the Supplementary Discussion. The particular binding energy and the charge state of Pt atoms at the steps depend on both the local step geometry and stoichiometry (Table 1). In the following, we show that a good agreement with all the experimental observations can only be achieved when considering segregation of Pt at steps in the presence of excess O (calculations denoted as O), while Pt segregation at stoichiometric steps (denoted S) exhibits significant discrepancies.

Pt at stoichiometric CeO$_2$(111) step edges. Pt segregation on step I-S yields Pt^{2+} species that are coordinated by four lattice O atoms in a characteristic PtO$_4$ planar unit (Fig. 3d). The PtO$_4$ unit is remarkably similar to that one proposed for Pt-doped ceria nanoparticles[4] and for surface reconstructions of Pd–ceria systems[29]. Instead, the different atomic structure of the step II-S edge prevents the formation of PtO$_4$ units, hinders the full Pt oxidation to Pt^{2+} and yields weakly oxidized Pt$^{\Delta+}$ species (Fig. 3e). Calculation results presented in Fig. 3d–g correspond to the Pt coverage at the steps 1/3 (1 Pt atom per 3 Ce step-edge atoms). For interpreting the capacity of the ceria step edges to accommodate a high density of Pt^{2+} ions, we calculate the adsorption of Pt at the ceria steps with increasing Pt coverage at the steps (Fig. 4), ranging from 1/3 to 1 (1 Pt atom per 1 Ce step-edge atom). On the step I-S, the maximum coverage of Pt^{2+} species is 2/3 (Fig. 4a). Higher Pt^{2+} coverages are unattainable and lead to nucleation of metallic Pt clusters, due to the large strain buildup resulting from long sequences of interconnected PtO$_4$ step units (Fig. 4b). On the step II-S, metallic Pt^0 species appear already for a coverage higher than 1/3 (Fig. 4c). Thus, on

Table 1 | Properties of Pt on CeO$_2$(111) obtained from DFT calculations.

Pt adsorption site	Binding energy (eV)	Formal charge	Bader charge (e)
Pt @ O vacancy	2.8	Pt$^{\Delta-}$	10.9
Pt @ CeO$_2$(111)	3.3	Pt$^{\Delta+}$	9.7
Pt$_6$ @ CeO$_2$(111)	4.4	Pt^0	9.9
Pt @ step I-S	5.0	Pt^{2+}	9.2
Pt @ step II-S	5.1	Pt$^{\Delta+}$	9.7
Pt @ step I-O	6.6	Pt^{2+}	8.6
Pt @ step II-O	6.7	Pt^{2+}	9.0
Pt bulk	5.5*	Pt^0	10.0

DFT, density functional theory; Pt, platinum; I-S and II-S, stoichiometric step types I and II; I-O and II-O, step types I and II with excess O atoms.
Results for the binding energies per atom, formal charges and Bader charges of Pt adatoms at the adsorption sites displayed in Fig. 3a–g. The binding energy for the Pt$_6$ cluster is the total binding energy divided by the number of Pt atoms. For reference, Pt bulk values are given in the last line.
*Bulk cohesive energy.

Pt step coverage

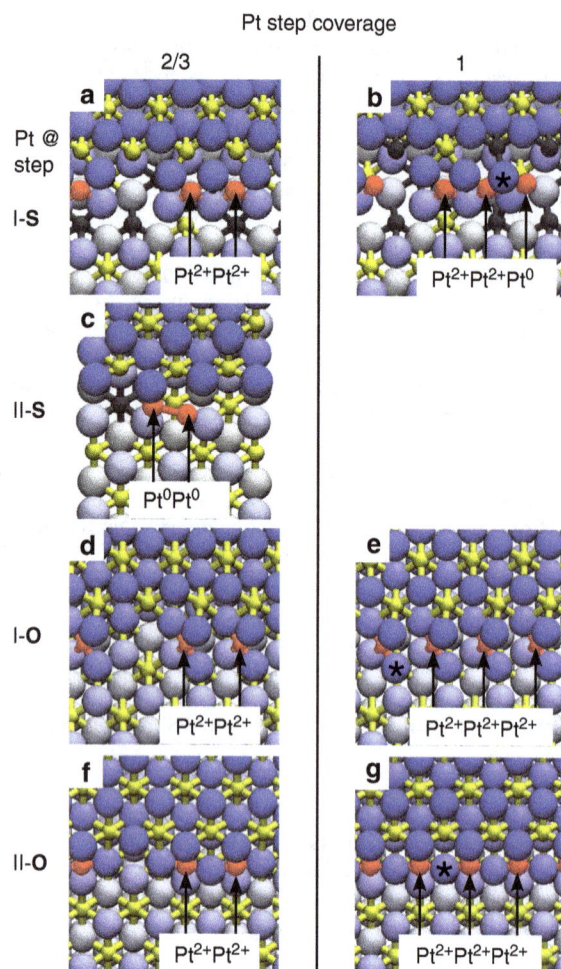

Figure 4 | Capacity of the CeO$_2$(111) step edges to accommodate Pt^{2+} ions obtained from DFT calculations. Calculated top views of the Pt binding to the steps I-S (**a,b**), step II-S (**c**), step I-O (**d,e**) and the step II-O (**f,g**) for Pt step coverage 2/3 (**a,c,f**) and 1 (**b,e,g**). At the step I-S, the limiting coverage of Pt^{2+} is 2/3 (**a**), additional Pt attaches to step edge as Pt^0 (**b**). At the step II-S, the Pt^{2+} coverage is 0. Pt atoms attach as weakly ionized Pt$^{\Delta+}$ and readily form metallic dimers (**c**) and clusters. On both steps I-O and II-O, excess oxygen can stabilize ionic Pt^{2+} at step edges as single ions appearing isolated or in groups up to 100% step coverage (**d-g**). The * symbol denotes the O atom removed to calculate the O vacancy formation energy reported in Table 2.

Table 2 | Minimum energy to remove an O atom from the CeO$_2$(111) surface obtained from DFT calculations.

O vacancy site	Pt step coverage	O vacancy formation energy (eV)
Pt @ step II-S	1/3	3.8
Pt @ step I-S	1/3	3.3
CeO$_2$(111)		2.5
step II-S	0	2.0
Pt @ step I-O	1/3	1.9
Pt @ step II-O	1/3	1.9
step I-S	0	1.8
Pt @ step I-O	1	1.8
Pt @ step II-O	1	1.7

DFT, density functional theory.
Results for steps I and II, and for Pt at steps I and II in the presence and absence of excess O. The energies for Pt @ steps were calculated by removing the O atom marked by the * symbol in Figs 3 and 4.

Pt^{2+} at CeO$_2$(111) step edges with excess O. Agreement between the theory and the experiment can be achieved when taking into account the step edges in the presence of an excess of O atoms. Irrespective of the local step geometry and Pt coverage at the steps, we find that excess O atoms readily bind to Pt at the ceria steps and drive a rearrangement of the step morphology forming ionized Pt^{2+} species incorporated in the planar PtO$_4$ moieties on both steps I and II (Fig. 3f,g). In the presence of excess of oxygen, Pt atoms bind stronger to the ceria step edges, with calculated binding energies up to 6.7 eV, which are higher than at the stoichiometric steps edges by ~1.6 eV, and which are also higher than the cohesive energy of bulk metallic Pt (Table 1). This condition, which determines the stability of the Pt^{2+} species at steps with respect to metallic Pt clusters, is fulfilled only in the presence of excess oxygen at the steps. The computed electronic structure and density of states of the PtO$_4$ moieties at the steps I-O and II-O (Supplementary Figs 5 and 6) confirm that the Pt$^0 \rightarrow$ Pt^{2+} oxidation results from the ionic Pt–O bond in the PtO$_4$ planar units, and that Ce^{3+} ions do not form in agreement with the experimental evidence (Fig. 2b). The calculated maximum coverage of Pt^{2+} at the steps I-O and II-O is 100% (Fig. 4e,g and Supplementary Table 2), as interconnected assemblies of the PtO$_4$ units can optimally fit the periodicities of both steps I and II at calculated Pt coverages at the step edges 1/3, 2/3 and 1 (Figs 3f,g and 4d–g). The presence of excess oxygen at steps therefore explains also the maximal Pt^{2+} ionization experimentally measured on the ceria-supported catalysts.

The stabilization of excess oxygen in the PtO$_4$ moieties by the Pt^{2+} ions suggests an oxygen source for redox reactions and hence provides a link between the presence of highly dispersed ionic Pt species on ceria and the increased redox reactivity of Pt^{2+}/CeO$_2$ single-atom catalysts[3]. The oxygen buffering capacity of ceria-based catalysts is associated with easy oxygen vacancy formation. We calculate the vacancy formation energy on the clean CeO$_2$(111) terrace, on the stoichiometric steps and on the Pt-decorated steps (Table 2). Compared with the energy of 2.5 eV calculated on the CeO$_2$(111) terrace, the energies required to remove an oxygen atom bound to Pt at the stoichiometric steps are 3.3 eV (from the PtO$_4$ unit at the step I-S) and 3.8 eV (from the PtO$_4$ unit at the step II-S; see Table 2). Pt segregation at the stoichiometric step edges yields the formation of strong Pt–O bonds and therefore hinders ceria O-buffering. Much lower energies are instead needed to remove the excess O incorporated in the PtO$_4$ units at steps I-O and II-O, where the O vacancy formation energy can be as low as 1.7 eV, lower or comparable to the values for the stoichiometric steps without Pt (2.0–1.8 eV;

samples with equal proportion of the stoichiometric steps I and II, *ab initio* calculations predict maximum Pt^{2+} coverage at the steps (≤33% of the step-edge sites) and maximum conversion of the Pt deposit to Pt^{2+} (≤33% of deposited Pt) that are well below the experimental values (50–80% of step-edge sites, up to 90% of deposited Pt, *cf.* Fig. 2a).

Most importantly, the calculations on the stoichiometric steps predict that Pt segregation, oxidation and the formation of the Pt^{2+} species are always accompanied by the reduction of surface Ce atoms from Ce^{4+} to Ce^{3+} (denoted in gray in Figs 3 and 4). The resulting concentration of the Ce^{3+} ions exceeds that of the Pt^{2+} ions by a factor of 2. This is in stark contrast with the resonant PES measurements on our samples showing that the concentration of Ce^{3+} is considerably lower than the concentration of Pt^{2+} after annealing the samples (Fig. 2b). This indicates that Pt is preferentially oxidized by other mechanisms than the Pt^0/Ce^{4+} redox couple.

Table 2)[30]. This indicates that the dispersed Pt^{2+} ions can enhance the oxygen storage capacity of ceria-based catalysts by assisting the reversible storage of excess O atoms.

The single-ion nature of Pt^{2+} in the PtO_4 units is preserved at all coverages, even when densely packed at the ceria step edges as interconnected PtO_4 units. Indeed, the Pt charge state, its local electronic structure and the O vacancy formation energy of the densely packed PtO_4 units are comparable to that of the isolated PtO_4 units at the steps (Supplementary Fig. 7 and Table 2). The PtO_4 units exhibit a large adaptability in stabilizing at different types of surface step edges, resulting in high effectiveness and capacity of the ceria surface to accommodate the Pt^{2+} ions. Regardless of the particular organization on the surface, the PtO_4 units are also always accessible to the reactants. Thus, the square-planar PtO_4 units carrying monodispersed Pt^{2+} ions can be considered elementary building blocks of single-atom Pt-ceria catalysts.

Discussion

On large-area samples, Pt^{2+} ions on ceria show exceptional reactivity with minimum Pt load in important applications: water–gas shift reaction[3], hydrogen oxidation on the anode of proton-exchange membrane fuel cell[11] and in the three-way catalyst converter[10]. In these applications, Pt^{2+} ions on ceria exhibit long-term stability under realistic reaction conditions of elevated temperatures and ambient pressure of reactant gases[3,31,32]. Pt^{2+} in large-area samples is routinely identified with PES. Complementary measurements with extended X-ray absorption fine structure (EXAFS)[33] and high-resolution transmission electron microscopy[3,4,31,32] confirm the absence of three-dimensional Pt or PtO_x clusters and, in agreement with the advanced PES measurements[34], identify Pt^{2+} as highly dispersed surface species on ceria[32].

Our present study identifies the stabilization of monodispersed Pt^{2+} ions with one particular defect site on the ceria surface—the monoatomic step edge—and excludes the stabilization of Pt^{2+} on the oxygen vacancies. Monodispersed Pt on ceria is observed to be effective in incorporating excess oxygen even in the unfavourable conditions of UHV experiment. Excess oxygen and Pt^{2+} arrange in the square-planar PtO_4 moieties decorating different types of the surface steps at coverages up to one PtO_4 per one step-edge Ce atom. The excess oxygen can be easily detached, indicating enhancement of the redox properties of ceria loaded with the Pt^{2+} ions. Adjusting the step density and the Pt load on the model $CeO_2(111)$ surface allows maximizing the coverage of Pt^{2+}, while suppressing the nucleation of metallic Pt^0 clusters. In the present experiment, we achieve surface coverage of Pt^{2+} 0.05 ML. A further increase of the completely monodispersed ionized Pt^{2+} coverage to 0.1 ML can be expected.

Step edges on ceria have been previously identified as preferred nucleation sites for supported metal clusters[20,21,27,28,35–37]. Our present study highlights the property of the step edges on ceria to provide specific structural and electronic environments for selective formation of monodispersed, thermally and chemically stable Pt^{2+} ions. The step edges represent intrinsic defects ubiquitously present on nanostructured ceria surfaces[4,38]; our results are thus applicable for the interpretation of the properties and the optimization of the Pt^{2+} load on large-area ceria supports[3,10,11]. More generally, the step edges may represent a common type of adsorption sites providing stabilization for monodispersed metal atoms and ions in any oxide-supported single-atom catalysts[15,39]. Our results therefore introduce important concepts of step reactivity[40] and step engineering[13,14] in understanding the stability, the activity and in designing new single-atom catalysts.

Methods

Experiment. The experiments were performed on surface science apparatuses in Surface Science Laboratory in Prague (STM, laboratory X-ray PES (XPS) with $hv = 1,487\,eV$ (Al Kα), low-energy electron diffraction) and at the Materials Science Beamline in Trieste (PES with $hv = 22$–$1,000\,eV$ (synchrotron), laboratory XPS with $hv = 1,487\,eV$ (Al Kα) and low-energy electron diffraction).

Preparation of the ceria substrates. The ceria layers and their Pt loading were prepared using the same procedures and parameters in both laboratories, and investigated by surface science methods *in situ* without exposing to air. The procedures and parameters of all samples are summarized in Supplementary Table 1. The ceria layers were prepared by deposition of Ce metal (Ce wire 99.9%, Goodfellow Cambridge Ltd) from Ta or Mo crucible heated by electron bombardment on clean Cu(111) substrate (MaTecK GmbH) in a background pressure of $5 \times 10^{-5}\,Pa$ of O_2 (5.0, Linde AG). The growth rate of CeO_2 was 6 ML per hour. Varying densities of 1 ML high steps on the prepared CeO_2 layers were obtained by growth of CeO_2 at constant substrate temperature 423 or 523 K (Method I in Supplementary Table 1) or linearly increasing substrate temperature from room temperature to 723 K (Method II in Supplementary Table 1)[16]. For experiments in Fig. 1, the ordered, fully oxidized layer of CeO_2 (Fig. 1a,b) and the ordered reduced layer of $CeO_{1.7}$ (Fig. 1d,e) were obtained by approach published in ref. 17 that yields the lowest step density. In this approach, first, fully reduced Ce_2O_3 layer is prepared by depositing metallic Ce on a CeO_2 layer and annealing in vacuum. Subsequently, the Ce_2O_3 layer is exposed to a controlled dose of O_2 at $5 \times 10^{-5}\,Pa$ and annealed to obtain desired stoichiometry $CeO_{1.7}$ or CeO_2 (Method III in Supplementary Table 1). The CeO_2 surface imaged in Fig. 1g,h was obtained by depositing 0.3 ML CeO_2 on the $CeO_2(111)$ substrate as in Fig. 1a, forming small ML-high islands. This homoepitaxy of CeO_2 on CeO_2 yields clearly arranged samples with high step density (Method IV in Supplementary Table 1).

Characterization of the ceria substrates. The thickness of the ceria layers was determined from the attenuation of the substrate $Cu\,2p_{3/2}$ XPS signal measured at $hv = 1,487\,eV$. For calculations, we used inelastic mean free path of electrons in CeO_2 11.2 Å. The thickness of the ceria layers was set between 20 and 40 Å or 7 and 12 ML with 1 ML corresponding to 3.1 Å, the distance between Ce(111) atomic planes of CeO_2. In this range of thickness, the coverage of the Cu substrate by ceria ranges between 97 and 100 % (ref. 17). For determining the density of 1 ML high steps, we use a semi-automated procedure when the first step outlines are marked in STM images manually. Step outlines are then mapped onto a properly scaled and rotated hexagonal mesh of surface Ce atoms. The atoms that are closest to the outlines are automatically identified as step-edge atoms and their density evaluated in ML. The error in determining the density of 1 ML high steps is estimated to be ± 10 % and is marked in Fig. 2a.

Preparation of the Pt deposit. Pt was deposited on the ceria layers from a Pt wire (99.99%, MaTecK GmbH) heated by electron bombardment. Pt was deposited on the sample surface at 300 K and subsequently stabilized by increasing the sample temperature to 700 K at the rate 2 K s^{-1}. Both Pt deposition and annealing proceeded in the UHV background pressure $5 \times 10^{-8}\,Pa$ or below. The thermal treatment supports the ionization of Pt to Pt^{2+}.

Characterization of the Pt deposit. The amount of Pt was calculated from the deposition time after calibrating the constant evaporation rate of the Pt evaporator. The evaporation rate was determined by a Quartz Crystal Microbalance and/or in a dedicated experiment from the thickness of 4-ML-thick Pt layers on $CeO_2(111)$/Cu(111) determined by attenuation of the substrate $Cu\,2p_{3/2}$ XPS signal measured at $hv = 1,487\,eV$. This dedicated experiment was used to correlate Pt evaporation rates between the two experimental apparatuses. For calculations, we used inelastic mean free path of electrons in Pt 8 Å. The fraction of Pt^{2+} after thermal treatment was determined by fitting the ionic Pt^{2+} and neutral Pt^0 component in the PES Pt 4f spectrum measured at $hv = 180\,eV$ (cf. Fig. 1c,f,i). The error in determining the Pt and Pt^{2+} amounts on the studied samples is ± 20 % and is marked in Fig. 2a. This error represents the calibration error of the Pt evaporation rate.

Resonant PES. Reduction of the ceria surface after deposition of Pt and thermal treatment was determined with resonant PES of Ce 4f state. We determine the so-called resonant enhancement ratio (RER) as defined in refs 41,42 from measurements of intensities of Ce^{3+} and Ce^{4+} components of valence-band resonant PES Ce 4f spectra of CeO_2 measured off-resonance ($hv = 115\,eV$) and on-resonance ($hv = 121.4\,eV$ for the Ce^{3+} component and $hv = 124.8\,eV$ for the Ce^{4+} component). The value of resonant enhancement ratio represents an upper estimate of the concentration of Ce^{3+} ions on the ceria surface and is plotted in Fig. 2b[17,42].

STM imaging. STM measurements were performed with commercial Pt–Ir tips (Unisoku). STM imaging of $CeO_2(111)$ and $Pt/CeO_2(111)$ films was available only via unoccupied states. We used sample voltages 2.5–3.5 V and tunnelling currents 25–75 pA.

Theory. All calculations were based on the DFT and were performed using the spin-polarized GGA + U approach[43], employing the Perdew–Burke–Ernzerhof exchange-correlation functional[44] and ultrasoft pseudopotentials[45]. The spin-polarized Kohn–Sham equations were solved with a plane-wave basis set and the Fourier representation of the charge density was limited by kinetic cutoffs of 40 and 320 Ry, respectively. The Quantum-ESPRESSO computer package was used in all the calculations[46]. In the Hubbard U term, the occupations of the f-orbitals were defined in terms of atomic wave function projectors and the value of the parameter U was set to 4.5 eV, following our previous studies[47,48].

Slab models. The ceria (111) surfaces were modelled with periodic (3×3) slabs being three CeO_2 ML thick and separated by more than 10 Å of vacuum in the direction perpendicular to the surface. The Brillouin zone was sampled at Gamma point. In the present work, we considered two low-energy ML-high steps that we label following ref. 25 as step I and step II. The edge of both step I and step II steps are oriented along the $[1\bar{1}0]$ direction. These surface steps were modelled with vicinal surfaces described with monoclinic periodic slabs separated by > 10 Å of vacuum in the direction perpendicular to the (111) terrace. The dimensions of the cells were 17.97×11.67 Å^2 along the $[1\bar{1}2]$ and $[1\bar{1}0]$ directions (step I) and 15.72×11.67 Å^2 along the $[1\bar{1}2]$ and $[1\bar{1}0]$ (step II). All the vicinal surfaces slabs included three CeO_2 ML. This thickness was shown to be sufficient to calculate the structural and thermodynamic properties of these steps[25]. The complete set of surface structures and systems considered in this work is listed in the Supplementary Note 1. All these systems were structurally optimized according to the Hellmann–Feynman forces. During the geometry optimization, the atomic positions of the lowermost CeO_2 ML were constrained, as well as those of the Ce atoms in the central ML, except for the Ce atoms below the step edge.

Energetics. Binding energies were computed as $1/N_{Pt} (E_{slab} + N_{Pt} E_{Pt} - E_{slab/Pt})$, where $E_{slab/Pt}$ is the total energy of the ceria slab containing N_{Pt} atoms of Pt, E_{slab} is the total energy of the corresponding relevant (stoichiometric, reduced or oxidized) Pt-free ceria slab and E_{Pt} is the total energy of a Pt atom in vacuum. The energies required to form an oxygen vacancy Ov were calculated as $(E_{slab/Ov} + \frac{1}{2} E_{O2} - E_{slab})$, where $E_{slab/Ov}$ and E_{slab} are the total energies of the ceria supercell with and without the O vacancy, respectively, whereas E_{O2} is the total energy of a gas-phase O_2 molecule compensated for the known overbinding predicted by (semi)local functionals for O_2.

References

1. Yang, X. et al. Single-atom catalysts: a new frontier in heterogeneous catalysis. Acc. Chem. Res. **46**, 1740–1748 (2013).
2. Flytzani-Stephanopoulos, M. & Gates, B. C. Atomically dispersed supported metal catalysts. Annu. Rev. Chem. Biomol. Eng. **3**, 545–574 (2012).
3. Fu, Q., Saltsburg, H. & Flytzani-Stephanopoulos, M. Active nonmetallic Au and Pt species on ceria-based water-gas shift catalysts. Science **301**, 935–938 (2003).
4. Bruix, A. et al. Maximum noble-metal efficiency in catalytic materials: atomically dispersed surface platinum. Angew. Chem. Int. Ed. **53**, 10525–10530 (2014).
5. Qiao, B. et al. Single-atom catalysis of CO oxidation using Pt$_1$/FeO$_x$. Nat. Chem. **3**, 634–641 (2011).
6. Novotný, Z. et al. Ordered array of single adatoms with remarkable thermal stability: Au/Fe$_3$O$_4$(001). Phys. Rev. Lett. **108**, 216103 (2012).
7. Parkinson, G. S. et al. Carbon monoxide-induced adatom sintering in a Pd-Fe$_3$O$_4$ model catalyst. Nat. Mater. **12**, 724–728 (2013).
8. Bliem, R. et al. Subsurface cation vacancy stabilization of the magnetite (001) surface. Science **346**, 1215–1218 (2014).
9. Li, F., Li, Y., Zeng, X. C. & Chen, Z. Exploration of high-performance single-atom catalysts on support M$_1$/FeO$_x$ for CO oxidation via computational study. ACS Catal. **5**, 544–552 (2015).
10. Hatanaka, M. et al. Ideal Pt loading for a Pt/CeO$_2$-based catalyst stabilized by a Pt-O-Ce bond. Appl. Catal. B Environ. **99**, 336–342 (2010).
11. Fiala, R. et al. Proton exchange membrane fuel cell made of magnetron sputtered Pt-CeO$_x$ and Pt-Co thin film catalysts. J. Power Sources **273**, 105–109 (2015).
12. Gong, X.-Q., Selloni, A., Batzill, M. & Diebold, U. Steps on anatase TiO$_2$(101). Nat. Mater. **5**, 665–670 (2006).
13. Barth, J. V., Costantini, G. & Kern, K. Engineering atomic and molecular nanostructures at surfaces. Nature **437**, 671–679 (2005).
14. Vang, R. T. et al. Controlling the catalytic bond-breaking selectivity of Ni surfaces by step blocking. Nat. Mater. **4**, 160–162 (2005).
15. Gong, X., Selloni, A., Dulub, O., Jacobson, P. & Diebold, U. Small Au and Pt clusters at the anatase TiO$_2$(101) surface: behavior at terraces, steps, and surface oxygen vacancies. J. Am. Chem. Soc. **130**, 370–381 (2008).
16. Dvořák, F. et al. Adjusting morphology and surface reduction of CeO$_2$(111) thin films on Cu(111). J. Phys. Chem. C **115**, 7496–7503 (2011).
17. Duchoň, T. et al. Ordered phases of reduced ceria as epitaxial films on Cu(111). J. Phys. Chem. C **118**, 357–365 (2014).
18. Zhou, Y., Perket, J. M. & Zhou, J. Growth of Pt nanoparticles on reducible CeO$_2$ (111) thin films: effect of nanostructures and redox properties of ceria. J. Phys. Chem. C **114**, 11853–11860 (2010).
19. Shao, X., Jerratsch, J.-F., Nilius, N. & Freund, H.-J. Probing the 4f states of ceria by tunneling spectroscopy. Phys. Chem. Chem. Phys. **13**, 12646–12651 (2011).
20. Lu, J.-L., Gao, H.-J., Shaikhutdinov, S. & Freund, H.-J. Morphology and defect structure of the CeO$_2$(111) films grown on Ru(0001) as studied by scanning tunneling microscopy. Surf. Sci. **600**, 5004–5010 (2006).
21. Zhou, Y. & Zhou, J. Interactions of Ni nanoparticles with reducible CeO$_2$(111) thin films. J. Phys. Chem. C **116**, 9544–9549 (2012).
22. Mullins, D. R. et al. Water dissociation on CeO$_2$(100) and CeO$_2$(111) thin films. J. Phys. Chem. C **116**, 19419–19428 (2012).
23. Bruix, A. et al. A new type of strong metal-support interaction and the production of H$_2$ through the transformation of water on Pt/CeO$_2$(111) and Pt/CeO$_x$/TiO$_2$(110) catalysts. J. Am. Chem. Soc. **134**, 8968–8974 (2012).
24. Negreiros, F. R. & Fabris, S. Role of cluster morphology in the dynamics and reactivity of subnanometer Pt clusters supported on ceria surfaces. J. Phys. Chem. C **118**, 21014–21020 (2014).
25. Kozlov, S. M., Viñes, F., Nilius, N., Shaikhutdinov, S. & Neyman, K. M. Absolute surface step energies: accurate theoretical methods applied to ceria nanoislands. J. Phys. Chem. Lett. **3**, 1956–1961 (2012).
26. Torbrügge, S., Cranney, M. & Reichling, M. Morphology of step structures on CeO$_2$(111). Appl. Phys. Lett. **93**, 073112 (2008).
27. James, T. E., Hemmingson, S. L. & Campbell, C. T. Energy of supported metal catalysts: from single atoms to large metal nanoparticles. ACS Catal. **5**, 5673–5678 (2015).
28. James, T. E., Hemmingson, S. L., Ito, T. & Campbell, C. T. Energetics of Cu adsorption and adhesion onto reduced CeO$_2$(111) surfaces by calorimetry. J. Phys. Chem. C **119**, 17209–17217 (2015).
29. Colussi, S. et al. Nanofaceted Pd-O sites in Pd-Ce surface superstructures: Enhanced activity in catalytic combustion of methane. Angew. Chem. Int. Ed. **48**, 8481–8484 (2009).
30. Kozlov, S. M. & Neyman, K. M. O vacancies on steps on the CeO$_2$(111) surface. Phys. Chem. Chem. Phys. **16**, 7823–7829 (2014).
31. Fiala, R. et al. Pt-CeO$_x$ thin film catalysts for PEMFC. Catal. Today **240**, 236–241 (2015).
32. Hatanaka, M. et al. Reversible changes in the Pt oxidation state and nanostructure on a ceria-based supported Pt. J. Catal. **266**, 182–190 (2009).
33. Nagai, Y. et al. Sintering inhibition mechanism of platinum supported on ceria-based oxide and Pt-oxide-support interaction. J. Catal. **242**, 103–109 (2006).
34. Matolín, V. et al. Platinum-doped CeO$_2$ thin film catalysts prepared by magnetron sputtering. Langmuir **26**, 12824–12831 (2010).
35. Zhou, J., Baddorf, A. P., Mullins, D. R. & Overbury, S. H. Growth and characterization of Rh and Pd nanoparticles on oxidized and reduced CeO$_x$(111) thin films by scanning tunneling microscopy. J. Phys. Chem. C **112**, 9336–9345 (2008).
36. Zhou, Y., Perket, J. M. & Zhou, J. Growth of Pt nanoparticles on reducible CeO$_2$(111) thin films: effect of nanostructures and redox properties of ceria. J. Phys. Chem. C **114**, 11853–11860 (2010).
37. Zhou, Y. & Zhou, J. Growth and sintering of Au – Pt nanoparticles on oxidized and reduced CeO$_x$(111) thin films by scanning tunneling microscopy. J. Phys. Chem. Lett. **1**, 609–615 (2010).
38. Sayle, T. X. T., Parker, S. C. & Sayle, D. C. Oxidising CO to CO$_2$ using ceria nanoparticles. Phys. Chem. Chem. Phys. **7**, 2936–2941 (2005).
39. Castellani, N. J., Branda, M. M., Neyman, K. M. & Illas, F. Density functional theory study of the adsorption of Au atom on cerium oxide: effect of low-coordinated surface sites. J. Phys. Chem. C **113**, 4948–4954 (2009).
40. Zambelli, T., Wintterlin, J., Trost, J. & Ertl, G. Identification of the 'active sites' of a surface-catalyzed reaction. Science **273**, 1688–1690 (1996).
41. Matolín, V. et al. Water interaction with CeO$_2$(111)/Cu(111) model catalyst surface. Catal. Today **181**, 124–132 (2012).
42. Mullins, D. R. The surface chemistry of cerium oxide. Surf. Sci. Rep. **70**, 42–85 (2015).
43. Cococcioni, M. & de Gironcoli, S. Linear response approach to the calculation of the effective interaction parameters in the LDA + U method. Phys. Rev. B **71**, 035105 (2005).
44. Perdew, J. P. J., Burke, K. & Ernzerhof, M. Generalized gradient approximation made simple. Phys. Rev. Lett. **77**, 3865–3868 (1996).
45. Vanderbilt, D. Soft self-consistent pseudopotentials in a generalized eigenvalue formalism. Phys. Rev. B **41**, 7892–7895 (1990).
46. Giannozzi, P. et al. QUANTUM ESPRESSO: a modular and open-source software project for quantum simulations of materials. J. Phys. Condens. Matter **21**, 395502 (2009).

47. Fabris, S., de Gironcoli, S., Baroni, S., Vicario, G. & Balducci, G. Taming multiple valency with density functionals: a case study of defective ceria. *Phys. Rev. B* **71**, 041102 (2005).

48. Fabris, S., Vicario, G., Balducci, G., De Gironcoli, S. & Baroni, S. Electronic and atomistic structures of clean and reduced ceria surfaces. *J. Phys. Chem. B* **109**, 22860–22867 (2005).

Acknowledgements

This work was supported by Czech Science Foundation (contract numbers 15-06759S and 13-10396S), and by the European Union via the FP7-NMP-2012 project chipCAT under contract number 310191 and the EU FP7 COST action CM1104. A.T. acknowledges the support of the Grant Agency of the Charles University, contract number 2048514. S.F. acknowledges the support provided by the Humboldt Foundation through a Friedrich Wilhelm Bessel Research Award. The high-performance computing resources were gratefully provided by ISCRA initiative of CINECA. CERIC-ERIC consortium is acknowledged for financial support.

Author contributions

F.D., A.T., M.V., T.S., I.M., J.M. and V.M. designed and performed the experiments. M.F.C., N.-D.T., F.R.N. and S.F. performed the DFT calculations. All authors interpreted the experimental and computational results. J.M., S.F., T.S., I.M. and V.M. wrote the manuscript. F.D., A.T., J.M., I.M., V.M. and S.F. provided funding.

Additional information

Competing financial interests: The authors declare no competing financial interests.

Micro-total envelope system with silicon nanowire separator for safe carcinogenic chemistry

Ajay K. Singh[1], Dong-Hyeon Ko[1], Niraj K. Vishwakarma[1], Seungwook Jang[1], Kyoung-Ik Min[1] & Dong-Pyo Kim[1]

Exploration and expansion of the chemistries involving toxic or carcinogenic reagents are severely limited by the health hazards their presence poses. Here, we present a micro-total envelope system (μ-TES) and an automated total process for the generation of the carcinogenic reagent, its purification and its utilization for a desired synthesis that is totally enveloped from being exposed to the carcinogen. A unique microseparator is developed on the basis of SiNWs structure to replace the usual exposure-prone distillation in separating the generated reagent. Chloromethyl methyl ether chemistry is explored as a carcinogenic model in demonstrating the efficiency of the μ-TES that is fully automated so that feeding the ingredients for the generation is all it takes to produce the desired product. Syntheses taking days can be accomplished safely in minutes with excellent yields, which bodes well for elevating the carcinogenic chemistry to new unexplored dimensions.

[1] National Center of Applied Microfluidic Chemistry, Department of Chemical Engineering, POSTECH (Pohang University of Science and Technology), Pohang 37673, Korea. Correspondence and requests for materials should be addressed to D.-P.K. (email: dpkim@postech.ac.kr).

It is not uncommon that carcinogenic reagents are needed for chemical syntheses. In fact, a number of carcinogenic reagents are utilized for the synthesis of drugs and other valuable products[1]. In certain syntheses, no alternative reagents are available. A case in point is chloromethylmethyl ether (CMME). It is one of the most popular chemical reagents and has drawn the attention of many chemists during the past decades in both academia and industry[2,3]. It has been broadly used in multi-step syntheses of drugs and natural products, including bactericides and pesticides, and functional chemicals, solvent for polymerization reactions, as well as acid-sensitive protecting groups for alcohols, phenols, thiols and carboxylic acids[4-9]. In particular, chloromethylation reaction is an important intermediate step to prepare anion exchange membrane for alkaline fuel cells, desalination and electro-dialysis applications[10].

The moisture sensitive CMME is highly carcinogenic and genotoxic by reacting spontaneously with nucleophilic DNA in the absence of enzyme, whereas even little exposure causes sore throat with fever and difficulty in breathing[11]. Despite the fact, there have been no clear-cut solutions to minimizing direct exposure to the carcinogenic reagent in CMME chemistry[2,12,13]. Therefore, scientific and industrial use of toxic CMME has been faced with serious safety issues in the synthesis, separation and transportation, which turned away many potential opportunities in this field. Thus, a safe and efficient chemical approach is needed to expand the scope of the carcinogenic chemistry, including the CMME chemistry, to new unexplored dimensions.

Continuous-flow microfluidic device has emerged as an efficient synthetic tool with attractive advantages such as high surface-to-volume ratio, and excellent mass and heat transfer, which leads to an enhancement in selectivity and a reduction in reaction time[14-25]. Recently, there have been several attempts to minimize the safety issues in the microfluidic processing of risky chemicals by separation through the embedded membranes[16,17,23,26-29]. However, the vulnerable polymer membrane as a physical barrier lowered diffusion rate and limited the operating conditions[30]. It is still challenging to demonstrate a zero exposure carcinogenic chemistry with excellent performance.

To realize the concept of a total process in this light, we present micro-total envelope system μ-TES), an automated total process that is enveloped totally from being exposed to carcinogen reagents. The μ-TES platform consisting of microfluidic devices enables in situ generation of the carcinogenic reagent, its separation from the reaction products, subsequent synthesis of the desired product with the carcinogenic reagent and decomposition of the unreacted carcinogenic reagent by quenching, separating the final desired product, all in a safe sequential manner. This μ-TES for CMME chemistry is depicted in Fig. 1, showing the four microfluidic sub-systems for generation, separation, reaction and quenching. The quenching part is elaborated in more detail later.

Results

In situ generation of CMME compound.
For in situ generation of CMME, hexanoyl chloride and dimethoxy methane as substrates were chosen among others[11]. A single by-product of non-volatile methyl hexanoate (boiling point 151 °C) was formed in this CMME generation because of its stoichiometric yield with extensive substrate availability[11]. Two reactants in separate syringes were injected into a polytetrafluoroethylene (PTFE) capillary microreactor through a T-mixer (T_1), and then passed through the tubing (id = 300 μm, length = 5.0 m) at 55 °C. Their flow rates were adjusted to maintain the 1:1 molar ratio stoichiometry (Supplementary Table 1). A back pressure regulator (BPR) was necessary to suppress the volatility of low boiling point dimethoxymethane (42 °C) reactant for a homogeneous liquid with no phase segregation. In general, a longer retention time promoted a higher CMME production. A residence time of 6 min at 40 p.s.i. was found to nearly complete the exchange reaction of dimethoxymethane to reach 98% yield of CMME (Supplementary Table 1). In contrast, a bulk CMME synthesis required 18 h of long reaction time[11]. The solvothermal-like condition in the capillary microreactor with intrinsic microfluidic advantages in mass and thermal transfer could be responsible for the accelerated synthesis.

Fabrication of membrane-free SiNW microseparator.
Routine purification of CMME by separation from reaction mixture invariably involves batch distillation under argon atmosphere, which poses safety and human health issues[11]. For the μ-TES, we developed a continuous-flow microfluidic device for safe and efficient purification of CMME from the in situ generated reaction mixture. This novel microseparator based on silicon nanowires (SiNWs) is shown in Fig. 2. The unique feature of this microseparator is that it does not require any membrane for the separation because of the definitive gas–liquid interface formed in the stable laminar flow because of the presence of SiNWs. This direct and side-by-side contact between gas and liquid phases that is rendered by simple feeding of contacting gas and liquid streams is most desirable for efficient diffusion kinetics for diverse gas separation tasks. It is definitely distinguished from the traditional gas–liquid segmented flow[31] and microchemical distillation systems[32] that required a low gas permeable membrane or a carrier gas (He or N_2) at elevated temperatures.

The microseparator is essentially a long serpentine tunnel imbedded in a solid block of a polydimethylsiloxane (PDMS) slab bonded to a silicon wafer patterned by photolithography[33]. The bottom part of the tunnel is formed in the chemically resistant PDMS slab, whereas the matching upper part is filled with the cone-shaped SiNWs clusters (Fig. 2a,d) fabricated by silver-assisted selective etching of silicon wafer[33-35]. The cone-shaped SiNWs were decorated with SiO_2 nanoparticles by a

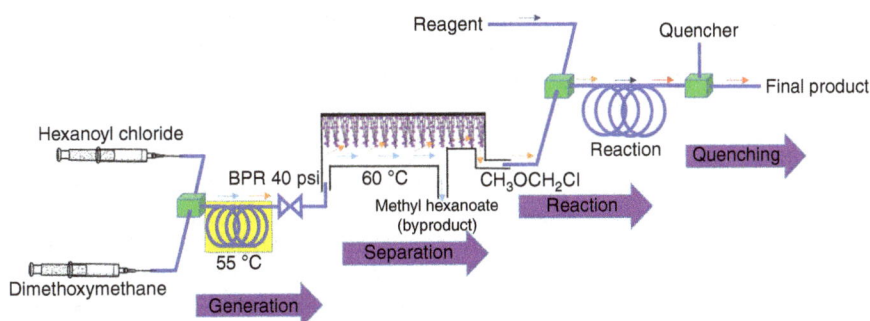

Figure 1 | Continuous-flow μ-TES. for safe utilization of carcinogenic chloromethyl methyl ether (CMME) via generation of CMME, separation, in situ consumption for forming a final product and quenching step.

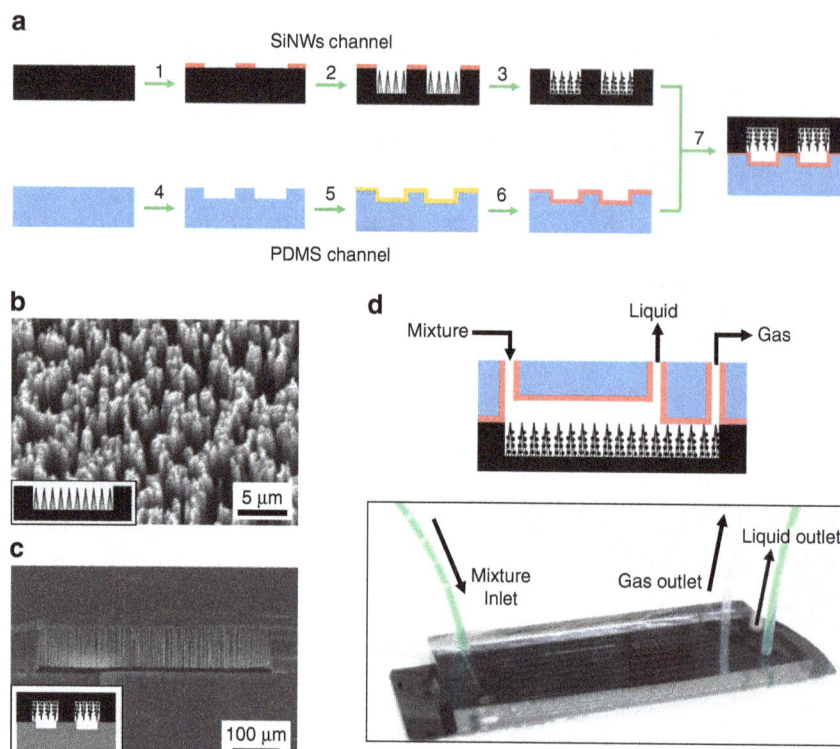

Figure 2 | Microseparator. (**a**) Schematic illustration of fabrication of membrane-free SiNW microseparator. Step 1: patterning of AZ photoresist protective layer on Si. Step 2: preparation of SiNW pattern. Step 3: fabrication of superamphiphobic SiNWs pattern. Step 4: spin-coating polyvinylsilazane (PVSZ) on plasma-treated PDMS channel. Step 5: ultraviolet curing of PVSZ-coated channel. Step 6: thermal bonding of PVSZ-coated PDMS channel on the SiNW channel. SEM images for (**b**) cross-sectional view of SiNW microseparator, (**c**) top view of cone-shaped SiNWs (100-300 nm in diameter and 75 μm in length). (**d**) Optical image of SiNWs microseparator, showing one inlet and two outlets.

sol-gel process to form a hierarchical structure with rough surface. They were then fluorinated to lower the surface energy and to enhance the chemical resistivity of the SiNWs surface (see Supplementary Methods).

This SiNWs surface was found to be superamphiphobic (both superhydrophobic and superoleophobic) as revealed by the high contact angles (CAs) of water (164°) on the surface, and of three organic solvents of dimethylsulphoxide (155°), hexadecane (120°) and methyl hexanoate (105.5°). The chemical and thermal stability of the prepared superamphiphobic SiNWs surface was confirmed by exposing to corrosive gases (either HCl or NH_3 vapour for 24 h) and thermal stress (300 °C for 1 h in air) and observing no change of CAs (Supplementary Fig. 6). The scanning electron microscopy (SEM) images in Fig. 2b,c and Supplementary Fig. 8 reveal that the diameter of Si nanowires is in the range of 100–300 nm and the length is ~75 μm, which can be controlled by varying the etching time. The SiO_2 nanoparticles decorating the nanowires are ~20 nm in diameter (see Supplementary Fig. 8C,D for details). It should be pointed out here that the SiO_2 nanotexturing on the smooth surface of cone-shaped SiNWs bundles with micrometre-scale intervals led to a hierarchically structured surface, which enhances the superamphiphobicity considerably.

To fabricate a membrane-free SiNW microseparator, a chemical-resistant polyvinylsilazane layer (ca 2–10 nm) was applied on the PDMS slab with the patterned channel structure to bond with the silicon slab, as reported elsewhere[16]. The membrane-free SiNW microseparator has one inlet for infusing a product mixture, and two outlets for separated gas and remaining liquid, respectively. The dimensions of the SiNWs channel are 500 μm in width, 70 μm in height and 43 cm in length (Fig. 2d)

and those of the PDMS channel are 500 μm, 15 μm and 40 cm, respectively, for the width, height and length.

Performance of *in situ* CMME separation. The gas–liquid mixture of the products generated in the generation part of the μ-TES squirts into the microseparator through the inlet tube in bursts of CMME gas plug followed by liquid plug containing by-products. One major role of the SiNWs in the microseparator is to enable a complete separation of gas from a gas and liquid mixture. Highly superamphiphobic nature of the nanowires ensures that the liquid from the reactor does not splash into the upper SiNWs channel so that a gas phase laminar flow is established in the SiNWs channel. The other is to collect only gaseous product evaporating from the liquid phase flow in the bottom PDMS channel by establishing a definitive gas–liquid interface between the two channels. To demonstrate the stable gas–liquid laminar flow with direct interface contact, a combination of dyed water (100 μl min^{-1}) and air (100 μl min^{-1}) were injected through individual inlets. No bubbling was observed in the liquid flow along a 40-cm microchannel, and no dyed water was observed in the gas outlet dipped into pure methyl hexanoate (see Supplementary Movie 1). The stable gas–liquid laminar flow was maintained under various flow rates of gas and liquids (methyl hexanoate) in the range of 0.5–100 μl min^{-1}. To the best of our knowledge, it is reported for the first time that a superamphiphobic surface enables handling of both liquid (organic solvent or water) and gas under dynamic flow conditions.

The separation efficiency for the generated CMME was optimized with respect to the channel dimensions, temperature and surface condition of Si nanowires. A low PDMS channel height (15 μm) relative to the height of SiNWs channel (75 μm) provided a short distance for the CMME to evaporate into the SiNWs channel. As shown in Table 1, the yield of CMME after the separation turned out to be 96% at 60 °C, which means that 98% of the CMME produced was separated because the reaction yield was 98%. Table 1 also clearly shows the beneficial effects of adding SiO_2 nanoparticles and making the surface fluorinated,

both of which render the structure more superamphiphobic. With the optimized conditions, the generated CMME was distilled quickly (3 min of residence time) at a rate of 17.5 μl min^{-1} with 98% separation efficiency. In the separation, the relatively non-volatile methyl hexanoate remains in the liquid phase.

Total organic processes by μ-TES. With the *in situ* generation of CMME and its separation from the reaction mixture established, it remains to demonstrate a safe CMME chemistry in the μ-TES. The protection of acid-sensitive phenol group by forming a heteroatom C–O bond was taken up first as a model reaction to utilize the purified CMME as a reactive electrophile in a serial process, which is suitable for synthesis of drug and selective functionalized chemistry[4–7]. For the reaction, a solution of phenol (1 M) in dichloromethane (DCM), and diisopropylethylamine (DIPEA) were separately introduced into PTFE tube through a T-shaped mixer by syringe pumps (Fig. 3). The mixed phenol and DIPEA solution was directly connected by a T-mixer (T2) into the out-flowing CMME (17.5 μl min^{-1}) from the microseparator to provide a molar ratio of 1:1.15 (phenol/CMME), which was passed through a PTFE tube (id = 500 μm, length = 560 cm). Temperature, reagent concentration and residence time (Supplementary Table 2) were optimized to form themethoxymethyl (MOM)-phenol product. Eventually, 99% yield of (methoxymethoxy)benzene (**2a**) was obtained in 4.7 min residence time at room temperature and ambient pressure. In

Table 1 | Continuous-flow separation performance of the membrane-free SiNW microseparator for the *in situ* synthesized CMME compound.

Entry	Hexanoyl chloride flow rate (μl min^{-1})	Dimethoxy methane flow rate (μl min^{-1})	Temperature (°C)	CH_3OCH_2Cl yield (%)*
1	36	22	60	96
2	36	22	55	70
3	36	22	50	2
4†	36	22	60	88
5‡	36	22	60	75

CMME, chloromethylmethyl ether; SiNW, silicon nanowire.
*Isolated yield is based on NMR analysis, retention time 3 min
†Without SiO_2 nanoparticle coating on SiNWs.
‡Without trichloro(1H, 1H, 2H, 2H-perfluorooctyl)silane coating.

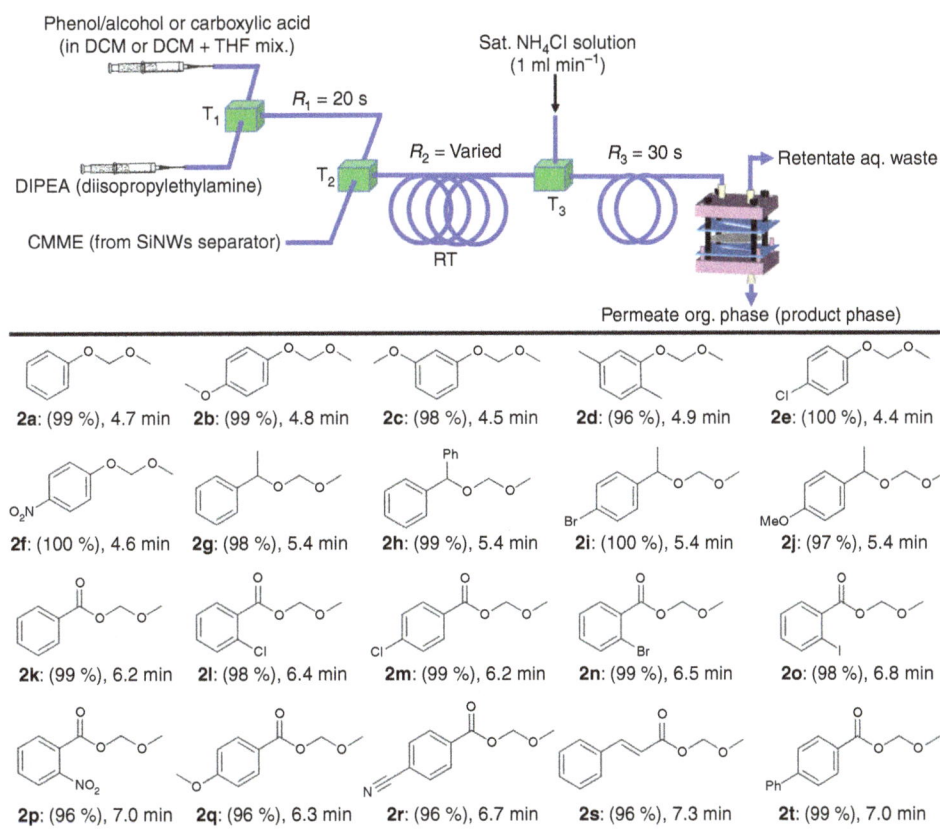

2a: (99 %), 4.7 min   2b: (99 %), 4.8 min   2c: (98 %), 4.5 min   2d: (96 %), 4.9 min   2e: (100 %), 4.4 min

2f: (100 %), 4.6 min   2g: (98 %), 5.4 min   2h: (99 %), 5.4 min   2i: (100 %), 5.4 min   2j: (97 %), 5.4 min

2k: (99 %), 6.2 min   2l: (98 %), 6.4 min   2m: (99 %), 6.2 min   2n: (99 %), 6.5 min   2o: (98 %), 6.8 min

2p: (96 %), 7.0 min   2q: (96 %), 6.3 min   2r: (96 %), 6.7 min   2s: (96 %), 7.3 min   2t: (99 %), 7.0 min

Figure 3 | Functional group protection Expanded reaction and quenching parts in Fig. 1 for the continuous-flow μ-TES for alkoxyalkylation reactions to obtain MOM group-protected phenol, alcohols and carboxylic acid products at room temperature and ambient pressure: automated serial steps of CMME generation (6 min), CMME separation (3 min), *in situ* consumption (time in the table) and quenching/separation step (30 s). (**2a–2f**: phenols (1 M in DCM, flow rate 180 μl min^{-1}, DIPEA (37 μl min^{-1})); alcohol products (**2g–2j**: alcohol (1 M in DCM, flow rate 170 μl min^{-1}), DIPEA (37 μl min^{-1})) and acid products (**2k, 2l** and **2s**: acid (1 M in DCM, flow rate 190 μl min^{-1}), DIPEA (37 μl min^{-1}); **2m, 2q** and **2t**: acid (1 M in DCM + THF mixture (20:1 ratio), flow rate 180 μl min^{-1}, DIPEA (37 μl min^{-1})); **2r**: acid (1 M in DCM + THF mixture (4:1 ratio), flow rate 180 μl min^{-1}, DIPEA (37 μl min^{-1})). Isolated yields in parenthesis by NMR analysis were averaged by conducting three experiments at least and the data error is within ±1.5%. THF, tetrahydrofuran.

contrast, the conventional batch process for the identical reaction showed lower yields (61–85%) even under excess use of CMME (1.5–2 equivalent) and long reaction time (24 h) (ref. 36) at lowered temperature[37]. With the intrinsically good characteristics of microreactor, the exothermic reaction of carcinogenic CMME with DIPEA and phenol could be conducted at room temperature.

For the total process strategy, a quenching step was devised to decompose the unreacted toxic CMME as well as to remove the reaction impurities (DIPEA, its salt DIPEA · HCl) in the product mixture. This continuous process of removing toxic and waste chemicals must contribute to reducing a tedious workup step[38] and accomplishing no exposure of carcinogenic CMME compound. In view of previous studies for quenching chemicals[2] and removing unwanted chemicals by extraction[16,17,23], an aqueous NH$_4$Cl saturated solution (flow rate = 1 ml min^{-1}) was infused through a T-junction to mix with the organic flowing out of the integrated capillary microreactor, forming organic-aqueous droplets for selective extraction of reaction wastes (excess DIPEA/their salt and excess of CMME) that preferentially dissolve in water (Fig. 3) (ref. 39). The extraction between liquid droplets was accomplished in the PTFE capillary (id = 500 μm, length = 320 cm, R_3 = 30 s, Supplementary Movie 2). Subsequently, the organic phase containing only the target product was separated from the aqueous phase using a PTFE membrane-embedded microseparator. The thin fluoropolymer-based PTFE membrane was preferentially wetted by the organic solvent (product) that permeated the membrane holes, whereas the non-wetting aqueous phase (waste) did not penetrate the membrane (details in Supplementary Methods) as reported in the literature[16,17,23]. Complete separation of the two phases was achieved at high flow rates (up to 300 μl min^{-1}).

Various protected phenol products were obtained with excellent yields over 96% (Fig. 3, **2a–2f**) in 13.9–14.4 min by the fully automated μ-TES (CMME generation (6 min), CMME separation (3 min), *in situ* consumption (4.4–4.9 min) and quenching/separation step (30 s)). In contrast, the traditional batch methods require 16 h to produce these substituted phenol products at 10.4 – 10.8 mmol h^{-1}. The identical methodology has been further applied to protecting alcohol group with excellent yields over 97% in 5.4 min of *in situ* consumption time (Fig. 3, **2g–2j**), whereas a batch process generally needed 16 h using excess (2.0–2.1 equivalent) CMME. In the case of protecting benzoic acid group, the reaction was nearly completed in only 6.2 min of *in situ* consumption time at room temperature (Fig. 3, **2k**). Insoluble DIPEA · HCl salt in DCM solvent caused tube clogging problem to an extent, which required the use of a co-solvent system (DCM + THF) and controlled concentration of the carboxylic acid reactant to dissolve the precipitate. Several substituted aromatic carboxylic acids (*o*-Cl, *p*-Cl, *o*-Br, *o*-I, *o*-NO2, *p*-MeO and *p*-CN) were successfully protected in excellent yields (**2l–2q**, 96–99%) within 6.2–7 min of *in situ* consumption time under the same system and the conditions for phenol group protection. Aliphatic- and biphenyl-based carboxylic acids (**2r** and **2s**) were also successfully protected with yields up to 96% (Fig. 3), which is superior to the conventional batch process that generally required several hours with additional cooling system.

Polymer chloromethylation by μ-TES. Moreover, an automated total process of CMME generation, separation, *in situ* consumption and quenching step was extended to the formation

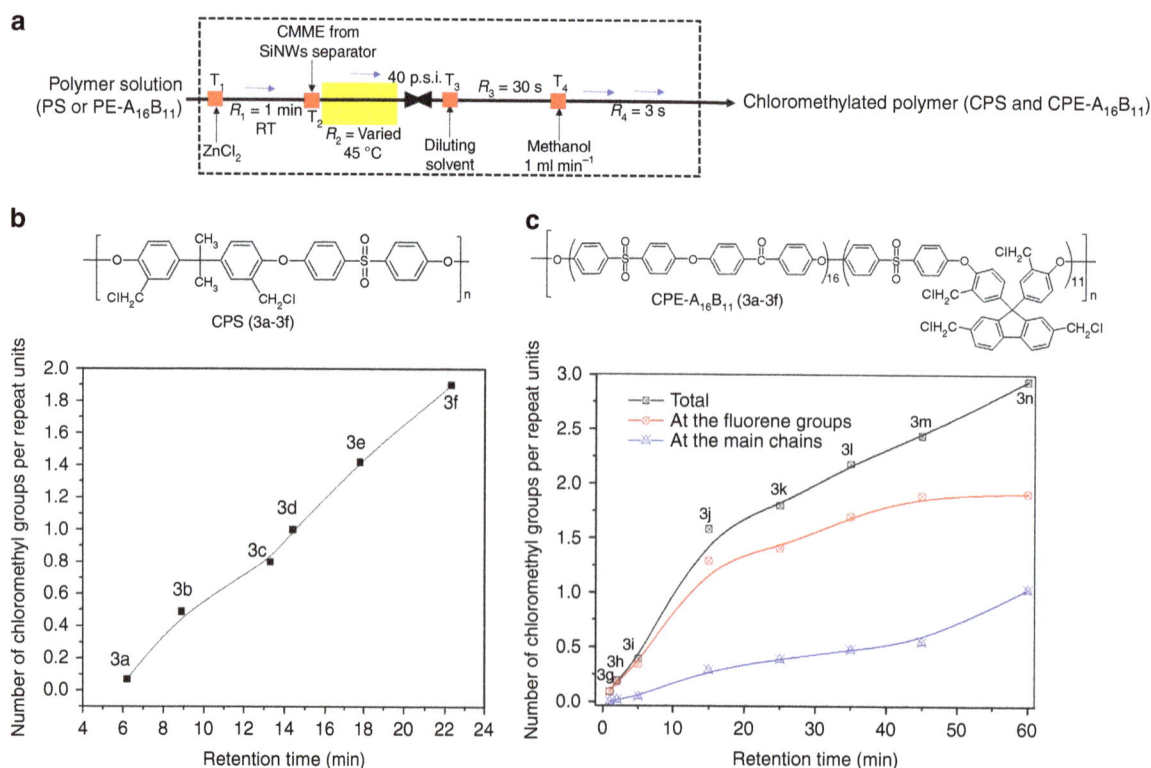

Figure 4 | Polymer functionalization. (**a**) Continuous-flow chloromethylation by an automated μ-TES (CMME generation (6 min), distilled separation (3 min), *in situ* consumption (times in the graph) and quenching/separation step (30 s)); (**b**) chloromethylation of polysulphone polymer (CPS) reaction condition: Polysulphone (1.2 gm in 10 ml THF solution, flow rate = 160 μl min^{-1}), ZnCl$_2$ in THF solution (0.05 M in THF, flow rate = 14 μl min^{-1}), CMME flow rate = 17.0 ± 0.5 μl min^{-1}); (**c**) PE-A$_{16}$B$_{11}$ polymer (1.0 g (fluorine unit = 1.03 mmol) in 20 ml TCE solution (flow rate 101 ml min^{-1}), ZnCl$_2$ in THF solution (0.25 M, flow rate = 11 μl min^{-1}), CMME flow rate = 17.0 ± 0.5 μl min^{-1}. The number of chloromethyl groups per repeat unit was measured by ^{1}H-NMR analysis and the data error is within ±1.5%.

of carbon–carbon bond via Friedel-craft reaction that is a significant route to functionalizing aromatic polymers or anion exchange resin through electrophilic substitution[13,40,41]. The degree of chloromethylation is very critical to the ion exchange capacity of polymer membrane, which is always challenging for the membrane scientists[42]. Polysulphone polymer (containing aromatic backbone) was chosen as a model substrate for Friedel-craft chloromethylation (Fig. 4). A solution of polysulphone (in THF) and $ZnCl_2$ (in THF) was mixed at a T-junction, which was then mixed at another T junction with the purified CMME from the membrane-free SiNW microseparator (see Fig. 4a and Supplementary Fig. 15 for more details). The reaction mixture was then introduced into a capillary microreactor system with endpoint BPR, which was submerged in a preheated oil bath at different temperatures (30–45 °C), and back pressures (40 p.s.i.). The reaction time of CMME ranged from 1 to 22 min with $ZnCl_2$ catalyst (0.25 M in THF). In general, the degree of chloromethylation (based on NMR analysis) gradually increased with longer reaction time (Fig. 4b, 3a–3e), eventually reaching 1.90 without gelation at the optimal conditions (40 p.s.i. BPR, 40 equivalent CMME concentration, 45 °C and 22 min chloromethylation, 3f) by preferably functionalizing the electron providing propane-2,2-diylgroup containing unit. In contrast, the conventional batch process yields a lower degree of chloromethylation (1.69) even with longer reaction time (90 min) at high temperature (75 °C) (ref. 20). The quenching/separation of the unreacted CMME by extraction through methanol-aqueous droplets was conducted in a continuous-flow device to reduce manpower and exposure of unreacted CMME chemicals.

Recently, aromatic multiblock copolymers have drawn interest as a new class of polymeric anion exchange membrane because the degree of chloromethylation could be highly improved[43]. The long process time over 144 h, however, could increase the risk of carcinogenic exposure to manpower. Therefore, the µ-TES was applied to chloromethylate a synthesized poly(arylene ether) block copolymer (PE-$A_{16}B_{11}$; details in Supplementary Methods)[43]. As shown in Fig. 4c and Supplementary Fig. 19, the degree of chloromethylation in the A unit of PE-$A_{16}B_{11}$ block copolymer could not exceed 1.0 for 60 min in situ reaction time at 45 °C with 80 equivalent of CMME, because electron-withdrawing ketone and sulfone groups were less reactive with electrophiles. In contrast, the fluorene groups of the B unit in the PE-$A_{16}B_{11}$ block copolymer were more reactive to be chloromethylated, considerably more at preferable positions (2 and 7) due to the absence of electron-withdrawing group[43].

Discussion

We have developed a µ-TES and established a zero exposure of carcinogenic reagent for CMME chemistry using the µ-TES. This autonomous serial process of CMME generation, self-purification, separation, reaction and quenching system does not require any additional workup and column chromatography, which completely remove the safety issues involving risky compounds in the chemical processes. In particular, a novel membrane-free SiNWs microseparator developed here has been shown to allow for the separation of low boiling chemicals by simple heating in a continuous-flow manner. This total process concept including the integrated system and procedure can be easily extended to other carcinogenic, explosive, toxic or noxious regents. More importantly, the system provided here would provide a safe and fast avenue for those potential workers who turned away or using alternative routes due to the safety and longevity issues in the areas of drug discovery, natural products, ion-exchange membranes, materials synthesis and biology.

Methods

General. Used material details given in Supplementary Methods. GC/MS spectrum was recorded by Agilent 5975C GC/MSD System (Agilent Technologies). ^{1}H NMR and ^{13}C NMR spectra were recorded on a Bruker 600 or 300. Field emission scanning electron microscope (FE-SEM) images were obtained using JSM-6700F. The synthesized product compounds were fully characterized by their Mass-spectra, ^{1}H and ^{13}C NMR data, by collecting for 30 min unless otherwise noted. Optimization yields were average of at least two experiments. Electron ionization mass spectra were recorded on 5675C VL MSD spectrometer (Agilent Technologies). CAs were measured using a SmartDrop (FemtoFab). The etched fraction of SiNWs was analysed by calculating the ratio of etched area to the whole area of SiNWs from the SEM image through Image J software.

Typical procedure for total organic process. A solution of substituted organic compound in DCM or THF + DCM mixture and excess DIPEA reactant were introduced to a T-mixer (T_1) of molar ratio of 1:1.15 to maintain reaction stoichiometry, and then passed through a PTFE tubing (id = 500 µm, length = varied) for varied retention time. Properly mixed organic compound and DIPEA solution were connected by another T-mixer (T_2) and directly to outlet of the membrane-free SiNW microseparator (see in Supplementary Methods in more details). Out-flowing CMME (17.5 µl min^{-1}) and the substrate were controlled to become molar ratio 1:1.15 (organic compound/CMME). The three components (phenol, DIPEA and out-flowing CMME) were mixed through T-junction (T_2; see Supplementary Fig. 10) and infused to PTFE tubing (id = 500 µm, length = varied) for varied time. Note that the tube length was varied with different retention times of chloromethylation. The organic group protection setup as aforementioned in the Supplementary Fig. 10 was connected to a quencher inlet for in situ decomposition of CMME and extraction removal of excess DIPEA/their salt. The saturated aqueous NH_4Cl quenching solution was merged to the DCM-based reaction mixture by T-junction to form aqueous-organic droplets (see Supplementary Fig. 11 for more details). Sufficient quenching and extraction from the above reaction mixture was observed at 1 ml min^{-1} flow rate of aqueous NH_4Cl. To separate organic phase containing the wanted product from the aqueous impurity phase, the additional PFFE membrane embedded separator was added to end of the integrated microreactor system, as similarly reported (see Supplementary Fig. 12 for more details).

Typical procedure for polymer chloromethylation. A solution of polymer (see Supplementary Methods in more details) was taken in one syringe and the solution of $ZnCl_2$ (varied) in THF was taken in another syringe (see Supplementary Figs 15 and 19 in more details). The two solutions were introduced to a T-mixer (T_1) in a flow rate (detail in Supplementary Methods), and then passed through PTFE tubing (id = 500 µm, length = varied) to mix for varied residence time. Next the reaction mixture was connected to outlet of CMME purified microreactor at a flow rate = 17.5 µl min^{-1} by another T-mixer (T_2) at 45 °C (Supplementary Figs 15 and 19 in more details). Three component mixture (polymer, $ZnCl_2$, CMME) was passed through a PTFE tubing (id = 500 µm, length = varied as mentioned in Supplementary Figs 15 and 19) for different retention times. The use of 40 p.s.i. BPR enhanced the degree of chloromethylation. Then controlled degree of chloromethylated polymer solution was connected with T-mixer (T_3), and THF/1,1,2,2-tetrachloroethane (TCE) solvent was added continuously (flow rate varied) and again passed through PTFE tubing (id = 500 µm, length = varied) to dilute the polymer solution for varied time. The properly diluted chloromethylated polymer solution was precipitated continuously by adding methanol (95% methanol and 5% water) solution (flow rate = 1 ml min^{-1}) through T-mixer (T_4) and passed through a PTFE tubing (id = 1,000 µm, length = 10 cm) to complete quenching of excess of $ZnCl_2$ and CMME during 3 s residence time with no clogging problem. Finally, the chloromethylated polymer was washed several times with de-ionized water to determine degree of chloromthylation by NMR analysis after drying.

References

1. Luch, A. Nature and nurture - lessons from chemical carcinogenesis. *Nat. Rev. Cancer* **5**, 113–125 (2005).
2. Berliner, M. A. & Belecki, K. Simple, rapid procedure for the synthesis of chloromethyl methyl ether and other chloro alkyl ethers. *J. Org. Chem.* **70**, 9618–9621 (2005).
3. Kim, J. *et al.* Catalytic polymerization of anthracene in a recyclable SBA-15 reactor with high iron content by a friedel–crafts alkylation. *Angew. Chem. Int. Ed.* **51**, 2859–2863 (2012).
4. Lazarski, K. E., Moritz, B. J. & Thomson, R. J. The total synthesis of isodon diterpenes. *Angew. Chem. Int. Ed.* **53**, 10588–10599 (2014).
5. Ito, H., Mitamura, Y., Segawa, Y. & Itami, K. Thiophene-based, radial π-conjugation: synthesis, structure, and photophysical properties of cyclo-1,4-phenylene-2′,5′-thienylenes. *Angew. Chem. Int. Ed.* **127**, 161–165 (2015).
6. Nishiuchi, T. & Iyoda, M. Bent π-conjugated systems composed of three-dimensional benzoannulenes. *Chem. Rec.* **15**, 329–346 (2015).

7. Yagi, A., Venkataramana, G., Segawa, Y. & Itami, K. Synthesis and properties of cycloparaphenylene-2,7-pyrenylene: a pyrene-containing carbon nanoring. *Chem. Commun.* **50**, 957–959 (2014).

8. Berkessel, A. *et al.* Umpolung by N-heterocyclic carbenes: generation and reactivity of the elusive 2,2-diamino enols (breslow intermediates). *Angew. Chem. Int. Ed.* **51**, 12370–12374 (2012).

9. Barl, N. M., Sansiaume-Dagousset, E., Karaghiosoff, K. & Knochel, P. Full functionalization of the 7-azaindole scaffold by selective metalation and sulfoxide/magnesium exchange. *Angew. Chem. Int. Ed.* **52**, 10093–10096 (2013).

10. Varcoe, J. R. *et al.* Anion-exchange membranes in electrochemical energy systems. *Energy Environ. Sci.* **7**, 3135–3191 (2014).

11. Linderman, R. J., Jaber, M. & Griedel, B. D. A simple and cost effective synthesis of chloromethyl methyl ether. *J. Org. Chem.* **59**, 6499–6500 (1994).

12. Barnes, D. M., Barkalow, J. & Plata, D. J. A facile method for the preparation of MOM-Protected carbamates. *Org. Lett.* **11**, 273–275 (2009).

13. Jasti, A., Prakash, S. & Shahi, V. K. Stable and hydroxide ion conductive membranes for fuel cell applications: chloromethylation and amination of poly(ether ether ketone). *J. Membr. Sci.* **428**, 470–479 (2013).

14. Nagaki, A., Ichinari, D. & Yoshida, J.-I. Three-component coupling based on flash chemistry. carbolithiation of benzyne with functionalized aryllithiums followed by reactions with electrophiles. *J. Am. Chem. Soc.* **136**, 12245–12248 (2014).

15. Poh, J.-S., Tran, D. N., Battilocchio, C., Hawkins, J. M. & Ley, S. V. A versatile room-temperature route to di- and trisubstituted allenes using flow-generated diazo compounds. *Angew. Chem. Int. Ed.* **54**, 7920–7923 (2015).

16. Basavaraju, K. C., Sharma, S., Maurya, R. A. & Kim, D.-P. Safe use of a toxic compound: heterogeneous OsO4 catalysis in a nanobrush polymer microreactor. *Angew. Chem. Int. Ed.* **52**, 6735–6738 (2013).

17. Sharma, S., Maurya, R. A., Min, K.-I., Jeong, G.-Y. & Kim, D.-P. Odorless isocyanide chemistry: an integrated microfluidic system for a multistep reaction sequence. *Angew. Chem. Int. Ed.* **52**, 7564–7568 (2013).

18. Becker, M. R. & Knochel, P. Practical continuous-flow trapping metalations of functionalized arenes and heteroarenes using TMPLi in the presence of Mg, Zn, Cu, or La halides. *Angew. Chem. Int. Ed.* **54**, 12501–12505 (2015).

19. Ley, S. V., Fitzpatrick, D. E., Ingham, R. J. & Myers, R. M. Organic synthesis: march of the machines. *Angew. Chem. Int. Ed.* **54**, 3449–3464 (2015).

20. Ingham, R. J. *et al.* A systems approach towards an intelligent and self-controlling platform for integrated continuous reaction sequences. *Angew. Chem. Int. Ed.* **54**, 144–148 (2015).

21. Gutmann, B., Cantillo, D. & Kappe, C. O. Continuous-Flow technology—a tool for the safe manufacturing of active pharmaceutical ingredients. *Angew. Chem. Int. Ed.* **54**, 6688–6728 (2015).

22. Gutmann, B., Elsner, P., Glasnov, T., Roberge, D. M. & Kappe, C. O. Shifting chemical equilibria in flow—efficient decarbonylation driven by annular flow regimes. *Angew. Chem. Int. Ed.* **126**, 11741–11745 (2014).

23. Maurya, R. A., Park, C. P., Lee, J. H. & Kim, D.-P. Continuous *in situ* generation, separation, and reaction of diazomethane in a dual-channel microreactor. *Angew. Chem. Int. Ed.* **50**, 5952–5955 (2011).

24. Singh, A. K. *et al.* Eco-efficient preparation of N-doped graphene equivalent and its application to metal free selective oxidation reaction. *Green Chem.* **16**, 3024–3030 (2014).

25. Fuse, S., Mifune, Y. & Takahashi, T. Efficient amide bond formation through a rapid and strong activation of carboxylic acids in a microflow reactor. *Angew. Chem. Int. Ed.* **53**, 851–855 (2014).

26. Elvira, K. S., i Solvas, X. C., Wootton, R. C. R. & deMello, A. J. The past, present and potential for microfluidic reactor technology in chemical synthesis. *Nat. Chem.* **5**, 905–915 (2013).

27. Tran, D. N., Battilocchio, C., Lou, S.-B., Hawkins, J. M. & Ley, S. V. Flow chemistry as a discovery tool to access sp2-sp3 cross-coupling reactions via diazo compounds. *Chem. Sci.* **6**, 1120–1125 (2015).

28. Gross, U., Koos, P., O'Brien, M., Polyzos, A. & Ley, S. V. A general continuous flow method for palladium catalysed carbonylation reactions using single and multiple tube-in-tube gas-liquid microreactors. *Eur. J. Org. Chem.* **2014**, 6418–6430 (2014).

29. Maurya, R. A., Min, K.-I. & Kim, D.-P. Continuous flow synthesis of toxic ethyl diazoacetate for utilization in an integrated microfluidic system. *Green Chem.* **16**, 116–120 (2014).

30. O'Brien, M., Baxendale, I. R. & Ley, S. V. Flow ozonolysis using a semipermeable Teflon AF-2400 membrane to effect gas – liquid contact. *Org. Lett.* **12**, 1596–1598 (2010).

31. Hartman, R. L., Sahoo, H. R., Yen, B. C. & Jensen, K. F. Distillation in microchemical systems using capillary forces and segmented flow. *Lab. Chip* **9**, 1843–1849 (2009).

32. Wootton, R. C. R. & deMello, A. J. Continuous laminar evaporation: micronscale distillation. *Chem. Commun.* 266–267 (2004).

33. Seo, J. *et al.* Switchable water-adhesive, superhydrophobic palladium-layered silicon nanowires potentiate the angiogenic efficacy of human stem cell spheroids. *Adv. Mater.* **26**, 7043–7050 (2014).

34. Walia, J., Dhindsa, N., Khorasaninejad, M. & Saini, S. S. Color generation and refractive index sensing using diffraction from 2D silicon nanowire arrays. *Small* **10**, 144–151 (2014).

35. De Volder, M. & Hart, A. J. Engineering hierarchical nanostructures by elastocapillary self-assembly. *Angew. Chem. Int. Ed.* **52**, 2412–2425 (2013).

36. Faujan, B. H. & Ahmad, J. M. B. A simple and clean method for methoxymethylation of phenols. *Pertanika* **12**, 71–78 (1989).

37. Jarowicki, K. & Kocienski, P. Protecting groups. *J. Chem. Soc. Perkin Trans* **1**, 2109–2135 (2001).

38. Wernerova, M. & Hudlicky, T. On the practical limits of determining isolated product yields and ratios of stereoisomers: reflections, analysis, and redemption. *Synlett* **2010**, 2701–2707 (2010).

39. *Acute Exposure Guideline Levels for Selected Airborne Chemicals* Vol 3 (The National Acadamic Press, 2013).

40. Wang, G., Weng, Y., Chu, D., Chen, R. & Xie, D. Developing a polysulfone-based alkaline anion exchange membrane for improved ionic conductivity. *J. Membr. Sci* **332**, 63–68 (2009).

41. Singh, A. K., Pandey, R. P. & Shahi, V. K. Fluorenyl phenolphthalein groups containing a multi-block copolymer membrane for alkaline fuel cells. *RSC Adv* **4**, 22186–22193 (2014).

42. Xu, W. *et al.* Highly stable anion exchange membranes with internal cross-linking networks. *Adv. Funct. Mater.* **25**, 2583–2589 (2015).

43. Tanaka, M. *et al.* Anion conductive block poly(arylene ether)s: synthesis, properties, and application in alkaline fuel cells. *J. Am. Chem. Soc.* **133**, 10646–10654 (2011).

Acknowledgements

We gratefully acknowledge the support from the National Research Foundation (NRF) of Korea grant funded by the Korean government (NRF-2008-0061983 and NRF-2014M1A8A1074940).

Author contributions

D.-P.K. conceived the project. A.K.S., D.-H.K., D.-P.K. designed the experiments. A.K.S., N.K.V. and D.-H.K. conducted the μ-TES process experiments. S.J. and K.-I.M. conducted the separator design experiments. N.K.V. conducted block copolymer synthesis. A.K.S. and D.P.K. wrote the paper. A.K.S. and N.K.V. contributed to analysis of data.

Additional information

Switchable friction enabled by nanoscale self-assembly on graphene

Patrick Gallagher[1], Menyoung Lee[1], Francois Amet[2,3], Petro Maksymovych[4], Jun Wang[4], Shuopei Wang[5], Xiaobo Lu[5], Guangyu Zhang[5], Kenji Watanabe[6], Takashi Taniguchi[6] & David Goldhaber-Gordon[1]

Graphene monolayers are known to display domains of anisotropic friction with twofold symmetry and anisotropy exceeding 200%. This anisotropy has been thought to originate from periodic nanoscale ripples in the graphene sheet, which enhance puckering around a sliding asperity to a degree determined by the sliding direction. Here we demonstrate that these frictional domains derive not from structural features in the graphene but from self-assembly of environmental adsorbates into a highly regular superlattice of stripes with period 4-6 nm. The stripes and resulting frictional domains appear on monolayer and multilayer graphene on a variety of substrates, as well as on exfoliated flakes of hexagonal boron nitride. We show that the stripe-superlattices can be reproducibly and reversibly manipulated with submicrometre precision using a scanning probe microscope, allowing us to create arbitrary arrangements of frictional domains within a single flake. Our results suggest a revised understanding of the anisotropic friction observed on graphene and bulk graphite in terms of adsorbates.

[1] Department of Physics, Stanford University, Stanford, California 94305, USA. [2] Department of Physics, Duke University, Durham, North Carolina 27708, USA. [3] Department of Physics and Astronomy, Appalachian State University, Boone, North Carolina 28608, USA. [4] Center for Nanophase Materials Sciences, Oak Ridge National Laboratory, Oak Ridge, Tennessee 37831, USA. [5] Institute of Physics, Chinese Academy of Sciences, Beijing 100190, China. [6] National Institute for Materials Science, 1-1 Namiki, Tsukuba 305-0044, Japan. Correspondence and requests for materials should be addressed to D.G.-G. (email: goldhaber-gordon@stanford.edu).

Nanometre-scale surface textures with long-range order often give rise to pronounced frictional anisotropy. These textures sometimes originate from crystal structures: periodic tetrahedral reversals in the antigorite lattice create nanoscale surface corrugations, which generate the anisotropic friction that governs certain seismic processes[1]. A large frictional anisotropy similarly arises for some quasicrystal intermetallics, whose surfaces are textured by atomic columns[2]. Rotationally aligned molecules also form ordered nanotextures with associated anisotropic friction, as observed in organic crystals[3]. In adsorbed organic films[4–6], rotational symmetry of the host surface permits multiple stable molecular orientations, yielding frictional domains with anisotropy along different axes.

From a technological standpoint, nanometre-scale systems with such multistability are appealing platforms for switches or memories. Bistable states in redox centres[7], rotaxane molecules[8] and iron clusters[9] can be addressed and switched using scanned probes, enabling dense information storage. Multistable nanotextures could find application in nanoelectromechanical systems if the friction-producing textures could be dynamically controlled, as in biomimetic tapes with magnetically actuated micropillars[10]. Existing schemes for tuning friction at submicrometre scales include Fermi level modulation in silicon[11] and mechanical oscillation of a sliding contact[12]—nonhysteretic techniques, which require maintenance of a voltage or oscillation, a disadvantage for circuitry.

In this study, we identify a friction-producing nanotexture that naturally forms on graphene exposed to laboratory air and exploit its multistability to hysteretically switch friction with submicrometre precision. Using high-resolution atomic force microscopy (AFM), we directly image a superlattice of nanoscale stripes on exfoliated graphene and we show that this striped nanotexture produces the anisotropic friction previously observed[13–15] on graphene monolayers. This nanotexture strongly resembles patterns of adsorbates observed on graphite[16–18] and we induce an apparently identical nanotexture on flakes of hexagonal boron nitride (hBN) using a thermal cycling procedure. Consistent with the adsorbate picture, we can rapidly and predictably reorient the frictional domains by scanning a probe tip along the flake in a chosen direction—a departure from nanoassembly techniques[19] such as dip-pen nanolithography[20] and nanografting[21], for which writing a different 'colour' requires submerging the sample in a different 'ink'.

Results

Superlattice of nanoscale stripes. To image friction, we measure the deflection (diving board motion) and torsion (axial twist) of a scanned AFM cantilever in light contact with the sample. The deflection signal primarily contains topographic information, while the meaning of the torsion signal depends on scan direction. For lateral scanning (motion perpendicular to cantilever axis; Fig. 1b lower panel), the torsion measures lateral tip-sample forces commonly interpreted as friction forces. In this 'friction-imaging' mode, tip-sample forces transverse to the scan direction result in deflection, contributing spurious topographic signals. When the cantilever is scanned longitudinally (Fig. 1c lower panel), the torsion signal directly measures tip-sample forces transverse to the scan direction. For an isotropic surface, this 'transverse force' signal is zero.

As reported previously[13,14], the friction signal of exfoliated monolayer graphene flakes on silicon dioxide reveals up to three distinct domains of friction despite a featureless topography signal (Fig. 1a,b). The domains vary in size from tens of nanometres to tens of micrometres and produce sharp contrast in transverse force, confirming their anisotropic character (Fig. 1c).

Tapping-mode AFM images taken with ultrasharp tips within the different domains (Fig. 1d) reveal periodic stripes along axes rotationally separated by 60° (angular orientation does not measurably vary within a given domain; see Supplementary Note 1). Within experimental error (typically ± 0.2 nm), stripe period (typically ~ 4 nm) does not change across a sample, although we have observed global changes in stripe period after thermal cycling (for example, from 4 to 6 nm; see Supplementary Note 2). Peak-to-trough stripe amplitude ranges between 10 and 100 pm, but strongly depends on tip conditions and oscillation parameters.

The observed frictional anisotropy of a given domain respects the symmetry of the stripe-superlattice. The friction signal approximately tracks the cosine of the angle between scan axis and stripes (Fig. 1e)—friction is maximized when the two are aligned—whereas the transverse force is zero when the stripes are perpendicular or parallel to the scan axis, as required by symmetry (Fig. 1f). In between these zeros, the transverse force changes sign so as to guide the sliding tip towards the low friction axis (lower panel in Fig. 1c). We conclude that the stripes in graphene produce the observed friction anisotropy, similar to friction-producing nanotextures in other systems[1–6].

The stripes are not unique to monolayer graphene on SiO_2. We observe stripe domains and anisotropic friction on graphene flakes up to 50 nm thick (the maximum thickness investigated) without change in stripe period and with minor change in magnitude of frictional anisotropy (Supplementary Note 3), as well as on graphene flakes on different substrates (Supplementary Note 4). Stripe domains can also form on exfoliated flakes of hBN on SiO_2: single crystals show at most three distinct domains of anisotropic friction (Fig. 2b,c), each characterized by a different orientation of stripes, whose typical period is ~ 4 nm (Fig. 2d). As for graphene, the friction signal is maximized when scanning along the stripes. However, whereas we observed stripes on nearly all graphene flakes as exfoliated, only occasionally did we observe stripes on hBN as exfoliated. We found that a cryogenic thermal cycle such as immersion in liquid nitrogen and subsequent removal to ambient conditions (Methods) would reliably produce stripes on hBN. A full understanding of the effect of thermal cycling is beyond the scope of our study; our limited variable-temperature AFM experiments found stripes to form on hBN on cooling from 300 to 250 K, although vacuum conditions probably influenced the evolution with temperature (Supplementary Note 5).

The behaviour of stripes on epitaxial heterostructures of graphene and hBN implies that stripes on both materials share a common origin. The nearly perfect rotational alignment between stacked lattices[22] results in a moiré pattern with lattice constant ~ 14 nm in regions where graphene has grown on the hBN (Fig. 3a). Despite this additional superstructure, stripes form on exposed layers of both graphene and hBN with no measurable difference in period and often appear to maintain phase across a graphene/hBN boundary. Furthermore, using the moiré pattern to infer lattice orientation[23] (Fig. 3b), we find that the stripes run along the armchair axes of both graphene and hBN in all 25 epitaxial heterostructures and 5 mechanically assembled heterostructures that we studied (Supplementary Note 6).

Previous studies have ascribed the anisotropic friction in monolayer graphene to periodic ripples in the graphene sheet induced by stress from the substrate[13–15]. Although our data confirm the presence of periodic structure, the extreme similarity of the stripes on graphene and hBN—materials with different bending stiffness and response to stress[24]—suggests that the stripes are adsorbates rather than features of the crystals themselves. The orientation of the stripes further rules out periodic ripples, which are expected to produce a high friction

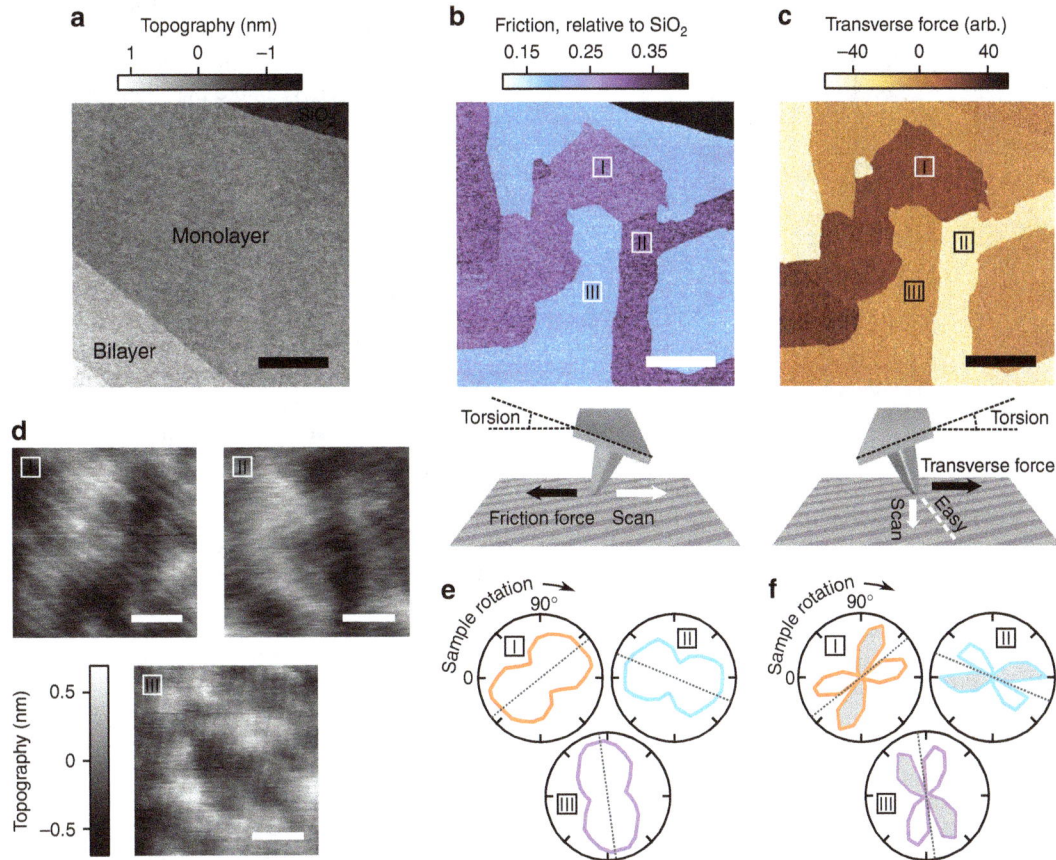

Figure 1 | Stripes on exfoliated graphene. (**a**) Contact mode topography scan of a graphene flake on silicon oxide, showing monolayer, bilayer and trilayer regions. Scale bar, 3 µm. (**b**) Simultaneously recorded friction signal (upper panel), showing three distinct domains of friction labeled I, II and III. Lower panel: cartoon of the friction imaging mode. The cantilever is scanned laterally and friction between the tip and sample produces the measured torsion of the cantilever. (**c**) Transverse force signal (upper panel) from the same region as in **b**, measured by recording the torsion while scanning the cantilever longitudinally (lower panel). Surface anisotropy pushes the tip towards the local 'easy' axis, creating a transverse force that twists the cantilever. (**d**) Tapping mode topography scans of the graphene monolayer, taken within each of the three domains. Each domain is characterized by stripes of period 4.3 ± 0.2 nm along one of three distinct axes rotationally separated by 60°. Scale bars, 20 nm. (**e**) Friction relative to SiO$_2$ for each domain as a function of clockwise sample rotation angle; zero degrees corresponds to the orientation shown in **a–c**. For each polar plot, the origin and circumference correspond to relative friction values of 0.15 and 0.4, respectively. Dotted lines indicate the sample rotations at which the stripes shown in **d** are parallel to the scan axis. The friction signal is approximately sinusoidal, with the highest friction produced when stripes are parallel to the scan axis. (**f**) Transverse force signal for each domain as a function of clockwise sample rotation angle. Unshaded and grey-shaded regions indicate positive and negative transverse signals, respectively. The origin of each polar plot is zero. The transverse signal for a given domain switches sign as the stripe axis rotates through the lateral axis.

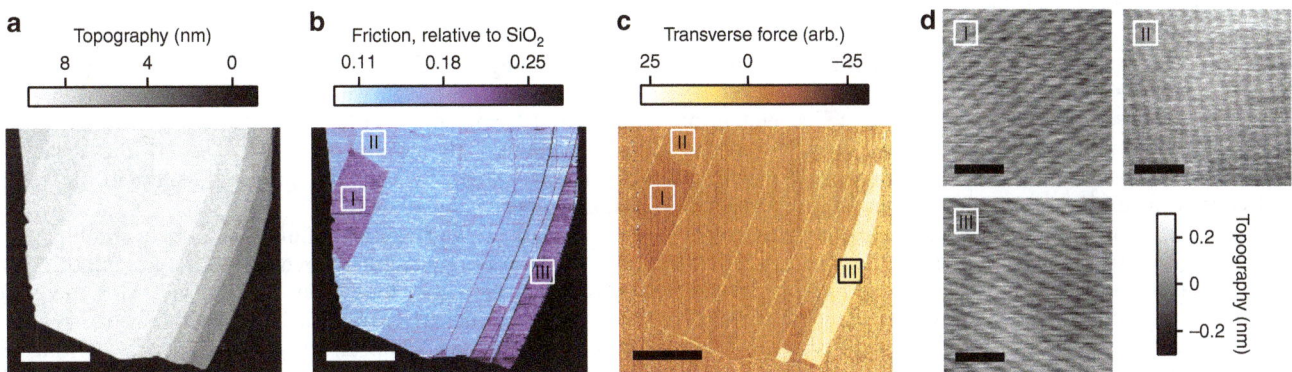

Figure 2 | Stripes on exfoliated hBN. (**a**) Contact mode topography scan of a terraced hBN flake, thickness 5–9 nm, after thermal cycling in liquid nitrogen. Scale bar, 5 µm. (**b,c**) Simultaneously recorded friction signal (**b**) and separately recorded transverse force signal (**c**) showing the presence of three distinct domains (I, II and III). The contrast between I and III is weak in friction, but strong in transverse force. (**d**) Tapping mode topography scans of the three domains, taken in the regions indicated in **b** and **c**. Each domain is characterized by stripes of period 4.7 ± 0.2 nm along one of three distinct axes rotationally separated by 60°. Scale bars, 20 nm.

a

b

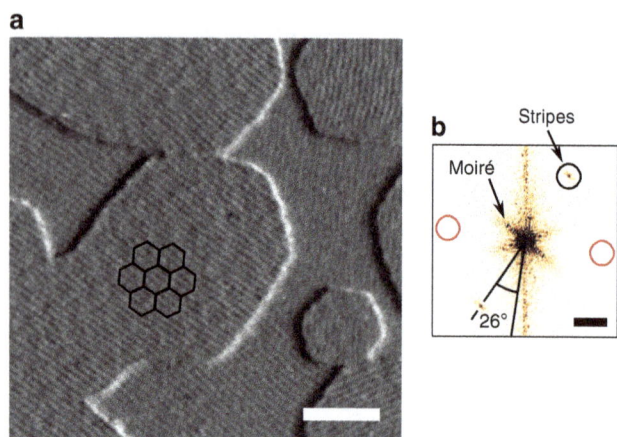

Figure 3 | Orientation of stripes on graphene and hBN. (**a**) Tapping mode topography image of graphene islands grown by van der Waals epitaxy on exfoliated hBN. The image has been differentiated along the horizontal axis for clarity. Graphene islands can be distinguished from the hBN surface by the presence of a moiré pattern, which is partially outlined in black for one of the grains. The sample surface is covered with stripes of period 4.3 ± 0.1 nm, oriented along one of three distinct axes rotationally separated by 60°. The stripe period is the same on graphene and hBN, and the stripes frequently appear to cross the graphene/hBN boundary without a phase slip. Scale bar, 50 nm. (**b**) Fast Fourier transform (FFT) of the topography signal used to produce **a**. The moiré pattern within the graphene grains appears as a sixfold-symmetric pattern with segments extending $\sim 70\,\mu m^{-1}$ from the origin; these protruding segments are parallel to the momentum-space moiré lattice vectors. The dominant stripe domain on graphene and hBN produces a pair of isolated points in the FFT, one of which is circled in black. The stripe axis is rotated 26 ± 4° from the moiré lattice vectors, indicating that the stripe axes are nearly aligned with the armchair axes of the graphene and hBN. The quoted angular precision reflects the width of the moiré peaks; we also expect a few-degree systematic error in the angular estimate, as a misalignment between graphene and hBN lattices of 0.1°—a reasonable expectation for van der Waals epitaxial heterostructures[23]—would rotate the moiré pattern by 4° with respect to the graphene lattice. The small area of nearly vertical stripes in **a** produces a pair of points, circled in red, which can barely be seen with this colourscale. Scale bar, 100 μm^{-1}.

axis perpendicular to the stripes[13,14] and a zigzag stripe axis[15,25]—both opposite to our findings. Periodic ripples have never been observed in scanning tunnelling microscopy (STM) of the graphene lattice and our STM data are no exception (Supplementary Note 7); on the other hand, adsorbates can be disturbed by the pressure of the STM tip under standard imaging conditions[26], perhaps explaining why the stripes that we observe have not previously been reported.

Various organic adsorbates are known to self-assemble into nanoscale stripes on graphite. Surfactant molecules, for instance, form stripes[17] whose 4–7 nm period is set by molecular length and Debye screening[27]; anisotropic van der Waals interactions align the stripes along the armchair axes[17]. Alkanes also produce armchair-aligned stripes of 4 nm period on graphite, where the period is again determined largely by molecular length[16]. Self-assembly of inorganic species has been reported as well: Lu et al.[18,28] observed crystallographically aligned stripes of 4 nm period on graphite submerged in water and correlated the stripes with the presence of dissolved nitrogen gas. Noting that gas enrichment at the interface between a hydrophobic surface and water is theoretically expected[29], Lu et al.[18,28] argued that the stripes were self-assembled columns of molecular nitrogen adsorbed to the graphite surface. Stripes of similar period were

later observed in ambient on multilayer epitaxial graphene[30,31]; following Lu et al.[18,28], these stripes were attributed to nitrogen adsorbates trapped at the interface between graphene and an ambient water layer. We note that these studies[18,28,30,31] do not provide a direct chemical analysis of the stripes to prove their nitrogen content. Why stripes should form instead of a homogeneous layer of nitrogen is also unexplained.

We propose that the stripes on graphene and hBN are self-assembled environmental adsorbates, in view of their similarity to stripes formed by adsorbates on graphitic surfaces[16–18,27,28,30,31] and their aforementioned dissimilarity to structural ripples, as well as our ability to manipulate the stripes by physical contact (see below). Although direct determination of the chemical makeup of the stripes is beyond the scope of our work, our data suggest that the species that self-assembles is airborne and ubiquitous in the laboratory, as stripes of uniform period fully cover our cleaved or annealed crystal surfaces that have not been exposed to any chemical processing (Methods). From this perspective, an interpretation in terms of nitrogen and water[18,28,30,31] or other common inorganic molecules is attractive. However, hydrocarbons are also plentiful in laboratory air (arising from, for instance, outgassing plastics or vacuum pump oil) and certain species could preferentially attach to graphene or hBN due to lattice match[16].

Recent work resolved nanoscale stripes in the transverse force response of bulk graphite and ascribed them to a novel puckering-induced stick–slip friction process[32]. These stripes produced domains of anisotropic friction[33] such as those on graphene and hBN. We suggest a reinterpretation of these data in terms of adsorbates, which would unify our understanding of anisotropic friction in graphite, graphene and hBN.

Manipulation of frictional domains. Adsorbates can sometimes be mechanically manipulated by AFM[19], raising the possibility of patterning friction on these materials. For monolayer graphene on SiO₂, scanning at the low normal force used for imaging (1 nN) often minimally affects the frictional domains, but scanning at high normal force (30 nN) reproducibly reorients the domains (Fig. 4a). We devised two standard approaches for domain manipulation (Fig. 4b). The 'brush stroke' consists of raster scanning a rectangular window at high normal force; we retract the tip after every line so that it only scans the sample in one direction. Brush strokes produce reproducible results—often a domain flop—that depend on the scan angle and the initial 'canvas' domain. For scan angles near the canvas stripe axis, the canvas switches to the domain with stripes next closest to the scan axis (Fig. 4c). Our second approach is to 'erase' the canvas domain within a rectangular scan window by rapid, back-and-forth scanning at high normal force. This mode destabilizes the domains within the scan window, leaving only the most stable domain, determined primarily by local strain and partly by scan axis. Although strain gradually varies across the flake (see discussion below), erasing still produces deterministic results within a specific region.

The brush stroke and eraser allow us to rapidly create patterns of friction with submicrometre precision. Without optimizing our procedure, creating a block letter 'S' 5 μm tall using the eraser took 16 min, whereas creating a 'U' using brush strokes took 36 min (Fig. 4d and Supplementary Movie 1). After writing, the pattern gradually decayed: here the 'S' widened, while the 'U' narrowed (Fig. 4e). We wrote the same pattern in different parts of the flake and found that whether a domain grew or shrank with time, and how rapidly it evolved, depended on position. The absence of other obvious symmetry-breaking mechanisms suggests that local strain induced by the substrate determines

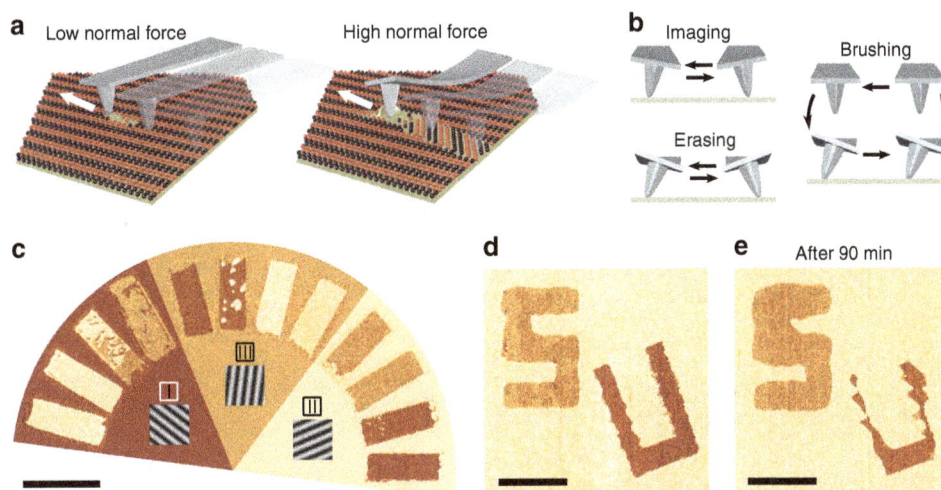

Figure 4 | Rewritable friction on monolayer graphene. (a) Cartoon illustrating the response of the striped adsorbates to the scanning tip. At low normal force, the tip minimally disturbs the stripes as it scans the surface and the stripe structure rapidly heals. At high normal force, the stripe structure is heavily disturbed, creating a new stripe domain in the wake of the scanning tip. **(b)** Summary of our scanning modes. For imaging, we rapidly scan the cantilever back and forth at low normal force, while slowly moving it in the direction perpendicular to the fast scan axis. The erasing mode is identical, but at high normal force. For a brush stroke, we raster-scan the cantilever such that the tip only moves in one direction when in contact with the sample. After scanning each line, we lift the cantilever, move it to the start of the next line and touch down again. **(c)** Domain switching as a function of scan angle on the monolayer flake studied in Fig. 1, rotated as in Fig. 1a–c. The image shown is a collage of 12 transverse force images, each taken after executing a single 3 μm by 1 μm brush stroke on a canvas composed initially of a single domain. For each canvas domain we show four brush strokes nearly parallel with the canvas stripes, where each brush stroke is directed radially outward from the origin of the semicircle. The brush strokes steer the canvas domain towards the domain whose stripes are next nearest the brush axis. Scale bar, 3 μm. **(d)** Transverse force image immediately after writing block letters 'S' and 'U' in domains III and I, respectively, on a canvas of domain II (same flake and orientation as in **a**). The block letter 'S' was written by 'erasing', whereas the 'U' was written using brush strokes. Scale bar, 3 μm. **(e)** Transverse force image of the same area, taken 90 min later. The 'S' (domain III) has expanded into the canvas, while the 'U' (domain I) has decayed.

the relative stability of the domains. Domain stability in turn determines the effective resolution of our patterning technique: although we can write crisp lines 100 nm wide in some parts of a flake, in other parts these features only persist for minutes before decaying to match the canvas domain. In addition, although we can pattern friction on several different monolayer graphene flakes, others show only weak response to both patterning modes described; the strain field in these flakes probably strongly favours the local canvas domain, making the canvas difficult to switch.

Whether patterning friction is possible on thicker crystals requires further investigation. Our first attempts indicate that domains can be rewritten with the eraser or brush stroke, although the resulting domains are not as sharp as on monolayer graphene. Proximity to the substrate could be stabilizing the stripes, allowing for more flexible control of domain shape. Our work underscores the major role played by adsorbates, rather than structural deformation, in determining friction on graphene and hBN—and perhaps on other layered materials, such as transition metal dichalcogenides. The periodic perturbation from the adsorbates might open gaps at the superlattice energy or modify the Fermi velocity in graphene[34], with measurable consequences for electronic properties of ultraclean graphene/hBN heterostructures[35].

Methods

Sample preparation. Flakes of graphene and hBN were prepared by mechanical exfoliation (3M Scotch 600 Transparent Tape or 3M Scotch 810 Magic Tape) under ambient conditions (40–60% relative humidity) on n-doped silicon wafers with 90 or 300 nm of thermal oxide. The substrates were not exposed to any chemical processing following thermal oxidation. For graphene exfoliation, we used bulk crystals of both Kish graphite (Sedgetech, USA) and highly oriented pyrolitic graphite (HOPG ZYA, SPI Supplies, USA) and observed no differences in superlattice phenomena between samples produced using different graphite sources or tapes. For hBN exfoliation, we used bulk crystals provided by Kenji Watanabe and Takashi Taniguchi. We also prepared graphene flakes on other

substrates (Supplementary Note 4), including SU-8 epoxy (MicroChem, USA), 200 nm of Au(111) on mica (Phasis, Switzerland) and 5 nm of Pt (electron-beam evaporation) on magnesium oxide (MTI, USA).

We prepared epitaxial graphene heterostructures on oxidized silicon substrates by mechanical exfoliation of hBN followed by graphene growth at 500 °C by a remote plasma-enhanced chemical vapour deposition process described previously[22]. We also mechanically assembled heterostructures of graphene on hBN using both wet[36] and dry[37] transfer methods. Polymer residues from the assembly process were removed by annealing samples in a tube furnace for 4 h at 500 °C under continuous flow of oxygen (50 sccm) and argon (500 sccm); before removal to air, we allowed the samples to cool (5–10 °C min^{-1}), to below 100 °C under the same flow of oxygen and argon.

Thermal cycling. We found stripes to appear on our samples after thermal cycling to liquid nitrogen temperatures or below using a variety of methods. Most commonly, and specifically for the sample shown in Fig. 2, we immersed the sample in liquid nitrogen for 1–5 min and then removed it to atmosphere, and blew off the condensation with dry air. This procedure would almost always produce stripes on graphene, hBN or graphene/hBN heterostructures. In other cases, we loaded the sample in the vacuum chamber of a cryostat—either a cryogen-free dilution refrigerator or a Quantum Design PPMS—and thermal cycled to a base temperature between 25 mK and 100 K. Cooling and warming rates varied between 1 and 30 K min^{-1}. We warmed up the samples under various atmospheres including moderate vacuum, helium gas or nitrogen gas; in all of these cases (over ten different samples cycled in the dilution refrigerator or PPMS) we found stripes on every flake or heterostructure (totaling several tens) that we checked. The epitaxial heterostructure in Fig. 3 was not cycled to low temperature: the sample displayed stripes in AFM with no further processing following removal from the growth furnace. Some of our assembled heterostructures (Supplementary Note 6) required a low-temperature thermal cycle to produce stripes after the oxygen/argon anneal, although in other cases we observed stripes without cryogenic treatment.

AFM and STM measurements. All images shown in Figs 1–4 were taken with a Park XE-100 AFM under ambient conditions (40–60% relative humidity) except for Fig. 4d,e, which were taken in 10% relative humidity by flooding the chamber of the XE-100 with dry air. (We observed no significant difference in domain mutability or evolution between 10 and 50% relative humidity.) To resolve the stripes in tapping mode, we used sharp silicon probes (MikroMasch Hi'Res-C15/Cr-Au) with a nominal tip radius of 1 nm, a typical resonant frequency of 265 kHz and a typical cantilever Q of 400. See Supplementary Note 8 for a detailed interpretation of the tapping mode topography signal.

For measurements in contact mode, we used silicon probes (MikroMasch HQ:NSC19/Al BS-15) with a nominal tip radius of 8 nm and a typical resonant frequency of 65 kHz. We used a normal force setpoint of 1 nN for all friction and transverse force imaging scans shown, with scan rates $\sim 10\,\mu m\,s^{-1}$. See Supplementary Note 8 for a discussion of the friction imaging mechanism. For domain manipulation we used a normal force setpoint of 30 nN. For brush strokes we used scan rates $\sim 30\,\mu m\,s^{-1}$, whereas for erasing we used scan rates $\sim 300\,\mu m\,s^{-1}$.

When imaging friction or transverse force, we collected torsion data for both forward-moving and backward-moving scans. To eliminate offsets in the friction and transverse force signals for Fig. 1e,f, we subtracted backward images from forward images and divided by two. All friction or transverse force images shown are just the forward scan, with any torsion offset eliminated by subtracting the average of forward and backward torsion values on SiO$_2$.

To study stripe formation with changing temperature (Supplementary Note 5), we used an Omicron varible-temperature AFM/STM operating in ultrahigh vacuum (UHV; 8×10^{-11} mbar). Samples were not baked in UHV before experiments. The sample stage was cooled by a copper braid attached to a cold sink held at low temperature by continuous flow of liquid nitrogen; by this method, we achieved a base temperature of 110 K. We used the same sharp probes as for ambient AFM (MikroMasch Hi'Res-C15/Cr-Au). In UHV, the cantilever Q reached 5,000, which significantly restricted scan speed for tapping mode; we therefore used on-resonance frequency-modulation mode, imaging at a typical frequency shift of -30 Hz. For all images, we applied a DC tip-sample bias to nullify the contact potential difference.

STM measurements (Supplementary Note 7) were carried out under ambient conditions using the Park XE-100. We prepared our tip by mechanically cutting a Pt/Ir wire and scanning the sample at high bias voltages until we achieved atomic resolution of the graphene lattice.

Error bars and lateral calibration. All values quoted for moiré period and angular orientation are extracted from the fast Fourier transform of the AFM images. AFM images of all heterostructures described in this study are corrected for thermal drift by performing an affine transformation to produce regular moiré hexagons (we used the free software Gwyddion, available at gwyddion.net). All error bars reflect the full width at half maximum of the peaks in the fast Fourier transform; for instance, 12.0 ± 0.5 nm means that the full width at half maximum of the peak maps to 1 nm in real space. The lateral scale of the Park XE-100 was calibrated by measuring the moiré period of graphene/hBN heterostructures grown by van der Waals epitaxy, in which the graphene and hBN lattices are nearly perfectly aligned, and defining this period (averaged over several samples) to be 13.6 nm. This definition corresponds to the assumption made in Supplementary Note 6 that the lattice constants for hBN and graphene are $a_{hBN} = 0.25$ nm and $a_{graphene} = a_{hBN}/1.018$. The lateral scale of the Omicron variable-temperature AFM was calibrated to the lateral scale of the Park XE-100 by measuring the moiré pattern of the same sample in both systems.

References

1. Campione, M. & Capitani, G. C. Subduction-zone earthquake complexity related to frictional anisotropy in antigorite. *Nat. Geosci.* **6**, 847–851 (2013).
2. Park, J. Y. *et al.* High frictional anisotropy of periodic and aperiodic directions on a quasicrystal surface. *Science* **309**, 1354–1356 (2005).
3. Bluhm, H., Schwarz, U. D., Meyer, K.-P. & Wiesendanger, R. Ansiotropy of sliding friction on the triglycine sulfate (010) surface. *Appl. Phys. A* **61**, 525–533 (1995).
4. Overney, R. M., Takano, H., Fujihara, M., Paulus, W. & Ringsdorf, H. Ansiotropy in friction and molecular stick-slip motion. *Phys. Rev. Lett.* **72**, 3546–3549 (1994).
5. Last, J. A. & Ward, M. D. Electrochemical annealing and friction anisotropy of domains in epitaxial molecular films. *Adv. Mater.* **8**, 730–733 (1996).
6. Liley, M. *et al.* Friction anisotropy and asymmetry of a compliant monolayer induced by a small molecular tilt. *Science* **280**, 273–275 (1998).
7. Gittins, D. I., Bethell, D., Schiffrin, D. J. & Nichols, R. J. A nanometre-scale electronic switch consisting of a metal cluster and redox-addressable groups. *Nature* **408**, 67–69 (2000).
8. Cavallini, M. *et al.* Information storage using supramolecular surface patterns. *Science* **299**, 531 (2003).
9. Loth, S., Baumann, S., Lutz, C. P., Eigler, D. M. & Heinrich, A. J. Bistability in atomic-scale antiferromagnets. *Science* **335**, 196–199 (2012).
10. Northen, M. T., Greiner, C., Arzt, E. & Turner, K. L. A gecko-inspired reversible adhesive. *Adv. Mater.* **20**, 3905–3909 (2008).
11. Park, J. Y., Ogletree, D. F., Thiel, P. A. & Salmeron, M. Electronic control of friction in silicon pn junctions. *Science* **313**, 186 (2006).
12. Socoliuc, A. *et al.* Atomic-scale control of friction by actuation of nanometer-sized contacts. *Science* **313**, 207–210 (2006).
13. Choi, J. S. *et al.* Friction anisotropy-driven domain imaging on exfoliated monolayer graphene. *Science* **333**, 607–610 (2011).
14. Choi, J. S. *et al.* Facile characterization of ripple domains on exfoliated graphene. *Rev. Sci. Instr.* **83**, 073905 (2012).
15. Choi, J. S. *et al.* Correlation between micrometer-scale ripple alignment and atomic-scale crystallographic orientation of monolayer graphene. *Sci. Rep.* **4**, 7263 (2014).
16. McGonigal, G. C., Bernhardt, R. H. & Thomson, D. J. Imaging alkane layers at the liquid/graphite interface with the scanning tunneling microscope. *App. Phys. Lett.* **57**, 28–30 (1990).
17. Manne, S. & Gaub, H. E. Molecular organization of surfactants at solid-liquid interfaces. *Science* **270**, 1480–1482 (1995).
18. Lu, Y.-H., Yang, C.-W. & Hwang, I.-S. Molecular layer of gaslike domains at a hydrophobic water interface observed by frequency-modulation atomic force microscopy. *Langmuir* **28**, 12691–12695 (2012).
19. Tseng, A. *Tip-Based Nanofabrication: Fundamentals and Applications* (Springer, 2011).
20. Piner, R. D., Zhu, J., Xu, F., Hong, S. & Mirkin, C. A. 'Dip-pen' nanolithography. *Science* **283**, 661–663 (1999).
21. Xu, S. & Liu, G.-Y. Nanometer-scale fabrication by simultaneous nanoshaving and molecular self-assembly. *Langmuir* **13**, 127–129 (1997).
22. Yang, W. *et al.* Epitaxial growth of single-domain graphene on hexagonal boron nitride. *Nat. Mater.* **12**, 792–797 (2013).
23. Tang, S. *et al.* Precisely aligned graphene grown on hexagonal boron nitride by catalyst free chemical vapor deposition. *Sci. Rep.* **3**, 2666 (2013).
24. Singh, S. K., Neek-Amal, M., Costamagna, S. & Peeters, F. M. Thermomechanical properties of a single hexagonal boron nitride sheet. *Phys. Rev. B* **87**, 184106 (2013).
25. Ma, T., Li, B. & Chang, T. Chirality- and curvature-dependent bending stiffness of single layer graphene. *Appl. Phys. Lett.* **99**, 201901 (2011).
26. Magonov, S. & Whangbo, M.-H. *Surface Analysis with STM and AFM: Experimental and Theoretical Aspects of Image Analysis* (Wiley, 2008).
27. Wanless, E. J. & Ducker, W. A. Organization of sodium dodecyl sulfate at the graphite-solution interface. *J. Phys. Chem.* **100**, 3207–3214 (1996).
28. Lu, Y.-H., Yang, C.-W. & Hwang, I.-S. Atomic force microscopy study of nitrogen molecule self-assembly at the HOPG-water interface. *Appl. Surf. Sci.* **304**, 56–64 (2014).
29. Dammer, S. M. & Lohse, D. Gas enrichment at liquid-wall interfaces. *Phys. Rev. Lett.* **96**, 206101 (2006).
30. Wastl, D. S. *et al.* Observation of 4 nm pitch stripe domains formed by exposing graphene to ambient air. *ACS Nano* **7**, 10032–10037 (2013).
31. Wastl, D. S., Weymouth, A. J. & Giessibl, F. J. Atomically resolved graphitic surfaces in air by atomic force microscopy. *ACS Nano* **8**, 5233–5239 (2014).
32. Rastei, M. V., Heinrich, B. & Gallani, J. L. Puckering stick-slip friction induced by a sliding nanoscale contact. *Phys. Rev. Lett.* **111**, 084301 (2013).
33. Rastei, M. V., Guzmán, P. & Gallani, J. L. Sliding speed-induced nanoscale friction mosaicity at the graphite surface. *Phys. Rev. B* **90**, 041409 (2014).
34. Park, C.-H., Yang, L., Son, Y.-W., Cohen, M. L. & Louie, S. G. Anisotropic behaviours of massless Dirac fermions in graphene under periodic potentials. *Nat. Phys.* **4**, 213–217 (2008).
35. Dean, C. R. *et al.* Boron nitride substrates for high-quality graphene electronics. *Nat. Nanotechnol.* **5**, 722–726 (2010).
36. Amet, F., Williams, J. R., Watanabe, K., Taniguchi, T. & Goldhaber-Gordon, D. Insulating behavior at the neutrality point in single-layer graphene. *Phys. Rev. Lett.* **110**, 216601 (2013).
37. Wang, L. *et al.* One-dimensional electrical contact to a two-dimensional material. *Science* **342**, 614–617 (2013).

Acknowledgements

We gratefully acknowledge Byong-man Kim and Ryan Yoo of Park Systems for verifying the presence of stripes in our samples using their Park NX-10 AFM. We thank Daniel Wastl for carefully reading our manuscript and for encouraging us to re-examine whether the stripes we observed were caused by periodic structural ripples or self-assembled adsorbates. We thank Trevor Petach and Arthur Barnard for other helpful discussions. Sample fabrication and ambient AFM/STM were performed at the Stanford Nano Shared Facilities with support from the Air Force Office of Science Research, Award Number FA9550-12-1-02520. Variable-temperature AFM studies were conducted at the Center for Nanophase Materials Sciences, which is a DOE Office of Science User Facility; our use of the facility was supported by the Center for Probing the Nanoscale, an NSF NSEC, under grant PHY-0830228. S.W., X.L. and G.Z. acknowledge support from the National Basic Research Program of China (Program 973) under grant 2013CB934500, the National Natural Science Foundation of China under grants 61325021 and 91223204, and the Strategic Priority Research Program (B) of the Chinese Academy of Sciences under grant XDB07010100. K.W. and T.T. acknowledge support from the Elemental Strategy Initiative conducted by the MEXT (Japan). T.T. acknowledges support from JSPS Grant-in-Aid for Scientific Research under grants 262480621 and 25106006.

Author contributions

P.G. identified the stripes, performed all experiments and wrote the paper. P.G. and F.A. fabricated the assembled heterostructures. M.L., F.A. and D.G.-G. discussed data and experimental directions, and assisted in writing the paper. P.M. and J.W. supported the variable-temperature AFM measurements. S.W., X.L. and G.Z. grew the epitaxial graphene/hBN heterostructures. K.W. and T.T. grew the bulk hBN crystals.

Additional information

Competing financial interests: The authors declare no competing financial interests.

Topological phase transitions and chiral inelastic transport induced by the squeezing of light

Vittorio Peano[1], Martin Houde[2], Christian Brendel[1], Florian Marquardt[1,3] & Aashish A. Clerk[2]

There is enormous interest in engineering topological photonic systems. Despite intense activity, most works on topological photonic states (and more generally bosonic states) amount in the end to replicating a well-known fermionic single-particle Hamiltonian. Here we show how the squeezing of light can lead to the formation of qualitatively new kinds of topological states. Such states are characterized by non-trivial Chern numbers, and exhibit protected edge modes, which give rise to chiral elastic and inelastic photon transport. These topological bosonic states are not equivalent to their fermionic (topological superconductor) counterparts and, in addition, cannot be mapped by a local transformation onto topological states found in particle-conserving models. They thus represent a new type of topological system. We study this physics in detail in the case of a kagome lattice model, and discuss possible realizations using nonlinear photonic crystals or superconducting circuits.

[1]Institute for Theoretical Physics, University of Erlangen-Nürnberg, Staudtstr. 7, 91058 Erlangen, Germany. [2]Department of Physics, McGill University, 3600 rue University, Montreal, Quebec, Canada H3A 2T8. [3]Max Planck Institute for the Science of Light, Günther-Scharowsky-Straße 1/Bau 24, 91058 Erlangen, Germany. Correspondence and requests for materials should be addressed to V.P. (email: Vittorio.Peano@fau.de).

Waves are not only ubiquitous in physics, but the behaviour of linear waves is also known to be very generic, with many features that are independent of the specific physical realization. This has traditionally allowed us to transfer insights gained in one system (for example, sound waves) to other systems (for example, matter waves). That strategy has even been successful for more advanced concepts in the field of wave transport. One important recent example of this kind is the physics of topological wave transport, where waves can propagate along the boundaries of a sample, in a one-way chiral manner that is robust against disorder scattering. While first discovered for electron waves, this phenomenon has by now also been explored for a variety of other waves in a diverse set of systems, including cold atoms[1], photonic systems[2] and more recently phononic systems[3-9].

In the case of topological wave transport, the connection between waves in different physical implementations can actually be so close that the calculations turn out to be the same. In particular, if we are dealing with matter waves moving in a periodic potential, the results do not depend on whether they are bosons or fermions, as long as interactions do not matter. The single-particle wave equation to be solved happens to be exactly the same. This has allowed to envision and realize photonic analogues of quantum-Hall effect[10-18], the spin Hall effect[19-22], Floquet topological insulators[23,24] and even Majorana-like modes[25]. More generally, the well-known classification of electronic band structures based on the dimensionality and certain generalized symmetries[26] directly applies to photonic systems provided that the particle number is conserved. As we now discuss, this simple correspondence will fail in the presence of squeezing.

Consider the most general quadratic Hamiltonian describing photons in a periodic potential in the presence of parametric driving:

$$\hat{H} = \sum_{\mathbf{k},n} \varepsilon_n[\mathbf{k}]\hat{b}_{\mathbf{k},n}^\dagger \hat{b}_{\mathbf{k},n} + \sum_{\mathbf{k},n,n'} \left(\lambda_{nn'}[\mathbf{k}]\hat{b}_{\mathbf{k},n}^\dagger \hat{b}_{-\mathbf{k},n'}^\dagger + \text{h.c.} \right). \quad (1)$$

The first term describes a non-interacting photonic band structure, where $\hat{b}_{\mathbf{k},n}$ annihilates a photon with quasimomentum $\mathbf{k}$ in the n-th band. The remaining two-mode squeezing terms are induced by parametric driving and do not conserve the excitation number. As we discuss below, they can be controllably realized in a number of different photonic settings. While superficially similar to pairing terms in a superconductor, these two-mode squeezing terms have a profoundly different effect in a bosonic system, as there is no limit to the occupancy of a particular single-particle state. They can give rise to highly entangled ground states, and even to instabilities.

Given these differences, it is natural to ask how anomalous pairing terms can directly lead to topological phases of light. In this work, we study the topological properties of two-dimensional photonic systems described by Equation (1), in the case where the underlying particle-conserving band structure has no topological structure, and where the parametric driving terms do not make the system unstable. We show that the introduction of particle non-conserving terms can break time-reversal symmetry (TRS) in a manner that is distinct from having introduced a synthetic gauge field, and can lead to the formation of bands having a non-trivial pattern of (suitably defined) quantized Chern numbers. This in turn leads to the formation of protected chiral edge modes: unlike the particle-conserving case, these modes can mediate a protected inelastic (but still coherent) scattering mechanism along the edge (that is, a probe field injected into the edge of the sample will travel along the edge, but emerge at a different frequency). In general, the topological phases we find here are distinct both from those obtained in the particle-conserving case, and from those found in topological super-conductors. We also discuss possible realizations of this model using a nonlinear photonic crystal or superconducting microwave circuits. Finally, we discuss the formal analogies and crucial differences between the topological phases of light investigated here and those recently proposed for other kinds of Bogoliubov quasiparicles[27-31] (see Discussion section).

Results

Kagome lattice model. For concreteness, we start with a system of bosons on a kagome lattice (Fig.1),

$$\hat{H}_0 = \sum_{\mathbf{j}} \omega_0 \hat{a}_{\mathbf{j}}^\dagger \hat{a}_{\mathbf{j}} - J\sum_{\langle \mathbf{j},\mathbf{j}'\rangle} \hat{a}_{\mathbf{j}}^\dagger \hat{a}_{\mathbf{j}'} \quad (2)$$

(we set $\hbar = 1$). Here we denote by $\hat{a}_{\mathbf{j}}$ the photon annihilation operator associated with lattice site $\mathbf{j}$, where the vector site index has the form $\mathbf{j} = (j_1, j_2, s)$. $j_1 j_2 \in Z$ labels a particular unit cell of the lattice, while the index $s = A, B, C$ labels the element of the sublattice. $\langle \mathbf{j}, \mathbf{j}' \rangle$ indicates the sum over nearest neighbours, and J is the (real valued) nearest-neighbour hopping rate; ω_0 plays the role of an onsite energy. As there are no phases associated with the hopping terms, this Hamiltonian is time-reversal symmetric and topologically trivial. We chose the kagome lattice because it is directly realizable both in quantum optomechanics[5] and in arrays of super-conducting cavity arrays[13,16]; it is also the simplest model where purely local parametric driving can result in a topological phase.

We next introduce quadratic squeezing terms to this Hamiltonian that preserve the translational symmetry of the lattice and that are no more non-local than our original,

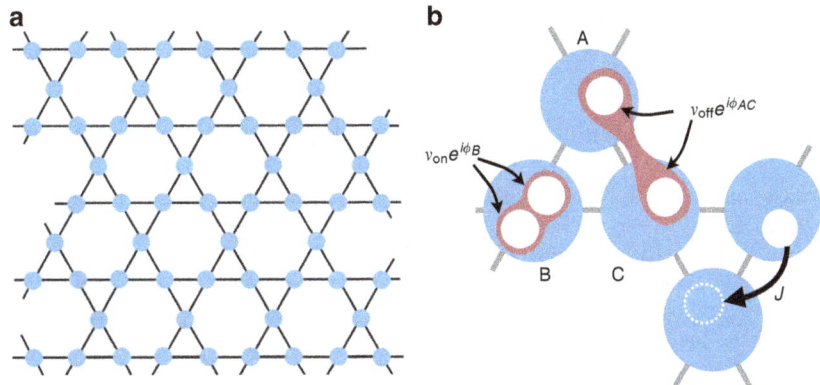

Figure 1 | Setup figure. (**a**) An array of nonlinear cavities forming a kagome lattice. (**b**) Photons hop between nearest-neighbour sites with rate J. Each cavity is driven parametrically leading to the creation of photon pairs on the same lattice site (rate ν_{on}) and on nearest-neighbour sites (rate ν_{off}). A spatial pattern of the driving phase is imprinted on the parametric interactions, breaking the time-reversal symmetry (but preserving the $C3$ rotational symmetry).

nearest-neighbour hopping Hamiltonian:

$$\hat{H}_{\mathrm{L}} = -\frac{1}{2}\left[v_{\mathrm{on}} \sum_{\mathbf{j}} e^{i\phi_s} \hat{a}_{\mathbf{j}}^{\dagger}\hat{a}_{\mathbf{j}}^{\dagger} + v_{\mathrm{off}} \sum_{\langle \mathbf{j},\mathbf{j'}\rangle} e^{i\phi_{ss'}} \hat{a}_{\mathbf{j}}^{\dagger}\hat{a}_{\mathbf{j'}}^{\dagger}\right] + \mathrm{h.c.} \quad (3)$$

Such terms generically arise from having a nonlinear interaction with a driven auxiliary pump mode (which can be treated classically) on each site, see for example, ref. 32. As we discuss below, the variation in phases in $\hat{H}_{\mathrm{L}}$ from site to site could be achieved by a corresponding variation of the driving phase of the pump. Note that we are working in a rotating frame where this interaction is time independent, and thus ω_0 should be interpreted as the detuning between the parametric driving and the true onsite (cavity) frequency ω_{cav} (that is, $\omega_0 = \omega_{\mathrm{cav}} - \omega_{\mathrm{L}}/2$, where the parametric driving is at a frequency ω_{L}). The parametric driving can cause the system to become unstable; we will thus require that the onsite energy (that is, parametric drive detuning) ω_0 be sufficiently large that each parametric driving term is non-resonant enough to ensure stability. If one keeps ω_0 fixed, this means that the parametric driving amplitudes v_{on}, v_{off} will be limited to some fraction of ω_0 (the particular value of which depends on J, Supplementary Note 1).

For a generic choice of phases in the parametric driving Hamiltonian of Equation (3), it is no longer possible to find a gauge where $\hat{H} = \hat{H}_0 + \hat{H}_{\mathrm{L}}$ is purely real when expressed in terms of real-space annihilation operators: hence, even though the hopping Hamiltonian $\hat{H}_0$ corresponds to strictly zero flux, the parametric driving can itself break TRS. In what follows, we will focus for simplicity on situations where time reversal and particle conservation are the only symmetries broken by the parametric driving: they will maintain the inversion and $\mathcal{C}3$ rotational symmetry of the kagome lattice. We will also make a global gauge transformation so that v_{off} is purely real, while $v_{\mathrm{on}} = |v_{\mathrm{on}}|e^{i\varphi_v}$. In this case, the only possible choices for the ϕ phases have the form $(\phi_A,\phi_B,\phi_C) = (\phi_{AB},\phi_{BC},\phi_{CA}) = \pm(0,\delta,2\delta)$ with $\delta = 2\pi m_v/3$, where m_v is an integer and is the vorticity of the parametric driving phases. We stress that these phases (and hence the sign of the TRS breaking) are determined by the phases of the pump modes used to generate the parametric interaction.

Gap opening and non-trivial topology. $\hat{H}_0$ is the standard tight-binding kagome Hamiltonian for zero magnetic field, and does not have band gaps: the upper and middle bands touch at the symmetry point $\mathbf{\Gamma} \equiv (0,0)$, whereas the middle and lower bands touch at the symmetry points $K = (2\pi/3,0)$ and $K' = (\pi/3,\pi/(3)^{1/2})$ where they form Dirac cones (Fig. 2a).

Turning on the pairing terms, the Hamiltonian $\hat{H} = \hat{H}_0 + \hat{H}_{\mathrm{L}}$ can be diagonalized in the standard manner as $\hat{H} = \sum_{n,\mathbf{k}} E_n[\mathbf{k}]\hat{\beta}_{n,\mathbf{k}}^{\dagger}\hat{\beta}_{n,\mathbf{k}}$, where the $\hat{\beta}_{n,\mathbf{k}}$ are canonical bosonic annihilation operators determined by a Bogoliubov transformation of the form (see Methods section):

$$\hat{\beta}_{n,\mathbf{k}}^{\dagger} = \sum_{s=A,B,C} u_{n,\mathbf{k}}[s]\hat{a}_{\mathbf{k},s}^{\dagger} - v_{n,\mathbf{k}}[s]\hat{a}_{-\mathbf{k},s}. \quad (4)$$

Here $\hat{a}_{\mathbf{k},s}$ are the annihilation operators in quasimomentum space, and $n = 1,2,3$ is a band index; we count the bands by increasing energy. The photonic single-particle spectral function now shows resonances at both positive and negative frequencies, $\pm E_n[k]$, corresponding to particle- and hole-type bands, Fig. 2d. Because of the TRS breaking induced by the squeezing terms, the band structure described by $E_n[\mathbf{k}]$ now exhibits gaps, Fig. 2b; furthermore, for a finite sized system, one also finds edge modes in the gap, Fig. 2d.

The above behaviour suggests that the parametric terms have induced a non-trivial topological structure in the wavefunctions

of the band eigenstates. To quantify this, we first need to properly identify the Berry phase associated with a bosonic band eigenstate in the presence of particle non-conserving terms. For each $\mathbf{k}$, the Bloch Hamiltonian $\hat{H}_{\mathbf{k}}$ corresponds to the Hamiltonian of a multi-mode parametric amplifier. Unlike the particle-conserving case, the ground state of such a Hamiltonian is a multi-mode squeezed state with non-zero photon number; it can thus have a non-trivial Berry's phase associated with it when $\mathbf{k}$ is varied, Supplementary Note 2. The Berry phase of interest for us will be the difference of this ground state Berry phase and that associated with a single quasiparticle excitation. One finds that the resulting Berry connection takes the form

$$\boldsymbol{\mathcal{A}}_n = i\langle \mathbf{k}, n|\hat{\sigma}_z \boldsymbol{\nabla}_k|\mathbf{k}, n\rangle. \quad (5)$$

Here the six vector of Bogoliubov coefficients $|\mathbf{k},\ n\rangle \equiv (u_{n,\mathbf{k}}[A], u_{n,\mathbf{k}}[B], u_{n,\mathbf{k}}[C], v_{n,\mathbf{k}}[A], v_{n,\mathbf{k}}[B], v_{n,\mathbf{k}}[C])$ plays the role of a singe-particle wavefunction, and $\hat{\sigma}_z$ acts in the particle-hole space, associating $+1$ to the u components and -1 to the v components, see Methods section. These effective wavefunctions obey the symplectic normalization condition

$$\langle \mathbf{k}, n|\hat{\sigma}_z|\mathbf{k}, n'\rangle = \sum_s u_{n,\mathbf{k}}^*[s]u_{n',\mathbf{k}}[s] - v_{n,\mathbf{k}}^*[s]v_{n',\mathbf{k}}[s] = \delta_{n,n'}. \quad (6)$$

Having identified the appropriate Berry connection for a band eigenstate, the Chern number for a band n is then defined in the usual manner:

$$C_n = \frac{1}{2\pi}\int_{BZ} (\boldsymbol{\nabla}\times\boldsymbol{\mathcal{A}}_n)\cdot\hat{z}. \quad (7)$$

The definition in Eq. (5) agrees with that presented in ref. 27 and (in one-dimension) ref. 29; standard arguments[27] show that the C_n are integers with the usual properties. We note that, as for superconductors, breaking the $U(1)$ (particle-conservation) symmetry remains compatible with a first-quantized picture after doubling the number of bands. The additional hole bands are connected to the standard particle bands by a particle–hole symmetry; see Methods section. In bosonic systems, the requirement of stability generally implies that particle and hole bands can not touch; this is true for our system. Thus, the sum of the Chern numbers over the particle bands (with $E > 0$) must be zero, and there cannot be any edge states with energies below the lowest particle bulk band (or in particular, at zero energy); Supplementary Note 1.

In the special case where we only have onsite parametric driving (that is, $v_{\mathrm{off}} = 0, v_{\mathrm{on}} \neq 0$), the Chern numbers can be calculated analytically (Supplementary Note 3). They are uniquely fixed by the pump vorticity. If $m_v = 0$, we have TRS and the band structure is gapless, while for $m_v = \pm 1$, $\boldsymbol{C} = (\mp 1, 0, \pm 1)$. This set of topological phases also occurs in a particle-number conserving model on the kagome lattice with a staggered magnetic field, that is, the Oghushi–Murakami–Nagaosa (OMN) model of the anomalous quantum-Hall effect[33,34].

In the general case, where we include offsite parametric driving, entirely new phases appear. We have computed the Chern numbers here numerically, using the approach of ref. 35. In Fig. 3a, we show the topological phase diagram of our system, where J/ω_0 and m_v are held fixed, while the parametric drive strengths $v_{\mathrm{on}}, v_{\mathrm{off}}$ are varied. Different colours correspond to different triplets $\boldsymbol{C} \equiv (C_1, C_2, C_3)$ of the band Chern numbers, with grey and dark-grey corresponding to the two phases already present in the OMN model. Strikingly, a finite off-diagonal coupling v_{off} generates a large variety of phases which are not present in the OMN model, including phases having bands with $|C_n| > 1$. The border between different topological phases represent topological phase transitions, and correspond to parameter values where a pair of bands touch at a particular symmetry

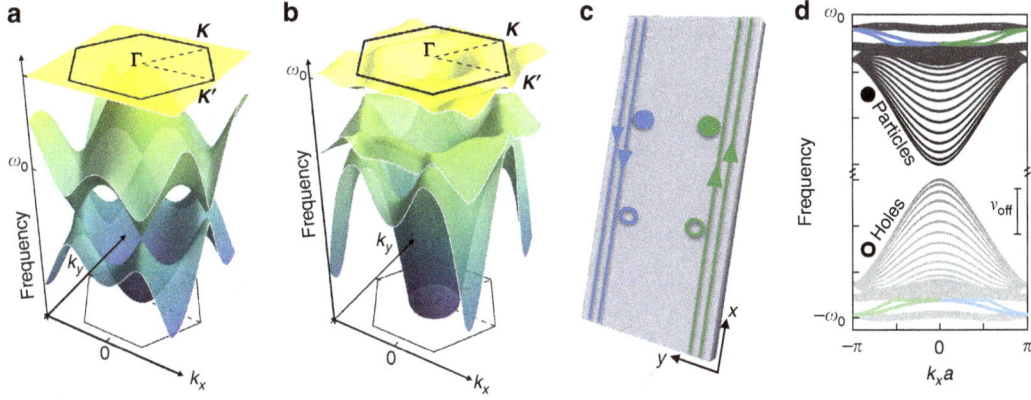

Figure 2 | Topological Band structure. (**a,b**) 3D plots of the bulk band structure. The hexagonal Brillouin zone is also shown. (**a**) In the absence of parametric driving, neighbouring bands touch at the rotational symmetry points **K**, **K′** and **Γ**. (**b**) The parametric driving opens a gap between subsequent bands. For the chosen parameters, there is a global band gap between the second and third band. (**d**) Hole and particle bands, $\pm E_m[k_x]$, in a strip geometry (sketched in **c**). The line intensity is proportional to the weight of the corresponding resonance in the photon spectral function, Supplementary Note 1. The edge states localized on the right (left) edge, plotted in green (blue), have positive (negative) velocity. Parameters: Hopping rate $J = 0.02\omega_0$ (ω_0 is the onsite frequency); (**b,d**), the parametric couplings are $v_{\mathrm{on}} = -0.085\omega_0$ and $v_{\mathrm{off}} = 0.22\omega_0$.

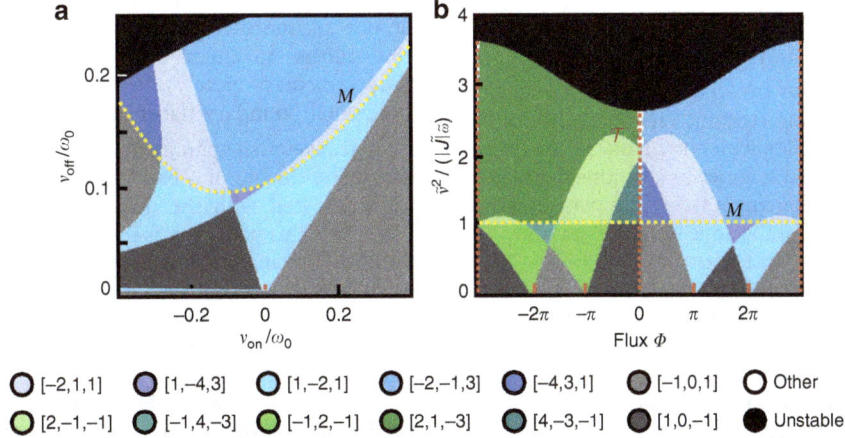

| ○ [−2,1,1] | ● [1,−4,3] | ● [1,−2,1] | ● [−2,−1,3] | ● [−4,3,1] | ● [−1,0,1] | ○ Other |
| ○ [2,−1,−1] | ● [−1,4,−3] | ● [−1,2,−1] | ● [2,1,−3] | ● [4,−3,−1] | ● [1,0,−1] | ● Unstable |

Figure 3 | Symplectic Topological phase diagrams. (**a**)Topological phase diagram for the parametrically driven kagome lattice model. The y (x) axis corresponds to the strength of the onsite parametric drive v_{on} (offsite parametric drive v_{off}), and different colours correspond to different triplets $\mathbf{C} = (C_1, C_2, C_3)$ of Chern numbers for the three bands of the model. Note that only the grey and dark-grey phases are found in the particle-conserving version of our model with a staggered field. We have fixed the hopping rate $J/\omega_0 = 0.02$, and the vorticity of the pump $m_v = 1$. (**b**) Same phase diagram, but now plotted in terms of the effective flux Φ and effective parametric drive $\tilde{v}$ experienced by α quasiparticles.

point; we discuss this further below. Via a standard bulk-boundary correspondence (Supplementary Note 4), the band Chern numbers for a particular phase determine the number of protected edge states that will be present in a system with a boundary; as usual, the number of edge states in a particular bandgap is obtained by summing the Chern numbers of lower-lying bands. We discuss these edge states in greater detail in a following subsection. Finally, the black regions in the phase diagram indicate regimes of instability, which occur when the parametric driving strength becomes too strong.

Dressed-state picture. To gain further insight into the structure of the topological phases found above, it is useful to work in a dressed-state basis that eliminates the local parametric driving terms from our Hamiltonian. We thus first diagonalize the purely local terms in the Hamiltonian; for each lattice site **j** we have

$$\hat{H}_j = \omega_0 \hat{a}_j^\dagger \hat{a}_j - \frac{1}{2}\left[v_{\mathrm{on}} e^{i\phi_j} \hat{a}_j^\dagger \hat{a}_j^\dagger + \mathrm{h.c.}\right] = \tilde{\omega}\hat{\alpha}_j^\dagger \hat{\alpha}_j \quad . \quad (8)$$

Here $\tilde{\omega} = \sqrt{\omega_0^2 - v_{\mathrm{on}}^2}$, and the annihilation operators $\hat{\alpha}_j$ are given by a local Bogoliubov (squeezing) transformation $\hat{\alpha}_j = e^{i\phi_j}e^{-i\varphi_v/2}(\cosh(r)a_j - e^{i\phi_j}e^{i\varphi_v}\sinh(r)\hat{a}_j^\dagger)$, where the squeezing factor r is

$$r = \frac{1}{4}\ln\left[\frac{\omega_0 + v_{\mathrm{on}}}{\omega_0 - v_{\mathrm{on}}}\right]. \quad (9)$$

On a physical level, the local parametric driving terms attempt to drive each site into a squeezed vacuum state with squeeze parameter r; the $\hat{\alpha}_j$ quasiparticles correspond to excitations above this reference state. Note that we have included an overall phase factor in the definition of the $\hat{\alpha}_j$, which will simplify the final form of the full Hamiltonian.

In this new basis of local quasiparticles, our full Hamiltonian takes the form

$$\hat{H} = \sum_j \tilde{\omega}\hat{\alpha}_j^\dagger \hat{\alpha}_j - \sum_{\langle j,l \rangle} \tilde{J}_{jl}\hat{\alpha}_j^\dagger \hat{\alpha}_l - \left(\frac{\tilde{v}}{2}\sum_{\langle j,l \rangle} \hat{\alpha}_j^\dagger \hat{\alpha}_l^\dagger + \mathrm{h.c.}\right). \quad (10)$$

The transformation has mixed the hopping terms with the non-local parametric terms: The effective counter-clockwise hopping matrix element is

$$\tilde{J}_{jl} = Je^{i\delta} + e^{3i\delta/2}\left[2J\cos\left(\frac{\delta}{2}\right)\sinh^2 r + v_{\text{off}}\sinh 2r\cos\left(\frac{\delta}{2} + \varphi_v\right)\right],$$

(11)

and the magnitude of the effective non-local parametric driving is

$$|\tilde{v}| = |v_{\text{off}}e^{-i(\delta/2 + \varphi_v)} + 2v_{\text{off}}\cos(\delta/2 + \varphi_v)\sinh^2 r + J\sinh 2r\cos(\delta/2)|.$$

(12)

Note that the phase of $\tilde{v}$ can be eliminated by a global gauge transformation, and hence it plays no role; we thus take $\tilde{v}$ to be real in what follows.

Our model takes on a much simpler form in the new basis: the onsite parametric driving is gone, and the non-local parametric driving is real. Most crucially, the effective hoppings can now have spatially varying phases, which depend both on the vorticity of the parametric driving in $\hat{H}_L$ (through δ), and the magnitude of the onsite squeezing (through r). In this transformed basis, the effective hopping phases are the only route to breaking TRS. Our model has thus been mapped onto the standard OMN model for the anomalous quantum-Hall effect, with an additional (purely real) nearest-neighbour two-mode squeezing interaction. In the regime where the parametric interactions between the $\hat{\alpha}$ quasiparticles are negligible (Supplementary Note 3), the complex phases correspond in the usual manner to a synthetic gauge field (that is, the effective flux Φ piercing a triangular plaquette would be $\Phi = 3\arg\tilde{J}$). In other words, the squeezing creates a synthetic gauge field for Bogoliubov quasiparticles. However, in the presence of substantial parametric interaction between $\hat{\alpha}$ quasiparticles, the parameter Φ can not be interpreted anymore as a flux: a flux of 2π can not be eliminated by a gauge transformation because the complex phases reappear in the parametric terms. In that case, only a periodicity of 6π in Φ is retained, since that corresponds to having trivial hopping phases of 2π.

Understanding the topological structure of this transformed Hamiltonian is completely sufficient for our purposes: one can easily show that the Chern number of a band is invariant under any local Bogoliubov transformation, hence the Chern numbers obtained from the transformed Hamiltonian in Equation (8) will coincide exactly with those obtained from the original Hamiltonian in Equation (3). We thus see that the topological structure of our system is controlled completely by only three dimensionless parameters: the flux Φ (associated with the hopping phases), the ratio $|\tilde{v}/\tilde{J}|$, and the ratio $\tilde{\omega}/|\tilde{J}|$.

The topological phase diagram for the effective model is shown in Fig. 3b. Again, one sees that as soon as the effective non-local parametric drive $\tilde{v}$ is non-zero, topological phases distinct from the standard (particle-conserving) OMN model are possible. The sign of the parametric pump vorticity m_v determines the sign of the effective flux Φ, c.f. Equation (11). As such, the right half of Fig. 3b (corresponding to $\Phi > 0$) is a deformed version of the phase diagram of the original model for pump vorticity $m_v = 1$, as plotted in Fig. 3a. Changing the sign of m_v (and hence Φ) simply flips the sign of all Chern numbers, Supplementary Note 3.

Our effective model provides a more direct means for understanding the boundaries between different topological phases. Most of these are associated with the crossing of bands at one or more high-symmetry points in the Brillouin zone; this allows an analytic calculation of the phase boundary (Supplementary Note 3). Perhaps most striking in Fig. 3b is the horizontal boundary (labelled $\mathcal{M}$), occurring at a finite value of the effective offsite parametric drive, $\tilde{v} \approx \sqrt{\tilde{J}\tilde{\omega}}$. This boundary is set by the closing of a band gap at the M points; as these points are associated with the decoupling of one sublattice from the other two, this boundary is insensitive to the flux Φ. Similarly, the vertical line labelled $\mathcal{T}$ denotes a line where the system has TRS, and all bands cross at the symmetry points K, K' and Γ. The case of zero pump vorticity $m_v = 0$ (not shown) is also interesting. Here the effective flux Φ depends on the strength of the parametric drivings, but is always constrained to be 0 or 3π. This implies that the effective Hamiltonian has TRS, even though the original Hamiltonian may not (that is, if Im $v_{\text{off}} \neq 0$, the original Hamiltonian does not have TRS). For $m_v = 0$, the parametric

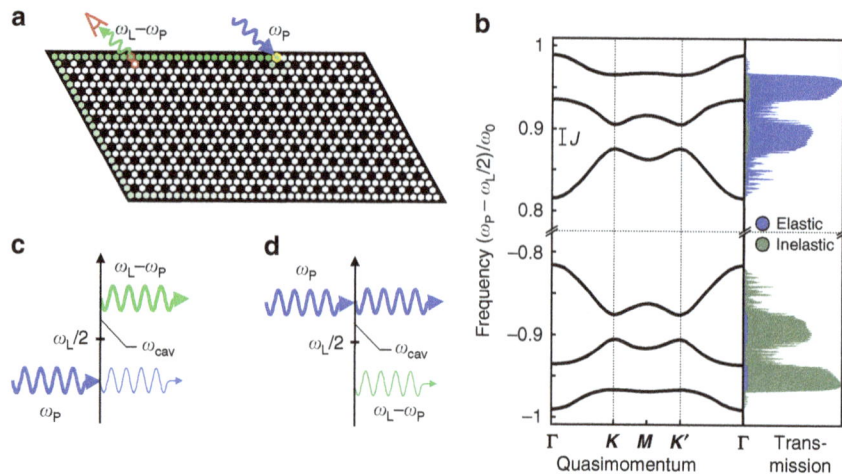

Figure 4 | Topologically protected transport in a finite system. (a) A probe beam at frequency ω_p inside the bulk band gap is focused on a site (marked in yellow) at the edge of a finite sample. The probability map of the light transmitted inelastically at frequency $\omega_L - \omega_p$ (where ω_L is the frequency of the drive tone applied to the auxiliary pump modes) clearly shows that the transport is chiral. (b) The elastic and inelastic transmission probability to a pair of sites along the edges (indicated in red in **a**) is plotted in blue and green, respectively. A cut through the bulk bands is shown to the left. (**c,d**) Sketch of the relevant scattering processes and energy scales. The inelastic (elastic) transmission has a larger rate when the light is injected in the hole (particle) band gap. Parameters: Hopping rate $J = 0.02\omega_0$ (ω_0 is the onsite frequency), parametric couplings $v_{\text{on}} = 0.4\omega_0$ and $v_{\text{off}} = 0.02\omega_0$, optical decay rate $\kappa = 0.001\omega_0$. (**a**) $\omega_p - \omega_L/2 = 0.95\omega_0$.

drivings do not open any band gap and the Chern numbers are not well defined.

Edge states and transport. Despite their modified definition, the Chern numbers associated with our Bogoliubov bands still guarantee the existence of protected chiral edge modes in a system with boundaries via a standard bulk-boundary correspondence, see Supplementary Note 4. These states can be used to transport photons by exciting them with an auxiliary probe laser beam, which is focused on an edge site and at the correct frequency. The lack of particle-number conservation manifests itself directly in the properties of the edge states: along with the standard elastic transmission they can also mediate inelastic scattering processes. In terms of the original lab frame, light injected at a frequency ω_p can emerge on the edge at frequency $\omega_L - \omega_p$, where ω_L is the frequency of the laser parametrically driving the system. This is analogous to the idler output of a parametric amplifier. Here both signal and idler have a topologically protected chirality.

Shown in Fig. 4 are the results of a linear response calculation describing such an experiment, applied to a finite system with corners. We incorporate a finite photon decay rate κ in the standard input–output formalism, see Methods section. Narrow-band probe light inside a topological band gap is applied to a site on the edge, and the resulting inelastic transmission probabilities to each site on the lattice are plotted, Fig. 4a. One clearly sees that the probe light is transmitted in a unidirectional way along the edge of the sample, and is even able to turn the corner without significant backscatter. The corresponding elastic transmission (not shown) is also chiral and shows the same spatial dependence. In Fig. 4b we show the elastic and inelastic transmissions to the sites indicated in red (rescaled by the overall transmission, $1 - R$ where R is the reflection probability at the injection site) as a function of the probe frequency ω_p. By scanning the laser probe frequency, one can separately address particle and hole band gaps. The relative intensity of the inelastic scattering component is highly enhanced when the probe beam is inside a hole band gap, see also the sketches in Fig. 4c,d. When the parametric interaction between the $\hat{\alpha}$ quasiparticles is negligible, the ratio of elastic and inelastic transmissions depends only on the squeezing factor r, (c.f. Equation (9)), see Methods section.

Physical realization. Systems of this type could be implemented in photonic crystal coupled cavity arrays[36] fabricated from nonlinear optical χ^2 materials[37–39]. The array of optical modes participating in the transport would be supplemented by pump modes (resonant with the pump laser at twice the frequency). One type of pump mode could be engineered to be spatially co-localized with the transport modes (ν_{on} processes), while others could be located in-between (ν_{off}). The required periodic phase pattern of the pump laser can be implemented using spatial light modulators or a suitable superposition of several laser beams impinging on the plane of the crystal. One method for realizing the required kagome lattice of defect cavities was discussed in ref. 5. Optomechanical systems offer another route towards generating optical squeezing terms[40,41], via the mechanically induced Kerr interaction, and this could be exploited to create an optomechanical array with a photon Hamiltonian of the type discussed here. Alternatively, these systems can be driven by two laser beams to create phononic squeezing terms[42]. A fourth alternative consists in superconducting microwave circuits of coupled resonators, where Josephson junctions can be embedded to introduce χ^2 and higher order nonlinearities, as demonstrated in refs 43,44. Kagome lattices of superconducting resonators have recently been implemented[45].

Discussion

Before concluding, it is worthwhile to discuss the connections between our work and other recent studies. A Hamiltonian of the general form of Equation (1) arises naturally in the mean-field description of a Bose-condensed phase. In this setting, the anomalous pairing terms describe the interactions with the condensate treated at the mean-field level. A few recent studies have proposed to take advantage of these interactions to selectively populate topological edge states[28,30] or, closer to our study, to induce novel topological phases. These include a study of a magnonic crystal[27], as well as general Bose–Einstein condensates in one-dimension[29] and in two-dimensions[31].

There are some crucial differences between the above studies and our work. In our case, Equation (1) describes the real particles of our system, not quasiparticles defined above some background. This difference is not just a question of semantics: in our case, topological effects can directly be seen by detecting photons, whereas in refs 29,31, one would need to isolate the contribution of a small number of Bogoliubov quasiparticles sitting on a much larger background of condensed particles. In addition, in our work the pairing terms in Equation (1) are achieved by driving the system, implying that negative and positive frequencies are clearly physically distinguished (that is, they are defined relative to a non-zero pump frequency). This is at the heart of the topologically protected inelastic scattering mechanism we describe, and is something that is not present in previous studies.

Our work opens the door to a number of interesting new directions. On the more practical side, one could attempt to exploit the unique edge states in our system to facilitate directional, quantum-limited amplification. On the more fundamental level, one could use insights from the corresponding disorder problem[46] and attempt to develop a full characterization of particle non-conserving bosonic topological states that are described by quadratic Hamiltonians. This would then be a counterpart to the classification already developed for fermionic systems[26].

Methods

Bogoliubov transformation and first-quantized picture. We find the normal mode decompositions leading to the band structures in Fig. 2 and the topological phase diagrams in Fig. 3 by introducing a first-quantized picture. Since the relevant Hamiltonians do not conserve the excitation number, this is only possible after doubling the degrees of freedom. This is achieved by grouping the annihilation operators with quasimomentum $\mathbf{k}$ and the creation operators with quasimomentum $-\mathbf{k}$ in the $2N$ vector of operators $\hat{\Psi}_\mathbf{k} = (\hat{a}_{\mathbf{k}1}, \ldots, \hat{a}_{\mathbf{k}N}, \hat{a}^\dagger_{-\mathbf{k}1}, \ldots, \hat{a}^\dagger_{-\mathbf{k}N})$ (where N is the unit cell dimension), and by casting the second quantized Hamiltonian $\hat{H}$ in the form

$$\hat{H} = \frac{1}{2} \sum_\mathbf{k} \hat{\Psi}^\dagger_\mathbf{k} \hat{h}(\mathbf{k}) \hat{\Psi}_\mathbf{k}. \tag{13}$$

The $2N \times 2N$ hermitian matrix $\hat{h}(\mathbf{k})$ plays the role of a single-particle Hamiltonian and is referred to as the Bogoliubov de Gennes Hamiltonian. By definition of the normal modes $\hat{H} = \sum_{\mathbf{k},n} E_n[\mathbf{k}] \hat{\beta}^\dagger_{n,\mathbf{k}} \hat{\beta}_{n,\mathbf{k}}$, we have $[\hat{H}, \hat{\beta}^\dagger_{n,\mathbf{k}}] = E_n[\mathbf{k}] \hat{\beta}^\dagger_{n,\mathbf{k}}$. By plugging into the above equation the Bogoliubov ansatz Equation (4) one immediately finds

$$\hat{h}(\mathbf{k})|\mathbf{k}_n\rangle = E_n[\mathbf{k}] \hat{\sigma}_z |\mathbf{k}_n\rangle. \tag{14}$$

Likewise, from $[\hat{H}, \hat{\beta}_{n,-\mathbf{k}}] = -E_n[-\mathbf{k}] \hat{\beta}_{n,-\mathbf{k}}$ one finds

$$\hat{h}(\mathbf{k})(\mathcal{K}\hat{\sigma}_x |-\mathbf{k}_n\rangle) = -E_n[-\mathbf{k}] \hat{\sigma}_z (\mathcal{K}\hat{\sigma}_x |-\mathbf{k}_n\rangle). \tag{15}$$

Here $\mathcal{K}$ denotes the complex conjugation and the matrix $\hat{\sigma}_x$ exchanges the u's and the v's Bogoliubov coefficients,

$$\hat{\sigma}_x = \begin{pmatrix} 0 & \mathbb{1}_N \\ \mathbb{1}_N & 0 \end{pmatrix}.$$

Thus, the spectrum of the $2N$ matrix $\hat{\sigma}_z \hat{h}(\mathbf{k})$ is formed by the set of $2N$ eigenenergies $E_n[\mathbf{k}]$ (belonging to the particle bands) and $-E_n[-\mathbf{k}]$ (belonging to the hole bands). *Vice versa*, to calculate the eigenenergies $E_n[\mathbf{k}]$ and $-E_n[-\mathbf{k}]$ and the vector of Bogoliubov coefficients in Equation (4), we have to solve the eigenvalue problem

$$\hat{\sigma}_z \hat{h}(\mathbf{k})|m\rangle = \lambda_m|m\rangle. \tag{16}$$

The solutions we are interested in should also display the symplectic orthonormality relations Equation (6).

We note in passing that so far we have implicitly assumed that the normal mode decomposition is possible. However, this is not always the case. When the matrix $\hat{\sigma}_z \hat{h}(\mathbf{k})$ has any complex eigenvalue, the Hamiltonian is unstable. Moreover, at the border of the stable and unstable parameter regions, the matrix $\hat{\sigma}_z \hat{h}(\mathbf{k})$ is not diagonalizable. The Supplementary Note 1 contains a stability analysis of our specific model.

In the stable regime of interest here the matrix $\hat{\sigma}_z \hat{h}(\mathbf{k})$ is diagonalizable and all its eigenvalues are real. In this case, its eigenvectors $|m\rangle$ can be chosen to be mutually $\hat{\sigma}_z$ orthogonal. In addition, there are exactly N positive (negative) norm eigenvectors. Thus, it is always possible to enforce the symplectic orthonormality relations Equation (6) by identifying the (appropriately normalized) positive and negative norm solutions with $|\mathbf{k}_n\rangle$ and $\mathcal{K}\hat{\sigma}_x|-\mathbf{k}_n\rangle$, respectively. The corresponding eigenvalues are then to be identified with $E_n[\mathbf{k}]$ (particle band structure) and $-E_n[-\mathbf{k}]$ (hole band structure), respectively

Particle–hole symmetry. The Bogoliubov de Gennes Hamiltonian has the generalized symmetry $\mathcal{C}^\dagger \hat{h}(\mathbf{k})\mathcal{C} = -\hat{h}(-\mathbf{k})$ where the charge conjugation operator C is anti-unitary and $C^2 = \mathbb{1}_{2N}$. Thus, our system represents the bosonic analogue of a superconductor in the Class D of the standard topological classification. This is a simple consequence of the doubling of the degrees of freedom in the single-particle picture. It simply reflects that the set of ladder operators $\hat{\beta}_{n,\mathbf{k}}^\dagger$ and $\hat{\beta}_{n,-\mathbf{k}}$ calculated from $\hat{h}(\mathbf{k})$ are the adjoint of the set of operators $\hat{\beta}_{n,\mathbf{k}}$ and $\hat{\beta}_{n,-\mathbf{k}}^\dagger$ calculated from $\hat{h}(-\mathbf{k})$.

Details of the transport calculations. In our transport calculations we have included photon decay. We adopt the standard description of the dissipative dynamics of photonic systems in terms of the Langevin equation and the input–output theory[47], for each site:

$$\dot{\hat{a}}_j = i[\hat{H}, \hat{a}_j] - \kappa \hat{a}_j/2 + \sqrt{\kappa}\hat{a}_j^{(\text{in})}. \tag{17}$$

In practice, we consider an array of detuned parametric amplifiers with intensity decay rate κ and add to the standard description of each parametric amplifier the inter-cell coherent coupling described in the main text. The last term describes the influence of the input field $\hat{a}_j^{(\text{in})}$ injected by an additional probe drive including also the environment vacuum fluctuations. The field $\hat{a}_j^{(\text{out})}$ leaking out of each cavity at site j is given by the input–output relations

$$\hat{a}_j^{(\text{out})} = \hat{a}_j^{(\text{in})} - \sqrt{\kappa}\hat{a}_j. \tag{18}$$

The above formulas give an accurate description of a photonic system where the intrinsic losses during injection and inside the system are negligible. Intrinsic photon absorption can be incorporated by adding another decay channel to the equation for the light field. It reduces the propagation length but does not change qualitatively the dynamics.

In Fig. 3, we show the probabilities $T_E(\omega, l, j)$ and $T_I(\omega, l, j)$ that a photon injected on site j with frequency $\omega_{\text{in}} = \omega + \omega_L/2$ is transmitted elastically (at frequency $\omega + \omega_L/2$) or inelastically (at frequency $\omega_L/2 - \omega$) to site l where it is detected. From the Kubo formula and the input–output relations we find

$$T_E(\omega, l, j) = \left|\delta_{lj} - i\kappa \tilde{G}_E(\omega, l, j)\right|^2, \tag{19}$$

$$T_I(\omega, l, j) = \kappa^2 \left|\tilde{G}_I(\omega, l, j)\right|^2. \tag{20}$$

Here $\tilde{G}_{E/I}(\omega, l, j)$ are the elastic and inelastic components of the Green's function in frequency space,

$$\tilde{G}_E(\omega, l, j) = -i\int_{-\infty}^{\infty} dt\,\Theta(t)\left\langle\left[\hat{a}_l(t), \hat{a}_j^\dagger(0)\right]\right\rangle e^{i\omega t}, \tag{21}$$

$$\tilde{G}_I(\omega, l, j) = -i\int_{-\infty}^{\infty} dt\,\Theta(t)\left\langle\left[\hat{a}_l^\dagger(t), \hat{a}_j^\dagger(0)\right]\right\rangle e^{i\omega t}. \tag{22}$$

In a N site array with single-particle eigenstates $|n\rangle = (u_n[1], \dots, u_n[N], v_n[1], \dots, v_n[N])^T$, the Green's functions read

$$G_E(\omega, l, j) = \sum_n \frac{u_n[l]u_n^*[j]}{\omega - E[n] + i\kappa/2} - \frac{v_n^*[l]v_n[j]}{\omega + E[n] + i\kappa/2}, \tag{23}$$

$$G_I(\omega, l, j) = \sum_n \frac{v_n[l]u_n^*[j]}{\omega - E[n] + i\kappa/2} - \frac{u_n^*[l]v_n[j]}{\omega + E[n] + i\kappa/2}. \tag{24}$$

We note that for a probe field inside the bandwidth of the particle (hole) sector but far detuned from the hole (particle) sector, only the first (second) term of the

summand in Equations (23) and (24) is resonant. Thus, as expected, the inelastic scattering is comparatively larger when the probe field is in the hole band gap.

It is easy to estimate quantitatively the relative intensities of elastically and inelastically transmitted light when the parametric interaction of the $\hat{\alpha}$ Bogoliubov quasiparticles is small (the regime where Φ can be interpreted as a synthetic gauge field experienced by the Bogoliubov quasiparticles). In this case, it is straightforward to show that $|v_n[j]/u_n[j]| \approx \tanh r$ independent of the eigenstate n and the site j. By putting together Equations (20, 23, 24) and neglecting the off-resonant terms we find that for $\omega, \tilde{\omega} \gg |\tilde{J}|, \kappa, |\omega - \tilde{\omega}|$,

$$T_I(\omega, l, j) \approx (\tanh r)^2 T_E(\omega, l, j) \approx T_I(-\omega, l, j) \approx (\coth r)^2 T_E(-\omega, l, j).$$

These analytical formulas agree quantitatively with the numerical results shown in Fig. 4b (note that in Fig. 4b the transmission at the output sites is rescaled by the overall transmission, $\sum_{l \neq j} T_I(\omega, l, j) + T_E(\omega, l, j)$).

References

1. Goldman, N., Juzeliunas, G., Öhberg, P. & Spielman, I. B. Light-induced gauge fields for ultracold atoms. *Rep. Prog. Phys.* **77**, 126401 (2014).
2. Lu, L., Joannopoulos, J. D. & Soljacic, M. Topological photonics. *Nat. Photon.* **8**, 821–829 (2014).
3. Prodan, E. & Prodan, C. Topological phonon modes and their role in dynamic instability of microtubules. *Phys. Rev. Lett.* **103**, 248101 (2009).
4. Kane, C. L. & Lubensky, T. C. Topological boundary modes in isostatic lattices. *Nat. Phys.* **10**, 39–45 (2013).
5. Peano, V., Brendel, C., Schmidt, M. & Marquardt, F. Topological phases of sound and light. *Phys. Rev. X* **5**, 031011 (2015).
6. Yang, Z. *et al.* Topological acoustics. *Phys. Rev. Lett.* **114**, 114301 (2015).
7. Süsstrunk, R. & Huber, S. D. Observation of phononic helical edge states in a mechanical topological insulator. *Science* **349**, 47–50 (2015).
8. Paulose, J., Chen, B. G. & Vitelli, V. Topological modes bound to dislocations in mechanical metamaterials. *Nat. Phys.* **11**, 153–156 (2015).
9. Nash, L. M. *et al.* Topological mechanics of gyroscopic metamaterials. *Proc. Natl Acad. Sci. USA* **112**, 14495–14500 (2015).
10. Haldane, F. D. M. & Raghu, S. Possible realization of directional optical waveguides in photonic crystals with broken time-reversal symmetry. *Phys. Rev. Lett.* **100**, 013904 (2008).
11. Raghu, S. & Haldane, F. D. M. Analogs of quantum-hall-effect edge states in photonic crystals. *Phys. Rev. A* **78**, 033834 (2008).
12. Wang, Z., Chong, Y., Joannopoulos, J. D. & Soljacic, M. Observation of unidirectional backscattering-immune topological electromagnetic states. *Nature* **461**, 772–775 (2009).
13. Koch, J., Houck, A. A., Le Hur, K. & Girvin, S. M. Time-reversal-symmetry breaking in circuit-QED-based photon lattices. *Phys. Rev. A* **82**, 043811 (2010).
14. Umucallar, R. O. & Carusotto, I. Artificial gauge field for photons in coupled cavity arrays. *Phys. Rev. A* **84**, 043804 (2011).
15. Fang, K., Yu, Z. & Fan, S. Realizing effective magnetic field for photons by controlling the phase of dynamic modulation. *Nat. Photon.* **6**, 782–787 (2012).
16. Petrescu, A., Houck, A. A. & Le Hur, K. Anomalous Hall effects of light and chiral edge modes on the Kagomé lattice. *Phys. Rev. A* **86**, 053804 (2012).
17. Tzuang, L. D., Fang, K., Nussenzveig, P., Fan, S. & Lipson, M. Non-reciprocal phase shift induced by an effective magnetic flux for light. *Nat. Photon.* **8**, 701–705 (2014).
18. Schmidt, M., Kessler, S., Peano, V., Painter, O. & Marquardt, F. Optomechanical creation of magnetic fields for photons on a lattice. *Optica* **2**, 635–641 (2015).
19. Hafezi, M., Demler, E. A., Lukin, M. D. & Taylor, J. M. Robust optical delay lines with topological protection. *Nat. Phys.* **7**, 907–912 (2011).
20. Khanikaev, A. B. *et al.* Photonic topological insulators. *Nat. Mater.* **12**, 233–239 (2012).
21. Hafezi, M., Mittal, S., Fan, J., Migdall, A. & Taylor, J. M. Imaging topological edge states in silicon photonics. *Nat. Photon.* **7**, 1001–1005 (2013).
22. Mittal, S. *et al.* Topologically robust transport of photons in a synthetic gauge field. *Phys. Rev. Lett.* **113**, 087403 (2014).
23. Kitagawa, T. *et al.* Observation of topologically protected bound states in photonic quantum walks. *Nat. Commun.* **3**, 882 (2012).
24. Rechtsman, M. C. *et al.* Topological creation and destruction of edge states in photonic graphene. *Phys. Rev. Lett.* **111**, 103901 (2013).
25. Bardyn, C.-E. & Imamoğlu, A. Majorana-like modes of light in a one-dimensional array of nonlinear cavities. *Phys. Rev. Lett.* **109**, 253606 (2012).
26. Ryu, S., Schnyder, A. P., Furusaki, A. & Ludwig, A. W. W. Topological insulators and superconductors: tenfold way and dimensional hierarchy. *New J. Phys.* **12**, 065010 (2010).
27. Shindou, R., Matsumoto, R., Murakami, S. & Ohe, J. Topological chiral magnonic edge mode in a magnonic crystal. *Phys. Rev. B* **87**, 174427 (2013).
28. Barnett, R. Edge-state instabilities of bosons in a topological band. *Phys. Rev. A* **88**, 063631 (2013).
29. Engelhardt, G. & Brandes, T. Topological bogoliubov excitations in inversion-symmetric systems of interacting bosons. *Phys. Rev. A* **91**, 053621 (2015).

30. Galilo, B., Lee, D. K. K. & Barnett, R. Selective population of edge states in a 2d topological band system. *Phys. Rev. Lett.* **115**, 245302 (2015).

31. Bardyn, C.-E., Karzig, T., Refael, G. & Liew, T. C. H. Chiral bogoliubov excitations in nonlinear bosonic systems. *Phys. Rev. B* **93**, 020502 (2016).

32. Gerry, C. C. & Knight, P. L. *Introductory Quantum Optics* (Cambridge University Press (2005).

33. Ohgushi, K., Murakami, S. & Nagaosa, N. Spin anisotropy and quantum hall effect in the *kagomé* lattice: chiral spin state based on a ferromagnet. *Phys. Rev. B* **62**, R6065–R6068 (2000).

34. Green, D., Santos, L. & Chamon, C. Isolated flat bands and spin-1 conical bands in two-dimensional lattices. *Phys. Rev. B* **82**, 075104 (2010).

35. Fukui, T., Hatsugai, Y. & Suzuki, H. Chern numbers in discretized brillouin zone: Efficient method of computing (spin) hall conductances. *J. Phys. Soc. Jpn.* **74**, 1674–1677 (2005).

36. Notomi, M., Kuramochi, E. & Tanabe, T. Large-scale arrays of ultrahigh-q coupled nanocavities. *Nat. Photon.* **2**, 741–747 (2008).

37. Mookherjea, S. & Yariv, A. Coupled resonator optical waveguides. *IEEE J. Quantum Elec.* **8**, 448 (2002).

38. Eggleton, B. J., Luther-Davies, B. & Richardson, K. Chalcogenide photonics. *Nat. Photon.* **5**, 141–148 (2011).

39. Dahdah, J., Pilar-Bernal, M., Courjal, N., Ulliac, G. & Baida, F. Near-field observations of light confinement in a two dimensional lithium niobate photonic crystal cavity. *J. Appl. Phys.* **110**, 074318 (2011).

40. Safavi-Naeini, A. H. *et al.* Squeezed light from a silicon micromechanical resonator. *Nature* **500**, 185–189 (2013).

41. Purdy, T. P., Yu, P. L., Peterson, R. W., Kampel, N. S. & Regal, C. A. Strong Optomechanical Squeezing of Light. *Phys. Rev. X* **3**, 031012 (2013).

42. Kronwald, A., Marquardt, F. & Clerk, A. A. Arbitrarily large steady-state bosonic squeezing via dissipation. *Phys. Rev. A* **88**, 063833 (2013).

43. Bergeal, N. *et al.* Analog information processing at the quantum limit with a Josephson ring modulator. *Nat. Phys.* **6**, 296–302 (2010).

44. Abdo, B., Kamal, A. & Devoret, M. Nondegenerate three-wave mixing with the Josephson ring modulator. *Phys. Rev. B* **87**, 014508 (2013).

45. Underwood, D. L., Shanks, W. E., Koch, J. & Houck, A. A. Low-disorder microwave cavity lattices for quantum simulation with photons. *Phys. Rev. A* **86**, 023837 (2012).

46. Gurarie, V. & Chalker, J. T. Bosonic excitations in random media. *Phys. Rev. B* **68**, 134207 (2003).

47. Clerk, A. A., Devoret, M. H., Girvin, S. M., Marquardt, F. & Schoelkopf, R. J. Introduction to quantum noise, measurement, and amplification. *Rev. Mod. Phys.* **82**, 1155–1208 (2010).

Acknowledgements

V.P., C.B., and F.M. acknowledge support by an ERC Starting Grant OPTOMECH, by the DARPA project ORCHID, and by the European Marie-Curie ITN network cQOM. M.H. and A.A.C. acknowledge support from NSERC.

Author contributions

V.P., A.A.C. and F.M. contributed to the conceptual development of the project and interpretation of results. Calculations and simulations were done by V.P., M.H. and C.B.

Additional information

Multiscale deformations lead to high toughness and circularly polarized emission in helical nacre-like fibres

Jia Zhang[1,*], Wenchun Feng[2,*], Huangxi Zhang[1], Zhenlong Wang[1], Heather A. Calcaterra[2], Bongjun Yeom[2], Ping An Hu[1] & Nicholas A. Kotov[2]

Nacre-like composites have been investigated typically in the form of coatings or free-standing sheets. They demonstrated remarkable mechanical properties and are used as ultrastrong materials but macroscale fibres with nacre-like organization can improve mechanical properties even further. The fiber form or nacre can, simplify manufacturing and offer new functional properties unknown yet for other forms of biomimetic materials. Here we demonstrate that nacre-like fibres can be produced by shear-induced self-assembly of nanoplatelets. The synergy between two structural motifs—nanoscale brick-and-mortar stacking of platelets and microscale twisting of the fibres—gives rise to high stretchability (>400%) and gravimetric toughness (640 J g^{-1}). These unique mechanical properties originate from the multiscale deformation regime involving solid-state self-organization processes that lead to efficient energy dissipation. Incorporating luminescent CdTe nanowires into these fibres imparts the new property of mechanically tunable circularly polarized luminescence. The nacre-like fibres open a novel technological space for optomechanics of biomimetic composites, while their continuous spinning methodology makes scalable production realistic.

[1] Key Laboratory of Micro-systems and Micro-structures Manufacturing, Ministry of Education, Harbin Institute of Technology, Harbin 150080, China. [2] Department of Chemical Engineering, University of Michigan, Ann Arbor, Michigan 48109-2136, USA. * These authors contributed equally to this work. Correspondence and requests for materials should be addressed to P.A.H. (email: hupa@hit.edu.cn) or to N.A.K. (email: kotov@umich.edu).

Realization of materials with high toughness combined with other properties is one of the key challenges for both load-bearing and functional materials[1]. This materials science challenge can often be addressed using biomimetic design taking naturally occurring composites that have been optimized over long evolutionary periods as inspiration. The 'brick–and–mortar' layered design of nacre, with alternating layers of inorganic platelets and biopolymers, inspired biomimetic research for several decades[2,3]. The materials architecture with alternating layers of hard inorganic components and soft organic polymers effectively arrests the propagation of cracks. The process has been replicated using a large variety of inorganic components, including clay[4-7], Al_2O_3 (refs 1,8) and layered double hydroxides[5], combined with various organic polymers including poly(vinyl alcohol) (PVA)[5], polyelectrolytes[6] and chitosan[7]. The layered biomimetic nanomaterials are much tougher than their inorganic and organic components alone, and often reveal exceptionally high strength and stiffness[1,9]. Further improvement of toughness in biomimetic nanocomposites is restricted, however, by the low strains (ε) of composite materials, especially when the volume fraction of the stiff inorganic phase is high. The nacre-like composites typically show strains $<5\%$ (ref. 10). In fact, the problem of low stretchability is quite general and observed for a variety of nanocomposites including graphene ribbons ($\varepsilon = 6\%$)[11].

The combination of two structural motifs at different scales, specifically nanoscale and microscale in this case, is designed to simultaneously increase both the stretchability and toughness of a composite[12,13]. Indeed, a stretchable graphene film combined with ripples and yarns exhibited improved tensile strain of 30% (ref. 12) and 76% (ref. 13), respectively. This inspired our search for methods to create nacre-like composites with multiscale structural motifs and evaluate their mechanical properties, which we expected to be quite unique as well as technologically valuable. Here we demonstrate that it is possible to transform flat nacre films into fibres that combine layered nanoscale and spiral microscale structural motifs. The resulting fibres can sustain longitudinal strains as high as 414%. This is 10–1,000 times higher than typical biomimeticaly designed layered composites and other fibre-like nanocomposites. The nacre-like fibres display an unusually high gravimetric toughness of $\sim 640\,\mathrm{J\,g^{-1}}$, which significantly exceeds those of natural nacre ($\sim 1\,\mathrm{J\,g^{-1}}$)[9], dragline silk (165 J g^{-1})[14], graphene (17 J m^{-3})[13], Kevlar(KM2) (78 J g^{-1})[14] and some of the best examples of composited single-wall carbon nanotube (SWNT) fibres (570–970 J g^{-1})[15-17]. Such unusual mechanical properties are attributed to multiscale deformation combining both the sliding of nanoscale platelets and unravelling of microscale spiral curls. Moreover, the described process of fibre spinning and strain-induced particle self-organization enables continuous scalable production of this material[18,19]. Furthermore, the helical patterns of the multiscale deformations causes circularly polarized luminescence (CPL) to be emitted at $\sim 575\,\mathrm{nm}$ when cadmium telluride (CdTe) nanowires are incorporated into PVA/$CaCO_3$ fibres. The high stretchability of the fibres allows the wide-range modulation of the luminescence dissymmetry ratio (g_{lum}) and the same is expected for many other optically active materials and different wavelengths. This novel optomechanical property of the fibres highlights the emergence of novel possibilities for engineering chiral nanomaterials that may be useful for remote monitoring of materials' strains.

Results

Preparation of $CaCO_3$ and graphene nanoplatelets. To prepare the nacre-like fibres we used two types of inorganic 'building blocks': one is platelets of $CaCO_3$ (vaterite) synthesized from calcium chloride and ethylene glycol by a hydrothermal method, and the other is graphene-based nanosheets (G) made by electrochemical exfoliation of highly oriented pyrolitic graphite. The vaterite plateletes had diameters and thicknesses of 4–20 μm and 100–500 nm (Supplementary Fig. 1), respectively; these dimensions are very similar to the microplatelets of the aragonite inorganic phase in seashell nacre[9]. G nanosheets displayed diameters and thicknesses of 5–45 μm and 1–5 nm, respectively (Supplementary Fig. 2a–c). $CaCO_3$ and G were chosen a pair of building blocks for the nacre-like composites because of their low density of defects[18], (Supplementary Fig. 2d,e) and they are known to display tensile strength and stiffness higher than the two-dimensional (2D) nanocarbon materials prepared from other oxidation methods, such as Hummers' method. For the polymeric component in both types of nanocomposites we used PVA, which exhibits affinity to, and forms stable dispersions with, both $CaCO_3$ and G[20,21].

Spinning of nacre-like fibres. The fibres were obtained by a combination of drawing and twisting the fibre in its wet state (Fig. 1). This technique was similar to those previously reported for making threads from cotton[22] as well as SWNTs[17,23], graphene[13] and Kevlar. Nevertheless, the fibres obtained in this study displayed essential morphological and functional differences compared with other fibres obtained from nanoscale components. The most important differences are that the platelet morphology of the inorganic components, rheological properties of the polymer–platelet mixture, and the addition of twisting during fibre processing imparts a multiscale organization to the fibres, leading to new deformation modalities and considerable improvement of mechanical performance.

Now we consider the fabrication process and different structural motifs in the fibres of the PVA/G composite. A stable, aqueous dispersion of G and PVA (Fig. 1, first step) was continuously extruded from a syringe, solidified by dehydration in an ethanol bath and drawn out using godet rollers. Similar to doctor blading, the sheer generated during twist drawing in the wet state results in self-organization of the nanosheets into the nacre-like brick-and-mortar architecture. The nanoscale layering was observed for all conditions tested, whereas the appearance of structural motifs at larger scale was depended on the content of organic phase in the composite. When the weight fraction of PVA was $<30\,\mathrm{wt\%}$ or $>80\,\mathrm{wt\%}$ (Supplementary Fig. 3a,h), the fibres showed helical ridges on their surfaces with a helical length periodicity of 220–300 μm (Supplementary Fig. 3a,h). When the PVA weight fraction was within this range, we observed additional coiling of the fibre, which resulted in a helical twist with a pitch of 100–200 μm (Supplementary Fig. 3b,d) and sub helical ridges of 0.92–20 μm (Supplementary Fig. 3e,g,i,j). Nearly perfect coiled fibres with all three structural motifs were obtained for PVA with weight fractions between 54 and 80 wt% (Supplementary Fig. 3c,f). The sub helical ridges on the surface were between 250 and 600 nm (Supplementary Fig. 3f).

Similar morphology of the fibres with nanoscale layering and two helical structured motifs can also be seen for PVA/$CaCO_3$ composites (Supplementary Fig. 4). In this case, the sub helical ridges with a wavelength of about 6–12 μm were observed when the weight fraction of PVA was $<50\,\mathrm{wt\%}$ and $>85\,\mathrm{wt\%}$ (Supplementary Fig. 4a,e,c,g), while uniform 95–105 μm coils appear for PVA fractions between 50 and 85 wt% (Supplementary Fig. 4b,f). Furthermore, the wavelengths of the helical ridges on the coils were calculated to be in the range of 4.5–10 μm (Supplementary Fig. 4f).

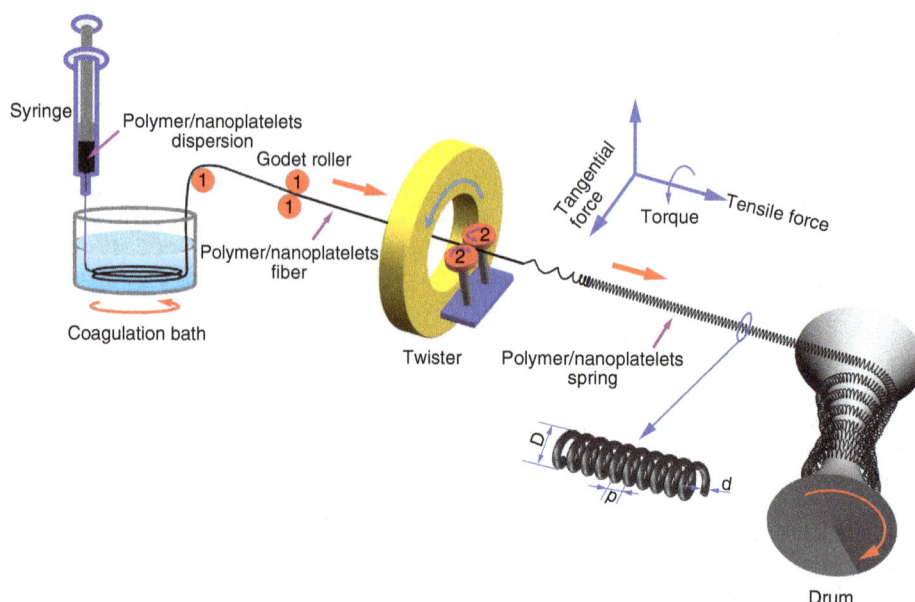

Figure 1 | Schematic diagram of the two-step formation of nacre-like composite fibre. First step: a belt-shaped nacre-like fibre was obtained via a wet-spinning process. Second step: The polymer/nanoplatelets fibre was twist spun.

Morphology evolution during twist spinning. To better understand the mechanical performance of the products, we investigated how the microscale morphology evolved after drawing them out of the coagulation bath. After extrusion of the PVA/G composite from a 400 μm diameter nozzle, a belt-like fibre (Fig. 2a), ~225 μm in diameter, was formed; its nanoscale organization was characterized by disordered packing of the nanosheets (Fig. 2g). After twisting, the fibres became thinner (diameters ~120 μm) and nearly circular in cross-section (Fig. 2b,h). Additional twisting further decreased the fibre diameters to ~110 μm, and resulted in the formation of uniform, spring-like coils (Fig. 2c).

A similar progression of fibre shapes can be seen for the PVA/CaCO$_3$ composite (Supplementary Fig. 5). Unlike the PVA/CaCO$_3$ composite, the surface of belt-like fibres from G sheets had a non-uniform orientation of platelets (Fig. 2d). After the twist spinning, the fibres showed helical corrugations along the fibre axis with periodicity of 250–600 nm (Fig. 2e,f). The torque applied during spinning of the wet fibre resulted in considerable improvement of platelet alignment (Fig. 2j–l). Enhanced alignment of the platelets was also confirmed using synchrotron small-angle X-ray scattering (S–SAXS). 2D scattering patterns accompanied with the profile of scattering intensity ($I \times \mathbf{q}^2$) as a function of scattering vector ($\mathbf{q}$) ($\mathbf{q} = 4\pi\sin\theta/\lambda$) and profile of scattering intensity (I) as a function of azimuthal angle (φ) are given in Fig. 2m–o and Supplementary Fig. 6, respectively. Since PVA scatters X-rays weakly[24], the diffraction peaks originated from the G. The elliptical S–SAXS pattern and absence of sharp scattering peak (Fig. 2m and Supplementary Fig. 6d) obtained for belt-like fibres before twisting indicated poor alignment of nanosheets[20,21]. The nearly perfect nacre-like layering in the transversal direction of the fibre indicated by the sharp scattering peak (Fig. 2n) was observed after twisting. This was accompanied by sharp peaks at 154° and 329°, as shown in Supplementary Fig. 6d, which highlighted the strong alignment of platelets in the twisting fibre. Further twisting led to the formation of a coiled fibre and its longitudinal direction alignment (Supplementary Fig. 6c) resulted in an absence of a sharp peak in the 2D scattering pattern (Fig. 2o). The monotonous intensity drop with no visible peak (Supplementary Fig. 6e) is characteristic of the uniform dispersion of G in the PVA matrix for all of the samples[20,21], which is both essential and non-trivial for composites with high content of G[25-27]. For both of the PVA/G and PVA/CaCO$_3$ composites, the nanoscale structure of the fibres was analogous to the nacre-like materials previously reported[5,6,8,9,21,22], with the exception of additional circular bending of the inorganic phase that becomes progressively more pronounced with increased coiling (Fig. 2j–l).

Mechanical properties of nacre-like fibres. Mechanical properties of the fibres were assessed using the stress–strain test (Fig. 3a). The tensile strength was 270 ± 30 MPa, which was comparable to other nacre-like materials made from clay, graphene or graphene oxide (GO). Herein, the tensile stress was calculated from the normalization of the applied force with a cross–section of microfibre after fracture. More importantly, elongation to fracture, that is, maximum tensile strain (ε), was as high as $330 \pm 60\%$; this compares favourably to tensile strains of natural nacre ($\varepsilon = 2\%$)[9], dragline silk ($\varepsilon = 31-47\%$)[14], graphene composites ($\varepsilon = 1.6-76\%$)[13,28], Kevlar(KM2) ($\varepsilon = 4.15-4.89\%$)[14] and SWNT composite fibres ($\varepsilon = 104-430\%$)[15,16]. The area under the stress-strain curve was used to calculate toughness, which was determined to be 460.3 ± 42 MJ m^{-3} and 168.1 ± 18.2 MJ m^{-3} for G (Fig. 3a) and vaterite fibres (Supplementary Fig. 7), respectively. In contrast, for an equivalent PVA fraction of 66 wt%, belt-like and circular fibres dissipated only 46.3 ± 5.4 MJ m^{-3} and 118.2 ± 9.7 MJ m^{-3}, respectively, before fracture. Because the density of our fibres is ~0.84 g cm^{-3}, the weight-specific toughness of the material was calculated to be 548.0 ± 50 J g^{-1}. For comparison, tensile strength, ε and toughness for 'flat' graphene paper obtained by vacuum assisted filtration (VAF)[29] without polymer were reported to be 41.7 ± 4 MPa, $1.34 \pm 0.07\%$ and 0.23 ± 0.03 J g^{-1}, respectively; the same values for PVA/graphene nanocomposites prepared via layer-by-layer assembly[29] were 143.1 ± 11.2 MPa, $0.051 \pm 0.013\%$ and 6.1 MJ m^{-3}, respectively. We also prepared a PVA/G composite using the VAF method with 66 wt% PVA (Supplementary Fig. 8). It showed tensile strength, ε and toughness of 65.8 ± 12 MPa, $95.5 \pm 8\%$ and 48.6 ± 6 J g^{-1}, respectively. The maximum strain sustained by the nacre-like PVA/G fibre is ~246 times higher than that of the neat G film or

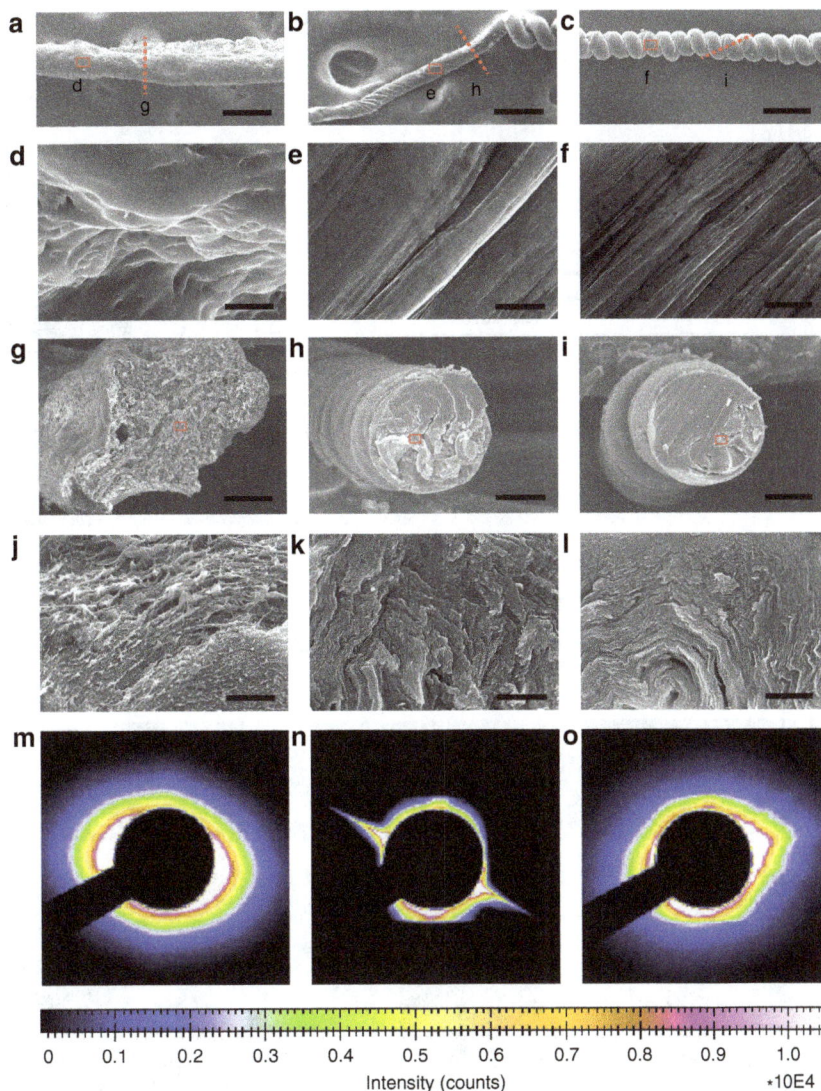

Figure 2 | Characterizations of the microscale morphology evolution of PVA/G composite fibre with PVA content of 66 wt% during spinning process. Structural changes of PVA/G composite fibres during the production process: (**a**) belt-like fibre after drawing from the coagulation bath, (**b**) circular fibre after initial twist spinning, (**c**) spring-like fibres after additional twist spinning, (**g–i**) SEM images of cross section of the three types of fibre represented by dashed lines shown in **a–c**, (**j–l**) SEM images of cross–section marked in red boxes in **g–i**, (**m–o**) show S–SAXS scattering patterns of the PVA/G fibre with PVA content of 66%, for belt-like fibre (**m**) in **a**, twisted fibre (**n**) in **b** and coiled fibre (**o**) in **c**. All the specimens were vertically fixed. Scale bars in **a,b,c** are 250 μm, **d,e,f,j,k,l** are 2.5 μm, (**g,h,i**) are 50 μm, respectively.

approximately 3.5–4 times higher than that of nacre-like flat PVA/G composite film (VAF) with the same G loading. The toughness of spring-like PVA/G fibre is ∼2,380 and 11 times higher than that of the G paper and 'flat' 66% PVA/G composite film made by VAF, respectively. Similarly, outstanding mechanical properties were also obtained for PVA/CaCO$_3$ fibres (Supplementary Fig. 7a), which showed a strain of 200.6 ± 14%, and toughness of 107.1 ± 11.6 J g^{-1}, with a fibre density of ∼1.57 g cm^{-3}.

A cyclic stretching test with ε = 20% was used to investigate the strain-hardening processes of the nacre-like fibres (Supplementary Fig. 9a). For the first to 50th cycles, the PVA/G fibres exhibited considerable plastic deformation typical of many biological and synthetic materials[30,31]. After the 50th cycle, the deformation became mostly elastic (Fig. 4). The tensile stress for ε = 20% saturates at ∼39 MPa for the 80th cycle (Fig. 4 and Supplementary Fig. 9b), which is about 1.18 times its initial value. Strain hardening was accompanied by tighter packing of the nanosheets and an increase of their curvature (Supplementary

Fig. 9c–f). After being pre-stretched, the fibre still showed the visible hysteresis between the loading and unloading curve with energy dissipation of 0.13 J g^{-1} (Supplementary Fig. 10), indicating viscoelastic behaviour of the nacre-like fibre. Notably, the spring constant (Supplementary Fig. 11) after strain hardening was 76.8 ± 1.8 N m^{-1}, twice as much as conventional steel-based tension springs and carbon nanocoils[32].

It is informative to compare the mechanical properties and toughness of the obtained nacre-like fibres with other well-known materials, such as dragline silk (165 J g^{-1})[14] and Kevlar(KM2) (78 J g^{-1})[14], as well as additional human-made fibres exemplified by various carbon nanotube- and graphene-based composites (Supplementary Table 1). Note that while nacre is known for its toughness of ∼1 J g^{-1}, the twisted PVA/CaCO$_3$ composite fibres are over ∼107 times tougher.

It is also instructive to compare the mechanical properties of the prepared twisted fibres with those made from neat PVA using the same apparatus. The maximum strain and toughness of neat PVA fibre without G are 102.6 ± 14% and 13.6 ± 3.8 J g^{-1},

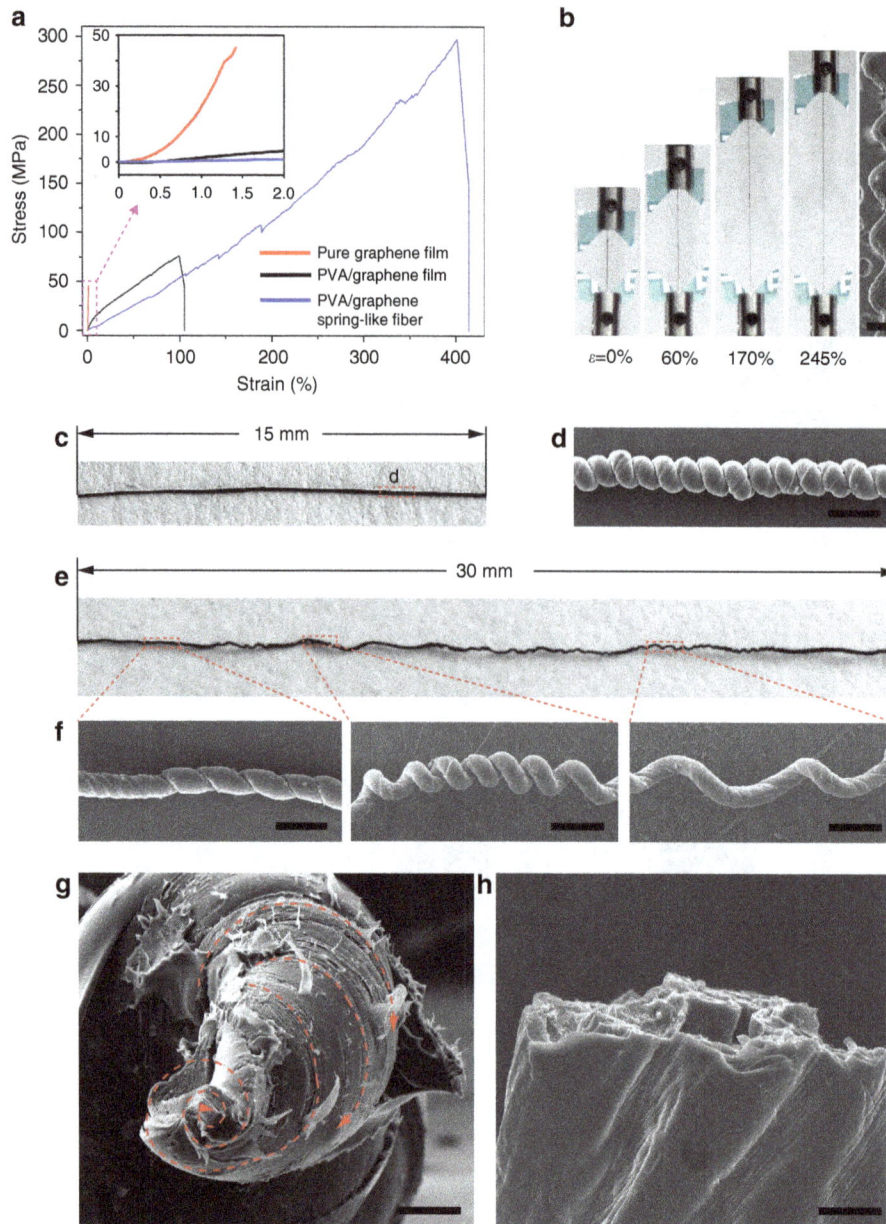

Figure 3 | Mechanical properties and characterizations of spring-like PVA/G fibre with PVA content of 66 wt%. (**a**) A typical tensile stress–strain curve of neat G film, PVA/G film, and spring-like fibre, (**b**) Photography of the fibre at strains of 0%, 60%, 170% and 245%, respectively. The right panel shows the SEM image of extended loops after breaking, bar, 250 µm, (**c,e**) Photographs of a 15-mm-long spring-like fibre before and after being stretched to 200%, showing a partial opening of loops, (**d,f**) Enlarged SEM images taken from different parts of the fibre, (**g,h**) Typical tensile fracture surface SEM image of spring-like fibre and belt-like fibre, the dash arrows in **g** show the progressive cone-shaped surface. Scale bars in **d** and **f** are 250 µm, **g** and **h** are 50 µm, repsectively.

respectively. This fact is remarkable because parent polymeric materials typically exhibit higher stretchability than their resulting composites. Both vaterite and G composites display greater strains than their parent polymers. The fibre's mechanical properties for different mass fractions of PVA peak at about 65% PVA (Supplementary Fig. 12).

The fibres can be made by a continuous mode, forming rolls of nacre-like fibrous material (Fig. 5) with identical morphology over the entire fibre, exemplified by a 4 cm long fibre, displaying uniform diameter and pitch over its entire length (Supplementary Fig. 11b).

Circularly polarized emission of helical fibres. Luminescent helical fibres were prepared by incorporating CdTe nanowires into PVA/CaCO$_3$ composites during the spinning process (Methods section). The content of CdTe nanowires was about 0.9 wt%, as measured by thermogravimetric analysis (Supplementary Fig. 15), and the fibres retained their original morphology and stretchability. Both left- and right-handed helical fibres can be produced through controlling the spinning direction (Fig. 6b inset). Strong CPL of these fibres was observed under the illumination of non-polarized light. Shifting the excitation wavelength by 20 nm (from 358 to 338 nm) did not change the peak maxima of the corresponding photoluminescence spectra (Supplementary Fig. 16), which demonstrated that Raman scattering was not a major contributor to these chiroptical properties and confirmed the photoluminescence origin of the observed spectra when the fibres are irradiated with CPL light.

Consecutive measurements of the same fibres yielded identical g_{lum} spectra (Supplementary Fig. 17), confirming that these fibres are relatively optically robust when irradiated.

The luminescence dissymmetry ratio (g_{lum}) was calculated as

$$g_{lum} = \frac{\Delta I}{\frac{1}{2}I} = \frac{I_L - I_R}{\frac{1}{2}(I_L + I_R)}$$

where I_L and I_R represent left- and right-handed CPL, respectively. Fibres of opposite handedness (Fig. 6a) exhibit almost mirror-imaged g_{lum} spectra along the x axis with peak maxima of similar intensities: 3.7×10^{-3} (580 nm) for the right-handed fibres and 4.0×10^{-3} (572 nm) for the left-handed fibres (Fig. 6b). The value of g_{lum} is comparable to that observed for other semiconductor materials such as cysteine-capped CdSe nanoparticles[33] (3×10^{-3} and 4×10^{-3}) and CdS nanoparticles templated in a protein nanocage[34] (4.4×10^{-3}), though our macroscale twisting process is considerably simpler.

CPL activity is associated with the helicity of the fibres. An effective validation of this mechanism would be modulation of CPL activity by fibre helicity changes induced by a mechanical

deformation operation such as stretching. Indeed, the g_{lum} spectra of luminescent spring fibres show a strong dependence of CPL attenuation on the fibre elongation (Fig. 6c,d and Supplementary Fig. 18). As the fibres are stretched from their original length to 1.5 and 2 times their original length, the g_{lum} at peak maxima is decreased from 3.7×10^{-3} to 2.8×10^{-3} and finally to 1.3×10^{-3} (Fig. 6c,d). This mechanical manipulation of CPL emission is efficient in a broad illumination wavelength range of 400–900 nm (Fig. 6d), showing that smaller g_{lum} values can be obtained from the increasingly stretched fibres.

Discussion

Previous versions of coiled fibres obtained by spinning and scrolling carbon nanotubes and GO films dissipated energy predominantly via macroscale deformation of the coils and displayed toughness of $28.7 \, J \, g^{-1}$ and $17 \, J \, m^{-3}$, respectively[13,35]. Other deformation modalities involving, for instance, nanoscale deformations, might be expected but were not investigated. This may be partially related to less pronounced structural changes than those observed in Figs 2 and 3, and the related inability to use SAXS to visualize the structural evolution under strain.

The unusually high toughness for nacre-like fibres in Figs 1 and 2 is attributed to multiscale deformation of the fibres on strain that involves both sliding/reorganization of the nanoscale sheets and microscale deformation of the coils. Indeed, opening of the coils and decreasing coil density can be seen as the strain increases from 0 to 60%, 170% and 245% (Fig. 3b). Various degrees of coil opening along the length of the fibre (Fig. 3c–f) suggests somewhat unequal strain along the fibre during stretching.

The combination of nanoscale and microscale deformations allows the material to experience strong intermolecular interactions between the nanosheets and the polymers, while uniformly distributing the stress along the length of the fibre. While the former mechanism is possible for 'flat' nacre-like composites, the latter, or uniform stress distribution, is not. Minor defects result in stress concentration in the small volumes of the material and consequent failure. Other processes to achieve high elongation and thereby toughness include using elastic polymers or

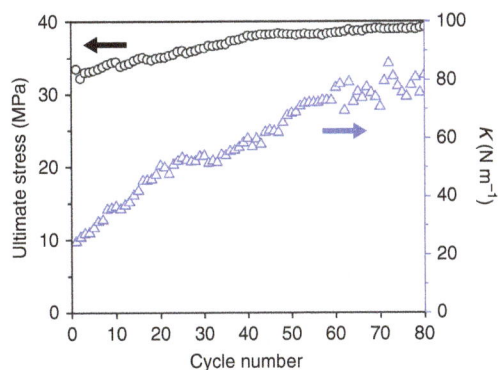

Figure 4 | Tensile testing. Tensile stress and spring constant (k) of PVA/G fibre at $\varepsilon = 20\%$ as a function of the cycle number during cyclic tensile testing.

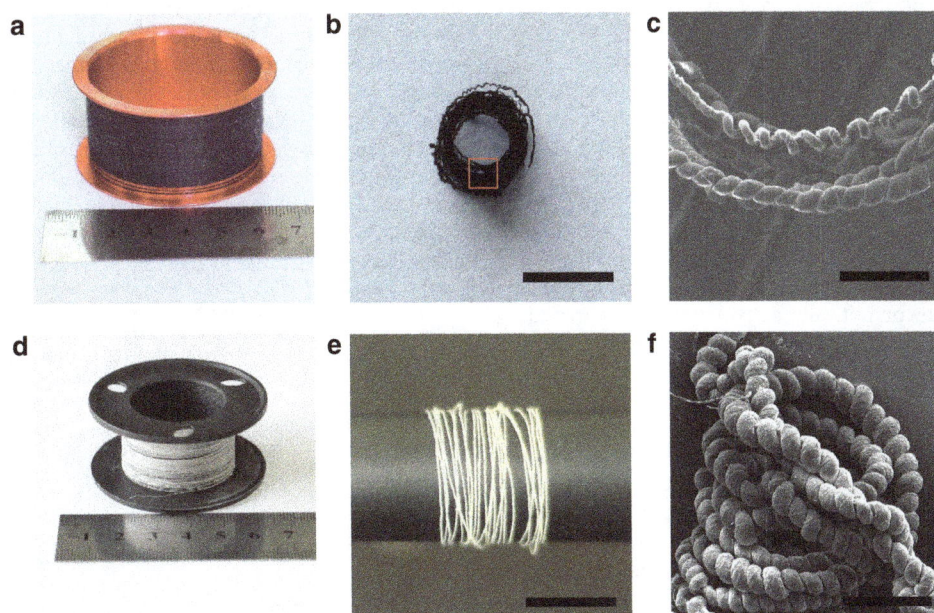

Figure 5 | Characterization of a coil of PVA/G and PVA/CaCO₃ fibre. (**a**) Photograph of PVA/G fibre wound on a reel, (**b**) Top view: a coil of spring-like PVA/G fibre, (**c**) SEM image of coiled spring-like PVA/G fibre marked in box in **b**, (**d**) Photograph of PVA/CaCO₃ fibre wound on a reel, (**e**) A coil of spring-like PVA/CaCO₃ on a roll, (**f**) SEM image of a coil of PVA/CaCO₃ fibre. Scale bars in **b** and **e** are 4 mm, **c** and **f** are 0.5 mm, respectively.

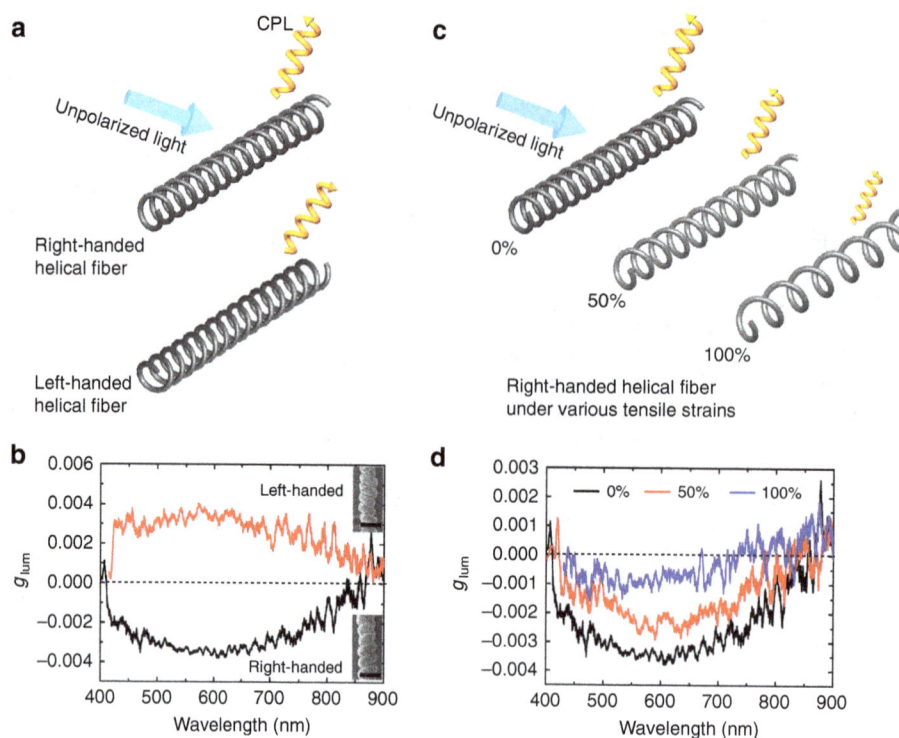

Figure 6 | Circularly polarized luminescence characteristics of PVA/CaCO₃/CdTe helical fibres. (**a**) Schematic illustration of different circularly polarized luminescence (CPL) emissions of left- and right-handed helical fibres under illumination of unpolarized light, (**b**) g_{lum} spectra of left- and right-handed helical fibres. Inset SEM images show a representative left-handed helical fibre (top) and a right-handed helical fibre (bottom), scale bar, 200 µm, (**c**) Schematic illustration of CPL activity from luminescent helical fibres at different tensile strains of 0, 50 and 100%, (**d**) g_{lum} spectra of right-handed helical fibres at various stretching states ($\epsilon = 0$, 50, 100%).

composites[17,36], taking advantage of the intrinsic elasticity of polymers[36] or interconnected networks and chemical bonds of constituents[17]. While yielding high-performance materials, the mechanism of energy dissipation in these fibres is different from that described for fibres with G, CaCO₃ and CdTe.

Distribution of stress over the extended volume of the material in helical fibres can be seen from the conical shape of the fracture surface (Fig. 3g). The cracks apparently do not propagate perpendicularly to the surface but do so parallel to the length of the fibre, which dissipates more energy. In contrast, the fracture surface in the belt-like fibre (Fig. 3h) is perpendicular to the fibre axis. Consequently, tensile strength, strechability and toughness of the helical fibres are about 4, 4 and 11.5 times higher than those of belt-like fibres, respectively (Fig. 3a). Similar tensile deformation behaviour can also be observed in the PVA/CaCO₃ composite fibre (Supplementary Fig. 7b,c).

The same multiscale deformations result in a novel chiroptical functionality: CPL. Circular asymmetry of emission is possible but difficult to obtain for optically active compounds including quantum dots[33], nanofibres[37], liquid crystals[38] and molecular luminophores[39]. In our case, CPL is not only strong in intensity and simple to realize, but it can also be effectively modulated by the macroscopic structures of the fibres including handedness and pitch. When the handedness of the host fibres was changed from left-handed to right-handed, there was a corresponding sign inversion of CPL from positive to negative (Fig. 6). The mechanism of g_{lum} modulation by deformations can be associated with strain-dependent variation of pitch of the fibre resulting in the transfer of chirality to smaller scales affecting the three-dimensional geometry of the nanowires in the matrix and efficiency of emission of left and right circularly polarized photons. Mechanical manipulation of CPL can be compared with other methods of its modulation including phase transition of

liquid crystals[38], varying aggregation states[37,40], temperature[38], chemical stimuli[41] and solvent[42].

We demonstrated that multiscale deformations in nacre-like fibres have important ramifications for both mechanical and optical properties of the nacre-like fibres. One of them is exceptional toughness while the other one is emergence of optical asymmetry for circularly polarized photons. Since strain-induced particle self-organization can be realized for a variety of nano-scale components[43,44], the method of fibre preparation can be generalized to other inorganic nanomaterials for mechanically tunable CPL-active composites and enables a novel techno-logically significant method to manipulate the chiroptical activity of materials that can be utilized among other applications for the stand-off assessment of mechanical strains.

Methods

Electrochemical exfoliation of graphene-based nanosheets. Highly oriented pyrolytic graphite (HOPG, $1 \times 1 \times 0.3$ cm) was employed as the electrode and source of graphene for electrochemical exfoliation. The anode was a block of HOPG, gripped and inserted into the electrolyte. A Pt plate served as the grounded electrode, and was placed parallel to the HOPG with a separation of about 4 cm. The electrolyte was prepared by taking 2.6 ml fuming sulfuric acid and diluting in 100 ml deionized water and then a 30% potassium hydroxide solution was added drop by drop to neutralize the sulfuric acid until the pH = ~1.2. The electro-chemical exfoliation process was carried out by applying DC bias on the HOPG electrode (from −10 V to +10 V). When the HOPG was completely exfoliated, G nanosheets were collected by vacuum filtration and rinsed at least three times with deionized water. After drying, the G was re-dispersed in dimethyl formamide (DMF) solution by gentle sonication. Finally, the suspension was centrifuged at 2,500 r.p.m. for 3 min to remove blocked graphite. The centrifuged suspension was later used for further characterization.

Characterization of low defect graphene-based nanosheets. Supplementary Figure 2a shows an scanning electron microscopy (SEM) image of G made by electrochemical exfoliation of HOPG on a SiO₂/Si substrate. The sheet sizes range from 5 to 45 µm. Supplementary Figure 2b shows a typical atomic force microscopy

(AFM) image of G sheet on a SiO_2/Si substrate with a thickness of ~ 2.8 nm (Supplementary Fig. 2c). We have measured tens of Gs and their thicknesses ranged from 1 to 5 nm, indicating that the number of layers in G is less than five.

Gs show smaller density of defects than conventionally used GO prepared by modified Hummers' method[19]. We have compared the Raman spectra of G sample with that of the original graphite and GO sample[19] (Supplementary Fig. 2d). The Raman spectrum of the G sample is very different from that of a GO sample. The appearance of D band indicates that there are still defects (disorder sp^3 structure, oxygen-contained functional groups) in G plane or at edges, which may originate from some oxidation of diluted sulfuric acid and positive electrical potential. The intensity ratio of D band over G band (I_D/I_G) was calculated to be 0.08, for the G is incomparable to negligible defect signal in graphite but much smaller than that of GO sample ($I_D/I_G = 1.12$). According to Tuinstra and Koenig[45], I_D/I_G varies inversely with the crystallite size (L_a) in graphite: L_a (nm) $= 2.4 \times 10^{-10} \times \lambda^4 (I_D/I_G)^{-1}$, where λ is the Raman excitation wavelength. Using this relationship, the crystallite sizes (graphitic domain size) of GO and as-prepared G are 18.3 ± 2.0 and 97.4 ± 43.8 nm, respectively. Moreover, the I_{2D}/I_G has been proved to be related to the degree of sp^2 structure in graphite, herein, the ratio of 2D band over G band (I_{2D}/I_G) for the graphene sample is significantly larger than that of GO sample. Furthermore, we have measured the thermogravimetric curves of the G, GO, pure PVA and PVA/G samples (Supplementary Fig. 2e). The GO sample shows a weak thermal stability, the mass loss starts once heating begins and is dramatic until 250 °C, which is similar to the results reported by Stankovich et al[46]. The mass loss in GO sample is attributed to pyrolysis of oxygen groups such as –OH, –O–, –COOH and so on, the higher defect density present in GO, the lower starting temperature for rapid mass loss and vice versa. Otherwise, a decrease in defect density would improve the thermal stability of G. The weight decreases slowly while the temperature is above 250 °C and approaches to 40% at 800 °C. But for the G, there is continuous mass loss in the whole range of heating and keeps $\sim 91\%$ mass at 800 °C. Notably, G possesses higher thermal stability than the GO indicating a lower defect in the G sheet, also. The small changes in weight of the G can be attributed to the release of absorbed water in the surface of the nanosheets.

Synthesis of $CaCO_3$ nanoplatelets. The $CaCO_3$ nanoplatelets were prepared as follows: 2 g of $CaCl_2$ and 2 g of urea were added into 6.0 ml of deionized water and stirred vigorously. Then 0.44 g of hexadecyl–trimethyl–ammonium bromide was added into the mixture, followed by adding 50 ml of ethylene glycol and 200 μl of Tween 20. The entire solution was transferred to a Teflon-lined stainless steel autoclave, and then was stirred for an additional 30 min. Finally, the autoclave was placed inside an electric oven at 140 °C for 4 h. After the reaction, the autoclave was cooled, and the suspension was centrifuged at 6,000 r.p.m. for 5 min. The centrifugation and redispersion cycles were repeated twice with deionized water and ethanol, respectively. The precipitate was oven dried at 60 °C for 12 h.

Supplementary Figure 1a shows the typical SEM image of as-synthesized $CaCO_3$ nanoplatelets, the lateral size is in the range of 4–20 μm with several to tens of layers. X-ray diffraction pattern (Supplementary Fig. 1b) shows that the products are mainly vaterite with a trace amount of aragonite and calcite (JCPDS 721652, lattice constants $a_0 = 4.99$ Å and $c_0 = 17.00$ Å, space group $R\bar{3}c$)[47]. The thickness of a single $CaCO_3$ nanosheet is in the range of 100–500 nm with the average of ~ 320 nm (Supplementary Fig. 1c and d), which is in the same scale as that for natural nacre[4,5,9].

Synthesis of TGA–CdTe NPs. $Cd(ClO_4)_2 \cdot 6H_2O$ (0.985 g) and thioglycolic acid (TGA, 0.392 ml) were dissolved in 125 ml of deionized water, followed by adjusting the pH to 11.2 with 1 M NaOH. This solution was placed in a three-neck round-bottom flask and purged with N_2 for 30 min. H_2Te gas (generated by reacting 0.13 g Al_2Te_3 with 25 ml of 0.5M H_2SO_4) was slowly passed through the solution. The solution was then allowed to reflux under N_2 at 100 °C for 120 min to obtain the TGA–CdTe NPs (Supplementary Fig. 13).

Preparation of TGA–CdTe nanowires. 15 ml of TGA–CdTe NP solution was mixed with 22.5 ml of methanol, yielding red precipitation in the solution. After centrifuging at 1,500 r.p.m. for 3 min, the supernatant was removed and the pre-cipitation was re-dispersed in deionized water and pH adjusted to 9 by the addition of 0.1 M HCl solution. The solution was aged in the dark at room temperature and nanowires (Supplementary Fig. 14) were formed after 6 days of aging.

Preparation of nacre-like PVA/G composite fibres. For preparation of PVA/G composite fibres, dried and powdered G was re-dispersed in deionized water by gentle sonication; the concentration was 10 mg ml^{-1}. PVA aqueous solution with a mass fraction of 5 wt% was prepared from dissolved PVA (molecular weight = $\sim 205,000$, hydrolysis $\leq 89\%$, Aladin) in deionized water. For preparing 66 wt% PVA/G mixture, 2.7 ml 5 wt% PVA aqueous solution was added to 7.3 ml G suspension (10 mg ml^{-1}) and stirred at 600 r.p.m. for 12 h. Then the mixture was concentrated by heating until a final mixture volume of 7 ml. Other ratios of PVA in the mixture was regulated by adding different volumes of 5 wt% aqueous PVA. For the wet-spinning process, first the concentrate was loaded into a 50 ml plastic syringe with a spinning nozzle (diameters = 100, 200 and 400 μm), and injected

into a container with 300 ml coagulation bath (for example, ethanol with 5 wt% NaOH, acetone, Na_2SO_4) by a syringe pump at speed of 0.05–3 ml min^{-1}. After coagulation for 0.5–5 min, the fibres were drawn out and transported by godet roller 1. For the second step, the PVA/G fibre was fixed and fed into the central axis of the twister by godet rollers 1 and 2. Then, the fibre was twisted into a spring-like shape and collected by the drum. Godet roller 2 was fixed on the twister and rotated with it, which kept the fibre in the central axis of the twister throughout the process. Usually, as-spinning fibre began to transfer to coiling fibre after twisting 33–36 circles per cm, then another 70–85 circles per cm twisting was needed to form a complete loop. Finally, the average density of loops was in the range of 10–25 loops per mm. We have prepared a series of mass fraction of PVA/G solutions (for example, 0, 4.4, 11 30, 54, 66, 80, 90 and 100 wt%) for spinning fibres. For comparison, G composite films with various PVA content were prepared by VAF.

Preparation of $PVA/CaCO_3$ composite fibres. The spinning process is the same as that of PVA/G detailed in Fig. 1. In brief, 0.4 g dried $CaCO_3$ nanoplatelets were added in 3.5 ml of ethanol and 5.0 ml of deionized water and stirred for 20 min. Then, 8 g of 5 wt% of aqueous solutions of PVA was added into the $CaCO_3$ solution and stirred for 24 h. The solution was centrifuged at 3,500 r.p.m. for 3 min. The sediment was re-dispersed in a little deionized water, and stirred vigorously (the content of $CaCO_3$ was about 0.2–0.4 g ml^{-1}). Then the mixture was injected into 5 wt% of ethanol solution of sodium hydroxide (the total volume was 200 ml) and a fibre was formed. To curdle the fibre, it was kept for 15 min in a coagulation bath. The wet fibre was dried at 60 °C for 1 h under vacuum. Twist-spinning process was similar to that for the PVA/G spring fibre.

Preparation of $PVA/CaCO_3/CdTe$ composite fibres. The spinning process is the same as that of $PVA/CaCO_3$ fibre detailed in Fig. 1. In brief, 0.4 g dried $CaCO_3$ nanoplatelets were added in 3.5 ml of ethanol and 5.0 ml of deionized water and stirred for 20 min. Then, 8 g of 5 wt% of aqueous solutions of PVA and 5 ml as-prepared CdTe nanowire solution was added into the above solution and stirred for 24 h. The solution was centrifuged at 3,500 r.p.m. for 3 min. The sediment was re-dispersed in a little deionized water, and stirred vigorously (the content of CdTe is about 0.9 wt% thermogravimetric analysis curve in Supplementary Fig. 15). Then the mixture was injected into pure ethanol solution and a fibre was formed. To curdle the fibre, it was kept for 15 min in a coagulation bath. Twist-spinning process was similar to that in the PVA/G spring fibre.

Characterization. SEM images were obtained using a Hitachi SU–8000 with accelerating voltage of 15 kV. AFM images were acquired in the tapping mode with a commercial multimode Nanoscope IIIa (Veeco Co., Ltd.). Raman spectra were collected using a LabRAM XploRA laser Raman spectroscope (HORIBA Jobin Yvon Co., Ltd.) using a 532-nm laser with incident power of ~ 1 mW. The G samples used in SEM, AFM and Raman characterizations were transferred to a silicon substrate that had a 300 nm thermal oxide layer. Thermogravimetric analysis was conducted in a nitrogen atmosphere from room temperature to 800 °C with a heating rate of 20 °C min^{-1}. About 5 mg of each sample was used for the tests. SAXS analyses were carried out at the Shanghai Synchrotron Radiation Facility, using a fixed wavelength of 1.24 Å, a sample-to-detector distance of 5,180 mm and an exposure time of 10 s. The scattering patterns were collected on a CCD camera. The area of the incident X-ray spot was 400×600 μm^2. Mechanical tests were carried out using the tensile tester equipped with a single column (Agilent Technologies T150UTM). For the test, each end of the fibre was first fixed on a paper with a cut window and then the paper was cut from two sides to free the fibre. For tension tests, the upper grip moved at a constant strain rate of 2.7×10^{-3} s^{-1}. All of the mechanical tests were carried out at a relative humidity of 50% and room temperature. All of the data were reported with a 95% confidence level. CPL studies: JASCO CPL–300 was used for CPL measurements. Emission intensity (arbitrary units), represented by the direct current voltage collected by the instrument, was measured along with CPL (unit: mdeg). Typical CPL settings were as follows: excitation wavelength = 358 nm, data pitch = 0.1 nm, scanning speed = 50 nm min^{-1}, digital integration time = 8 s and 10 accumulations. The luminescence dissymmetry ratio g_{lum} was calculated using the following equation: $g_{lum} = 0.00006978 \times$ CPL/emission. As-prepared spring fibres (coagulated in ethanol) incorporating CdTe nanowires were used for CPL studies. Multiple strands of the original spring fibres were tethered onto the sample holder so that they aligned vertically along the centre of the irradiation beam. The original fibres were slowly stretched after a 30-s immersion in a mixed solvent system of water and ethanol with a volume ratio of 1:9, followed by re-attachment to the sample holder for subsequent CPL measurements.

References

1. Shim, B. S. et al. Multiparameter structural optimization of single-walled carbon nanotube composites: toward record strength, stiffness, and toughness. ACS Nano **3**, 1711–1722 (2009).

2. Munch, E. *et al*. Tough, bio–inspired hybrid materials. *Science* **322**, 1516–1520 (2008).

3. Huang, Z. W. & Li, X. D. Order–disorder transition of aragonite nanoparticles in nacre. *Phys. Rev. Lett.* **109**, 025501 (2012).

4. Tang, Z. Y., Kotov, N. A., Magonov, S. & Ozturk, B. Nanostructured artificial nacre. *Nat. Mater.* **2**, 413–418 (2003).

5. Podsiadlo, P. *et al*. Ultrastrong and stiff layered polymer nanocomposites. *Science* **318**, 80–83 (2007).

6. Walther, A. *et al*. Supramolecular control of stiffness and strength in lightweight high-performance nacre-mimetic paper with fire-shielding properties. *Angew. Chem. Int. Ed.* **49**, 6448–6453 (2010).

7. Yao, H.-B., Tan, Z.-H., Fang, H.-Y. & Yu, S.-H. Artificial nacre-like bionanocomposite films from the self-assembly of chitosan–montmorillonite hybrid building blocks. *Angew. Chem. Int. Ed.* **49**, 10127–10131 (2010).

8. Bonderer, L. J., Studart, A. R. & Gauckler, L. J. Bioinspired design and assembly of platelet reinforced polymer films. *Science* **319**, 1069–1073 (2008).

9. Jackson, A. P. *et al*. The mechanical design of nacre. *Proc. R Soc. Lond. B* **234**, 415–440 (1988).

10. Bae, S. H. *et al*. Graphene-based transparent strain sensor. *Carbon* **51**, 236–242 (2013).

11. Li, X. *et al*. Stretchable and highly sensitive graphene-on-polymer strain sensors. *Sci. Rep.* **2**, 870–875 (2012).

12. Wang, Y. *et al*. Super-elastic graphene ripples for flexible strain sensors. *ACS Nano* **5**, 3645–3650 (2011).

13. Cruz–Silva, R. *et al*. Super-stretchable graphene oxide macroscopic fibers with outstanding knotability fabricated by dry film scrolling. *ACS Nano* **8**, 5959–5967 (2014).

14. Vollrath, F. & Knight, D. P. Liquid crystalline spinning of spider silk. *Nature* **410**, 541–548 (2001).

15. Dalton, A. B. *et al*. Super-tough carbon-nanotube fibers. *Nature* **423**, 703–703 (2003).

16. Miaudet, P. *et al*. Hot-drawing of single and multiwall carbon nanotube fibers for high toughness and alignment. *Nano Lett.* **5**, 2212–2215 (2005).

17. Shin, M. K. *et al*. Synergistic toughening of composite fibers by self-alignment of reduced graphene oxide and carbon nanotubes. *Nat. Commun.* **3**, 650–657 (2012).

18. Sun, C.-Y. *et al*. High-quality thin graphene films from fast electrochemical exfoliation. *ACS Nano* **5**, 2332–2339 (2011).

19. Wang, L., Xu, L., Kuang, H., Xu, C. & Kotov, N. A Dynamic nanoparticle assemblies. *Acc. Chem. Res.* **45**, 1916–1926 (2012).

20. Hu, X. Z., Xu, Z., Liu, Z. & Gao, C. Liquid crystal self-templating approach to ultrastrong and tough biomimic composites. *Sci. Rep.* **3**, 2374–2381 (2013).

21. Kou, L. & Gao, C. Bioinspired design and macroscopic assembly of poly(vinyl alcohol)-coated graphene into kilometers-long fibers. *Nanoscale* **5**, 4370–4378 (2013).

22. Zhang, M., Atkinson, K. R. & Baughman, R. H. Multifunctional carbon nanotube yarns by downsizing an ancient technology. *Science* **306**, 1358–1361 (2004).

23. Vigolo, B. *et al*. Macroscopic fibers and ribbons of oriented carbon nanotubes. *Science* **290**, 1331–1334 (2000).

24. Stribeck, N., Zeinolebadi, A., Fakirov, S., Bhattacharyya, D. & Botta, S. Extruded blend films of poly(vinyl alcohol) and polyolefins: common and hard-elastic nanostructure evolution in the polyolefin during straining as monitored by SAXS. *Sci. Technol. Adv. Mater.* **14**, 035006 (2013).

25. Kamal, T. *et al*. An *in situ* simultaneous SAXS and WAXS survey of PEBAX® nanocomposites reinforced with organoclay and POSS during uniaxial deformation. *Polymer* **53**, 3360–3367 (2012).

26. Potts, J. R., Dreyer, D. R., Bielawski, C. W. & Ruoff, R. S. Graphene-based polymer nanocomposites. *Polymer* **52**, 5–25 (2011).

27. Mamedov, A. A. *et al*. Molecular design of strong SWNT/polyelectrolyte multilayers composites. *Nat. Mater.* **1**, 190–194 (2002).

28. Dong, Z. L. *et al*. Facile fabrication of light, flexible and multifunctional graphene fiber. *Adv. Mater.* **24**, 1856–1861 (2012).

29. Zhu, J., Zhang, H. & Kotov, N. A. Thermodynamic and structural insights into nanocomposites engineering by comparing two materials assembly techniques for graphene. *ACS Nano* **7**, 4818–4829 (2013).

30. Gardel, M. L. *et al*. Elastic behavior of cross-linked and bundled actin networks. *Science* **304**, 1301–1305 (2004).

31. Schmoller, K. M., Fernández, P., Arevalo, R. C., Blair, D. L. & Bausch, A. R. Cyclic hardening in bundled actin network. *Nat. Commun.* **1**, 134–143 (2010).

32. Chen, X. Q., Zhang, S. L., Dikin, D. A., Ding, W. Q. & Ruoff, R. S. Mechanics of carbon nanocoil. *Nano Lett.* **3**, 1299–1304 (2003).

33. Tohgha, U. *et al*. Ligand induced circular dichroism and circularly polarized luminescence in CdSe quantum dots. *ACS Nano* **7**, 11094–11102 (2013).

34. Naito, M., Iwahori, K., Miura, A., Yamane, M. & Yamashita, I. Circularly polarized luminescent CdS quantum dots prepared in a protein nanocage. *Angew. Chem. Int. Ed.* **49**, 7006–7009 (2010).

35. Shang, Y. Y. *et al*. Super-stretchable spring-like carbon nanotube ropes. *Adv. Mater.* **24**, 2896–2900 (2012).

36. Chen, T., Hao, R., Peng, H. S. & Dai, L. M. High-performance, stretchable, wire–shaped supercapacitors. *Angew. Chem. Int. Ed.* **53**, 1–6 (2014).

37. Kumar, J. *et al*. Circularly polarized luminescence in chiral aggregates: dependence of morphology on luminescence dissymmetry. *J. Phys. Chem. Lett.* **5**, 316–321 (2014).

38. Ye, Q., Zhu, D., Zhang, H., Lu, X. & Lu, Q. Thermally tunable circular dichroism and circularly polarized luminescence of tetraphenylethene with two cholesterol pendants. *J. Mater. Chem. C* **3**, 6997–7003 (2015).

39. Coughlin, F. J. *et al*. Synthesis, separation, and circularly polarized luminescence studies of enantiomers of iridium(III) luminophores. *Inorg. Chem.* **47**, 2039–2048 (2008).

40. Tsumatori, H., Nakashima, T. & Kawai, T. Observation of chiral aggregate growth of perylene derivative in opaque solution by circularly polarized luminescence. *Org. Lett.* **12**, 2362–2365 (2010).

41. Maeda, H. *et al*. Chemical-stimuli-controllable circularly polarized luminescence from anion–responsive π–conjugated molecules. *J. Am. Chem. Soc.* **133**, 9266–9269 (2011).

42. Van Delden, R. A. *et al*. Remarkable solvent-dependent excited-state chirality: a molecular modulator of circularly polarized luminescence. *J. Am. Chem. Soc.* **125**, 15659–15665 (2003).

43. Kim, Y. *et al*. Recon?gurable chiroptical nanocomposites with chirality transfer from the macro- to the nanoscale. *Nat. Mater.* doi:10.1038/nmat4525 (2016).

44. Kim, Y. *et al*. Reconfigurable chiroptical nanocomposites with macro-to-nano chirality transfer. *Nat. Mater.* http://dx.doi.org/10.1038/nmat4525 (2016).

45. Tuinstra, F. & Koenig, J. L. Raman spectrum of graphite. *J. Chem. Phys.* **53**, 1126–1130 (1970).

46. Stankovich, S. *et al*. Synthesis of graphene-based nanosheets via chemical reduction of exfoliated graphite oxide. *Carbon* **45**, 1558–1565 (2007).

47. Li, X. Q. & Zeng, H. C. Calcium carbonate nanotablets: bridging artificial to natural nacre. *Adv. Mater.* **24**, 6277–6282 (2012).

Acknowledgements

We acknowledge financial support from National Natural Science Foundation of China (NSFC, No. 51502059, 61172001, 21373068, 61390502), the National Basic Research Program of China (2013CB632900), the Foundational Research Funds for the Central Universities (NO. HIT. NSRIF. 201641), the Self-Planned Task of State Key Laboratory of Robotics and System (HIT) (NO. SKLRS201509B) and the China Postdoctoral Science Foundation Grant (NO. 2015M570285). This work was partially supported by the Center for Photonic and Multiscale Nanomaterials (C-PHOM) funded by the National Science Foundation (NSF) Materials Research Science and Engineering Center program DMR 1120923; NSF EFRI–ODISSEI: Multiscale Origami for Novel Photonics, Energy Conversion, NSF–1240264. We thank the University of Michigan's EMAL for its assistance with electron microscopy, and for the NSF grant #DMR-9871177 for funding of the JEOL 2010F analytical electron microscope used in this work. We thank Professor F. G. Bian, Dr X. H. Li, and Dr F. Tian in the SSRF (BL16B1 experimental station) for S–SAXS characterization. We thank Dr J. Q. Yang for the analysis of S–SAXS data.

Author contributions

P.A.H. and N.A.K. supervised the project. J.Z. carried out fibre fabrication, characterization and testing. W.F. carried out the synthesis and characterization of CdTe NPs and NWs. W.F. and H.A.C. performed the CPL experiments. B.Y. carried out fibre testing and characterization. H.X.Z. synthesized CaCO₃ and fabricated PVA/CaCO₃ fibre. N.A.K. and P.A.H. analysed the data and co-wrote the paper. All of the authors discussed the results and commented on the manuscript.

Additional information

Competing financial interests: The authors declare no competing financial interests.

Three-dimensional structural dynamics and fluctuations of DNA-nanogold conjugates by individual-particle electron tomography

Lei Zhang[1,2], Dongsheng Lei[1], Jessica M. Smith[3,4,5], Meng Zhang[1], Huimin Tong[1], Xing Zhang[1,2], Zhuoyang Lu[1,6,7,8], Jiankang Liu[6,7,8], A. Paul Alivisatos[3,4,5,9] & Gang Ren[1]

DNA base pairing has been used for many years to direct the arrangement of inorganic nanocrystals into small groupings and arrays with tailored optical and electrical properties. The control of DNA-mediated assembly depends crucially on a better understanding of three-dimensional structure of DNA-nanocrystal-hybridized building blocks. Existing techniques do not allow for structural determination of these flexible and heterogeneous samples. Here we report cryo-electron microscopy and negative-staining electron tomography approaches to image, and three-dimensionally reconstruct a single DNA-nanogold conjugate, an 84-bp double-stranded DNA with two 5-nm nanogold particles for potential substrates in plasmon-coupling experiments. By individual-particle electron tomography reconstruction, we obtain 14 density maps at ∼2-nm resolution. Using these maps as constraints, we derive 14 conformations of dsDNA by molecular dynamics simulations. The conformational variation is consistent with that from liquid solution, suggesting that individual-particle electron tomography could be an expected approach to study DNA-assembling and flexible protein structure and dynamics.

[1] The Molecular Foundry, Lawrence Berkeley National Laboratory, Berkeley, California 94720, USA. [2] Department of Applied Physics, School of Science, Xi'an Jiaotong University, Xi'an 710049, China. [3] Materials Sciences Division, Lawrence Berkeley National Laboratory, Berkeley, California 94720, USA. [4] Department of Chemistry, University of California, Berkeley, California 94720, USA. [5] Department of Materials Science, University of California, Berkeley, California 94720, USA. [6] Center for Mitochondrial Biology and Medicine, The Key Laboratory of Biomedical Information Engineering of the Ministry of Education, Xi'an Jiaotong University, Xi'an 710049, China. [7] School of Life Science and Technology, Xi'an Jiaotong University, Xi'an 710049, China. [8] Frontier Institute of Science and Technology, Xi'an Jiaotong University, Xi'an 710049, China. [9] Kavli Energy NanoScience Institute, University of California, Berkeley, California 94720, USA. Correspondence and requests for materials should be addressed to G.R. (email: gren@lbl.gov).

Organic–inorganic-hybridized nanocrystals are a valuable class of new materials that are suitable for addressing many emerging challenges in biological and material sciences[1,2]. Nanogold and quantum dot conjugates have been used extensively as biomolecular markers[3,4], whereas DNA base pairing has directed the self-assembly of discrete groupings and arrays of organic and inorganic nanocrystals in the formation of a network solid for electronic devices and memory components[5]. Discretely hybridized gold nanoparticles conjugated to DNA were developed as a molecular ruler to detect sub-nanometre distance changes via plasmon-coupling-mediated variations in dark-field light scattering[3,6]. For many of these applications, it is desirable to obtain nanocrystals functionalized with discrete numbers of DNA strands[7,8]. In all of these circumstances, the soft components can fluctuate, and the range of these structural deviations have not previously been determined with a degree of rigour that could help influence the future design and use of these assemblies.

Conformational flexibility and dynamics of the DNA-nanogold conjugates limit the structural determination by X-ray crystallography, nuclear magnetic resonance spectroscopy and single-particle cryo-electron microscopic (cryo-EM) reconstruction because they do not crystallize, are not sufficiently small for nuclear magnetic resonance studies and cannot be classified into a limited number of classes for single-particle EM reconstruction. In addition, three-dimensional (3D) structure averaged from tens of thousands of different macromolecular particles obtained without prior knowledge of the macromolecular structural flexibility could result in an absence of flexible domains upon using the single-particle reconstruction method, for example, two ankyrin repeated regions of TRPV1 were absent in its atomic resolution 3D density map[9].

A fundamental experimental solution to reveal the structure of a flexible macromolecule should be based on the determination of each individual macromolecule's structure[10]. Electron tomography (ET) provides high-resolution images of a single object from a series of tilted viewing angles[11]. ET has been applied to reveal the 3D structure of a cell section and an individual bacterium at nanometre-scale resolution[12]. However, reconstruction from an individual macromolecule at an intermediate resolution (1–3 nm) remains challenging due to small molecular weight and low image contrast. Although, the first 3D map of an individual macromolecule, a fatty acid synthetase molecule, was reconstructed from negative-staining (NS) ET by Hoppe et al.[13], serious doubts have been raised regarding the validity of this structure[14], as this molecule received a radiation dose hundreds of times greater than the reported damage threshold[15]. Recently, we investigated the possibility based on simulated and real experimental NS and cryo-ET images[10]. We showed that a single-protein 3D structure at an intermediate resolution (1–3 nm) is potentially achieved using our proposed individual-particle ET (IPET) method[10,16–18]. IPET, an iterative refinement process using automatically generated dynamic filters and soft masks, requires no pre-given initial model, class averaging or lattice, but can tolerate small tilt errors and large-scale image distortion via decreasing the reconstruction image size to reduce the negative effects on 3D reconstruction. IPET allows us to obtain a 'snapshot' single-molecule 3D structure of flexible proteins at an intermediate resolution, and can be even used to reveal the macromolecular dynamics and fluctuation[17].

Here we use IPET, cryo-EM and our previously reported optimized NS (OpNS)[19,20] techniques to investigate the morphology and 3D structure of hybridized DNA-nanogold conjugates. These conjugates were self-assembled from a mixture of two monoconjugates, each consisting of 84-bp single-stranded DNA and a 5-nm nanogold particle. The dimers were separated by anion-exchange high-performance liquid chromatography (HPLC) and agarose gel electrophoresis as potential substrates in plasmon-coupling experiments. By OpNS-ET imaging and IPET 3D reconstruction, we reconstruct a total of 14 density maps at a resolution of ~2 nm from 14 individual double-stranded DNA (dsDNA)-nanogold conjugates. Using these maps as constraints, we derive 14 conformations of dsDNA by projecting a standard flexible dsDNA model onto the observed maps using molecular dynamics (MD) simulations. The variation of the conformations was largely consistent with that from liquid solution, and suggests that the IPET approach provides a most complete experimental determination of flexibility and fluctuation range of these directed nanocrystal assemblies to date. The general features revealed by this experiment can be expected to occur in a broad range of DNA-assembled nanostructures and flexible proteins.

Results

Cryo-EM and NS images of DNA-nanogold conjugates.
The sample of HPLC-purified 84-bp dsDNA (molecular mass = ~52 kDa) and two 5-nm nanogold conjugates (Supplementary Fig. 1) were examined by two methods, that is, cryo-EM (native buffer, vitreous ice, no staining) and OpNS[19,20]. Cryo-EM, a method often used to prevent artefacts induced by fixatives and stains, can be used to study protein structures under near-native conditions. However, small proteins (<100 kDa) are challenging to be visualized or 3D reconstructed. NS, a historical method, can be used for high-contrast imaging of small proteins through heavy metal salts coating the proteins surfaces[20,21]. However, heavy metal reaction can also cause a potential artefact by conventional protocols, including rouleaux formation of lipoproteins[19,22]. We used cryo-EM as a control, investigated the parameters related to lipoprotein rouleaux artefact[19] and reported an OpNS protocol[19,22]. The OpNS protocol has been tested by structure-known proteins, including 53-kDa cholesteryl ester transfer protein[23], GroEL and proteasomes[20], and the flexible proteins with structure partially known, including immunoglobulin (Ig)-G1 antibody and its peptide conjugates[16,17]. The surrounding heavy metal atoms provide more electron scattering and radiation damage resistance than the protein light atoms. In this study, we used cryo-EM and OpNS methods to examine the same sample under -178 °C and room temperature, respectively, (Fig. 1).

Survey cryo-EM and OpNS-EM micrographs (Fig. 1a,e) showed that each pair of nanogold particles was near one another. A statistical analysis of 1,032 nanogold particles from cryo-EM showed that the particles had a diameter of ~63.5 ± 6.7 Å (mean ± s.d.) and a peak population (~25.1%) diameter of 63.6 ± 1.0 Å (black solid line in Fig. 1c). This measurement is consistent with those from OpNS, that is, 606 nanogold particles from NS micrographs showed a diameter of ~63.0 ± 6.4 Å (mean ± s.d.) and a peak population (~25.5%) diameter of 62.8 ± 1.0 Å (blue dashed line in Fig. 1c).

To quantitatively identify whether the pairs of nanogold particles are strongly linked together by a statistical method, Pearson's correlation coefficients (PCC)[24] were used. The PCC, $r_{x,x}$ (defined as $r_{x,x} = \sum_{i=1}^{n} \left(x_i^a - \bar{x}^a \right)\left(x_i^b - \bar{x}^b \right) / \sqrt{ \sum_{i=1}^{n} \left(x_i^a - \bar{x}^a \right)^2 \sum_{i=1}^{n} \left(x_i^b - \bar{x}^b \right)^2 }$ for the x axis coordinates of two objects, a and b) and $r_{y,y}$ for the y axis coordinates were 0.9996 and 0.9976 for cryo-EM, and 0.9984 and 0.9983 for NS-EM results, respectively. The coefficients corresponded well with previous transmission electron microscopy (TEM) observations of the same sample in liquid solution (that is, $r_{x,x} = 0.934$ and $r_{y,y} = 0.943$; ref. 24). The high PCC suggests that the pair of nanogold particles are strongly linked together[24].

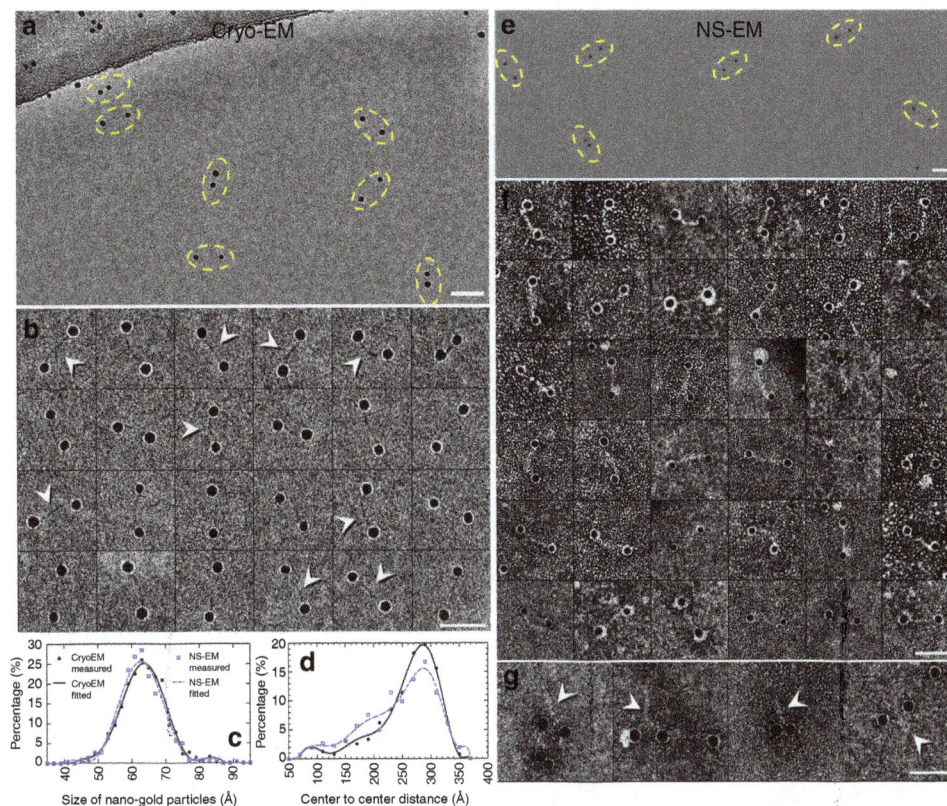

Figure 1 | Cryo-EM and OpNS-EM images of dsDNA-nanogold conjugates. (**a**) Cryo-EM images (vitreous buffer, no staining) of 5-nm nanogold particles conjugated to 84-bp dsDNA via a 5′-thiol linker. Pairs of nanogold were marked by yellow dashed ovals. (**b**) 24 representative cryo-EM images of the particles of DNA-nanogold conjugates. The polygonal-shaped areas are the nanogold particles, which were bridged by a fibre-shaped density (high contrast densities were indicated by arrows), ∼20–30 nm in length and ∼2 nm in width. The surfaces of the nanogold particles were coated with a layer of extraneous PEG for surface protection. (**c**) Histogram of the geometric diameters of 1,032 nanogold particles from cryo-EM images and 606 nanogold particles from NS images. (**d**) Histogram of the DNA lengths measured from 516 conjugates from cryo-EM and 303 conjugates from NS. The centre-to-centre length was measured between the centres of each nanogold particle pair. (**e**) NS images and (**f**) 36 representative NS images of the particles. The polygonal-shaped nanogold particles were bridged by a fibre-shaped density, and their surfaces were coated with a layer of extraneous PEG for surface protection. (**g**) A few pairs of nanogold particles were significantly closer in distance to each other, whereas their bridging fabric-like densities were thicker (indicated by arrows), likely due to the supercoiling of the dsDNA. Scale bars = 30 nm.

Zoom-in images of 24 representative cryo-EM particle pairs (Fig. 1b) and 36 representative NS particle pairs (Fig. 1f) showed that the polygonal-shaped nanogold particles were bridged by an ∼2-nm-width fibre-shaped density. A statistical analysis of the distances among the 516 pairs of cryo-EM nanogold particles yielded a length of 255.3 ± 48.7 Å (mean ± s.d., measured from the centre to centre of the nanogold particles) with a peak population (∼19.7%) distance of 286.4 ± 10 Å (black solid line in Fig. 1d). The distances are similar to that from 303 pairs of NS particles, that is, 245.5 ± 62.6 Å (mean ± s.d.) with a peak population (∼15.6%) distance of 287.0 ± 10 Å (blue dashed line in Fig. 1d). Moreover, both lengths were consistent with those measured from liquid solution by small-angle X-ray scattering (SAXS), that is, 28–30 nm (ref. 25) and a standard model of 84-bp dsDNA (∼2 nm wide and ∼30 nm long).

In addition, several pairs of nanogold particles presented abnormally closer to one another in both cryo-EM and NS. The higher-contrast NS images showed that their fibre-shaped bridging densities appeared thicker, but had lengths ranging from ∼20 to 30 nm, thus seeming similar to those of the full-length dsDNA (Fig. 1g). We suspected that these particles may be formed by two conjugates, in which each conjugate lost one nanogold, but met and formed a supercoil via their two dsDNAs. Since the mass is only 52 kDa above that of the regular conjugates, these particles are too small to be identified or isolated by our filtration.

3D reconstruction of an individual dsDNA-nanogold conjugate. To obtain a 3D structure of the dsDNA-nanogold conjugates, we used the IPET technique[10] rather than the conventional single-particle reconstruction method or sub-volume averaging method because these conjugates were not guaranteed to share the same structure (DNA is naturally flexible and dynamic in structure). IPET is used to obtain the *ab initio* 3D structure of an individual particle imaged from a series of tilt angles (Fig. 2a).

Although the DNA portion in the cryo-EM images could be barely visible under a dose of ∼20e$^-$ Å^{-2}, beyond this dose limitation, the contrasts rapidly disappeared, which prevents us from a full ET data acquisition (∼80 micrographs) and further 3D reconstruction. Thus, we used OpNS-EM for 3D reconstruction.

Although the signal-to-noise ratios (SNRs) of the nanogold portion of ET images (from −60° to +60° at 1.5° increments) were only ∼0.19 to ∼0.41 with an average of ∼0.31, the overall shape of dsDNA was still visible (Fig. 2b and Supplementary Movie 1). After contrast transfer function (CTF) correction, the tilt images were iteratively aligned to a global centre before achieving a final *ab initio* 3D reconstruction (Fig. 2b). During the iterations, SNR of the dsDNA portion gradually increased up to ∼2.44. The final 3D showed an overall handcuff shape (Fig. 2c) at a resolution of ∼14.7 Å, measured based on a Fourier shell correlation (FSC) analysis (details are provided in the Methods section; black line in Fig. 2e). To avoid the potential

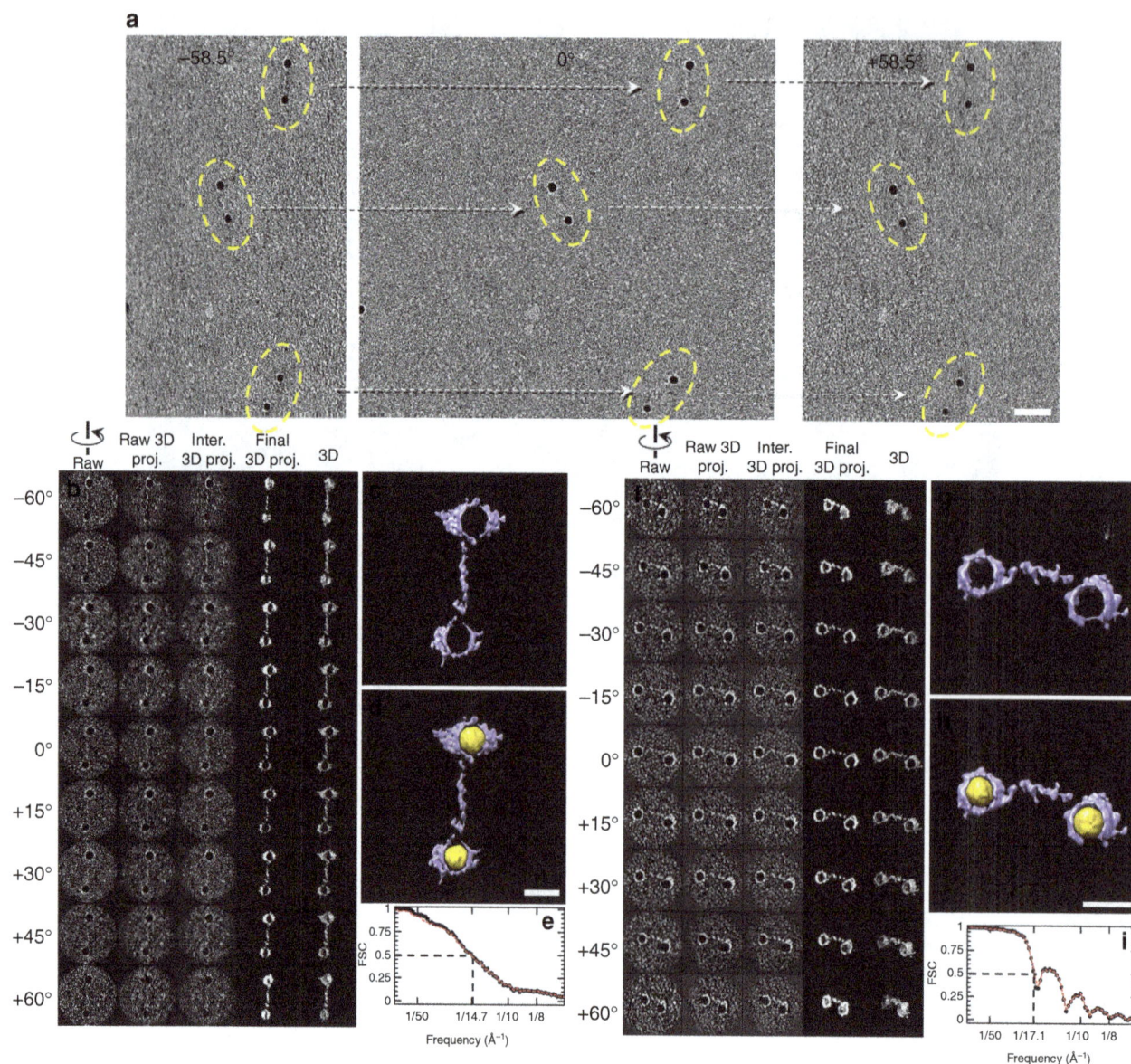

Figure 2 | 3D reconstruction of two representative DNA-nanogold conjugates by IPET. (**a**) The OpNS samples of DNA-nanogold conjugates were imaged using ET from a series of tilt angles (from −60° to +60° at 1.5° intervals). Three targeted particles (yellow circled) with their orthogonal views are indicated by the linked dashed arrows in the three selected ET tilt micrographs. The relative tilt angles are indicated in each image, and the axis of the tilt is vertical to the images. (**b**) Nine representative tilt images of the first targeted individual particle are displayed in the first column from the left (SNR of DNA portion: ∼0.31). Using IPET, the tilt images (after CTF correction) were gradually aligned to a common centre for 3D reconstruction via an iterative refinement process. The projections of the intermediate and final 3D reconstructions at the corresponding tilt angles are displayed in the next four columns according to their corresponding tilt angles. (**c**) Final IPET 3D density map of the targeted individual particle (SNR of DNA portion: ∼2.44). (**d**) The final 3D density map and its overlaid 3D density maps (final map in blue and its reversed map in gold) indicated the overall conformation of the DNA-nanogold conjugates. (**e**) The FSC analyses under including (black line) and excluding (red line) nanogold portions (two density maps reconstructed from odd and even numbers of tilt images) revealed that the resolutions of the IPET 3D density map were both ∼14.7 Å. (**f–i**) The 3D density map of a second individual DNA-nanogold conjugate was reconstructed from the tilt images (SNR of DNA portion: ∼0.56) using IPET. The FSC analysis showed that the 3D reconstruction resolution (SNR of DNA portion: ∼3.26) was ∼17.1 Å. Scale bars = 20 nm (**a**) and 10 nm (**d,h**).

overestimation of resolution by nanogolds instead of correctly reflecting the resolution of the dsDNA, we masked out the dsDNA portion only to repeat the FSC analyses. The FSC curve was nearly identical to that for nanogolds (red versus black lines in Fig. 2e), suggesting that the resolution is not overestimated by nanogolds.

The nanogold particle surface appeared with a layer of densities, which is possibly thiolated short-chain polyethylene glycol (PEG) molecules used to stabilize the particles against aggregation at high ionic strength (Fig. 2c). Considering that

the nanogold particles were in opposite image contrast to the dsDNA, we reversed the image contrast of the final 3D (coloured in gold) and overlaid it with original 3D to display both the dsDNA and nanogold particles in a same 3D map (Fig. 2d). This overlaid map showed the nanogold particles with diameters of ∼73.0 and ∼72.0 Å bridged by a high-density fabric with dimensions of ∼242.0 Å × ∼18.0 Å × ∼18.0 Å (Fig. 2d). The nanogold particle surfaces were surrounded by irregularly shaped densities, the PEG surface protection layer.

Although the 3D resolution was insufficient to determine the structure of the dsDNA, the overall shape could be used as a constraint to flexibly dock a standard structure of an 84-bp dsDNA into it to achieve a dsDNA conformation. By satisfying both the best fit to the density map and the chemical minimal energy requirements, we obtained a dsDNA conformation via gently bending the straight dsDNA structure into the density map using the CHARMM force field for all MD simulations[26,27] (Fig. 3a and Supplementary Movie 1).

By repeating the above process, we reconstructed 3D from second individual dsDNA-nanogold conjugate (Fig. 2f–i). The representative tilt images showed that the dsDNA was still visible (Fig. 2f), and the SNR of dsDNA portions ranged from ~0.41 to ~0.62, with an average of 0.56. Through IPET reconstruction, the tilt images were aligned to a global centre (Fig. 2f), and the SNR was increased up to ~3.26 (Fig. 2g). The final 3D reconstruction at ~17.1-Å resolution (based on the FSC analyses with and without the nanogold portion conditions; Fig. 2i)

showed a handcuff shape. The overlaid density map (the final 3D map with its reversed map, coloured in gold) showed that the nanogold particles had diameters of ~71.0 and ~65.0 Å bridged by a fabric-like density with dimensions of ~191.0 Å × ~20.0 Å × ~20.0 Å (Fig. 2h). Again, the nanogold particles were surrounded by irregularly shaped densities of a PEG surface protection layer. By flexibly docking the standard dsDNA structure into the bridging portion and following MD simulation for energy minimization, a second conformation of dsDNA was obtained (Fig. 3b).

3D reconstructions of another 12 DNA-nanogold conjugates. Through particle-by-particle 3D reconstructions, additional 12 conjugates were reconstructed using IPET (Fig. 3, Supplementary Figs 2–13 and Supplementary Table 1). The selected projections of the final 3D reconstructions showed a fabric DNA density between two nanogold particles (Fig. 3c). The overlaid density

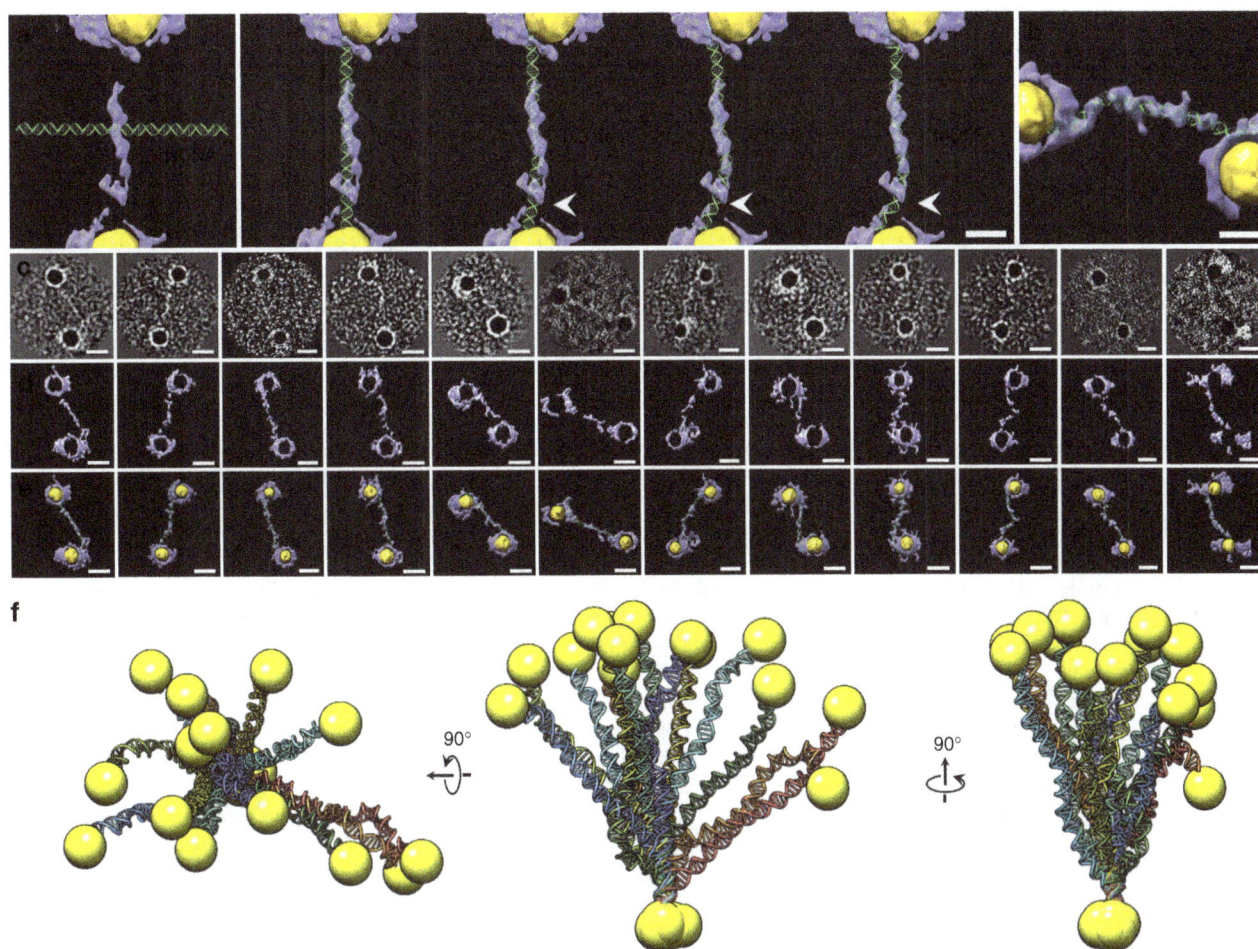

Figure 3 | 3D conformations of 14 dsDNA structures. Conformations obtained by flexibly fitting the dsDNA model onto the EM density maps using targeted MD simulations. (**a**) The final density map provided a constraint for the TMD simulations to achieve a new DNA conformation. Four snapshot images during the TMD simulation illustrate the process of flexibly docking the DNA model into the IPET density map to achieve a new conformation of DNA (arrow indicated the great conformational changing portion of DNA). During this process, the DNA conformation was allowed to change its structure while maintaining its chemical geometry and bonds with local energy minimization. (**b**) The final conformation of the second dsDNA structure was obtained from the second density map by following the same processes. (**c-e**) Gallery of 12 additional conformations from the 3D density maps of an additional 12 DNA-nanogold conjugates reconstructed using IPET. (**c**) Selected projections of the 3D density map of each individual DNA-nanogold conjugate. (**d**) Final 3D density maps of the individual DNA-nanogold conjugates. (**e**) The overlaid density maps of the final 3D reconstruction (grey) and its reversed contrast map (gold) revealed the overall conformation of the DNA-nanogold conjugates. A standard 84-bp dsDNA structure was flexibly docked into each density map to achieve the new conformations via TMD simulations. (**f**) Conformational flexibility and dynamics of the DNA-nanogold conjugate. Fourteen conformations of the DNA-nanogold conjugates were aligned together based on their first 14 bp. The distribution of dsDNA is shown from three orthogonal views. Scale bars = 5 nm (**a,b**) and 10 nm (**c-e**).

maps (reversed maps are coloured in gold; Fig. 3d,e) also confirmed that the nanogold particles were connected by densities attributable to dsDNA. Although up to ~30% of the fabric-like densities in those maps (Fig. 3e) could not be fully observed under the selected contour levels owing to various factors, such as uneven staining, image noise and reconstruction errors, these defects have a limited effect on the spacing distribution and the overall shape determination of the dsDNA due to its connectivity. The resolutions of ~20 Å still allowed us to flexibly dock the dsDNA model to obtain 12 additional DNA conformations via MD simulations (Fig. 3e and Supplementary Figs 2c–13c).

Statistical analyses of DNA-nanogold conjugates structures. Aligning the 14 conformations of dsDNA along their first 14 bp yielded a distribution in a shape of a bundle of flowers (Fig. 3f). Considering that the 14 conformations is insufficient to reveal the full 3D distribution of DNA conformations, only 1D distribution analysis was conducted, that is, the nanoparticle size and DNA length. The histogram of the nanogold particle sizes measured from the 3D reconstructions (the measurement method is described in Methods section) showed that the geometric mean of the nanoparticle diameters was 65.7 ± 5.0 Å, which is similar to the diameter measured from the 2D images of 606 nanogold particles (62.8 ± 1.0 Å at peak population, ~25.5%; Fig. 1d and Supplementary Table 1). The average distance measured from the 3D reconstructions (the measurement method is described in Methods section) was 291.1 ± 31.9 Å (mean ± s.d.), which was longer than the mean distance measured from the 2D images (245.5 ± 62.6 Å), but was similar to the distance of 287.0 ± 10.0 Å at the peak population (~15.6%; Fig. 1d). Approximately 69% of the distances was shorter than the peak distance, whereas ~31% were longer (Supplementary Table 1). This uneven distribution of the distance around the distance of the peak population is likely owing to the portion of dsDNA that formed a supercoiled structure but were not 3D reconstructed or measured. These variations reflected the conformational flexibility and dynamics of the dsDNA between the two conjugated nanogold particles.

Although the particles were flash-fixed by stain heavy metals and attached to a substrate (carbon film, which may cause certain artefacts, such as a preferred orientation, flatness and an uneven staining distribution), the statistical analyses showed that the measured lengths from 303 dimers were highly similar to the same sample measured in solution (Supplementary Table 2; refs 24,25). In detail, the distances measured using SAXS from the same sample in solution were ~280 Å on average and ~320 Å at the peak population[25], which were ~10–15% longer than those measured from our 2D images (Fig. 1d). Considering that the length measured from 2D images corresponded to the projection distance of the 3D length in solution, the 2D projection is naturally shorter than the length in 3D by a factor of $\pi/4$ under an isotropic distribution assumption condition. Based on the solution measured distances, their corresponding 2D projection distances should be ~220 Å ($\pi/4 \times 280$ Å) on average and ~250 Å ($\pi/4 \times 320$ Å) at the peak population, which were ~10–12% shorter than our measurement from 2D images. Our measured lengths (~246 and ~287 Å) were between the 3D lengths (~280 and ~320 Å) and the 2D projection lengths (~220 and ~250 Å), suggesting that our particles have a certain preferred orientation to the carbon film. The mean distance measured in solution via *in situ* TEM was ~180 Å (ref. 24), which was ~26% shorter than the mean measured from our 2D images. Although the short-distance views of the targeted conjugate was specifically chosen for easy tracking and imaging in the *in situ* experiment, the mean value was still close to the error bar range in our measurement.

Analyses of bending energy of DNA-nanogold conjugates. The conformations of dsDNA provide an opportunity to study the DNA bending energy. The bending energy can be calculated based on a simple worm-like chain (WLC) model[25,28,29], in which the energy of dsDNA is simplified to the bending potential while ignoring the distortion potential at the single-base pair level. The calculation of the bending energy depends on the local bending angles of each base pair, which is governed by a single parameter for the mechanical response, that is, the persistence length. Although the persistence length could be extracted by measuring the tangent correlation function for a WLC, the length of our 84-bp DNA here is too short to derive a persistence length. Thus, the widely used persistence length of 50 nm was used to compute the bending energy as this length is a favourable parameter to describe the dsDNA conformational statistics for constructs composed of tens of base pairs[25].

To measure the local bending angle (Fig. 4a), each of two nearby base pairs was first fitted with a small standard cylinder with a fixed length and width (Fig. 4b,c). Two types of angles can be defined to represent the bending angle of the base pair, that is, the angle θ_i, formed between centre-to-centre directions of three nearby cylinders (Fig. 4d), and the angle φ_i, formed between the two axes of nearby cylinders (Fig. 4e). The energies of each base pair were calculated and summed to represent the total energy of this dsDNA conformation. The energy distribution of the 14 conformations showed that the averaged bending energies were ~116 and ~169 kcal mol^{-1} based on two types of angle definition (Fig. 4f and Supplementary Table 2). The average bending energies were ~2–3 times higher than the bending energy calculated based on a theoretical WLC model prediction on 84-bp DNA (~50 kcal mol^{-1} at room temperature)[25], suggesting that the 14 DNA conformations were more flexible than the prediction.

Parameters influence the length and bending energy of DNA. Above analyses showed two definition types of bending angles, that is, θ_i and φ_i, which could cause a ~50% difference on the computed bending energy from the experimental DNA conformations. Considering that the EM observed DNA is more flexible than predicted, it is worthy to evaluate whether other factors may also influence the measurement of the DNA length and bending energy, such as the fluctuation of distal ends, DNA modelling methods, noise bias in the EM density map, initial model bias and temperature. Additional analyses were performed as the following: (i) the bending energy based on the central 42 bp was recalculated. This was because 84-bp DNA is relatively short and the ends of the chain exhibit greater fluctuations than the middle portion. The results obtained from the calculation showed that the bending energy is ~15% less than when using all 84 bp (Supplementary Table 2), suggesting that DNA is still more flexible than the WLC prediction. (ii) The initial model of DNA was obtained by manually fitting a standard DNA model to each EM configuration. Considering that the manual operation may lead to kinks, which are difficult to be repaired by MD simulation, we used another method to generate a smooth curved model of DNA whose structure was as close as possible to the canonical B-form DNA double helix (details are in Methods section). Before conducting any simulation, we calculated its bending energy, which is only ~10% of the above bending energy, and only ~20% of the WLC prediction, suggesting that the smooth DNA is more stiff than the WLC prediction, and EM configurations. (iii) After submitting this smooth model for energy minimization using the nanoscale molecular dynamics version 2 (NAMD2) software package[30], we found that the bending energy was increased ~4–5 times, thus nearly approaching the WLC

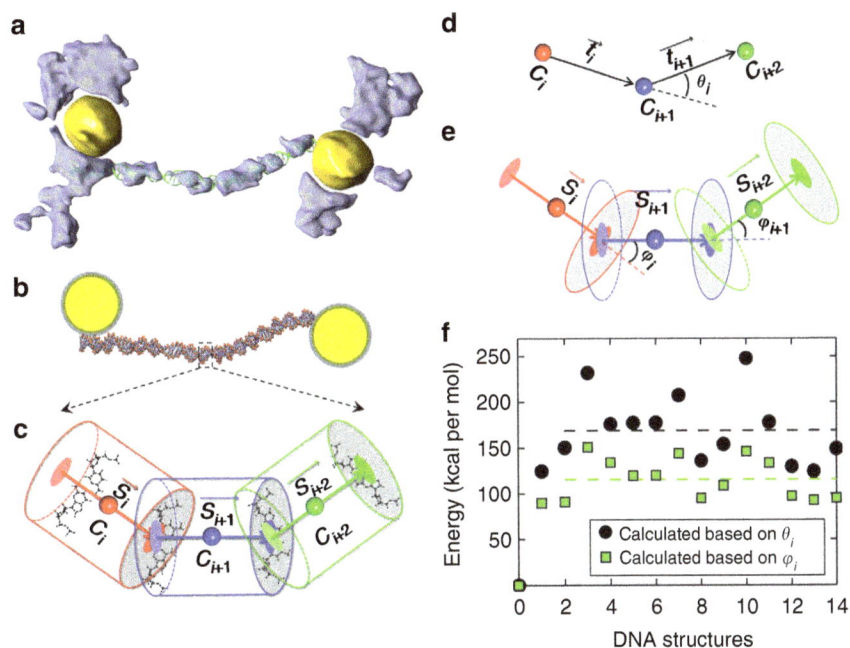

Figure 4 | Bending energy distribution of dsDNA. (**a**) dsDNA conformation was obtained by fitting the standard dsDNA model into the IPET 3D density map. (**b**) Schematic model illustrating that the nanogold interacts with the dsDNA and that the dsDNA contains kink regions that carry bending elasticity. (**c**) Cylinder model illustrating the bending angles between two connected base pairs. The cylinder is defined by the two consecutive dsDNA base pairs. (**d**) The bending angle can be presented by the angle θ_i, formed by the centres of three consecutive cylinders, or (**e**) by the angle φ_i, formed by the centre axes of two consecutive cylinders. (**f**) Based on the two types of measured angles, θ_i and φ_i, the bending energies for each DNA conformation were calculated and plotted based on a simple WLC model. The averaged bending energies from the two types of angles are indicated by the dashed lines. The bending energy of the standard DNA model was also calculated and indicated as structure no. 0 as a control.

prediction. (iv) However, after submitting the model to further MD simulations for only 0.1 ns, the calculated energy jumped up to $\sim 80\%$ of the energy from the EM configuration, thereby confirming that the DNA is more flexible than the WLC prediction. This result suggests that the MD simulations may be the key role to increase the bending energy and result in a more flexible DNA model. (v) To evaluate how the noise in the density map may influence the bending energy, we conducted MD flexible fitting (MDFF) based on the smooth model to constrain its structure with the EM density map. The calculated bending energy immediately jumped up to an even higher level, that is, $\sim 10\%$ more than that from the EM configurations. This test suggests that the MDFF and noise in the EM density could be critical in causing the DNA model to be more flexible than it should be. (vi) To avoid a potential influence to the energy from the given initial models, a standard and straight model of DNA was used for MD simulations under the same condition, that is, under 0.15 M physiological salt solution, temperature of 298 K and a pressure of 1 atm. Using NAMD2 for 20 ns of simulation without any constraint for DNA conformational changes, the equilibration of DNA in a box of $\sim 347.3\,\text{Å} \times \sim 93.4\,\text{Å} \times \sim 50.8\,\text{Å}$ was monitored using the root mean square deviation by visual MD (VMD)[31] (Supplementary Fig. 14a). The bending energies (Supplementary Fig. 14b) and the distances between the two distal ends (Supplementary Fig. 14c) revealed that the system became nearly balanced after 8 ns. The statistical analyses of the bending energies of the DNA in its last 10-ns simulations showed that the average energy was $99.1 \pm 10.9\,\text{kcal mol}^{-1}$ with a peak population ($\sim 7.1\%$) energy of $97.4 \pm 1.0\,\text{kcal mol}^{-1}$ based on the bending angles of the φ_i calculation, whereas the average energy was $152.1 \pm 16.1\,\text{kcal mol}^{-1}$ with a peak population ($\sim 4.6\%$) energy of $151.3 \pm 1.0\,\text{kcal mol}^{-1}$ based on the bending angles of the θ_i calculation (Supplementary Fig. 14b,d). This energy is

surprisingly similar to that of the EM configuration (~ 10–15% lower) (Supplementary Table 2). The statistical analyses of the length between the two distal ends of equilibrated DNA in the last 10 ns of the simulations showed that the average length was $268.5 \pm 2.1\,\text{Å}$ with a peak population ($\sim 18.5\%$) length of $267.9 \pm 0.5\,\text{Å}$ (Supplementary Fig. 14c,e). The distance is similar to the length at the peak population measured from TEM images. (vii) To evaluate how temperature influences the bending energy measurement using MD simulations, the above processes were repeated under a higher temperature, that is, 310 K instead of 298 K (Supplementary Fig. 15). The bending energies calculated under the higher temperature in the last 10 ns showed that the average energy was $120.8 \pm 14.7\,\text{kcal mol}^{-1}$ with a peak population ($\sim 6.1\%$) energy of $115.1 \pm 1.0\,\text{kcal mol}^{-1}$ based on the bending angles of the φ_i calculation, whereas the average energy was $177.9 \pm 15.6\,\text{kcal mol}^{-1}$ with a peak population ($\sim 5.1\%$) energy of $177.5 \pm 1.0\,\text{kcal mol}^{-1}$ based on the bending angles of the θ_i calculation (Supplementary Fig. 15b,d). The energy is increased by $\sim 20\%$ from those under 298 K and becomes more similar to that of the EM configuration, suggesting that the temperature is related, but not critical. The statistical analyses on the length showed that the averaged length was $269.5 \pm 2.9\,\text{Å}$ with a peak population ($\sim 13.3\%$) length of $271.1 \pm 0.5\,\text{Å}$ (Supplementary Fig. 15c,e), which is similar to those measured from the EM configurations and 298 K MD simulation, suggesting that the length is insensitive to the temperature. (viii) To further confirm that the length is insensitive to bending energy, one EM configuration with the DNA length of $\sim 241.0\,\text{Å}$ was performed by MD simulations under a length constrain (Supplementary Fig. 16). This length is close to the mean length of the DNA portion estimated from solution using SAXS[25]. The bending energy in the last 10 ns was $105.8 \pm 10.7\,\text{kcal mol}^{-1}$ with a peak population ($\sim 7.3\%$) energy of $102.7 \pm 1.0\,\text{kcal mol}^{-1}$

based on the bending angles of the φ_i calculation, whereas the average energy was 163.6 ± 17.4 kcal mol^{-1} with a peak population ($\sim 5.0\%$) energy of 158.4 ± 1.0 kcal mol^{-1} based on the bending angles of the θ_i calculation (Supplementary Fig. 16b,c). This energy is similar to that of the other EM configurations and simulations, suggesting that the length cannot reflect the bending energy and flexibility of DNA.

Above tests showed, although EM density maps contain noise, that the bending energy can be influenced limitedly by the initial models and the EM map configuration. The similar bending energies calculated from different MD simulations and initial models suggested that the DNA is more flexible than WLC prediction. However, we cannot exclude those MD simulations that may result in DNA, which presents more flexibility than WLC prediction.

Discussion

Although the direct imaging of dsDNA has been previously reported using heavy metal shadowing[32,33] and NS methods[34–36], to the best of our knowledge, the 3D structure of an individual dsDNA strand has not previously been achieved. It has been thought that individual dsDNA would be destroyed under the high energy of the electron beam before a 3D reconstruction, or even a 2D image, is able to be achieved. Our NS tilt images showing fibre-shaped dsDNA bridging two conjugated nanogold particles demonstrated that the dsDNA can in fact be directly visualized using EM, which is consistent with the recently reported single-molecule DNA sequencing technique via TEM[36]. The resolutions of our density maps ranged from ~ 14 to ~ 23 Å, demonstrating that an intermediate-resolution 3D structure can be obtained for each individual macromolecule. This capability is consistent with our earlier report of a ~ 20-Å resolution 3D reconstruction of an individual IgG1 antibody using the same approach[16,17].

Notably, a total dose of $\sim 2,000 e^- $ Å^{-2} used in our ET data acquisition is significantly above the limitation conventionally used in cryo-EM (~ 80–$100 e^-$ Å^{-2}), which can be suspected to have certain artefact from radiation damage. In cryo-EM, the radiation damage could cause sample bubbling, deformation and knockout effects; in NS, only the knockout phenomena is often observed, in which the protein is surrounded by heavy atoms that were kicked out by electron beam. Since the sample was coated with heavy metal atoms and were dried in air, the bubbling and deformation phenomena were not usually observed. The heavy metal atoms that coat the surface of the biomolecule can provide a much higher electron scattering than from a biomolecule only inside lighter atoms. The scattering is sufficiently high to provide enough image contrast at our 120-kV high tension; thus, a further increase to the scattering ability by reducing the high tension to 80 kV may not be necessary for this NS sample. In addition, the heavy atoms can provide more radiation resistance and allow the sample to be imaged under a higher dose condition. The exact dose limitation for NS is still unknown. The radiation damage related artefact in NS samples is knockout, which could reduce the image contrast and lower the tilt image alignment accuracy and 3D reconstruction resolution. In our study, a total dose of $2,000 e^-$ Å^{-2} did not cause any obvious knockout phenomenon, but provides a sufficiently high contrast for the otherwise barely visible DNA conformations in each tilt series. The direct confirmation of visible DNA in each tilt image is essentially important to us to validate each 3D reconstruction, especially considering this relatively new approach.

Our 3D reconstruction algorithm used an *ab initio* real-space reference-projection match iterative algorithm to correct the centres of each tilt images, in which the equal tilt angle step for

3D reconstruction of a low contrast and asymmetric macromolecule was used. This method is different from recently reported Fourier-based iterative algorithm, termed equally sloped tomography, in which the pseudo-polar fast Fourier transform, the oversampling method and internal lattice of a targeted nanoparticle are used to achieve 3D reconstruction at atomic resolution[37].

It is generally challenging to achieve visualization and 3D reconstruction on an individual, small and asymmetric macromolecule by other conventional methods; our method demonstrated its capability for 3D reconstruction of 52 kDa 84-bp dsDNA through these studies: IgG1 antibody 3D structural fluctuation[17], peptide-induced conformational changes on flexible IgG1 antibody[5], floppy liposome surface binding with 53-kDa proteins[38], all of which suggest that this method could be used to serve the community as a novel tool for studying flexible macromolecular structures, dynamics and fluctuations of proteins, and for catching the intermediate 3D structure of protein assembling.

DNA-based self-assembling materials have been developed for use in materials science and biomedical research, such as DNA origami designed for targeted drug delivery. The structure, design and control require feedback from the 3D structure, which could validate the design hypothesis, optimize the synthesis protocol and improve the reproducible capability, while even providing insight into the mechanism of DNA-mediated assembly.

Methods

Synthesis of DNA-nanogold conjugates. Conjugation of DNA-nanogold was synthesized according to a previously published procedure[1]. In brief, ~ 5.5-nm nanogold particles were stabilized via exchanging with bis-(p-sulfonatophenyl) phenylphosphine. DNA sequences modified with a 5′-thiol moiety were purified via polyacrylamide gel electrophoresis. DNA thiolated at the 5′ end was re-suspended in buffer (10 mM Tris pH 8 and 0.5 mM EDTA). Nanogold particles and DNA were combined at a stoichiometric ratio of 1:2 in the presence of a reducing agent. The formed monoconjugates were separated using anion-exchange HPLC, and the fractions were concentrated using an Amicon Ultra spin filter, MW = 100,000 (EMD Millipore Corp., Billerica, MA, USA). Twenty microlitres of nanogold monoconjugates, each containing complementary strands of DNA, were combined stoichiometrically, as determined by absorption at 520 nm, and were allowed to react overnight at room temperature[7]. The final conjugates were purified from unreacted monoconjugates via agarose gel electrophoresis.

Sequences. 84-Base DNA sequence. 5′-thiol-CCGGCGGCCCAGGTGTA TCAGTGTTCGTTGCAAGCTCCAACATCTGAGTACCACGCATACTAT ACTTGAAATATCCGCGCCCGG-3′.
84-Base DNA complement. 5′-thiol-CCGGGCGCGGATATTTCAAGTATA GTATGCGTGGTACTCAGATGTTGGAGCTTGCAACGAACACTGATACAC CTGGGCCGCCGG-3′.

Preparation of cryo-EM and OpNS-EM specimens. The cryo-EM specimens were prepared as described previously[39]. In brief, an aliquot ($\sim 4 \mu l$) of DNA-nanogold sample at a concentration of $\sim 20 \mu g$ ml^{-1} was placed on a glow-discharged lacey carbon film-coated copper grid (Cu-200LC, Pacific Grid-Tech, San Francisco, CA, USA). The samples were blotted with filter paper from both sides at $\sim 90\%$ humidity and 4 °C with a Leica EM GP rapid-plunging device (Leica, Buffalo Grove, IL, USA) and then flash-frozen in liquid ethane. The flash-frozen grids were transferred into liquid nitrogen for storage. The NS specimens were prepared by OpNS protocol as described previously[19,22]. In brief, an aliquot ($\sim 4 \mu l$) of DNA-nanogold sample at a concentration of $\sim 20 \mu g$ ml^{-1} was placed on a thin carbon-coated 200-mesh copper grid (Cu-200CN, Pacific Grid-Tech; CF200-Cu, Electron Microscopy Sciences, Hatfield, PA, USA) that had been glow-discharged. After ~ 1 min of incubation, the excess solution was blotted with filter paper. The grid was then washed with water and stained with 1% (w/v) uranyl formate on Parafilm before air-drying with nitrogen[19,22].

TEM un-tilted data acquisition and imaging process. Both cryo-EM and OpNS samples were examined using a Zeiss Libra 120 Plus TEM (Carl Zeiss NTS) operating at 120 kV high tension with 20 eV in-column energy filtering at -178 °C (for cryo-EM) and room temperature (for NS). A Gatan 915 cryo-holder was used. The micrographs were acquired using a Gatan UltraScan 4K × 4K CCD under a magnification of 20–125 kx (each pixel of the micrographs corresponded to ~ 0.59 to ~ 0.094 nm in specimens) under a dose of ~ 5–$15 e^-$ Å^{-2} (for cryo-EM) and

~40–90e$^-$ Å^{-2} (for NS-EM), and defocus of 0.2–3 µm (for cryo-EM) and <1.0 µm (for NS-EM). The defocus and astigmatism of each micrograph were examined using EMAN *ctfit* software[40] after the X-ray speckles were removed. The micrographs with distinguishable drift effects were excluded. The remaining micrographs were filtered using a Gaussian boundary high-pass filter within a resolution range of ~200–500 nm.

The statistical analysis of the diameter of the nanogold particles was based on the geometric mean[22] of 1,032 cryo-EM nanogold particles and 606 NS nanogold particles. Each particle was measured based on two perpendicular directions in which one of the diameters was measured along the longest direction of the particle[22]. The DNA lengths were measured by calculating the distance between the centres of two nanogolds (centre-to-centre distance). The lengths of 516 pairs from cryo-EM images and 303 pairs from NS images were measured and then submitted for histogram analyses by using Matlab software.

ET data acquisition and image pre-process. The TEM holder was tilted at angles ranging from −60° to +60° at 1.5° increments and controlled using Gatan tomography software that was pre-installed in the microscopes (Zeiss Libra 120 TEM). The TEM was operated under 120 kV with a 20 eV energy filter. The tilt series were acquired using a Gatan Ultrascan 4K × 4K CCD camera under low-defocus conditions (<1.0 µm) and a magnification of 125 kx (0.094 nm per pixel). The total electron doses were ~1,500–2,500e$^-$ Å^{-2}. The micrographs were initially aligned together using the IMOD software package[41]. The CTF was then corrected using TOMOCTF[42]. The tilt series of the particles in square windows of 512 × 512 pixels (~48 nm) were semi-automatically tracked and windowed using IPET software[10].

IPET 3D reconstruction. In the IPET reconstruction process[10], a tilt series of CTF-corrected images containing a single DNA-nanogold particle (in size of ~48 nm) was directly back-projected into an *ab initio* 3D density map as an initial model based on its corresponding goniometer tilt angles. The refinement was started using this initial model to align each tilt image via translational alignment to the projections of the initial model. During the refinement, automatically generated filters and a circular-shaped mask with a Gaussian boundary were sequentially applied to the tilt images and references to increase the alignment accuracy[10]. The resolution was defined based on FSC in which the aligned images were split into two halves based on an odd- or even-numbered index to generate two 3D reconstructions for computing their FSC curve against the spatial frequency shells in Fourier space. The frequency at which the FSC curve first falls to a value of 0.5 was used to represent the resolution of the IPET 3D density map. All of the IPET density maps presented in the figures were low-pass filtered to 16–20 Å. The DNA portion SNR in each 3D map was calculated using the equation $SNR_y = (I_s - I_b)/N_b$, where I_s is the average power inside the particle, I_b is the average power outside the particle (nanogold area was excluded), and N_b is the s.d. of the noise calculated from the background s.d. outside the particle (nanogold area was excluded). The particle area was defined using a particle-shaped mask generated from the IPET final 3D with low-pass filtering to ~25–30 Å and set as three times its molecular weight. This same method was used to calculate the 2D SNR. The 2D mask was generated based on the 3D projection at each tilt angle.

Conformation of DNA. A standard model of the 84-bp double-helix B-form DNA (dsDNA) was generated using the online DNA sequence to structure tool[43]. By comparing the standard model to the density map, we aligned the best-fit positions onto the EM density map envelope using Chimera[44]. Using the best-fit positions as the target markers, we drove the DNA conformational change of the model to agree with the DNA portion in the 3D density map with a pre-calculated force via a targeted MD (TMD) simulation[45]. After 0.6 ns of TMD simulation, the DNA was moved onto the density map, and we further equilibrated the DNA conformation in a water solution at room temperature via a MDFF simulation under the constraint of the original density map while protecting the secondary structure. Both TMD and MDFF simulations were performed using NAMD2 software package from the University of Illinois at Urbana-Champaign[30].

A method to generate a smooth DNA model to match the EM configuration was conducted as follows: the DNA portion between two nanogolds was first marked out by a series of consecutive 3D points. Labelled points were automatically fitted with a smooth quadratic or cubic Bézier curve by GraphiteLifeExplorer software[46]. A DNA model was generated to be as close as possible to the curve with a canonical DNA helix. For comparison, the model was also submitted for refinement by an energy minimization with NAMD2 at 0.15-M salt concentrations. The energy minimization was last for 30,000 steps and followed by 0.1-ns all-atom simulation. The model was also constrained by the electron density map using MDFF. All the MD simulations were performed using CHARMM force field[26,27].

Calculation of dsDNA bending energy. To calculate the DNA bending energy, we measured the local bending angles, as illustrated in Fig. 4c–e. In detail, the two nearby bases of the standard DNA model were rigidly aligned to the derived dsDNA conformations using VMD[31]. Based on the centres (C_i) and centre directions (indicated as a vector $\vec{S_i}$) of the aligned standard DNA bases, whose

orientations are represented by a small cylinder, the two types of DNA bending angles were measured, that is, the angle θ_i formed by the centres of three consecutive cylinders and the angles φ_i between the central axes of two consecutive cylinders. Therefore, the DNA bend energy E_{bend} was calculated according to $E_{bend} = k_b T \frac{l_p}{d} \sum_i (1 - \cos \theta_i)$ or $E_{bend} = k_b T \frac{l_p}{d} \sum_i (1 - \cos \varphi_i)$, where l_p is the dsDNA persistence length (50 nm) and d is the distance between the base pairs (3.4 Å; ref. 25).

MD simulation of dsDNA bending energy and length in solution. The MD simulation analyses were performed according to the following three steps: (i) a standard model of 84-bp dsDNA was embedded into a cubic box that extended at least 15 Å away from the DNA surface. The box contained a total of 49,679 TIP3P water molecules, 307 Na$^+$ atoms and 141 Cl$^-$ atoms to simulate the 0.15-M physiological salt concentration and was constructed using VMD[31]. The system, which contained a total of 154,813 atoms, was subjected to energy minimization via 20,000 steps to remove the atomic clashes using the NAMD2. The DNA backbone atoms were fixed in the first 10,000 steps. In the second 10,000 steps, the DNA backbone atoms were constrained under a force constant of 5 kcal mol^{-1} Å^{-2}. The energy-minimized system was subsequently heated from 0 to 298 K over 120 ps to initiate the 20-ns all-atom MD simulations. The systems under 1 atm and 298 K reached equilibrium after 10 ns; the DNA backbone constraints were removed after the first 0.2 ns of the simulations. During the simulations, the temperature was maintained via Langevin dynamics[47] with a damping coefficient of 5 ps, and the 1 atm pressure was maintained using the Langevin piston Nose–Hoover method[48] with a piston period of 100 fs and a decay rate of 50 fs. Periodic boundary conditions and a cutoff distance of 12 Å for van der Waals' interactions were applied, and the Particle–Mesh Ewald method[49] with a grid spacing of <1 Å was used to compute the long-range electrostatic interactions. (ii) The above system was submitted again under a higher balanced temperature of 310 K instead of 298 K. (iii) The 11th dsDNA conformation derived from the previous flexible docking was embedded into a cubic box containing 52,093 TIP3P water molecules, 313 Na$^+$ atoms and 147 Cl$^-$ atoms. The criterion to select this conformation as a representative conformation of the dsDNA model was a length of ~241.0 Å, which was close to the mean length of the DNA estimated from solution by SAXS[25]. The system, which contained 162,067 atoms, was subjected to energy minimization via 10,000 steps to remove the atomic clashes using NAMD2. The energy-minimized system was subsequently heated from 0 to 293 K over 60 ps to initiate the 20-ns all-atom MD simulation, in which the end-to-end distance of this DNA was constrained at 241.0 Å under a force constant of 50 kcal mol^{-1} Å^{-2}. This system under 1 atm and 293 K reached equilibrium after 10 ns of simulation. In the above processes, the bending energies and length of the dsDNA conformations in the last 10 ns were submitted to the statistical analyses.

References

1. Alivisatos, A. P. *et al.* Organization of 'nanocrystal molecules' using DNA. *Nature* **382**, 609–611 (1996).
2. Mirkin, C. A., Letsinger, R. L., Mucic, R. C. & Storhoff, J. J. A DNA-based method for rationally assembling nanoparticles into macroscopic materials. *Nature* **382**, 607–609 (1996).
3. Elghanian, R., Storhoff, J. J., Mucic, R. C., Letsinger, R. L. & Mirkin, C. A. Selective colorimetric detection of polynucleotides based on the distance-dependent optical properties of gold nanoparticles. *Science* **277**, 1078–1081 (1997).
4. Wu, X. *et al.* Immunofluorescent labeling of cancer marker Her2 and other cellular targets with semiconductor quantum dots. *Nat. Biotechnol.* **21**, 41–46 (2003).
5. Zheng, J. *et al.* Two-dimensional nanoparticle arrays show the organizational power of robust DNA motifs. *Nano Lett.* **6**, 1502–1504 (2006).
6. Barrow, S. J., Wei, X., Baldauf, J. S., Funston, A. M. & Mulvaney, P. The surface plasmon modes of self-assembled gold nanocrystals. *Nat. Commun.* **3**, 1275 (2012).
7. Claridge, S. A., Liang, H. W., Basu, S. R., Frechet, J. M. & Alivisatos, A. P. Isolation of discrete nanoparticle-DNA conjugates for plasmonic applications. *Nano Lett.* **8**, 1202–1206 (2008).
8. Jones, M. R., Seeman, N. C. & Mirkin, C. A. Nanomaterials. Programmable materials and the nature of the DNA bond. *Science* **347**, 1260901 (2015).
9. Liao, M., Cao, E., Julius, D. & Cheng, Y. Structure of the TRPV1 ion channel determined by electron cryo-microscopy. *Nature* **504**, 107–112 (2013).
10. Zhang, L. & Ren, G. IPET and FETR: experimental approach for studying molecular structure dynamics by cryo-electron tomography of a single-molecule structure. *PLoS ONE* **7**, e30249 (2012).
11. Milne, J. L. & Subramaniam, S. Cryo-electron tomography of bacteria: progress, challenges and future prospects. *Nat. Rev. Microbiol.* **7**, 666–675 (2009).
12. Komeili, A., Li, Z., Newman, D. K. & Jensen, G. J. Magnetosomes are cell membrane invaginations organized by the actin-like protein MamK. *Science* **311**, 242–245 (2006).
13. Hoppe, W., Gassmann, J., Hunsmann, N., Schramm, H. J. & Sturm, M. Three-dimensional reconstruction of individual negatively stained yeast fatty-acid

synthetase molecules from tilt series in the electron microscope. *Hoppe-Seylers Z Physiol Chem* **355**, 1483–1487 (1974).

14. Frank, J. *Electron Tomography, Methods for Three-Dimensional Visualization of Structures in the Cell* (Springer, 2006).

15. Unwin, P. N. & Henderson, R. Molecular structure determination by electron microscopy of unstained crystalline specimens. *J. Mol. Biol.* **94**, 425–440 (1975).

16. Tong, H. *et al.* Peptide-conjugation induced conformational changes in human IgG1 observed by optimized negative-staining and individual-particle electron tomography. *Sci. Rep.* **3**, 1089 (2013).

17. Zhang, X. *et al.* 3D structural fluctuation of IgG1 antibody revealed by individual particle electron tomography. *Sci. Rep.* **5**, 9803 (2015).

18. Lu, Z. *et al.* Calsyntenin-3 molecular architecture and interaction with neurexin 1alpha. *J. Biol. Chem.* **289**, 34530–34542 (2014).

19. Zhang, L. *et al.* An optimized negative-staining protocol of electron microscopy for apoE4 POPC lipoprotein. *J. Lipid Res.* **51**, 1228–1236 (2010).

20. Rames, M., Yu, Y. & Ren, G. Optimized negative staining: a high-throughput protocol for examining small and asymmetric protein structure by electron microscopy. *J. Vis. Exp.* **90**, e51087 (2014).

21. Ohi, M., Li, Y., Cheng, Y. & Walz, T. Negative staining and image classification - powerful tools in modern electron microscopy. *Biol. Proced. Online* **6**, 23–34 (2004).

22. Zhang, L. *et al.* Morphology and structure of lipoproteins revealed by an optimized negative-staining protocol of electron microscopy. *J. Lipid Res.* **52**, 175–184 (2011).

23. Zhang, L. *et al.* Structural basis of transfer between lipoproteins by cholesteryl ester transfer protein. *Nat. Chem. Biol.* **8**, 342–349 (2012).

24. Chen, Q. *et al.* 3D motion of DNA-Au nanoconjugates in graphene liquid cell electron microscopy. *Nano Lett.* **13**, 4556–4561 (2013).

25. Mastroianni, A. J., Sivak, D. A., Geissler, P. L. & Alivisatos, A. P. Probing the conformational distributions of subpersistence length DNA. *Biophys. J.* **97**, 1408–1417 (2009).

26. Foloppe, N. & MacKerell, A. D. All-atom empirical force field for nucleic acids: I. Parameter optimization based on small molecule and condensed phase macromolecular target data. *J. Comput. Chem.* **21**, 86–104 (2000).

27. MacKerell, A. D. & Banavali, N. K. All-atom empirical force field for nucleic acids: II. Application to molecular dynamics simulations of DNA and RNA in solution. *J. Comput. Chem.* **21**, 105–120 (2000).

28. Bustamante, C., Smith, S. B., Liphardt, J. & Smith, D. Single-molecule studies of DNA mechanics. *Curr. Opin. Struct. Biol.* **10**, 279–285 (2000).

29. Bustamante, C., Bryant, Z. & Smith, S. B. Ten years of tension: single-molecule DNA mechanics. *Nature* **421**, 423–427 (2003).

30. Kale, L. *et al.* NAMD2: Greater scalability for parallel molecular dynamics. *J. Comput. Phys.* **151**, 283–312 (1999).

31. Humphrey, W., Dalke, A. & Schulten, K. VMD: visual molecular dynamics. *J. Mol. Graph.* **14**, 33–38 (1996).

32. Hall, C. E. Method for the observation of macromolecules with the electron microscope illustrated with micrographs of DNA. *J. Biophys. Biochem. Cytol.* **2**, 625–628 (1956).

33. Beer, M. Electron microscopy of unbroken DNA molecules. *J. Mol. Biol.* **3**, 263–266 (1961).

34. Zobel, C. R. & Beer, M. Electron stains. I. Chemical studies on the interaction of DNA with uranyl salts. *J. Biophys. Biochem. Cytol.* **10**, 335–346 (1961).

35. Beer, M. & Zobel, C. R. Electron stains. II: Electron microscopic studies on the visibility of stained DNA molecules. *J. Mol. Biol.* **3**, 717–726 (1961).

36. Bell, D. C. *et al.* DNA base identification by electron microscopy. *Microsc. Microanal.* **18**, 1049–1053 (2012).

37. Scott, M. C. *et al.* Electron tomography at 2.4-angstrom resolution. *Nature* **483**, 444–447 (2012).

38. Zhang, M. *et al.* HDL surface lipids mediate CETP binding as revealed by electron microscopy and molecular dynamics simulation. *Sci. Rep.* **5**, 8741 (2015).

39. Jones, M. K. *et al.* Assessment of the validity of the double superhelix model for reconstituted high density lipoproteins: a combined computational-experimental approach. *J. Biol. Chem.* **285**, 41161–41171 (2010).

40. Ludtke, S. J., Baldwin, P. R. & Chiu, W. EMAN: semiautomated software for high-resolution single-particle reconstructions. *J. Struct. Biol.* **128**, 82–97 (1999).

41. Kremer, J. R., Mastronarde, D. N. & McIntosh, J. R. Computer visualization of three-dimensional image data using IMOD. *J. Struct. Biol.* **116**, 71–76 (1996).

42. Fernandez, J. J., Li, S. & Crowther, R. A. CTF determination and correction in electron cryotomography. *Ultramicroscopy* **106**, 587–596 (2006).

43. Arnott, S., Campbell-Smith, P. J. & Chandrasekaran, R. *In Handbook of Biochemistry and Molecular Biology*, Vol II (ed. Fasman G.P.) 3rd edn, 411–422 (CRC Press, Cleveland, 1976).

44. Pettersen, E. F. *et al.* UCSF Chimera—a visualization system for exploratory research and analysis. *J. Comput. Chem.* **25**, 1605–1612 (2004).

45. Schlitter, J., Engels, M. & Kruger, P. Targeted molecular dynamics: a new approach for searching pathways of conformational transitions. *J. Mol. Graph.* **12**, 84–89 (1994).

46. Hornus, S., Levy, B., Lariviere, D. & Fourmentin, E. Easy DNA modeling and more with GraphiteLifeExplorer. *PLoS ONE* **8**, e53609 (2013).

47. Grest, G. S. & Kremer, K. Molecular dynamics simulation for polymers in the presence of a heat bath. *Phys. Rev. A* **33**, 3628–3631 (1986).

48. Feller, S. E., Zhang, Y. H., Pastor, R. W. & Brooks, B. R. Constant-pressure molecular-dynamics simulation - the Langevin Piston method. *J. Chem. Phys.* **103**, 4613–4621 (1995).

49. Darden, T., York, D. & Pedersen, L. Particle mesh Ewald - an N.Log(N) method for Ewald sums in large systems. *J. Chem. Phys.* **98**, 10089–10092 (1993).

Acknowledgements

We thank Drs Qian Chen and Phillip Geissler for their discussion and comments, and Mr Matthew Rames for editing. This material is based upon work supported by the National Science Foundation under grant DMR-1344290. Work at the Molecular Foundry was supported by the Office of Science, Office of Basic Energy Sciences of the US Department of Energy under contract no. DE-AC02-05CH11231. G.R. is supported by the National Heart, Lung, and Blood Institute of the National Institutes of Health (no. R01HL115153) and the National Institute of General Medical Sciences of the National Institutes of Health (no. R01GM104427). L.Z. is partly supported by National Natural Science Foundation of China (no. 11504287) and Young Talent Support Plan of Xi'an Jiaotong University; Z.L. and J.L. are partly supported by the National Basic Research Program of Ministry of Science and Technology, China (no. 2015CB553602).

Author contributions

This project was initiated and designed by J.M.S., P.A. and G.R., and refined by L.Z. and G.R. J.S. prepared the conjugates. L.Z., H.T., Z.L. and G.R. prepared TEM samples and/or acquired the data. L.Z. and G.R. processed the data, and L.Z. solved the IPET 3D structures. X.Z., M.Z. and L.Z. docked the model, and D.L. measured the angles and computed/analysed the energies of dsDNA in solution by MD simulations. L.Z., J.M.S., D.L., P.A., J.L. and G.R. interpreted and manipulated the structures. G.R. drafted the initial manuscript, which was revised by L.Z., J.M.S., X.Z., D.L., Z.L., H.T., M.Z., J.L. and P.A.

Additional information

Accession codes: TEM 3D density maps of 14 DNA-nanogold conjugates are available from the EM data bank as EMDB IDs 2948–2961.

A highly active and stable hydrogen evolution catalyst based on pyrite-structured cobalt phosphosulfide

Wen Liu[1], Enyuan Hu[2], Hong Jiang[3], Yingjie Xiang[4], Zhe Weng[1], Min Li[5], Qi Fan[1], Xiqian Yu[2], Eric I. Altman[5] & Hailiang Wang[1]

Rational design and controlled synthesis of hybrid structures comprising multiple components with distinctive functionalities are an intriguing and challenging approach to materials development for important energy applications like electrocatalytic hydrogen production, where there is a great need for cost effective, active and durable catalyst materials to replace the precious platinum. Here we report a structure design and sequential synthesis of a highly active and stable hydrogen evolution electrocatalyst material based on pyrite-structured cobalt phosphosulfide nanoparticles grown on carbon nanotubes. The three synthetic steps in turn render electrical conductivity, catalytic activity and stability to the material. The hybrid material exhibits superior activity for hydrogen evolution, achieving current densities of $10\,mA\,cm^{-2}$ and $100\,mA\,cm^{-2}$ at overpotentials of $48\,mV$ and $109\,mV$, respectively. Phosphorus substitution is crucial for the chemical stability and catalytic durability of the material, the molecular origins of which are uncovered by X-ray absorption spectroscopy and computational simulation.

[1] Department of Chemistry and Energy Sciences Institute, Yale University, 520 West Campus Drive, West Haven, Connecticut 06511, USA. [2] Chemistry Department, Brookhaven National Laboratory, Upton, New York 11973, USA. [3] Beijing National Laboratory for Molecular Sciences, College of Chemistry and Molecular Engineering, Peking University, Beijing 100871, China. [4] Department of Mechanical Engineering and Materials Science, Yale University, 520 West Campus Drive, West Haven, Connecticut 06511, USA. [5] Department of Chemical and Environmental Engineering, Yale University, 520 West Campus Drive, West Haven, Connecticut 06511, USA. Correspondence and requests for materials should be addressed to H.W. (email: hailiang.wang@yale.edu).

With the rising concern over energy crisis and environmental pollution, there has been a growing need to replace fossil fuels with clean and sustainable energy carriers. Molecular hydrogen, with its high energy density and non-polluting characteristics, has been regarded as one of the most promising new fuels[1,2]. There are various methods to produce hydrogen, among which water splitting driven by electricity generated from renewable energy sources is an attractive way to support the future hydrogen economy. Efficient electrolytic hydrogen generation relies heavily on active, durable and affordable catalysts to accelerate the kinetics[3–5]. Although platinum (Pt) has been widely acknowledged as the most active catalyst for the hydrogen evolution reaction (HER), this precious metal is scarce in the earth's crust and expensive for large scale applications.

In pursuit of inexpensive replacements for Pt as the HER electrocatalyst in acidic solutions, numerous inorganic materials based on non-precious transition metals, including sulfides[6–8], selenides[9,10], phosphides[11–13] and others[14,15] have been explored. Extensive effort has been devoted to improving the HER catalytic activity of molybdenum disulfide (MoS_2) by identifying and exposing active sites[6,16], as well as enhancing electron conduction through nanostructuring[17], shape control[6,18], phase engineering[19,20], doping[21,22], intercalation[23] and hybridization[17,24,25]. Recently a ternary cobalt phosphosulfide pyrite-type structure (CoPS) has been reported with superior catalytic activity for HER[26]. Cobalt disulfide (CoS_2) is another material of interest[27,28]. However, the existing HER catalyst materials based on non-noble metal elements are still less satisfactory in terms of both activity and stability, which calls for further material structure innovation to realize efficient and cost effective HER catalysis.

Here we report a design and synthesis of a highly active and stable HER electrocatalyst material consisting of pyrite-structured cobalt phosphosulfide (CoS|P) nanoparticles anchored on carbon nanotubes (CNTs). The material architecture is built by a three-step chemical synthesis: Strong interactions with CNTs and particle size control are first established by the selective growth of cobalt(II,III) oxide (Co_3O_4) nanoparticles on CNTs; High catalytic activity for HER is then rendered by conversion of Co_3O_4 to CoS_2 nanoparticles; Good chemical stability and catalytic durability are lastly obtained from substituting some of the sulfur with phosphorus. The unique material structure directly enables some of the highest HER catalytic performance among all Co-based catalyst materials. In 0.5 M H_2SO_4, the CoS|P/CNT hybrid exhibits a negligible onset overpotential and a Tafel slop of 55 mV per decade with an exchange current density

of 1.14 mA cm^{-2}. At a mass loading of 1.6 mg cm^{-2}, the hybrid material requires overpotentials of only 48 mV and 109 mV to reach stable catalytic current densities of 10 mA cm^{-2} and 100 mA cm^{-2} respectively, representing one of the few most active non-Pt catalysts for HER. Phosphorus substitution in the pyrite structure is a critical step that renders chemical stability and catalytic durability to the CoS|P/CNT hybrid material. Density functional theory (DFT) calculations confirm the structural stability of pyrite CoS|P and suggest stronger metal–ligand bonding as a contributor to the improved stability, which is supported by X-ray absorption spectroscopy data.

Results

Sequential synthesis of CoS|P/CNT. The synthetic strategy for the CoS|P/CNT hybrid material structure involves three steps, as illustrated in Fig. 1. In the first step, Co_3O_4 nanoparticles were directly and selectively grown onto mildly oxidized multi-wall CNTs (see Supplementary Methods for CNT oxidation) by a hydrolysis reaction of cobalt acetate at 80 °C. An ethanol/water mixed solvent was used to slow down the hydrolysis reaction, facilitate interactions between the Co^{2+} and the functional groups on the CNT surface, and ensure selective nucleation and growth of Co_3O_4 nanoparticles on CNTs. $NH_3 \cdot H_2O$ was added into the reaction system to coordinate the Co^{2+} and thus further reduce the hydrolysis rate to limit the size of the resulting Co_3O_4 nanoparticles and optimize their distribution on the CNTs. Scanning electron microscopy (SEM), transmission electron microscopy (TEM) and X-ray diffraction (XRD) characterizations revealed the product structure as spinel structured Co_3O_4 (PDF#43-1003) nanoparticles with an average size of 5–10 nm anchored on CNTs (Supplementary Fig. 1).

The second step is a hydrothermal reaction at 200 °C to convert Co_3O_4 into CoS_2. Thioacetamide (CH_3CSNH_2) was used as a slow-release S precursor, which reacted with water to gradually generate the H_2S reactant and therefore allowed for facile chemical conversion from oxide to sulfide without damaging the material morphology and hybrid structure. SEM and TEM imaging showed nanoparticles with an average size of 10–20 nm attached to CNTs (Supplementary Fig. 2a,b). The nanoparticles were confirmed to be pyrite-structured CoS_2 (PDF#41–1471) by XRD (Supplementary Fig. 2d). Lattice fringes of the CoS_2 nanoparticles on CNTs were recorded with high-resolution TEM (Supplementary Fig. 2c). Energy dispersive spectroscopy (EDS) under scanning TEM (STEM) mode was used to map the distributions of elements of interest in the material structure. It is clear from the result that the

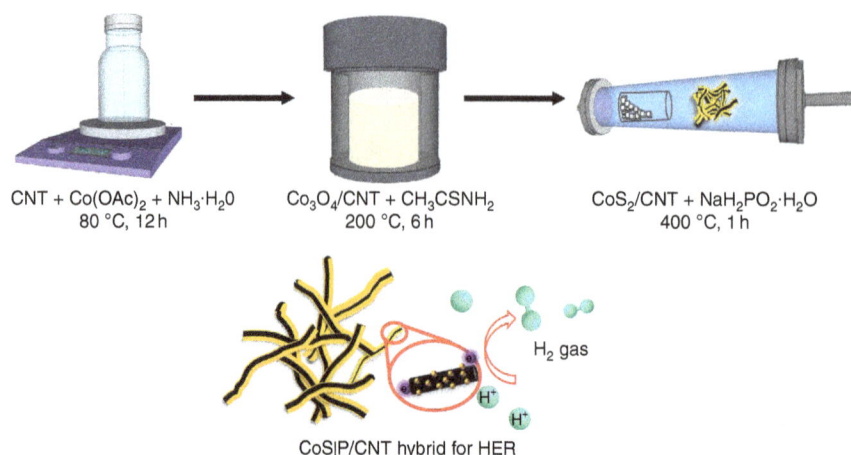

CNT + Co(OAc)$_2$ + NH$_3\cdot$H$_2$O   Co$_3$O$_4$/CNT + CH$_3$CSNH$_2$ CoS$_2$/CNT + NaH$_2$PO$_2\cdot$H$_2$O
80 °C, 12 h   200 °C, 6 h   400 °C, 1 h

H$_2$ gas

CoS|P/CNT hybrid for HER

Figure 1 | Schematic illustration of the sequential synthesis of the CoS|P/CNT hybrid material for HER catalysis. The CoS|P/CNT is synthesized through three steps including hydrolysis, hydrothermal sulfurization and solid/gas-phase phosphorization.

nanoparticles consist of Co and S, and they are well-anchored on the CNT surface (Supplementary Fig. 3).

The third step features a solid/gas-phase reaction at 400 °C to introduce P into the CoS_2 structure. $NaH_2PO_2 \cdot H_2O$ was used as a precursor to generate the PH_3 reactant via thermal decomposition. The PH_3 then reacted with the CoS_2 nanoparticles on CNTs to form the final CoS|P/CNT hybrid material. SEM and TEM characterizations confirmed the microstructure of nanoparticles with an average size of 10–20 nm on CNTs (Fig. 2a,b), suggesting that the substitution process had negligible influence on the material morphology and nanoparticle size. XRD measurement of the material generated a diffraction pattern characteristic of a pyrite structure with almost identical lattice parameters as CoS_2 (Fig. 2c), indicating that the substitution process happened in the form of P replacing S, which did not alter the crystal structure or lattice parameters due to the very similar atomic sizes of S and P. The lattice fringes of the CoS|P nanoparticles were imaged by high-resolution TEM (Fig. 2d). The interplanar spacing of 0.277 nm corresponding to the (200) crystallographic planes of the pyrite structure is in consistency with the XRD result. The high-resolution TEM images also excluded the existence of core-shell structured nanoparticles.

Structural and chemical analysis. To further understand the structure of the CoS|P/CNT hybrid material, STEM-EDS characterization was performed to gain elemental composition and distribution information. The EDS maps of Co, S and P overlap quite well (Fig. 3a,b), suggesting P has been uniformly doped into the crystal structure of nanoparticles. It is also evident from the EDS mapping that the CoS|P nanoparticles are closely anchored onto the CNT surfaces. The average atomic ratio of P/S in the nanoparticles was calculated to be ~ 1.0 from the EDS spectrum (Fig. 3c and Supplementary Fig. 4), giving a $\sim 50\%$ substitution of the S sites. Raman spectroscopy was used to probe chemical bonding information in the CoS|P nanoparticles. For the CoS_2/CNT, two peaks were observed at 284 and 385 cm^{-1}

(Fig. 3d, blue trace). These Raman peaks could be attributed to the characteristic E_g and A_g vibrational modes corresponding to the in-phase stretching and pure libration of the S–S dumbbells in the pyrite structure[29–31]. For the CoS|P/CNT, the peaks slightly shifted to higher wavenumber (Fig. 3d, red trace) as a result of partial phosphorus substitution for sulfur and possible formation of P–S dumbbells. The characteristic Raman pattern also excludes formation of CoP[32].

X-ray photoelectron spectroscopy (XPS) was employed to investigate the surface composition and oxidation states of the catalysts. For the CoS_2/CNT, Co $2p_{3/2}$ and $2p_{1/2}$ core level peaks were observed at binding energies of 778.8 eV and 794.0 eV, respectively, together with satellite features (Fig. 3e), which match literature results on CoS_2 (ref. 28). The Co $2p$ core level spectrum of the CoS|P/CNT hybrid is very similar to that of the CoS_2/CNT, with binding energies of Co $2p_{3/2}$ and $2p_{1/2}$ peaks at 779.2 eV and 794.1 eV, respectively (Fig. 3e). The negligible change in the Co $2p$ spectrum verifies that the oxidation state of Co is not affected by P substitution. The S $2p$ core level spectrum of the CoS_2/CNT shows lower binding energy components at 162.6 eV and 163.9 eV (S $2p_{3/2}$ and $2p_{1/2}$) attributed to sulfide species as well as higher binding energy components at 168.5 eV and 169.6 eV (S $2p_{3/2}$ and $2p_{1/2}$), respectively, characteristic of sulfate species (Fig. 3f). Existence of the sulfate components suggests that the CoS_2/CNT catalyst is slightly oxidized on surface. Interestingly, no sulfate features were found in the S $2p$ core level spectrum of the CoS|P/CNT (Fig. 3f), indicating P doping could prevent CoS_2 from oxidation under ambient conditions. The P $2p$ core level spectrum of the CoS|P/CNT displays two peak regions (Fig. 3g), with one centred at the binding energy of 129.3 and 130.1 eV (P $2p_{3/2}$ and $2p_{1/2}$), which can be assigned to phosphorus anions, and the other at 133.6 and 134.4 eV (unresolved doublet) characteristic of phosphate-like P. The existence of the high oxidation state P could be ascribed to surface oxidation under ambient conditions as often observed for metal phosphide materials[33–35]. Furthermore, a surface P/S ratio of ~ 1.0 was derived from the XPS results, close to the bulk P/S ratio measured

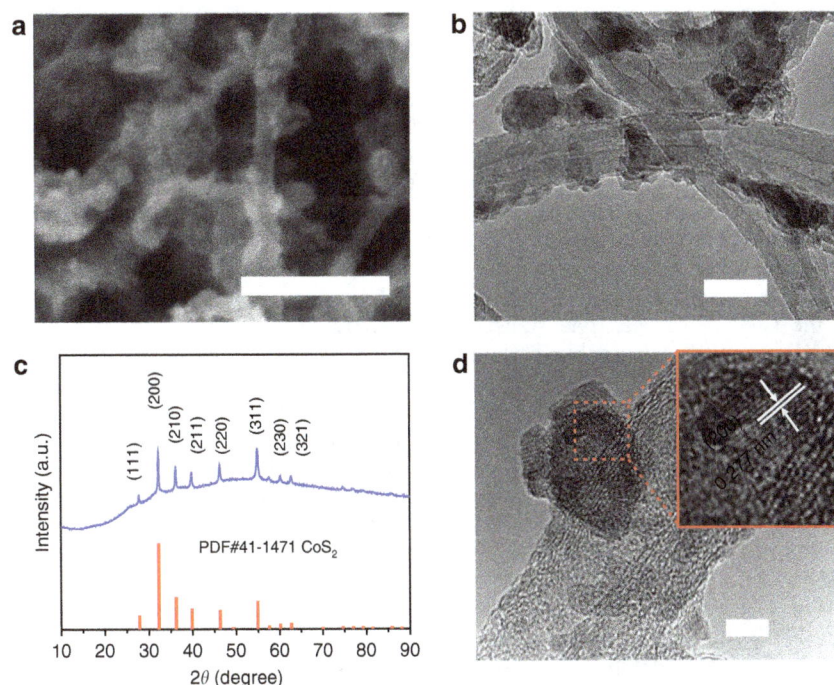

Figure 2 | Structural characterizations of the CoS|P/CNT hybrid material. (**a**) SEM image of CoS|P/CNT; Scale bar, 200 nm. (**b**) Low-magnification TEM image of CoS|P/CNT showing nanoparticles attached to CNTs; Scale bar, 20 nm. (**c**) XRD pattern of CoS|P/CNT as compared with the pyrite-phase CoS_2 standard (PDF#41-1471). (**d**) High-resolution TEM image showing the (200) lattice fringes of pyrite-phase CoS|P; Scale bar, 5 nm.

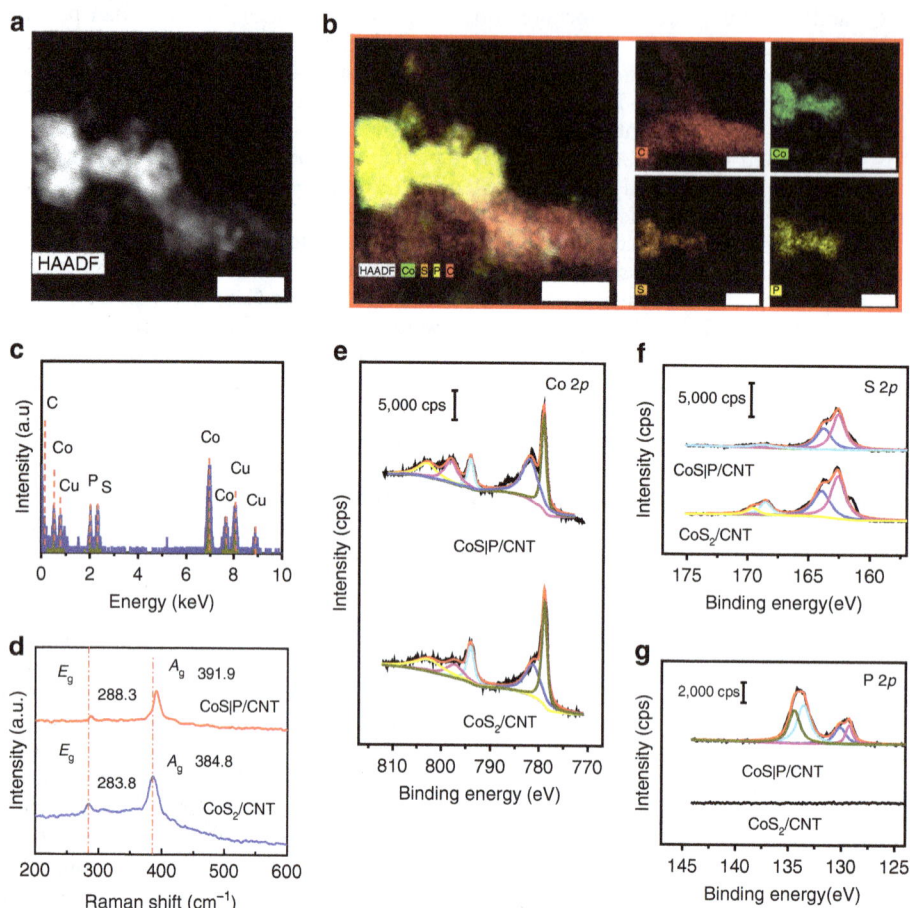

Figure 3 | Composition and chemical analysis for the CoS|P/CNT hybrid material. (a) STEM image recorded by a high-angle annular dark field (HAADF) detector showing CoS|P nanoparticles attached on CNTs. Scale bar, 20 nm. (b) STEM-EDS mapping of CoS|P/CNT catalyst showing the distributions of Co (green), P (yellow) and S (orange) within the nanoparticles closely attached to C (red). Scale bar, 20 nm. (c) EDS spectrum of CoS|P/CNT. (d) Raman spectra of CoS_2/CNT and CoS|P/CNT. (e-g) Co 2p, S 2p and P 2p core level XPS spectra of CoS|P/CNT and CoS_2/CNT.

by EDS. This again corroborates that P is uniformly distributed within the nanoparticles rather than forming a shell-like structure on the original CoS_2 nanoparticles.

Electrocatalytic hydrogen evolution. HER electrocatalytic activity of the CoS|P/CNT hybrid material was assessed in 0.5 M H_2SO_4 aqueous solution. Figure 4a shows the polarization curve of the CoS|P/CNT hybrid material as compared with a benchmark Pt/C catalyst at a scan rate of $5 \, \text{mV s}^{-1}$. With a mass loading of $1.6 \, \text{mg cm}^{-2}$, the CoS|P/CNT electrode showed a negligible onset overpotential versus the reversible hydrogen electrode (RHE). The catalytic current density increased rapidly with further cathodic polarization to $10 \, \text{mA cm}^{-2}$, $20 \, \text{mA cm}^{-2}$ and $100 \, \text{mA cm}^{-2}$ at overpotentials of 48 mV, 65 mV and 109 mV, respectively (Fig. 4a). An exchange current density of $1.14 \, \text{mA cm}^{-2}$ and a Tafel slope of 55 mV per decade were derived from the polarization curve (Fig. 4b), suggesting a different HER mechanism from Pt which showed a 30 mV per decade Tafel slope indicative of a Volmer–Tafel mechanism[36]. Such performance represents arguably higher HER catalytic activity than any other Co-based catalyst materials reported in the literature, placing our material at the top of all existing noble-metal-free HER catalyst materials working in acidic media (Supplementary Table 1). Our catalyst also showed high durability for HER catalysis. The initial current density ($\sim 45 \, \text{mA cm}^{-2}$) maintained after 24 h of continuous hydrogen

production (Fig. 4c). Stability for 100 h of HER catalysis was also confirmed (Supplementary Fig. 5). We also performed 2,000 cycles of cyclic voltammetry between 0.25 and -0.12 V versus RHE on the CoS|P/CNT catalyst. The polarization curves showed negligible shift during the test (Fig. 4d). Chronoamperometry in combination with gas chromatography and mass spectrometry (MS) revealed Faradic efficiency of $\sim 100\%$ for H_2 (Supplementary Fig. 6).

Discussion

The excellent HER catalytic activity and durability of the CoS|P/CNT hybrid material is a direct outcome of its unique material structure imparted by the three designed chemical reaction steps. The first step of reaction grows Co_3O_4 nanoparticles on CNTs, which builds the strong electrical and chemical coupling between the nanoparticles and CNTs[37,38]. Consequently, nanoparticles are anchored on the highly conductive CNT network, which can rapidly transport electrons from external circuit to nanoparticle/electrolyte interface for hydrogen evolution. This step also brings size control of the nanoparticles, which is essential for increasing the electrochemically active surface area and reducing the electron diffusion length within each nanoparticle.

The second step of reaction converts the Co_3O_4 nanoparticles to CoS_2 nanoparticles, through which highly active sites for HER catalysis are created. The CoS_2/CNT material is already as active as the final CoS|P/CNT. At a mass loading of $0.8 \, \text{mg cm}^{-2}$, HER

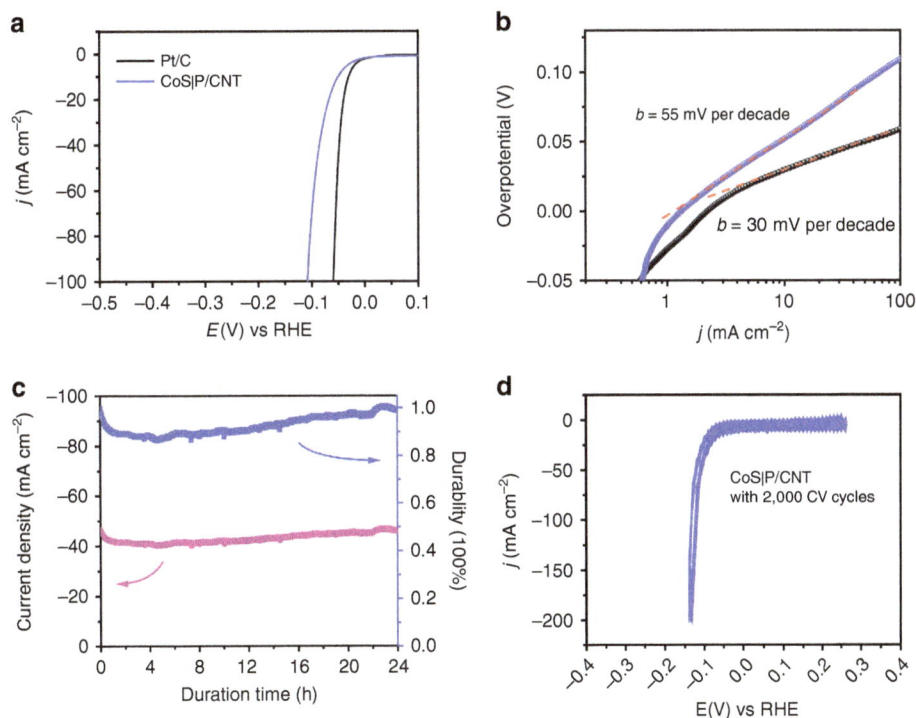

Figure 4 | Electrocatalytic hydrogen evolution over the CoS|P/CNT catalyst. (**a**) Polarization curves for HER on the CoS|P/CNT hybrid and a commercial Pt/C catalyst at 5 mV s^{-1}. The catalyst mass loading was 0.4 mg cm^{-2} for Pt/C catalyst. (**b**) Tafel plots for the CoS|P/CNT and Pt/C catalysts derived from the polarization curves in **a**. (**c**) Chronoamperometric response ($j \sim t$ curve) recorded on the CoS|P/CNT electrode at a constant overpotential of 95 mV with iR compensation. (**d**) CV test between 0.25 and − 0.12 V versus RHE at a scan rate of 100 mV s^{-1} for 2,000 cycles. The catalyst mass loading of CoS|P/CNT was 1.6 mg cm^{-2} unless otherwise noted.

current density of 10 mA cm^{-2} was reached at an overpotential of 61 mV for the CoS$_2$/CNT as compared with 64 mV for the CoS|P/CNT under the same condition (Supplementary Fig. 8a). This already represents one of the most active cobalt chalcogenide HER catalyst materials[10,28,39–42]. The sequential synthetic method, namely oxide growth followed by conversion to sulfide, is responsible for the superior catalytic activity, as corroborated by our control experiment (Supplementary Fig. 7).

In spite of the excellent activity as a result of size control and CNT hybridization, the intrinsic instability of CoS$_2$ in strong acid is exacerbated. The CoS$_2$/CNT material is extremely unstable during HER catalysis in 0.5 M H$_2$SO$_4$. Under constant potential operation, the current density decreased drastically by 70% in less than 30 min (Fig. 5a). Such a dramatic deterioration was accompanied by substantial dissolution of the CoS$_2$ active phase into the electrolyte. A Co concentration of ∼2.1 p.p.m. in the electrolyte was measured by inductively coupled plasma MS (ICP-MS) after 30 min of HER catalysis (Fig. 5b), corresponding to about 17.5% of the CoS$_2$ having been dissolved. The concentration of dissolved Co gradually increased to ∼2.9 p.p.m. over 20 h of HER operation. The CoS$_2$/CNT material is also sensitive to air and moisture. After being stored in ambient condition for 2 weeks, the CoS$_2$ nanoparticles were completely oxidized to CoSO$_4 \cdot 7$H$_2$O (PDF#16-04872), which is soluble in water and HER inactive (Fig. 5c).

The third step of reaction (P substitution) is critical to the chemical stability and catalytic durability of the final material (Fig. 5 and Supplementary Fig. 8). No phase change was detected by XRD for the CoS|P/CNTs material after being stored in ambient condition for 2 weeks (Fig. 5c), suggesting significantly improved chemical stability to oxygen and moisture. This is consistent with our XPS results that P substitution could mitigate sulfate formation on CoS$_2$ surface. The CoS|P/CNTs catalyst was

able to sustain a current density of ∼10 mA cm^{-2} during 20 h of continuous HER operation (Fig. 5a), substantially more stable than the CoS$_2$/CNT catalyst under working conditions. In consistency, the amount of Co leaching into the electrolyte was much lower than that for the CoS$_2$/CNT catalyst. The Co concentration in electrolyte gradually reached ∼0.6 p.p.m. within 12 h and did not increase significantly in the following 8 h (Fig. 5b). To visually demonstrate the stability difference between the CoS$_2$/CNT and CoS|P/CNT in the electrolyte, we conducted a colorimetric comparison (Supplementary Methods) after the CoS$_2$/CNT and CoS|P/CNT hybrids were soaked in 0.5 M H$_2$SO$_4$ for 2 h. The Nitrite R salt is widely used as a colour indicator to detect cobalt in solution as it can form a red coloured complex with Co^{2+} ions. The solution in which the CoS$_2$/CNT was soaked exhibited an orange colour (Fig. 5d), clearly showing a considerable amount of Co dissolved into the acid. In contrast, the solution in which the CoS|P/CNT was soaked remained almost the same colour as the blank control, confirming greatly improved stability of the CoS|P/CNT catalyst against acid corrosion. Our control experiment clearly verified that it is indeed the P substitution rather than the high-temperature annealing process that renders the excellent catalytic durability (Supplementary Fig. 9).

It is worth emphasizing here that the P substitution step is only for improving material stability and catalytic durability, but not for enhancing catalytic activity. This is different from the recent study where the CoPS is much more active than the CoS$_2$ with similar morphology[26]. It is also noted that our CoS|P is likely in a different structure than the recently reported CoPS. The CoPS is a distinct ternary pyrite-type phase featuring P–S bonds without any S–S bonds as in the CoS$_2$ pyrite structure. The corresponding XRD and Raman peaks significantly shifted as a result of the shrunk lattice parameters. In contrast, our CoS|P is more likely to

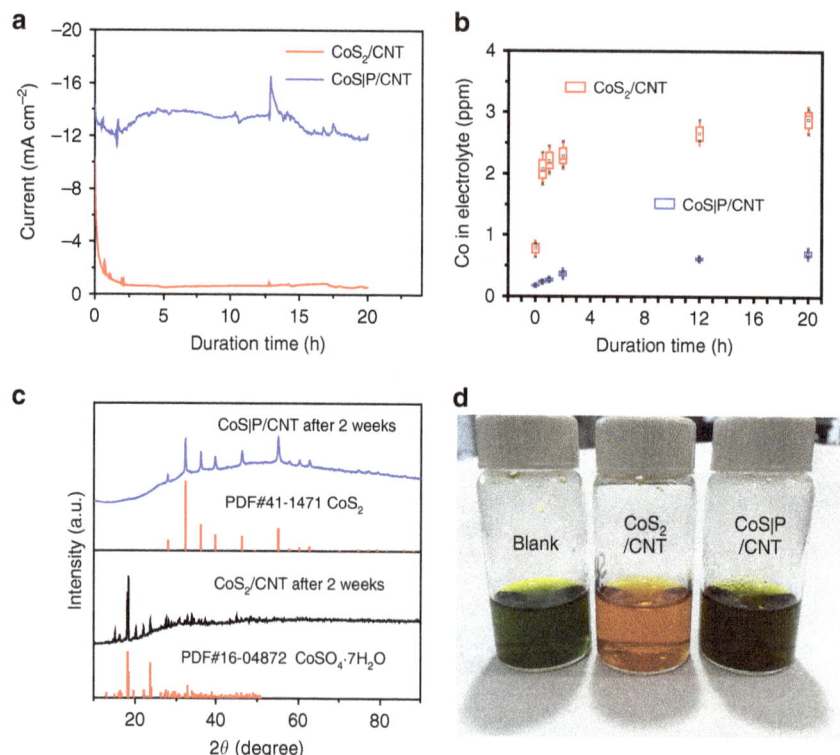

Figure 5 | Comparison of chemical stability and catalytic durability between CoS$_2$/CNT and CoS|P/CNT. (**a**) Typical chronoamperometric responses ($j \sim t$ curves) of the CoS$_2$/CNT and CoS|P/CNT catalysts driving hydrogen evolution at the overpotential of 77 mV without iR compensation for 20 h in 0.5 M H$_2$SO$_4$ solution. About 0.4 mg of each catalyst was loaded on a carbon fibre paper with 0.5 cm^2 of active area. The sharp current fluctuations were caused by the sampling of electrolyte during the electrolysis process. (**b**) Box plots (median and quartiles) representing the concentrations of Co dissolved in 20 ml of electrolyte as the HER catalysis proceeds. The vertical whiskers represent the s.d. The statistics are derived from at least three independent measurements. (**c**) XRD patterns of CoS$_2$/CNT and CoS|P/CNT after 2 weeks of storage in ambient conditions. (**d**) Colorimetric comparison of the CoS$_2$/CNT and CoS|P/CNT hybrids soaked in 0.5 M H$_2$SO$_4$ solution for 2 h; Nitrite R salt was used as the colour indicator.

be a CoS$_2$ pyrite structure (no lattice parameter shrinking) with some of the S atoms randomly substituted by P atoms. The structural difference is a direct outcome of the different synthetic methods adopted. The CoPS is prepared by converting Co(OH)(CO$_3$)$_{0.5}$·XH$_2$O with pre-formed P$_x$S$_y$, therefore the P–S bonding is exclusive and CoPS is the only available composition. In this study, CoS|P is derived from CoS$_2$ so that the P/S ratio in the CoS$_{2-x}$P$_x$/CNT materials can be readily tuned between 0 and 1 while keeping the mother structure and lattice parameters unchanged (Supplementary Fig. 10). As we gradually increase the substitution level of P, the influence on HER activity is negligible but the catalytic durability improves drastically (Supplementary Fig. 11).

To probe the molecular origins of the structural stability of CoS|P/CNT, we performed X-ray absorption near edge structure (XANES) spectroscopy measurements. The absorption edges of the Co K-edge spectra of CoS$_2$/CNT and CoS|P/CNT lie close to each other (Fig. 6a), suggesting similar oxidation states of Co in both materials, which match the XPS results. The relatively small pre-edge peaks of both spectra suggest that the Co ions invariably reside in octahedral coordination environment (Fig. 6c). It is noted that the CoS|P/CNT spectrum shows a shoulder peak at ~7,717 eV. Similar features have been observed in other systems and are attributed to the covalency effect[43]. This correlates well with the pre-edge absorption intensity difference between the CoS|P/CNT and CoS$_2$/CNT (inset of Fig. 6a). For both CoS|P and CoS$_2$, the lowest unoccupied molecular orbitals are the anti-bonding e_g^* orbitals[44]. These orbitals are due to hybridization between transition metal 3d and ligand 3p orbitals and are thus very sensitive to covalency. The stronger

pre-edge adsorption of CoS|P/CNT compared with that of CoS$_2$/CNT suggests more p features in the Co 3d state and therefore stronger covalency for the CoS|P compound. Stronger covalency between the transition metal and ligands in CoS|P/CNT is also supported by the S K-edge XANES spectra. As shown in Fig. 6b, the CoS|P/CNT spectrum exhibits a stronger peak at ~2,469 eV than the CoS$_2$/CNT, corresponding to higher probability of 1s to e_g^* transition as a result of stronger covalency of the metal–ligand bonds in CoS|P[45].

To further rationalize the experimental findings, we performed first-principles DFT calculations for CoS$_{2-x}$P$_x$ ($x=0$, 0.5, 1.0, 1.5 and 2) with both the cubic (pyrite) and monoclinic structures. The results for the relative stability of the two phases are shown in Fig. 6d. We found that CoS$_2$ is more stable in the cubic structure than in the monoclinic structure, but only with a marginal energy difference of about 0.03 eV per formula unit (f.u.). In contrast, CoP$_2$ is substantially more stable in the monoclinic structure by about 0.6 eV per f.u. When S is substituted by P, the phosphosulfides would be more stable in the monoclinic structure within a large range of x if we assume a linear dependence of the relative stability of the two phases on the composition. However, first-principles calculations clearly suggest that CoS$_{2-x}$P$_x$ is more stable in the cubic structure for $x \leq 1.0$, which agrees well with our experimental results. The remarkable stability of the cubic structure can be attributed to the fact that cubic-phase CoS$_2$ and CoP$_2$ have very similar optimized equilibrium volume (V_0) (41.6 versus 40.1 Å^3 per f.u.). In comparison, the V_0 difference between monoclinic-phase CoS$_2$ and CoP$_2$ is much larger (42.8 versus 39.2 Å^3 per f.u.). Substitution of S by P would induce much less strain in the cubic than in the monoclinic structure.

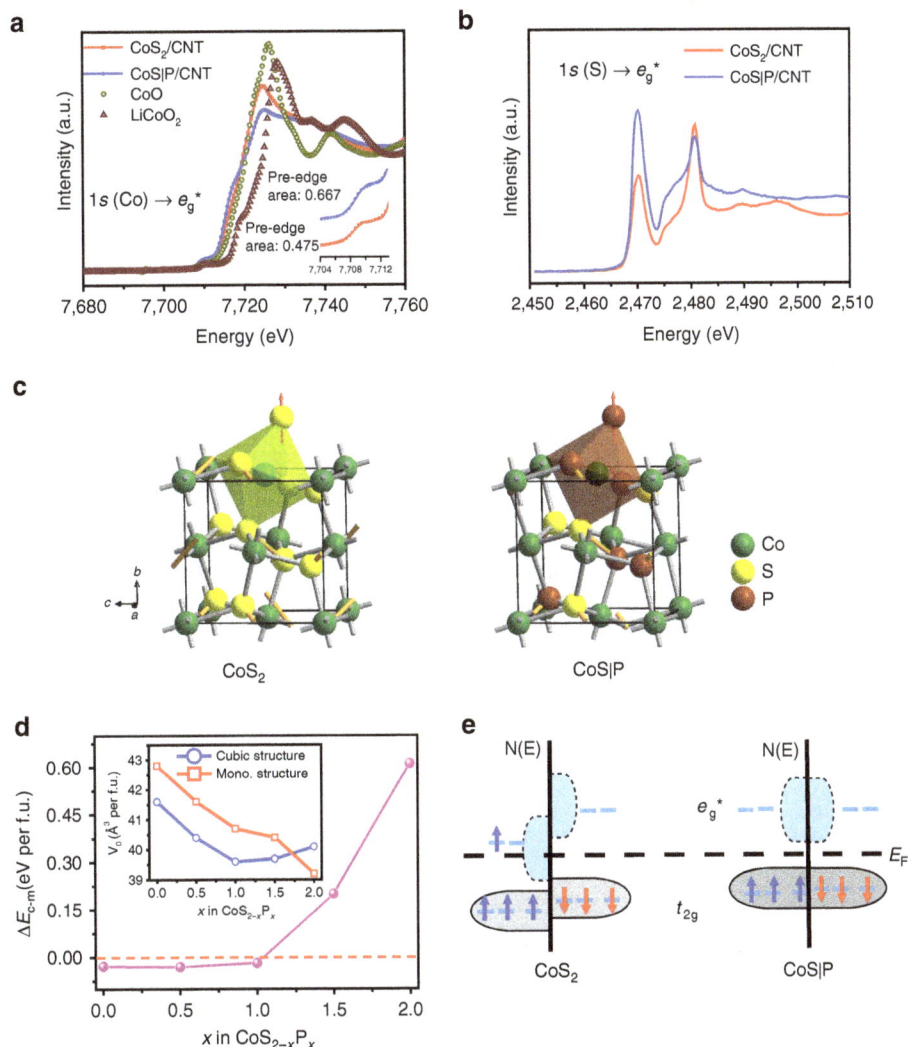

Figure 6 | XANES spectra of CoS|P/CNT and structural stability discussion based on first-principles calculations. (a) Cobalt K-edge XANES spectra of CoS$_2$/CNT and CoS|P/CNT compared with Co^{II} (CoO) and Co^{III} (LiCoO$_2$) standards. Pre-edge features correspond to transition from 1s (Co) orbital to e_g^* anti-bonding state, which is metal 3d and ligand 3p hybrid orbitals. The pre-edge features were fitted by two pseudo-Voigt functions with the results shown in the inset graph. **(b)** Sulfur K-edge XANES spectra of CoS$_2$/CNT and CoS|P/CNT. **(c)** Structure illustration of pyrite-phase CoS$_2$ and CoS|P (CoS$_{2-x}$P$_x$, $x = 1$), each with a representative coordination polyhedron. **(d)** The energy difference per formula unit (f.u.) between the cubic and monoclinic phases as a function of the P substitution extent obtained from DFT calculations. The inset shows the equilibrium volume per f.u. in the cubic and monoclinic phases as a function of the P substitution extent. **(e)** Conceptual energy level diagrams of the frontier molecular orbitals for pyrite-phase CoS$_2$ and CoS|P derived from the calculated electronic structures (Supplementary Fig. 12). CoS$_2$ is magnetic at room temperature. As a result spin-up and spin-down electrons have different energy levels. CoS|P (CoS$_{2-x}$P$_x$, $x = 1$) is non-magnetic.

Therefore, the pyrite-structured CoS$_{2-x}$P$_x$ is stable within a quite wide range of P doping levels ($x \leq 1.0$).

First-principles calculations can also provide insight into why incorporation of P improves the chemical stability of pyrite-structured CoS$_2$. From projected density of states analysis (Supplementary Fig. 12), we see that P substitution significantly influences the nature of chemical bonding between Co and S/P. In the pyrite structure, each Co atom is coordinated in an octahedral ligand field (Fig. 6c), and therefore the 3d orbitals are split into t_{2g} and e_g^* manifolds that are of non-bonding and anti-bonding characteristics, respectively. As qualitatively demonstrated in Fig. 6e, the highest occupied states in CoS$_2$ are of anti-bonding nature, which is the origin of the instability. When half of the S atoms are replaced by P, which has fewer valence electrons, the anti-bonding e_g^* orbitals are depleted, which strengthens the chemical bonding between Co and ligands and thus enhances the chemical stability of the material. This is in good agreement

with our XANES results. In conclusion, we demonstrate a novel structure design and synthesis of a highly active and stable HER catalyst material based on pyrite-structured CoS|P. The sequential synthetic strategy we adopt imparts electrical conductivity, catalytic activity and stability to the material. The CoS|P/CNT catalyst exhibits arguably the highest catalytic activity among all non-noble metal based catalysts. P substitution is critical to chemical stability and catalytic durability of the material. The molecular origins are rationalized by spectroscopy characterization and computational modelling.

Methods

Material synthesis. The CoS|P/CNT hybrid was prepared through a three-step method. In the first step (synthesis of Co$_3$O$_4$/CNT), 4 mg of mildly oxidized CNTs (the CNTs were oxidized following a modified Hummers method as described in Supplementary Information) were dispersed in 14 ml of ethanol by sonication for 1 h. Then, 0.8 ml of 0.2 M cobalt acetate aqueous solution and 0.6 ml of NH$_4$OH

(30%) were added to the suspension. The hydrolysis reaction was kept at 80 °C in oil bath with stirring for 12 h. After that, the product was collected by centrifuge. The precipitate was then washed with ethanol and DI water. The resulting Co_3O_4/CNT was lyophilized. To prepare CoS_2/CNT, 20 mg of Co_3O_4/CNT was dispersed in 20 ml of DI water by sonication for 40 min, followed by the addition of 0.75 ml of 1 M thioacetamide solution. After that, the reaction mixture was transferred to a 40 ml autoclave for hydrothermal reaction at 200 °C for 6 h. The resulting product was collected by centrifugation and repeatedly washed with DI water. The CoS_2/CNT hybrid was then freeze-dried. In the third step, 5 mg of CoS_2/CNT and 100 mg of $NaH_2PO_2 \cdot H_2O$ were placed at two separate positions in a ceramic crucible with the $NaH_2PO_2 \cdot H_2O$ at the upstream side. The samples were heated at 400 °C for 1 h with Ar gas flowing at 200 s.c.c.m. The final product contains about 60 wt% of CoS|P and 40 wt% of CNTs.

Electrochemical measurements. To prepare catalyst ink, 1 mg of CoS|P/CNT was mixed with 190 µl of water, 50 µl of ethanol and 10 µl of 5 wt% Nafion solution by sonication for 1 h. Subsequently, 50–200 µl of the catalyst ink was drop-dried onto a carbon fibre paper (Spectracarb 2050 A from Fuel Cell Store) to cover an area of $0.5 cm^2$ (0.4–1.6 mg cm^{-2}). The electrode was further heated at 90 °C in vacuum for 2 h. HER catalytic measurements were performed with a CHI 760D electrochemistry workstation (CH instruments, USA). A conventional three electrode cell configuration was employed. A saturated calomel electrode (SCE) was used as the reference electrode, and a graphite rod was used as the counter electrode. 0.5 M H_2SO_4 solution was used as electrolyte. Linear sweep voltammetry was recorded at a scan rate of 5 mV s^{-1}. All the polarization curves were iR-corrected. The reference electrode was calibrated against the RHE as shown in Supplementary Fig. 13. All the potentials reported in our work were converted according to E (versus RHE) = E (versus SCE) + 0.278 V.

First-principles calculations. All DFT calculations were conducted by using the Vienna *ab initio* simulation package (VASP) suite that is based on the projector-augmented wave (PAW) approach and the plane wave basis set[46]. The structures of $CoS_{2-x}P_x$ (x = 0, 0.5, 1.0, 1.5 and 2), including lattice constants and internal coordinates, in both the cubic (pyrite) and monoclinic phases, were optimized using the Perdew–Burke–Ernzerhof[47] approximation for the exchange-correlation functional with a plane wave energy cutoff of 400 eV, which is about 1.5 times of the default cutoff value such that the Pulay stress problem can be avoided. To account for the possible strong correlation effects in the Co 3d electrons, the energy difference between the cubic and monoclinic phases was calculated using the PBE plus the Hubbard U correction (PBE + U) approach in the rotationally invariant scheme (LDAUTYPE = 1)[48] with U = 4.5 eV and J = 0.5 eV. We used the special quasi-random structure approach[49] implemented in the ATAT code[50] to model the random substitution of S by P in $CoS_{2-x}P_x$ using a supercell of 24 atoms.

References

1. Turner, J. A. Sustainable hydrogen production. *Science* **305**, 972–974 (2004).
2. Mazloomi, K. & Gomes, C. Hydrogen as an energy carrier: Prospects and challenges. *Renew. Sust. Energ. Rev.* **16**, 3024–3033 (2012).
3. Yang, J. & Shin, H. S. Recent advances in layered transition metal dichalcogenides for hydrogen evolution reaction. *J. Mater. Chem. A* **2**, 5979–5985 (2014).
4. Faber, M. S. & Jin, S. Earth-abundant inorganic electrocatalysts and their nanostructures for energy conversion applications. *Energy Environ. Sci.* **7**, 3519–3542 (2014).
5. Zou, X. & Zhang, Y. Noble metal-free hydrogen evolution catalysts for water splitting. *Chem. Soc. Rev.* **44**, 5148–5180 (2015).
6. Jaramillo, T. F. *et al.* Identification of active edge sites for electrochemical H_2 evolution from MoS_2 nanocatalysts. *Science* **317**, 100–102 (2007).
7. Laursen, A. B., Kegnaes, S., Dahl, S. & Chorkendorff, I. Molybdenum sulfides-efficient and viable materials for electro - and photoelectrocatalytic hydrogen evolution. *Energy Environ. Sci.* **5**, 5577–5591 (2012).
8. Voiry, D. *et al.* Enhanced catalytic activity in strained chemically exfoliated WS_2 nanosheets for hydrogen evolution. *Nat. Mater.* **12**, 850–855 (2013).
9. Wang, H. *et al.* $MoSe_2$ and WSe_2 nanofilms with vertically aligned molecular layers on curved and rough surfaces. *Nano Lett.* **13**, 3426–3433 (2013).
10. Kong, D., Wang, H., Lu, Z. & Cui, Y. $CoSe_2$ nanoparticles grown on carbon fibre paper: an efficient and stable electrocatalyst for hydrogen evolution reaction. *J. Am. Chem. Soc.* **136**, 4897–4900 (2014).
11. McEnaney, J. M. *et al.* Amorphous molybdenum phosphide nanoparticles for electrocatalytic hydrogen evolution. *Chem. Mater.* **26**, 4826–4831 (2014).
12. Jiang, P. *et al.* A cost-effective 3D hydrogen evolution cathode with high catalytic activity: FeP nanowire array as the active phase. *Angew. Chem. Int. Ed.* **53**, 12855–12859 (2014).
13. Liu, Q. *et al.* Carbon nanotubes decorated with CoP nanocrystals: a highly active non-noble-metal nanohybrid electrocatalyst for hydrogen evolution. *Angew. Chem. Int. Ed.* **53**, 6710–6714 (2014).

14. Li, Y. H. *et al.* Local atomic structure modulations activate metal oxide as electrocatalyst for hydrogen evolution in acidic water. *Nat. Commun.* **6**, 8064 (2015).
15. Yan, H. J. *et al.* Phosphorus-modified tungsten nitride/reduced graphene oxide as a high-performance, non-noble-metal electrocatalyst for the hydrogen evolution reaction. *Angew. Chem. Int. Ed.* **54**, 6325–6329 (2015).
16. Kong, D. *et al.* Synthesis of MoS_2 and $MoSe_2$ films with vertically aligned layers. *Nano Lett.* **13**, 1341–1347 (2013).
17. Li, Y. G. *et al.* MoS_2 nanoparticles grown on graphene: an advanced catalyst for the hydrogen evolution reaction. *J. Am. Chem. Soc.* **133**, 7296–7299 (2011).
18. Yang, Y., Fei, H., Ruan, G., Xiang, C. & Tour, J. M. Edge-oriented MoS_2 nanoporous films as flexible electrodes for hydrogen evolution reactions and supercapacitor devices. *Adv. Mater.* **26**, 8163–8168 (2014).
19. Voiry, D. *et al.* Conducting MoS_2 nanosheets as catalysts for hydrogen evolution reaction. *Nano Lett.* **13**, 6222–6227 (2013).
20. Morales-Guio, C. G. & Hu, X. Amorphous molybdenum sulfides as hydrogen evolution catalysts. *Acc. Chem. Res.* **47**, 2671–2681 (2014).
21. Kibsgaard, J. & Jaramillo, T. F. Molybdenum phosphosulfide: an active, acid-stable, earth-abundant catalyst for the hydrogen evolution reaction. *Angew. Chem. Int. Ed.* **53**, 14433–14437 (2014).
22. Deng, J. *et al.* Triggering the electrocatalytic hydrogen evolution activity of the inert two-dimensional MoS_2 surface via single-atom metal doping. *Energy Environ. Sci.* **8**, 1594–1601 (2015).
23. Wang, H. *et al.* Electrochemical tuning of vertically aligned MoS_2 nanofilms and its application in improving hydrogen evolution reaction. *Proc. Natl Acad. Sci. USA* **110**, 19701–19706 (2013).
24. Wang, T. *et al.* Electrochemically fabricated polypyrrole and MoS_x copolymer films as a highly active hydrogen evolution electrocatalyst. *Adv. Mater.* **26**, 3761–3766 (2014).
25. Gao, M. R. *et al.* An efficient molybdenum disulfide/cobalt diselenide hybrid catalyst for electrochemical hydrogen generation. *Nat. Commun.* **6**, 5982 (2015).
26. Caban-Acevedo, M. *et al.* Efficient hydrogen evolution catalysis using ternary pyrite-type cobalt phosphosulphide. *Nat. Mater.* **14**, 1245–1251 (2015).
27. Zhang, H. *et al.* Highly crystallized cubic cattierite CoS_2 for electrochemically hydrogen evolution over wide pH range from 0 to 14. *Electrochim. Acta* **148**, 170–174 (2014).
28. Faber, M. S. *et al.* High-performance electrocatalysis using metallic cobalt pyrite CoS_2 micro- and nanostructures. *J. Am. Chem. Soc.* **136**, 10053–10061 (2014).
29. Anastassakis, E. Light scattering and ir measurements in XS_2 pryite-type compounds. *J. Chem. Phys.* **64**, 3604 (1976).
30. Lyapin, S. *et al.* Raman studies of nearly half-metallic ferromagnetic CoS_2. *J. Phys. Condens. Matter* **26**, 396001 (2014).
31. Shadike, Z., Cao, M. H., Ding, F., Sang, L. & Fu, Z. W. Improved electrochemical performance of CoS_2-MWCNT nanocomposites for sodium-ion batteries. *Chem. Commun.* **51**, 10486–10489 (2015).
32. Ma, L. B. *et al.* CoP nanoparticles deposited on reduced graphene oxide sheets as an active electrocatalyst for the hydrogen evolution reaction. *J. Mater. Chem. A* **3**, 5337–5343 (2015).
33. Jiang, P. *et al.* Synthesis of FeP_2/C nanohybrids and their performance for hydrogen evolution reaction. *J. Mater. Chem. A* **3**, 499–503 (2015).
34. Pu, Z., Liu, Q., Asiri, A. M. & Sun, X. Tungsten phosphide nanorod arrays directly grown on carbon cloth: a highly efficient and stable hydrogen evolution cathode at all pH values. *ACS Appl. Mater. Interfaces* **6**, 21874–21879 (2014).
35. Xiao, M., Miao, Y., Tian, Y. & Yan, Y. Synthesizing Nanoparticles of Co-P-Se compounds as electrocatalysts for the hydrogen evolution reaction. *Electrochim. Acta* **165**, 206–210 (2015).
36. Saraby-Reintjes, A. Kinetic criteria for the mechanism of the hydrogen evolution reaction. *Electrochim. Acta* **31**, 251–254 (1986).
37. Fan, Q., Liu, W., Weng, Z., Sun, Y. & Wang, H. Ternary hybrid material for high-performance lithium-sulfur battery. *J. Am. Chem. Soc.* **137**, 12946–12953 (2015).
38. Weng, Z. *et al.* Metal/oxide interface nanostructures generated by surface segregation for electrocatalysis. *Nano Lett.* **15**, 7704–7710 (2015).
39. Kong, D., Cha, J. J., Wang, H., Lee, H. R. & Cui, Y. First-row transition metal dichalcogenide catalysts for hydrogen evolution reaction. *Energy Environ. Sci.* **6**, 3553–3558 (2013).
40. Peng, S. *et al.* Cobalt sulfide nanosheet/graphene/carbon nanotube nanocomposites as flexible electrodes for hydrogen evolution. *Angew. Chem. Int. Ed.* **53**, 12594–12599 (2014).
41. Zhang, H. *et al.* A metallic CoS_2 nanopyramid array grown on 3D carbon fibre paper as an excellent electrocatalyst for hydrogen evolution. *J. Mater. Chem. A* **3**, 6306–6310 (2015).

42. Liu, Q. *et al.* CoSe₂ nanowires array as a 3D electrode for highly efficient electrochemical hydrogen evolution. *ACS Appl. Mater. Interfaces* **7**, 3877–3881 (2015).

43. Gyu Kim, M., Sang Cho, H. & Hyun Yo, C. Fe K-edge X-Ray absorption (XANES/EXAFS) spectroscopic study of the nonstoichiometric SrFe₁₋ₓSnₓO₃₋ᵧ system. *J. Phys. Chem. Solids* **59**, 1369–1381 (1998).

44. Bither, T. A., Bouchard, R. J., Cloud, W. H., Donohue, P. C. & Siemons, W. J. Transition metal pyrite dichalcogenides high-pressure synthesis and correlation of properties. *Inorg. Chem.* **7**, 2208–2220 (1968).

45. Sugiura, C. Sulfur-K X-ray absorption-spectra of FeS, FeS₂, and Fe₂S₃. *J. Chem. Phys.* **74**, 215–217 (1981).

46. Kresse, G. & Furthmuller, J. Efficiency of *ab initio* total energy calculations for metals and semiconductors using a plane-wave basis set. *Comput. Mater. Sci.* **6**, 15–50 (1996).

47. Perdew, J. P., Burke, K. & Ernzerhof, M. Generalized gradient approximation made simple. *Phys. Rev. Lett.* **77**, 3865–3868 (1996).

48. Liechtenstein, A. I., Anisimov, A. I. & Zaanen, J. Density-functional theory and strong interactions: orbital ordering in mott-hubbard insulators. *Phys. Rev. B Condens. Matter* **52**, R5467–R5470 (1995).

49. Wei, S., Ferreira, L. G., Bernard, J. E. & Zunger, A. Electronic properties of random alloys: special quasirandom structures. *Phys. Rev. B Condens Matter* **42**, 9622–9649 (1990).

50. van de Walle, A. Multicomponent multisublattice alloys, nonconfigurational entropy and other additions to the Alloy Theoretic Automated Toolkit. *Calphad J.* **33**, 266–278 (2009).

Acknowledgements

The work is partially supported by the Yale University and the Global Innovation Initiative from Institute of International Education. The work at BNL was supported by the US Department of Energy, the Assistant Secretary for Energy Efficiency and Renewable Energy, Office of Vehicle Technologies under Contract Number DE-SC0012704. We acknowledge technical support from the scientists at beamlines 9-BM-B and 12-BM-B of APS (ANL), supported by the U.S. DOE under Contract No. DE-AC02-06CH11357. M.L. and E.I.A. acknowledge the support of the US Department of Energy through Basic Energy Sciences grant DE-FG02-98ER14882 and the use of facilities supported by the National Science Foundation through the Yale Materials Research Science and Engineering Center (Grant No. MRSEC DMR-1119826). H. J. acknowledges the financial support of National Natural Science Foundation of China (Projects No. 1373017 and 21321001). We thank Prof. Fei Wei (Tsinghua University) for providing the CNTs. We appreciate acquisition of XPS spectra by Baowen Li (CMCM IBS Center, the Ulsan National University of Science and Technology).

Author contributions

W.L. and H.W. conceived the research. W.L. and Y.X. synthesized the materials and performed the electrochemical measurements. E.H. performed the XANES measurements. H.J. performed the DFT calculations. W.L., Z.W., M.L. and Q.F. performed the material characterizations. All authors discussed and analysed the data. W.L., E.H., H.J. and H.W. wrote the paper. All authors discussed and commented on the manuscript.

Additional information

25

Multifunctional hydrogel nano-probes for atomic force microscopy

Jae Seol Lee[1,*], Jungki Song[1,*], Seong Oh Kim[2], Seokbeom Kim[1], Wooju Lee[1], Joshua A. Jackman[2], Dongchoul Kim[1], Nam-Joon Cho[2,3] & Jungchul Lee[1]

Since the invention of the atomic force microscope (AFM) three decades ago, there have been numerous advances in its measurement capabilities. Curiously, throughout these developments, the fundamental nature of the force-sensing probe—the key actuating element—has remained largely unchanged. It is produced by long-established microfabrication etching strategies and typically composed of silicon-based materials. Here, we report a new class of photopolymerizable hydrogel nano-probes that are produced by bottom-up fabrication with compressible replica moulding. The hydrogel probes demonstrate excellent capabilities for AFM imaging and force measurement applications while enabling programmable, multifunctional capabilities based on compositionally adjustable mechanical properties and facile encapsulation of various nanomaterials. Taken together, the simple, fast and affordable manufacturing route and multifunctional capabilities of hydrogel AFM nano-probes highlight the potential of soft matter mechanical transducers in nanotechnology applications. The fabrication scheme can also be readily utilized to prepare hydrogel cantilevers, including in parallel arrays, for nanomechanical sensor devices.

[1] Department of Mechanical Engineering, Sogang University, 35 Baekbeom-ro (Sinsu-dong), Mapo-gu, Seoul 04107, South Korea. [2] School of Materials Science and Engineering and Centre for Biomimetic Sensor Science, Nanyang Technological University, 50 Nanyang Drive, Singapore 637553, Singapore. [3] School of Chemical and Biomedical Engineering, Nanyang Technological University, 62 Nanyang Drive, Singapore 637459, Singapore. * These authors contributed equally to this work. Correspondence and requests for materials should be addressed to N.-J.C. (email: njcho@ntu.edu.sg) or to J.L. (email: jayclee@sogang.ac.kr).

Acompelling motivation in surface science is the development of nanoscale probes to seamlessly interrogate the molecular-level properties of material interfaces. This goal envisions probes which both engage and respond to the local material environment. To this end, scanning probe microscopy[1,2] techniques have revolutionized our experimental capabilities for surface metrology. Atomic force microscopy (AFM)[3] is one of the most successful scanning probe microscopy techniques and employs a cantilever-mounted tip to probe atomic details of a surface. When the tip approaches a surface, the cantilever deflection is influenced by atomic interactions between the tip and sample. Depending on the application and sample properties, the AFM probe design can be varied for optimal sensing—common parameters to adjust include mechanical properties of the cantilever such as spring constant[4] and resonance frequency[5] as well as the tip geometry[6]. In the case of conventional silicon-based probes, mechanical characteristics of the probe are mainly controlled by geometrical dimensions of the cantilever. Indeed, there is a narrow tuning range of the elastic modulus, and tip geometries are typically confined to cones or pyramids with fixed aspect ratios stemming from limited recipes for materials etching. Another option is to functionalize the tip to improve imaging performance or enable a specific application. In such cases, the tip can be treated with a covalent surface modification (for example, functional groups by silane or thiol chemistry[7]), which have a high-aspect ratio nanomaterial (for example, carbon nanotube[8]) attached to the tip apex, or be fabricated from an unconventional material (for example, hydrophobic acrylate and epoxy blend[9] or SU8 photoplastic[10]). Silicon-based probes remain the standard technology in the field for both the cantilever and tip components due to advanced fabrication capabilities, including tip designs with small radii of curvature for high-resolution performance. At the same time, the current emphasis on silicon-based technologies has limited efforts to discover new promising materials or fabrication methods for further innovating AFM probes.

With a growing range of AFM nanotechnology applications[11,12], the need for developing multifunctional AFM probes is paramount, motivating the exploration of new material compositions and fabrication strategies[13]. In particular, there is significant opportunity to develop AFM probes that go beyond two-dimensional surface functionalization and have three-dimensional (3D) programmable features with compositionally tunable properties. To realize this goal, molecular self-assembly offers key advantages over conventional microfabrication for materials programming, that is, imparting functionality through modular combinations of polymeric, organic and inorganic nanomaterials. Photopolymerizable hydrogels are an excellent class of candidate material with tunable mechanical properties[14,15], vast functionalization possibilities[16–18] and flexibility to encapsulate nanomaterials of varying size[19,20]. Furthermore, the soft and compliant nature of hydrogels could be useful for soft matter and biological AFM applications, for which sample wear and damage is a major challenge[21]. Hence, hydrogels have strong merits to be explored as a material for fabricating AFM probes.

Realizing the potential of hydrogel composites as nanoscale actuating probes for AFM applications[5,22,23] would also represent a significant technical advance for nanomechanical sensors in general. The functional features of AFM probes inspired the creation of the field of cantilever-based nanomechanical sensing, which involves highly sensitive detection of biological and chemical analytes among other application possibilities[24,25]. As with AFM probes, tipless cantilever sensors are typically fabricated from silicon-based materials; however, there has been strong interest in exploring polymeric materials[26,27] with lower mechanical stiffness as an alternative, often superior option for high-sensitivity detection[25,28]. In some cases, nanoparticles have been incorporated into the polymeric cantilevers for multifunctional applications[29]. To date, the cantilevers have been composed of relatively stiff, hydrophobic polymers, while softer hydrogels have been explored as a surface coating option to improve the stability and reusability of SU8 polymeric cantilevers[30] as well as for ion sensing[31]. Hydrogel-based cantilevers with lower, and widely tunable, mechanical stiffnesses could enable more sensitive detection capabilities while also providing an improved tool for nanomechanical measurements on soft matter systems. These features would be especially advantageous if a simple and reproducible fabrication scheme could be employed for producing the hydrogel cantilevers. Taken together, all these points highlight the potential significance of developing hydrogel-based cantilevers, both in the context of fully integrated AFM probes as well as for cantilever-based nanomechanical sensor devices.

Towards this goal, we explore the design, fabrication and testing of multifunctional hydrogel AFM probes. From a fabrication perspective, the design of hydrogel AFM probes represents a particularly challenging feature because it involves manufacturing and integrating hydrogel cantilevers and tips—the nanoscale geometrical features of the latter are also important. The learnings from work in this direction can be directly applied to fabricating tipless cantilevers as well. Experimentally, our findings indicate that hydrogel AFM probes enable stable and reproducible force, and imaging measurements of various inorganic and biological samples in air and liquid conditions. The combination of fabrication possibilities and multifunctional versatility afforded by the hydrogel probes offers the first demonstration of how 3D materials design strategies can be utilized for AFM applications, thereby showing the potential of soft matter to provide superior and versatile actuators for surface metrology. The hydrogel nano-probes developed in this work have the potential to be extended to cantilever-based nanomechanical sensing applications in both single measurement and array configurations.

Results

Fabrication of hydrogel AFM probes. Hydrogel materials are readily fabricated by a molecular self-assembly process that involves light-sensitive polymerization reactions[32]. A wide variety of materials can form hydrogels and an even greater range of materials can be encapsulated within hydrogels[33]. While hydrogels are routinely fabricated on the microscale, hydrogel fabrication on the nanoscale requires a delicate approach based on high precision tuning of the tip dimensions to enable various scanning probe applications (Fig. 1a). To fabricate hydrogel AFM probes, we took advantage of a bottom-up strategy that uses aqueous conditions, which contrasts with the typical etching of silicon probes that involves harsh treatments and organic solvents[34]. As presented in Fig. 1b, the strategy involves fabrication of the tipless hydrogel cantilever, which is first prepared by ultraviolet light-induced curing of a pre-polymer solution introduced into the cantilever beam mould. The tipless hydrogel cantilever then makes contact with a tip mould filled with pre-polymer solution, followed by a second round of ultraviolet light exposure which cures the hydrogel in the tip mould and results in firm attachment between the cantilever and tip. Before the second ultraviolet light exposure, the hydrogel-filled tip mould can be optionally deformed to apply compressive strains which facilitate tunable tip sharpness and aspect ratio. After tip fabrication, a metal reflective coating is then added onto the top portion of the cantilever.

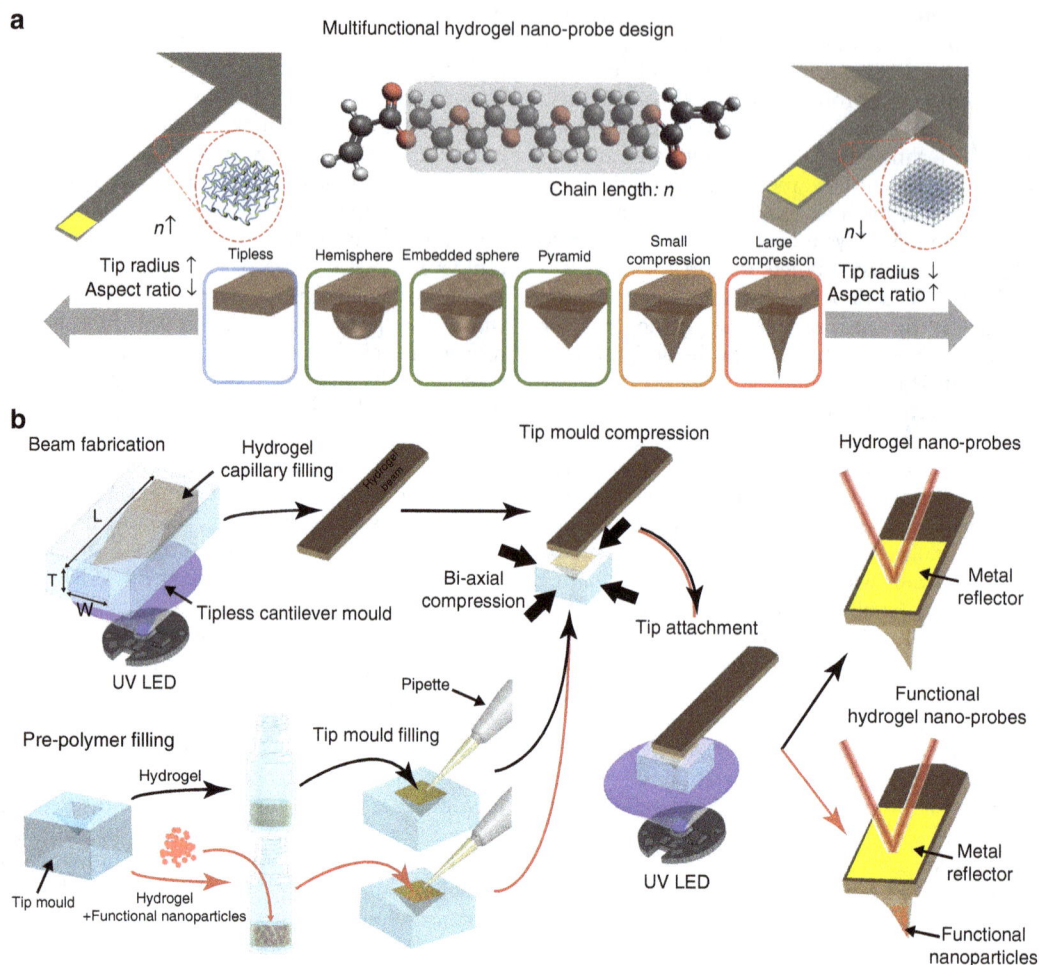

Figure 1 | Design and fabrication of hydrogel AFM nano-probes. (**a**) Conceptual schematic for beam and tip geometry to tune probe characteristics for various AFM applications. Capillary driven filling of ultraviolet curable hydrogels into the engineered beam mould facilitates fabrication of tipless hydrogel cantilevers with tunable spring constants over several orders of magnitude by varying the MW of the monomer composition and geometrical dimensions of the cantilever. Nanoscale dimensions of the tip shape can also be tuned by preparing different tip moulds or by applying deformation to a given mould. When the tip mould is compressed, the tip becomes sharper and the tip aspect ratio typically increases. (**b**) Fabrication method for hydrogel AFM probes. A tipless hydrogel cantilever is first prepared by ultraviolet curing of the pre-polymer solution introduced into the cantilever beam mould. The tipless hydrogel cantilever then makes contact with a tip mould filled with pre-polymer solution without or with encapsulated functional elements, followed by a second round of ultraviolet exposure which cures the hydrogel in the tip mould and results in firm attachment between the cantilever and tip. Before the second ultraviolet exposure, the hydrogel-filled tip mould can be optionally deformed by applying bi-axial compressive strains to facilitate tunable tip sharpness and aspect ratio.

The detailed fabrication set-up is presented in Fig. 2a. A hydrogel cantilever was fabricated by using a poly-dimethylsiloxane (PDMS) replica mould inside a curing set-up (Supplementary Fig. 1). The pre-polymer solution containing poly(ethylene glycol) diacrylate (PEG-DA; molecular weight (MW) of $250 \, g \, mol^{-1}$) was filled into the mould and the solution was cured by applying ultraviolet light (Supplementary Fig. 1). To functionalize the hydrogel cantilever with a tip, simultaneous hydrogel tip fabrication and attachment was performed (Fig. 2b). The steps entailed filling a tip mould with PEG-DA pre-polymer solution, contacting the cantilever base with the tip mould, and applying ultraviolet light to cure the hydrogel tip (Fig. 2b and Supplementary Figs 2 and 3). This last step cross-linked the PEG-DA monomers in the tip mould and also led to cross-linking between PEG-DA monomers in the tip and cantilever base. Hence, an integrated hydrogel AFM probe was created that possesses a cantilever base and tip. Control of the tip geometry and sharpness was established by using a compression jig that is included in the fabrication set-up (Fig. 2c). Sequential uniaxial and bi-axial compression enables precise tip

shapes to be formed. The resulting hydrogel cantilevers had controllable length and exhibited flatness in ambient conditions (Fig. 2d). Representative scanning electron microscopy images further demonstrate the fabrication of hydrogel tips with various shapes, including embedded spheres, hemispheres and pyramids (Fig. 2e). For pyramidal tips, the tip radius was estimated by circular fitting of the tip apex region (Supplementary Fig. 4). Importantly, hydrogel tips with sub 20 nm radii were achievable with the compressible replica moudling approach. Taken together, the characterization studies indicate that the tip radius and aspect ratio can be precisely tuned for different applications (Fig. 2f and Supplementary Figs 5 and 6). As PEG-DA hydrogels are optically transparent, the detector side of the cantilever was partially covered by a reflective metal (Ti/Au; 20/100 nm) coating for AFM operation (Supplementary Fig. 7) and the cantilevers remained flat in air and water (Supplementary Fig. 8). For more precise registration of the tip, the area and position of the metal coating can be adjusted to allow the scanning tip to be visually accessible (Supplementary Fig. 9).

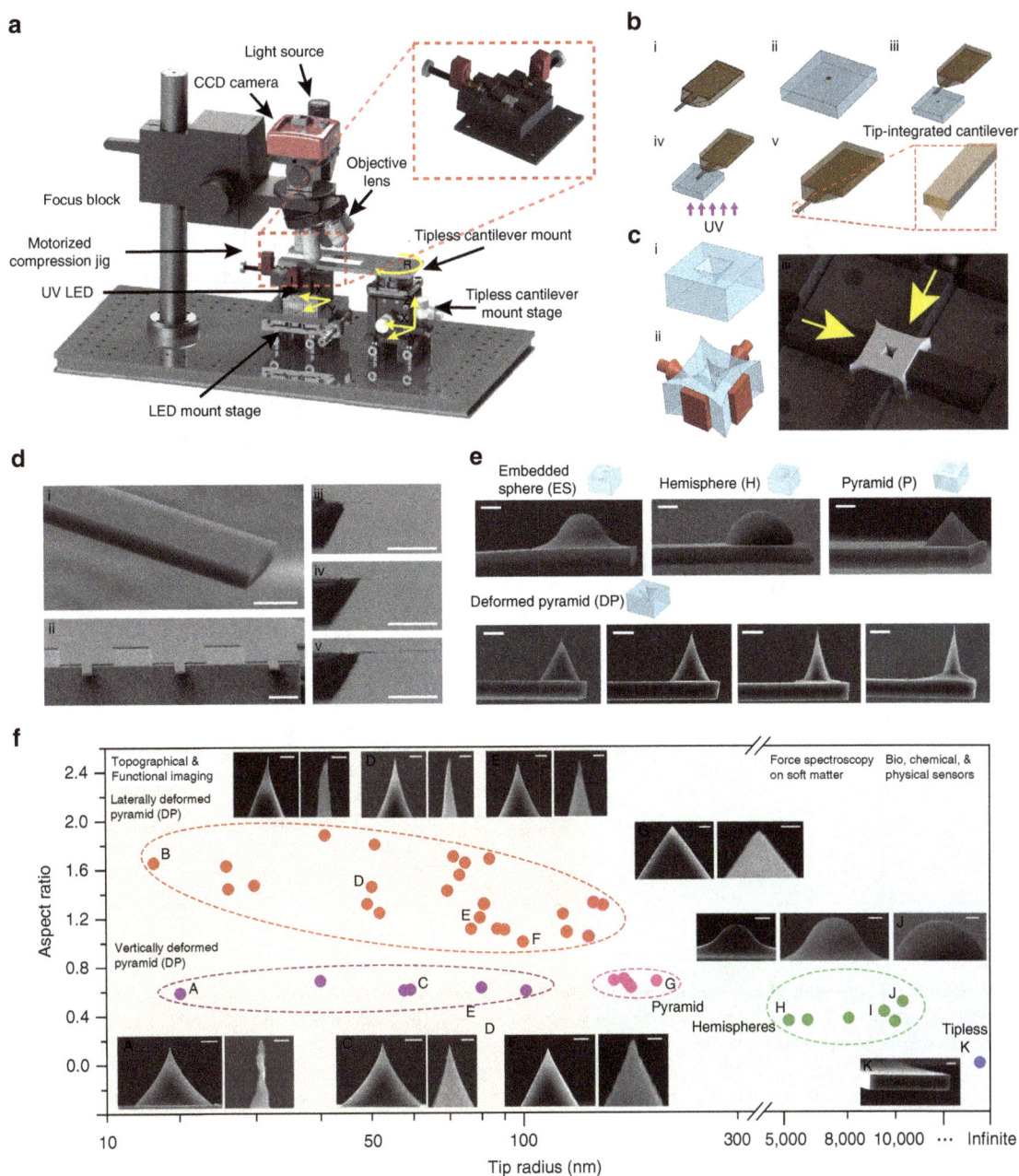

Figure 2 | Replica moudling strategy to fabricate hydrogel nano-probes. (**a**) Experimental set-up for fabricating hydrogel nano-probes, which consists of a custom-built microscope, a ultraviolet light-emitting diode (LED), and a compression jig for tip shape tuning. (**b**) Tip integration process (i) Fabricated tipless hydrogel cantilever, (ii) Fill the pre-polymer solution into a tip mould (in this case, an inverted pyramid although the design can vary), (iii) Approach the tipless hydrogel cantilever towards the hydrogel-filled tip mould and make contact between the cantilever and tip mould, (iv) Cure hydrogel tip with ultraviolet light, (v) Separate the tip-integrated hydrogel cantilever from tip mould). (**c**) Tip shape tuning via compression replica moudling (i) PDMS pyramidal tip mould, (ii) bi-axial compression to the tip mould and 3D schematic (iii) for bi-axial compression of the tip mould. (**d**) Scanning electron and optical micrographs of a single tipless hydrogel cantilever (i) and its array (ii). Tipless hydrogel cantilevers with lengths of (iii) 300, (iv) 500 and (v) 700 μm, respectively. Scale bars are 20, 200 and 500 μm, for (i), (ii) and (iii–v), respectively. (**e**) Scanning electron micrographs showing different tips integrated on hydrogel cantilevers—embedded sphere (ES), hemisphere (H), pyramid (P) and deformed pyramid (DP). All scale bars are 10 μm. (**f**) Summary of tip radii and aspect ratios of various deformed pyramidal tips. Scale bars are 5 μm for overall views, 200 nm for the zoom-ins of A and B, and 1 μm for the other zoom-ins (C–G).

Tunable mechanical force sensing. A major advantage of hydrogel materials is that molecular characteristics can be easily customized by adjusting factors such as monomer size[35], polymer weight fraction[36] and curing condition[37]. Depending on the PEG-DA MW, the mechanical properties of the cantilever are tunable. With increasing MW, the cantilevers become more flexible, as indicated by trends in the relative deflection (Fig. 3a). Likewise, with increasing ultraviolet light dose in the curing step,

the elastic modulus of PEG-DA MW 250 cantilevers in air increases[38] (Fig. 3b). To characterize the mechanical properties of hydrogel cantilevers in water, stylus forces were calibrated and transient polymer swelling was first assessed as a function of PEG-DA MW (Supplementary Figs 10 and 11). Under steady-state conditions, hydrated PEG-DA hydrogels with higher average MW monomers exhibited greater swelling (Fig. 3c and Supplementary Fig. 12). Accordingly, the elastic

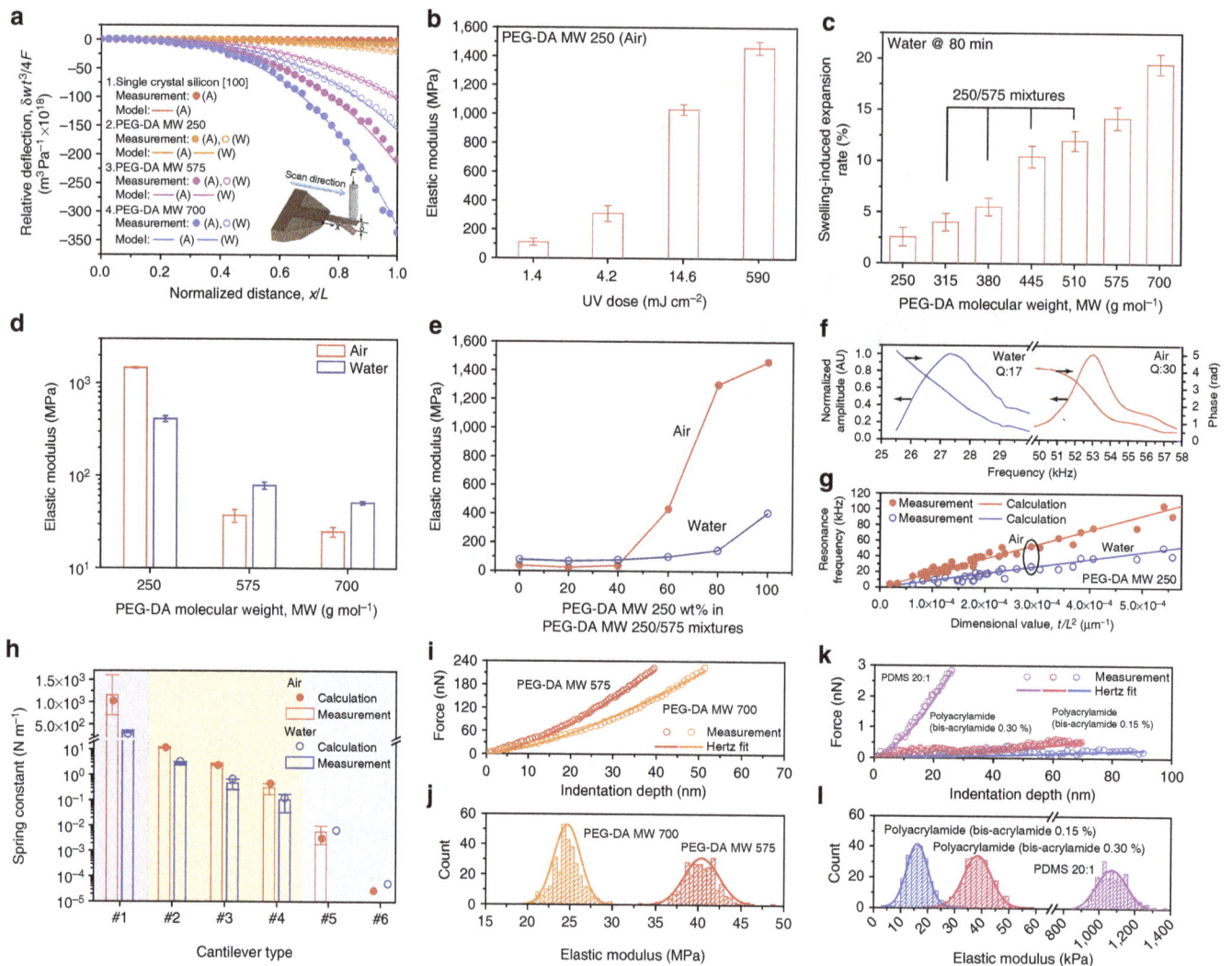

Figure 3 | Tunable mechanical properties of hydrogel nano-probes for force-sensing applications. (**a**) Relative deflection of hydrogel cantilevers as a function of normalized displacement. 'A' and 'W' represent air and water, respectively. (**b**) Elastic moduli of PEG-DA MW 250 hydrogel cured under different ultraviolet doses. Error bars represent s.d. of the mean with $N = 3$ measurements. (**c**) Swelling-induced expansion of PEG-DA hydrogels in water. (**d**) Elastic moduli of PEG-DA MW 250, 575 and 700 in air and water. Error bars represent s.d. of the mean with $N = 3$ measurements. (**e**) Elastic moduli of PEG-DA MW 250/575 mixtures in air and water as a function of wt% of PEG-DA MW 250. (**f**) Resonance spectra of a hydrogel cantilever in air and water (length, width and thickness: 222, 50 and 15 μm). (**g**) Resonance frequencies of various hydrogel cantilevers with different dimensions in air and water. The two data points enclosed by the ellipse correspond to the measurement values for the hydrogel cantilever shown in **f**. A total of 53 and 35 measurements were completed in air and water, respectively. (**h**) Experimental and theoretical spring constant values for hydrogel cantilevers in air and water (length, width and thickness: 500, 350 and 100 μm (#1); 220, 50 and 19 μm (#2); 400, 50 and 20 μm (#3); 375, 50 and 11 μm (#4); 505, 50 and 11 μm (#5); and 1500, 35 and 5 μm (#6)). Cantilevers #1–4 were fabricated using PEG-DA MW 250, and cantilevers #5 and #6 were fabricated using PEG-DA MW 700. Error bars represent s.d. of the mean with $N = 3$ measurements. (**i**) Force-indentation curves on PEG-DA hydrogel slabs in air with Hertzian curve fits. (**j**) Elastic moduli of different MW PEG-DA samples in air. (**k**) Force-indentation curves on different polymeric substrates in water with Hertzian curve fits. (**l**) Elastic moduli of PDMS 20:1 and polyacrylamide with different bis-acrylamide concentrations in water.

modulus of PEG-DA MW 575 and 700 cantilevers increased in water due to greater stress among the swollen, cross-linked polymer chains[39,40] (Fig. 3d). On the other hand, the elastic modulus of hydrated PEG-DA MW 250 cantilevers decreased by 30%. In this latter case, the lack of swelling is advantageous for AFM applications. Parylene coating is an alternative and optional step to partial metal coating to prevent issues resulting from water absorption (Supplementary Fig. 13). To finely adjust the elastic modulus across a wide range (~30 MPa to 1.5 GPa), cantilevers were also prepared from PEG-DA MW 250 and 575 mixtures (Fig. 3e and Supplementary Fig. 14). Widely tunable and comparably low elastic moduli of hydrogel cantilevers are in stark contrast with those of silicon-based cantilevers which

exhibit a more limited tuning range that typically relies on crystallinity[41].

Next, we investigated the resonance spectra of a PEG-DA MW 250 cantilever in air and water. The fundamental resonance frequencies in air and water were 53 and 27 kHz, respectively. The quality factors in air and water were 30 and 17, respectively (Fig. 3f). By adjusting the geometrical dimensions of the PEG-DA MW 250 cantilever, we varied the resonance frequency between 4 and 105 kHz in air and between 2 and 51 kHz in water, respectively (Fig. 3g). By adjusting the cantilever dimensions and PEG-DA MW, the spring constant was also varied between 0.000027 and 1,022 N m^{-1} in air, and between 0.000053 and 280 N m^{-1} in water (Fig. 3h and Supplementary Table 1). In particular, the spring constants of some hydrogel cantilevers in water were specifically tuned to match those of commercial silicon cantilevers (ranging from 0.09 to $2 \, N \, m^{-1}$) to compare contact mode AFM imaging performance, as described below. The mechanical force-sensing capabilities of the hydrogel cantilevers were determined by force-displacement measurements[42] (Supplementary Method, Supplementary Figs 15–17, and Supplementary Table 2). As shown in Fig. 3i–l, the hydrogel cantilever is well-suited for measuring the elastic moduli of various hard and soft polymeric substrates in cases where the hydrogel tip deformation is negligibly small (see Supplementary Discussion, Supplementary Fig. 18 and Supplementary Table 3). The measured elastic moduli of all substrates were in good agreement with results obtained by stylus measurements or reference values (Supplementary Table 4). Collectively, the experiments indicate that the bottom-up fabrication approach permits excellent control over the mechanical properties of hydrogel cantilevers and the spring constant can be readily tuned over at least eight orders of magnitude from ~ 0.00003 to $\sim 1,000 \, N \, m^{-1}$ which is a wider tuning range than that offered by commercial silicon-based cantilevers (Supplementary Fig. 19). It is therefore feasible to employ integrated hydrogel probes in AFM applications on both soft and hard substrates as well as for nanomechanical sensing applications[43–46].

Deflection and resonance stabilities. To investigate the stability of hydrogel contact and noncontact probes, we measured time-dependent variations of the static deflection and resonance frequency of hydrogel probes in different surrounding

media conditions (Supplementary Discussion and Supplementary Fig. 20a,b). Measurement results were plotted as Allan deviations (Fig. 4a,b) and s.d.'s (Fig. 4c,d). Both the Allan deviations and the s.d.'s of the static deflection and resonance frequency indicate that hydrogel probes are most stable in water. High ionic strength conditions in phosphate-buffered saline (PBS) solution caused the hydrogel probes to be marginally less stable than devices operating in water[47]. Resonance frequencies measured in PBS also differed from measurements conducted in water (Supplementary Fig. 20b). For benchmarking, the stability characteristics of hydrogel probes were further compared with those of commercial silicon probes which had similar spring constants or resonance frequency values (Supplementary Discussion and Supplementary Fig. 20c,d). In air, the hydrogel probes are slightly less stable than the silicon probes for static operation when the fluctuation of static deflection is directly compared (Supplementary Fig. 20c). Once the different probe lengths are taken into account, the hydrogel probes appear to be more stable than the silicon probes (Fig. 4c). On the other hand, in water and PBS conditions, hydrogel probes are more stable than silicon probes for both static and dynamic operation in terms of both absolute and relative figures (Fig. 4c,d and Supplementary Fig. 20c,d). Given these experimental findings that demonstrate hydrogel probes have stable deflection and resonance stabilities in air and liquid conditions, we next explored the potential of hydrogel probes for AFM applications.

AFM performance of hydrogel probes. To initially check the feasibility of hydrogel probes for AFM applications, calibration

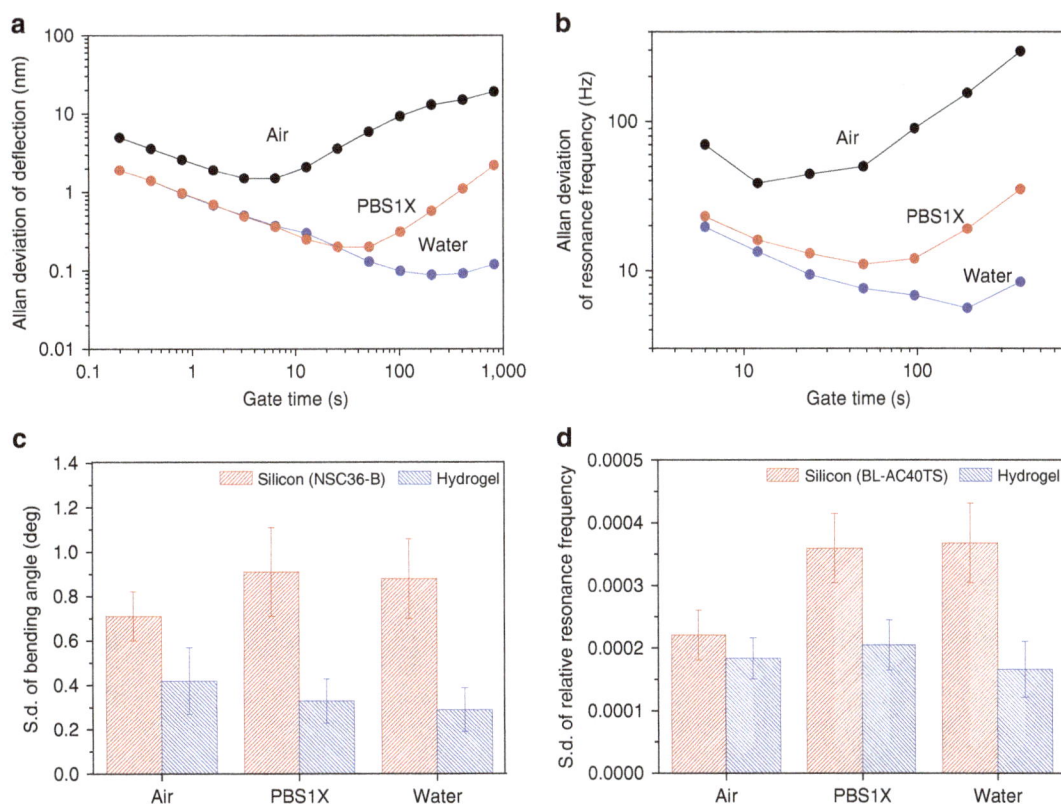

Figure 4 | Stability of hydrogel probes in different environmental conditions. (a,b) Allan deviation of the static deflection of a hydrogel probe (length, width and thickness: 240, 50 and 10 µm) **(a)** and Allan deviation of the resonance frequency of another hydrogel probe (length, width and thickness: 220, 50 and 20 µm) **(b)** in air, water and PBS solution. **(c,d)** S.d. of the bending angles of three silicon (NSC36-B, Mikromasch) and three hydrogel probes with similar spring constants **(c)** and s.d. of the relative resonance frequencies of three silicon (BL-AC40TS, Olympus) and three hydrogel probes with similar resonance frequencies in air, water and PBS solution. Error bars represent s.d. of the average from three measurements for each device ($N = 9$ measurements for each probe material).

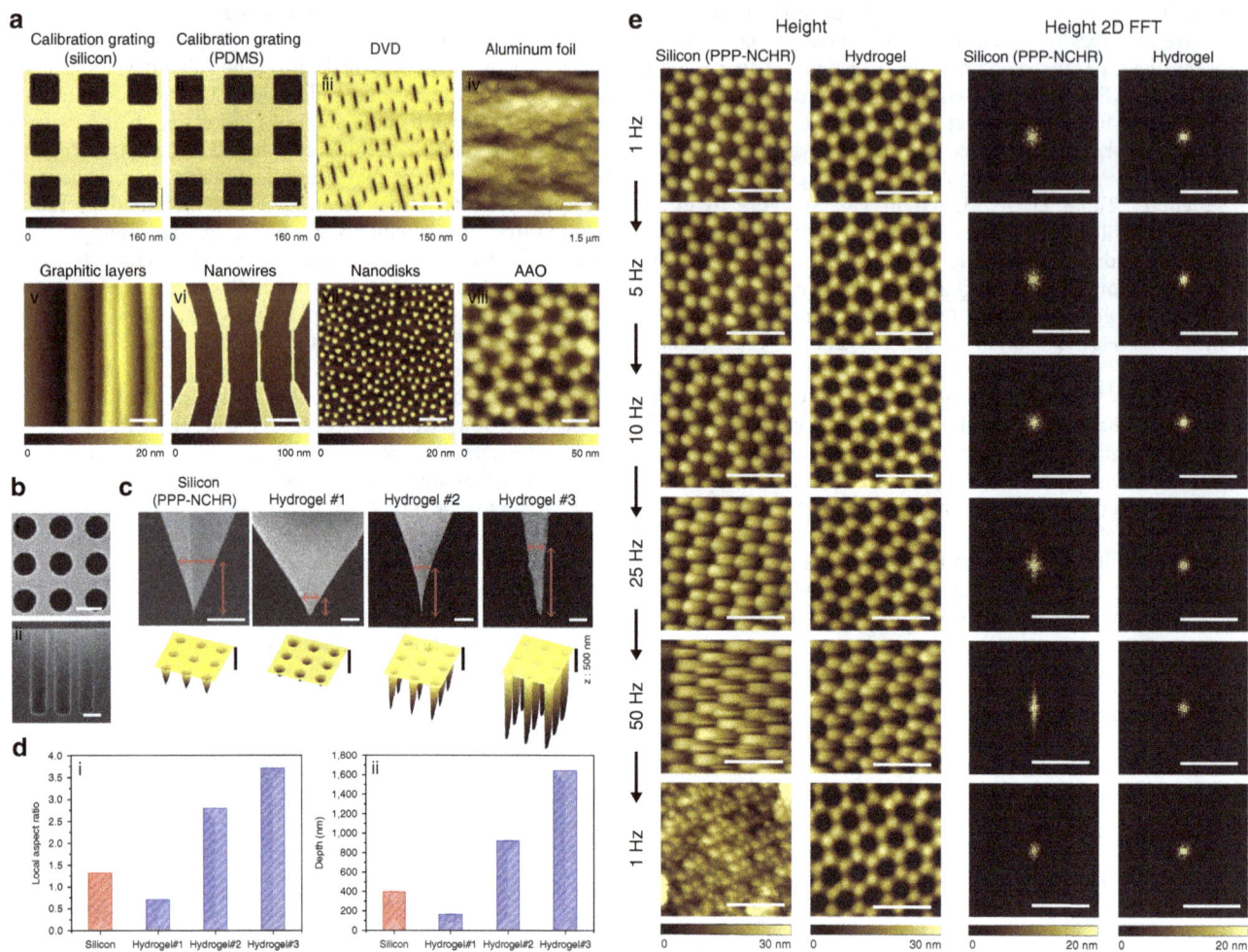

Figure 5 | Noncontact mode imaging performance of hydrogel probes and comparison with silicon probes. (**a**) Noncontact mode height images of calibration gratings (silicon and PDMS replica), digital versatile disk media with written data bits, nanowires, aluminium foil, graphitic layers, nanodisks and an AAO sample (70 nm pore diameter). Scale bars are 5, 5, 5, 10, 1, 5, 0.4 and 0.1 μm for (i–viii), respectively. (**b**) Top (i) and side (ii) view of SEM images of cylindrical pores with 1 μm diameter, 1.4 μm pitch and 5 μm depth. The side view was taken after the sample was cleaved. Scale bars are 1 μm. (**c**) Scanning electron microscopy (SEM) images showing tip apex regions of commercial silicon (PPP-NCHR, Nanosensors) and fabricated hydrogel (length, width and thickness: 200, 50 and 20 μm) probes and corresponding 3D representations of noncontact height images for the deep cylindrical pores. All scale bars are 1 μm. (**d**) Local aspect ratio (i) and maximum depth (ii) measured with silicon and hydrogel probes. (**e**) Height images for another AAO sample (35 nm pore diameter) taken by commercial silicon (PPP-NCHR, Nanosensors) and fabricated hydrogel (length, width and thickness: 170, 50 and 20 μm) probes at 1, 5, 10, 25, 50 and 1 Hz, respectively. Height images are 2D fast Fourier transformed (2D FFT) to quantitatively compare image distortions and artifacts. Scale bars for height and 2D FFT images are 100 nm and 200 μm^{-1}, respectively. The same colour bars are used for height and 2D FFT images, respectively.

gratings were imaged in contact and noncontact modes with fabricated hydrogel and commercial silicon probes (Fig. 5a, i–ii and Supplementary Fig. 21). Hydrogel probes were then employed to image various samples including digital versatile disk media with data written, aluminium foil, graphitic layers, nanowires, nanodisks and anodized aluminium oxide (AAO) samples (Fig. 5a, iii–viii). Hydrogel probes successfully resolved ~1–2 nm steps in graphitic layers and 70 nm diameter pores in AAO. Once a softer hydrogel nano-probe (Supplementary Fig. 22) was employed in contact mode imaging of graphitic layers, enhanced friction contrast was obtained (Supplementary Fig. 23). In addition, a silicon sample with a deep cylindrical pore array was prepared by deep reactive ion etching and imaged by three fabricated hydrogel probes with local aspect ratio of 0.7, 2.79 and 3.71 (Fig. 5b). A commercial silicon probe with local aspect ratio of 1.32 was tested as well (Fig. 5c). Here, the local aspect ratio is defined as the ratio of tip height to tip width corresponding to the pore diameter represented with vertical and

horizontal arrows in the top panel of Fig. 5c. Three-dimensional representations of noncontact mode height images indicate that hydrogel probes with a deformed pyramidal tip outperform the silicon probe while the hydrogel probe with a regular non-deformed pyramidal tip underperforms the silicon probe (Fig. 5c). The local aspect ratio of each probe and the corresponding maximum depth measured are shown in Fig. 5d. The depth measured by the hydrogel probe with local aspect ratio of 3.71 is ~4.1 times deeper than that measured by the silicon probe. Once the tip mould is initially replicated from a specific tip shape and then bi-axially compressed, the local aspect ratio always increases. Therefore, hydrogel probes show excellent demonstrated potential to outperform silicon probes in the case of imaging deep pores and trenches.

In addition, the advantage of a relatively low quality factor for high-speed imaging with hydrogel probes was investigated. A hydrogel probe with resonance frequency of 141 kHz and quality factor of 35.1 was used to image an AAO sample with

35 nm diameter pores at various scan rates. For benchmarking, a commercial silicon probe with resonance frequency of 322 kHz and quality factor of 589 was employed (Fig. 5e). Both hydrogel and silicon probes exhibited tip sharpness of ~ 10 nm and feedback control parameters for hydrogel and silicon probes were individually optimized at a 1-Hz scan rate, and then maintained for other subsequently higher scan rates. The hydrogel probe maintained similar imaging quality at all scan rates while the silicon probe produced degraded image qualities at scan rates higher than 25 Hz. Height images obtained by the silicon probes exhibited much greater distortions and artifacts at scan rates above 25 Hz. To quantitatively compare the distortions, all height images were two-dimensional fast Fourier transformed (FFT) using the Hanning window to derive the modulus—the absolute value of the complex Fourier coefficient. There were hexagonal bright spots at the centre in common for both probes at all scan rates. While the symmetry of the bright spots was maintained for the hydrogel probe, it was broken and significantly stretched along the slow scan axis for the silicon probe. The superior performance of the hydrogel probes can be explained by the cantilever bandwidth, B, which is proportional to the ratio of the resonance frequency, f_0, to the quality factor, Q. The f_0/Q ratios are 4.02 and 0.55 kHz for hydrogel and silicon probes, respectively. The lower noise observed with the hydrogel probe may be attributed to the higher f_0/Q of the hydrogel probe by an order of magnitude because a higher f_0/Q is known to be beneficial for high-speed imaging[48]. After imaging at 50 Hz, the AAO sample was imaged again at 1 Hz with both probes to check for any damage. Interestingly, only the hydrogel probe offered image quality similar to that of the first 1-Hz scan. The silicon probe tip was significantly damaged while the hydrogel probe only underwent regular wear (Supplementary Fig. 24). While the imaging performance of the silicon probe could possibly be improved by using one with a softer cantilever[49], the result nevertheless demonstrates that the hydrogel AFM probe is useful for high-speed imaging applications.

We next investigated the imaging performance of hydrogel probes in liquid conditions. First, changes in tip radius induced by physical contact and swelling were shown to be negligible for typical liquid imaging conditions (Supplementary Fig. 25). Experiments with human MRC-5 fibroblast cells and HeLa cells attested to the biological imaging capabilities of the hydrogel AFM probe (Fig. 6 and Supplementary Figs 26–28). The imaging performance of hydrogel and silicon AFM probes on MRC-5 fibroblast cell surfaces was directly compared in parallel experiments. For these experiments, the silicon probe (BL-AC40TS, Olympus) had a spring constant of 2 N m^{-1} and a pyramidal tip with a 10-nm radius of curvature, while the hydrogel probe had a cantilever with a 1-N m^{-1} spring constant and a pyramidal tip with a sub 30 nm radius of curvature. Feedback control parameters were individually optimized for each cantilever used. Comprehensive experiments were performed at increasing scan rates (1–10 Hz) at an identical contact force of 3 nN, and demonstrated that hydrogel and silicon probes had similar performances up to 5 Hz. However, at a 10-Hz rate, only the hydrogel probe maintained consistent image quality while significant distortions are observed in the image obtained by the silicon probe specifically designed for high-speed imaging in liquids (Fig. 6a,b). High-speed imaging for MRC-5 fibroblast cells in noncontact mode performed with both silicon and hydrogel probes also showed the outperformance of hydrogel probes over silicon probes (Supplementary Fig. 29).

Another set of cellular imaging experiments was also conducted to compare image quality[21] by hydrogel and silicon probes at a high contact force. We prepared a MRC-5 fibroblast cell sample and imaged the same region of the cellular plasma membrane sequentially with a hydrogel, silicon and again hydrogel probe (Fig. 6c,d). Fine details of the cell surface topography were revealed by the hydrogel probe in both experiments, with appreciably lower spatial resolution afforded by the silicon probe even though the same contact force of 300 nN was applied to both probes. It should be stressed that the low spatial resolution obtained with the silicon probe was not due to sample damage by the hydrogel probe because the second scan with the hydrogel probe showed comparable results to the first scan with the hydrogel probe. Next, the cells were imaged in PBS solution under dynamic temperature cycling between 34.5 and 40.4 °C and the results were comparable to those obtained on the same cell surface in measurements conducted at a fixed temperature of around 36 °C (Supplementary Fig. 30). Lastly, human fibroblast cells were continuously imaged in PBS solution with contact mode for 4 h to investigate the long-term stability of hydrogel probes in a practical setting (Supplementary Fig. 31). The image quality was maintained without any evidence of degradation or contamination. The hydrogel-metal interface was durable enough to function properly after 50 h immersion in water (Supplementary Fig. 32).

Durability test and tip recovery. Hydrogel probes with tip radii of ~ 30 nm were employed to investigate wear characteristics in air and PBS solution by repeated imaging on silicon and PDMS gratings (Supplementary Methods and Supplementary Figs 33 and 34). Scanning electron microscopy images show the tip regions of five hydrogel probes before and after wear tests (Fig. 7a). When a hydrogel probe was used to image a hard silicon grating, the tip wear was significantly reduced in noncontact mode (Fig. 7b). Tip wear in noncontact mode is attributed to intermittent hammering impact[50,51] while tip wear in contact mode is attributed to constant sliding friction[52,53]. In addition, tip wear during imaging in air occurs more quickly than that observed during imaging in PBS solution (Fig. 7b). The liquid surrounding the hydrophilic tip may act as a lubricant and/or coolant to slow down wear. When a hydrogel probe was used to image a soft PDMS grating, tip wear was significantly decelerated compared to imaging on the hard silicon sample (Fig. 7b). Nonetheless, with any operation mode and imaging condition, wear of the hydrogel tip cannot be completely prevented. One can readily replace worn hydrogel probes with new hydrogel probes through fast and cost-effective fabrication means. We applied oxygen plasma ashing[54] to re-sharpen worn hydrogel probes. Worn hydrogel probes with tip radii of 130 and 340 nm were regenerated to have sub 30 nm tip radii after oxygen plasma ashing (Fig. 7c). This post-treatment step is applicable to re-sharpening of hydrogel probe tips degraded during any AFM application.

Multifunctional capabilities via materials encapsulation. Beyond imaging capabilities, hydrogel AFM probes enable multifunctional features based on materials encapsulation[55,56]. Taking advantage of the bottom-up fabrication approach, we incorporated various nanomaterials into the aqueous pre-polymer solution before photopolymerization. By tuning the hydrogel mesh network dimensions[57], different nanomaterials can be encapsulated indefinitely or exhibit controlled release profiles. Due to the modular design of the hydrogel tip and cantilever, the hydrogel material in both parts can be varied to encapsulate different materials (Supplementary Fig. 35 and Supplementary Table 5). As demonstrated in Fig. 8a, this approach can be applied to fluorescent dyes (Rhodamine B), quantum dots (cadmium telluride, CdTe), magnetic nanoparticles (cobalt, Co) and plasmonic nanoparticles (gold, Au) across a wide range of sizes

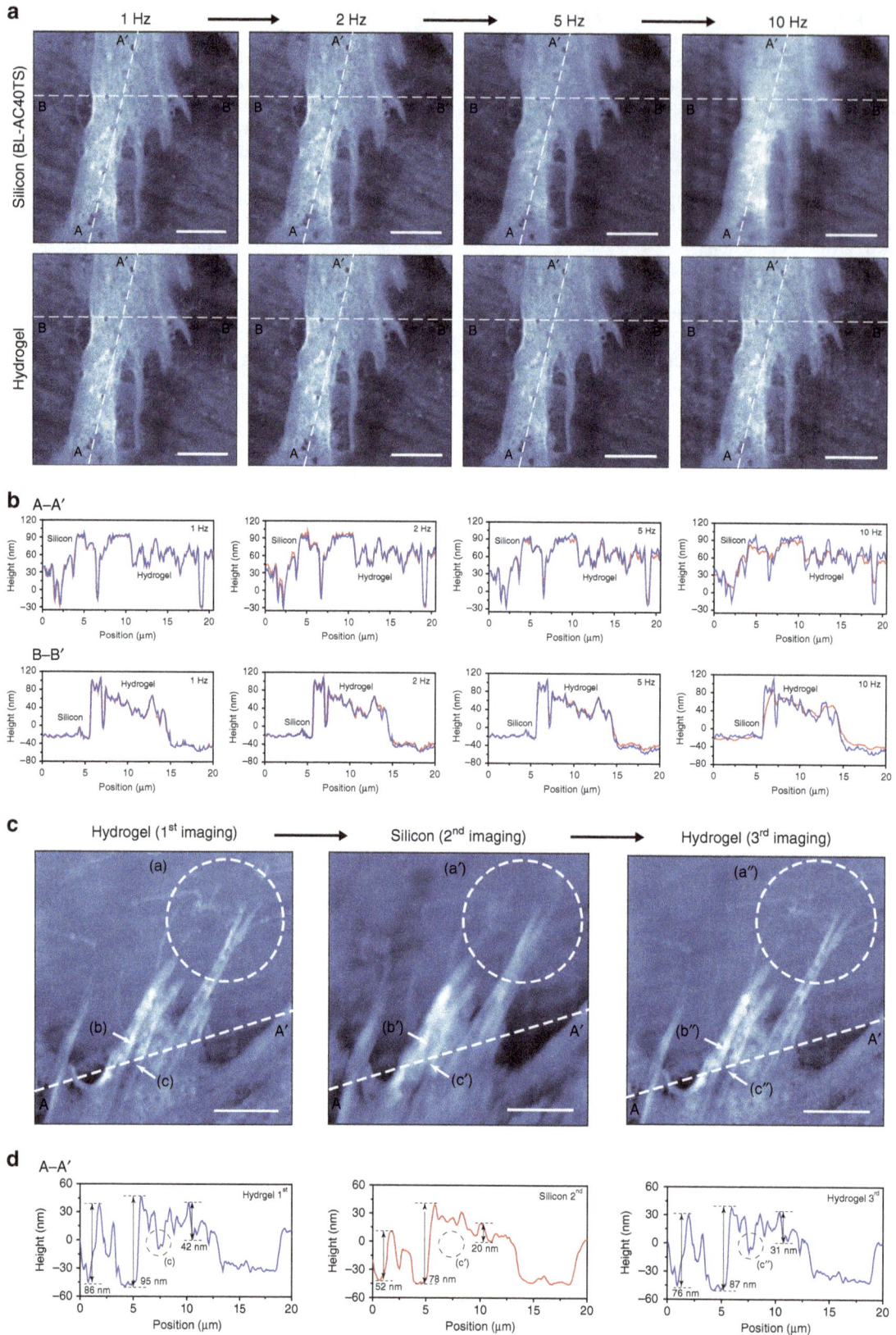

Figure 6 | Contact mode imaging results in liquid using hydrogel probes and performance comparison with silicon probes. (**a**) Contact mode height images, with fixed 3 nN contact forces, for human fibroblast cells (MRC-5) in PBS imaged by a commercial silicon probe (first row) and a fabricated hydrogel probe (second row) at various scan rates. All scale bars are 5 μm. (**b**) Height profiles along the A–A′ and B–B′ lines in **a**. (**c**) Contact mode height images, with fixed 300 nN contact force, for MRC-5 cells in PBS sequentially imaged by using hydrogel (first), silicon (second) and hydrogel (third) probes at a scan rate of 1 Hz. All scale bars are 5 μm. (**d**) Height profiles along the A–A′ line in **c**.

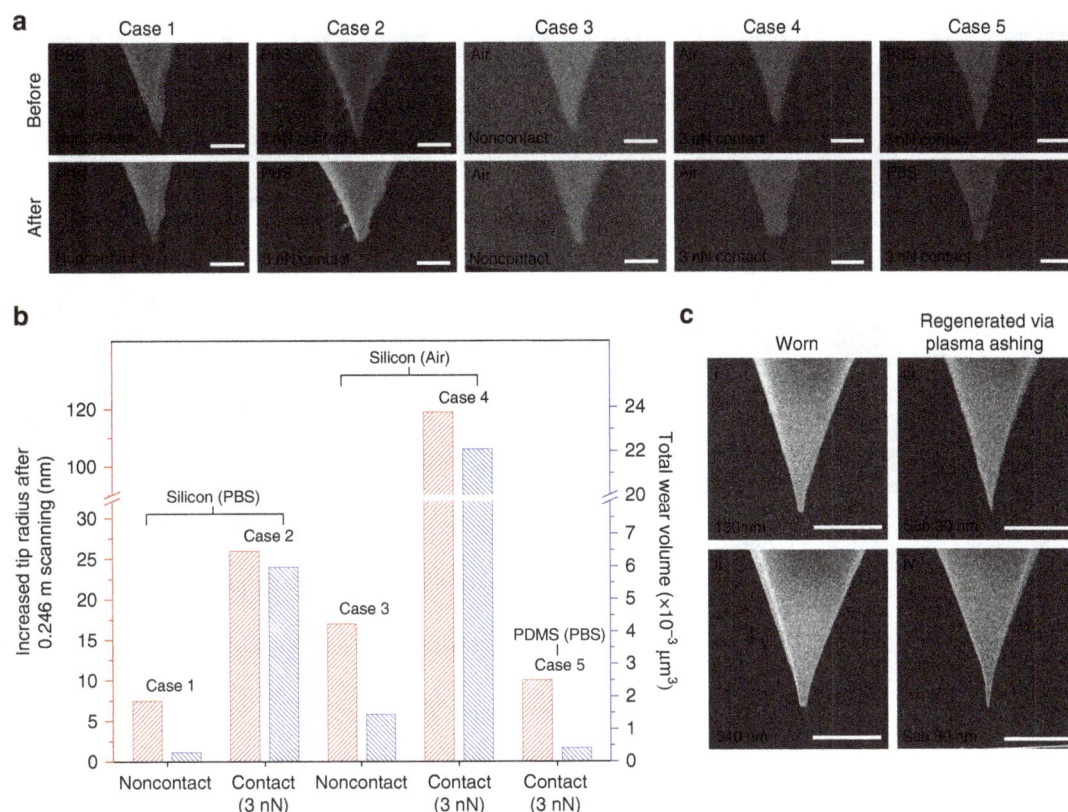

Figure 7 | Wear characteristics of hydrogel nano-probes and regeneration via oxygen plasma ashing. (**a**) Scanning electron microscopy (SEM) images of hydrogel probe tips taken before and after wear tests. Imaging mode, condition and surrounding media are indicated on the left side of each micrograph. A hard silicon calibration grating sample was used for test cases 1–4, and a PDMS calibration grating sample replicated from the silicon grating was used for test case 5. All scale bars are 500 nm. (**b**) Increased tip radius and total wear volume after scanning a 0.246-m length (12 frame imagings with a scan size of $40 \times 40\ \mu m^2$) for each operating condition. (**c**) SEM images of worn hydrogel nano-probe tips before (i, ii) and after (iii, iv) oxygen plasma ashing. Hydrogel nano-probes with tip radii of 130 and 340 nm were re-sharpened to have sub 30 nm tip radii after plasma ashing. All scale bars are 5 μm.

from 1.6 to 50 nm diameter. The approach enables nanomaterials to come into close proximity to target interfaces (for example, cell surfaces) in a highly controlled manner, and the hydrogel network minimizes nanomaterial toxicity[58] and related issues. With different loading methods, the nanomaterials were encapsulated throughout the entire tip (Fig. 8a, i, iii, v and vii) or into the tip apex only (Fig. 8a, ii, iv, vi and viii). This control was achieved by filling the mould completely with pre-polymer solution loaded with nanomaterial, or only partial filling with the former and then filling the remaining volume in the tip mould with pre-polymer solution without loaded nanomaterial. Following the tip attachment process as described above, the tips were integrated with the cantilevers to form the functional probes. Using this approach, several nanomaterials could also be sequentially encapsulated within the AFM tip, demonstrating excellent spatial control (Fig. 8b).

To explore the multifunctional capabilities of hydrogel probes, we focused on materials encapsulation as well as several AFM applications: local heating via induction, temperature sensing via quantum dot fluorescence quenching and swelling-mediated materials delivery. In all the experiments described below, the tips were composed of PEG-DA of varying MW and had a pyramidal tip shape. With embedded magnetic Co nanoparticles in PEG-DA MW 250, out-of-plane bending was induced by a permanent magnet (Supplementary Fig. 36) and localized inductive heating was induced by an externally applied alternating current magnetic field[59] (Fig. 8c and Supplementary Fig. 37). The temperature rise was indirectly confirmed by a decrease in the fundamental resonance frequency of the cantilever in water (Fig. 8d,e) and could also be detected in air (Supplementary

Fig. 38). To confirm that the temperature rise is due to the encapsulated Co nanoparticles, the same experiment was also performed using an identical hydrogel probe without embedded Co nanoparticles and there was a negligible signal response. With embedded CdTe quantum dots in PEG-DA MW 700, local temperature sensing[60] was enabled when the tip contacted an indium tin oxide-coated glass substrate under Joule heating (Fig. 8f). Fluorescence emission intensities decreased and the peak wavelength increased as the temperature at the contact point increased in water (Fig. 8g,h) or air (Supplementary Fig. 39) where embedded CdTe quantum dots remained stable due to their diameters larger than mesh sizes of PEG-DA monomers (Supplementary Fig. 40). Supplementary Figure 41 summarizes temperature sensitivities of CdTe quantum dots embedded hydrogel nano-probes. Importantly, local induction heating and *in situ* temperature sensing could be performed simultaneously in water or air using a multifunctional tip loaded with both Co and CdTe nanoparticles (Fig. 8j,k and Supplementary Figs 42 and 43).

The encapsulation and controlled release of Rhodamine B dye from a PEG-DA MW 575 tip was also investigated (Fig. 8l). When the hydrogel tip is in ambient conditions, the encapsulated dye remains stable but diffuses away over time in aqueous conditions. The rate of release depends on properties of the encapsulated molecule along with the polymer mesh size, and is controllable. Dye encapsulation can take place during fabrication through inclusion in the pre-polymer solution, or after fabrication if the tip is immersed in a molecular ink depot[39] (Supplementary Fig. 44a). A single hydrogel probe can be used for multiple rounds of materials loading and delivery (Fig. 8m). Vice versa, a dye-loaded tip can make contact with a polymeric substrate

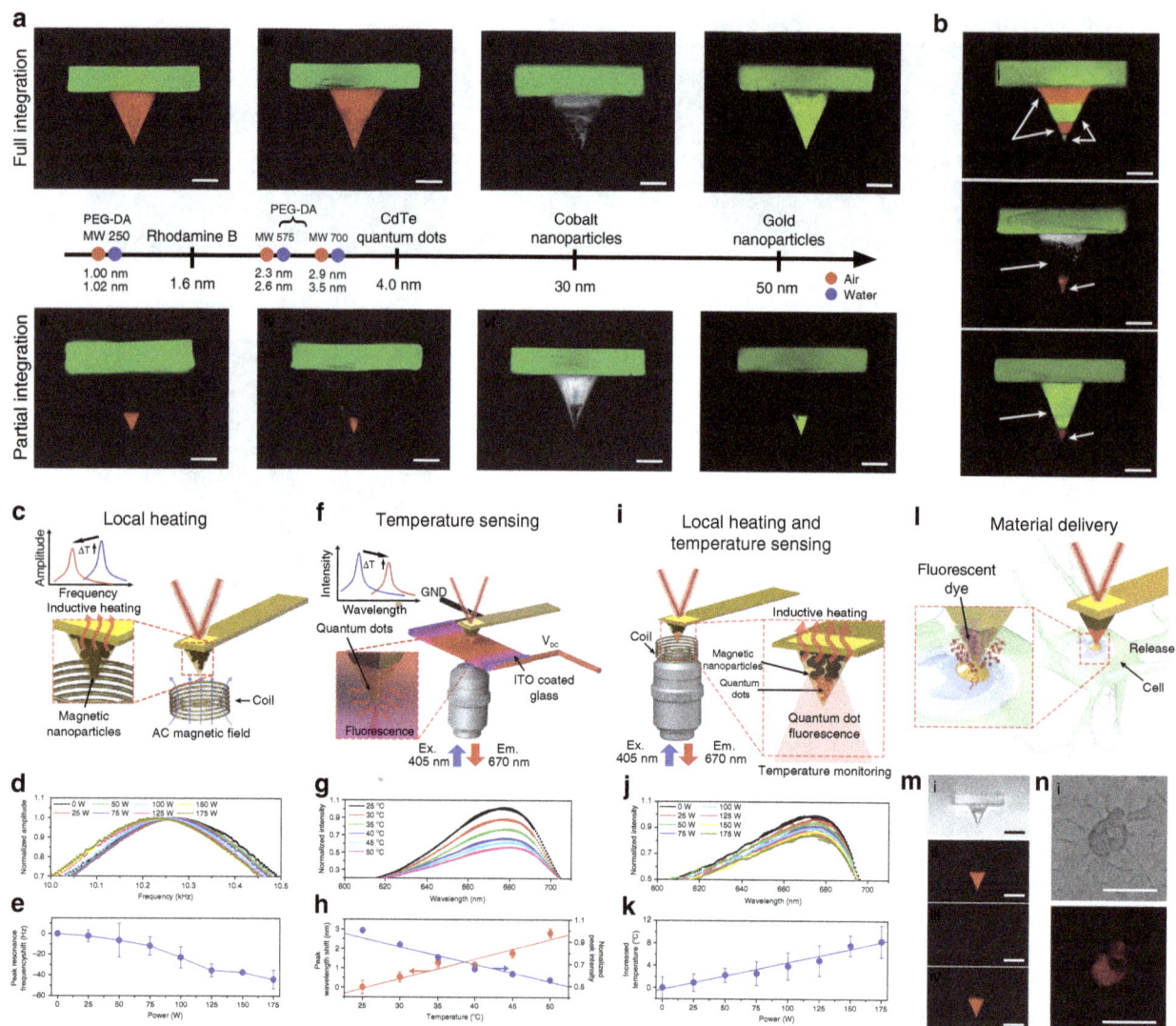

Figure 8 | Programming of multifunctional hydrogel nano-probes. (**a**) Hydrogel nano-probes with integrated functional nanomaterials achieved through materials encapsulation. Depending on the encapsulated nanomaterial, the tip structure is composed of PEG-DA with varying MW as specified: (i, ii) Rhodamine B dye (MW 575), (iii, iv) cadmium telluride (CdTe) quantum dots (MW 700), (v, vi) cobalt (Co) nanoparticles (MW 250) and (vii, viii) gold (Au) nanoparticles (MW 700). The mesh sizes of PEG-DA monomers with different MWs are presented. All scale bars are 10 μm. (**b**) Multifunctional hydrogel nano-probes with sequential loading of FITC and Rhodamine B (i), CdTe quantum dots and Co nanoparticles (ii), and CdTe quantum dots and gold nanoparticles (iii). All scale bars are 10 μm. (**c**) Schematic for local inductive heating. (**d**) Normalized amplitude spectra of a hydrogel nano-probe with a Co nanoparticle-embedded tip under induction heating in water. (**e**) Resonance frequency shift as a function of the input power. (**f**) Schematic for local temperature sensing via fluorescence quenching of quantum dots. (**g**) Fluorescence spectra from a hydrogel nano-probe with a CdTe quantum dot-embedded tip in contact with an ITO-coated glass substrate under Joule heating in water. (**h**) Peak wavelength shift and the normalized peak intensity as a function of the temperature at the tip contact point. (**i**) Schematic for a dual function hydrogel nano-probe for local heating and temperature sensing. (**j**) Fluorescence spectra from a hydrogel nano-probe with sequentially embedded CdTe quantum dots and Co nanoparticles under induction heating. (**k**) Temperature increase as a function of the power applied to the induction coil. (**l**) Schematic for localized materials delivery. (**m**) Bright-field and fluorescence microscopy images of a hydrogel nano-probe before (i) and after (ii) the first loading as well as after the first delivery (iii) and after the second loading (iv), respectively. All scale bars are 20 μm. (**n**) Bright-field optical microscopy image (i) and fluorescence microscopy image (ii) for a breast cancer cell (MCF-7) demonstrating localized delivery of Rhodamine B dye. Scale bars are 10 μm. All error bars represent the s.d. of the mean with $N = 3$ measurements.

and, with sufficient contact force and time, locally deliver the dye molecules. Localized materials delivery was optimally achieved with high dye concentration and short contact time (Supplementary Fig. 44b). The experimental results are consistent with simulations of the delivered Rhodamine B concentration as a function of the contact time and molecular diffusion properties in the system (Supplementary Figs 45 and 46). Extending this concept to cellular applications, localized delivery was demonstrated on the surface of live human breast cancer cells at 37 °C (Fig. 8n). A gentle contact force of 1–2 nN

was applied in conjunction with a relatively blunt tip of 150 nm curvature radius to permit effective materials delivery without tip penetration into the cell membrane. This demonstration validates the potential of localized materials delivery with hydrogel tips which offer size-controlled delivery of loaded materials based on a tunable mesh size through the direct contact area. Such aspects distinguish the proposed swelling-mediated materials delivery from dip-pen nanolithography[61]. Altogether, the application examples highlight the range of possibilities for programmable and multifunctional design of hydrogel AFM probes. Looking forward, it should also be stressed that bulk fabrication of hydrogel probes at the wafer scale is possible. Towards this goal, preliminary data supports that hydrogel nano-probes obtained by our PDMS-based fabrication scheme are free of silicone oil contaminants[62] (Supplementary Fig. 47), can be produced in parallel by a slit type shadow mask (Supplementary Fig. 48), and the corresponding production costs are forecast to be appreciably lower than the production costs for silicon probes (Supplementary Fig. 49).

Discussion

Herein, we have demonstrated the bottom-up fabrication of hydrogel AFM probes to enable a wide range of stiffness tuning and nanoscale tip geometries. Importantly, by encapsulating functional nanomaterials into the hydrogel tip, we also demonstrated the programmable and multifunctional capabilities of hydrogel probes, including local heating, temperature sensing and materials delivery among other possibilities. The developments in this study set new precedents for AFM probe design, and a number of innovations following this idea can be envisioned. In particular, hydrogel cantilevers may facilitate high-speed imaging capabilities due in part to lower elastic moduli and quality factors which make the imaging tip compliant and effectively absorb mechanical noise during scanning. Given the breadth of hydrogel research and functionalization possibilities, there is also strong motivation to develop the hydrogel probe as a platform for various applications and materials encapsulation strategies. A key advantage is the simplicity of bottom-up fabrication with straightforward photopolymerization reactions. The fabrication steps in our scheme are widely used in polymer chemistry and materials science, accessible to many researchers and generalizable to many different polymer choices. Considering that silicon probe fabrication strategies are highly specialized and offer limited customization, the opportunities afforded by hydrogel AFM probes are compelling to further explore. Without a doubt, silicon probes have played a critical role in AFM instrumentation and will continue to do so. Our contention is not that hydrogel probes will replace silicon probes. Both types of probes have important roles and complement one another. Rather, hydrogel probes are a harbinger of the potential that lies ahead with the convergence of soft matter and nanotechnology. Such potential can also be readily extended to fabricating hydrogel cantilevers for nanomechanical sensor devices, including in parallel array configurations for high-throughput applications.

Methods

Hydrogel preparation. Aqueous pre-polymer solutions were prepared using PEG-DA monomers (Sigma Aldrich), with phenylbis(2,4,6-trimethylbenzoyl) phosphine oxide (Sigma Aldrich) as the photoinitiator. The PEG-DA:photoinitiator weight ratio was 98:2. In most experiments, only one type of PEG-DA monomer was used with the average MW varied between 250 and 700 g mol^{-1}. In some experiments, PEG-DA mixtures of the MW 250 and MW 575 monomers were prepared to finely tune the average MW. After preparation, the aqueous pre-polymer solutions were magnetically stirred for 24 h and kept in brown bottles wrapped in aluminium foil to prevent light exposure before experiment.

Hydrogel cantilever fabrication. Replica moudling was employed to fabricate hydrogel cantilevers (Supplementary Fig. 1). First, an SU8 cantilever master mould was fabricated. SU8-2015 and SU8-2005 photoresists (Microchem) were spin-coated onto 4-inch silicon wafers for 30 s (5,000 and 3,000 rpm for 10 and 15 μm target thicknesses, respectively, with 2015 type; and 3,000 rpm for 5 μm target thickness with 2005 type), and then soft-baked for 4 min at 105 °C. The photoresists were next exposed to a 365-nm i-line ultraviolet source (20 mW cm^{-2}; MA6, Karl Suss) for 15 and 10 s for 2015 and 2005 types, respectively. A commercial cantilever handle (PPP-NCHR, Nanosensors) was used to define the cantilever handle mould. A PDMS pre-polymer mixture (weight ratio 10:1; Sylgard 184, Dow Corning) was poured onto the two master moulds and cured on a hot plate at 100 °C for 60 min. The replica from the top cantilever mould was cut to have an inlet for injecting PEG-DA pre-polymer

solution and aligned on top of the other replica from the bottom handle mould. The aqueous pre-polymer solution (98:2 weight ratio of PEG-DA and photoinitiator) was pipetted near the inlet and introduced into the aligned replica mould assembly via capillary action and cured by using an ultraviolet light-emitting diode source (CBT-90-ultraviolet-C31-M400-22, Luminus Devices) for 5 s (ultraviolet dose: 590 mJ cm^{-2}). After the top cantilever replica mould was removed, the cured PEG-DA cantilever was cut to the desired length and separated from the bottom handle replica mould, which was facilitated by an oxygen inhibition layer between the PDMS and PEG-DA hydrogel.

Hydrogel tip integration. A 4-inch single crystal silicon wafer was thermally oxidized to grow a 600-nm-thick silicon dioxide film. The oxide layer was patterned in a reactive ion etcher with a photoresist mask (PR1-1000 A, Futurrex) to leave exposed circular regions (10–20 μm in diameter). This exposed silicon region on the oxide-patterned wafer was etched anisotropically in 30 wt% aqueous potassium hydroxide solution at 70 °C for 15 min to form negative pyramids or etched isotropically in a mixture of hydrofluoric, nitric and acetic acids (HNO$_3$:HF:CH$_3$COOH = 95:2:3 v v^{-1}v^{-1}), at room temperature for 5 min resulting in negative hemispheres. Then, the oxide layer was removed in 49% hydrofluoric acid and the wafer was cleaned with piranha solution, a mixture of sulfuric acid and hydrogen peroxide (H$_2$SO$_4$:H$_2$O$_2$ = 3:1 v v^{-1}, 10 min, 100 °C) and deionized water. 20 ml of the photoresist (PR1-2000 A, Futurrex) mixed with 1 μl of a polystyrene microsphere suspension (20 μm in diameter, 4.55 × 10^7 particles per ml, Polysciences Inc.) was spin-coated (1,000 rpm/15 s-3,000 rpm/ 20 s-1,000 rpm/10 s) onto a 4-inch silicon wafer, and then baked on a hot plate at 120 °C for 1 min to embed the microspheres into a thin layer of photoresist. The wafer with negative pyramids or hemispheres and the wafer with embedded microspheres was silanized under vacuum conditions (−80 kPa) for 30 min by 95% trichlorosilane (tridecafluoro-1,1,2,2-tetrahydrooctyl, Gelest) before PDMS replica moulding. PDMS pre-polymer was poured onto silanized wafers and cured on a hot plate at 95 °C for 40 min. The cured PDMS substrate was peeled off and cut for preparing embedded sphere moulds (Supplementary Fig. 2a). The cured PDMS substrate with positive hemispheres or pyramids was silanized to be used as the master mould. Once the PDMS pre-polymer was poured and cured on the silanized PDMS substrate, the final PDMS substrate was peeled off and cut for preparing hemisphere (H; Supplementary Fig. 2b) and pyramid (P) moulds (Supplementary Fig. 2c). The fabrication process for pyramid moulds is similar to that of hemisphere moulds except for the shape of the etched silicon. The P mould can be deformed by using a custom-made compression jig (Fig. 2a) to control the aspect ratio and radius of curvature of the pyramidal tip. Embedded sphere and H tip moulds are typically used without deformation. Each tip mould placed on top of the ultraviolet light-emitting diode was filled with PEG-DA pre-polymer solution. After a hydrogel cantilever contacted the PEG-DA filled tip mould, ultraviolet exposure was applied to cure the tip. In the case of deformed pyramid moulds, lateral or vertical compression was applied by using piezomotors (Picomotor #8303, Newport) within the strain range of 10–14% and 2–6%, respectively. If the empty mould is compressed first and then filled with hydrogel, the apex of the deformed pyramid mould may not be completely filled. Excess PEG-DA pre-polymer solution on tip mould compression was guided along the cantilever via capillary action over the handle. No special care is necessary when the cured tip integrated on the cantilever is separated from the mould due to the volume contraction of the cured PEG-DA hydrogel and the oxygen inhibition layer.

Cell preparation. Human breast cancer cells (MCF-7, Sigma), human fibroblast cells (MRC-5, Japanese Collection of Research Bioresources Cell Bank; JCRB Cell Bank) and HeLa cells (ATCC CCL-2, ATCC) were used in experiments. Cells were maintained in Eagle's Minimum Essential Medium (ATCC 30-2003, ATCC) supplemented with 10% foetal bovine serum (SV30160.03IR, Hyclone, Thermo Fisher Scientific) and 1% penicillin/streptomycin (10378016, Invitrogen Life Technique) at 37 °C in a humidified atmosphere containing 5% CO$_2$. Cells were harvested for subculture post trypsinization, washed and re-suspended at 1 × 10^5 ml^{-1} concentration in a normal growth medium. For AFM experiments, the cell concentration was adjusted to ~1 × 10^3 ml^{-1}, and then 1 ml of the cell suspension was seeded on a 35-mm culture dish. Cells were washed by PBS for 5 min and fixed by 4% paraformaldehyde before AFM experiment.

Oxygen plasma ashing of worn hydrogel tip. Worn hydrogel nano-probes with various tip radii were oxygen plasma-ashed by reactive ion etching (RIE 80 plus, Oxford Instrument) with a forward power of 43 W, oxygen flow rate of 6 s.c.c.m., and chamber pressure of 30 mTorr. With these processing parameters, an average etch rate of ~10 nm min^{-1} was obtained. Considering the tip radius of a worn hydrogel nano-probe and the average etch rate, the processing time was determined for each hydrogel nano-probe.

References

1. Bottomley, L. A. Scanning probe microscopy. *Anal. Chem.* **70**, 425–476 (1998).
2. Binnig, G. & Rohrer, H. Scanning tunneling microscopy. *IBM J. Res. Dev.* **44**, 279–293 (2000).

3. Binnig, G., Quate, C. F. & Gerber, C. Atomic force microscope. *Phys. Rev. Lett.* **56**, 930–933 (1986).

4. Bull, M. S., Sullan, R. M. A., Li, H. & Perkins, T. T. Improved single molecule force spectroscopy using micromachined cantilevers. *ACS Nano* **8**, 4984–4995 (2014).

5. Li, M., Tang, H. X. & Roukes, M. L. Ultra-sensitive NEMS-based cantilevers for sensing, scanned probe and very high-frequency applications. *Nat. Nanotechnol.* **2**, 114–120 (2007).

6. Wilson, N. R. & Macpherson, J. V. Carbon nanotube tips for atomic force microscopy. *Nat. Nanotechnol.* **4**, 483–491 (2009).

7. Butt, H.-J., Cappella, B. & Kappl, M. Force measurements with the atomic force microscope: Technique, interpretation and applications. *Surf. Sci. Rep.* **59**, 1–152 (2005).

8. Dai, H., Hafner, J. H., Rinzler, A. G., Colbert, D. T. & Smalley, R. E. Nanotubes as nanoprobes in scanning probe microscopy. *Nature* **384**, 147–150 (1996).

9. Kim, J. M. & Muramatsu, H. Two-photon photopolymerized tips for adhesion-free scanning-probe microscopy. *Nano Lett.* **5**, 309–314 (2005).

10. Genolet, G. *et al.* Soft, entirely photoplastic probes for scanning force microscopy. *Rev. Sci. Instrum.* **70**, 2398–2401 (1999).

11. Dufrêne, Y. F., Martínez-Martín, D., Medalsy, I., Alsteens, D. & Müller, D. J. Multiparametric imaging of biological systems by force-distance curve-based AFM. *Nat. Methods* **10**, 847–854 (2013).

12. Müller, D. J. & Dufrêne, Y. F. Atomic force microscopy as a multifunctional molecular toolbox in nanobiotechnology. *Nat. Nanotechnol.* **3**, 261–269 (2008).

13. Hafner, J., Cheung, C.-L., Woolley, A. & Lieber, C. Structural and functional imaging with carbon nanotube AFM probes. *Prog. Biophys. Mol. Biol.* **77**, 73–110 (2001).

14. Sun, J.-Y. *et al.* Highly stretchable and tough hydrogels. *Nature* **489**, 133–136 (2012).

15. Imran, A. B. *et al.* Extremely stretchable thermosensitive hydrogels by introducing slide-ring polyrotaxane cross-linkers and ionic groups into the polymer network. *Nat. Commun.* **5**, 5124 (2014).

16. Gou, M. *et al.* Bio-inspired detoxification using 3D-printed hydrogel nanocomposites. *Nat. Commun.* **5**, 3774 (2014).

17. Lee, H., Kim, J., Kim, H., Kim, J. & Kwon, S. Colour-barcoded magnetic microparticles for multiplexed bioassays. *Nat. Mater.* **9**, 745–749 (2010).

18. Purcell, B. P. *et al.* Injectable and bioresponsive hydrogels for on-demand matrix metalloproteinase inhibition. *Nat. Mater.* **13**, 653–661 (2014).

19. Lee, W., Cho, N.-J., Xiong, A., Glenn, J. S. & Frank, C. W. Hydrophobic nanoparticles improve permeability of cell-encapsulating poly (ethylene glycol) hydrogels while maintaining patternability. *Proc. Natl Acad. Sci. USA* **107**, 20709–20714 (2010).

20. Xu, F. *et al.* Release of magnetic nanoparticles from cell-encapsulating biodegradable nanobiomaterials. *ACS Nano* **6**, 6640–6649 (2012).

21. Ivanovska, I. L., Miranda, R., Carrascosa, J. L., Wuite, G. J. & Schmidt, C. F. Discrete fracture patterns of virus shells reveal mechanical building blocks. *Proc. Natl Acad. Sci. USA* **108**, 12611–12616 (2011).

22. Novak, P. *et al.* Nanoscale live-cell imaging using hopping probe ion conductance microscopy. *Nat. Methods* **6**, 279–281 (2009).

23. Bourlon, B., Wong, J., Mikó, C., Forró, L. & Bockrath, M. A nanoscale probe for fluidic and ionic transport. *Nat. Nanotechnol.* **2**, 104–107 (2007).

24. Carrascosa, L. G., Moreno, M., Álvarez, M. & Lechuga, L. M. Nanomechanical biosensors: a new sensing tool. *TrAc Trends Anal. Chem.* **25**, 196–206 (2006).

25. Boisen, A., Dohn, S., Keller, S. S., Schmid, S. & Tenje, M. Cantilever-like micromechanical sensors. *Rep. Prog. Phys.* **74**, 036101 (2011).

26. Calleja, M. *et al.* Highly sensitive polymer-based cantilever-sensors for DNA detection. *Ultramicroscopy* **105**, 215–222 (2005).

27. Calleja, M., Tamayo, J., Nordström, M. & Boisen, A. Low-noise polymeric nanomechanical biosensors. *Appl. Phys. Lett.* **88**, 113901 (2006).

28. Nordström, M. *et al.* SU-8 cantilevers for bio/chemical sensing; fabrication, characterisation and development of novel read-out methods. *Sensors* **8**, 1595–1612 (2008).

29. Gammelgaard, L., Rasmussen, P. A., Calleja, M., Vettiger, P. & Boisen, A. Microfabricated photoplastic cantilever with integrated photoplastic/carbon based piezoresistive strain sensor. *Appl. Phys. Lett.* **88**, 113508 (2006).

30. Zhang, Y., Kim, H. H., Kwon, B. H. & Go, J. S. Polymeric cantilever sensors functionalized with multiamine supramolecular hydrogel. *Sens. Actuator B Chem.* **178**, 47–52 (2013).

31. Bashir, R., Hilt, J., Elibol, O., Gupta, A. & Peppas, N. Micromechanical cantilever as an ultrasensitive pH microsensor. *Appl. Phys. Lett.* **81**, 3091–3093 (2002).

32. Wichterle, O. & Lim, D. Hydrophilic gels for biological use. *Nature* **185**, 117–118 (1960).

33. West, J. L. & Hubbell, J. A. Photopolymerized hydrogel materials for drug delivery applications. *React. Polym.* **25**, 139–147 (1995).

34. Albrecht, T., Akamine, S., Carver, T. & Quate, C. Microfabrication of cantilever styli for the atomic force microscope. *J. Vac. Sci. Technol. A* **8**, 3386–3396 (1990).

35. Temenoff, J. S., Athanasiou, K. A., Lebaron, R. G. & Mikos, A. G. Effect of poly (ethylene glycol) molecular weight on tensile and swelling properties of oligo (poly (ethylene glycol) fumarate) hydrogels for cartilage tissue engineering. *J. Biomed. Mater. Res.* **59**, 429–437 (2002).

36. Davis, T. P. & Huglin, M. B. Studies on copolymeric hydrogels of N-vinyl-2-pyrrolidone with 2-hydroxyethyl methacrylate. *Macromolecules* **22**, 2824–2829 (1989).

37. Cvetkovic, C. *et al.* Three-dimensionally printed biological machines powered by skeletal muscle. *Proc. Natl Acad. Sci. USA* **111**, 10125–10130 (2014).

38. Bae, M., Gemeinhart, R. A., Divan, R., Suthar, K. J. & Mancini, D. C. Fabrication of poly (ethylene glycol) hydrogel structures for pharmaceutical applications using electron beam and optical lithography. *J. Vac. Sci. Technol. B* **28**, C6P24–C26P29 (2010).

39. Anbergen, U. & Oppermann, W. Elasticity and swelling behaviour of chemically crosslinked cellulose ethers in aqueous systems. *Polymer (Guildf).* **31**, 1854–1858 (1990).

40. Butler, M. F., Clark, A. H. & Adams, S. Swelling and mechanical properties of biopolymer hydrogels containing chitosan and bovine serum albumin. *Biomacromolecules* **7**, 2961–2970 (2006).

41. Hopcroft, M., Nix, W. D. & Kenny, T. W. What is the Young's Modulus of Silicon? *J. Microelectromech. Syst.* **19**, 229–238 (2010).

42. Plodinec, M. *et al.* The nanomechanical signature of breast cancer. *Nat. Nanotechnol.* **7**, 757–765 (2012).

43. Ruz, J., Tamayo, J., Pini, V., Kosaka, P. & Calleja, M. Physics of nanomechanical spectrometry of viruses. *Sci. Rep.* **4**, 6051 (2014).

44. Shekhawat, G. S. & Dravid, V. P. Nanomechanical sensors: bent on detecting cancer. *Nat. Nanotechnol.* **8**, 77–78 (2013).

45. Arlett, J., Myers, E. & Roukes, M. Comparative advantages of mechanical biosensors. *Nat. Nanotechnol.* **6**, 203–215 (2011).

46. Calleja, M., Kosaka, P. M., San Paulo, Á. & Tamayo, J. Challenges for nanomechanical sensors in biological detection. *Nanoscale* **4**, 4925–4938 (2012).

47. Sağlam, D., Venema, P., de Vries, R. & van der Linden, E. The influence of pH and ionic strength on the swelling of dense protein particles. *Soft Matter* **9**, 4598–4606 (2013).

48. Adams, J. D. *et al.* Harnessing the damping properties of materials for high-speed atomic force microscopy. *Nat. Nanotechnol* **11**, 147–151 (2015).

49. Brown, B. P., Picco, L., Miles, M. J. & Faul, C. F. Opportunities in high-speed atomic force microscopy. *Small* **9**, 3201–3211 (2013).

50. Vahdat, V. *et al.* Atomic-scale wear of amorphous hydrogenated carbon during intermittent contact: a combined study using experiment, simulation, and theory. *ACS Nano* **8**, 7027–7040 (2014).

51. Vahdat, V., Grierson, D. S., Turner, K. T. & Carpick, R. W. Mechanics of interaction and atomic-scale wear of amplitude modulation atomic force microscopy probes. *ACS Nano* **7**, 3221–3235 (2013).

52. Mo, Y., Turner, K. T. & Szlufarska, I. Friction laws at the nanoscale. *Nature* **457**, 1116–1119 (2009).

53. Vargonen, M., Yang, Y., Huang, L. & Shi, Y. Molecular simulation of tip wear in a single asperity sliding contact. *Wear* **307**, 150–154 (2013).

54. Jung, B. J. *et al.* Fabrication of 15 nm curvature radius polymer tip probe on an optical fiber via two-photon polymerization and O2 plasma ashing. *Curr. Appl. Phys.* **13**, 2064–2069 (2013).

55. Delgado, M., Spanka, C., Kerwin, L. D., Wentworth, P. & Janda, K. D. A tunable hydrogel for encapsulation and controlled release of bioactive proteins. *Biomacromolecules* **3**, 262–271 (2002).

56. Ashley, G. W., Henise, J., Reid, R. & Santi, D. V. Hydrogel drug delivery system with predictable and tunable drug release and degradation rates. *Proc. Natl Acad. Sci. USA* **110**, 2318–2323 (2013).

57. Wu, Y., Joseph, S. & Aluru, N. Effect of cross-linking on the diffusion of water, ions, and small molecules in hydrogels. *J. Phys. Chem. B* **113**, 3512–3520 (2009).

58. Major, J., Treharne, R., Phillips, L. & Durose, K. A low-cost non-toxic post-growth activation step for CdTe solar cells. *Nature* **511**, 334–337 (2014).

59. Mohr, R. *et al.* Initiation of shape-memory effect by inductive heating of magnetic nanoparticles in thermoplastic polymers. *Proc. Natl Acad. Sci. USA* **103**, 3540–3545 (2006).

60. Liang, R. *et al.* A temperature sensor based on CdTe quantum dots–layered double hydroxide ultrathin films via layer-by-layer assembly. *Chem. Commun.* **49**, 969–971 (2013).

61. Piner, R. D., Zhu, J., Xu, F., Hong, S. & Mirkin, C. A. "Dip-pen" nanolithography. *Science* **283**, 661–663 (1999).

62. Lo, Y.-S. *et al.* Organic and inorganic contamination on commercial AFM cantilevers. *Langmuir.* **15**, 6522–6526 (1999).

Acknowledgements

This research was supported by the Commercializations Promotion Agency for R&D Outcomes (COMPA) (2015K000127) and the National Research Foundation of Korea (NRF) funded by the Korea government (MSIP; NRF-2013R1A1A1076080 and NRF-2014R1A2A1A11053283), the National Research Foundation of Singapore

(NRF-NRFF2011-01 and NRF-CRP10-2012-07) and the National Medical Research Council (NMRC/CBRG/0005/2012).

Author contributions

N.-J.C. and J.L. developed the idea; J.S.L. and J.S. fabricated and characterized the cantilever probes; J.S.L., J.S., S.O.K. and S.K. performed the experiments; J.A.J., N.-J.C., and J.L. were involved in study design and data interpretation; W.L. and D.K. performed the numerical analysis; J.A.J., N.-J.C., and J.L. wrote the manuscript. All authors discussed the results and commented on the manuscript.

Additional information

Competing financial interests: The authors declare no competing financial interests.

26

Flexible single-layer ionic organic–inorganic frameworks towards precise nano-size separation

Liang Yue[1], Shan Wang[1], Ding Zhou[1], Hao Zhang[1], Bao Li[1] & Lixin Wu[1,2]

Consecutive two-dimensional frameworks comprised of molecular or cluster building blocks in large area represent ideal candidates for membranes sieving molecules and nano-objects, but challenges still remain in methodology and practical preparation. Here we exploit a new strategy to build soft single-layer ionic organic–inorganic frameworks via electrostatic interaction without preferential binding direction in water. Upon consideration of steric effect and additional interaction, polyanionic clusters as connection nodes and cationic pseudorotaxanes acting as bridging monomers connect with each other to form a single-layer ionic self-assembled framework with 1.4 nm layer thickness. Such soft supramolecular polymer frameworks possess uniform and adjustable ortho-tetragonal nanoporous structure in pore size of 3.4–4.1 nm and exhibit greatly convenient solution processability. The stable membranes maintaining uniform porous structure demonstrate precisely size-selective separation of semiconductor quantum dots within 0.1 nm of accuracy and may hold promise for practical applications in selective transport, molecular separation and dialysis systems.

[1] State Key Laboratory of Supramolecular Structure and Materials, College of Chemistry, Jilin University, Changchun 130012, PR China. [2] Key Laboratory of Natural Resources of Changbai Mountain and Functional Molecules, Yanbian University, Yanji 133002, PR China. Correspondence and requests for materials should be addressed to L.W. (email: wulx@jlu.edu.cn) or to H.Z. (email: hao_zhang@jlu.edu.cn).

Ultrathin porous membranes[1] have received increasing attention over recent years because of their superiorities, for example, permeation[2,3], selective transport[4] and molecular and nano-object sieving[5] in the fields of chemical engineering[6], biomedicine[7], environment and energy[8] and materials science[9]. Single-layer two-dimensional (2D) polymers[10–13] with identical porosity and good processability represent a type of ideal porous membranes, but the big challenges still remain in structural design and synthetic methodology. Currently, in order to integrate discrete subunits into 2D connection, building block monomers must be specially designed with rigid planar structure[10] or capable of preorganization into 2D mesostructure assisted by crystallization[11–13], solid surface[14,15] or interface[16] support. These methods often bring about inevitable limitations, such as monomer shape and crosslinking reaction, structural damage in single-layer exfoliation and transferability. Therefore, a facile and general method to construct flexible single-layer 2D polymer in solution is still highly desired. As an applicable choice, supramolecular self-assembly[17–22] based on intermolecular interactions has been proved to be a promising approach to integrate bridging monomers with connection nodes for 2D supramolecular polymer. Through coordination interaction, Kim's group prepared 2D polyrotaxane network with large cavities and channels, demonstrating a viable approach for modular porous solids[23,24]. Furthermore, Li and his coworkers developed single-layer 2D honeycomb supramolecular organic framework connected by inclusion interaction[25]. However, owing to the inevitably strong interlayer and weak in-layer interactions, a new synthetic methodology for single-layer soft frameworks in all desired area via a facile and rapid procedure is highly desired.

Polyoxometalates[26–29] (POMs) are a kind of discrete metal-oxide clusters with precise chemical composition and structure. The rich architectures, uniform morphologies and multiple negative charges make POMs outstanding candidates for self-assembly[30–40]. When the POM cluster is considered as connection node and the electrostatic interaction between POM and cationic linker is applied as binding force, the construction of ionic organic–inorganic frameworks (IOIFs) in supramolecular network can be expected in principle. Based on the rational design of building block and ionic connection, the regular porosity can be predicted in supramolecular polymers. In addition, the non-direction preference of ionic bonding can bring flexibility for the self-assembled framework. Thus, the integration of identical porosity and convenient processability can be predicted to exist in the IOIFs. The tough barrier from the non-saturation and non-preferential direction of ionic bond needs to be overcome. Herein, we report a new strategy to construct free-standing single-layer 2D supramolecular polymer frameworks by means of synergistic ionic self-assembly of cationic α-cyclodextrin (CD)-based pseudorotaxane bridging sticks with the anionic $PW_{11}VO_{40}^{4-}$ (PWV^{4-}) cluster nodes in water. Because negative charges of the POM are delocalized and moveable[41–43], cationic groups can spontaneously adjust their locations around the POM depending on steric effect and additional interaction. Therefore, the introduced CD provides steric guidance and lateral hydrogen bonding for the 2D arrangement of cationic sticks around polyanionic node. The cluster with four negative charges acts as both the crosslinker to bind with cations and the capper to lock pseudorotaxane sticks. Given the large-scale single-layer nanoporous sheets and convenient solution processability, ultrafiltration membranes (pore size 3.4–4.1 nm) are prepared through simple filtration under slightly reduced pressure. Interestingly, the prepared nanoporous membrane not only maintains the regular porosity, but also stands without supporting. More significantly, the

membrane can realize the pinpoint size-selective separation of semiconductor quantum dots (QDs) in accuracy of 0.1 nm, just encountering a quick filtration under reduced pressure.

Results

Construction of bridging stick with bulky cationic heads. The bolaform cationic molecules (Azo-Tr/TeEG·2Br) comprised of two cationic azobenzene (Azo) groups connecting with a tri-/tetra-ethylene glycol (Tr/TeEG) spacer were synthesized and their chemical structures were characterized by ^{1}H NMR and electrospray ionization mass spectra (ESI-MS) (Supplementary Figs 1–10). The reasons we designed such a stick molecule are that the two cationic heads on both sides can perform as a stick to bind with polyanions (Fig. 1), and the Azo group can recognize with CD to block the possible aggregation and control the steric adaptation. The spacer is used to adjust the pore size and flexibility of as-prepared self-assemblies. We firstly examined host–guest interaction between Azo-TrEG·2Br and CD by ^{1}H NMR spectra in D_2O (Fig. 2, Supplementary Fig. 11 and Supplementary Table 1). Upon addition of host CDs, all proton signals of guest bolaform Azo-TrEG·2Br shift downfield. The largest shift (0.82 p.p.m.) for proton H(f) in Azo group is observed, while the signals of H(j) and H(k) belonging to spacer chain shift downfield (0.14 and 0.07 p.p.m.) slightly, indicating that only Azo groups are included in CDs. The plot of inclusion amount calculated by relative integral area of H(d) versus the molar ratio of CD to Azo-TrEG·2Br suggests that over 96% Azo-TrEG·2Br form pseudorotaxane in 2:1 stoichiometry (Azo-TrEG@CD·2Br). The 2D NOESY spectrum of CD and Azo-TrEG·2Br mixture (molar ratio 2:1) in D_2O (Supplementary Fig. 12a) displays four NOE correlation signals between H(3) and H(5) in CD cavity and H(e–g) in the Azo group. Specifically, proton H(3) correlates with H(f) and H(g) at (8.37, 3.75 p.p.m.) and (7.78, 3.75 p.p.m.), while proton H(5) associates with H(e) and H(f) at (7.84, 3.58 p.p.m.) and (8.37, 3.58 p.p.m.), further identifying that Azo group is included in CD in a slightly acclivitous orientation. Moreover, the ESI-MS of CD and Azo-TrEG·2Br mixture clearly reveals formation of Azo-TrEG@CD·2Br as a [2+1] inclusion complex (Supplementary Fig. 13).

End-capping and crosslinking reaction by ionic bond. Then, upon the addition of 0.5 equivalent of PWV^{4-} into the

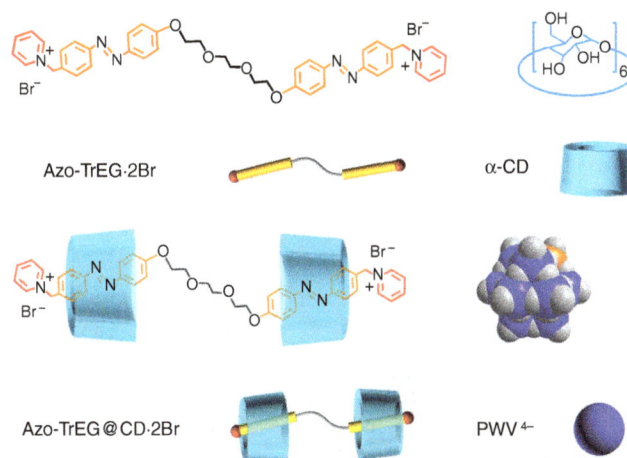

Figure 1 | Schematic representation chemical structures and representations. Cationic bolaform molecule bearing two guest groups (Azo-TrEG·2Br), non-ionic host molecule (α-CD), pseudorotaxane unit (Azo-TrEG@CD·2Br) and POM polyanionic cluster (PWV[4−]).

Azo-TrEG@CD · 2Br solution, the cationic heads of the bridging stick linker can be tethered by polyanionic cluster capper via electrostatic interaction, yielding an organic–inorganic supramolecular rotaxane through electrostatic interaction. By referring to the chemical shifts of isolated Azo-TrEG@CD · 2Br, all proton signals of pyridinium head group broaden and shift downfield, for instance of 0.05 p.p.m. movement for H(c) shown in Fig. 2a, implying the braced electrostatic interaction. The X-ray photoelectron spectra (Supplementary Fig. 14) show that the counterions of Azo-TrEG@CD · 2Br have been substituted by PWV^{4-} cluster in the final complex after separation from solution. Combined these results with elemental analysis (Supplementary Table 2), a full charge neutralization between Azo-TrEG@CD^{2+} and PWV^{4-} is affirmed, thus yielding a tightly bound electrostatic complex with Azo-TrEG@CD^{2+} units. ^{31}P NMR spectra of PWV^{4-} (Supplementary Fig. 15) show no change in chemical shift after the substitution of counterions, implying the stable cluster's structure in [Azo-TrEG@CD][PWV]. Meanwhile, the proton signals of Azo groups still closing to their included state support the inclusion interaction maintaining in the terminal locked polyrotaxane. However, compared with Azo-TrEG@CD · 2Br alone, only three NOE correlation signals appear in 2D NOESY NMR spectrum of

[Azo-TrEG@CD][PWV] (Supplementary Fig. 12b). The proton correlations of H(e) with H(5), and H(g) with H(3) retain constant. Instead of correlations of H(f) with H(3) and H(5), a new signal at (6.07, 3.62 p.p.m.) that is assigned to the correlation of H(d) and H(5) appears. These definite correlations reveal that the electrostatic combination to the cationic head results in a large tilting of Azo group in CD cavity, which compresses the outside methylene group next to pyridyl head deeply into CD cavity. The tilting angle corresponding to the normal of CD cavity is estimated to be around 23.4° (Supplementary Fig. 16). Interestingly, from proton chemical shifts of CD molecule after mixing with PWV^{4-} (Supplementary Fig. 17), we observe the interaction between the narrow ring of CD and PWV^{4-} existing in aqueous solution. Therefore, the acclivitous recognition model is conducive to decrease the distance and strengthen the interaction between the narrow ring of CD and cluster. The crystal structure analysis on the interaction between CD and a similar POM to the present one in the recent reported result highly supports this assignment[44].

We further used circular dichroism spectroscopy to characterize inclusion interaction in POM-locked polyrotaxane. The induced spectrum of isolate Azo-TrEG@CD · 2Br (Fig. 3a) shows a positive Cotton effect corresponding to $\pi - \pi^{\star}$ transition of Azo group at 360 nm and a negative Cotton effect ascribing to $n - \pi^{\star}$ transition of the same group at 440 nm, indicating the strong inclusion interaction. The spectral feature indicates Azo group in approximately parallels to the normal of CD cavity[45,46]. Upon addition of PWV^{4-} (0.1–0.7 equivalent), both intensities of positive and negative Cotton effects gradually decrease. As is

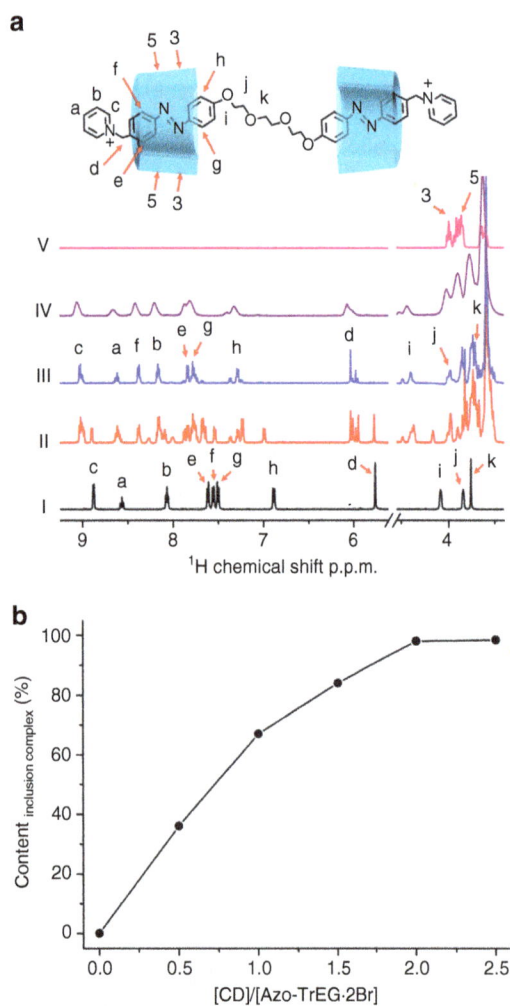

Figure 2 | ^{1}H NMR spectra. (a) Chemical shift of Azo-TrEG · 2Br upon addition of 0 (I), 1.0 (II), 2.0 (III) eq. CDs; Azo-TrEG@CD · 2Br upon addition of 0.5 equivalent PWV^{4-} (IV); and isolated CD (V) (D$_2$O, 25 °C), **(b)** plot of relative content of inclusion complex calculated from integral area of H(d) versus the molar ratio of CD to Azo-TrEG · 2Br.

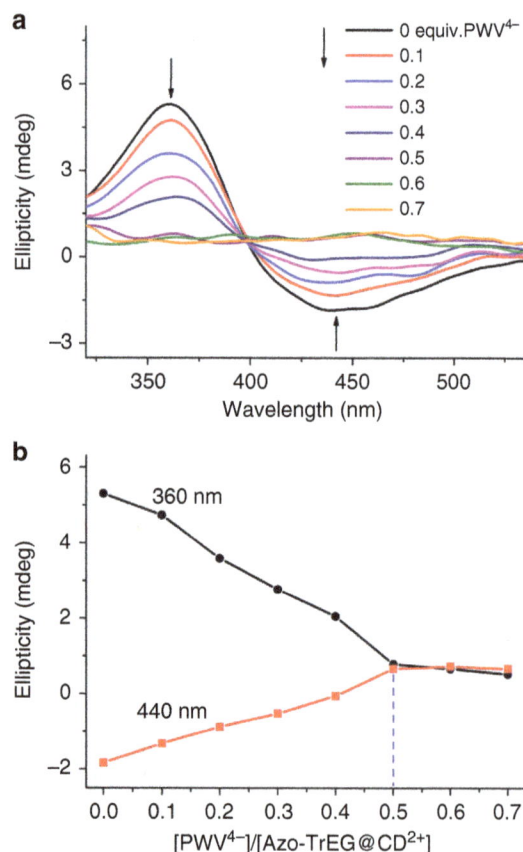

Figure 3 | Circular dichroism spectra (CDS). (a) Azo-TrEG@CD · 2Br aqueous solution upon addition of PWV^{4-} (from 0 to 0.7 equivalents), and **(b)** intensity plots of Cotton signals at 360 and 440 nm versus the molar ratio of PWV^{4-} to Azo-TrEG@CD^{2+}.

known, the deviation of transition moment for guest molecule from normal axis of CD cavity leads to the weakness of induced chirality until to silence[46,47]. The acclivitous state of Azo group in CD cavity further certifies its source from electrostatic interaction, which is in good accordance with the preponderant mode inferred from 2D NOESY spectrum. Thus, the model can be used to trace the binding ratio between PWV^{4-} cluster and Azo-TrEG@CD^{2+} by simply monitoring induced Cotton signals of Azo groups. The intensity plots of Cotton effects at 360 and 440 nm versus the molar ratio of PWV^{4-} to Azo-TrEG@CD^{2+} illustrate a definite 1:2 stoichiometry at the turning point when the charges of PWV^{4-} are fully neutralized by Azo-TrEG@CD^{2+} (Fig. 3b). Considering the multiple binding sites in above building components, a supramolecular ionic framework becomes apparently dominant in limited possibilities of connection geometry.

Synergistic ionic self-assembly for 2D framework. In an attempt to assemble 2D frameworks by using the present system via ionic interaction, steric effect and other additional interaction, such as hydrogen bond, must be combined. In other case of POM-based ionic self-assembly system, the negative charges of POMs are widely accepted to be delocalized though there was no a precisely theoretical calculation due to the difficulty for solution system[41]. As an indirect example, the DFT calculation indicates that the electrons move towards contacting site when a POM adsorbs on graphene surface[42]. The molecular dynamics simulation for POMs in aqueous solution revealed that the paired ions moved freely within the region bounded by Bjerrum's length[43]. Thus, the delocalized negative charges on a POM catch its counterions anywhere around the cluster surface. But the distribution of negative charges corresponding to binding sites is affected or dominated by steric effect and additional interaction among those trapped cationic components dynamically. As an example, the lateral van der Waals interaction was found to trigger the cationic groups around each POM from a mean distribution to a separated state of hydrophilic and hydrophobic components. This means that organic cations accumulate on opposite sides or even equatorial plane of one POM, depending on the packing fashion propelled by the interfacial energy[48–52]. Therefore, under driven by proper addition interaction and steric effect, the preferential distribution of negative charges on POM, directed by counterions, becomes possible.

In light of this, PWV^{4-} cluster performing the node in construction of 2D IOIF upon simply modulating binding angle between cationic bridging sticks surrounding it becomes rational. Based on similar understanding, the non-centrosymmetric PWV^{4-} cluster does not affect the movement of negative charges since those self-assembled structures of POM complexes in solutions were proved to be independent on the symmetry of POM clusters[53]. As for detailed estimation on steric adaptation between CD-shielded cations and PWV^{4-} cluster, firstly, the POM has enough space to accommodate four CD-shielded cations packing in one plane, which is the prerequisite for 2D framework (complemented structural analysis shown in Supplementary Figs 18,19). Secondly, the calculated distance between neighbouring CD shields in a tetragonal fashion is ~0.23 nm (Supplementary Fig. 19b), in perfect agreement with the reported distance of hydrogen bond[54,55], indicating the favourable lateral interaction between the narrow rings of neighbouring CD-shielded cations. Of course, besides the planar square framework, the steric tetrahedron binding style is also possible theoretically. However, suppose the tetrahedron structure was preferential, the distance between CD shields would increase to ~0.58 nm (Supplementary Fig. 20), too far for the

formation of hydrogen bonding between neighbouring CD shields. So, the efficient hydrogen bonds of CD-shielded cations around each POM assists for the 2D framework.

Characterizations for 2D IOIF structure. Neither Azo-TrEG@CD·2Br nor PWV^{4-} alone shows Tyndall phenomenon in aqueous solution, while a distinct light-scattering effect emerges in their 2:1 mixture solution (Supplementary Fig. 21), implying the generation of ionic self-assembly. We further characterized the structure by using microscopic techniques. Atomic force microscopic (AFM) image demonstrates that the supramolecular architectures appear as very thin sheets with irregular shape at the beginning (Supplementary Fig. 22). After a while, micrometre-scale sheets can be observed (Fig. 4) but the thickness maintains constant, indicating stepwise 2D growing of the architectures. Based on the thickness analysis in a large area, the average height is at 1.43 – 1.48 nm, in perfect agreement with the interlayer spacing (1.49 nm) measured by powder XRD of freeze-dried sample (Supplementary Fig. 23a). Considering the accord of this value with the diameter (1.46 nm) of CD ring[56], a single-layer 2D assembling structure can be rationally inferred.

Transmission electron microscopic (TEM) measurements further demonstrate the micrometre-scale sheets of singly layered ionic supramolecular assembly (Fig. 5a). The wrinkled and folded edges (Supplementary Fig. 24) illustrate the high flexibility and free-standing feature of the single-layer assembly. The energy-dispersive X-ray spectral analysis of the observed sheets points out the presence of tungsten element (Supplementary Fig. 25), confirming the inorganic clusters existing in the sheets.

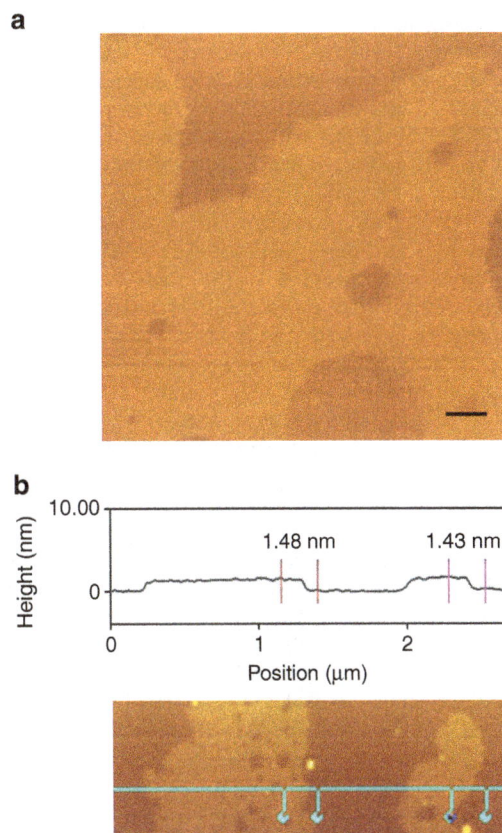

Figure 4 | AFM image. (a) Tapping mode image and (b) height profile analysis of [Azo-TrEG@CD][PWV] architecture spreading on mica. Scale bar, 500 nm (a).

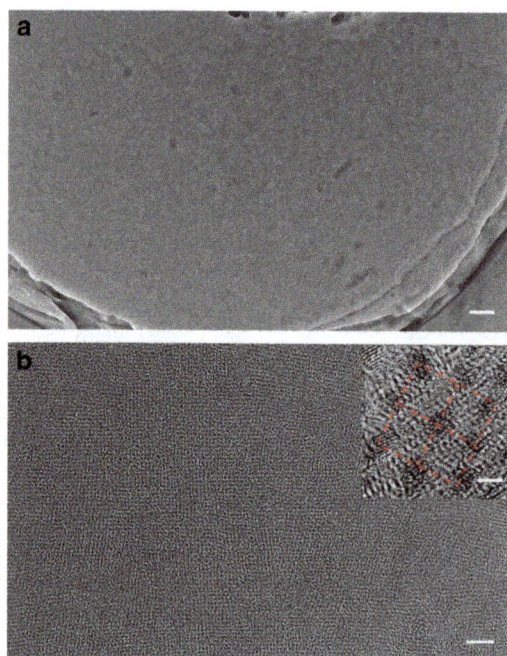

Figure 5 | TEM images. Single-layer self-assembly of [Azo-TrEG@CD] [PWV] IOIF in: (**a**) wide and (**b**) amplified scale, while the insert presents a high resolution TEM image taken from **b**. Scale bars, 100 nm (**a**) 20 nm (**b**) 2 nm (inset in **b**).

Because of the electron density contrast between organic and inorganic components, in the case without addition of any staining agents, the inorganic clusters can be well discerned as dark spots locating at the nodal points. The formed long-range uniform orthogonal mesh structure with edge length ~ 3.7 nm in the single-layer sheet is observed definitely (Fig. 5b). This fantastic framework structure is further identified by XRD measurement (Supplementary Fig. 23b and Supplementary Table 3). The diffraction pattern of sample film prepared by filtration of IOIF self-assembly solution exhibits nine diffraction peaks, which can be perfectly indexed into an in-layer lamellar structure with 3.7 nm of spacing, in perfect accord with the value estimated from high resolution TEM image. This value is also in good agreement with the ideal length calculated from planar square framework (3.8–4.6 nm) after considering the utmost shrinkage and stretching of TrEG spacer inserted in bolaform cation (Supplementary Fig. 26). These results strongly point out that the square framework exists in the single-layer 2D architecture comprised of Azo-TrEG@CD^{2+} bridging stick and PWV^{4-} node via electrostatic interaction, as illustrated in Fig. 6. It is worth pointing out that the size heterogeneity of PWV^{4-} and Azo-TrEG@CD^{2+} and the flexibility of IOIF can weaken the interlayer interactions, thus leading to dominant single-layer frameworks in water. Besides, the hydrophilicity of the building blocks facilitates the stable dispersion of single-layer IOIF in water.

To demonstrate the role of α-CD in synergistic ionic self-assembly for 2D framework structure, we repeat the preparation of single layered self-assembly by using β-CD to replace α-CD. As predicted, we do not observe any expected layer assemblies, but instead, nubbly and distorted bulk aggregations are found (Supplementary Fig. 27). This difference supports above analysis that the size matching and the lateral interaction between neighbouring CD molecules drives the tetragonal distribution of α-CD-shielded cations around POM. Because the larger size of β-CD, the used POM cannot provide enough space for four β-CD-shielded cations packing in the same planar

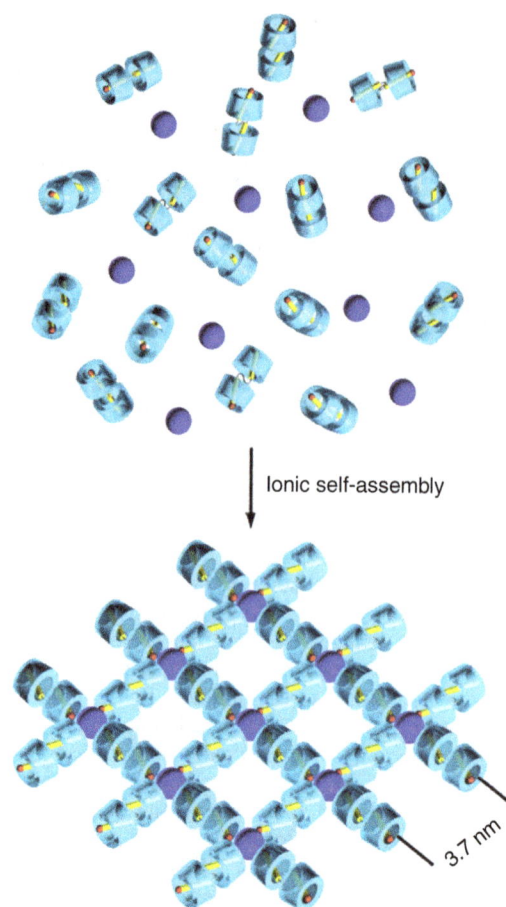

Figure 6 | Schematic representation of mechanism. Possible process for spontaneous formation of 2D supramolecular framework via ionic self-assembly of cationic pseudorotaxane unit (Azo-TrEG@CD^{2+}) and POM cluster (PWV^{4-}).

style (Supplementary Fig. 28) as that of α-CD. Consequently, a steric distribution becomes unavoidable.

Characterizations and nano-size separation of IOIF membrane. The uniform mesh-like structure and the flexibility of the single-layer frameworks in aqueous solution exhibit the processing feature and simplify the fabrication of nanoporous membrane to a facile suction filtration under a slightly reduced pressure. Typically, the Tr-membrane (0.25 mg cm^{-2}, 2 cm in diameter) was prepared from filtration of [Azo-TrEG@CD][PWV] solution (0.04 mg ml^{-1}, 20 ml) on a supporting polycarbonate filter under the vacuum pressure of $-2,000$ Pa, as shown in Fig. 7a. After drying at 40 °C for 48 h and suffering the dissolution of polycarbonate filter in chloroform, a free-standing transparent IOIF membrane is obtained with rough thickness $\sim 0.43 \mu$m (Fig. 7b and Supplementary Fig. 29b). No obvious defects or pinholes (Supplementary Fig. 30) have been found, displaying intactness and good mechanical stability. With this method, in general, the area of the IOIF membranes is not restricted and the membranes can be prepared independent of supporting substrates' shape and size. Considering the torsion-induced shrinking of TrEG chain, the possible mesh size for the prepared Tr-membranes is estimated in the region from 2.4 to 3.4 nm (Supplementary Fig. 31). These specific features including the regular porosity and flexibility, especially the convenience in preparation provide applicable separation capability for bigger size objects that general MOF materials could not conduct.

To demonstrate the capacity of the supramolecular IOIF in size-selective separation, three kinds of small molecules, rhodamine B, xylenol orange (pH = 4.0 and 7.9) and the mixture of α-, β- and γ-CDs, which are positive, negative and nonionic at different pH conditions while maintaining the size less than 2 nm, were chosen as filtered chemical objects. The rhodamine B and xylenol orange filtrates were detected by UV–vis spectra (Supplementary Fig. 32), and CD mixture filtrate was detected by matrix-assisted laser desorption/ionization time-of-flight (MALDI-TOF) mass spectra (Supplementary Fig. 33). As predicted, all three kinds of small molecules, regardless of positive charge, negative charge or non-charge, can pass through the membrane without obvious quantity loss. According to the size matching, the result indicates a fluently passing ability for organic molecules smaller than the diameter of estimated square pores.

QDs are one of typical competitive luminescent probes for bioimaging and emitting materials for illumination and display[57]. Because the emission properties of QDs are highly dependent on size and distribution, the screen and separation by sizes from the ensembles are significant for further functional optimization. The capability of prepared IOIF membrane for the separation of

QDs is evaluated. Two differently sized CdTe QDs modified with 1-thioglycerol (TG) in aqueous solution are applied. The smaller one (TG-1) has an average particle size $D = 3.3$ nm in diameter with green emission ($\lambda_{max} = 533$ nm) and the bigger one (TG-2) has the size $D = 4.4$ nm in diameter with red emission ($\lambda_{max} = 611$ nm) after considering the thickness of modified surface layer. The mixture solution of TG-1 and TG-2 was filtered through Tr-membrane and the filtrate was monitored by emission spectrum. Interestingly, the luminescent photographs (Fig. 8a,b, inset) using to monitor the filtration process show that the orange luminescence of QDs mixture solution turns green after the filtration, indicating that only smaller size QDs passed through the Tr-membrane. Accompanying by a 10 nm blue-shifting ($\lambda_{max} = 523$ nm) of the emission band of the filtrate, the full-width at half-maximum (FWHM) narrows from 41.4 to 39.1 nm. Apparently, this spectral change corresponds to particle size having decreased to ~3.0 nm, while the larger size part of TG-1 and whole TG-2 are blocked by the supramolecular polymer membrane, as supported by red luminescence on filtered membrane (Fig. 8a,c, inset). By washing the filtered membrane with water, the collected residue QDs display two emission bands at 546 and 598 nm, further confirming that the larger sized part of TG-1 and whole TG-2 in QD ensembles have been successfully separated off. We also characterized the QDs in filtrate by TEM and indeed observed QDs with smaller and narrower size distribution than those without encountering filtration (Fig. 8d,e). These results verify that the critical separation size is in 3.0 − 3.3 nm, very close to the estimated mesh dimension in prepared Tr-membrane. After drying at 40 °C for 48 h, the Tr-membrane has the same filtration capability as the fresh made one (Supplementary Fig. 34), indicating the maintained mesh structures and IOIF membrane strength after drying treatment.

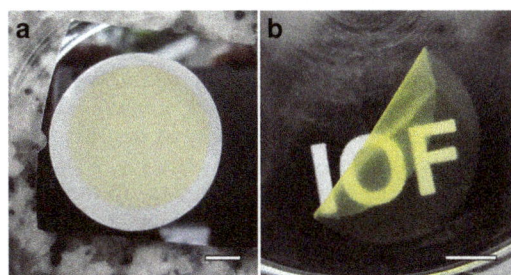

Figure 7 | Digital photographs. (**a**) The as-prepared IOIF membrane on polycarbonate filter and (**b**) the isolated IOIF membrane obtained through drying in oven at 40 °C for 48 h and dissolving polycarbonate supporting filter in chloroform. Scale bars, 0.5 cm (**a**,**b**).

Mesh-size modulation and separation efficiency. The pore of prepared single-layer framework is adjustable and can be modulated by simply increasing the length of flexible spacer:

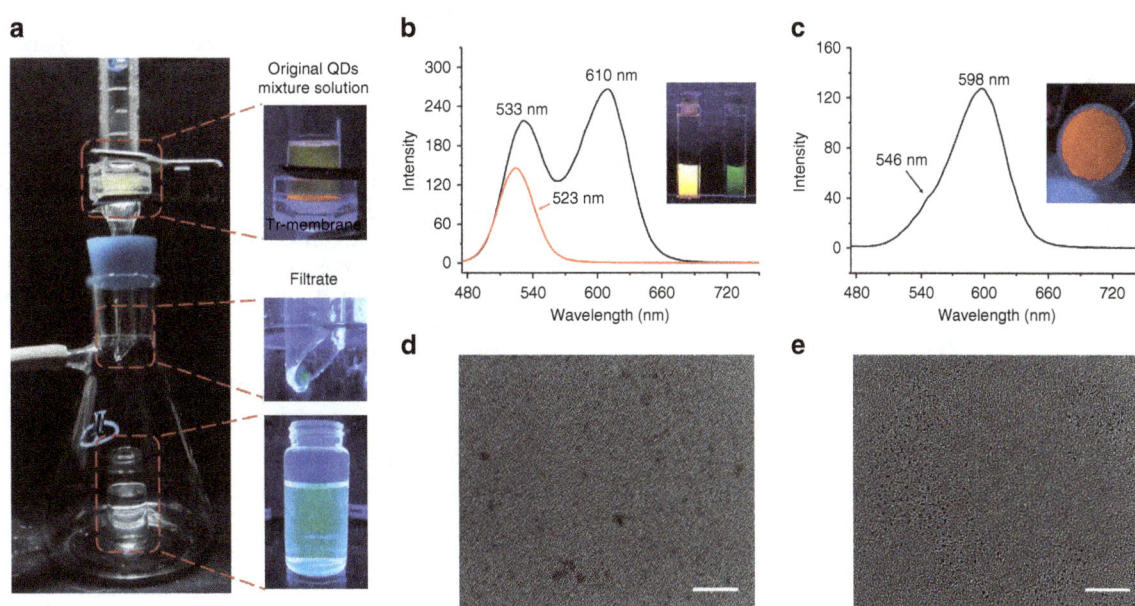

Figure 8 | Precise nano-size-selective separation. (**a**) Digital photographs of installation and corresponding amplifications of main parts under 365-nm light irradiation for examination of QDs mixture solution of TG-1 ($\lambda_{max} = 533$ nm and $D = 3.3$ nm) and TG-2 ($\lambda_{max} = 611$ nm and $D = 4.4$ nm) passing through Tr-membrane. Luminescent spectra of (**b**) initial QDs mixture solution (black line) and filtrate (red line), where the inset presents corresponding photographs of original solution (left) and filtrate (right) under 365-nm light; (**c**) residual QDs washed out from filtrated Tr-membrane, where the inset is the corresponding photograph of residual QDs under 365-nm light. TEM images of QDs from (**d**) initial mixture solution and (**e**) filtrate. Scale bars, 20 nm (**d**,**e**).

TeEG. The obtained [Azo-TeEG@CD][PWV] displays the whole characteristics of [Azo-TrEG@CD][PWV] on framework structure and filtration property except the increased pore size (2.4 − 4.1 nm) after considering the shrinkage and stretching of TeEG chain (Supplementary Figs 35–41 and Supplementary Table 2). The TG-1 passes through Te-membrane in both cases of isolated solution (Supplementary Fig. 42) and its mixture with TG-2 (Fig. 9a). However, TG-2 is still impossible to penetrate through the mesh of Te-membrane. The critical value estimated from the filtration cutoff for the Te-membrane should be in the region between sizes of TG-1 and TG-2. More interestingly, we can combine the two IOIF membranes bearing different mesh dimensions for selective size optimization of QDs mixture. For example, for the mixture solution of TG-1 and TG-2, we firstly filtered out the larger sized TG-2 by using Te-membrane, and then conducted another filtration of the filtrate by using Tr-membrane. Thus, the QDs showing luminescence at 543 nm with the dimension between two cutoff sizes of Tr- and Te-membranes are obtained by washing the residue on Tr-membrane (Fig. 9b). This simple combination of filtration membranes provides more precise classification of QDs. By taking isolated TG-1 as reference, the separation experiment indicates that the smaller size part of TG-1 can also be further separated by using Tr-membrane (Supplementary Fig. 43). The spectral analysis of the residual QDs blocked by Tr-membrane indicates the larger-size composition of TG-1 in diameter of 3.6 nm (λ_{max} = 542 nm). With this fast and facile filtration, we provide an alternative analysis method for size distribution of QDs in addition to high-resolution TEM.

Generally, the aforementioned separation is independent of the surface species stabilizing QDs because when 3-mercapto-propionic acid-modified QDs in size of 4.0 (MPA-1) and 4.8 nm (MPA-2) are used in the filtration, we also obtain identical

separation effect (Supplementary Figs 44–46). By evaluating the filtration of MPA-1 and its mixture of TG-2 with Te-membrane, we can definitely deduce the practical critical value of separation in a much precise region of 3.9 − 4.0 nm (Supplementary Fig. 45).

The separation efficiency of QDs is carried out by evaluating relative fluorescence intensity before and after the filtration. As shown in Table 1, for Te-membrane, because the size of TG-1 (3.3 nm) is much smaller than the mesh size of Te-membrane, up to 93.4% of TG-1 pass through the filter (Supplementary Fig. 42). Considering few larger sized QDs having been blocked on the IOIF membrane, the filtration displays very high efficiency for QDs in smaller size. In the case of mixture of TG-1 and TG-2, because the TG-2 in larger size are impossible to pass through the filter, the residue attached on the membrane obstructs the passing channel of TG-1, generating a decreased separation efficiency to 81.3% (Fig. 9a). For Tr-membrane, only the smaller size part of TG-1 penetrate through the membrane, leading to decreased separation efficiency of 76.3% and 73.4% for isolated TG-1 (Supplementary Fig. 43a) and its mixture with TG-2 (Fig. 8b), respectively. For other separations, Te-membrane performs 80.4% separation efficiency for MPA-1 alone (Supplementary Fig. 45a) and 71.2% for the mixture of MPA-1 and TG-2 (Supplementary Fig. 45b), while Tr-membrane displays 68.8% separation efficiency for TG-1 mixing with MPA-2 (Supplementary Fig. 46b). Because of the emission of MPA-1 locating at 523 nm, it is difficult to evaluate the separation efficiency of Tr-membrane for TG-1 and MPA-1 mixture accurately by analysing relative fluorescence intensity at the same wavelength before and after filtration (Supplementary Fig. 46a). In general, the separation efficiency is highly affected by the thickness of IOIF membrane. We prepared Tr-membranes with different thickness by varying the concentration of sample solution (Supplementary Fig. 29). With the concentration decreasing, the thickness becomes thinner and uneven. When the thickness is less than 200 nm, the separation of QDs mixture becomes incomplete because of the possible defects in the membrane formation upon suction, which results in leakage of some larger sized QDs into the filtrate. But, with the thickness increasing, the separation of QDs mixture solution performs well. The emission band and FWHM in the spectra of filtrates show no obvious change except a gradual decrease of separation efficiency (Supplementary Fig. 47).

Durability and reusability. To demonstrate the durability of prepared IOIF membranes, 20 time-separation experiments (20 ml for each separation, 400 ml in total) for TG-1 and TG-2 mixture solution were consecutively carried out by using the same Tr-membrane. After the continuous separation, a slight decrease of separation efficiency is found but there is no significant change by examining the emission band and FWHM of the filtrate (Supplementary Fig. 48 and Supplementary Table 4), implying the good durability of the IOIF membranes. The burst strength of

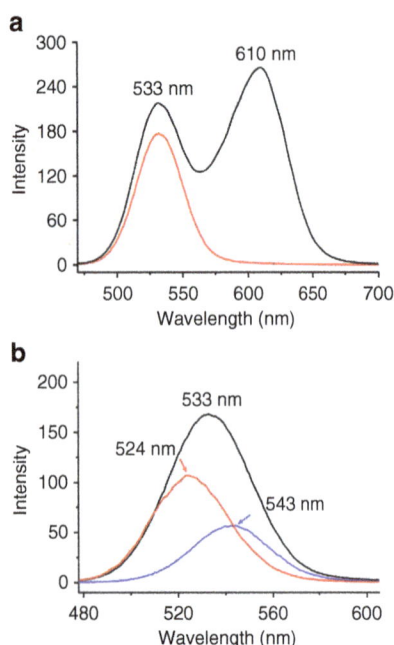

Figure 9 | Two-step separation. Fluorescent spectra of (**a**) initial QDs mixture solution of TG-1 (λ_{max} = 533 nm, D = 3.3 nm) and TG-2 (λ_{max} = 611 nm, D = 4.4 nm) (black line), and the filtrate (red line) passing through Te-membrane; and (**b**) the filtrate obtained from the first filtration of original TG-1 and TG-2 mixture solution through Te-membrane (λ_{max} = 533 nm, black line), the filtrate obtained from the second filtration through Tr-membrane (λ_{max} = 524 nm, red line), and the residual QDs washed out from the filtrated Tr-membrane (λ_{max} = 543 nm, blue line).

Table 1 | Summary of separation efficiency of Tr/Te-membranes for QDs solutions.

	QDs separation efficiency[*]	
	Tr-membrane	Te-membrane
TG-1	76.3%	93.4%
TG-1 + TG-2	73.4%	81.3%

QDs, quantum dots.
[*]The separation efficiency was calculated from comparison of relative fluorescence intensity at emission wavelength 523 and 533 nm (for Tr and Te membranes, respectively) before and after the filtration.

Tr-membrane (0.25 mg cm^{-2}, ~0.43 µm) spreading on a support bearing average pore of 220 nm in diameter is 1.49~1.66 MPa with the test area of 3.8×10^{-6} cm^2 (Supplementary Figs 49,50), revealing high stability of the IOIF membranes. Furthermore, after washing out the residual QDs, we re-dispersed the filtered Tr-membrane in water. After sonication for a while (Supplementary Fig. 51), a new Tr-membrane can be prepared by filtrating the re-dispersed solution and used to separate QDs mixture solution of TG-1 and TG-2. A similar separation capability as that of fresh one is observed (Supplementary Fig. 52), demonstrating the reusability of the IOIF membranes.

Discussion

Overall, we have created a facile and convenient strategy for the fabrication of free-standing single-layer 2D IOIFs. In contrast to the known driving forces, electrostatic interaction was employed as the main binding force for the supramolecular hybrid frameworks. Through the synergetic ionic self-assembly of cationic pseudorotaxane unit as bridging stick and anionic PWV^{4-} as capper and node in aqueous solution, the obtained IOIF supramolecular polymer effectively overcomes the interlayer interaction and structure stiffness. The prepared single-layer framework not only integrates inorganic clusters and rotaxane-type units into branched polymeric architectures but also possesses highly ordered nanoporous mesh and good solution processability. This type of soft framework structure constructed by ionic bonding without preferential direction and its distinctive properties such as stability and plasticity offer an unprecedented opportunity to fabricate ultrafiltration membrane towards precisely screening nanoparticles. Because of the facile methodology and uniform nanoporous structure, the membranes prepared with 2D IOIF assemblies hold steady promise for practical applications in selective transport, molecular separation and dialysis systems. Meanwhile, the synergetic ionic self-assembly strategy can be hopefully extended to diverse IOIFs for advanced supramolecular materials.

Methods

Materials and instruments. The general chemicals, α-, β-, γ-CD (α-, β-, γ-CD), rhodamine B and xylenol orange, are the products of TCI Chemicals (China) Pvt. Ltd. Other chemicals and solvents were purchased from Beijing Chemical Reagent Industry and used as received. Acetonitrile was dried over P$_2$O$_5$ and distilled prior to use. N,N-dimethylformamide (DMF) was dried with CaH$_2$ for several days and distilled before using. Doubly distilled water (Milli-Pore 18.2 MΩ cm^{-1}) was used in the experiment. The polycarbonate membranes with 0.2-µm pore size are the product of Whatman Filters (a GE Healthcare brand). The polyanionic cluster PWV^{4-} was prepared following the published procedure[58]. Four kinds of CdTe QDs were prepared and characterized following the published procedures[59].

^{1}H NMR spectra were recorded on a Bruker AVANCE 500 MHz spectrometer while the chemical shifts were corrected by the solvent value ($\delta = 4.79$ p.p.m. for D$_2$O and $\delta = 2.50$ p.p.m. for DMSO-d_6). UV–vis spectra were carried out on a spectrometer (Varian CARY 50 Probe). The fluorescent spectra were carried out using the spectrophotometer (Shimadzu RF-5301PC). Circular dichroism spectroscopy were performed on a Bio-Logic MOS-450 spectropolarimeter in water with step size of 1 nm and speed of 5 nm s^{-1} at 25 °C. AFM images were taken with a SPA-300HV (Seiko, Japan) under ambient conditions. AFM was operated in the tapping mode with an optical readout using Si cantilevers. Scanning electron microscope images were acquired on a JEOL FESEM 6700F field emission scanning electron microscope (JEOL, Japan). TEM and HRTEM were conducted on a JEOL JEM 2010 electron microscope under an accelerating voltage of 200 kV without staining and energy-dispersive X-ray spectrum was collected on a FEI Tecnai F20 microscope operating at an accelerating voltage of 200 kV. XRD data were recorded on a Rigaku SmartLab X-ray diffractometer using Cu Kα$_1$ radiation at wavelength of 1.542 Å. MALDI-TOF mass spectra was recorded on an autoflex MALDI-TOF/TOF (Bruker, Germany) mass spectrometer equipped with a nitrogen laser (337 nm, 3 ns pulse). The ESI-MS/MS spectra were carried out by POEMS inductively coupled plasma mass spectrometer (TJA, USA). X-ray photoelectron spectra was carried out on an ESCALAB 250 spectrometer with a monochromic X-ray source (Al Kα line, 1,486.6 eV) and the charging shift was corrected by the binding energy of C(1 s) at 284.6 eV. Inductively coupled plasma

atomic emission spectrometry (ICP-AES) was carried out on the PerkinElmer Optima 3300DV (PerkinElmer, USA). Organic element analysis was carried out on the vario MACRO cube CHNS (Elementar, Germany).

Synthesis of Azo-Tr/TeEG · 2Br. The bolaform cationic organic compounds were synthesized according to the route in Supplementary Fig. 1. Azo compound **a**, sulfonated compound of glycol **b**, coupling product **c** was prepared according to the literatures[60,61]. Compound **c** bearing TrEG or TeEG chain (1 g, 1.44 or 1.35 mmol) was added to a 10 ml round bottom flask containing 5 ml of DMF and 2 ml of pyridine. The reaction mixture was heated under stirring in an oil bath at 100 °C for 48 h. Then the reaction solution was cooled to room temperature and transferred dropwise to 500 ml of ethyl acetate. The yielded precipitate was collected by filtration and washed with ethyl acetate (1,000 ml) for three times to give the product (0.93 g, 76% for Azo-TrEG · 2Br; 0.90 g, 74% for Azo-TeEG · 2Br). ^{1}H NMR (500 MHz, D$_2$O) for Azo-TrEG · 2Br: δ 8.87 p.p.m. (d, $J = 5.6$ Hz, 4H, Ar-H), 8.57 (t, $J = 7.9$ Hz, 2H, Ar-H), 8.07 (t, 4H, Ar-H), 7.62 (d, $J = 8.4$ Hz, 4H, Ar-H), 7.57 (d, $J = 9.0$ Hz, 4H, Ar-H), 7.51 (d, $J = 8.5$ Hz, 4H, Ar-H), 6.92 (d, $J = 9.0$ Hz, 4H, Ar-H), 5.75 (s, 4H, CH$_2$), 4.10 (t, 4H, CH$_2$), 3.86 (t, 4H, CH$_2$) and 3.77 (s, 4H, CH$_2$). ^{13}C NMR (125 MHz, DMSO-d_6): δ 161.7, 152.8, 146.3, 146.2, 145.0, 136.4, 130.1, 128.7, 124.9, 123.0, 115.3, 70.1, 68.9, 67.8 and 62.9. ESI-MS (m/z) [M]$^{2+}$: calculated for C$_{42}$H$_{42}$N$_6$O$_4$: 347.4; found: 347.4. ^{1}H NMR (500 MHz, DMSO-d_6) for Azo-TeEG · 2Br: δ 9.26 p.p.m. (d, $J = 5.7$ Hz, 4H, Ar-H), 8.67 (t, $J = 7.8$ Hz, 2H, Ar-H), 8.22 (t, $J = 7.1$ Hz, 4H, Ar-H), 7.98 − 7.80 (m, 8H, Ar-H), 7.72 (d, $J = 8.3$ Hz, 4H, Ar-H), 7.17 (d, $J = 8.9$ Hz, 4H, Ar-H), 5.97 (s, 4H, CH$_2$), 4.21 (t, 4H, CH$_2$), 3.79 (t, 4H, CH$_2$) and 3.65 − 3.50 (m, 8H, CH$_2$). ^{13}C NMR (126 MHz, DMSO-d_6) δ 162.1, 152.8, 146.63, 145.57, 145.4, 136.8, 130.4, 129.0, 125.3, 123.3, 115.6, 70.4, 70.3, 69.3, 68.2 and 63.3. ESI-MS (m/z) [M]$^{2+}$: calculated for C$_{44}$H$_{46}$N$_6$O$_5$: 369.2; found: 369.6.

Preparation of Azo-Tr/TeEG@CD · 2Br and [Azo-Tr/TeEG@CD][PWV]. In a typical procedure, Azo-TrEG · 2Br (4.5 mg, 5.3 µmol) and α-CD (10.2 mg, 10.5 µmol) were mixed and dissolved in 60 ml water under sonication for 0.5 h to generate pseudorotaxane unit Azo-TrEG@CD · 2Br. The final framework assembly of [Azo-TrEG@CD][PWV] (0.17 mg ml^{-1}) was constructed by simply mixing the as-prepared Azo-TrEG@CD · 2Br and PWV^{4-} solution that is prepared by dissolving PWV^{4-} (7.6 mg, 2.6 µmol) in 60 ml water according to certain stoichiometry.

Preparation of IOIF membrane. The IOIF membranes were prepared by a simple filtration of sample solutions containing single-layer IOIF self-assemblies through a commercial filter with even dispersed pores in certain size under vacuum pressure of − 2,000 Pa. After washing with water, the prepared nanofiltration membrane has the same size as the effective filtration area of supporting substrate used in the installation. In a typical procedure, 20 ml as-prepared [Azo-TrEG@CD][PWV] solution (0.04 mg ml^{-1}) was filtered over a supporting filter (Whatman Nuclepore Track-Etched Polycarbonate Membrane; effective filtration area: 3.14 cm^2; pore size: 200 nm) under the preset vacuum pressure. Water (10 ml) was subsequently used to wash the membrane. Under the same vacuum pressure, the IOIF membranes with various thicknesses were prepared by using 20 ml sample solution with concentrations: 0.02, 0.04, 0.06 and 0.08 mg ml^{-1}, and corresponding thicknesses are measured to be 0.20~0.35, 0.43, 1.43 and 2.19 µm, respectively.

Filtration of small organic molecules. The solutions of dyes, rhodamine B and xylenol orange, were prepared at the concentration of 0.42 and 0.48 mM, respectively. In the case of CD mixture, each of concentrations of α-CD, β-CD and γ-CD was maintained at 5.1 mM. In all filtration experiments, the volume of the original solutions was 20 ml, the vacuum pressure was about − 5,000 Pa, and the flux was set at 15.3 m^3 m^{-2} h^{-1} bar^{-1}.

Size-selective separation of CdTe QDs. In all QDs filtration experiments, the volume of QD solutions for each filtration was maintained at 20 ml, the vacuum pressure was about − 5,000 Pa, and the flux was ~15.3 m^3 m^{-2} h^{-1} bar^{-1}. The concentrations of QDs were estimated as follows: 0.05 mM for TG-1, 0.10 mM for TG-2, 0.02 mM for MPA-1, 0.10 mM for MPA-2, 0.05 and 0.10 mM for the mixture of TG-1 and TG-2, 0.05 and 0.02 mM for the mixture of TG-1 and MPA-1, 0.05 and 0.10 mM for the mixture of TG-1 and MPA-2, and 0.02 and 0.10 mM for the mixture of MPA-1 and TG-2, respectively, according to their absorbance. By putting the filtered Tr-membrane into 10 ml water and undergoing an oscillation for a while, the residual QDs could be washed out from the membrane (Supplementary Fig. 51). During this process, it is hard to maintain the membrane intact. The QDs mixing with a few fragments of Tr-membrane in solution could be separated by centrifugation (1,500 r.p.m., 3 min).

Sample preparation and measurement for TEM. We used a thin copper ring to spread a thin layer of [Azo-Tr/TeEG@CD][PWV] solution (0.02 mg ml^{-1}) like a soap film and cast it on a copper grid. The procedure was repeated three times for one sample to ensure enough amounts of samples being attached on the copper grid. During the measurement, longer time exposure to the high-energy electron beam will destroy the framework structure, resulting in less ordered structure. The HRTEM images were tracked within a quite short time. In order to obtain a clear high contrast image, we used a smart camera technique for image collection, which was based on a continuous acquisition of images to get high image contrast. Because of the heating disturbance of the sample during the electron beam irradiation, the image superposition brought slight ghosting phenomenon, which causes the size of inorganic clusters seemed larger than their ideal dimension.

Sample preparation for AFM measurement. The sample for AFM measurement was prepared by a dip-coating technique. First, we dipped a mica wafer quickly into the as-prepared [Azo-Tr/TeEG@CD][PWV] solution with the concentration of $0.02\,mg\,ml^{-1}$, and then slowly withdrew at a constant speed of $1\,mm\,min^{-1}$. During the process, the IOIF assembly was attached on the surface of the mica in a single layer.

Sample preparation for XRD measurement. The powdered sample was prepared by the lyophilization of the [Azo-TrEG@CD][PWV] solution ($0.17\,mg\,ml^{-1}$, 120 ml) and then grinding it into powder. The film sample was prepared by the filtration of [Azo-TrEG@CD][PWV] solution ($0.18\,mg\,ml^{-1}$, 120 ml) over a supporting filter (Whatman Nuclepore Track-Etched Polycarbonate Membrane; effective filtration area: $3.14\,cm^2$; pore size: 200 nm), drying in oven at 40 °C for 48 h.

Particle size calculation of CdTe QDs. The core diameter (D_c) of CdTe QDs was calculated according to the following published equation[62]: $D_c = (9.8127 \times 10^{-7})\lambda^3 - (1.7147 \times 10^{-3})\lambda^2 + (1.0064)\lambda - 194.84$, where λ (nm) is the maximum wavelength corresponding to the first excitonic absorption peak of QDs. The full diameter of surface stabilized CdTe QDs (average diameter D) was calculated by summing the molecular length of ligands (l) and the calculated D_c value, where the molecular length of ligand stabilizer is estimated ~0.46 nm for TG and ~0.65 nm for MPA, simulated by ChemBio 3D (12.0 version). Thus, the diameters (D) of four QDs were calculated to be 3.3 nm for TG-1, 4.4 nm for TG-2, 4.0 nm for MPA-1 and 4.8 nm for MPA-2.

References

1. Cheng, W., Campolongo, M. J., Tan, S. J. & Luo, D. Freestanding ultrathin nano-membranes via self-assembly. *Nano Today* **4**, 482–493 (2009).
2. Boukhvalov, D. W., Katsnelson, M. I. & Son, Y. W. Origin of anomalous water permeation through graphene oxide membrane. *Nano Lett.* **13**, 3930–3935 (2013).
3. Celebi, K. *et al.* Ultimate permeation across atomically thin porous graphene. *Science* **344**, 289–292 (2014).
4. Shen, J. *et al.* Membranes with fast and selective gas-transport channels of laminar graphene oxide for efficient CO_2 capture. *Angew. Chem. Int. Ed.* **54**, 578–582 (2015).
5. Mi, B. Graphene oxide membranes for ionic and molecular sieving. *Science* **343**, 740–742 (2014).
6. Gaborski, T. R. *et al.* High-performance separation of nanoparticles with ultrathin porous nanocrystalline silicon membranes. *ACS Nano* **4**, 6973–6981 (2010).
7. Raaijmakers, M. J. T. *et al.* Enzymatically active ultrathin pepsin membranes. *Angew. Chem. Int. Ed.* **54**, 5910–5914 (2015).
8. Shi, Z. *et al.* Ultrafast separation of emulsified oil/water mixtures by ultrathin free-standing single-walled carbon nanotube network films. *Adv. Mater.* **25**, 2422–2427 (2013).
9. Striemer, C. C., Gaborski, T. R., McGrath, J. L. & Fauchet, P. M. Charge- and size-based separation of macromolecules using ultrathin silicon membranes. *Nature* **445**, 749–753 (2007).
10. Baek, K. *et al.* Free-standing, single-monomer-thick two-dimensional polymers through covalent self-assembly in solution. *J. Am. Chem. Soc.* **135**, 6523–6528 (2013).
11. Kissel, P. *et al.* A two-dimensional polymer prepared by organic synthesis. *Nat. Chem.* **4**, 287–291 (2012).
12. Kissel, P., Murray, D. J., Wulftange, W. J., Catalano, V. J. & King, B. T. A nanoporous two-dimensional polymer by single-crystal-to-single-crystal photopolymerization. *Nat. Chem.* **6**, 774–778 (2014).
13. Kory, M. J. *et al.* Gram-scale synthesis of two-dimensional polymer crystals and their structure analysis by X-ray diffraction. *Nat. Chem.* **6**, 779–784 (2014).
14. Xu, L. *et al.* Surface-confined single-layer covalent organic framework on single-layer graphene grown on copper foil. *Angew. Chem. Int. Ed.* **53**, 9564–9568 (2014).
15. Bebensee, F. *et al.* A surface coordination network based on copper adatom trimers. *Angew. Chem. Int. Ed.* **53**, 12955–12959 (2014).
16. Bauer, T. *et al.* Synthesis of free-standing, monolayered organometallic sheets at the air/water interface. *Angew. Chem. Int. Ed.* **50**, 7879–7884 (2011).
17. Lehn, J. M. Toward self-organization and complex matter. *Science* **295**, 2400–2403 (2002).
18. De Greef, T. F. *et al.* Supramolecular polymerization. *Chem. Rev.* **109**, 5687–5754 (2009).
19. Krieg, E., Weissman, H., Shirman, E., Shimoni, E. & Rybtchinski, B. A recyclable supramolecular membrane for size-selective separation of nanoparticles. *Nat. Nanotechnol.* **6**, 141–146 (2011).
20. Wang, C., Wang, Z. & Zhang, X. Amphiphilic building blocks for self-assembly: from amphiphiles to supra-amphiphiles. *Acc. Chem. Res.* **45**, 608–618 (2012).
21. Harada, A., Kobayashi, R., Takashima, Y., Hashidzume, A. & Yamaguchi, H. Macroscopic self-assembly through molecular recognition. *Nat. Chem.* **3**, 34–37 (2011).
22. Yamaguchi, H. *et al.* Photoswitchable gel assembly based on molecular recognition. *Nat. Commun.* **3**, 603 (2012).
23. Whang, D. & Kim, K. Polycatenated two-dimensional polyrotaxane net. *J. Am. Chem. Soc.* **119**, 451–452 (1997).
24. Lee, E., Heo, J. & Kim, K. A three-dimensional polyrotaxane network. *Angew. Chem. Int. Ed.* **39**, 2699–2701 (2000).
25. Zhang, K. D. *et al.* Toward a single-layer two-dimensional honeycomb supramolecular organic framework in water. *J. Am. Chem. Soc.* **135**, 17913–17918 (2013).
26. Pope, M. T. & Müller, A. Polyoxometalate chemistry: an old field with new dimensions in several disciplines. *Angew. Chem. Int. Ed. Engl.* **30**, 34–48 (1991).
27. Long, D. L., Burkholder, E. & Cronin, L. Polyoxometalate clusters, nanostructures and materials: from self assembly to designer materials and devices. *Chem. Soc. Rev.* **36**, 105–121 (2007).
28. Dolbecq, A., Dumas, E., Mayer, C. R. & Mialane, P. Hybrid organic-inorganic polyoxometalate compounds: from structural diversity to applications. *Chem. Rev.* **110**, 6009–6048 (2010).
29. Proust, A. *et al.* Functionalization and post-functionalization: a step towards polyoxometalate-based materials. *Chem. Soc. Rev.* **41**, 7605–7622 (2012).
30. Lopez, X., Carbo, J. J., Bo, C. & Poblet, J. M. Structure, properties and reactivity of polyoxometalates: a theoretical perspective. *Chem. Soc. Rev.* **41**, 7537–7571 (2012).
31. Izzet, G. *et al.* Cyclodextrin-induced auto-healing of hybrid polyoxometalates. *Angew. Chem. Int. Ed.* **51**, 487–490 (2012).
32. He, P., Xu, B., Wang, P. P., Liu, H. & Wang, X. A monolayer polyoxometalate superlattice. *Adv. Mater.* **26**, 4339–4344 (2014).
33. Long, D. L., Tsunashima, R. & Cronin, L. Polyoxometalates: building blocks for functional nanoscale systems. *Angew. Chem. Int. Ed.* **49**, 1736–1758 (2010).
34. Liu, T., Diemann, E., Li, H., Dress, A. W. M. & Müller, A. Self-assembly in aqueous solution of wheel-shaped Mo_{154} oxide clusters into vesicles. *Nature* **426**, 59–62 (2003).
35. Ishii, Y., Takenaka, Y. & Konishi, K. Porous organic-inorganic assemblies constructed from keggin polyoxometalate anions and calix[4]arene-Na$^+$ complexes: structures and guest-sorption profiles. *Angew. Chem. Int. Ed.* **43**, 2702–2705 (2004).
36. Yin, P. *et al.* Chiral recognition and selection during the self-assembly process of protein-mimic macroanions. *Nat. Commun.* **6**, 6475 (2015).
37. Wei, H. *et al.* Tunable, luminescent, and self-healing hybrid hydrogels of polyoxometalates and triblock copolymers based on electrostatic assembly. *Chem. Commun.* **50**, 1447–1450 (2014).
38. Du, D., Qin, J., Li, S., Su, Z. & Lan, Y. Recent advances in porous polyoxometalate-based metal-organic framework materials. *Chem. Soc. Rev.* **43**, 4615–4632 (2014).
39. Miras, H. N., Yan, J., Long, D. L. & Cronin, L. Engineering polyoxometalates with emergent properties. *Chem. Soc. Rev.* **41**, 7403–7430 (2012).
40. Miras, H. N., Vilà-Nadal, L. & Cronin, L. Polyoxometalate based open-frameworks (POM-OFs). *Chem. Soc. Rev.* **43**, 5679–5699 (2014).
41. Polarz, S., Smarsly, B. & Antonietti, M. Colloidal organization and clusters: self-assembly of polyoxometalate-surfactant complexes towards three-dimensional organized structures. *ChemPhysChem* **7**, 457–461 (2001).
42. Wen, S. *et al.* Theoretical investigation of structural and electronic propertyies of $[PW_{12}O_{40}]^{3-}$ on graphene layer. *Dalton Trans.* **41**, 4602–4607 (2012).
43. Leroy, F., Miró, P., Poblet, J. M., Bo, C. & Ávalos, J. B. Keggin polyoxoanions in aqueous solution: ion pairing and its effect on dynamic properties by molecular dynamics simulations. *J. Phys. Chem. B* **112**, 8591–8599 (2008).
44. Wu, Y. *et al.* Complexation of polyoxometalates with cyclodextrins. *J. Am. Chem. Soc.* **137**, 4111–4118 (2015).
45. Harata, K. & Uedaira, H. The circular dichroism spectra of the β-cyclodextrin complex with naphthalene derivatives. *Bull. Chem. Soc. Jpn* **48**, 375–378 (1975).
46. Kodaka, M. A general rule for circular dichroism induced by a chiral macrocycle. *J. Am. Chem. Soc.* **115**, 3702–3705 (1993).
47. Liu, Y., Zhao, Y. L., Zhang, H. Y. & Song, H. B. Polymeric rotaxane constructed from the inclusion complex of β-cyclodextrin and 4,4′-dipyridine by coordination with nickel(II) ions. *Angew. Chem. Int. Ed.* **42**, 3260–3263 (2003).

48. Ito, T., Sawada, K. & Yamase, T. Crystal structure of bis(dimethyldioctade-cylammonium) hexamolybdate: a molecular model of langmuir-blodgett films. *Chem. Lett.* **32**, 938–939 (2003).

49. Nyman, M. *et al.* Comparative study of inorganic cluster-surfactant arrays. *Chem. Mater.* **17**, 2885–2895 (2005).

50. Ito, T. & Yamase, T. Inorganic-organic hybrid layered crystal composed of polyoxomolybdate and surfactant with π electrons. *Chem. Lett.* **38**, 370–371 (2009).

51. Yang, Y., Wang, Y., Li, H., Li, W. & Wu, L. Self-assembly and structural evolvement of polyoxometalate-anchored dendron complexes. *Chem. Eur. J.* **16**, 8062–8071 (2010).

52. Nisar, A., Lu, Y., Zhuang, J. & Wang, X. Polyoxometalate nanocone nanoreactors: magnetic manipulation and enhanced catalytic performance. *Angew. Chem. Int. Ed.* **50**, 3187–3192 (2011).

53. Wang, Y., Li, W. & Wu, L. Organic-inorganic hybrid supramolecular gels of surfactant-encapsulated polyoxometalates. *Langmuir* **25**, 13194–13200 (2009).

54. Liu, Y., Zhao, Y., Zhang, H., Yang, E. & Guan, X. Binding ability and assembly behavior of β-cyclodextrin complexes with 2,2'-dipyridine and 4,4'-dipyridine. *J. Org. Chem.* **69**, 3383–3390 (2004).

55. Ghosh, P., Maity, A., Das, T., Dash, J. & Purkayastha, P. Modulation of small molecule induced architecture of cyclodextrin aggregation by guest structure and host size. *J. Phys. Chem. C* **115**, 20970–20977 (2011).

56. Saenger, W. *et al.* Structures of the common cyclodextrins and their larger analogues-beyond the doughnut. *Chem. Rev.* **98**, 1787–1802 (1998).

57. Rosi, N. L. & Mirkin, C. A. Nanostructures in biodiagnostics. *Chem. Rev.* **105**, 1547–1562 (2005).

58. Domaille, P. J. The 1- and 2-dimensional tungsten-183 and vanadium-51 NMR characterization of isopolymetalates and heteropolymetalates. *J. Am. Chem. Soc.* **106**, 7677–7687 (1984).

59. Zhou, D. *et al.* Simple synthesis of highly luminescent water-soluble CdTe quantum dots with controllable surface functionality. *Chem. Mater.* **23**, 4857–4862 (2011).

60. Chen, X. *et al.* Light-controllable reflection wavelength of blue phase liquid crystals doped with azobenzene-dimers. *Chem. Commun.* **49**, 10097–10099 (2013).

61. Commins, P. & Garcia-Garibay, M. A. Photochromic molecular gyroscope with solid state rotational states determined by an azobenzene bridge. *J. Org. Chem.* **79**, 1611–1619 (2014).

62. Yu, W. W., Qu, L., Guo, W. & Peng, X. Experimental determination of the extinction coefficient of CdTe, CdSe, and CdS nanocrystals. *Chem. Mater.* **15**, 2854–2860 (2003).

Acknowledgements

We acknowledge financial support from National 973 Program (2013CB834503), NSFC (21574057, 91227110 and 21221063), Changbai Mountain Scholar of Jilin Province and Ministry of Education of China (20120061110047).

Author contributions

L.W. conceived the idea and performed the data interpretation. L.Y. designed and carried out the experiments, performed the data interpretation and wrote the manuscript. S.W. and B.L. synthesized polyoxometalate and performed AFM characterization. D.Z. and H.Z. provided CdTe QDs and helpful suggestions in nano-size separation. All authors contributed to the revising of the manuscript.

Additional information

27

Photoresponse of supramolecular self-assembled networks on graphene–diamond interfaces

Sarah Wieghold[1,2,*], Juan Li[2,3,*], Patrick Simon[3,4], Maximilian Krause[1,2], Yuri Avlasevich[5], Chen Li[5], Jose A. Garrido[3,4], Ueli Heiz[1,2], Paolo Samori[6], Klaus Müllen[5], Friedrich Esch[1,2], Johannes V. Barth[2,3] & Carlos-Andres Palma[2,3]

Nature employs self-assembly to fabricate the most complex molecularly precise machinery known to man. Heteromolecular, two-dimensional self-assembled networks provide a route to spatially organize different building blocks relative to each other, enabling synthetic molecularly precise fabrication. Here we demonstrate optoelectronic function in a near-to-monolayer molecular architecture approaching atomically defined spatial disposition of all components. The active layer consists of a self-assembled terrylene-based dye, forming a bicomponent supramolecular network with melamine. The assembly at the graphene–diamond interface shows an absorption maximum at 740 nm whereby the photoresponse can be measured with a gallium counter electrode. We find photocurrents of 0.5 nA and open-circuit voltages of 270 mV employing 19 mW cm^{-2} irradiation intensities at 710 nm. With an *ex situ* calculated contact area of $9.9 \times 10^2 \mu m^2$, an incident photon to current efficiency of 0.6% at 710 nm is estimated, opening up intriguing possibilities in bottom-up optoelectronic device fabrication with molecular resolution.

[1] Chemie-Department, Technische Universität München, Lichtenbergstraße 4, Garching 85748, Germany. [2] Catalysis Research Center, Technische Universität München, Ernst-Otto-Fischer-Straße 1, Garching 85748, Germany. [3] Physik-Department, Technische Universität München, James-Franck-Strasse 1, Garching 85748, Germany. [4] Walter Schottky Institut, Technische Universität München, Am Coulombwall 4, Garching 85748, Germany. [5] Max Planck Institute for Polymer Research, Ackermannweg 10, Mainz 55128, Germany. [6] ISIS & icFRC, Université de Strasbourg & CNRS, 8 allée Gaspard Monge, Strasbourg 67000, France. * These authors contributed equally to this work. Correspondence and requests for materials should be addressed to K.M. (email: muellen@mpip-mainz.mpg.de) or to F.E. (email: friedrich.esch@ch.tum.de) or to J.V.B. (email: jvb@tum.de) or to C.-A.P. (email: c.a.palma@tum.de).

Elemental crystals, interfaces and (macro)molecules have become integral parts of modern semiconductor devices which are shaping twenty-first century technology. Thus far, the first generation of organic semiconductor devices, for example, light-emitting diodes[1] and photovoltaics[2], are based on (macro)molecules processed with a wide range of methods in an effort to make them compatible with the capabilities of the semiconductor industry. These methods consist mainly of thin film technologies, including spin-coating[3], sublimation[1], printing[4], crystallization[5] and self-assembly[6–8] fabrication procedures or combinations thereof[9]. However, the aforementioned strategies have been often centred around the fabrication and the tuning of the properties of the (macro) molecular active components in thin films[10]. Consequently, spatial orientation of both the active layer's components and their interfaces are not usually known with atomic precision. Thus, *a posteriori* characterization techniques are required and the final absolute atomic-scale spatial constitution of the device as a whole is rarely reproducible. Although an atomically precise layout of the constituting components is not always required for a device's function, it may be critical for its optimization[11]. For instance, sensitizing interfaces with molecules was early recognized[12] as efficient means to photovoltaic charge generation. By optimizing the sensitizer surface area, this strategy became technologically viable[13,14]. Similarly, by improving chemical precision over donor and acceptor polymers[15], organic solar cell efficiencies grew rapidly[16]. Thus, it is clear that for a transition to a second generation of organic device engineering, their constituents must be fabricated not only with high interfacial and chemical control but also with exquisite spatio-temporal heteromolecular precision, where the absolute location of different molecular components is mastered and precisely known *a priori*. So far, device elements approaching such molecular precision employ single-molecule configurations[17,18], which are not yet ready to be implemented for large-area technological applications. One strategy for large-area, artificial molecularly precise device fabrication is to grow architectures from the bottom-up[19], at interfaces with solutions[20] or under vacuum. Supramolecular hydrogen bonded[21,22], metal-organic[23,24] or covalent[25–27] multi-component surface-confined networks provide a route to precisely organize different building blocks relative to each other in two dimensions (2D). These 2D surface assemblies can be engineered[28,29] with increasing level of prediction[30,31] to afford device functionalities ontop of specific substrates. In addition, 2D networks can act as templates for the growth of three-dimensional networks[32,33], paving the road towards vertical heteromolecular control via monolayer-by-monolayer growth. Such architectures may present precise interpenetrated morphologies[34] with ideal nanoporous and columnar order, which have long been considered optimal configurations for organic solar cells[11,35].

Here we demonstrate the photoresponse of a bicomponent supramolecular network (Fig. 1a) on transparent, graphene-passivated H-C(100) diamond (GHD) and employing a gallium droplet as a counter electrode (Fig. 1b). The network is built with a chromophore, consisting of a terrylene diimide (TDI) derivative (**1**) and melamine (**2**). After initial molecular characterization by means of scanning tunnelling microscopy, we show generated photocurrents of (0.5 ± 0.2) nA and photovoltages of (270 ± 120) mV at 19 mW cm^{-2} irradiation intensities at 710 nm (uncertainties are s.d. in tenths of measurements on different sample preparations). We find incident photon to electron efficiencies (IPCE) at 710 nm of (0.6 ± 0.25)% when estimating the contact area *ex situ*, yielding photocurrent densities of (47 ± 5) μA cm^{-2}. Our work introduces bottom-up supramolecular network engineering on 2D materials for molecularly precise function with atomically defined interfaces.

Results

Chemical design and synthesis. Heteromolecular recognition between tritopic melamine and complementary ditopic linkers via hydrogen bond formation was introduced as means to achieve hexagonal supramolecular architectures[21,36]. By proper chemical design, the complementary linker can be engineered[31,37] for self-assembly with increasing level of predictability at the solid–liquid interface, thereby decreasing the content of poorly ordered and glassy phases. For instance, we have previously shown that peripheral substitution of the linker favours the formation of networks over tightly packed patterns[22]. Further, the linker can be imbued with functional properties. Rylene dyes are poly(peri-naphthalene)s[38] that show a high photostability[39], which make them preferred constituents for application in optoelectronic devices. Extending the π-system of the dye allows for tuning the optical properties and thus, shifting the absorption maximum to higher wavelengths. At the same time, rylenes are known to be amongst the most efficient organic absorbers[38,40]. Thus, rylenes are promising for the infrared regime with high transparency in the visible spectrum, for use in the next generation of facade and window building technology[41]. By introducing diimide terminations, the molecules acquire the supramolecular moiety necessary for triple hydrogen bond complementary recognition with the melamine cornerstone. Substituents at the bay position are used to improve the solubility[38] and favour porous network formation. Thus a novel TDI, bearing NH groups in the imide structure, is synthesized according to Fig. 2 (see Supplementary Methods for details). The presence of NH groups makes the solubility poor and purification of the rylene dyes rather difficult. Therefore, we use bulky 2,6-diisopropylphenyl groups in the starting one-pot reaction of perylene monoimide **3** and naphthalene monoimide **4** to create the soluble compound **5**. After bromination and phenoxylation steps, tetra(*t*-octylphenoxy) substituted **7**, showing outstanding solubility[42], can be used in the following. Hydrolysis of the imide groups under basic conditions, results in bis-anhydride **8** which is reacted with ammonium acetate to afford the target **1**. In both cases the presence of four voluminous *t*-octylphenoxy groups made purification and processing possible. Comparing with soluble perylene diimide (PDI) analogues, TDIs have higher absorption coefficients, which render TDI a better light-harvesting efficiency[43]. Moreover, the additional naphthalene unit in the TDI structure separates the tetraphenoxy groups and makes the TDI molecules less twisted[44]. Thus, improved planarity and increased absorptivity[38,40] render the engineered TDI supramolecular assemblies advantageous over shorter PDI assemblies[21,45].

Scanning tunnelling microscopy characterization. The successful formation of highly regular 2D supramolecular networks between **1** and **2** has been investigated *ex situ* (on a model substrate) and *in situ* (in the device configuration) by means of scanning tunnelling microscopy. Self-assembly experiments involved applying a mixture of **1** + **2** in 1,2,4-tri-chlorobenzene (TCB) with 1–5% of dimethylsulfoxide (DMSO), first on highly ordered pyrolytic graphite (HOPG, Fig. 3a,c) and then on transparent platforms made by graphene transferred to hydrogenated H-C(100) diamond (GHD). In the scanning tunnelling microscope (STM) images, molecular features can be resolved. Figure 3a shows how the observed hexagonal supramolecular features perfectly match the expected chemical structure in Fig. 3b. The molecular geometry was optimized with the

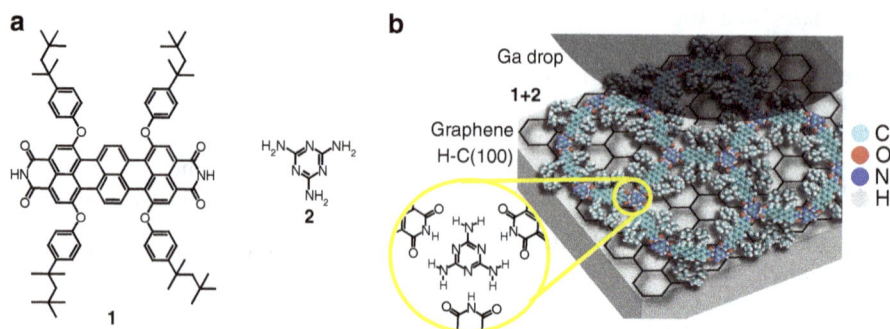

Figure 1 | Chemical structures of the molecules and setup. (a) Structure of TDI tetracarboxylic acid derivative (**1**) and melamine (**2**) (**b**). Schematic drawing of the photoresponse device setup including the ideal representation of the **1** + **2** mixture yielding a [**1**₃**2**₂]ₙ hexagonal supramolecular network via hydrogen bonds (yellow circle).

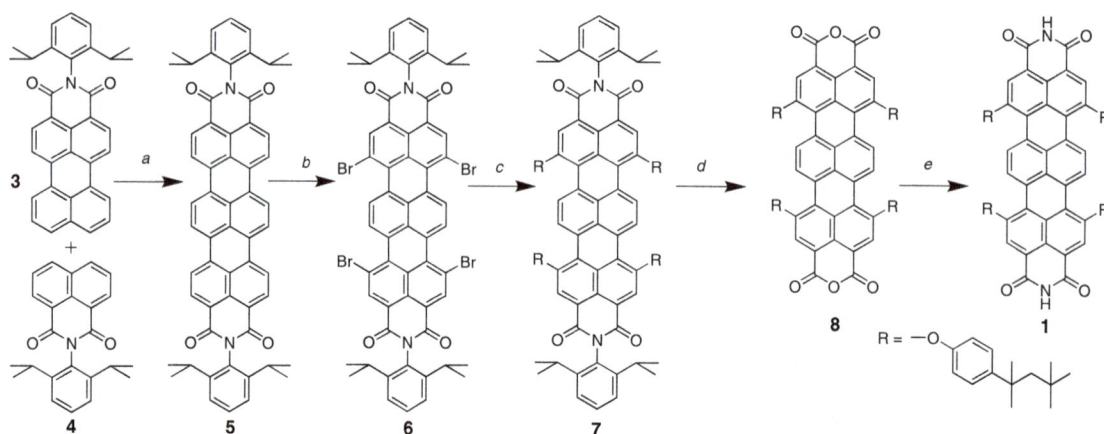

Figure 2 | Synthesis of novel TDI (1). (a) Base-induced fusion of naphthaleneimide and peryleneimide: diazabicyclo[4.3.0]non-5-ene, *t*-BuONa, diglyme, 130 °C, 3 h, 42%; (b) tetrabromination of TDI: Br₂, chloroform, reflux, 12 h, 75%; (c) phenoxylation of TDI: 4-(1,1,3,3-tetramethylbutyl)phenol, K₂CO₃, N-methylpyrrolidone, 80 °C, 8 h, 86%; (d) base-induced hydrolysis of bisimide into bis-anhydride: KOH, KF, 2-methyl-2-butanol, reflux, 53%; and (e) NH-imidization of terrylene bis-anhydride: ammonium acetate, propionic acid, reflux, 20%.

MMFF molecular force field. The 2D fast Fourier transform in Fig. 3c provides evidence for the regularity in the monolayer supported on HOPG. At GHD, only nanocrystalline hexagonal domains with sizes of tens of nanometres are monitored (Fig. 3d). Defects and impurities on GHD make the extended growth of crystalline bicomponent networks challenging, as discussed below. The experimental unit cell parameters for the network formation on GHD amount to $a = (3.9 \pm 0.2)$ nm and $b = (4.0 \pm 0.2)$ nm and $a,b = 62 \pm 1°$, which are in perfect agreement with the values measured on HOPG.

UV–visible measurements. With an atomically flat and transparent platform such as GHD, capable of supporting the bimolecular 2D self-assembly, optical spectral properties of crystalline supramolecular layers can now be investigated. Thus, UV–vis absorption measurements were performed to distinguish between changes in the spectrum of **1** upon hydrogen bond recognition with **2**. The absorption spectra of (5 ± 1) μl of the pristine **1** in a 12 μM TCB solution and the pure TCB solvent on the diamond supported graphene surface are shown in Fig. 4a. Two distinct absorption peaks at 665 and 735 nm can be distinguished. This corresponds to a bathochromic shift of 46 and 66 nm with respect to UV–vis measurements in solution (Supplementary Fig. 1). Note that the TCB solvent used for the drop-casting shows no absorption in that wavelength range. The absorption peak-to-baseline signal of **1**, 0.012 at (735 ± 2) nm (see Supplementary

Fig. 1 for absolute absorption units), is indicative of approximately a monolayer of **1** when compared with the absorbance of monolayer perylene tetracarboxylic anhydride on graphene[46], ~0.007 at 702 nm. The molar attenuation coefficient of perylene tetracarboxylic anhydrides and PDIs[47] of $\sim 5 \times 10^4$ M^{-1} cm^{-1} is half the one of analogue TDI derivatives[48], close to 1×10^5 M^{-1} cm^{-1}, at the respective absorbance maxima. When (5 ± 1) μl solutions of **1** and **2** with a concentration ratio of 12 μM:8 μM are applied to the sample, a fivefold reduction of the signal of **1** is observed along with a bathochromic shift of the absorbance maximum to (740 ± 5) nm (Fig. 4b, uncertainties are s.d. between four different preparations). At least a 1.5-fold reduction in the absorbance is expected when comparing the molecular density of molecules of **1** (0.31 nm^{-1}) with that of molecules of **1** in the **1** + **2** network (0.21 nm^{-1}). Incidentally, a fourfold reduction of the absorbance maximum is also prominent in π-stacks of perylenes[49], in part due to specific surface reduction[50,51]. Because the reduction of the absorbance is not an effect of variations in the drop-cast solution volume or concentration (see Supplementary Fig. 1), we suggest it is the combined effect of a looser packing (increased unit cell) of **1** + **2** and its aforementioned π-stacking[33].

Photoresponse characteristics. To preserve pristine molecular interfaces aiming at molecular precision, a gallium droplet has been used to soft-contact the supramolecular network. Current–

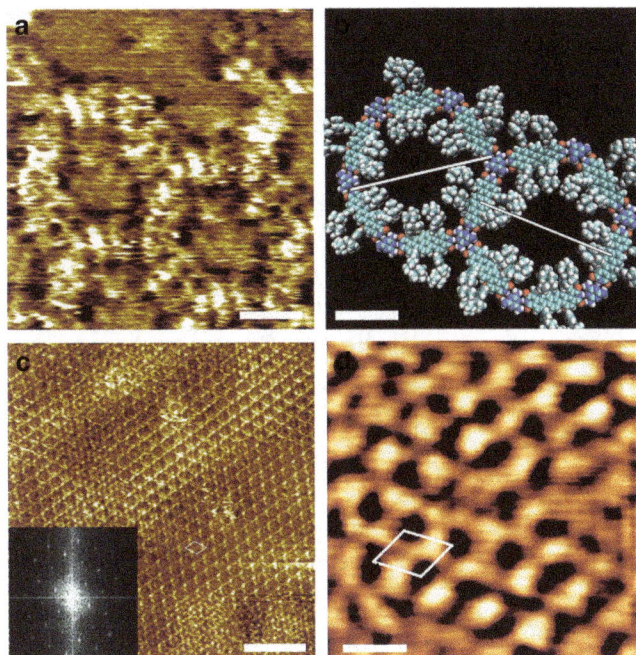

Figure 3 | STM images showing the assembly of 1 + 2 on graphite and graphene-passivated H-C(100) diamond substrates. (a) STM constant current image of **1 + 2** (16 µM: 10 µM) and underlying HOPG interface. Scale bar, 2 nm **(b)**. Molecular model minimized by the MMFF force field, the hexagonal size corresponding to the theoretical unit cell is $a = b = 4.2$ nm and the pore diameter is $d = 4.6$ nm. Scale bar, 2 nm **(c)**. STM large-area constant height image (12 µM:8 µM). (inset) 2D fast Fourier transform showing the high crystallinity of the assembly on HOPG. Unit cell $a = (4.1 \pm 0.2)$ nm, $b = (4.3 \pm 0.2)$ nm and $a,b = 65 \pm 2°$ Scale bar 20 nm **(d)**. Gaussian-filtered STM constant current image of **1 + 2** on GHD. Unit cell $a = (3.9 \pm 0.2)$ nm and $b = (4.0 \pm 0.2)$ nm and $a,b = 62 \pm 2°$. Area 13.8 nm^2. Tunneling parameters: average tunneling current $(I_t) = 20$ pA, sample voltage $(V_t) = 300$ mV. Scale bar, 5.5 nm.

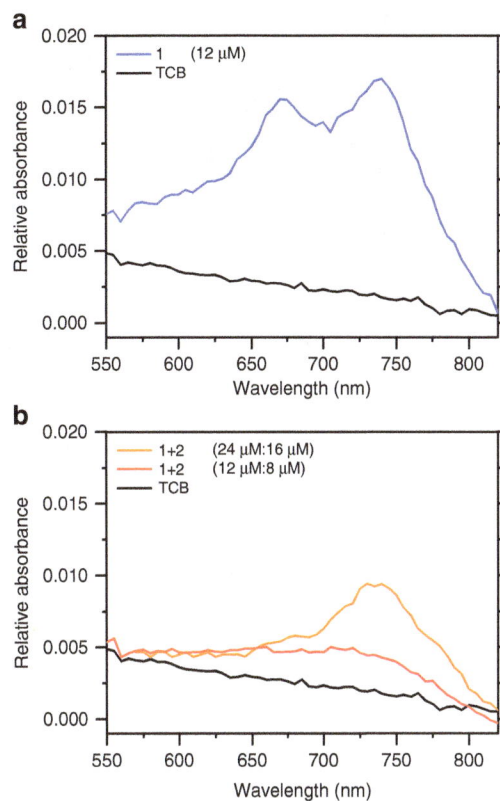

Figure 4 | UV-vis spectroscopy on graphene-passivated H-C(100) diamond. (a) 1 (12 µM) and pure TCB baseline spectra. **(b). 1 + 2** (12 µM:8 µM) and pure TCB baseline absorbance spectra. **1 + 2** (24 µM:16 µM) is also shown as additional evidence of strong absorbance reduction upon complexation with **2**.

voltage (*IV*) measurements have been used to characterize the device element's photoresponse. The top contact was fabricated by letting a liquid gallium droplet cover a blunt tungsten tip and slowly cool at room temperature (Fig. 5a). Gallium alloys have been widely used as a replacement for mercury for creating macroscopic device contacts with molecular layers[52,53]. Our strategy consists of approaching the gallium-coated tip to the substrate and stabilizing it to a current setpoint of 2 nA at 100 mV, with the help of a modified STM. Contact and wetting of the GHD substrate by the gallium was inferred by fluorescence spectroscopy (see Methods and Supplementary Fig. 2) and by measuring the current during approach of the electrode to the surface (Supplementary Fig. 3). In addition, stepwise increase of the setpoint current shows stable semiconductor characteristics up to 100 nA (Supplementary Fig. 4). The large-area *IV* measurements were performed on the **1 + 2** on GHD and bare GHD substrates. The *IV* spectra of GHD (Fig. 5c, black dotted line) and **1 + 2** on GHD (Fig. 5c, blue dotted line) were recorded with a forward sweep of 200 mV s^{-1}. For obtaining the photoresponse characteristics, the current was recorded under illumination by a red light-emitting diode (LED) ($\lambda = 710$ nm) with a measured power of 19 mW cm^{-2}. When bare GHD was employed, no photoresponse was observed (Fig. 5c, black line). Conversely, the illuminated *IV* curve of the 12 µM:8 µM **1 + 2** on GHD monolayer photoresponse exhibits characteristic features (Fig. 5c, red line). Under illumination, a finite current flows at

zero bias voltage, the short-circuit current I_{SC}. An average short-circuit current of $I_{SC} = (0.5 \pm 0.2)$ nA and open-circuit voltage of $V_{oc} = (270 \pm 120)$ mV were measured by illuminating the system with monochromatic light of 710 nm. Typical maximum and minimum values, from the average photoresponse characteristic, are also shown as shaded areas in Fig. 5c. In addition, *IV* curves were recorded by illuminating the **1 + 2** network on graphene with monochromatic light of 520 nm of 13 mW cm^{-2} (Fig. 5d), where **1** does not absorb light. Indeed, no photocurrent was generated under 520 nm light irradiation conditions where TDI does not absorb. Figure 5e,f depict the back illumination geometries employed. Figure 5g shows the stability of the photovoltage generated by the junctions as a function of light on–off cycles (employing functionalized gallium tips, see Methods). The data corresponds to ~25% of various measured junctions. In the remaining junctions, a different regime is observed, where a clear increase in the current with 710 nm irradiation occurs but neither open-circuit voltage nor short-circuit current are detected (Fig. 5h). Because current-distance spectroscopy reveals a clear exponential dependence of the current with the distance, these junctions do not physically contact the substrate. Hence, this non-contact regime is attributed to a photoexcitation effect, where additional tunnelling channels are opened upon light exposure. All in all the results show that the supramolecular network based on **1 + 2** molecules specifically generates a photoresponse at the designated wavelength. It is instructive to approximate the tunnelling contact area of the gallium droplet (~250 µm diameter) to estimate photoconversion efficiencies. By *ex situ* contact junctions with

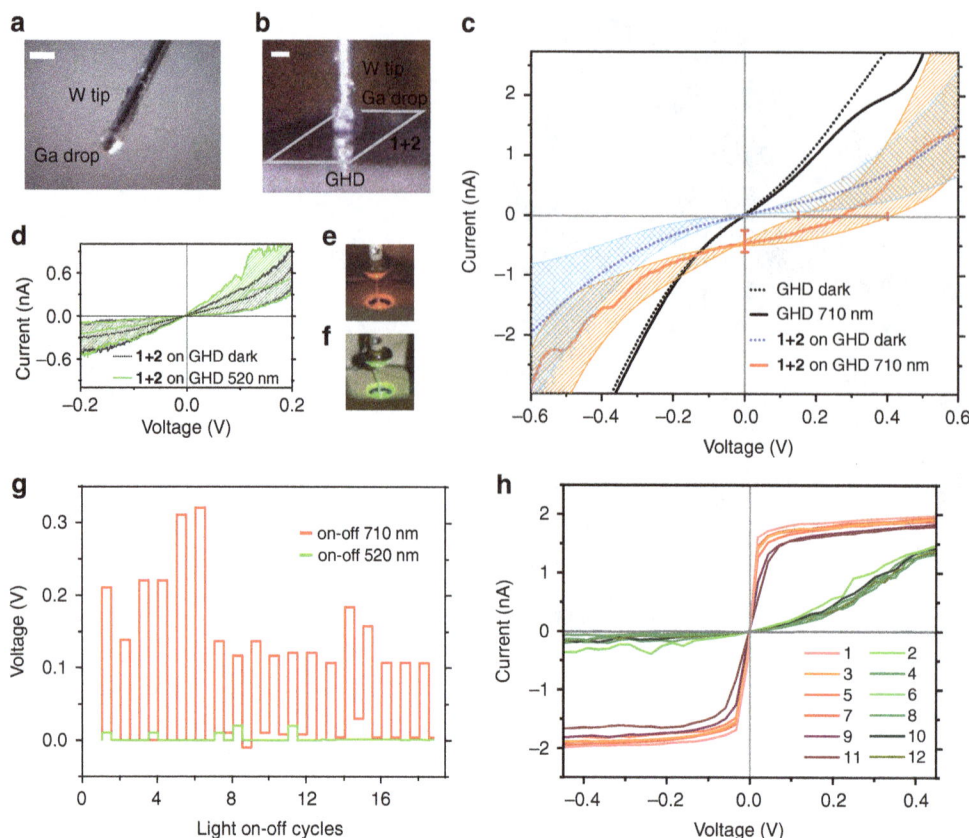

Figure 5 | Photoresponse of a supramolecular network optoelectronic device element. (**a**). Tungsten STM tip with a gallium droplet. (**b**) Gallium electrode in contact with the sample before the tunneling spectroscopy measurements. (**c**). Current–voltage characteristics (contact regime see main text) of pristine GHD (black lines) and of **1** + **2** on GHD before (blue lines) and after (red lines) irradiation at $\lambda = 710$ nm (**d**). Current–voltage characteristics before (black lines) and after (green lines) $\lambda = 520$ nm photon irradiation. Approach parameters: $I_t = 2$ nA and $V_t = 100$ mV. Striped areas indicate the maximum and minimum currents observed in more than ten curves in a single junction, while error bars indicate the standard deviation for five junctions in three samples and are reported in the main text as $I_{SC} = (0.5 \pm 0.2)$ nA, $V_{oc} = (270 \pm 120)$ mV (**e**). Back illumination geometry with $\lambda = 710$ nm (**f**). Back illumination geometry with $\lambda = 520$ nm. (**g**). Open-circuit voltage for consecutive on-off irradiation cycles at 710 nm (red lines), followed by on–off cycles at $\lambda = 520$ nm (green lines) of **1** + **2**. Every set of six consecutive cycles was recorded with different junction (sample, tip) preparations. The constant dark voltage background for each set was substracted (~ 0.05 V). For stability, these studies were performed with functionalized eutectic gallium-indium electrodes, see Methods. (**h**) Scanning tunnelling spectroscopy (non-contact regime, see main text) for consecutive irradiation cycles 1-12 at $\lambda = 710$ nm (red lines), followed by $\lambda = 520$ nm (green lines), showing drastic changes assigned to electron tunneling through photoexcited states. Setpoint parameters: $I_t = 1$ nA, $V_t = 300$ mV. For all panels, data shown was neither filtered nor averaged. Scale bars, 250 μm.

insulating fluorescence dyes (see Methods and Supplementary Fig. 2), an area of $(9.9 \pm 0.6) \times 10^2$ μm^2 can be estimated for the *in situ* experiment. This area is roughly 2% of the projected area under the gallium electrode $(49 \times 10^3$ μm^2, using a radius of 125 μm). With such estimation, current densities of 10^{-4} A cm^{-2} at 0.5 V can be derived from the *IV* spectra. These current densities are comparable to those reported by contacting -S-C$_{16}$H$_{33}$ self-assembled monolayers-metal interfaces of similar contact areas[53], indicating a soft-contact. Further, with the derived photocurrent density of $J_{SC} = (47 \pm 5)$ μA cm^{-2} a monochromatic IPCE of (0.6 ± 0.25)% can be estimated at 710 nm, 19 mW cm^{-2} irradiation intensities.

Discussion

We have fabricated surface-confined bicomponent assemblies on GHD based on a functional dye[38] absorbing at 740 nm. The active layer ideally consists of a self-assembled TDI-based supramolecular nanoporous network exhibiting nanocrystalline hexagonal order. Shorter diimide-based molecules like naphtalenes have been found to stack in the third dimension

through van der Waals face-to-face stacking[33]. Thus, the employed system presents an avenue towards molecularly precise three-dimensional devices. During the measurements, atomically flat and transparent all-carbon GHD served as the photoanode, while a gallium junction was used as top cathode electrode, respectively. For sake of maintaining molecular integrity at the interface, as observed by STM in Fig. 3, a soft gallium electrode is used as a contact. The photoresponse exhibited a three order of magnitude increase in the short-circuit current, (0.5 ± 0.2) nA, with respect to the dark current (Supplementary Fig. 5) and an open-circuit voltage of (270 ± 120) mV. It is worth mentioning that average short-circuit current and open-circuit voltage for the single component **1** were measured as (0.09 ± 0.01) nA and (160 ± 60) mV, respectively. The reported open-circuit voltage values are close to the energy level difference between the highest occupied molecular orbital (HOMO) of **1** and the work function of graphene. The electrode work function is 4.5 eV (ref. 54) for graphene and 4.3 eV (ref. 55) for gallium. The calculated first excitation of **1** occurs at 1.6 eV (775 nm) and the HOMO of **1** is 4.8 eV below the vacuum level (Supplementary Fig. 6). The correlation with the energy level difference might be coincidental, as the origins of the open-circuit voltage in excitonic solar cells are under intense discussion[56]. In our setup, a full molecular monolayer is guaranteed by employing concentrations and volumes, equivalent to 3.8×10^{13} molecules of **1** per substrate (and similar amount for **2**). Considering a substrate area of $5 \times 5 \, mm^2$ and three TDI molecules per **1 + 2** unit cell area of $13.8 \, nm^2$, a monolayer is formed with 5.4×10^{12} TDI molecules. The higher amount of molecules applied to the substrate was employed to compensate for ring stain effects when drying drop-casting solutions (see Supplementary Fig. 1). A peak UV–vis absorbance of 0.012 at $(735 \pm 2) \, nm$ for **1** provides additional evidence of a molecular monolayer. Upon **1 + 2** supramolecular network formation, absorbance is reduced, as expected because the assembly of a porous architecture entails a lower molecular surface density. In addition, formation of π-aggregate layers[49,50], reduces the resulting absorbance cross section, as well as hydrogen bonding[57] where charge transfer is likely to occur[58].

Our estimate on the tunnelling contact area allows us to elaborate on the technological implications of photoresponsive surface assemblies. Previously reported optimized photovoltaic devices of thin films of precursor molecule **7** blended with an organic acceptor[59], featured IPCEs[35] of 0.3% at 700 nm. These thin films were prepared by spin-coating solutions of $13 \, mg \, ml^{-1}$. The comparable estimate of an IPCE of $(0.6 \pm 0.25)\%$ at 710 nm for our system prepared by drop-casting a $5 \, \mu l$ of a solution 1,000 times more diluted ($0.015 \, mg \, ml^{-1}$ or approximately $12 \, \mu M$) suggests that the photovoltaic response of few monolayers of self-assembled molecular architectures could outperform the response of bulk spin-coated materials. This is in part due to high internal quantum yields for monolayer absorbers or high collection efficiencies at interfaces, as reported for C_{60}-porphyrin mixed self-assembled dyads[60], natural photosystem-I[18,61] and naturally occurring 2D crystals[62,63]. In our configuration, two molecular interfaces are formed, one between **1 + 2** on GHD and one between **1 + 2** and gallium oxide on gallium. The oxide[64] tunnelling barrier between gallium and the supramolecular assembly forms a blocking-layer that prevents the efficient collection of photogenerated electrons at the gallium electrode. We suggest that the resulting photoresponsive device element is hole-only, that is, photogenerated holes are readily collected at the graphene photoanode, while electrons have to tunnel to the gallium electrode. Hence, it is expected that the tuning of the work function and of the appropriate tunnelling junction material (for example, allowing hole-only and electron-only transport in the pertinent contacts)[65] will radically improve the efficiency of monolayer-thin organic devices.

In summary, bottom-up modular self-assembled networks and 2D materials grant access to device fabrication with molecular precision. We have shown the first macroscopic (μm scale) photoresponse characterization of a bicomponent supramolecular interfacial assembly. By *ex situ* estimations of the tunnelling area, our non-optimized device element configuration yields an IPCE (at 710 nm) as high as 0.6% in air, opening novel avenues towards tandem photovoltaics from monolayer-thin sensitizers. More importantly, we highlighted how, more than a decade since the introduction of interfacial bottom-up modular self-assembly[21,23,24] and *in situ* on-surface synthesis[25] for atomically precise fabrication, serious efforts are still required for molecularly precise device fabrication, with high throughput, large-area, dedicated analytical methods simply lacking. Our work motivates rapid progress in molecular engineered manufacturing and monolayer-by-monolayer molecular printing methods, which potentially grant access to exponential optimization of device performance.

Methods

CVD-grown graphene transfer and network formation. The experiments were performed under ambient conditions on CVD-grown graphene ($10 \times 10 \, mm$ on a copper foil, Graphene Platform, Japan) on hydrogenated diamond (Element Six, thickness: $500 \, \mu m$) surface. A high purity diamond C-(100) plate was cleaned with pure N-methyl-2-pyrrolidone (NMP) and isopropanol sonication treatment. Oxidation of the substrate was conducted with oxygen microwave plasma (TePla 100-E): 600 s at 200 W load coil power and 50 Pa oxygen pressure at a constant oxygen flow rate equivalent to $90 \, cm^3 \, min^{-1}$ (SCCM). To ensure a high quality conductive termination, hydrogenation of the diamond surface was performed in a quartz tube reactor (Seki Technotron Corp.) of a microwave-coupled ASTEX plasma system with three steps: 750 W and 50 mbar hydrogen pressure at a constant hydrogen flow rate of 100 SCCM at 700 °C for 15 min; 230 W, 10 mbar hydrogen pressure, 100 SCCM at 300 °C for 10 min; 0 W, 10 mbar hydrogen pressure, 100 SCCM at 35 °C for 30 min. The hydrogenated diamond was characterized via STM (Supplementary Fig. 7). After these treatments, the graphene layer was transferred[66,67] on diamond H-C(100). Melamine (Fluka, 52549, 99%) was used as received. Dimethyl sulfoxide (DMSO, Sigma-Aldrich, 99.9%) and anhydrous 1,2,4-trichlorobenzene (TCB, Sigma-Aldrich, 99.9%) were used as solvents without further purification. Uncertainty values were derived from the standard deviation of the balance's linearity (Sartorius CPA2245). The mother solutions in 10% DMSO and 90% in TCB were sonicated and heated to 80 °C and a dilution was prepared in TCB with a concentration ratio of **1 + 2** of $12 \, \mu M:8 \, \mu M$. The network was obtained by drop-casting $(5 \pm 1) \, \mu l$ of **1 + 2** on GHD.

Scanning tunnelling microscopy measurements. STM measurements (Agilent Technologies 5,100) were performed in constant current and constant height mode. The scanning tips were prepared by mechanically cutting a Pt/Ir wire (80:20%, Goodfellow, UK). STM data were analysed with the free WSxM software (Nanotec Electronica S.L., Spain) and the Gwyddion software[68]. All images, except Fig. 3c with Gaussian filtering, are shown with line-wise flattening to remove tilting effect of the substrate plane. The network structures were modelled by the Merck modular force field (MMFF)[69]. Supplementary Fig. 8 reports the STM data of the molecule of **1** in HOPG.

UV-visible absorption spectroscopy. UV–vis measurements were performed with a UV/VIS/NIR Spectrometer, Lambda 900 (Perkin Elmer). The spectra were recorded at the full spectra range (2,000–200 nm) with 5 nm data interval, 0.32 and 0.68 s integration time for UV–vis and NIR, respectively. For absorption measurements, $(5 \pm 1) \, \mu l$ of **1 + 2** ($12 \, \mu M:8 \, \mu M$) in 1,2,4-Trichlorobenzene (TCB, Sigma-Aldrich, 99.9%) were drop-casted on GHD and dried in air. The spectrum was averaged 10 times and is plotted versus the wavelength for each sample.

Photoresponse measurements. *IV* measurements were performed with a home-built STM described in detail elsewhere[70]. A blunt tungsten tip was dipped into a heated gallium droplet and directly mounted into the tip holder of the scanner. The gallium electrode was approached to the surface with approach parameters of 2 nA and 100 mV. The spectroscopy data were recorded with a forward sweep rate of $200 \, mV \, s^{-1}$ using a Femto pre-amplifier. For recording the on–off cycle photovoltages in Fig. 5g, EGaIn electrodes coated with alkyl thiols were employed for increased stability. The tungsten blunt tips were dipped in EGaIn (495425, Sigma-Aldrich) until a smooth coating was obtained and subsequently immersed in a pure solution of 1-dodecanethiol (471364, Sigma-Aldrich) for 15 min. These *IV* measurements were independently performed in an Agilent Technologies 5,100 using a logarithmic current amplifier to avoid current saturation. Before each single tunnelling spectroscopy measurement, the feedback vertical position of the electrode was regulated again to a tunnelling current of 1 nA and a voltage of 300 mV and turned off. The data were recorded with a forward sweep rate of $20 \, mV \, s^{-1}$. For the illumination, a 710 nm (30 mW, 18°, Roithner Lasertechnik, Austria) and 520 nm (9,600 mCd, 123 mW, 30°, Nichia, Japan) LED were used. The photovoltaic detection limit of our setup is 90 mV, calculated as trice the standard deviation of the dark voltage. The acquired data in Fig. 5c,d,g corresponds to stable contact junctions (no observable oscillations in the junction z axis piezo electric control nor identifiable non-contact tunnelling junction formation, see main text) among tenths of different area surveys on four different samples.

Fluorescence measurement of the Ga droplet contact area. A Ga droplet was prepared as employed for the photocurrent measurements. The droplet was brought into tunnelling contact with a thin film of fluorescent dye on HOPG surface with the same approach and current parameters used in the STM measurement. By employing a nm-thick film of an insulating dye Rhodamin B (Radiant Dyes, Wermelskirchen), a contact junction forms, implying physical contact with the monolayer. The physical contact of the gallium droplet with a thick film of Rhodamin B leads to a higher deformation the droplet, increasing the contact area in comparison to the photoresponse contacts. Therefore, this method accurately estimates an upper bound to the actual device element contact area and therefore, a minimum current density and efficiency. To form the nm-thick film, HOPG was spin-coated (480 r.p.m.) three times with $10 \, \mu l$ of a saturated solution of

Rhodamin B in acetone, and dried between each application. The tip was retracted from the surface, fixed to a microscope slide and imaged under fluorescence conditions with a fluorescence microscope (Leica DMI 3000B, Wetzlar). The image of the tip is shown in Supplementary Fig. 2 with enhanced contrast to make the contour of the tip visible (darker area). From the raw data (greyscale TIFF-image), the number of pixels with brightness higher than a certain threshold was extracted. From the length scale per pixel (known from calibration), the total area corresponding to the pixels was calculated. The threshold was set to a clear gap between brighter and darker pixels in the histogram and the corresponding area in the image was identified.

References

1. Tang, C. W. & VanSlyke, S. A. Organic electroluminescent diodes. *Appl. Phys. Lett.* **51,** 913–915 (1987).
2. Tang, C. W. 2-Layer organic photovoltaic cell. *Appl. Phys. Lett.* **48,** 183–185 (1986).
3. Burroughes, J. H. *et al.* Light-emitting diodes based on conjugated polymers. *Nature* **347,** 539–541 (1990).
4. Hebner, T. R., Wu, C. C., Marcy, D., Lu, M. H. & Sturm, J. C. Ink-jet printing of doped polymers for organic light emitting devices. *Appl. Phys. Lett.* **72,** 519–521 (1998).
5. deBoer, R. W. I., Gershenson, M. E., Morpurgo, A. F. & Podzorov, V. Organic single-crystal field-effect transistors. *Phys. Status Solidi A* **201,** 1302–1331 (2004).
6. Halik, M. & Hirsch, A. The potential of molecular self-assembled monolayers in organic electronic devices. *Adv. Mater.* **23,** 2689–2695 (2011).
7. Yamamoto, Y. *et al.* Photoconductive coaxial nanotubes of molecularly connected electron donor and acceptor layers. *Science* **314,** 1761–1764 (2006).
8. Wasielewski, M. R. Self-assembly strategies for integrating light harvesting and charge separation in artificial photosynthetic systems. *Acc. Chem. Res.* **42,** 1910–1921 (2009).
9. Briseno, A. L. *et al.* Patterning organic single-crystal transistor arrays. *Nature* **444,** 913–917 (2006).
10. Murphy, A. R. & Fréchet, J. M. J. Organic semiconducting oligomers for use in thin film transistors. *Chem. Rev.* **107,** 1066–1096 (2007).
11. Heeger, A. J. 25th anniversary article: bulk heterojunction solar cells: understanding the mechanism of operation. *Adv. Mater.* **26,** 10–27 (2014).
12. Vlachopoulos, N., Liska, P., Augustynski, J. & Grätzel, M. Very efficient visible-light energy harvesting and conversion by spectral sensitization of high surface-area polycrystalline titanium-dioxide films. *J. Am. Chem. Soc.* **110,** 1216–1220 (1988).
13. O'Regan, B. & Grätzel, M. A low-cost, high-efficiency solar-cell based on dye-sensitized colloidal TiO$_2$ films. *Nature* **353,** 737–740 (1991).
14. Hagfeldt, A., Boschloo, G., Sun, L. C., Kloo, L. & Pettersson, H. Dye-sensitized solar cells. *Chem. Rev.* **110,** 6595–6663 (2010).
15. Scharber, M. C. *et al.* Design rules for donors in bulk-heterojunction solar cells - towards 10% energy-conversion efficiency. *Adv. Mater.* **18,** 789–794 (2006).
16. Krebs, F. C., Espinosa, N., Hosel, M., Sondergaard, R. R. & Jorgensen, M. 25th anniversary article: rise to power-OPV-based solar parks. *Adv. Mater.* **26,** 29–38 (2014).
17. Park, H. *et al.* Nanomechanical oscillations in a single-C-60 transistor. *Nature* **407,** 57–60 (2000).
18. Gerster, D. *et al.* Photocurrent of a single photosynthetic protein. *Nat. Nanotechnol.* **7,** 673–676 (2012).
19. Palma, C.-A. & Samorì, P. Blueprinting macromolecular electronics. *Nat. Chem.* **3,** 431–436 (2011).
20. Rabe, J. P. & Buchholz, S. Commensurability and mobility in two-dimensional molecular patterns on graphite. *Science* **253,** 424–427 (1991).
21. Theobald, J. A., Oxtoby, N. S., Phillips, M. A., Champness, N. R. & Beton, P. H. Controlling molecular deposition and layer structure with supramolecular surface assemblies. *Nature* **424,** 1029–1031 (2003).
22. Palma, C.-A. *et al.* Tailoring bicomponent supramolecular nanoporous networks: phase segregation, polymorphism, and glasses at the solid-liquid interface. *J. Am. Chem. Soc.* **131,** 13062–13071 (2009).
23. Stepanow, S. *et al.* Steering molecular organization and host-guest interactions using two-dimensional nanoporous coordination systems. *Nat. Mater.* **3,** 229–233 (2004).
24. Barth, J. V. Molecular architectonic on metal surfaces. *Annu. Rev. Phys. Chem* **58,** 375–407 (2007).
25. Grill, L. *et al.* Nano-architectures by covalent assembly of molecular building blocks. *Nat. Nanotechnol.* **2,** 687–691 (2007).
26. Zwaneveld, N. A. *et al.* Organized formation of 2D extended covalent organic frameworks at surfaces. *J. Am. Chem. Soc.* **130,** 6678–6679 (2008).
27. Ciesielski, A. *et al.* Dynamic covalent chemistry of bisimines at the solid/liquid interface monitored by scanning tunnelling microscopy. *Nat. Chem.* **6,** 1017–1023 (2014).
28. Barth, J. V., Costantini, G. & Kern, K. Engineering atomic and molecular nanostructures at surfaces. *Nature* **437,** 671–679 (2005).
29. Ciesielski, A., Palma, C.-A., Bonini, M. & Samorì, P. Towards supramolecular engineering of functional nanomaterials: pre-programming multi-component 2D self-assembly at solid-liquid interfaces. *Adv. Mater.* **22,** 3506–3520 (2010).
30. Palma, C.-A., Cecchini, M. & Samorì, P. Predicting self-assembly. *Chem. Soc. Rev.* **41,** 3713–3730 (2012).
31. Whitelam, S. *et al.* Common physical framework explains phase behavior and dynamics of atomic, molecular, and polymeric network formers. *Phys. Rev. X* **4,** 011044–1 011044-12 (2014).
32. Blunt, M. O. *et al.* Guest-induced growth of a surface-based supramolecular bilayer. *Nat. Chem.* **3,** 74–78 (2011).
33. Li, J. *et al.* Three-dimensional bicomponent supramolecular nanoporous self-assembly on a hybrid all-carbon atomically flat and transparent platform. *Nano Lett.* **14,** 4486–4492 (2014).
34. Halls, J. J. M. *et al.* Efficient photodiodes from interpenetrating polymer networks. *Nature* **376,** 498–500 (1995).
35. Gunes, S., Neugebauer, H. & Sariciftci, N. S. Conjugated polymer-based organic solar cells. *Chem. Rev.* **107,** 1324–1338 (2007).
36. Zerkowski, J. A., Seto, C. T. & Whitesides, G. M. Solid-state structures of rosette and crinkled tape motifs derived from the cyanuric acid melamine lattice. *J. Am. Chem. Soc.* **114,** 5473–5475 (1992).
37. Palma, C.-A., Samorì, P. & Cecchini, M. Atomistic simulations of 2D bicomponent self-assembly: from molecular recognition to self-healing. *J. Am. Chem. Soc.* **132,** 17880–17885 (2010).
38. Weil, T., Vosch, T., Hofkens, J., Peneva, K. & Müllen, K. The rylene colorant family--tailored nanoemitters for photonics research and applications. *Angew. Chem. Int. Ed. Engl.* **49,** 9068–9093 (2010).
39. Wöll, D. *et al.* Polymers and single molecule fluorescence spectroscopy, what can we learn? *Chem. Soc. Rev.* **38,** 313–328 (2009).
40. Chen, L., Li, C. & Müllen, K. Beyond perylene diimides: synthesis, assembly and function of higher rylene chromophores. *J. Mater. Chem. C* **2,** 1938–1956 (2014).
41. Li, G., Zhu, R. & Yang, Y. Polymer solar cells. *Nat. Photon.* **6,** 153–161 (2012).
42. Nolde, F. *et al.* Synthesis and modification of terrylenediimides as high-performance fluorescent dyes. *Chem. Eur. J.* **11,** 3959–3967 (2005).
43. Avlasevich, Y., Li, C. & Müllen, K. Synthesis and applications of core-enlarged perylene dyes. *J. Mater. Chem.* **20,** 3814–3826 (2010).
44. Würthner, F. Bay-Substituted Perylene & Bisimides Twisted fluorophores for supramolecular chemistry. *Pure Appl. Chem.* **78,** 2341–2349 (2006).
45. Madueno, R., Raisanen, M. T., Silien, C. & Buck, M. Functionalizing hydrogen-bonded surface networks with self-assembled monolayers. *Nature* **454,** 618–621 (2008).
46. Huang, S. *et al.* Molecular selectivity of graphene-enhanced Raman scattering. *Nano Lett.* **15,** 2892–2901 (2015).
47. Würthner, F. *et al.* Preparation and characterization of regioisomerically pure 1,7-disubstituted perylene bisimide dyes. *J. Org. Chem.* **69,** 7933–7939 (2004).
48. Pschirer, N. G., Kohl, C., Nolde, T., Qu, J. Q. & Müllen, K. Pentarylene- and hexarylenebis(dicarboximide)s: near-infrared-absorbing polyaromatic dyes. *Angew. Chem. Int. Ed. Engl.* **45,** 1401–1404 (2006).
49. Chen, Z. J. *et al.* Photoluminescence and conductivity of self-assembled π-π stacks of perylene bisimide dyes. *Chem. Eur. J.* **13,** 436–449 (2007).
50. Spano, F. C. The spectral signatures of frenkel polarons in h- and j-aggregates. *Acc. Chem. Res.* **43,** 429–439 (2010).
51. Weigand, R., Rotermund, F. & Penzkofer, A. Aggregation dependent absorption reduction of indocyanine green. *J. Phys. Chem. A* **101,** 7729–7734 (1997).
52. Nijhuis, C. A., Reus, W. F., Siegel, A. C. & Whitesides, G. M. A molecular half-wave rectifier. *J. Am. Chem. Soc.* **133,** 15397–15411 (2011).
53. Nijhuis, C. A., Reus, W. F., Barber, J. R. & Whitesides, G. M. Comparison of SAM-based junctions with Ga$_2$O$_3$/Egaln top electrodes to other large-area tunnelling junctions. *J. Phys. Chem. C* **116,** 14139–14150 (2012).
54. Giovannetti, G. *et al.* Doping graphene with metal contacts. *Phys. Rev. Lett.* **101,** 026803–1 026803-4 (2008).
55. Srisonphan, S., Jung, Y. S. & Kim, H. K. Metal-oxide-semiconductor field-effect transistor with a vacuum channel. *Nat. Nanotechnol.* **7,** 504–508 (2012).
56. Gregg, B. A. Excitonic solar cells. *J. Phys. Chem. B.* **107,** 4688–4698 (2003).
57. Llanes-Pallas, A. *et al.* Engineering of supramolecular H-bonded nanopolygons via self-assembly of programmed molecular modules. *J. Am. Chem. Soc.* **131,** 509–520 (2009).
58. Reece, S. Y. & Nocera, D. G. Proton-coupled electron transfer in biology: results from synergistic studies in natural and model systems. *Annu. Rev. Biochem.* **78,** 673–699 (2009).
59. Gorenflot, J. *et al.* Detailed study of N,N '-(diisopropylphenyl)-terrylene-3,4:11,12-bis(dicarboximide) as electron acceptor for solar cells application. *Synth. Met.* **161,** 2669–2676 (2012).
60. Imahori, H. & Fukuzumi, S. Porphyrin- and fullerene-based molecular photovoltaic devices. *Adv. Funct. Mater.* **14,** 525–536 (2004).

61. Carmeli, I., Frolov, L., Carmeli, C. & Richter, S. Photovoltaic activity of photosystem I-based self-assembled monolayer. *J. Am. Chem. Soc.* **129**, 12352–12353 (2007).

62. Bernardi, M., Palummo, M. & Grossman, J. C. Extraordinary sunlight absorption and one nanometer thick photovoltaics using two-dimensional monolayer materials. *Nano Lett.* **13**, 3664–3670 (2013).

63. Lee, C. H. *et al.* Atomically thin P-N junctions with van Der Waals heterointerfaces. *Nat. Nanotechnol.* **9**, 676–681 (2014).

64. Čechal, J. *et al.* Characterization of oxidized gallium droplets on silicon surface: an ellipsoidal droplet shape model for angle resolved X-ray photoelectron spectroscopy analysis. *Thin Solid Films* **517**, 1928–1934 (2009).

65. Shieh, J. T. *et al.* The effect of carrier mobility in organic solar cells. *J. Appl. Phys.* **107**, 084503–1 084503–9 (2010).

66. Yan, Z. *et al.* Toward the synthesis of wafer-scale single-crystal graphene on copper foils. *ACS Nano* **6**, 9110–9117 (2012).

67. Li, X. *et al.* Large-area synthesis of high-quality and uniform graphene films on copper foils. *Science* **324**, 1312–1314 (2009).

68. Nečas, D. & Klapetek, P. Gwyddion: an open-source software for SPM data analysis. *Cent. Eur. J. Phys.* **10**, 181–188 (2012).

69. Halgren, T. A. Merck molecular force field. I. Basis, form, scope, parameterization, and performance of MMFF94. *J. Comput. Chem.* **17**, 490–519 (1996).

70. Wilms, M., Kruft, M., Bermes, G. & Wandelt, K. A new and sophisticated electrochemical scanning tunneling microscope design for the investigation of potentiodynamic processes. *Rev. Sci. Instrum.* **70**, 3641–3650 (1999).

Acknowledgements

This work was supported by the European Commission through the European Research Council projects MolArt (GA-247299) and SUPRAFUNCTION (GA-257305) and the Graphene Flagship (GA-604391), as well as by the DFG Excellence Clusters Nanosystems Initiative Munich (NIM) and Munich Center for Advanced Photonics (MAP). We thank Katharina Henneberg for the fluorescence measurements and Daniel Mosegui for discussions. J.L. thanks the China Scholarship Council for financial support. S.W. thanks Lea Nienhaus for additional data. We gratefully acknowledge the collaboration with Martino Saracino and Klaus Wandelt on the instrumentation development. P. Samorì acknowledges the Agence Nationale de la Recherche through the LabEx project Chemistry of Complex Systems (ANR-10-LABX-0026_CSC) and the International Center for Frontier Research in Chemistry (icFRC).

Author contributions

S.W. and J.L. contributed equally to the work. S.W., J.L., M.K and C.-A.P. performed the UV–vis and *IV* spectroscopy experiments. J.L. and C.-A.P. performed the STM experiments. Y.A., C.L. and K.M. synthesized the molecules. P. Samorì, F.E., J.V.B. and C.-A.P. planned the experiments. P. Simon, J.A.G. and U.H. developed experimental infrastructure. S.W., F.E., J.V.B. and C.-A.P. wrote the manuscript. All authors discussed the manuscript. K.M. and C.-A.P. supervised the research.

Additional information

Permissions

All chapters in this book were first published in NC, by Nature Publishing Group; hereby published with permission under the Creative Commons Attribution License or equivalent. Every chapter published in this book has been scrutinized by our experts. Their significance has been extensively debated. The topics covered herein carry significant findings which will fuel the growth of the discipline. They may even be implemented as practical applications or may be referred to as a beginning point for another development.

The contributors of this book come from diverse backgrounds, making this book a truly international effort. This book will bring forth new frontiers with its revolutionizing research information and detailed analysis of the nascent developments around the world.

We would like to thank all the contributing authors for lending their expertise to make the book truly unique. They have played a crucial role in the development of this book. Without their invaluable contributions this book wouldn't have been possible. They have made vital efforts to compile up to date information on the varied aspects of this subject to make this book a valuable addition to the collection of many professionals and students.

This book was conceptualized with the vision of imparting up-to-date information and advanced data in this field. To ensure the same, a matchless editorial board was set up. Every individual on the board went through rigorous rounds of assessment to prove their worth. After which they invested a large part of their time researching and compiling the most relevant data for our readers.

The editorial board has been involved in producing this book since its inception. They have spent rigorous hours researching and exploring the diverse topics which have resulted in the successful publishing of this book. They have passed on their knowledge of decades through this book. To expedite this challenging task, the publisher supported the team at every step. A small team of assistant editors was also appointed to further simplify the editing procedure and attain best results for the readers.

Apart from the editorial board, the designing team has also invested a significant amount of their time in understanding the subject and creating the most relevant covers. They scrutinized every image to scout for the most suitable representation of the subject and create an appropriate cover for the book.

The publishing team has been an ardent support to the editorial, designing and production team. Their endless efforts to recruit the best for this project, has resulted in the accomplishment of this book. They are a veteran in the field of academics and their pool of knowledge is as vast as their experience in printing. Their expertise and guidance has proved useful at every step. Their uncompromising quality standards have made this book an exceptional effort. Their encouragement from time to time has been an inspiration for everyone.

The publisher and the editorial board hope that this book will prove to be a valuable piece of knowledge for researchers, students, practitioners and scholars across the globe.

List of Contributors

Simone Meloni and Ursula Rothlisberger
Laboratoire de Chimie et Biochimie Computationnelles, ISIC, FSB-BCH, École Polytechnique Fédérale de Lausanne (EPFL), Lausanne CH-1015, Switzerland
National Competence Center of Research (NCCR) MARVEL — Materials' Revolution: Computational Design and Discovery of Novel Materials, Lausanne CH-1015, Switzerland

Thomas Moehl, Shaik Mohammed Zakeeruddin and Michael Graetzel
Laboratory of Photonics and Interfaces, ISIC, Swiss Federal Institute of Technology (EPFL), Lausanne CH-1015, Switzerland

Wolfgang Tress
Laboratory of Photonics and Interfaces, ISIC, Swiss Federal Institute of Technology (EPFL), Lausanne CH-1015, Switzerland
Group for Molecular Engineering of Functional Materials, ISIC-Valais, Swiss Federal Institute of Technology (EPFL), Lausanne CH-1015, Switzerland

Marius Franckevičius
Laboratory of Photonics and Interfaces, ISIC, Swiss Federal Institute of Technology (EPFL), Lausanne CH-1015, Switzerland
Center for Physical Sciences and Technology, Savanoriu̧ Avenue 231, Vilnius LT-02300, Lithuania

Michael Saliba, Yong Hui Lee, Peng Gao and Mohammad Khaja Nazeeruddin
Group for Molecular Engineering of Functional Materials, ISIC-Valais, Swiss Federal Institute of Technology (EPFL), Lausanne CH-1015, Switzerland

Saurabh Kumar Singh and Gopalan Rajaraman
Department of Chemistry, Indian Institute of Technology, Bombay Powai, Mumbai 400076, India

Hyeongwook Im, Taewoo Kim, Hyelynn Song and Jongho Choi
School of Mechanical and Aerospace Engineering, Seoul National University, Seoul 151-742, South Korea

Jae Sung Park
Institute of Advanced Machinery and Design, Seoul National University, Seoul 151-742, South Korea

Raquel Ovalle-Robles
Nano-Science & Technology Center, Lintec of America, Inc., Richardson, Texas 75081, USA

Hee Doo Yang
Department of NanoMechatronics Engineering, College of Nanoscience and Nanotechnology, Pusan National University, Busan 609-735, South Korea

Kenneth D. Kihm
Department of Mechanical, Aerospace and Biomedical Engineering, University of Tennessee, Knoxville, Tennessee 37996, USA

Ray H. Baughman
Alan G. MacDiarmid NanoTech Institute, University of Texas at Dallas, Richardson, Texas 75080, USA

Hong H. Lee
School of Chemical and Biological Engineering, Seoul National University, Seoul 151-744, South Korea

Tae June Kang
Department of Mechanical Engineering, INHA University, Incheon 22212, South Korea

Yong Hyup Kim
School of Mechanical and Aerospace Engineering, Seoul National University, Seoul 151-742, South Korea
Institute of Advanced Aerospace Technology, Seoul National University, Seoul 151-742, South Korea

Minghui Liu
Center for Molecular Design and Biomimetics, the Biodesign Institute at Arizona State University, Tempe, Arizona 85287, USA

Zhao Zhao, Yan Liu and Hao Yan
Center for Molecular Design and Biomimetics, the Biodesign Institute at Arizona State University, Tempe, Arizona 85287, USA
School of Molecular Sciences, Arizona State University, Tempe, Arizona 85287, USA

Jinglin Fu and Ting Zhang
Department of Chemistry, Center for Computational and Integrative Biology, Rutgers University-Camden, Camden, New Jersey 08102, USA

Soma Dhakal, Alexander Johnson-Buck and Nils G. Walter
Department of Chemistry, Single Molecule Analysis Group, University of Michigan, Ann Arbor, Michigan 48109, USA

Neal W. Woodbury
School of Molecular Sciences, Arizona State University, Tempe, Arizona 85287, USA
Center for Innovations in Medicine, the Biodesign Institute at Arizona State University, Tempe, Arizona 85287, USA

Chuanshou Wang, Jing Wang, Jinxing Zhang and Yu Tian
Department of Physics, Beijing Normal University, 100875 Beijing, China

Xiaoxing Ke
EMAT (Electron Microscopy for Materials Science), University of Antwerp, Groenenborgerlaan 171, B-2020 Antwerpen, Belgium
Institute of Microstructures and Properties of Advanced Materials, Beijing University of Technology, 100124 Beijing, China

Gustaaf Van Tendeloo
EMAT (Electron Microscopy for Materials Science), University of Antwerp, Groenenborgerlaan 171, B-2020 Antwerpen, Belgium

Jianjun Wang and Ce-Wen Nan
State Key Laboratory of New Ceramics and Fine Processing, School of Materials Science and Engineering, Tsinghua University, 100084 Beijing, China

Long-Qing Chen
State Key Laboratory of New Ceramics and Fine Processing, School of Materials Science and Engineering, Tsinghua University, 100084 Beijing, China
Department of Materials Science and Engineering, The Pennsylvania State University, University Park, Pennsylvania, 16802 Pennsylvania, USA

Renrong Liang
Tsinghua National Laboratory for Information Science and Technology, Institute of Microelectronics, Tsinghua University, 100084 Beijing, China

Zhenlin Luo
National Synchrotron Radiation Laboratory and CAS Key Laboratory of Materials for Energy Conversion, University of Science and Technology of China, 230026 Hefei, China

Di Yi, Ramamoorthy Ramesh
Department of Materials Science and Engineering, University of California, 94720 Berkeley, California, USA

Qintong Zhang and Xiu-Feng Han
Beijing National Laboratory of Condensed Matter Physics, Institute of Physics, Chinese Academy of Science, 100190 Beijing, China

Ulrike Böhm, Stefan W. Hell and Roman Schmidt
Department of NanoBiophotonics, Max Planck Institute for Biophysical Chemistry, Am Fassberg 11, Göttingen 37077, German

Yao Li
Department of Applied Physics, Stanford University, Stanford, California 94305, USA

Karel-Alexander N. Duerloo and Evan J. Reed
Department of Material Science and Engineering, Stanford University, Stanford, California 94305, USA

Kerry Wauson
Klipsch School of Electrical and Computer Engineering, New Mexico State University, Las Cruces, New Mexico 88003, USA

Adam Sweetman, Mohammad A. Rashid, Samuel P. Jarvis, Janette L. Dunn, Philipp Rahe and Philip Moriarty
School of Physics and Astronomy, University of Nottingham, Nottingham NG7 2RD, UK

Xiumei Geng, Weiwei Sun, WeiWu, Alaa Al-Hilo, Fumiya Watanabe and Tar-pin Chen
Department of Physics and Astronomy, University of Arkansas at Little Rock, 2801 South University Avenue, Little Rock, Arkansas 72204, USA

Benjamin Chen
Department of Physics, University at Buffalo, Buffalo, New York 14260, USA

Mourad Benamara
Institute for Nanoscale Materials Science and Engineering, University of Arkansas, Fayetteville, Arkansas 72701, USA

Hongli Zhu
Department of Mechanical and Industrial Engineering, Northeastern University, Boston, Massachusetts 02115, USA

Jingbiao Cui
Department of Physics and Materials Science, University of Memphis, Memphis, Tennessee 38152, USA

Bernd Gludovatz
Materials Sciences Division, Lawrence Berkeley National Laboratory, Berkeley, California 94720, USA

Anton Hohenwarter
Department of Materials Physics, Montanuniversität Leoben and Erich Schmid Institute of Materials Science, Austrian Academy of Sciences, Leoben 8700, Austria

Keli V.S. Thurston and Robert O. Ritchie
Materials Sciences Division, Lawrence Berkeley National Laboratory, Berkeley, California 94720, USA
Department of Materials Science and Engineering, University of California, Berkeley, California 94720, USA

Hongbin Bei
Materials Sciences and Technology Division, Oak Ridge National Laboratory, Oak Ridge, Tennessee 37831, USA

ZhenggangWu
Department of Materials Sciences and Engineering, University of Tennessee, Knoxville, Tennessee 37996, USA

Easo P. George
Materials Sciences and Technology Division, Oak Ridge National Laboratory, Oak Ridge, Tennessee 37831, USA
Department of Materials Sciences and Engineering, University of Tennessee, Knoxville, Tennessee 37996, USA

Xingjun Zhu
Institutes of Biomedical Sciences, Fudan University, 220 Handan Road, Shanghai 200433, China

Wei Feng, Jiachang Li, Min Chen and Yun Sun
Department of Chemistry, State Key Laboratory of Molecular Engineering of Polymers, Fudan University, 220 Handan Road, Shanghai 200433, China

Jian Chang and Yan-Wen Tan
Department of Physics, Fudan University, 220 Handan Road, Shanghai 200433, China

Fuyou Li
Department of Chemistry, State Key Laboratory of Molecular Engineering of Polymers, Fudan University, 220 Handan Road, Shanghai 200433, China
Collaborative Innovation Center of Chemistry for Energy Materials, Fudan University, 220 Handan Road, Shanghai 200433, China

Xiao Tao Geng, Dong-Eon Kim and Seungchul Kim
Max Planck Center for Attosecond Science, Max Planck POSTECH/KOREA Res. Initiative, Pohang, Gyeongbuk 376-73, South Korea
Department of Physics, Center for Attosecond Science and Technology (CASTECH), POSTECH, Pohang, Gyeongbuk 376-73, South Korea

Byung Jae Chun and Young-Jin Kim
School of Mechanical and Aerospace Engineering, Nanyang Technological University (NTU), 50 Nanyang Avenue, Singapore 639798, Singapore

Ji Hoon Seo and Kwanyong Seo
Department of Energy Engineering, Ulsan National Institute of Science and Technology (UNIST), Ulsan 689-798, South Korea

Hana Yoon
Energy Storage Department, Korea Institute of Energy Research (KIER), Daejeon 305-343, South Korea

Kentaro Watanabe
International Center for Materials Nanoarchitectonics, National Institute for Materials Science, 1-1 Namiki, Ibaraki 305-0044, Japan
Graduate School of Engineering Science, Osaka University, 1-3 Machikaneyama-cho, Osaka 560-8531, Japan

Takahiro Nagata, Seungjun Oh, Yutaka Wakayama and Takashi Sekiguchi
International Center for Materials Nanoarchitectonics, National Institute for Materials Science, 1-1 Namiki, Ibaraki 305-0044, Japan

János Volk
MTA EK Institute of Technical Physics and Materials Science, Konkoly Thege M. ut 29-33, Budapest 1121, Hungary

Yoshiaki Nakamura
Graduate School of Engineering Science, Osaka University, 1-3 Machikaneyama-cho, Osaka 560-8531, Japan

You-Chia Chang
Center for Photonics and Multiscale Nanomaterials, University of Michigan, 2200 Bonisteel Blvd., Ann Arbor, Michigan 48109, USA
Department of Physics, University of Michigan, 450 Church St, Ann Arbor, Michigan 48109, USA

Che-Hung Liu, Zhaohui Zhong and Theodore B. Norris
Center for Photonics and Multiscale Nanomaterials, University of Michigan, 2200 Bonisteel Blvd., Ann Arbor, Michigan 48109, USA
Department of Electrical Engineering and Computer Science, University of Michigan, 1301 Beal Avenue, Ann Arbor, Michigan 48109, USA

Chang-Hua Liu
Department of Electrical Engineering and Computer Science, University of Michigan, 1301 Beal Avenue, Ann Arbor, Michigan 48109, USA

Siyuan Zhang and Seth R. Marder
School of Chemistry and Biochemistry, Georgia Institute of Technology, 901 Atlantic Drive, Atlanta, Geogia 30332, USA

Evgenii E. Narimanov
School of Electrical and Computer Engineering and Birck Nanotechnology Center, Purdue University, 1205 West State Street, West Lafayette, Indiana 47907, USA

Ju-Hyung Kim
Surface and Interface Science Laboratory, RIKEN, 2-1 Hirosawa, Wako, Saitama 351-0198, Japan
Department of Chemical Engineering, Pukyong National University, 365 Sinseon-ro, Nam-gu, Busan 608-739, Republic of Korea
Center for Organic Photonics and Electronics Research (OPERA), Kyushu University, 744 Motooka, Nishi, Fukuoka 819-0395, Japan
Department of Advanced Materials Science, The University of Tokyo, 5-1-5 Kashiwanoha, Kashiwa, Chiba 277-8561, Japan

Jean-Charles Ribierre, Yu Seok Yang and Chihaya Adachi
Center for Organic Photonics and Electronics Research (OPERA), Kyushu University, 744 Motooka, Nishi, Fukuoka 819-0395, Japan

Maki Kawai
Department of Advanced Materials Science, The University of Tokyo, 5-1-5 Kashiwanoha, Kashiwa, Chiba 277-8561, Japan

Jaehoon Jung
Surface and Interface Science Laboratory, RIKEN, 2-1 Hirosawa, Wako, Saitama 351-0198, Japan
Department of Chemistry, University of Ulsan, 93 Daehak-ro, Nam-gu, Ulsan 44610, Republic of Korea

Takanori Fukushima
Chemical Resources Laboratory, Tokyo Institute of Technology, 4259 Nagatsuta, Midori-ku, Yokohama 226-8503, Japan

Yousoo Kim
Surface and Interface Science Laboratory, RIKEN, 2-1 Hirosawa, Wako, Saitama 351-0198, Japan

Anton Kuzyk and Simon Stoll
Max Planck Institute for Intelligent Systems, Heisenbergstrasse 3, D-70569 Stuttgart, Germany

Masayuki Endo
Institute for Integrated Cell-Material Sciences (WPI-iCeMS), Kyoto University, Yoshida-ushinomiyacho, Sakyo-ku, Kyoto 606-8501, Japan

Yangyang Yang and Hiroshi Sugiyama
Institute for Integrated Cell-Material Sciences (WPI-iCeMS), Kyoto University, Yoshida-ushinomiyacho, Sakyo-ku, Kyoto 606-8501, Japan

Department of Chemistry, Graduate School of Science, Kyoto University, Kitashirakawa-oiwakecho, Sakyo-ku, Kyoto 606-8502, Japan

Xiaoyang Duan and Na Liu
Max Planck Institute for Intelligent Systems, Heisenbergstrasse 3, D-70569 Stuttgart, Germany
Kirchhoff Institute for Physics, University of Heidelberg, Im Neuenheimer Feld 227, D-69120 Heidelberg, Germany

Alexander O. Govorov
Department of Physics and Astronomy, Ohio University, Athens, Ohio 45701, USA

Jianbo Yin, HuanWang, Lei Liao, Li Lin, Hailin Peng and Zhongfan Liu
Center for Nanochemistry, Beijing Science and Engineering Center for Nanocarbons, Beijing National Laboratory for Molecular Sciences, College of Chemistry and Molecular Engineering, Peking University, 202 Chengfu Road, Haidian District, Beijing 100871, China

Han Peng and Yulin Chen
Clarendon Laboratory, Department of Physics, University of Oxford, Parks Road, Oxford OX1 3PU, UK

Zhenjun Tan and Xiao Sun
Center for Nanochemistry, Beijing Science and Engineering Center for Nanocarbons, Beijing National Laboratory for Molecular Sciences, College of Chemistry and Molecular Engineering, Peking University, 202 Chengfu Road, Haidian District, Beijing 100871, China
Academy for Advanced Interdisciplinary Studies, Peking University, Beijing 100871, China

Ai Leen Koh
Stanford Nano Shared Facilities, Stanford University, Stanford, California 94305, USA

Filip Dvořák, Andrii Tovt, Mykhailo Vorokhta, Tomáš Skála, Iva Matolínová, Josef Mysliveček and Vladimír Matolín
Charles University in Prague, Faculty of Mathematics and Physics, V Holešovičkách 2, Prague 18000, Czech Republic

Matteo Farnesi Camellone
CNR-IOM DEMOCRITOS, Istituto Officina dei Materiali, Consiglio Nazionale delle Ricerche, Via Bonomea 265, Trieste 34136, Italy

Nguyen-Dung Tran and Stefano Fabris
CNR-IOM DEMOCRITOS, Istituto Officina dei Materiali, Consiglio Nazionale delle Ricerche, Via Bonomea 265, Trieste 34136, Italy

SISSA, Scuola Internazionale Superiore di Studi Avanzati, Via Bonomea 265, Trieste 34136, Italy

Ajay K. Singh, Dong-Hyeon Ko, Niraj K. Vishwakarma, Seungwook Jang, Kyoung-Ik Min and Dong-Pyo Kim
National Center of Applied Microfluidic Chemistry, Department of Chemical Engineering, POSTECH (Pohang University of Science and Technology), Pohang 37673, Korea

Patrick Gallagher, Menyoung Lee and David Goldhaber-Gordon
Department of Physics, Stanford University, Stanford, California 94305, USA

Francois Amet
Department of Physics, Duke University, Durham, North Carolina 27708, USA
Department of Physics and Astronomy, Appalachian State University, Boone, North Carolina 28608, USA

Petro Maksymovych and Jun Wang
Center for Nanophase Materials Sciences, Oak Ridge National Laboratory, Oak Ridge, Tennessee 37831, USA

Shuopei Wang, Xiaobo Lu and Guangyu Zhang
Institute of Physics, Chinese Academy of Sciences, Beijing 100190, China

Kenji Watanabe and Takashi Taniguchi
National Institute for Materials Science, 1-1 Namiki, Tsukuba 305-0044, Japan

Vittorio Peano and Christian Brendel
Institute for Theoretical Physics, University of Erlangen-Nürnberg, Staudtstr. 7, 91058 Erlangen, Germany

Martin Houde and Aashish A. Clerk
Department of Physics, McGill University, 3600 rue University, Montreal, Quebec, Canada H3A 2T8

Florian Marquardt
Institute for Theoretical Physics, University of Erlangen-Nürnberg, Staudtstr. 7, 91058 Erlangen, Germany
Max Planck Institute for the Science of Light, Günther-Scharowsky-Stra_e 1/Bau 24, 91058 Erlangen, Germany

Jia Zhang, Huangxi Zhang, Zhenlong Wang and Ping An Hu
Key Laboratory of Micro-systems and Micro-structures Manufacturing, Ministry of Education, Harbin Institute of Technology, Harbin 150080, China

Wenchun Feng, Heather A. Calcaterra, Bongjun Yeom and Nicholas A. Kotov
Department of Chemical Engineering, University of Michigan, Ann Arbor, Michigan 48109-2136, USA

Lei Zhang and Xing Zhang
The Molecular Foundry, Lawrence Berkeley National Laboratory, Berkeley, California 94720, USA
Department of Applied Physics, School of Science, Xi'an Jiaotong University, Xi'an 710049, China

Dongsheng Lei, Meng Zhang, Huimin Tong and Gang Ren
The Molecular Foundry, Lawrence Berkeley National Laboratory, Berkeley, California 94720, USA

Jessica M. Smith
Materials Sciences Division, Lawrence Berkeley National Laboratory, Berkeley, California 94720, USA
Department of Chemistry, University of California, Berkeley, California 94720, USA
Department of Materials Science, University of California, Berkeley, California 94720, USA

Zhuoyang Lu
The Molecular Foundry, Lawrence Berkeley National Laboratory, Berkeley, California 94720, USA
Center for Mitochondrial Biology and Medicine, The Key Laboratory of Biomedical Information Engineering of the Ministry of Education, Xi'an Jiaotong University, Xi'an 710049, China
School of Life Science and Technology, Xi'an Jiaotong University, Xi'an 710049, China
Frontier Institute of Science and Technology, Xi'an Jiaotong University, Xi'an 710049, China

Jiankang Liu
Center for Mitochondrial Biology and Medicine, The Key Laboratory of Biomedical Information Engineering of the Ministry of Education, Xi'an Jiaotong University, Xi'an 710049, China
School of Life Science and Technology, Xi'an Jiaotong University, Xi'an 710049, China
Frontier Institute of Science and Technology, Xi'an Jiaotong University, Xi'an 710049, China

A. Paul Alivisatos
Materials Sciences Division, Lawrence Berkeley National Laboratory, Berkeley, California 94720, USA
Department of Chemistry, University of California, Berkeley, California 94720, USA
Department of Materials Science, University of California, Berkeley, California 94720, USA
Kavli Energy NanoScience Institute, University of California, Berkeley, California 94720, USA

Wen Liu, Zhe Weng, Hailiang Wang and Qi Fan
Department of Chemistry and Energy Sciences Institute, Yale University, 520 West Campus Drive, West Haven, Connecticut 06511, USA

Enyuan Hu and Xiqian Yu
Chemistry Department, Brookhaven National Laboratory, Upton, New York 11973, USA

Hong Jiang
Beijing National Laboratory for Molecular Sciences, College of Chemistry and Molecular Engineering, Peking University, Beijing 100871, China

Yingjie Xiang
Department of Mechanical Engineering and Materials Science, Yale University, 520 West Campus Drive, West Haven, Connecticut 06511, USA

Eric I. Altman and Min Li
Department of Chemical and Environmental Engineering, Yale University, 520 West Campus Drive, West Haven, Connecticut 06511, USA

Jae Seol Lee, Jungki Song, Seokbeom Kim, Wooju Lee, Dongchoul Kim and Jungchul Lee
Department of Mechanical Engineering, Sogang University, 35 Baekbeom-ro (Sinsu-dong), Mapo-gu, Seoul 04107, South Korea

Seong Oh Kim and Joshua A. Jackman
School of Materials Science and Engineering and Centre for Biomimetic Sensor Science, Nanyang Technological University, = 50 Nanyang Drive, Singapore 637553, Singapore

Nam-Joon Cho
School of Materials Science and Engineering and Centre for Biomimetic Sensor Science, Nanyang Technological University,50 Nanyang Drive, Singapore 637553, Singapore
School of Chemical and Biomedical Engineering, Nanyang Technological University, 62 Nanyang Drive, Singapore 637459, Singapore

Liang Yue, Shan Wang, Ding Zhou, Hao Zhang and Bao Li
State Key Laboratory of Supramolecular Structure and Materials, College of Chemistry, Jilin University, Changchun 130012, PR China

Lixin Wu
Key Laboratory of Natural Resources of Changbai Mountain and Functional Molecules, Yanbian University, Yanji 133002, PR China

Sarah Wieghold, Maximilian Krause, Ueli Heiz and Friedrich Esch
Chemie-Department, Technische Universität München, Lichtenbergstra_e 4, Garching 85748, Germany
Catalysis Research Center, Technische Universität München, Ernst-Otto-Fischer-Stra_e 1, Garching 85748, Germany

Juan Li, Johannes V. Barth and Carlos-Andres Palma
Catalysis Research Center, Technische Universität München, Ernst-Otto-Fischer-Stra_e 1, Garching 85748, Germany
Physik-Department, Technische Universität München, James-Franck-Strasse 1, Garching 85748, Germany

Patrick Simon and Jose A. Garrido
Physik-Department, Technische Universität München, James-Franck-Strasse 1, Garching 85748, Germany
Walter Schottky Institut, Technische Universität München, Am Coulombwall 4, Garching 85748, Germany

Yuri Avlasevich, Chen Li and Klaus Müllen
Max Planck Institute for Polymer Research, Ackermannweg 10, Mainz 55128, Germany

Paolo Samorí
ISIS & icFRC, Universitéde Strasbourg & CNRS, 8 allée Gaspard Monge, Strasbourg 67000, France

Index

A

Aerogel Sheet Electrodes, 18-19, 25
Anisotropic Friction, 153-154, 156
Anodized Aluminium Oxide (aao), 202
Atomic Force Microscope (afm), 196
Atomic Structure, 53, 70-71, 141, 194

B

Bacteriorhodopsin, 125
Biotechnology, 27

C

Carbon Nanotube, 18, 20, 25-26, 169, 171, 175-176, 194, 208, 218
Carcinogenic Chemistry, 146-147
Catalytic Activity, 22, 27-28, 30, 187-188, 190-191, 193-194
Cathodoluminescence Imaging, 100
Chemical Synthesis, 68, 108, 152, 188
Chiral Inelastic Transport, 160
Chiroptical Function, 125
Chloromethyl Methyl Ether, 146-147, 151-152
Circularly Polarized Emission, 168, 172
Cobalt Phosphosulfide, 187-188
Cryogenic Applications, 76
Cryogenic Temperatures, 75-79, 81-82
Current-voltage Hysteresis, 1

D

Damage-tolerance, 75, 78-79
Density Functional Theory (dft), 54, 61, 188
Dna-nanogold Conjugates, 177-182, 184, 186
Dynamic Force Microscopy (dfm), 61

E

Electrocatalytic Hydrogen Production, 187
Electrochemical Thermal Energy, 18
Electron Tomography, 177, 185-186
Electronic Structures, 116-117, 120-123, 141, 193
Electrostatic Gating, 53-54, 57, 59
Energy Dissipation, 168, 171, 174

F

Facile Temperature, 83-84
Ferroelastic Switching, 36-38, 40-41, 43-44
Ferroelectric Effect, 1-2, 4, 8
Ferroelectric Polarization, 7-8, 36-37, 41-42, 44

F

Fluorescence Nanoscopy, 45, 50-51
Frequency Comb, 93-99
Frequency Metrology, 93, 99
Functional Biomaterials, 27
Functionalization, 17, 61, 65-66, 127, 150, 197, 207, 218

G

Giant Magnetic Anisotropy, 10
Grapheme, 111
Graphene Hyperbolic Metamaterials, 109
Graphene Photodetectors, 115, 131, 136
Graphene-diamond Interfaces, 220
Growth Sectors, 100-101, 103-107

H

Hydrogen Evolution Reaction, 59, 68-69, 73, 188, 194
Hyperbolic Dispersion, 109-110, 113-114

I

Intermolecular Potential, 61, 63
Ionic Polarization, 1, 4, 7-8

K

Kagome Lattice Model, 160-161, 163

L

Large Hyperfine Interactions, 10
Layered-perovskite Thin Film, 36
Local Carrier Concentrations, 100, 106

M

Magnetic Resonance, 91, 178
Mediumentropy Alloy, 75
Metal Interfaces, 116, 123, 224
Metal-dielectric Multilayers, 109
Metal-metal Raman Stretching, 68
Micro-total Envelope System, 146-147
Microscale Twisting, 168
Molecular Motion, 125-126, 128
Molecular Spectroscopy, 93
Molybdenum Disulfide Nanosheets, 68
Multiferroic Oxides, 36
Multiscale Deformations, 168-169, 174

N

Nano-size Separation, 210, 214, 219
Nanocaged Enzymes, 27, 30, 33

Nanoscale Self-assembly, 153

O

Optical Topological Transition, 109, 112-114

Optoelectronic Function, 220

Organic-inorganic Frameworks, 210-211

Orientational Dependence, 61

P

Perovskite Solar Cells, 1, 3, 7-9

Photoactivated Localization Microscopy (palm), 46

Photocurrent Generation, 131, 133-136

Photoelectron Spectroscopy, 69, 73, 138, 189

Photoresponse, 115, 131, 137, 220-225

Photothermal Ablation, 83, 87, 89, 92

Photothermal Therapy, 83-84, 86-89, 91-92

Plasmonic Nanosystem, 125-128

Plastic Instability, 75, 79

Polyanionic Surfaces, 27

Protease Digestion, 27

Q

Quadratic Squeezing, 161

S

Scanning Electron Microscopy (sem), 76, 101, 132, 135, 148, 174, 188, 202, 205

Scanning Tunneling Microscopy, 59, 66, 136, 144, 207

Semiconductor Nanorods, 100

Silicon Nanowire, 146, 149, 152

Single-molecule Magnets, 10-11, 16-17

Spatial Resolution, 83-84, 89, 100-101, 105-107, 133, 136, 203

Stimulated Emission Depletion (sted), 46

Structural Dynamics, 59, 177

Structural Integrity, 31, 116-119, 122

Structural Semiconductor, 53, 59

Supramolecular Carpet, 116-117

Surface Plasmon Resonance, 93, 99

Surface Step Decoration, 138

Switchable Friction, 153

T

Thermal Streams, 18

Topological Phase Transitions, 160, 162

Transition Metal Dichalcogenides (tmds), 54

Transmission Electron Microscopy (tem), 28, 126, 132

U

Ultra-high Vacuum (uhv), 139

Upconversion Nanocomposite, 83-84, 89

V

Van Hove Singularity, 131, 136

www.ingramcontent.com/pod-product-compliance
Lightning Source LLC
Chambersburg PA
CBHW080516200326
41458CB00012B/4226